C0-AVC-646

Signals and Signal Transduction Pathways in Plants

Signals and Signal Transduction Pathways in Plants

Edited by

Klaus Palme

Reprinted from Plant Molecular Biology, Vol. 26(5), 1994

Kluwer Academic Publishers
Dordrecht / Boston / London

A C.I.P. Catalogue record for this book is available from the Library of Congress

ISBN 0-7923-3364-0

Published by Kluwer Academic Publishers,
P.O. Box 17, 3300 AA Dordrecht, The Netherlands.

Kluwer Academic Publishers incorporates
the publishing programmes of
D. Reidel, Martinus Nijhoff, Dr W. Junk and MTP Press.

Sold and distributed in the U.S.A. and Canada
by Kluwer Academic Publishers,
101 Philip Drive, Norwell, MA 02061, U.S.A.

In all other countries, sold and distributed
by Kluwer Academic Publishers Group,
P.O. Box 322, 3300 AH Dordrecht, The Netherlands.

Printed on acid-free paper

Printed in the Netherlands

Contents

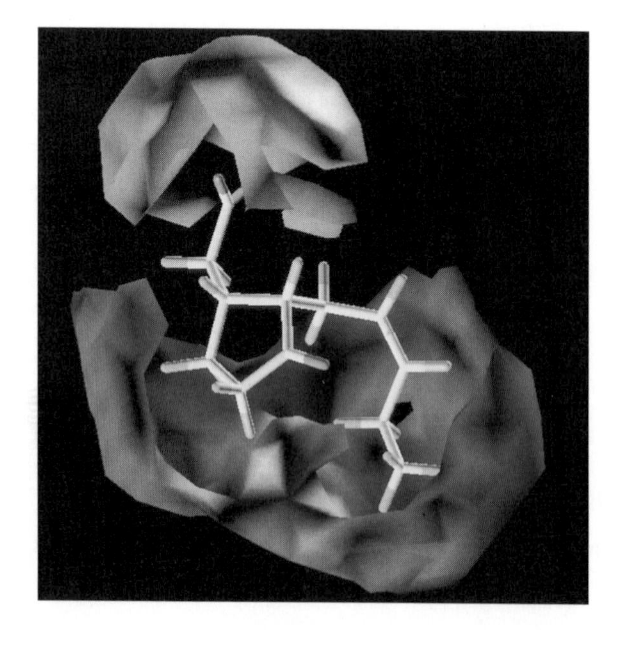

Photo cover

Molecular lipophilicity potential of 3R,7S-jasmonic acid. Blue regions are lipophilic, red regions are lipophobic. This model was computed by Patrick Gaillard, Pierre-Alain Carrupt and Nicolas Guex at the University of Lausanne. For details see article by Farmer, pages 1423–1437.

Plant Molecular Biology **26**: 1237, 1994.

Preface

Plants have remarkable strategies at their disposal to match their growth and development with the challenges and opportunities of their environment. To understand the mechanisms that underlie the perception of the environmental changes, developmental cues, hormones and other signals that elicit, activate and modulate the activity of plant signalling pathways at the molecular level, is a major challenge. What is the cellular structure and organization of these signalling pathways and how many such pathways are there in plant cells? What signals affect their activity and how widely is each of these signals used? How are the second messengers and their physiological consequences integrated? While most of these questions are just being started to be addressed with the help of the tools provided by molecular biology and genetics, we have already seen an explosive increase in information describing these signalling processes. Contributors to this special issue of *Plant Molecular Biology* highlight some of these important areas in plant biology and cell signalling. They deal with the control of plant cell proliferation and the molecular dissection of the roles of signal molecules in processes ranging from embryogenesis to flowering. New insights towards approaching an understanding of the physiology of the classical plant hormones are presented as well as an extensive discussion of other growth regulators such as lipo-chitin oligosaccharides and fatty acids. An integrated discussion of these themes together with reviews of current knowledge of plant receptor-like kinases, ion channels and transporters present a comprehensive current overview on plant signalling processes. The panoply of molecular tools and techniques as well as the blooming field of plant genetics is providing an exciting ground for major breakthroughs in this field. We hope that this journal-issue will convey some of the excitement felt among the plant researchers and that it will stimulate readers' interest in many aspects of plant biology.

KLAUS PALME
Guest-Editor

Plant Molecular Biology **26**: 1239–1270, 1994.
© 1994 *Kluwer Academic Publishers. Printed in Belgium.*

Transmembrane signalling in eukaryotes: a comparison between higher and lower eukaryotes

A. Lyndsay Drayer* and Peter J.M. van Haastert
*Department of Biochemistry, University of Groningen, Nijenborgh 4, Groningen, The Netherlands (*author for correspondence)*

Received 1 April 1994; accepted 26 April 1994

Key words: eukaryotes, cell surface receptors, second messengers, transmembrane signal transduction (pathway)

Abbreviations: C, catalytic subunit PKA; EGF, epidermal growth factor; FGF, fibroblast growth factor; GRK, G-protein coupled receptor kinase; $InsP_3$, inositol 1,4,5-trisphosphate; LAR, leucocyte common antigen-related; LCA, leucocyte common antigen; MAP kinase, mitogen-activated protein kinase; PDE, phosphodiesterase; PDGF, platelet-derived growth factor; PKA, cAMP-dependent protein kinase; PKC, protein kinase C; PKG, cGMP-dependent protein kinase; PLA_2, phospholipase A_2; PLC, phospholipase C; PLD, phospholipase D; $PtdInsP_2$, phosphatidylinositol 4,5-bisphosphate; PTP, protein tyrosine phosphatase; R, regulatory subunit PKA; SH domain, *src* homology domain; TCR, T cell receptor; TPA, 12-*O*-tetradecanoylphorbol-13-acetate

Introduction

All living organisms react to signals from their environment such as light, temperature, sound and chemical substances in order to survive and adapt to their surroundings. In multicellular organisms communication between cells is necessary to regulate growth and differentiation [3]. Neighbouring cells can communicate with each other by direct cell-cell contact through plasma-membrane-bound signalling molecules or through the formation of gap junctions [176]. Cells are able to communicate with other cells some distance away through secretion of chemical signals. The secreted signals can be soluble molecules such as most hormones and neurotransmitters; they bind to receptors on the cell surface of target cells. Hydrophobic signal molecules such as the steroid and thyroid hormones are able to pass the lipid bilayer of the plasma membrane and bind to specific proteins inside the cell [76]. The complex of hydrophobic hormone bound to its receptor protein is directly able to influence gene transcription by binding to specific DNA sequences [221].

This review discusses some of the mechanisms by which cells communicate with their surroundings through extracellular signals which bind to cell surface receptors. The receptors are able to transduce signals across the plasma membrane by activating intracellularly located proteins of the transmembrane signal transduction pathway. Many pathways were originally characterized in lower eukaryotes and were later shown to be present in higher organisms. Genetic analysis demonstrates that many components have been conserved during evolution and are shared by mammals and other vertebrates, invertebrates and microorganisms.

Cell surface receptors

The receptors are classified into different families based on structural similarities and modes of transduction. Transmitter-gated ion channels, such as the nicotinic acetylcholine, glycine and γ-aminobutyrate (GABA) receptors, are heteropentameric proteins surrounding a membrane pore. These receptors contain intrinsic channelling activity to allow the passage of ions across the cell membrane when activated by ligand binding [250, 100, 161]. The surface receptors that will be discussed here directly modulate the activity of proteins inside the cell upon binding of their ligand. G-protein-linked receptors have a putative structure containing seven transmembrane-spanning domains (Fig. 1A). Intracellularly the receptor couples to an intermediate protein to regulate an enzyme or ion channel. Within the family of catalytic receptors binding of ligand and generation of the intracellular signal are functions of one molecule. These receptors contain a single membrane-spanning domain. The catalytic receptors operate directly as enzymes through an intracellularly located domain with guanylyl, cyclase, tyrosine kinase or tyrosine phosphatase activity (Fig. 1B–D).

G-protein-coupled receptors

Proteins belonging to the G-protein-coupled receptor superfamily share two properties. First, binding of ligand to the receptor induces the activation of a heterotrimeric guanine nucleotide-binding protein, or G-protein. Through G-proteins these receptors are linked to intracellular effector enzymes such as adenylyl cyclase, phospholipase C, cGMP-dependent phosphodiesterase, and some ion channels. Second, the deduced amino acid sequences of cloned G-protein-coupled receptor genes predict a common

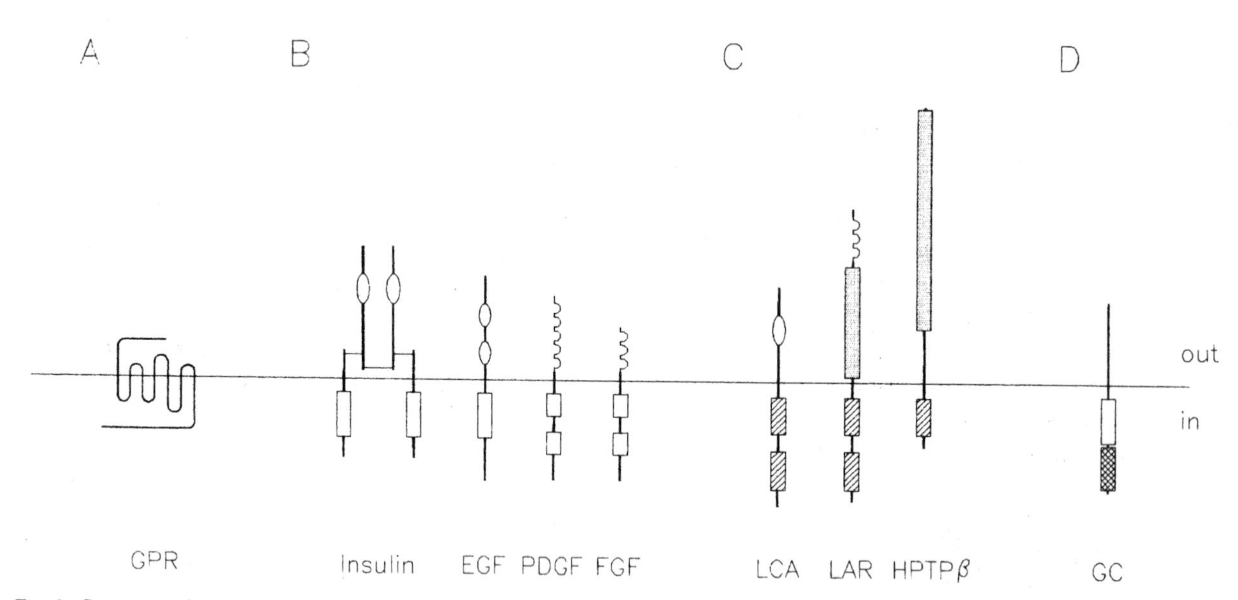

Fig. 1. Structure of cell surface receptor proteins. In the plasma membrane surface receptors the amino terminus is extracellular, whereas the carboxy terminus is cytoplasmic. Conserved regions: *open oval*, cystein rich domain; *open box*, protein tyrosine kinase catalytic domain; *half circle*, immunoglobuline-like region; *diagonally striped box*, protein tyrosine phosphatase domain; *filled box*, fibronectin type III repeat; *hatched box*, guanylyl cyclase catalytic domain. A. G-protein-linked receptors (GPR) contain seven hydrophobic stretches which are proposed to be embedded in the plasma membrane. The length of the extracellular domain varies great in the different receptors belonging to this family. B. Protein tyrosine kinase receptors. The insulin receptor and growth factor receptor subclasses PDGF (platelet-derived growth factor), EGF (epidermal grow factor) and FGF (fibroblast growth factor) have an intracellulary located protein tyrosine kinase domain. The kinase domains of the PDGF and EGF receptors contain an insertion sequence. C. Protein tyrosine phosphatase receptors. Structures of the leucocyte common antigen (LCA) receptor, LCA-related receptor (LAR), and human protein tyrosine phosphatase β (HPTPβ) receptor. HPTPβ receptor contains a single protein tyrosine phosphatase domain. D. Guanyly cyclase (GC) receptors contain an intracellularly located guanylyl cyclase catalytic domain and a protein tyrosine kinase-like domain.

structure consisting of seven hydrophobic transmembrane-spanning α helices (Fig. 1A). This structure is based on hydropathy plots, and is predicted to be similar to that of bacteriorhodopsin (which however does not bind a G-protein), identified by electron crystallography [112]. In the seven-membrane-spanning structure model the amino terminus is extracellular, the carboxyl terminus cytoplasmic, with the helices arranged counterclockwise in the membrane. Based on structural similarities and amino acid sequence homology G-protein-coupled receptors can be divided into different groups.

A homogeneous group is formed by G-protein-linked receptors in mammalian cells that are activated by small peptide ligands. This group consists of receptors for a family of related brain-gut peptides, such as secretin and vasoactive intensinal peptide, and receptors for calcium-regulating peptide hormones, such as calcitonin and parathyroid hormone [256]. These receptors show a high degree of similarity to each other in the hydrophobic membrane-spanning regions, but are not homologous to other G-protein-linked receptors (less than 12%). They contain a large (ca. 130 amino acids) amino-terminal domain with conserved cysteine residues.

Recently, a Ca^{2+}-sensing receptor gene from bovine and human parathyroid containing seven putative membrane-spanning domains was cloned [30]. The receptor is thought to regulate calcium homeostasis by sensing extracellular Ca^{2+} levels [230]. It shares similarity (30%) with a separate group of G-protein-linked receptors, the metabotropic glutamate receptors [204]. The sequences of both receptor types predict a very large putative extracellular domain (over 600 amino acids) containing conserved cysteine residues and a hydrophobic segment in the amino terminus. The Ca^{2+}-sensing receptor shows clusters of acidic amino acid residues possibly involved in calcium binding. The presence of other extracellular ion-sensing receptors, for example for K^+ and PO_4^{3-}, is assumed [30].

The main group of G-protein-linked receptors in mammalian cells bind diverse ligands: glycoproteins such as thyroid-stimulating and follicle-stimulating hormones, small organic molecules such as adrenaline and acetylcholine, hydrophobic compounds such as the cannabinoids, and the chromophore retinal. The putative extracellular amino-terminal domains of the receptors vary greatly in length. The receptors share a limited number of conserved amino acid sequences, mainly in the putative membrane-spanning regions [233]. Olfactory receptors were cloned using a polymerase chain reaction (PCR) strategy based on the assumption that the receptors would contain conserved transmembrane amino acid sequences [32]. A gustatory receptor, expressed in taste buds, is similar to the olfactory receptors (56% amino acid identity) and may be a taste receptor [1]. Through examination of features common to G-protein-linked receptors, regions for interaction between receptor and ligand, and receptor and G-protein are predicted [reviewed in 243]. A region with less sequence identity among this group of diverse receptors is the proposed third cytoplasmic loop between the fifth and sixth transmembrane regions; this region is thought to be involved in the interaction of the receptor with specific G-proteins. However, for rhodopsin a small loop is predicted, but it is capable of activating a variety of G-proteins. The membrane-spanning regions are proposed to form a ligand-binding pocket in the α- and β-adrenergic receptors, muscarinic acetylcholine receptors, and the receptors involved in vision, the opsins [166, 206, 207]. In the photoreceptor system, retinal is proposed to be surrounded by the transmembrane domains of the visual light pigment rhodopsin and activated when photons are absorbed. The three colour pigment receptor molecules expressed in cones are proposed to be arranged in a similar way. Visual pigments of the fruit fly *Drosophila* photoreceptor cells, R1–R8, were shown to share sequence homology with mammalian rhodopsin [217, 322]. Mammalian and *Drosophila* opsin genes are thus proposed to derive from a common ancestor.

The proposed topology of seven hydrophobic membrane-spanning domains in receptors linked to G-proteins is found in microorganisms. Conserved receptors are cAMP receptors of *Dictyos-*

telium discoideum and pheromone receptors of the yeasts *Saccharomyces cerevisiae* and *Schizosaccharomyces pombe*. The cellular slime mould *Dictyostelium discoideum* is an ideal system for studying signal transduction processes [63, 306]. Growth and cellular differentiation are entirely separated processes. The differentiation programme is induced by starvation during which cells communicate with each other by means of chemotactic signals. The four cAMP receptors involved in chemotaxis and development are expressed at distinct stages of *Dictyostelium* development [143, 129, 244]. In the budding yeast *S. cerevisiae* and the fission yeast *S. pombe* mating involves the fusion of two opposite cell types to produce a diploid cell [119, 312]. Each partner produces a peptide pheromone that is detected by surface receptors of the opposite type (Fig. 2). In *S. cerevisiae* the **a**- and α-mating factors are produced by **a** and α cell types, respectively. The **a** cell type produces a unique receptor protein encoded by the *STE2* gene, whereas the α cell type produces a receptor protein encoded by the *STE3* gene. Although the protein sequences encoded by *STE2* and *STE3* are predicted to have the same structure characteristic of the G-protein-coupled receptors, amino acid similarity between the two receptors is absent [33, 205, 105]. In *S. pombe* the two mating types are called h^+ and h^-; the pheromones secreted by these cells are called P-factor and M-factor, respectively. The deduced amino acid sequences of the *mam2* gene encoding the receptor for P-factor and the *map3* gene encoding the receptor for M-factor are very different [142, 287]. However, the Mam2 receptor of *S. pombe* shows significant homology (26% amino acid identity) with the α-factor receptor Ste2 of *S. cerevisiae*, and the Map3 receptor is homologous to the a-factor receptor Ste3. In *S. pombe* the mating factors are not only involved in the initiation of mating, as in *S. cerevisiae* but also in meiosis.

Receptor protein tyrosine kinases

Receptor tyrosine kinases contain an extracellular amino-terminal ligand-binding domain, a single putative membrane-spanning domain and intracellularly, a tyrosine kinase domain. The receptors belonging to this family can be divided

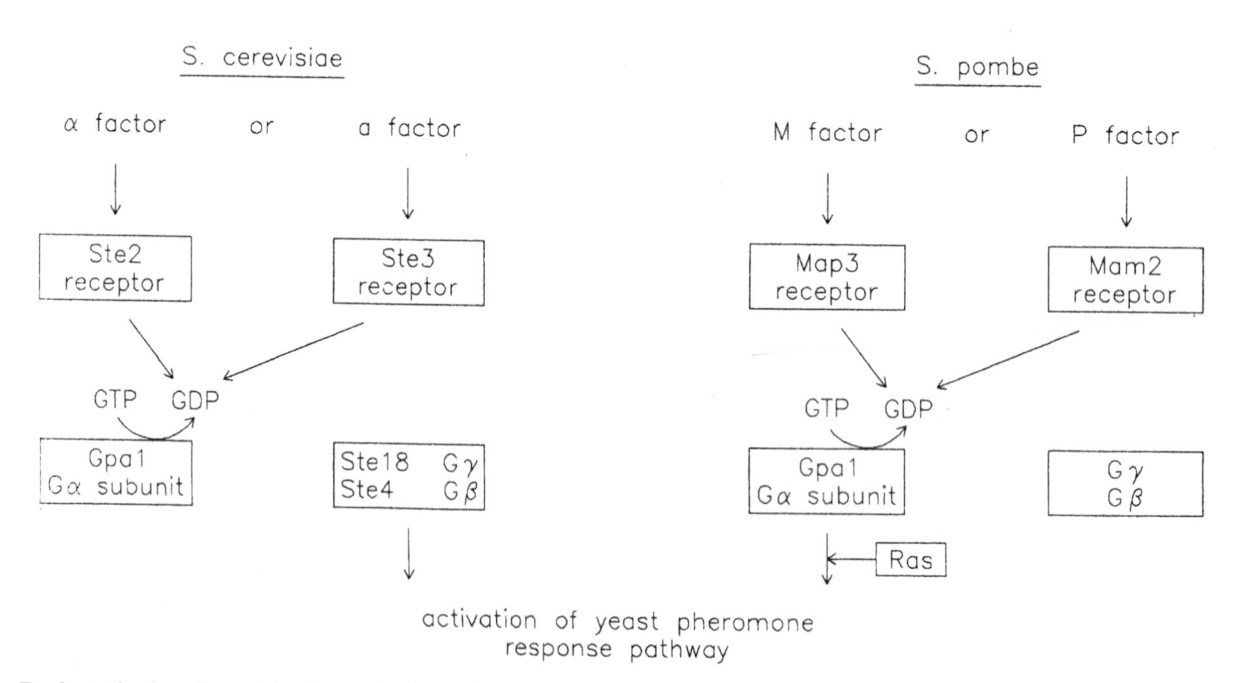

Fig. 2. Activation of yeast signal transduction pathways by heterotrimeric G-proteins. In *Saccharomyces cerevisiae* βγ subunits of G-proteins transduce the pheromone signal, whereas in *Schizosaccharomyces pombe* an α subunit and a Ras protein are involved.

into four subclasses based on sequence similarity and structural characteristics [302]: the epidermal growth factor (EGF) receptor, the insulin receptor, the platelet-derived growth factor (PDGF) and the fibroblast growth factor (FGF) receptor subfamilies (Fig. 1B). The EGF and insulin receptors share homologous cysteine repeats in their extracellular domains. The extracellular domains of PDGF receptors have an immunoglobulin-like structure. Members of the FGF receptor subclass contain three immunoglobulin-like sequence repeats, similar to the five repeats in the PDGF receptor. Binding of ligand to the extracellular domain induces receptor dimerization, which in turn is thought to induce the activation of the intracellular tyrosine kinase domain [reviewed in 246]. The monomeric ligand EGF, and the dimeric PDGF mediate dimerization of neighbouring receptors [111, 99]. The insulin receptor is a dimer, which is stabilized upon binding of its ligand. Activation of receptor tyrosine kinases induces cells to proliferate and differentiate in the case of the growth factor receptors, or to turn on metabolic pathways in the case of the insulin receptor [247].

The tyrosine kinase domain is the most conserved region between all receptor tyrosine kinase molecules. It contains the consensus sequence for ATP binding, resembling the catalytic subunit of cAMP-dependent protein kinases. Mutations in the ATP binding site result in the absence of protein tyrosine kinase activity and abolish all signal transduction processes [see references in 302]. The kinase domains of the PDGF and FGF receptor classes contain an insertion sequence of hydrophilic amino acid residues, dividing the kinase domain in half. The kinase insertion sequence is probably not involved in kinase activity, but is thought to play a role in receptor interactions with substrate proteins [138].

Autophosphorylation of the receptor plays an important role in both enzyme activation and interaction with the substrate protein [147]. The activated receptors transduce extracellular signals not only by phosphorylation, but also by direct protein-protein interactions. Many protein substrates of growth factor receptor tyrosine kinases contain the *src* homology domains SH2, a motif originally found in the cytoplasmic tyrosine kinases related to c-Src. The SH2 domains bind to autophosphorylated tyrosine residues on the receptors [reviewed in 233]. In the PDGF and EGF receptors phosphorylation sites are found in the non-catalytic regions of the cytoplasmic domains. Substrate proteins include phospholipase Cγ1, the phosphotyrosine phosphatase Syp, the GTPase-activating protein of Ras (GAP), and Gbr2 (growth factor receptor-bound protein 2), an adaptor protein involved in Ras signalling [139, 140, 184, 179, 4, 197]. The receptor tyrosine kinases are thus able to couple to many, diverse intracellular signalling pathways, each of the substrate proteins binding at specific phosphorylation sites within the receptor [38, 78, 267]. The activated insulin receptor phosphorylates an SH2-docking protein, named IRS1, on multiple tyrosine residues [280]. This intermediate protein is able to bind the SH2 domains of Grb2, Syp, and phosphatidylinositol 3-kinase [280, 223].

The receptor tyrosine kinase homologues identified in nonmammals control cell fate determination during development. In *Drosophila* the Sevenless receptor is required for photoreceptor cell R7 development [103], and the EGF receptor homologue DER affects the correct dorsoventral patterning of the eggshell and the embryo [232]. The structure of the torso receptor of *Drosophila* is reminiscent of the mammalian PDGF receptor, and involved in determination of terminal cell fates [268]. The Let-23 receptor tyrosine kinase from the nematode *Caenorhabditis elegans* is structurally similar to the mammalian EGF receptor, and is required for vulval induction [7]. The developmental pathways activated by receptor tyrosine kinases in invertebrates have been investigated genetically, and shown to involve signalling through Ras proteins (see below).

Receptor protein tyrosine phosphatases

The overall putative structure of receptor tyrosine phosphatases resembles that of receptor tyrosine kinases, with an extracellular amino-terminal domain, a single membrane-spanning domain, and intracellulary the catalytic domain [87]. The different receptor forms share conserved catalytic domains with the soluble protein tyrosine phosphatases; there is no similarity with serine/threonine phosphatases. After the first, soluble, protein tyrosine phosphatase, named PTP-1B, was purified from human placenta, part of its sequence was recognized to be similar to the tandem C-terminal homologous domains of the leucocyte common antigen (LCA), CD45 [41, 42]. CD45 is expressed as a major surface component on nucleated hematopoietic cells and contains high intrinsic phosphatase activity [300]. Its putative structure includes a cysteine-rich extracellular domain, a hydrophobic region that has the characteristics of a membrane-spanning structure and two intracellular tyrosine phosphatase domains (Fig. 1C). Only one tyrosine phosphatase domain contained *in vitro* enzymatic activity when expressed in *Escherichia coli* whereas the other, carboxyl-terminal domain, although showing no detectable catalytic activity, influenced substrate specificity [276]. CD45 is involved in regulating T-cell receptor (TCR) signalling, as CD45-deficient T-cells failed to proliferate or produce cytokines after induction with antigen [228, 309]. CD45 is proposed to link the TCR to intracellular signalling pathways such as phosphatidylinositol hydrolysis and calcium mobilization, by dephosphorylating a putative negative regulatory site of cytosolic protein tyrosine kinases [151, 152]. In particular, CD45 is proposed to activate the Lck and Fyn tyrosine kinases [198, 260]. As ligands for receptor tyrosine phosphatases are undefined, a chimaeric tyrosine phosphatase receptor was constructed, in which the extracellular and transmembrane domains of CD45 were replaced with those of the EGF receptor [62]. This chimaeric receptor was able to restore TCR signalling in CD45-deficient cells. Interestingly, addition of EGF ligand to these cells abolished all TCR mediated signalling, indicating that *in vivo* CD45 tyrosine phosphatase activity may be constitutively functional and negatively regulated by ligand binding.

Members of the family of receptor protein tyrosine phosphatases contain structural diverse amino-terminal domains (Fig. 1C). The extracellular domain of CD45 is heavily glycosylated and contains a cystein-rich region. The leucocyte common antigen-related (LAR) receptor, isolated by cross-hybridization to an LCA cDNA probe, is expressed on cells of epithelial origin [274]. The amino-terminal region of LAR and its *Drosophila* homologues DLAR and DPTP, contain immunoglobulin- and fibronectin type III-like domains, similar to domains found in neural cell adhesion molecules [275]. Thus, it is possible that LAR, DLAR and DPTP are cell adhesion receptors. Other receptor tyrosine phosphatases, mammalian HPTPα and HPTPε, contain very short external segments [153]. HPTPβ and *Drosophila* DPTP10D contain multiple fibronectin type III domains and are unique in that they have only one of the conserved intracellular tyrosine phosphatase domains [153, 295]. Mammalian HPTPζ contains a domain with homology to the enzyme carbonic anhydrase besides fibronectin type III repeats [12].

Apart from CD45 function, a definitive role for receptor protein tyrosine phosphatases in cellular signalling has not been established yet. The *in vivo* function of non-receptor tyrosine phosphatases has been established during recent years through genetic studies in lower eukaryotic organisms. Non-receptor protein tyrosine phosphatases appear to play a role in the development of *Dictyostelium* and *Drosophila*, and in cell-cycle regulation in *Schizosaccharomyces pombe*. The fission yeast *S. pombe* contains three tyrosine phosphatases, pyp1, pyp2 and pyp3, which play a role in the timing of the onset of mitosis through regulation of the cell-cycle cdc2 protein kinase [98]. Pyp1 and pyp2 phosphatases negatively regulate mitosis; deletion of both genes is lethal, whereas overexpression of these proteins results in a delay in mitotic onset [194, 216]. Deletion of the *pyp3* gene has no effect, indicating that it does not have an essential role in the cell cycle, but overproduction advances the onset of mitosis [193]. The differences in function of these phosphatases could be reflected in their structure; pyp1 and pyp2 contain a long amino-terminal extension preceding the catalytic domain, which is lacking in pyp3. *In vitro*, pyp3 was shown to directly activate cdc2 by removing an inhibitory phosphate of the kinase [reviewed in 158]. Pyp1 and pyp2 do not interact directly with cdc2, but positively regulate the activity of kinases involved in inhibiting cdc2 through phosphorylation.

In *Dictyostelium* the PTP1 tyrosine phosphatase is involved in regulating development [118]. Gene disruptants formed normal fruiting bodies, but the developmental process was accelerated. Cells overproducing PTP1 showed a delayed development, which was not completed to form normal fruiting bodies. In *Drosophila*, the *corckscrew* gene encodes a putative non-receptor tyrosine phosphatase containing two adjacent SH2 domains in its amino-terminal non-catalytic region [226]. It is a positive transducer of the signal generated by the receptor tyrosine kinase torso, possibly by binding to the activated torso receptor tyrosine kinase [227]. Corckscrew mammalian homologues are PTP1C, PTP1D, SH-PTP1 and SH-PTP2 in man, and Syp in mouse [258, 307, 89, 81]. Syp and PTP1D were shown to be regulated by binding to EGF and PDGF receptors through their SH2 domains. Binding of the phosphatases did not dephosphorylate the receptors, but the interaction with the activated receptor tyrosine kinases phosphorylates and thereby activates the phosphatases.

Guanylyl cyclases

Cyclic GMP is an ubiquitous mediator regulating cGMP-dependent ion channels, phosphodiesterases and protein kinases, involved in processes as diverse as renal and intestinal ion transport, maintenance of blood pressure, and light response in photoreceptor cells [125, 277]. Soluble and membrane-bound guanylyl cyclase enzymes have been cloned. The ligands that activate membrane-bound guanylyl cyclases are peptides [reviewed in 93]. In mammals guanylyl cyclase type A [45, 178] is activated by the atrial and brain natriuretic peptides ANP and BNP, guanylyl cyclase type B [40, 253] by type C natriuretic peptide, and guanylyl cyclase type C [254] is activated by enterotoxin of *E. coli* and by an endogenous peptide from the gut named guanylin. Guanylyl cyclase in sea urchin spermatozoa is stimulated by the peptides resact and speract secreted by sea urchin eggs [294, 266]. These peptides influence sperm motility. The ligand for the mammalian receptor cyclase, human retinal guanylyl cyclase, is as yet unidentified [263]. This cyclase could be responsible for the resynthesis of cGMP after phototransduction. Recoverin, a calcium sensor in vision, was initially proposed to be a soluble activator of retinal guanylyl cyclase [70, 159]. However, further experiments demonstrated that it was not recoverin that stimulates photoreceptor guanylyl cyclase, but an at present unidentified factor [121]. Calcium-bound recoverin is proposed to be involved in prolonging the photoresponse, possibly by blocking phosphorylation of activated rhodopsin [137].

Guanylyl cyclase receptors are predicted to be single membrane spanning proteins, analogous to receptor tyrosine kinases (Fig. 1D) [321]. The extracellular amino terminus is proposed to bind ligand. The cyclase catalytic domain is located at the carboxy terminus. Guanylyl cyclase receptors contain an intracellular domain homologous to protein kinases. The position of the kinase domain in the protein appears to be similar to that in receptor tyrosine kinases, and many of the amino acids highly conserved among the catalytic regions of protein kinases are present. However, no protein kinase activity has been detected in the guanylyl cyclase receptors. Deletion of the kinase-like domain resulted in a constitutive activation of guanylyl cyclase, proposing a role for this domain in the regulation of the catalytic domain of guanylyl cyclase receptors [44, 150]. The guanylyl cyclase catalytic domain shows a high degree of conservation with each of the catalytic domains of adenylyl cyclases. Dimerization of two cyclase catalytic domains is proposed to be required for catalytic activity, analogous to dimerization of receptor tyrosine kinases [93].

The soluble forms of guanylyl cyclase are not stimulated directly by hormones, but contain a prosthetic heme group and are regulated by nitric oxide, presumably through binding of nitric oxide to the heme group [reviewed in 248]. Mammalian soluble guanylyl cyclase isoforms are heterodimers with each subunit containing a cyclase catalytic domain similar to that of the membrane-bound form [see references in 46, 97]. Transfection experiments suggest that expression of both subunits is required for catalytic activity [202, 108, 109]. Unlike the membrane-bound forms, the soluble guanylyl cyclase do not contain a region homologous to protein kinases. A head-specific soluble guanylyl cyclase has been cloned from *Drosophila* [320]. Activation of this cyclase is thought to be involved in opening of the cGMP-dependent light-activated channel in invertebrate photoreceptor cells. Guanylyl cyclase in the protozoans *Paramecium* and *Tetrahymena* is tightly associated with the cytoskeleton and plays a role in cell motility. The genes have not yet been identified. The Ca^{2+} flux across the ciliary membranes of *Paramecium* and *Tetrahymena* stimulates guanylyl cyclase through direct binding with a Ca^{2+}/calmodulin complex [144, 251, 252]. In these organisms an increase in intracellular Ca^{2+} stimulates enzyme activity in the cilia. In *Dictyostelium*, guanylyl cyclase is activated by extracellular cAMP. The enzyme activity appears to be controlled through a cAMP surface receptor and G-protein pathway [127]. Like human retinal guanylyl cyclase, the *Dictyostelium* enzyme is inhibited by Ca^{2+} ions [303]. Cloning of the gene will have to reveal to which class *Dictyostelium* guanylyl cyclase belongs.

G-proteins

G-proteins are a superfamily of proteins that bind and hydrolyze GTP. They include proteins involved in protein synthesis, such as the elongation factor Tu (EF-Tu), and two classes of proteins that transduce signals; large heterotrimeric G-proteins and small proteins such as the proto-oncogene Ras [reviewed in 28]. Heterotrimeric G-proteins consist of α, β and γ subunits. Activated by binding of GTP, the protein dissociates; the β and γ subunits form a tightly associated complex, free from the α GTP-bound subunit. The crystal structures of proteins representative of each G-protein family have been reported [16, 210, 218]. Although bacterial EF-Tu, Ras (or p21, the products of the *ras* oncogenes) and $G_{t\alpha}$ (the α-subunit of the hetrotrimeric G-protein transducin) share less than 20% amino acid sequence identity, their GTPase domains demonstrate the same overall structure. In the GTPase core, five sequence regions conserved in all GTPases are found in loops that bind GTP and GDP. G-proteins function as molecular switches which cycle between the GTP-bound active form and the GDP-bound inactive form. The interconversion between the active and inactive state occurs by GTP hydrolysis, and from the inactive to the active state by nucleotide exchange. The interconversion of Ras from the GDP to the GTP bound state and vice versa is regulated by accessory proteins [reviewed in 24]. Guanine-nucleotide exchange factors (GEFs) mediate the replacement of GDP with GTP. Through the association of a GEF with the G-protein, GDP dissociates from the complex at an increased rate. Subsequently, the 'empty' Ras/GEF complex binds GTP, leading to the dissociation of GEF and activation of the Ras G-protein. GTPase-activating proteins (GAPs) accelerate the G-protein's intrinsic GTPase activity. In heterotrimeric G-proteins the situation is different: G-protein α subunits contain both intrinsic GEF and GAP activities. The 3-dimensional structure of $G_{t\alpha}$ predicts a unique α-helical domain, inserted in the GTPase core domain [210]. Together these domains surround the guanine nucleotide. The α-helical domain in the GTPase core has been proposed to be involved in GDP release as well as acting as a built-in GAP for G_{α}. In heterotrimeric G-proteins the bound GDP is inaccessible, unlike the situation in other G-proteins which bind the nucleotide in a partially exposed surface cleft. Activation of an appropriate G-protein-linked receptor is proposed to stimulate opening of the closed structure of G_{α} to allow exchange of GDP for GTP.

Heterotrimeric G-proteins

Binding of GTP to a heterotrimeric G-protein results in dissociation and the formation of an active α subunit and/or active $\beta\gamma$ subunit complex. The regions of the Gα-protein that interact with $\beta\gamma$ subunits, receptors and effectors have been reviewed recently by Conklin and Bourne [55]. A single receptor can activate multiple G-protein molecules thereby amplifying the external signal. Some α subunits possess specific residues that can be covalently modifided by bacterial toxins through ADP-ribosylation of the α subunit. Modification by pertussis toxin uncouples receptors from G-proteins thereby inhibiting signalling, whereas cholera toxin constitutively activates the G-protein by inhibiting GTPase activity [310, 305].

Amino acid sequence comparisons of the α subunits present in mammalian cells reveal four subfamilies after which the heterotrimeric G-protein is named: G_s, G_i, G_q and G_{12} [272, 113]. As G-protein were first identified functionally, names were assigned in accordance with the role they performed; for instance, members of the G_s family stimulated adenylyl cyclase, whereas G_i inhibited this enzyme. Other α subunits were isolated by genetic techniques; cellular functions are not always known. The α subunit of the G-protein involved in odorant signalling, G_{olf}, belongs to the G_s family [130]. It couples the odorant receptors with a distinct form of adenylyl cyclase. Structurally, the G-proteins involved in vertebrate vision, transducins (G_t), belong to the G_i family. Activated transducin binds to one of the two inhibitory γ subunits of a cGMP-dependent phosphodiesterase (PDE). The activated PDE increases the rate of cGMP hydrolysis which leads to closure of cGMP-gated cation channels in the plasma membrane [277]. G_{t1} is activated by rhodopsin in retinal rods, while G_{t2}, expressed in cones, is activated by cone opsins [170]. Gustducin, a member of the G_i family, is expressed specifically in taste buds and is presumably involved in taste sensing [188]. The function of $G_{\alpha o}$ also classified as a G_i protein, is unclear; it could be involved in regulating various ion channels [113]. Members of the G_q family stimulate a specific group of phosphoinositide-specific phospholipase C isoforms [292, 264, 22]. Functions for the G_{12} family have not been identified yet.

Structurally, the G-protein α_s, α_i, α_q and α_o subunits from *Drosophila* show a clear relation to the classes found in mammals [reviewed in 122]. The *concertina* gene is most similar to α subunits of mammalian G_{12}, and is required for normal gastrulation [222]. The functions of the other G-proteins remain to be identified. *Drosophila* α_q is expressed only in photoreceptor cells, where it is proposed to play a role in the activation of phospholipase C in phototransduction. In the nematode *C. elegans* G_α subunits homologous to α_s, a_q and α_o were found, but also three unique Gα proteins [175, 86, R.H.A. Plasterk, personal communication]. A cDNA cloned from *Xenopus laevis* oocytes is 89% identical to the α subunit of rat G_o [213].

Eight G_α genes have been identified in *Dictyostelium* that are expressed at distinct stages of the developmental cycle [234, 101, 314]. Outside of the proposed guanine nucleotide-binding domains, they bear no specific homology to a mammalian subtype. Some are essential for chemotaxis and development. Deletion of $G_\alpha2$, expressed at high levels during aggregation, resulted in cells which could not aggregate and in which all cAMP-receptor-mediated signalling was lost [155, 141]. $G_\alpha4$ is essential for proper development and spore production [102]. Deletion of $G_\alpha1$, expressed during vegetative growth and early development, did not result in a growth or developmental phenotype, suggesting it is not essential under normal conditions. It appears to play a role in cytokinesis, as cells overexpressing $G_\alpha1$ during the growth phase yield very large, multinucleated cells [155].

The budding yeast *S. cerevisiae* contains two G-protein α subunit genes, *GPA1* and *GPA2* [200, 201]. These α subunits have an extra region of 80–100 amino acids inserted near the amino-terminus compared to mammalian α subunits. Disruption of the *GPA1* gene leads to cell cycle arrest in G1 phase [195, 66]. Gpa1 plays a negative role in mating signal transduction, as in the absense of Gpa1 the pheromone response pathway is constitutively activated. Disruption of *GPA2* did not cause any obvious phenotype [201]. In the fission yeast *S. pombe* also two α subunits have been identified. Disruption of *S. pombe gpa1* resulted in viable, but sterile cells, indicating it is required in the developmental pathway for mating and sporulation (Fig. 2) [211]. *S. pombe* Gpa2 is involved in the monitoring of nutrition [123]. Disruption of the *gpa2* gene resulted in smaller cells, which grew slower than the wild type. cAMP levels were only one-third of wild-type level, and *gpa2* null cells did not produce cAMP in response to glucose stimulation, coupling Gpa2 to adenylyl cyclase.

The β and γ subunits of G-proteins form a tightly associated $\beta\gamma$ complex. The amino-terminus of the β subunit has been identified as an essential region by $\beta\gamma$ interaction, presumably through the formation of a structure called an α-helical coiled coil with the γ subunit [94]. The heterotrimer is associated with the plasma membrane, although none of the G-protein subunits contain potential membrane-spanning domains. Lipid modifications in the γ subunits (prenylation) serve to anchor the subunits to the membrane. The $\beta\gamma$ subunits were first merely thought to be inhibitors of G-protein activity and non-specific anchors. However, it is now evident that $\beta\gamma$ subunits are capable of activating effector proteins themselves [21]. In mammals $\beta\gamma$ subunits have been demonstrated to interact with some forms of adenylyl cyclase and phospholipase C, and stimulate opening of K^+ channels [36, 289, 177]. The activation of effector molecules can depend on the combined interaction of α and $\beta\gamma$ subunits, derived from two different G-proteins, as will later be discussed for activation of adenylyl cyclase.

In *Dictyostelium*, a single β subunit is expressed during its entire life-cycle [171]. This suggests the β subunit couples to all of the transiently expressed G-protein α subunits. Deletion of this gene does not influence growth, but inhibits chemotaxis and further development.

In the budding yeast *S. cerevisiae* $\beta\gamma$ subunits are involved in responses to mating factors (Fig. 2). Mutations in the *STE4* or *STE18* genes, encoding the β and γ subunits respectively, suppress the lethality of mutations in the *GPA1* GENE [311]. By $\beta\gamma$ subunits of *S. cerevisiae* participate in the activation of a mitogen-activated protein (MAP) kinase cascade, linking receptor activation to intracellular phosphorylation. In *S. pombe* however, interaction between the Gpa1 α subunit of a heterotrimeric G-protein and a Ras homologue is required for activation of a MAP-kinase cascade (discussed further in Ras signalling).

Ras signalling pathways

The superfamily of Ras-related G-proteins consists of small, monomeric proteins. Based on their structural and functional homology they are divided into four subfamilies [106]. Two subgroups containing Ras- and Rho-like proteins control extracellular signalling pathways. Ras-like proteins play a role in the control of normal and transformed cell growth and differentiation. The Rho-like proteins control signal pathways involved in the organization of the actin cytoskeleton. The Rab- and ARF-like (ADP-ribosylation factor) members of the Ras-related superfamily are involved in intracellular vesicle transport [25, 271, 257]. Recently, ARF was shown to activate phospholipase D [31, 54]. Therefore, assembly of coat proteins and regulation of the phospholipid content of membranes could both the mediated by ARF [131].

Ras genes have been identified in virtually all eukaryotic organisms examined. Post-translational modifications are necessary to localize Ras to the plasma membrane. Ras proteins act as key signal-transducing elements, coupling receptor-activated pathways to a cascade of protein kinases which regulate the activity of nuclear transcription factors. Biochemical studies in animal cells and genetic studies in lower eukaryotes have led to the identification of proteins involved in the activation of Ras by receptor tyrosine kinases. In addition to SH2 domains, many proteins involved in transducing signals contain a related conserved *src* homology domain, SH3 [147]. The number of SH2 and SH3 domains differs and some proteins only contain SH2 or SH3 domains. Adaptor proteins containing one SH2 and two SH3 domains regulate the specificity of protein-protein interactions from the activated receptor to the nucleotide exchange factor. In mammalian cells the adaptor protein Grb2 binds, via its SH3 domains, to a nucleotide exchange factor for Ras, named mSos [179, 43]. After cell stimulation the Grb2/mSos complex binds to phosphorylated receptor tyrosine kinases via the SH2 domain of Grb2. The result is translocation of mSos to the plasma membrane where it is now able to activate Ras (Fig. 3). Similar pathways for Ras activation were identified in *Drosophila* and *C. elegans* [reviewed in 247]. Sem-5 is the *C. elegans* homologue of Grb2 in the Let-23 tyrosine kinase pathway involved in vulval induction [51]. In *Drosophila* the adaptor protein Drk binds to the activated *Sevenless* receptor tyrosine kinases [265, 214]. Mammalian Sos is named after the *Drosophila* Son of sevenless (Sos) guanine nucleotide exchange factor [26].

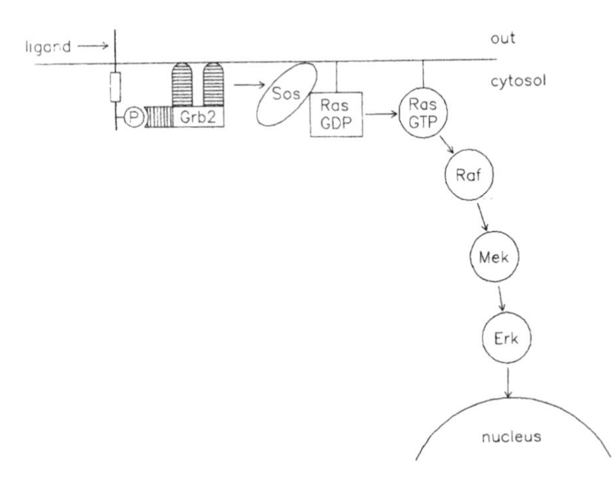

Fig. 3. Activation of Ras by protein tyrosine kinase growth factor receptors. Upon ligand activation, a growth factor receptor autophosphorylates tyrosine residues in its cytoplasmic domain. The adaptor protein Grb2 binds to a specific phosphorylated tyrosine in the receptor via its SH2 domain. The SH3 domains of Grb2 bind to a proline-rich, carboxyl-terminal region in Sos and target the complex to the plasma membrane, where Sos activates Ras by stimulating exchange of GDP for GTP. Ras activation leads to activation of nuclear transcription factors through the Raf, Mek and Erk protein kinases.

Downstream targets of Ras in mammalian cells are the mitogen-activated protein (MAP) kinases, Erk1 and 2 [27, 110]. MAP kinases were shown to be activated in response to growth factors, and are postulated to include nuclear transcription factors such as Fos and Jun as substrates [225, 96]. The MAP kinase cascade consists of MAP kinase kinase kinase (MAPKKK), MAP kinase kinase (MAPKK or MEK) and MAP kinase (Table 1). MAP kinases become activated only when both tyrosine and threonine residues are phosphorylated on a specific TEY sequence in the protein [2, 208]. This phosphorylation is catalyzed in mammals by Mek, a unique kinase with dual specificity for both tyrosine and serine/threonine [259, 56, 255]. In the Ras signalling pathway Mek is activated by phosphorylation on serine and threonine residues by Raf, a mammalian MAPKKK [61, 157]. The direct effector of Ras in this pathway remains to be identified, it is proposed to be Raf (Fig. 3). Besides Ras, protein kinase C was shown to activate Raf (see 'Phospholipid-dependent protein kinases').

A similar protein kinase cascade is activated during pheromone-induced mating in yeast cells (Table 1) [reviewed in 74]. In *S. cerevisiae* $\beta\gamma$ subunits of G-proteins participate in the activation of MAP kinases instead of Ras. The $\beta\gamma$ subunits activate, probably directly, the protein kinase Ste20 [163]. Ste20 mediates activation of a MAPKKK, Ste11, which in turn activates Ste7, the yeast homologue of mammalian Mek1. Activation of the *S. cerevisiae* MAP kinases Fus3 and Kss1 required tyrosine and threonine phosphorylation, which is achieved through Ste7 activity [75]. In the fission yeast *S. pombe* the pathway is activated by the Gpa1 α subunit, in combination

Table 1. Components of the MAP kinase cascade. The MAP kinase cascade is conserved from yeasts to vertebrates (see text for abbreviations).

	Mammals	*S. cerevisiae*	*S. pombe*
Activator	Ras/PKC	$G\beta\gamma$ + Ste20	$G\alpha$ + Ras
MAPKKK	Raf	Ste11	Byr2
MAPKK	Mek	Ste7	Byr1
MAPK	Erk1, 2	Fus3, Kss1	Spk1

with a Ras homologue, instead of $\beta\gamma$ subunits [211]. The protein kinases Byr2, Byr1 and Spk1 of *S. pombe* are structurally related to Ste11, Ste7 and Fus3/Kss1, respectively (Table 1). A mammalian homologue of Ste11/Byr2 called Mekk has been identified, which is able to catalyse the phosphorylation of Mek [160]. Therefore it is possible that heterotrimeric G-proteins also stimulate the MAP kinase pathway in mammals.

Dictyostelium contains at least two *Ras* genes encoding proteins homologous to mammalian Ras, expressed at different stages of development [237, 240]. *DdrasG* is expressed during growth and early development, *Ddras* during multicellular development. *Dictyostelium* cells overexpressing a mutant Ddras protein, Ddras-Thr^{12}, show an abnormal development, forming multiple tips during culmination [238]. The phospholipid turnover in these cell is increased which is due to increased phosphatidylinositol kinase activity, leading to increased levels of inostiol 1,4,5-trisphosphate in a compartment with a high metabolic turnover [304].

S. cerevisiae contains two closely related *RAS* genes [57, 231]. Although neither are essential genes by themselves, deletion of both genes is lethal, indicating that some Ras function is required for cell growth [133, 290]. The activity of Ras is regulated by yeast proteins encoded by CDC25 [35, 29], a GDP/GTP exchange protein homologous in its carboxyl-terminal domain to Sos, and *IRA1* AND *IRA2*, encoding GTPase-activating proteins [286]. Ras in *S. cerevisiae* was shown to activate adenylyl cyclase [297]. *S. pombe* has a single *ras* gene which encodes a protein similar in size to the mammalian Ras proteins, and does not contain the large inserts found in the C-terminal regions of *S. cerevisiae* Ras proteins [90].

Effectors catalysing the formation of second messengers

Upon stimulation of cell surface receptors, the activity of second messenger-generating enzymes, adenylyl cyclase and PLC, is regulated. Adenylyl cyclase catalyses the formation of cAMP, a ubiquitous intracellular second messenger in many cell types. cAMP plays an important role in signal transduction through its activation of cAMP-dependent protein kinase and regulation of ion channel functions [3]. It is firmly established that many hormones and neurotransmitters stimulate the hydrolysis of PtdIns(4,5)P_2 to form the second messenger Ins(1,4,5)P_3 and diacylglycerol by activating phosphoinositide-specificic phospholipase C enzymes [19]. Diacylglycerol is also generated through hydrolysis of other phospholipids, mainly phosphatidylcholine (PC) [173]. Hormone-induced breakdown of PC can be achieved through activation of a variety of phospholipases, namely PC-specific PLC, phospholipase D (PLD) and phospholipase A_2 (PLA_2) [reviewed in 52]. Activation of cytosolic PLA_2 is involved in generation of arachidonic acid which is further metabolized to eicosanoids; its role in intracellular signalling is still unclear. PC hydrolysis by PLC yields diacylglycerol, while hydrolysis by PLD yields phosphatidic acid (PA) which can be converted to diacylglycerol by PA phosphohydrolase. Activation of phospholipase-induced PC breakdown can be achieved by different mechanisms: by G-proteins, by protein kinase C (PKC), by Ca^{2+}, and by growth factor receptors [reviewed in 77]. No sequences for PLD enzymes have been reported yet. A sequence for cytosolic PLA_2 from mammalian cells reveals a calcium regulatory domain present in some isoforms of PKC (see 'Phospholipid-dependent protein kinases') [50].

Adenylyl cyclase

In mammalian cells six distinct membrane bound adenylyl cyclase enzymes have been identified and their genes cloned [154, 80, 11, 92, 135]. All share considerable sequence homology and based on hydropathy plots a similar topology has been predicted. The proposed structure of the mammalian membrane-bound forms consists of twelve transmembrane-spanning regions, divided in two domains of six spans each (Fig. 4A; [154]). It

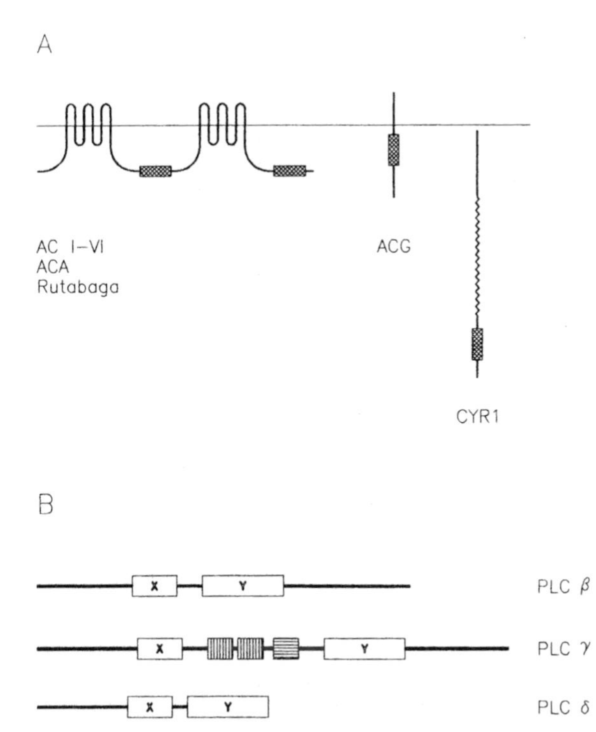

Fig. 4. Effector enzymes adenylyl cyclase and phospholipase C. Conserved regions: *hatched box*, adenylyl cyclase catalytic domain; *zig zag line*, repeated domain similar to domain in human glycoproteins; *X* and *Y box*, proposed catalytic domains in PLC; *vertically striped box*, SH2 domain; *horizontally striped box*, SH3 domain. A. Mammalian adenylyl cyclase types I to VI (AC I–VI), *Dictyostelium* adenylyl cyclase expressed during aggregation (ACA) and *Drosophila* Rutabaga contain two domains with six hydrophobic stretches each, which are proposed to traverse the plasma membrane. *Dictyostelium* adenylyl cyclase expressed during germination (ACG) contains a single membrane-spanning domain. Yeast adenylyl cyclase (CYR1) does not contain potential membrane-spanning domains. B. Phosphoinositide-specific phospholipase C enzymes PLC β, γ and δ contain two conserved domains, X and Y. In PLC γ regulatory SH domains are located between domains X and Y.

resembles the structure of various plasma membrane channels and transporters, although no evidence for channel activity has been found in adenylyl cyclase. The catalytic core consists of two large cytoplasmic domains; one located between the two membrane-spanning clusters, the other at the carboxyl end of the enzym. These domains within the protein sequence are similar to one another, and to the catalytic domains of guanylyl cyclases. The mammalian adenylyl cyclase subtypes are all activated by $G_s\alpha$ subunits, but differ in their response to regulation by Ca^{2+}/calmodulin and $\beta\gamma$ subunits of heterotrimeric G proteins [reviewed in 48]. Type I and III adenylyl cyclase are stimulated by Ca^{2+}/calmodulin. Type I can be stimulated directly by Ca^{2+} and calmodulin, or indirectly by stimulation of muscarinic receptors that mediate mobilization of intracellular Ca^{2+}. Type III, the adenylyl cyclase found in olfactory neurons [11], requires activation of the enzyme by G-protein coupled receptors before stimulation by the Ca^{2+}/calmodulin complex can occur [47]. The α_s and α_i subunits of G proteins were initially classified as stimulating or inhibiting adenylyl cyclase, respectively. Now it is clear that $\beta\gamma$ subunits can differentially regulate adenylyl cyclase activity [289]. In the presence of activated $G_{s\alpha}$, addition of $\beta\gamma$ subunits inhibited type I adenylyl cyclase activity, enhanced stimulation of type II and IV adenylyl cyclase activity, but did not affect the activity other types [288]. *In vivo* the $\beta\gamma$ subunits of G_i proteins were proposed to mediate these effects [79]. Thus, this appears to represent a mechanism for cross-talk between signalling pathways.

The *Drosophila* adenylyl cyclase *rutabaga* gene is most similar to the mammalian type I enzyme in sequence and Ca^{2+}/calmodulin responsiveness [169]. In flies this adenylyl cyclase is involved in learning and memory processes; it is expressed in the mushroom bodies of the fly brain. Expression of *rutabaga* in mammalian cells shows its activity is regulated by endogenous G proteins and calmodulin [169].

In *Dictyostelium* the adenylyl cyclase expressed during aggregation, ACA, is responsible for receptor and G protein-regulated adenylyl cyclase activity [229]. The proposed structure resembles that of mammalian membrane-bound adenylyl cyclases. Cells with a disrupted *aca* gene fail to aggregate, but are not affected in growth or chemotaxis. The proposed topology of a second adenylyl cyclase, ACG, expressed during germination, consists of a large extracellular domain connected to a single transmembrane-spanning domain and one cytoplasmic catalytic domain at the carboxyl terminus (see Fig. 4A). Although this structure resembles that of membrane-bound guanylyl cyclases, ACG was shown to contain adenylyl and not guanylyl cyclase activity [229]. ACG expressed in cells with a disrupted *aca* gene secrete cAMP constitutively and cAMP production is not regulated by surface receptors.

S. cerevisiae adenylyl cyclase, encoded by the *CYR1* gene, is required for cell growth [186]. It resembles the above discussed adenylyl cyclases only in its carboxyl terminal catalytic domain, and lacks potential transmembrane-spanning domains (see Fig. 4A; [134]). CYR1 contains multiple copies of a 23 amino acid repeating unit similar to a repeat found in human glycoproteins. Budding yeast adenylyl cyclase is not regulated by heterotrimeric G proteins, but stimulated by Ras proteins. *S. pombe* adenylyl cyclase is similar to *CYR1* of the budding yeast [317]. Its structure predicts the 23-amino acid repeats and a carboxyl-terminal adenylyl cyclase domain, but *S. pombe* adenylyl cyclase misses an amino-terminal segment present in *S. cerevisiae* adenylyl cyclase. The Gpa2 α subunit of a heterotrimeric G-protein regulates cAMP production in *S. pombe* as cells defective Gpa2 fail to produce cAMP in response to glucose stimulation [123].

Phosphoinositide-specific phospholipase C

The enzyme phospholipase C (PLC) generates two second messengers upon phosphatidylinositol 4,5-bisphosphate ($PtdInsP_2$) hydrolysis: membrane-bound diacylglycerol and water-soluble inositol 1,4,5-trisphosphate ($InsP_3$). Through its activation of PKC, diacylglycerol is involved in regulating protein phosphorylation, while $InsP_3$ regulates intracellular Ca^{2+} [209, 273, 19]. Calcium, contained within intracellular stores, is released to the cytosol when $InsP_3$ binds to its receptor (see below).

The mammalian phospholipase C enzyme family specific for hydrolysing polyphosphatidylinositols is divided into three classes based upon sequence conservation, each class containing isoforms [see references in 53, 239]. All groups contain two conserved amino acid domains, named X and Y, which are thought to form the catalytic site (Fig. 4B). PLC-β isoforms contain a large carboxyl-terminal domain after the conserved Y region which is involved in PLC activation by G-protein α subunits [219, 315]. PLC-γ isoforms contain the *src* homology domains SH2 and SH3 located in between the X and Y domains [279, 270]. This region is not essential for $PtdInsP_2$ hydrolysing activity [73], but the SH2 domains target PLC to tyrosine phosphorylated sequences in growth factor receptors, as described previously. The SH3 domain is involved in targeting the enzyme to cytoskeletal components [13]. PLC-δ does not contain a large carboxyl-terminal sequence, nor the SH domains. The amino-terminal domain of PLC-δ was shown to form part of a $PtdInsP_2$-binding site proposed to bind the enzyme to the membrane surface during hydrolysis [236, 49]. On the protein level the existence of a PLC-α isoform has been implicated. However, a sequence claimed to encode PLC-α [15], with no sequence similarity to known PLC isoforms, was in fact shown to be protein disulphide isomerase [269]. The PLC-α proteins, with a low molecular weight, could be proteolytic fragments derived from other PLCs.

The activity of the three PLC classes are regulated differently, although all depend on the presence of Ca^{2+}. The PLC-γ family is activated by tyrosine phosphorylation, as discussed above. In a reconstitution system with purified proteins, m1 muscarinic receptor and G-proteins of the G_q family were sufficient to stimulate PLC-β_1 [17]. Furthermore, with this system it was also shown that addition of PLC-β_1 stimulated hydrolysis of G_q-bound GTP. Thus, this PLC isoform serves as a GAP for the G-protein that mediates its activation [18]. Another α subunit of the G_q family, $G_{\alpha 16}$, was the most efficient in activating PLC-β_2 [164]. G-protein $\beta\gamma$ subunits increased PLC-β_1, -β_2 and -β_3 activity two-, four- and eight-fold, respectively [39, 220]. In PLC-β_2 regions for $\beta\gamma$ interaction were shown to be located in the amino-terminal region of the PLC protein, and are thus separate from Gα-activating regions [316]. PLC-δ_1 activity was increased two-fold after stimulation with $\beta\gamma$ subunits [220]. In permeabilized HL60 cells, G-protein-stimulated PLC activity depended on the presence of a cytosolic compound identified as phosphatidylinositol transfer protein, involved in transporting the lipid [293].

In Swiss 3T3 cells, diacylglycerol levels in the nucleus increased and $PtdInsP_2$ levels decreased upon stimulation with insulin-like growth factor 1 [67]. PLC activity in the nucleus has been reported, as have other enzymes involved in the inositide cycle [68, 69, 185]. However, immunologically only the PLC-β, isoform could be detected in the nucleus. Therefore, the regulation of PLC in the nucleus remains unclear, as PLC-β has been reported to be regulated by G-proteins, not by growth factor receptors.

Activation of PLC activity has proven to be not essential for induction of DNA synthesis in mammalian cell lines [114, 196]. The role of PLC in several invertebrates and microorganisms has well been studied. In *Drosophila* PLC is implicated to be involved in phototransduction [23, 249]. No receptor potential A (*norpA*) mutants fail to respond to light stimulation. These mutants contain no PLC activity in the head region. The *norpA* gene encodes a protein with similarity to bovine retinal PLC-β [82]. A second PLC in *Drosophila*, also of the β type, is expressed in the central nervous system [262]. *Xenopus* contains a PLC protein with 64% identity to the mammalian PLC-β_3 subtype [181]. Injection of oocytes with PLC antisense oligonucleotides significantly reduced receptor stimulated Cl^- currents. The microorganisms *S. cerevisiae* and *Dictyostelium* contain a PLC-δ like sequence [318, 71]. Deletion of PLC in yeast resulted in viable cells which however were retarded in cell growth when growing conditions were not optimal [88, 224]. In *Dictyostelium* deletion of the DdPLC gene resulted in cells containing no measurable PLC activity [72]. Surprisingly, the cells were able to grow and develop. Analysis of the $InsP_3$ concentration revealed that levels in cells with a deleted *plc* gene were only slightly lower (20%) than in wild-type cells. As there are no indications for other PLCs, alternative pathways for synthesizing $InsP_3$ have to be considered.

Proteins activated by second messengers

Cytosolic protein kinases are divided into different families, based on their primary structure and functional studies. We will discuss two types regulated by second messengers: protein kinase C and cyclic nucleotide-dependent protein kinases. The number of identified genes belonging to the protein kinase family has risen dramatically during the past years [120]. Sequence alignments have identified a conserved catalytic core of about 260 residues shared by all protein kinases [107]. For many, only the deduced protein sequences have been reported, with no known protein function yet. The structure of the catalytic domain is derived from the crystal structure of cAMP-dependent protein kinase [145]. It consists of a bilobed structure with an ATP-binding site on the smaller lobe, and a substrate-binding site on the larger lobe. Transfer of γ-phosphate of ATP from the protein kinase to a serine, threonine or tyrosine residue of the substrate is considered to alter the conformation, and thus the activity, of the substrate protein.

Several ion channels are activated by second messengers. The cGMP-gated cation channel of rod photoreceptor cells and the cAMP-gated channel from olfactory neurons contain a carboxyl-terminal cyclic-nucleotide-binding region, similar to the tandem repeat in cGMP-dependent protein kinase and cAMP-dependent protein kinase, respectively [136, 65]. The ryanodine receptor and the $InsP_3$ receptor are intracellular calcium channels regulated by Ca^{2+} and $InsP_3$. The $InsP_3$ receptor is conserved as genes encoding similar proteins have been found in mammals, *Xenopus* and *Drosophila*.

Cyclic nucleotide-dependent protein kinases

The intracellular second messengers cGMP and cAMP regulate physiological functions by activation of specific family of serine/threonine protein kinases, the cyclic nucleotide-dependent protein kinases. The cAMP-dependent protein kinase (PKA) plays a role in the expression of a large number of genes and has been shown to regulate the function of ion channels and the activity of metabolic enzymes. The cGMP-dependent protein kinase (PKG) is involved in platelet aggregation and the relaxation of hormonally contracted smooth muscle. Through phosphorylation of calcium channels, PKG regulates cytosolic calcium levels, although the mechanism by which PKG induces a reduction in the calcium concentration is still unresolved [189]. Although they bind different cyclic nucleotides, PKA and PKG show considerable similarity in amino acid sequence, especially in their catalytic domains (70% homology), and their cyclic nucleotide-binding domains. Each of the cyclic nucleotide-binding domains of PKA and PKG shows similarity in amino acid sequence (20%) to the cAMP-binding domain of the catabolite gene activator protein (CAP) from *E. coli* [187].

In mammalian cells, inactive PKA is a tetramer, consisting of two regulatory (R) subunits and two catalytic (C) subunits (R_2C_2). In the absence of cAMP, the R subunit functions to inhibit the activity of the C subunit. The C subunit is proposed to recognize the R subunit in a manner comparable to the recognition of protein substrates, each R subunit binding to one C subunit at its catalytic site. Upon cAMP binding, the holoenzyme dissociates into R_2 dimers and active monomeric C subunits. Three different mammalian genes for the C subunit of PKA have been identified, C_α, C_β and C_γ, which are 80% similar to each other [14]. The conserved protein kinase catalytic domain comprises most of the PKA C subunit, from amino acid residue 40 to 300 (see Fig. 5A; total length ca. 350 amino acids). Two different types of mammalian R subunits, R_I and R_{II}, have been found [296, 284]. The R subunit is composed of an amino-terminal dimerization domain, a pseudosubstrate domain which is able to bind the protein kinase inhibitor PKI [146] and is involved in interaction with the C subunit, and at the carboxy terminus two tandem cAMP-binding domains (Fig. 5A).

Three *S. cerevisiae* catalytic subunit genes with predicted structural and amino acid sequence similarity to mammalian PKA catalytic subunits have been identified, *TPK1*, *TPK2* and *TPK3* [299]. Gene disruption experiments have demonstrated that the proteins have overlapping functions, as the presence of at least one C subunit is sufficient for normal growth. Haploid spores lacking all three genes are able to germinate, but grow extremely slowly [299]. *S. cerevisiae* contains one gene for a PKA regulatory subunit encoded by the *BCY1* gene homologous to mammalian isoforms [298, 37]. *BCY1* expressed in *E. coli* was shown to bind 2 mol of cAMP per mole of R monomer [128].

In contrast to mammalian and yeast PKAs, the *Dictyostelium* holoenzyme is a dimer (RC) [58]. The R subunit of *dictyostelium* PKA does not contain the amino-terminal domain suggested to be required for dimerization (see Fig. 5A; [199]). The pseudosubstrate domain is conserved, and *Dictyostelium* R subunits can form holoenzymes with mammalian C subunits [167]. Two potential cAMP-binding sites were found in the R subunit sequence, which was surprising since binding experiments indicated the presence of only one cAMP-binding site [64, 59]. The carboxyl-terminal cAMP-binding domain is predicted to bind cAMP with low affinity [199]. *Dictyostelium* contains one catalytic subunit gene *pkaC* [34, 183] which encodes a protein that is twice as large as mammalian PKA C subunits [6]. Cells in which the *pkaC* gene has been disrupted do not aggregate, while cells overexpressing PKA C are accelerated in their development [183, 5].

Comparison of the predicted structures for PKA and PKG enzymes suggest that early in eukaryotic evolution the fusion of separate genes for a regulatory and a catalytic domain resulted in the formation of a chimaeric PKG gene [285]. Purified bovine PKG was shown to exist as a homodimer, consisting of two identical subunits [172, 313]. Amino acid sequence comparison of PKG genes from mammalian cells and *Drosophila* predict a polypeptide containing an N-terminal dimerization domain, two cGMP-binding domains and the catalytic domain (see Fig. 5A; [132, 116]). Dimerization probably occurs via a hydrophobic leucine/isoleucine-zipper motif, which is not found in the dimerization domain of

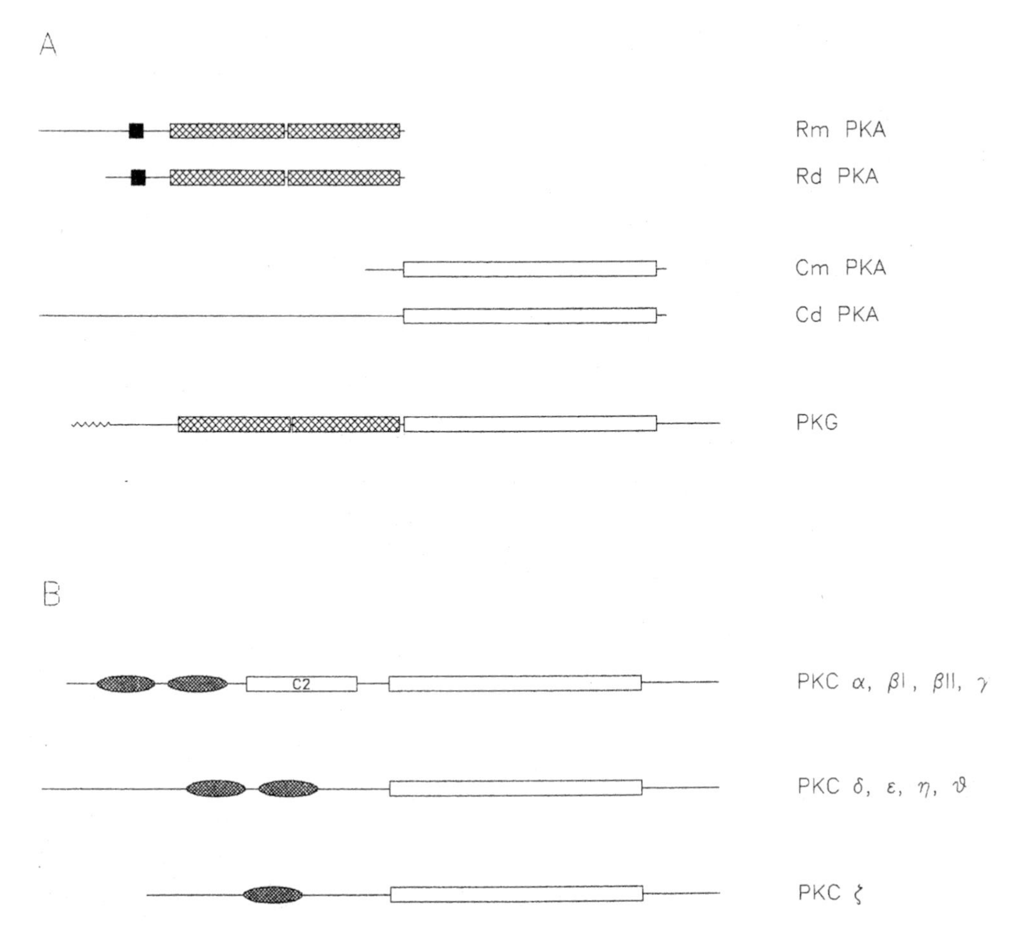

Fig. 5. Comparison of structures of cyclic nucleotide activated protein kinases and protein kinase C. Conserved regions: *open box*, protein kinase catalytic domain; *closed box*, pseudosubstrate domain; *hatched box*, cyclic nucleotide binding domain; *zig zag line*, leucine/ isoleucine-zipper motif; *filled oval*, zink-finger-like domain; *C2*, Ca^{2+}-activated domain. A. The regulatory subunit of *Dictyostelium* (Rd) cAMP-dependent protein kinase (PKA) misses an amino-terminal dimerization domain present in mammalian R (Rm) subunits. Compared to the mammalian catalytic (Cm) subunit of PKA, the catalytic subunit of *Dictyostelium* (Cd) contains an amino-terminal extension. The regulatory subunits of PKA contain a pseudosubstrate domain for binding to the C subunit. In cGMP-dependent protein kinase (PKG) the cyclic nucleotide-binding and catalytic domains are located in the same molecule. B. Protein kinase C (PKC) isoforms α, βI, βII and γ contain a calcium regulatory sequence, missing in the other isoforms. PKC ζ has only one zinc-finger-like domain.

mammalian PKAs [10]. Mammalian cells contain two distinct classes of PKG genes that encode different PKG proteins, showing homology in the catalytic and cGMP-binding domains, but no homology in the amino terminus [301].

PKG enzymes purified from the lower eukaryotes *Paramecium*, *Tetrahymena* and *Dictyostelium* are found in the monomeric form [reviewed in 116]. No sequences for these PKG enzymes have been reported yet. The presence of monomeric cGMP and cAMP kinases in some unicellular organisms suggests that dimerization of the kinases occurred later in evolution.

Phospholipid-dependent protein kinases

Protein kinase C (PKC) was originally characterized as a phospholipid-diacylglycerol-Ca^{2+}-dependent protein kinase [reviewed in 209]. Its assumed role in tumorigenesis arises through the fact that PKC proteins serve as receptors for phorbol ester tumour promoters. Tumour-promoting phorbol esters such as 12-*O*-tetradecanoylphorbol-13-acetate (TPA) directly activate PKC *in vitro* and are used as analogues of the physiological PKC activator, diacylglycerol. Diacylglycerol produced by hydrolysis of phospholipids by phosphoinositide-specific PLC is involved in transient activation of PKC [173]. However, sustained activation of PKC is necessary for cell proliferation and differentiation [8]. Diacylglycerol produced from phosphatidylcholine breakdown mediated by other signal-activated phospholipases, notably phospholipase D, is proposed to be involved in the long-term activation of PKC [77].

Molecular cloning and biochemical analysis have revealed the presence of twelve PKC isoforms in mammalian tissues at present [see references in 209, 9, 60]. All require the presence of phospholipid for activation, but differ in their need for calcium and diacylglycerol. The conventional isoforms, PKC-α, -βI, -βII and -γ, are activated by Ca^{2+} and DAG. The amino-terminal half of PKC proteins contain the regulatory domain consisting of a tandem repeat of a zinc-finger-like sequence and, in the contential PKC isoforms, a second conserved domain (C2) (see Fig. 5B). The regulatory domain also contains a strech of basic amino acids which appears to be an inhibitory pseudosubstrate domain [17]. The carboxyl-terminal region is the catalytic domain, consisting of the ATP-binding site and sequences similar to other protein kinases. The novel class of PKCs, PKC-δ, -ε, -η and -θ, contain the cysteine-rich zinc-finger-like sequences, but not the C2 domain (Fig. 5B). Novel PKC isoforms are activated by DAG and the tumour-promoting phorbol esters, but are insensitive to Ca^{2+}, proposing a role for the C2 domain in Ca^{2+} activation [148, 212]. Regulation of the atypical PKC ζ has not yet been fully established, but this isoform is insensitive to Ca^{2+}, DAG and phorbol esters [175, 9]. The amino-terminal domain of PKC-ζ contains only a single zinc-finger-like sequence [25]. PKC-ζ has been shown to be activated by the phospholipids PS and phosphatidylinositol 3,4,5-trisphosphate (PIP_3) [203]. Three new PKC isoforms have been cloned, but their sequences have not been reported yet.

Phorbol ester binding to PKC appears to be directly involved in down-regulating PKC activity as conventional and novel PKCs were degraded after prolonged treatment with phorbol ester, while PKC-ζ protein levels remained unchanged [308, 174]. The presence of multiple PKC isotypes has led to the suggestion that they perform specific functions within the cell. PKC-γ and -ε expression is restricted to the central nervous tissues, PKC-η to the lung and skin, while other PKCs are expressed in many different cell types.

Phorbol ester stimulation of cells results in the increased transcription of many genes and phosphorylation of a variety of nuclear proteins. PKC-α was shown to directly phosphorylate and activate the Raf kinase [149]. Raf-1 is involved in activating nuclear transcription factors through MEK and MAP kinase, as discussed previously. Thus PKC provides an alternative route besides *ras* to activate oncogene class transcription factors. Upon stimulation, translocation of specific PKC isozymes to the nucleus was demonstrated in different cell lines; translocation was shown to coincide with an increase in nuclear PKC activity [67, 115, 162]. Translocation of PKC to the nucleus is suggested to be a second, direct way for PKC involvement in regulation of nuclear proteins.

In the nematode *Caenorhabditis elegans* strains were isolated which were resistant to the phorbol ester tumour promotor TPA. These strains were mutated in the *tpa-1* gene. The predicted *tpa-1* protein contains the kinase catalytic domain and the two zinc-finger-like structures, but lacks the C2 domain in its amino-terminal region [282]. *Drosophila* contains three PKC genes [241, 245]. Two are transcribed predominantly in brain tissue, one encoding a conventional, the other a novel PKC-like protein. The third PKC gene is expressed specifically in photoreceptor cells [245]. Interestingly, it contains the potential calcium regulatory site and is proposed to function in *Drosophila* phototransduction adaptation.

A PKC-like enzyme activity was demonstrated

to be present in the micro-organism *Dictyostelium* [180]. The myosin heavy-chain kinase of *Dictyostelium* is a member of the PKC family [235]. In the regulatory domain it contains the basic pseudosubstrate domain and the two cystein-rich sequences. In the C2 region it is only 10% similar to the putative Ca^{2+} regulatory domain. A gene, *PKC1*, encoding a PKC enzyme closely related to mammalian conventional PKC has been isolated from *S. cerevisiae*. PKC1 is implicated to play a role in the cell division cycle [169]. Deletion of *PKC1* resulted in recessive lethality. The nonviable spores do germinate but are blocked in further growth.

Inositol 1,4,5-trisphosphate receptors

As mentioned above, $InsP_3$ produced by phospholipase C-mediated hydrolysis of $PtdInsP_2$ is able to release calcium from intracellular stores. $InsP_3$ binds to specific receptors that are coupled to calcium channels located in the endoplasmic reticulum (see Fig. 6; [190, 242]). Reconstitution of purified $InsP_3$-binding protein into liposomes showed that calcium was conducted as a function of ligand binding, suggesting that the same molecule mediates both $InsP_3$ binding and the release of Ca^{2+} [83, 182]. Cloning of the $InsP_3$ receptor from mammalian cerebellum revealed a primary structure containing several putative membrane-spanning domains and a striking homology with another intracellular calcium channel, the ryanodine receptor of the sarcoplasmic reticulum [91, 190, 283].

The $InsP_3$ receptor and the ryanodine receptor

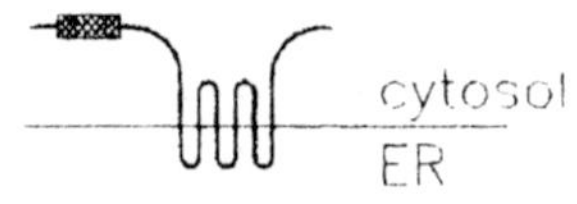

Fig. 6. Structure of the $InsP_3$ receptor. The $InsP_3$ receptor is an intracellular calcium channel responsible for mobilizing stored calcium. Four subunits combined form a functional ion channel. The exact number of membrane-spanning domains remains to be determined; here six are proposed as in a number of other ion channels. The amino-terminal $InsP_3$-binding domain (*hatched box*) and carboxyl terminus are located in the cytosol.

show the greatest stretches of amino acid identity in their carboxyl termini, which contain the putative membrane-spanning domains. Their tertiary structures are predicted to be similar as both receptor types are homotetrameric proteins, in which the carboxyl terminus of each receptor subunit is suggested to form a part of the calcium channel [95]. The $InsP_3$ receptor contains an amino-terminal cytoplasmic domain with an $InsP_3$-binding site (see Fig. 6; [191]). $InsP_3$ receptor sequences isolated from mammalian cells indicate the presence of distinct receptor subtypes [278]. Expression of the amino-terminal region of two different receptor types from rat brain demonstrated similar specificity for $InsP_3$, but different affinities [278]. Binding of $InsP_3$ results in a conformational change of the receptor, which is proposed to result in opening of the channel. The receptor is proposed to traverse the endoplasmic reticulum membrane an even number of times, although the precise number remains to be determined. The region between the ligand-binding domain and the calcium channel is proposed to be involved in regulation of the $InsP_3$ receptor. This region contains putative PKA phosphorylation sites. PKA-induced phosphorylation of the $InsP_3$ receptor did not influence $InsP_3$ binding to the receptor, but decreased $InsP_3$-induced calcium release [281]. Autophosphorylation could also play a role as the receptor was shown to function as a protein kinase, phosphorylating both exogenous substrates and the receptor itself [85]. ATP promotes $InsP_3$-induced calcium release, and ATP-binding sites are predicted in the regulatory domain of $InsP_3$ receptors [84, 182].

In *Drosophila* and *Xenopus* $InsP_3$ receptors homologous to those in mammals were identified. The *Drosophila* $InsP_3$ receptor shares 57% amino acid sequence identity with its mouse homologue [319]. The *Drosophila* $InsP_3$ receptor sequence however does not indicate it is a substrate for PKA. mRNA localization and $InsP_3$ binding to membrane preparations show that the receptor is expressed highly in the leg and thorax region and, interestingly, in the retina and antenna. As $InsP_3$ is thought to act as a second messenger in *Drosophila* vision, the $InsP_3$ receptor described here is proposed to play a role in depolarization of photoreceptor cells through $InsP_3$-induced calcium release [319]. The *Xenopus* $InsP_3$ receptor shows 90% amino acid sequence identity with its mouse homologue. It contains ATP-binding sites and PKA phosphorylation sites in between the ligand-binding amino terminus and calcium channel domain in the carboxyl terminus [156]. *Xenopus* egg activation, measured by $InsP_3$-responsive cortical contraction, was inhibited in eggs microinjected with $InsP_3$ receptor antisense oligonucleotides. Immunocytochemical localization experiments suggest a role for the $InsP_3$ receptor in the formation and propagation of Ca^{2+} waves during fertilization [156].

The $InsP_3$-induced calcium release measured in some studies indicates co-operative binding of $InsP_3$ to each of the four binding sites of the tetrameric receptor, while other studies indicate that binding of one $InsP_3$ molecule is sufficient to open the channel [reviewed in 291]. Calcium is not released gradually in response to increased $InsP_3$ concentration, but a fixed proportion is released (quantal release), with the remainder becoming accessible at higher $InsP_3$ concentrations [261]. Besides $InsP_3$, Ca^{2+} itself also modulates channel opening [124]. Increased concentrations of cytosolic calcium first stimulate its own release, which is then followed by an inhibitory effect [192]. Thus calcium is capable of both positive and negative feedback regulation of the $InsP_3$ receptor, and may be involved in regulation of calcium oscillations and waves in the cell [reviewed in 20].

Discussion and conclusions

Comparisons between higher and lower eukaryotes show that many processes involved in signal transduction have been evolutionary conserved. Pathways for activation of intracellular proteins through heterotrimeric G-proteins and Ras proteins are found in micro-organisms, invertebrates and vertebrates. Stimulation of tyrosine kinase pathways results in phosphorylation of proteins and the formation of active protein complexes; these reactions occur probably in all eukaryotes. Some proteins such as G-proteins, adenylyl cyclase and phospholipase C are remarkably well conserved in diverse organisms, whereas only the overall structure is conserved in other proteins, as is the case for the seven transmembrane segments in G-protein-linked surface receptors.

Although many proteins involved in transducing signals are present in virtually all eukaryotes, their connection in pathways may form specialized systems. Divergent pathways exist, as is the case for photoreceptor activation and activation of protein kinase cascades through Ras and heterotrimeric G-proteins. In mammalian rod cells, rhodopsin activates a specialized G protein, G_t, which transduces the signal to PDE in order to regulate cGMP levels. However, in *Drosophila*, rhodopsin appears to activate PLC. The signal is most likely transduced by G_q, which also mediates activation of PLC in mammalian systems. A second example of divergence is Ras. Ras proteins in mammalian cells are stimulated via receptor tyrosine kinases and activate a MAP kinase cascade. A Ras protein together with the α subunit of a heterotrimeric G-protein is involved in activating the MAP kinase pathway in *S. pombe*. In the yeast *S. cerevisiae* Ras has a completely different role as an activator of adenylyl cyclase; the MAP kinase cascade is activated by G-protein $\beta\gamma$ subunits.

Cellular effects are often the result of the interaction of multiple signal transduction pathways (cross-talk). G-proteins are able to regulate some forms of adenylyl cyclase and phospholipase C through both α and $\beta\gamma$ subunits. These subunits can be derived from different G-proteins, activated through independent surface receptors. Thus the effect of one ligand can be either enhanced or down regulated by the effect of another. Furthermore, PLC can be activated by two independent pathways, one coupled to tyrosine kinase-linked receptors, the other to heterotrimeric G-proteins. On the other hand, the MAP kinase activation pathway serves as a convergence point at which multiple signalling pathways meet. One of the proteins involved, Mek, can be activated by different protein kinases, Raf and

Mek kinase. These in turn are activated by multiple proteins, including Ras, PKC and possibly heterotrimeric G-proteins.

Feedback regulation of proteins plays an important role in controlling signal transduction and signal termination. G-protein-linked receptors demonstrate diminished responsiveness to their ligand by an uncoupling process during which the receptor is phosphorylated. This can be achieved by second-messenger-activated kinases such as PKA and PKC, or by the recently characterized G-protein coupled receptor kinases (GRKs) [126, 165]. β-Adrenergic receptor kinase activity is enhanced by $\beta\gamma$ subunits [104]. Hereby G-protein $\beta\gamma$ subunits negatively regulate the receptors' ability to activate G-proteins. Negative feedback regulation by phosphorylation of receptors by PKA appears to be a major mechanism for desensitization of adenylyl cyclase.

The mechanisms of signal transduction in plants are just beginning to be elucidated. It will be interesting to learn which pathways that are conserved between mammals and eukaryotic micro-organisms, are also present in plants. Knowledge of unique plant signal transduction cascades will be especially informative to discriminate between the general principles of sensory transduction and those used to provide specialized cells.

Acknowledgement

We thank Hubèr Timmermans for preparing the figures.

References

1. Abe K, Kusakabe Y, Tanemura K, Emori Y, Arai S: Primary structure and cell-type specific expression of a gustatory G protein-coupled receptor related to olfactory receptors. J Biol Chem 268: 12033–12039 (1993).
2. Ahn NG, Seger R, Bratlien RL, Diltz CD, Tonks NK, Krebs EG: Multiple components in an epidermal growth factor-stimulated protein kinase cascade – *in vitro* activation of a myelin basic protein/microtubule-associated protein 2 kinase. J Biol Chem 266: 4220–4227 (1991).
3. Alberts B, Bray D, Lewis J, Raff M, Roberts K, Watson JD: Molecular Biology of the Cell, pp. 681–726. Garland Publishing, New York/London (1989).
4. Anderson D, Koch CA, Grey L, Ellis C, Moran MF, Pawson T: Binding of SH2 domains of phospholipase Cγ1, GAP, and Src to activated growth factor receptors. Science 250: 979–982 (1990).
5. Anjard C, Pinaud S, Kay RR, Reymond CD: Overexpression of DdPK2 protein kinase causes rapid development and affects the intracellular cAMP pathway of *Dictyostelium discoideum*. Development 115: 785–790 (1992).
6. Anjard C, Etchebehere L, Pinaud S, Veron M, Reymond CD: An unusual catalytic subunit for the cAMP-dependent protein kinase of *Dictyostelium discoideum*. Biochemistry 32: 9532–9538 (1993).
7. Aroian RV, Koga M, Mendel JE, Ohshima Y, Sternberg PW: The *let-23* gene necessary for *Caenorhabditis elegans* vulval induction encodes a tyrosine kinase of the EGF receptor subfamily. Nature 348: 693–699 (1990).
8. Asaoka Y, Oka M, Yoshida K, Nishizuka Y: Metabolic rate of membrane-permeant diacylglycerol and its relation to human resting T-lymphocyte activation. Proc Natl Acad Sci USA 88: 8681–8685 (1991).
9. Asaoka Y, Nakamura S, Yoshida K, Nishizuka Y: Protein kinase C, calcium and phospholipid degradation. Trends Biochem Sci 17: 414–417 (1992).
10. Atkinson RA, Saudek V, Huggins JP, Pelton JT: ^{1}H NMR and circular dichroism studies of the N-terminal domain of cyclic GMP dependent protein kinase: a leucine/isoleucine zipper. Biochemistry 30: 9387–9395 (1991).
11. Bakalyar HA, Reed RR: Identification of a specialized adenylyl cyclase that may mediate odorant detection. Science 250: 1403–1405 (1990).
12. Barnea G, Grumet M, Sap J, Margolis RU, Schlessinger J: Close similarity between receptor-linked tyrosine phosphatase and rat brain proteoglycan. Cell 76: 205 (1993).
13. Bar-Sagi D, Rotin D, Batzer A, Mandiyan V, Schlessinger J: SH3 domains direct cellular localization of signalling molecules. Cell 74: 83–91 (1993).
14. Beebe SJ, Øyen O, Sandberg M, Frøysa A, Hansson V, Jahnsen T: Molecular cloning of a tissue-specific protein kinase (C_γ) from human testis – representing a third isoform for the catalytic subunit of cAMP-dependent protein kinase. Mol Endocrinol 4: 465–474 (1990).
15. Bennet CF, Balcarek JM, Varrichio A, Crooke ST: Molecular cloning and complete amino-acid sequence of form-I phosphoinositide-specific phospholipase C. Nature 334: 268–270 (1988).
16. Berchtold H, Reshetnikova L, Reiser COA, Schirmer NK, Sprinzi M, Hilgenfeld R: Crystal structure of active elongation factor Tu reveals major domain rearrangements. Nature 365: 126–132 (1993).
17. Bernstein G, Blank JL, Smrcka AV, Higashijima T,

Sternweis PC, Exton JH, Ross EM: Reconstitution of agonist-stimulated phosphatidylinositol 4,5-bisphosphate hydrolysis using purified m1 muscarinic receptor, Gq/11, and phospholipase Cβ1. J Biol Chem 267: 8081–8088 (1992).
18. Bernstein G, Blank JL, Jhon DK, Exton JH, Rhee SG, Ross EM: Phospholipase Cβ1 is a GTPase-activating protein for Gq/11, its physiologic regulator. Cell 70: 411–418 (1992).
19. Berridge MJ, Irvine RF: Inositol phosphates and cell signalling. Nature 341: 197–205 (1989).
20. Berridge MJ: Inositol trisphosphate and calcium signalling. Nature 361: 315–325 (1993).
21. Birnbaumer L: Receptor-to-effector signalling through G proteins: roles for βγ dimers as well as α subunits. Cell 71: 1069–1072 (1992).
22. Blank JL, Ross AH, Exton JH: Purification and characterization of two G-proteins that activate the β1 isozyme of phosphoinositide-specific phospholipase C. J Biol Chem 266: 18206–18216 (1991).
23. Bloomquist BT, Shortridge RD, Schneuwly S, Perdew M, Montell C, Steller H, Rubin G, Pak WL: Isolation of a putative phospholipase C gene of *Drosophila*, *norpA*, and its role in phototransduction. Cell 54: 723–733 (1988).
24. Boguski MS, McCormick F: Proteins regulating Ras and its relatives. Nature 366: 643–654 (1993).
25. Boman AL, Taylor TC, Melançon P, Wilson KL: A role for ADP-ribosylation factor in nuclear vesicle dynamics. Nature 358: 512–51 (1992).
26. Bonfini L, Karlovich CA, Dasgupta C, Banerjee U: The Son of sevenless gene product: a putative activator of Ras. Science 25: 603–606 (1992).
27. Boulton TG, Nye SH, Robbins DJ, Ip NY, Radziejewska E, Morgenbesser SD, DePinho RA, Panayotatos N, Cobb MH, Yancopoulos GD: ERKs: a family of protein-serine/threonine kinases that are activated and tyrosine phosphorylated in response to insulin and NGF. Cell 65: 663–675 (1991).
28. Bourne HR, Sanders DA, McCormick F: The GTPase superfamily: conserved structure and molecular mechanism. Nature 349: 117–127 (1991).
29. Broek D, Toda T, Michaeli T, Levin L, Birchmeier C, Zoller M, Powers S, Wigler M: The *Saccharomyces cerevisiae* CDC25 gene product regulates the RAS/adenylate cyclase pathway. Cell 48: 789–799 (1987).
30. Brown EM, Gamba G, Riccardi D, Lombardi M, Butters R, Kifor O, Sun A, Hediger MA, Lytton J, Hebert SC: Cloning and characterization of an extracellular Ca^{2+}-sensing receptor from bovine parathyroid. Nature 366: 575–580 (1993).
31. Brown HA, Gutowski S, Moomaw CR, Slaughter C, Sternweis PC: ADP-ribosylation factor, a small GTP-dependent regulatory protein, stimulates phospholipase D activity. Cell 75: 1137–1144 (1993).
32. Buck L, Axel R: A novel multigene family may encode odorant receptors: a molecular basis for odor recognition. Cell 65: 175–187 (1991).
33. Burkholder AC, Hartwell LH: The yeast α-factor receptor; structural properties deduced from the sequence of the STE2 gene. Nucl Acids Res 13: 8463–8475 (1985).
34. Bürki E, Anjard C, Scholder JC, Reymond CD: Isolation of two genes encoding putative protein kinases regulated during *Dictyostelium discoideum* development. Gene 102: 57–65 (1991).
35. Camonis JH, Kakeline M, Bernard G, Garreau H, Boy-Marcotte E, Jacquet M: Characterization, cloning and sequence analysis of the CDC25 gene which controls the cyclic AMP level of *Saccharomyces cerevisiae*. EMBO J 5: 375–380 (1986).
36. Camps M, Hou C, Sidiropoulos D, Stock JB, Jakobs KH, Gierschik P: Stimulation of phospholipase C by guanine-nucleotide-binding protein βγ subunits. Eur J Biochem 206: 821–831 (1992).
37. Cannon JF, Tatchell K: Characterization of *Saccharomyces cerevisiae* genes encoding subunits of cyclic AMP-dependent protein kinase. Mol Cell Biol 7: 2653–2663 (1987).
38. Cantley LC, Auger KR, Carpenter C, Duckworth B, Graziani A, Kapeller R, Soltoff S: Oncogenes and signal transduction. Cell 64: 281–302 (1991).
39. Carozzi A, Camps M, Gierschik P, Parker PJ: Activation of phosphatidylinositol lipid-specific phospholipase C β3 by G-protein βγ subunits. FEBS Lett 315: 340–342 (1993).
40. Chang MS, Lowe DG, Lewis M, Hellmiss R, Chen E, Goeddel DV: Differential activation by atrial and brain natriuretic peptides of two different receptor guanylate cyclases. Nature 341: 68–72 (1989).
41. Charbonneau H, Tonks NK, Walsh KA, Fischer EH: The leukocyte common antigen (CD45): a putative receptor-linked protein tyrosine phosphatase. Proc Natl Acad Sci USA 85: 7182–7186 (1988).
42. Charbonneau H, Tonks NK, Kumar S, Diltz CD, Harrylock M, Cool DE, Kregs EG, Fischer EH, Walsh KA: Human placenta protein-tyrosine-phosphatase: amino acid sequence and relationship to a family of receptor-like proteins. Proc Natl Acad Sci USA 86: 5252–5256 (1989).
43. Chardin P, Camonis J, Gale WL, Van Aelst L, Schlessinger J, Wigler MH, Bar-Sagi D: Human Sos1: a guanine nucleotide exchange factor for Ras that binds to GRB2. Science 260: 1338–1343 (1993).
44. Chinkers M, Garbers DL: The protein kinase domain of the ANP receptor is required for signalling. Science 245: 1392–1394 (1989).
45. Chinkers M, Garbers DL, Chang MS, Lowe DG, Chin H, Goeddel DV, Schulz S: A membrane form of guanylate cyclase is an atrial natriuretic peptide receptor. Nature 338: 78–83 (1989).
46. Chinkers M, Garbers DL: Signal transduction by guanylyl cyclases. Annu Rev Biochem 60: 553–575 (1991).

47. Choi EJ, Xia Z, Storm DR: Stimulation of the type III olfactory adenylyl cyclase by calcium and calmodulin. Biochemistry 31: 6492–6498 (1992).
48. Choi EJ, Xia Z, Villacres EC, Storm DR: The regulatory diversity of the mammalian adenylyl cyclases. Curr Opin Cell Biol 5: 269–273 (1993).
49. Cifuentes ME, Honkanen L, Rebecchi MJ: Proteolytic fragments of phosphoinositide-specific phospholipase Cδ1 – catalytic and membrane binding properties. J Biol Chem 268: 11586–11593 (1993).
50. Clark JD, Lin LL, Kriz RW, Ramesha CS, Sultzman LA, Lin AY, Milona N, Knopf JL: A novel arachidonic acid-selective cytosolic PLA_2 contains a Ca^{2+}-dependent translocation domain with homology to PKC and GAP. Cell 65: 1043–1051 (1991).
51. Clark SG, Stern MJ, Horvitz HR: C. elegans cell-signalling gene sem-5 encodes a protein with SH2 and SH3 domains. Nature 356: 340–344 (1992).
52. Cockroft S: G-protein-regulated phospholipases C, D and A_2-mediated signalling in neutrophils. Biochim Biophys Acta 1113: 135–160 (1992).
53. Cockroft S, Thomas GMH: Inositol-lipid-specific phospholipase C isoenzymes and their differential regulation by receptors. Biochem J 288: 1–14 (1992).
54. Cockroft S, Thomas GMH, Fensome A, Geny B, Cunningham E, Gout I, Hiles I, Totty NF, Truong O, Hsuan JJ: Phospholipase D: a downstream effector of ARF in granulocytes. Science 263: 523–526 (1994).
55. Conklin BR, Bourne HR: Structural elements of Galpha subunits that interact with G protein $\beta\gamma$, receptors and effectors. Cell 73: 631–641 (1993).
56. Crews CM, Alessandrini A, Erikson RL: The primary structure of MEK, a protein kinase that phosphorylates the ERK gene product. Science 258: 478–480 (1992).
57. DeFeo-Jones D, Scolnick EM, Koller R, Dhar R: Ras-related gene sequences identified and isolated from *Saccharomyces cerevisiae*. Nature 306: 707–709 (1983).
58. De Gunzburg J, Veron M: A cAMP-dependent protein kinase is present in differentiating *Dictyostelium discoideum* cells. EMBO J 1: 1063–1068 (1982).
59. De Gunzburg J, Part D, Guiso N, Veron M: An unusual adenosine 3′5′-phosphate dependent protein kinase from *Dictyostelium discoideum*. Biochemistry 23: 3805–3812 (1984).
60. Dekker LV, Parker PJ: Protein kinase C – a question of specificity. Trends Biochem Sci 19: 73–77 (1994).
61. Dent P, Haser W, Haystead TAJ, Vincent LA, Robers TM, Sturgill TW: Activation of mitogen-activated protein kinase kinase by v-Raf in NIH 3T3 cells and *in vitro*. Science 257: 1404–1407 (1992).
62. Desai DM, Sap J, Schlessinger J, Weiss A: Ligand-mediated negative regulation of a chimeric transmembrane receptor tyrosine phosphatase. Cell 73: 541–554 (1993).
63. Devreotes P: *Dictyostelium discoideum*: a model system for cell-cell interactions and development. Science 245: 1054–1058 (1989).
64. de Wit RJW, Arents JC, van Driel R: Ligand binding properties of the cytoplasmic cAMP-binding protein of *Dictyostelium discoideum*. FEBS Lett 145: 150–154 (1982).
65. Dhallan RS, Yau K, Schrader KA, Reed RR: Primary structure and functional expression of a cyclic nucleotide-activated channel from olfactory neurons. Nature 347: 184–187 (1990).
66. Dietzel C, Kurjan J: The yeast SCG1 gene: a Gα-like protein implicated in the a and α-factor response pathway. Cell 50: 1001–1010 (1987).
67. Divecha N, Banfić H, Irvine RF: The polyphosphoinositide cycle exists in the nuclei of Swiss 3T3 cells under the control of a receptor (for IGF-1) in the plasma membrane, and stimulation of the cycle increases nuclear diacylglycerol and apparently induces translocation of protein kinase C to the nucleus. EMBO J 10: 3207–3214 (1991).
68. Divecha N, Banfic H, Irvine RF: Inositides and the nucleus and inositides in the nucleus. Cell 74: 405–407 (1993).
69. Divecha N, Rhee SG, Letcher AJ, Irvine RF: Phosphoinositide signalling enzymes in rat liver nuclei: phosphoinositidase C isoform β1 is specifically, but not predominantly, located in the nucleus. Biochem J 289: 617–620 (1993).
70. Dizhoor AM, Ray S, Kumar S, Niemi G, Spencer M, Brolley D, Walsh KA, Philipov PP, Hurley JB, Stryer L: Recoverin: a calcium sensitive activator of retinal rod guanylate cyclase. Science 251: 915–918 (1991).
71. Drayer AL, van Haastert PJM: Molecular cloning and expression of a phosphoinositide-specific phospholipase C of *Dictyostelium discoideum*. J Biol Chem 267: 18387–18392 (1992).
72. Drayer AL, van der Kaay J, Mayr GW, van Haastert PJM: Role of phospholipase C in *Dictyostelium*: formation of inositol 1,4,5-trisphosphate and normal development in cells lacking phospholipase C activity. EMBO J 13: 1601–1609 (1994).
73. Emori Y, Homma Y, Sorimachi H, Kawasaki H, Nakanisho O, Suzuki K, Takenawa TA: A second type of rat phosphoinositide-specific phospholipase C containing a src-related sequence not essential for phosphoinositide-hydrolyzing activity. J Biol Chem 264: 21885–21890 (1989).
74. Errede B, Levin DE: A conserved kinase cascade for MAP kinase activation in yeast. Curr Opin Cell Biol 5: 254–260 (1993).
75. Errede B, Gartner A, Zhou Z, Nasmyth K, Ammerer G: Map kinase-related FUS3 from *Saccharomyces cerevisiae* is activated by Ste7 *in vitro*. Nature 362: 261–267 (1993).
76. Evans RM: The steroid and thyroid hormone receptor superfamily. Science 240: 889–895 (1988).

77. Exton JH: Signalling through phosphatidylcholine breakdown. J Biol Chem 265: 1–4 (1990).
78. Fantl WJ, Escobedo JA, Martin GA, Turck CW, Cel Rosario M, McCormick F, Williams LT: Distinct phosphotyrosines on a growth factor receptor bind to specific molecules that mediate different signalling pathways. Cell 69: 413–423 (1992).
79. Federman AD, Conklin BR, Schrader KA, Reed RR, Bourne HR: Hormonal stimulation of adenylyl cyclase through G_i-protein $\beta\gamma$ subunits. Nature 356: 159–161 (1992).
80. Feinstein PG, Schrader KA, Bakalyar HA, Tang WJ, Krupinski J, Gilman AG, Reed RR: Molecular cloning and characterization of a Ca^{2+}/calmodulin-insensitive adenylyl cyclase from rat brain. Proc Natl Acad Sci USA 88: 10173–10177 (1991).
81. Feng GS, Hui CC, Pawson T: SH2-containing phosphotyrosine phosphatase as a target of protein-tyrosine kinases. Science 259: 1607–1607 (1993).
82. Ferreira PA, Shortridge RD, Pak WL: Distinctive subtypes of bovine phospholipase C that have preferential expression in the retina and high homology to the *norpA* gene product of *Drosophila*. Proc Natl Acad Sci USA 90: 6042–6046 (1993).
83. Ferris CD, Huganir RL, Supattapone S, Snyder SH: Purified inositol 1,4,5-trisphosphate receptor mediates calcium flux in reconstituted lipid vesicles. Nature 342: 87–89 (1989).
84. Ferris CD, Huganir RL, Snyder SH: Calcium flux mediated by purified inositol 1,4,5-trisphosphate receptor in reconstituted lipid vesicles is allosterically regulated by adenine nucleotides. Proc Natl Acad Sci USA 87: 2147–2151 (1990).
85. Ferris CD, Cameron AM, Bredt DS, Huganir RL, Snyder SH: Autophosphorylation of inositol 1,4,5-trisphosphate receptors. J Biol Chem 267: 7036–7041 (1992).
86. Fino Silva I, Plasterk RHA: Characterization of a G-protein α-subunit gene from the nematode *Caenorhabditis elegans*. J Mol Biol 215: 483–487 (1990).
87. Fischer EH, Charbonneau H, Tonks NK: Protein tyrosine phosphatases: a diverse family of intracellular and transmembrane enzymes. Science 253: 401–406 (1991).
88. Flick JS, Thorner J: Genetic and biochemical characterization of a phosphatidylinositol-specific phospholipase C in *Saccharomyces cerevisiae*. Mol Cell Biol 13: 5861–5876 (1993).
89. Freeman RM, Plutzky J, Neel BG: Identification of a human src homology 2-containing protein-tyrosine-phosphatase: a putative homolog of *Drosophila* corkscrew. Proc Natl Acad Sci USA 89: 11239–11243 (1992).
90. Fukui Y, Kaziro Y: Molecular cloning and sequence analysis of a ras gene from Schizosaccharomyces pombe. EMBO J 4: 687–691 (1985).
91. Furuichi T, Yoshikawa S, Niyawaki A, Wada K, Maeda N, Mikoshiba K: Primary structure and functional expression of the inositol 1,4,5-trisphosphate-binding protein P_{400}. Nature 342: 32–38 (1989).
92. Gao B, Gilman AG: Cloning and expression of a widely distributed (type IV) adenylyl cyclase. Proc Natl Acad Sci USA 88: 10178–10182 (1991).
93. Garbers DL: Guanylyl cyclase receptors and their endocrine, paracrine, and autocrine ligands. Cell 71: 1–4 (1992).
94. Garritsen A, van Galen PJM, Simonds WF: The N-terminal coiled-coil domain of β is essential for γ association: a model for G-protein $\beta\gamma$ subunit interaction. Proc Natl Acad Sci USA 90: 7706–7710 (1993).
95. Gill DL: Receptor kinships revealed. Nature 342: 16–18 (1989).
96. Gille H, Sharrocks AD, Shaw PE: Phosphorylation of transcription factor $p62^{TCF}$ by MAP kinase stimulates ternary complex formation at c-fos promotor. Nature 358: 414–417 (1992).
97. Giuili G, Scholl U, Bulle F, Guellaën G: Molecular cloning of the cDNAs coding for the two subunits of soluble guanylyl cyclase from human brain. FEBS Lett 304: 82–88 (1992).
98. Gould KL, Nurse P: Tyrosine phosphorylation of the fission yeast $cdc2^+$ protein kinase regulates entry into mitosis. Nature 342: 39–45 (1989).
99. Greenfield C, Hils I, Waterfield MD, Federwisch W, Wollmer A, Blundell TL, McDonald N: EGF binding induces a conformational change in the external domain of its receptor. EMBO J 8: 4155–4124 (1989).
100. Grenningloh G, Rienitz A, Schmitt B, Methfessel C, Zensen M, Beyreuther K, Gundelfinger ED, Betz H: The strychnine-binding subunit of the glycine receptor shows homology with nicotinic acetylcholine receptors. Nature 328: 215–220 (1987).
101. Hadwiger JA, Wilkie TM, Stratmann M, Firtel RA: Identification of Dictyostelium G_α genes expressed during multicellular development. Proc Natl Acad Sci USA 88: 8213–8217 (1991).
102. Hadwiger JA, Firtel RA: Analysis of $G_\alpha 4$, a G-protein subunit required for multicellular development in *Dictyostelium*. Genes Devel 6: 38–49 (1992).
103. Hafen E, Basler K, Edstroem JE, Rubin GM: Sevenless, a cell-specific homeotic gene of *Drosophila*, encodes a putative transmembrane receptor with a tyrosine kinase domain. Science 236: 55–63 (1987).
104. Haga K, Haga T: Activation by G protein $\beta\gamma$ subunits of agonist- or light-dependent phosphorylation of muscarinic acetylcholine receptors and rhodopsin. J Biol Chem 267: 2222–2227 (1992).
105. Hagen DC, McCaffrey G, Sprague Jr GF: Evidence the yeast STE3 gene encodes a receptor for the peptide pheromone a factor: gene sequence and implications for the structure of the presumed receptor. Proc Natl Acad Sci USA 83: 1418–1422 (1986).

106. Hall A: Ras-related proteins. Curr Opin Cell Biol 5: 265–268 (1993).
107. Hanks SK, Quinn AM, Hunter T: The protein kinase family: conserved features and deduced phylogeny of the catalytic domains. Science 241: 42–51 (1988).
108. Harteneck C, Koesling D, Söling A, Schultz G, Böhme E: Expression of soluble guanylyl cyclase – catalytic activity requires two enzyme subunits. FEBS Lett 272: 221–223 (1990).
109. Harteneck C, Wedel B, Koesling D, Malkewitz J, Böhme E, Schultz G: Molecular cloning and expression of a new α-subunit of soluble guanylyl cyclase – interchangeability of the α-subunits of the enzyme. FEBS Lett 292: 217–222 (1991).
110. Hattori S, Fukuda M, Yamashita T, Nakamura S, Gotoh Y, Nishida E: Activation of mitogen-activated protein kinase and its activator by ras in intact cells and in a cell-free system. J Biol Chem 267: 20346–20351 (1992).
111. Heldin CH, Ernlund A, Rorsman C, Rönnstrand L: Dimerization of the B-type platelet-derived growth factor receptors occurs after ligand binding and is closely associated with receptor kinase activation. J Biol Chem 264: 8905–8912 (1989).
112. Henderson R, Baldwin JM, Ceska TA, Zemlin F, Beckmann E, Downing KH: Model for the structure of bacteriorhodopsin based on high-resolution electron cryo-microscopy. J Mol Biol 213: 899–929 (1990).
113. Hepler JR, Gilman AG: G proteins. Trends Biochem Sci 17: 383–387 (1992).
114. Hill TD, Dean NM, Mordan LJ, Lau AF, Kanemitsu MY, Boynton AL: PDGF-induced activation of phospholipase C is not required for induction of DNA synthesis. Science 248: 1660–1663 (1990).
115. Hocevar BA, Fields AP: Selective translocation of βII-protein kinase C to the nucleus of human promyelocytic (HL60) leukemia cells. J Biol Chem 266: 28–33 (1991).
116. Hofmann F, Dostmann W, Keilbach A, Landgraf W, Ruth P: Structure and physiological role of cGMP-dependent protein kinase. Biochim Biophys Acta 1135: 51–60 (1992).
117. House C, Kemp B: Protein kinase C contains a pseudosubstrate prototope in its regulatory domain. Science 238: 1726–1728 (1987).
118. Howard PK, Sefton BM, Firtel RA: Analysis of a spatially regulated phosphotyrosine phosphatase identifies tyrosine phosphorylation as a key regulatory pathway in *Dictyostelium*. Cell 71: 637–647 (1992).
119. Hughes DA, Yamamoto M: Ras and signal transduction during sexual differentiation in the fission yeast *Schizosaccharomyces pombe*. In: Kurjan J, Taylor BL (eds) Signal Transduction: Prokaryotic and Simple Eukaryotic Systems, pp. 123–146. Academic Press, New York (1993).
120. Hunter T: A thousand and one protein kinases. Cell 50: 823–829 (1987).
121. Hurley JB, Dizhoor AM, Ray S, Stryer L: Recoverin's role: conclusion withdrawn. Science 260: 740 (1993).
122. Hurley JB: G proteins of *Drosophila melanogaster*. In: Kurjan J, Taylor BL (eds) Signal Transduction: Prokaryotic and Simple Eukaryotic Systems, pp. 377–389. Academic Press, New York (1993).
123. Isshiki T, Mochizuki N, Maeda T, Yamamoto M: Characterization of a fission yeast gene, gpa2, that encodes a Gα subunit involved in the monitoring of nutrition. Genes Devel 6: 2455–2462 (1992).
124. Iino M, Endo M: Calcium-dependent immediate feedback control of inositol 1,4,5-trisphosphate-induced Ca^{2+} release. Nature 360: 76–78 (1992).
125. Inagami T: Atrial natriuretic factor. J Biol Chem 264: 3043–3046 (1989).
126. Inglese J, Freedman NJ, Koch WJ, Lefkowitz RJ: Structure and mechanism of the G protein-coupled receptor kinases. J Biol Chem 268: 23735–23738 (1993).
127. Janssens PMW, de Jong CCC, Vink AA, van Haastert PJM: Regulatory properties of magnesium-dependent guanylate cyclase in *Dictyostelium discoideum* membranes. J Biol Chem 264: 4329–4335 (1989).
128. Johnson KE, Cameron S, Toda T, Wigler M, Zoller MJ: Expression in Escherichia coli of BCY1, the regulatory subunit of cyclic AMP-dependent protein kinase from *Saccharomyces cerevisiae* – purification and characterization. J Biol Chem 262: 8636–8642 (1987).
129. Johnson RL, Saxe CL, Gollop R, Kimmel AR, Devreotes PN: Identification and targeted gene disruption of cAR3, a cAMP receptor subtype expressed during multicellular stages of *Dictyostelium* development. Genes Devel 7: 273–282 (1992).
130. Jones DT, Reed RR: G_{olf}: an olfactory neuron specific-G protein involved in odorant signal transduction. Science 244: 790–795 (1989).
131. Kahn RA, Yucel JK, Malhotra V: ARF signalling: a potential role for phospholipase D in membrane traffic. Cell 75: 1045–1048 (1993).
132. Kalderon D, Rubin GM: cGMP-dependent protein kinase genes in *Drosophila*. J Biol Chem 264: 10738–10748 (1989).
133. Katoaka T, Powers S, McGill C, Fasano O, Strathern J, Broach J, Wigler M: Genetic analysis of yeast RAS1 and RAS2 genes. Cell 37: 437–445 (1984).
134. Kataoka T, Broek D, Wigler M: DNA sequence and characterization of the *S. cerevisiae* gene encoding adenylate cyclase. Cell 43: 493–505 (1985).
135. Katsushika S, Chen L, Kawabe J, Nilakantan R, Halnon NJ, Homcy CJ, Ishikawa Y: Cloning and characterization of a sixth adenylyl cyclase isoform: types V and VI constitute a subgroup within the mammalian adenylyl cyclase family. Proc Natl Acad Sci USA 89: 8744–8778 (1992).
136. Kaupp UB, Niidome T, Tanabe T, Terada S, Bonigk W, *et al.*: Primary structure and functional expression from

complementary DNA of the rod photoreceptor cyclic GMP-gated channel. Nature 342: 762–766 (1989).
137. Kawamura S: Rhodopsin phosphorylation as a mechanism of cyclic GMP phosphodiesterase regulation by S-modulin. Nature 362: 855–857 (1993).
138. Kazlauskas A, Cooper JA: Autophosphorylation of the PDGF receptor in the kinase insert region regulates interactions with cell proteins. Cell 58: 1121–1132 (1989).
139. Kazlauskas A, Kashishian A, Cooper JA, Valius M: GTPase-activating protein and phosphatidylinositol 3-kinase bind to distinct regions of the platelet-derived growth factor receptor β subunit. Mol Cell Biol 12: 2534–2544 (1992).
140. Kazlauskas A, Feng GS, Pawson T, Valius M: The 64 kD protein that associates with the PDGF receptor subunit via tyrosine 1009 is the SH2 containing phosphotyrosine phosphatase Syp. Proc Natl Acad Sci USA 90: 6939–6942 (1993).
141. Kesbeke F, Snaar-Jagalska BE, van Haastert PJM: Signal transduction in Dictyostelium fgdA mutants with a defective interaction between surface cAMP receptor and a GTP-binding regulatory protein. J Cell Biol 197: 521–528 (1988).
142. Kitamura K, Shimoda C: The *Schizosaccharomyces pombe mam2* gene encodes a putative pheromone receptor which has a significant homology with the *Saccharomyces cerevisiae* Ste2 protein. EMBO J 10: 3743–3751 (1991).
143. Klein PS, Sun TJ, Saxe CL, Kimmel AR, Johnson RJ, Devreotes PN: A chemoattractant receptor controls development in *Dictyostelium discoideum*. Science 241: 1467–1472 (1988).
144. Klump S, Kleefeld G, Schultz JE: Calcium/calmodulin-regulated guanylate cyclase of the excitable ciliary membrane from *Paramecium*. J Biol Chem 258: 12455–12549 (1983).
145. Knighton DR, Zheng J, Ten Eyck LF, Ashford VA, Xuong N, Taylor SS, Sowadski JM: Crystal structure of the catalytic subunit of cyclic adenosine monophosphate-dependent protein kinase. Science 254: 407–414 (1991).
146. Knighton DR, Zheng J, Ten Eyck LF, Xyong N, Taylor SS, Sowadski JM: Structure of a peptide inhibitor bound to the catalytic subunit of cyclic adenosine monophosphate-dependent protein kinase. Science 253: 414–420 (1991).
147. Koch CA, Anderson D, Moran MF, Ellis C, Pawson T: SH2 and SH3 domains: elements that control interactions of cytoplasmic signaling molecules. Science 252: 668–674 (1991).
148. Koide H, Ogita K, Kikkawa U, Nishizuka Y: Isolation and characterization of the ε subspecies of protein kinase C from rat brain. Proc Natl Acad Sci USA 89: 1149–1153 (1992).
149. Koch W, Heidecker G, Kochs G, *et al.*: Protein kinase C α activates RAF-1 by direct phosphorylation. Nature 364: 249–252 (1993).
150. Koller KJ, De Sauvage FJ, Lowe DG, Goeddel DV: Conservation of the kinaselike regulatory domain is essential for activation of the natriuretic peptide receptor guanylyl cyclases. Mol Cell Biol 12: 2581–2590 (1992).
151. Koretzky GA, Picus J, Thomas ML, Weiss A: Tyrosine phosphatase CD45 is essential for coupling T-cell antigen receptor to the phosphatidylinositol pathway. Nature 346: 66–68 (1990).
152. Koretzky GA, Picus J, Schultz T, Weiss A: Tyrosine phosphatase CD45 is required for T-cell antigen receptor and CD2-mediated activation of a protein tyrosine kinase and interleukin 2 production. Proc Natl Acad Sci USA 88: 2037–2041 (1991).
153. Krueger NX, Streuli M, Saito H: Structural diversity and evolution of human receptor-like protein tyrosine phosphatases. EMBO J 9: 3241–3252 (1990).
154. Krupinski J, Coussen F, Bakalyar HA, Tang WJ, Feinstein PG, Orth K, Slaughter C, *et al.*: Adenylyl cyclase amino acid sequence: possible channel- or transporter-like structure. Science 244: 1558–1564 (1989).
155. Kumagai A, Pupillo M, Gundersen R, Miake-Lye R, Devreotes PN, Firtel RA: Regulation and function of Gα protein subunits in *Dictyostelium*. Cell 57: 265–275 (1989).
156. Kume S, Muto A, Aruga J, Nakagawa T, Michikawa Y, Furuichi T, Nakade S, Okano H, Mikoshiba K: The *Xenopus* IP_3 receptor: structure, function, and localization in oocytes and eggs. Cell 73: 555–570 (1993).
157. Kyriakis JM, App H, Zhang X, Banerjee P, Brautigan DL, Rapp UR, Avruch J: Raf-1 activates MAP kinase-kinase. Nature 358: 417–421 (1992).
158. Labib K, Nurse P: Bring on the phosphatases. Curr Biol 3: 164–166 (1993).
159. Lambrecht HG, Koch KW: A 26 kD calcium binding protein from bovine rod outer segments as modulator of photoreceptor guanylate cyclase. EMBO J 10: 793–798 (1991).
160. Lange-Carter Ca, Pleiman CM, Gardener AM, Blumer KJ, Johnson GL: A divergence in the MAP kinase regulatory network defined by MEK kinase and Raf. Science 260: 315–319 (1993).
161. Langosch D, Thomas L, Betz H: Conserved quaternary structure of ligand-gated ion channels: the postsynaptic glycine receptor is a pentamer. Proc Natl Acad Sci USA 85: 7394–7398 (1988).
162. Leach KL, Ruff VA, Jarpe MB, Adams LD, Fabbro D, Raben DM: α-Thrombin stimulates nuclear diglyceride levels and differential nuclear localization of protein kinase C isozymes in IIC9 cells. J Biol Chem 267: 21816–21822 (1992).
163. Leberer E, Dignard D, Harcus D, Thomas DY, Whiteway M: The protein kinase homologue Ste20p is

required to link the yeast pheromone response G-protein $\beta\gamma$ subunits to downstream signalling components. EMBO J: 11: 4815–4824 (1992).

164. Lee CH, Park D, Wu D, Rhee SG, Simon MI: Members of the $G_q\alpha$ subunit gene family activate phospholipase C-β isozymes. J Biol Chem 267: 16044–16047 (1992).
165. Lefkowitz RJ: Protein-coupled receptor kinases. Cell 74: 409–412 (1993).
166. Lefkowitz RJ, Caron MG: Adrenergic receptors – models for the study of receptors coupled to guanine nucleotide regulatory proteins. J Biol Chem 263: 4993–4996 (1988).
167. Leichtling BH, Spitz E, Rickenberg HV: A cAMP-binding protein from Dictyostelium discoideum regulates mammalian protein kinase. Biochem Biophys Res Common Biophys 100: 515–522 (1981).
168. Levin DR, Fields FO, Kuniswawa R, Bishop JM, Thorner J: A candidate protein kinase C gene, PKC1, is required for the S. cerevisiae cell cycle. Cell 62: 213–224 (1990).
169. Levin LR, Han PL, Hwang PM, Feinstein PG, Davis RL, Reed RR: The *Drosophila* learning and memory gene *rutabaga* encodes a Ca^{2+}/calmodulin-responsive adenylyl cyclase. Cell 68: 479–489 (1992).
170. Lerea CL, Somers DE, Hurley JB, Klock IB, Bunt-Milam AH: Identification of specific transducin α subunits in retinal rod and cone photoreceptors. Science 234: 77–80 (1986).
171. Lilly P, Wu L, Welker DL, Devreotes PNA: G-protein β subunit is essential for *Dictyostelium* development. Genes Devel 7: 986–995 (1993).
172. Lincoln TM, Thompson M, Cornwell TL: Purification and characterization of two forms of cyclic GMP-dependent protein kinase from bovine aorta. J Biol Chem 263: 17632–17637 (1988).
173. Liscovitch M: Crosstalk among multiple signal-activated phospholipases. Trends Biochem Sci 17: 393–399 (1992).
174. Liyanage M, Frith D, Livneh E, Stabel S: Protein kinase C group B members PKC-δ, -ε, -ζ, and PKC-L(η) – comparison of properties of recombinant protein *in vitro* and *in vivo*. Biochem J 283: 781–787 (1992).
175. Lochrie MA, Mendel JE, Sternberg PW, Simon MI: Homologous and unique G protein alpha subunits in the nematode *Caenorhabditis elegans*. Cell Regul 2: 135–154 (1991).
176. Loewenstein WR: The cell-to-cell channel of gap junctions. Cell 48: 725–726 (1987).
177. Logothetis DE, Kurachi Y, Galper J, Neer EJ, Clapham DE: The $\beta\gamma$ subunits of GTP-binding proteins activate the muscarinic K^+ channel in heart. Nature 325: 321–326 (1987).
178. Lowe DG, Chang MS, Hellmiss R, Chen E, Singh S, Garbers DL, Goeddel DV: Human atrial natriuretic peptide receptor defines a new paradigm for second messenger signal transduction. EMBO J 8: 1377–1384 (1989).
179. Lowenstein EJ, Daly RJ, Batzer AG, Li W, Margolis B, Lammers R, Ullrich A, Skolnik EY, Bar-Sagi D, Schlessinger J: The SH2 and SH3 domain-containing protein GRB2 links receptor tyrosine kinases to ras signalling. Cell 70: 431–442 (1992).
180. Ludérus MEE, van der Most RG, Otte AP, van Driel R: A protein kinase C-related enzyme activity in *Dictyostelium discoideum*. FEBS Lett 253: 71–75 (1989).
181. Ma HW, Blitzer RD, Healy ED, Premont RT, Landau EM, Iyengar R: Receptor-evoked Cl-current in Xenopus Oocytes is mediated through a β-type phospholipase C. J Biol Chem 268: 19915–19918 (1993).
182. Maeda N, Kawasaki T, Nakade S, Yokota N, Taguchi T, Kasai M, Mikoshiba K: Structural and functional characterization of inositol 1,4,5-trisphosphate receptor channel from mouse cerebellum. J Biol Chem 266: 1109–1116 (1991).
183. Mann SKO, Yonemoto WM, Taylor SS, Firtel RA: DdPK3, which plays essential roles during *Dictyostelium* development, encodes the catalytic subunit of cAMP-dependent protein kinase. Proc Natl Acad Sci USA 89: 10701–10705 (1992).
184. Margolis B, Li N, Koch A, Mohammadi M, Hurwitz D, Ullrich A, Zilberstein A, Pawson T, Schlessinger J: The tyrosine phosphorylated carboxy terminus of the EGF receptor is a binding site for GAP and PLCγ. EMBO J 9: 4375–4380 (1990).
185. Martelli AM, Gilmour RS, Bertagnolo V, Neri LM, Manzoli L, Cocco L: Nuclear localization and signalling activity of phosphoinositidase Cβ in Swiss 3T3 cells. Nature 358: 242–245 (1992).
186. Matsumoto K, Uno I, Oshima Y, Ishikawa T: Isolation and characterization of yeast mutants deficient in adenylate cyclase and cAMP-dependent protein kinase. Proc Natl Acad Sci USA 79: 2355–2359 (1982).
187. McKay DB, Weber IT, Steitz TA: Structure of catabolite gene activator protein at 2.9-A resolution. J Biol Chem 257: 9518–9524 (1982).
188. McLaughlin SK, McKinnon PJ, Margolskee RF: Gustducin is a taste-cell-specific G protein closely related to the transducins. Nature 357: 563–569 (1992).
189. Méry PF, Lohmann SM, Walter U, Fischmeister R: Ca^{2+} current is regulated by cyclic GMP-dependent protein kinase in mammalian cardiac myocytes. Proc Natl Acad Sci USA 88: 1197–1201 (1991).
190. Mignery GA, Südhof TC, Takei K, De Camilli P: Putative receptor for inositol 1,4,5-trisphosphate similar to ryanodine receptor. Nature 342: 192–195 (1989).
191. Mignery GA, Südhof TC: The ligand binding site and transduction mechanism in the inositol-1,4,5-triphosphate receptor. EMBO J 9: 3893–3898 (1990).
192. Mignery GA, Johnston PA, Sudhof TC: Mechanism of Ca2+ inhibition of inositol 1,4,5-trisphosphate (InsP3)

binding to the cerebellar InsP3 receptor. J Biol Chem 267: 7450–7455 (1992).
193. Millar JBA, Lenaers G, Russel P: Pyp3 PTPase acts as a mitotic inducer in fission yeast. EMBO J 11: 4933–4941 (1992).
194. Millar JBA, Russel P, Dixon JE, Guan KL: Negative regulation of mitosis by two functionally overlapping PTPases in fission yeast. EMBO J 11: 4943–4952 (1992).
195. Miyajima I, Nakafuku M, Nakayama N, Brenner C, Miyajima A, Kaibuchi AK, Arai K, Kaziro Y, Matsumoto K: GPA1, a haploid-specific essential gene, encodes a yeast homolog of mammalian G protein which may be involved in mating factor signal transduction. Cell 50: 1011–1019 (1987).
196. Mohammadi M, Dionne Ca, Li W, Li N, Spivak T, Honegger AM, Jaye M, Schlessinger J: Point mutation in FGF receptor eliminates phosphatidylinositol hydrolysis without affecting mitogenesis. Nature 358: 681–684 (1992).
197. Moran MF, Koch CA, Anderson D, Ellis C, England L, Martin GS, Pawson T: Src homology region 2 domains direct protein-protein interactions in signal transduction. Proc Natl Acad Sci USA 87: 8622–8626 (1990).
198. Mustelin T, Coggeshall KM, Altman A: Rapid activation of the T-cell tyrosine protein kinase $pp56^{lck}$ by the CD45 phosphotyrosine phosphatase. Proc Natl Acad Sci USA 86: 6302–6306 (1989).
199. Mutzel R, Lacombe ML, Simon MN, De Gunzburg J, Veron M: Cloning and cDNA sequence of the regulatory subunit of cAMP-dependent protein kinase from *Dictyostelium discoideum*. Proc Natl Acad Sci USA 84: 6–10 (1987).
200. Nakafuku M, Itoh H, Nakamura S, Kaziro Y: Occurrence in *Saccharomyces cerevisiae* of a gene homologous to the cDNA coding for the subunit of mammalian G proteins. Proc Natl Acad Sci USA 84: 2140–2144 (1987).
201. Nakafuku M, Obara T, Kaibuchi K, Miyajima I, Miyajima A, Itoh H, Nakamura S, Arai K, Matsumoto K, Kaziro Y: Isolation of a second yeast *Saccharomyces cerevisiae* gene (GPA2) coding for guanine nucleotide-binding regulatory protein: studies on its structure and possible functions. Proc Natl Acad Sci USA 85: 1374–1378 (1988).
202. Nakane M, Arai K, Saheki S, Kuno T, Buechler W, Murad F: Molecular cloning and expression of cDNAs coding for soluble guanylate cyclase from rat lung. J Biol Chem 265: 16841–16845 (1990).
203. Nakanishi H, Brewer KA, Exton JH: Activation of the ζ isozyme of protein kinase C by phosphatidylinositol 3,4,5-trisphosphate. J Biol Chem 268: 13–16 (1993).
204. Nakanishi S: Molecular diversity of glutamate receptors and implications for brain function. Science 258: 597–603 (1992).
205. Nakayama N, Miyajima A, Arai K: Nucleotide sequences of STE2 and STE3, cell type-specific sterile genes from *Saccharomyces cerevisiae*. EMBO J 4: 2643–2648 (1985).
206. Nathans J, Hogness DS: Isolation and nucleotide sequence of the gene encoding human rhodopsin. Proc Natl Acad Sci USA 81: 4851–4855 (1984).
207. Nathans J, Thomas D, Hogness DS: Molecular genetics of human color vision: the genes encoding blue, green, and red pigments. Science 232: 193–202 (1986).
208. Nishida E, Gotoh Y: The map kinase cascade is essential for diverse signal transduction pathways. Trends Biochem Sci 18: 128–131 (1993).
209. Nishizuka Y: The molecular heterogeneity of protein kinase C and its implications for cellular regulation. Nature 224: 661–665 (1988).
210. Noel JP, Hamm HE, Sigler PB: The 2.2 A crystal structure of transducin-α complexed with GTPγS. Nature 366: 654–663 (1993).
211. Obara T, Nakafuku M, Yamamota M, Kaziro Y: Isolation and characterization of a gene encoding a G-protein α subunit from *Schizosaccharomyces pombe*: involvement in mating and sporulation pathways. Proc Natl Acad Sci USA 88: 5877–5881 (1991).
212. Ogita K, Miyamoto S, Yamaguchi K, Koide H, Fumisawa N, Kikkawa U, Sahara S, Fukami Y, Nishizuka Y: Isolation and characterization of δ-subspecies of protein kinase C from rat brain. Proc Natl Acad Sci USA 89: 1592–1596 (1992).
213. Olate J, Jorquera H, Purcell P, Codina J, Birnbaumer L, Allende JE: Molecular cloning and sequence determination of a cDNA coding for the α-subunit of a G_0-type protein of *Xenopus laevis* oocytes. FEBS Lett 244: 188–192 (1989).
214. Olivier JP, Raabe T, Henkemeyer M, Dickson B, Mbamalu G, Margolis B, Schlessinger J, Hafen E, Pawson T: A *Drosophila* SH2-SH3 adaptor protein implicated in coupling the sevenless tyrosine kinase to an activator of Ras guanine nucleotide exchange, Sos. Cell 73: 179–191 (1993).
215. Ono Y, Fujii T, Ogita K, Kikkawa U, Igarashi K, Nishizuka Y: Protein kinase C ζ subspecies from rat brain: its structure, expression, and properties. Proc Natl Acad Sci USA 86: 3099–3103 (1989).
216. Ottillie S, Chernoff J, Hannig F, Hoffman CS, Erikson RL: The fission yeast genes $pyp1^{+}$ and $pyp2^{+}$ encode protein tyrosine phosphatases that negatively regulate mitosis. Mol Cell Biol 12: 5571–5580 (1992).
217. O'Tousa JE, Baehr W, Martin RL, Hirsh J, Pak WL, Applebury ML: The Drosophila ninaE gene encodes an opsin. Cell 40: 839–850 (1985).
218. Pai EF, Krengerl U, Petsko GA, Goody RS, Kabsch W, Wittinghofer A: Refined crystal structure of the triphosphate conformation of H-ras p21 at 1.35 A resolution: implications for the mechanisms of GTP hydrolysis. EMBO J 9: 2351–2359 (1990).
219. Park D, Jhon DY, Lee CW, Ryu SH, Rhee SG: Removal of the carboxyl-terminal region of phospholipase

C-β1 by calpain abolishes activation by Gαq. J Biol Chem 268: 3710–3714 (1993).
220. Park D, Jhon DY, Lee CW, Lee KH, Rhee SG: Activation of phospholipase C isozymes by G protein $\beta\gamma$ subunits. J Biol Chem 268: 4573–4576 (1993).
221. Parker MG: Steroid and related receptors. Curr Opin Cell Biol 5: 499–504 (1993).
222. Parks S, Wieschaus E: The *Drosophila* gastrulation gene *concertina* encodes a Gα-like protein. Cell 64: 447–458 (1991).
223. Pawson T, Schlessinger J: SH2 and SH3 domains. Curr Biol 3: 434–442 (1993).
224. Payne WE, Fitzgerald-Hayes M: A mutation in PLC1, a candidate phosphoinositide-specific phospholipase C gene from *Saccharomyces cerevisiae*, causes aberrant mitotic chromosome segregation. Mol Cell Biol 13: 4351–4364 (1993).
225. Pelech SL, Sanghere JS: Mitogen-activated protein kinases: versatile transducers for cell signalling. Trends Biochem Sci 17: 233–238 (1992).
226. Perkins LA, Larsen I, Perrimon N: Corkscrew encodes a putative protein tyrosine phosphatase that functions to transduce the terminal signal from the receptor tyrosine kinase torso. Cell 70: 225–236 (1992).
227. Perrimon N: The torso receptor protein-tyrosine kinase signalling pathway: an endless story. Cell 74: 219–222 (1993).
228. Pingel JT, Thomas ML: Evidence that the leukocyte-common antigen is required for antigen-induced T lymphocyte proliferation. Cell 58: 1055–1065 (1989).
229. Pitt GS, Milona N, Borleis J, Lin KC, Reed RR, Devreotes PN: Structurally distinct and stage-specific adenylyl cyclase genes play different roles in *Dictyostelium* development. Cell 69: 305–315 (1992).
230. Pollak MR, Brown EM, Chou YHW, Hebert SC, Marx SJ, Steinmann B, Levi T, Seidman CE, Seidman JG: Mutations in the human Ca^{2+}-sensing receptor gene cause familial hypocalciuric hypercalcemia and neonatal severe hyperparathyroidism. Cell 75: 1297–1303 (1993).
231. Powers S, Kataoka T, Fasano O, Goldfarb M, Broach J, Wigler M: Genes in *S. cerevisiae* encoding proteins with domains homologous to the mammalian ras proteins. Cell 36: 607–612 (1984).
232. Price JV, Clifford RJ, Schupbach T: The maternal ventralizing locus torpedo is allelic to faint little ball, an embryonic lethal, and encodes the *Drosophila* EGF receptor homolog. Cell 56: 1085–1092 (1989).
233. Probst WC, Snyder LA, Schuster DI, Brosius J, Sealfon SC: Sequence alignment of the G-protein coupled receptor superfamily. DNA Cell Biol 11: 1–20 (1992).
234. Pupillo M, Kumagai A, Pitt GS, Firtel RA, Devreotes PN: Multiple α subunits of guanine nucleotide-binding proteins in *Dictyostelium*. Proc Natl Acad Sci USA 86: 4892–4896 (1989).
235. Ravid S, Spudich JA: Membrane-bound Dictyostelium myosin heavy chain kinase: a developmentally regulated substrate-specific member of the protein kinase C family. Proc Natl Acad Sci USA 89: 5877–5881 (1992).
236. Rebecchi M, Peterson A, McLaughlin S: Phosphoinositide-specific phospholipase C-δ1 binds with high affinity to phospholipid vesicles containing phosphatidylinositol 4,5-bisphosphate. Biochemistry 31: 12742–12747 (1992).
237. Reymond CD, Gomer RH, Medhy MC, Firtel RA: Developmental regulation of a Dictyostelium gene encoding a protein homologous to mammalian ras protein. Cell 39: 141–148 (1984).
238. Reymond CD, Gomer RH, Nellen W, Theibert A, Devreotes P, Firtel RA: Phenotypic changes induced by a mutated ras gene during the development of *Dictyostelium* transformants. Nature 323: 340–343 (1986).
239. Rhee SG, Choi KD: regulation of inositol phospholipid-specific phospholipase C isozymes. J Biol Chem 267: 12393–12396 (1992).
240. Robbins SM, Williams JG, Jermyn KA, Spiegelman GB, Weeks G: Growing and developing *Dictyostelium* cells express different ras genes. Proc Natl Acad Sci USA 86: 938–942 (1989).
241. Rosenthal A, Rhee L, Yadegari R, Paro R, Ullrich A, Goeddel DV: Structure and nucleotide sequence of a Drosophila melanogaster protein kinase C gene. EMBO J 6: 433–441 (1987).
242. Satoh T, Ross CA, Villa A, Supattapone S, Pozzan T, Snyder SH, Meldolesi J: The inositol 1,4,5-trisphosphate receptor in cerebellar Purkinje cells: quantitative immunogold labeling reveals concentration in an ER subcompartment. J Cell Biol 111: 615–624 (1990).
243. Savarese TM, Fraser CM: *In vitro* mutagenesis and the search for structure-function relationships among G protein-coupled receptors. Biochem J 283: 1–19 (1992).
244. Saxe CL, Ginsburg GT, Louis JM, Johnson Rn, Devreotes PN, Kimmel AR: cAR2, a prestalk cAMP receptor required for normal tip formation and late development of Dictyostelium discoideum. Genes Devel 7: 262–272 (1992).
245. Schaeffer E, Smith D, Mardon G, Quinn W, Zuker C: Isolation and characterization of two new *Drosophila* protein kinase C genes, including one specifically expressed in photoreceptor cells. Cell 57: 403–412 (1989).
246. Schlessinger J, Ullrich A: Growth factor signalling by receptor tyrosine kinases. Neuron 9: 383–391 (1992).
247. Schlessinger J: How receptor tyrosine kinases activate ras. Trends Biochem Sci 18: 273–275 (1993).
248. Schmidt HHHW, Lohmann SM, Walter U: The nitric oxide and cGMP signal transduction system: regulation and mechanism of action. Biochim Biophys Acta 1178: 153–175 (1993).
249. Schneuwly S, Burg MG, Lending C, Perdew MH, Pak WL: Properties of photoreceptor-specific phospholipase C encoded by the *norpA* gene of *Drosophila melanogaster*. J Biol Chem 266: 24314–24319 (1991).
250. Schofield PR, Darlison MG, Fujita N, Burt DR,

Stephenson FA, Rodriguez H, Rhee LM, Ramachandran J, Reale V, Glencorse TA, Seeburg PH, Barnard EA: Sequence and functional expression of the $GABA_A$ receptor shows a ligand-gated receptor super-family. Nature 328: 221–227 (1987).

251. Schultz JE, Schönefeld U, Klumpp S: Calcium/calmodulin-regulated guanylate cyclase and calcium-permeability in the ciliary membrane from *Tetrahymena*. Eur J Biochem 137: 89–94 (1983).
252. Schultz JE, Pohl T, Klumpp S: Voltage-gated Ca^{2+} entry into *Paramecium* linked to intraciliary increase in cyclic GMP. Nature 322: 271–273 (1986).
253. Schulz S, Singh S, Bellet RA, Singh G, Tubb DJ, Chin H, Garbers DL: The primary structure of a plasma membrane guanylate cyclase demonstrates diversity within this new receptor family. Cell 58: 1155–1162 (1989).
254. Schulz S, Green CK, Yuen PST, Garbers DL: Guanylyl cyclase is a heat-stable enterotoxin receptor. Cell 63: 941–948 (1990).
255. Seger R, Seger D, Lozeman FJ, Ahn NG, Graves LM, Campbell JS, Ericsson L, Harrylock M, Jensen AM, Krebs EG: Human T-cell mitogen-activated protein kinase kinases are related to yeast signal transduction kinases. J Biol Chem 267: 25628–25631 (1992).
256. Segre GV, Goldring SR: Receptors for secretin, calcitonin, parathyroid hormone (PTH)/PTH-related peptide, vasoactive intestinal peptide, glucagonlike peptide 1, growth hormone-releasing hormone, and glucagon belong to a newly discovered G-protein linked receptor family. Trends Endocrinol Metab 4: 309–314 (1993).
257. Serafini T, Orci L, Amherdt M, Brunner M, Kahn RA, Rothman JE: ADP-ribosylation factor is a subunit of the coat of Golgi-derived COP-coated vesicles: a novel role for a GTP-binding protein. Cell 67: 239–253 (1991).
258. Shen SH, Bastien L, Posner BI, Chretien P: A protein-tyrosine phosphatase with sequence similarity to the SH2 domain of the protein-tyrosine kinases. Nature 352: 736–739 (1991).
259. Shirakabe K, Gotoh Y, Nishida E: A mitogen-activated protein (MAP) kinase activating factor in mammalian mitogen-stimulated cells is homologous to *Xenopus* M phase MAP kinase activator. J Biol Chem 267: 16685–16690 (1992).
260. Shiroo M, Goff L, Biffen M, Shivnan E, Alexander D: CD45 tyrosine phosphatase-activated $p59^{fyn}$ couples the T cell antigen receptor to pathways of diacylglycerol production, protein kinase C activation and calcium influx. EMBO J 11: 4887–4897 (1993).
261. Short AD, Klein MG, Schneider MF, Gill DL: Inositol 1,4,5-trisphosphate-mediated quantal Ca^{2+} release measured by high resolution imaging of Ca^{2+} within organelles. J Biol Chem 268: 25887–25983 (1993).
262. Shortridge RD, Yoon J, Lending CR, Bloomquist BT, Perdew MH, Pak WL: A *Drosophila* phospholipase C gene that is expressed in the central nervous system. J Biol Chem 266: 12474–12480 (1991).
263. Shyjan AW, de Sauvage FJ, Gillett NA, Goeddel DV, Lowe DG: Molecular cloning of a retina-specific membrane guanylyl cyclase. Neuron 9: 727–737 (1992).
264. Smercka AV, Hepler JR, Brown KO, Sternweis PC: Regulation of polyphosphoinositide-specific phospholipase C activity by purified Gq. Science 251: 804–807 (1991).
265. Simon MA, Dodson GS, Rubin GM: An SH3-SH2-SH3 protein is required for $p21^{Ras1}$ activation and binds to Sevenless and Sos proteins *in vitro*. Cell 73: 169–177 (1993).
266. Singh S, Lowe DG, Thorpe DS, Rodriguez H, Kuang WJ, Dangott L, Chinkers M, Goeddel DV, Garbers DL: Membrane guanylate cyclase is a cell-surface receptor with homology to protein kinases. Nature 334: 708–710 (1988).
267. Songyang Z, Shoelson SE, Chaudhuri M, Gish G, Pawson P, Haser WG, King F, Roberts T, Ratnofsky S, Lechleider RJ, *et al.*: SH2 domains recognize specific phosphopeptide sequences. Cell 72: 767–778 (1993).
268. Sprenger F, Stevens LM, Müsslein-Volhard C: The *Drosophila* gene *torso* encodes a putative receptor tyrosine kinase. Nature 338: 478–483 (1989).
269. Srivastava SP, Fuchs JA, Holtzman JL: The reported cDNA sequence for phospholipase C alpha encodes protein disulfide isomerase, isozyme Q-2 and not phospholipase C. Biochem Biophys Res Common 193: 971–978 (1993).
270. Stahl ML, Ferenz CR, Kelleher KL, Kriz RW, Knopf JL: Sequence similarity of phospholipase C with the non-catalytic region of *src*. Nature 332: 269–272 (1988).
271. Stearns T, Willingham MC, Botstein D, Kahn RA: ADP-ribosylation factor is functionally and physically associated with the Golgi comples. Proc Natl Acad Sci USA 87: 1238–1242 (1990).
272. Strathmann MP, Simon MI: $G\alpha 12$ and $G\alpha 13$ subunits define a fourth class of G protein α subunits. Proc Natl Acad Sci USA 88: 5582–5586 (1991).
273. Streb H, Irvine RF, Berridge MJ, Schulz I: Release of Ca^{2+} from a nonmitochondrial intracellular store in pancreatic acinar cells by inositol-1,4,5-trisphosphate. Nature 306: 67–68 (1983).
274. Streuli M, Krueger NX, Hall LR, Schlossman SF, Saito H: A new member of the immunoglobulin superfamily that has a cytoplasmic region homologous to the leukocyte common antigen. J Exp Med 168: 1523–1530 (1988).
275. Streuli M, Krueger NX, Tsai AYM, Saito H: A family of receptor-linked protein tyrosine phosphatases in humans and *Drosophila*. Proc Natl Acad Sci USA 86: 8698–8702 (1989).
276. Streuli M, Krueger NX, Thai T, Tang M, Saito H: Distinct function roles of the two intracellular phosphatase like domains of the receptor-linked protein tyrosine

phosphatases LCA and LAR. EMBO J 9: 2399–2407 (1990).
277. Stryer L: Visual excitation and recovery. J Biol Chem 266: 10711–10714 (1991).
278. Südhof TC, Newton CL, Archer III BT, Ushkaryov YA, Mignery GA: Structure of a novel InsP3 receptor. EMBO 10: 3199–3206 (1991).
279. Suh PG, Ryu SH, Moon KH, Suh HW, Rhee SG: Inositolphospholipid-specific phospholipase C: complete cDNA and protein sequences and sequence homology to tyrosine kinase-related oncogene products. Proc Natl Acad Sci USA 85: 5419–5423 (1988).
280. Sun XJ, Rothenberg P, Kahn CR, Backer JM, Araki E, Wilden PA, Cahill DA, Goldstein BJ, White MF: Structure of the insulin receptor substrate IRS-1 defines a unique signal transduction protein. Nature 352: 73–77 (1991).
281. Supattapone S, Danoff SK, Theibert A, Joseph SK, Steiner J, Snyder SH: Cyclic AMP-dependent phosphorylation of a brain inositol trisphosphate receptor decreases its release of calcium. Proc Natl Acad Sci USA 85: 8747–8750 (1988).
282. Tabuse Y, Nishiwaki K, Miwa J: Mutations in a protein kinase C homolog confer phorbol ester resistance on *Caenorhabditis elegans*. Science 243: 1713–1716 (1989).
283. Takeshima H, Nishimura S, Matsumoto T, Ishida H, Kangawa K, Minamino N, Matsuo H, Ueda M, Hanaoka M, Hirose T, Numa S: Primary structure and expression from complementary DNA of skeletal muscle ryanodine receptor. Nature 339: 439–445 (1989).
284. Takio K, Smith SB, Krebs EG, Walsh KA, Titani K: Amino acid sequence of the regulatory subunit of bovine type II adenosine cyclic 3′5′-phosphate dependent protein kinase. Biochemistry 23: 4200–4206 (1984).
285. Takio K, Wade RD, Smith SB, Krebs EG, Walsh KA, Titani K: Guanosine cyclic 3′5′-phosphate dependent protein kinase, a chimeric protein homologous with two separate protein families. Biochemistry 23: 4207–4218 (1984).
286. Tanaka K, Nakafuku M, Satoh T, Marshall MS, Gibbs JB, Matsumoto K, Kaziro Y, Toh-e: *S. cerevisiae* genes IRA1 and IRA2 encode proteins that may be functionally equivalent to mammalian ras GTPase activating protein. Cell 60: 803–807 (1990).
287. Tanaka K, Davey J, Imai Y, Yamamoto M: *Schizosaccharomyces pombe* map3$^+$ encodes the putative M-factor receptor. Mol Cell Biol 10: 4303–4313 (1993).
288. Tang W, Gilman AG: Type-specific regulation of adenylyl cyclase by G protein $\beta\gamma$ subunits. Science 254: 1500–1503 (1991).
289. Tang W, Gilman AG: Adenylyl cyclases. Cell 70: 869–872 (1992).
290. Tatchell K, Chaleff DT, DeFeo-Jones D, Scolnick EM: Requirement of either of a pair of ras-related genes of *Saccharomyces cerevisiae* for spore viability. Nature 309: 523–527 (1984).
291. Taylor CW, Marshall ICB: Calcium and inositol 1,4,5-trisphosphate receptors: a complex relationship. Trends Biochem Sci 17: 403–407 (1992).
292. Taylor SJ, Chae HZ, Rhee SG, Exton JH: Activation of the β1 isozyme of phospholipase C by a subunits of the Gq class of G proteins. Nature 350: 516–518 (1991).
293. Thomas GMH, Cunningham E, Fensome A, Ball A, Totty NF, Truong O, Hsuan JJ, Cockroft S: An essential role for phosphatidylinositol transfer protein in phospholipase C-mediated inositol lipid signalling. Cell 74: 919–928 (1993).
294. Thorpe DS, Garbers DL: The membrane form of guanylate cyclase – homology with a subunit of the cytoplasmic form of the enzyme. J Biol Chem 264: 6545–6549 (1989).
295. Tian SS, Tsoulfas P, Zinn K: Three receptor-linked protein-tyrosine phosphatases are selectively expressed on central nervous system axons in the *Drosophila* embryo. Cell 67: 675–685 (1991).
296. Titani K, Sasagawa T, Ericsson LH, Kumar S, Smith SB, Krebs EG, Walsh KA: Amino acid sequence of the regulatory subunit of bovine type I adenosine cyclic 3′5′-phosphate dependent protein kinase. Biochemistry 23: 4193–4199 (1984).
297. Toda T, Uno I, Ishikawa T, Powers S, Kataoka T, Broek D, Cameron S, Broach J, Matsumoto K, Wigler M: In yeast, RAS proteins are controlling elements of adenylate cyclase. Cell 40: 27–36 (1985).
298. Toda T, Cameron S, Sass P, Zoller M, Scott JD, McMullen B, Hurwitz M, Krebs EG, Wigler M: Cloning and characterization of BCY1, a locus encoding a regulatory subunit of the cyclic AMP-dependent protein kinase in Saccharomyces cerevisiae. Mol Cell Biol 7: 1371–1377 (1987).
299. Toda T, Cameron S, Sass P, Zoller M, Wigler M: Three different genes in *S. cerevisiae* encode the catalytic subunits of the cAMP-dependent protein kinase. Cell 50: 277–287 (1987).
300. Tonks NK, Charbonneau H, Diltz CD, Fischer EH, Walsh KA: Demonstration that leukocyte common antigen CD45 is a protein tyrosine phosphatase. Biochemistry 27: 8695–8701 (1988).
301. Uhler MD: Cloning and expression of a novel cyclic GMP-dependent protein kinase from mouse brain. J Biol Chem 268: 13586–13591 (1993).
302. Ullrich A, Schlessinger J: Signal transduction by receptors with tyrosine kinase activity. Cell 61: 203–212 (1990).
303. Valkema R, van Haastert PJM: Inhibition of receptor-stimulated guanylyl cyclase by intracellular calcium ions in *dictyostelium discoideum* cells. Biochem Biophys Res Commun 186: 263–268 (1992).
304. van der Kaay J, Draijer R, van Haastert PJM: Increased conversion of phosphatidylinositol to phosphatidylinositol phosphate in Dictyostelium cells expressing a

mutated ras gene. Proc Natl Acad Sci USA 87: 9197–9201 (1990).

305. van Dop C, Tsubokawa M, Bourne HR, Ramachandran J: Amino acid sequence of retinal transducin at the site ADP-ribosylated by cholera toxin. J Biol Chem 259: 696–698 (1984).
306. van Haastert PJM: Signal transduction and the control of development in *Dictyostelium discoideum*. Devel Biol 1: 159–167 (1990).
307. Vogel W, Lammers R, Huang J, Ullrich A: Activation of a phosphotyrosine phosphatase by tyrosine phosphorylation. Science 259: 1611–1614 (1993).
308. Ways DK, Cook PP, Webster C, Parker PP: Effect of phorbol esters on protein kinase C-ζ. J Biol Chem 267: 4799–4805 (1992).
309. Weaver CT, Pingel JT, Nelson JO, Thomas ML: $CD8^+$ T-cell clones deficient in the expression of the CD45 protein tyrosine phosphatase have impaired responses to T-cell receptor stimuli. Mol Cell Biol 11: 4415–4422 (1991).
310. West RE, Moss J, Vaughan M, Liu T, Liu TY: Pertussis toxin-catalyzed ADP-ribosylation of transducin. J Biol Chem 260: 14428–14430 (1985).
311. Whiteway M, Hougan L, Dignard D, Thomas DY, Bell L, Saari GC, Grant FJ, O'Hara P, MacKay VL: The STE4 and STE18 genes of yeast encode potential β and γ subunits of the mating factor receptor-coupled G protein. Cell 56: 467–477 (1989).
312. Whiteway M, Errede B: Signal transduction pathway for pheromone response in Saccharomyces cerevisiae. In: Kurjan J, Taylor BL (eds) Signal Transduction: Prokaryotic and simple Eukaryotic Systems, pp. 189–237. Academic Press, New York (1993).
313. Wolfe L, Corbin JD, Francis SH: Characterization of a novel isozyme of cGMP-dependent protein kinase from bovine aorta. J Biol Chem 264: 7734–7741 (1989).
314. Wu L, Devreotes PN: *dictyostelium* transiently expresses eight distinct G-protein α-subunits during its developmental program. Biochem Biophys Res Commun 179: 1141–1147 (1991).
315. Wu D, Jiang H, Katz A, Simon MI: Identification of critical regions on phospholipase C-β1 required for activation by G-proteins. J Biol Chem 268: 3704–3709 (1993).
316. Wu D, Katz A, Simon MI: Activation of phospholipase C β2 by the α and $\beta\gamma$ subunits of trimeric GTP-binding protein. Proc Natl Acad Sci USA 90: 5297–5301 (1993).
317. Yamawaki-Kataoka Y, Tamaoki T, Choe HR, Tanaka H, Kataoka T: Adenylate cyclase in yeast: a comparison of the genes from *Schizoaccharomyces pombe* and *Saccharomyces cerevisiae*. Proc Natl Acad Sci USA 86: 5693–5697 (1989).
318. Yoko-o T, Matsui Y, Yagisawa H, Nojima H, Uno I, Toh-e A: The putative phosphoinositide-specific phospholipase C gene, PLC1, of the yeast *Saccharomyces cerevisiae* is important for cell growth. Proc Natl Acad Sci USA 90: 1804–1808 (1993).
319. Yoshikawa S, Tanimura T, Miyawaki A, Nakamura M, Yuzaki M, Furuichi T, Mikoshiba K: Molecular cloning and characterization of the inositol,4,5-trisphosphate receptor in *Drosophila melanogaster*. J Biol Chem 267: 16613–16619 (1992).
320. Yoshikawa S, Miyamoto I, Aruga J, Furuichi T, Okano H, Mikoshiba K: Isolation of a *Drosophila* gene encoding a head-specific guanylyl cyclase. J Neurochem 60: 1570–1573 (1993).
321. Yuen PST, Garbers DL: Guanylyl cyclase-linked receptors. Annu Rev Neurosci 15: 193–225 (1992).
322. Zuker CS, Cowman AF, Rubin GM: Isolation and structure of rhodopsin gene from *D. melanogaster*. Cell 40: 851–858 (1985).

Plant Molecular Biology **26**: 1271–1287, 1994.

Signal transduction in the sexual life of *Chlamydomonas*

Lynne M. Quarmby
HeartCell Laboratory of Chlamydomonas Research, Department of Anatomy & Cell Biology, Emory University School of Medicine, Atlanta, GA 30322, USA

Received 8 March 1994; accepted 26 April 1994

Key words: *Chlamydomonas*, calcium, cAMP, adenylyl cyclase, gametogenesis, mating response

Abstract

Several signal transduction pathways play important roles in the sexual life cycle of *Chlamydomonas*. Nitrogen deprivation, perhaps sensed as a drop in intracellular [NH_4^+], triggers a signal transduction pathway that results in altered gene expression and the induction of the gametogenic pathway. Blue light triggers a second signalling cascade which also culminates in gene induction and completion of gametogenesis. New screens have uncovered several mutants in these pathways, but so far we know little about the biochemical events that transduce the environmental signals of nitrogen deprivation and blue light into the changes in gene transcription that produce gametes.

Cell-cell contact of mature, complementary gametes elicits a number of responses that prepare the cells for fusion. Contact is sensed by the agglutinin-mediated cross-linking of flagellar membrane proteins. An increase in [cAMP] couples protein cross-linking to the mating responses. In *C. reinhardtii* the cAMP signal appears to be generated by the sequential stimulation of as many as 3 distinct adenylyl cyclase activities. Although the molecular mechanisms of adenylyl cyclase activations are poorly understood, Ca^{2+} may play a role. Most of the mating responses appear to be triggered by a cAMP-dependent protein kinase, but here too, Ca^{2+} may play a role. Numerous mutants are facilitating studies of the signalling pathways that trigger the mating responses.

Cell fusion triggers another series of events that culminate in the expression of zygote specific genes. The mature zygote is sensitive to a light signal which stimulates the expression of genes whose products are essential for germination. The signal transduction pathways that trigger zygospore formation and germination are ripe for investigation in this experimentally powerful system.

Introduction

The unicellular green alga, *Chlamydomonas* can be found frantically swimming hither and thither in ponds and puddles, propelled by a synchronized ciliary-type beat of its two flagella. Sensing both the direction and the intensity of a light source, these active cells accumulate where the light is most appropriate for photosynthesis. In the vegetative phase, *Chlamydomonas* cells are haploid and mitotic. Nutrient starvation and a light signal stimulate vegetative cells to differentiate into gametes which, at a glance, do not look much different than the vegetative cells. The gametes are highly motile and phototactic, but are metabollically adapted to survive prolonged periods of nitrogen starvation. Concurrent with the metabolic adaptions to low nitrogen, differentiation pathways, at least one of which is triggered by blue light, produce cells that are competent to mate. Different species of *Chlamydomonas* exhibit differences in mating behaviour, but all species have two genetically determined mating types: plus (mt^+) and minus (mt^-). An encounter between two gametes of opposite mating type, but of the same species, triggers a series of events leading to fusion and the production of a quadraflagellate zygote. Fusion initiates another differentiation pathway, and the meiotic zygote resorbs its four flagella and becomes encased in a thick wall. In this condition, the species is able to survive not only nitrogen starvation, but also desiccation. Ultimately, four vegetative cells hatch from the zygospore. For a superb overview of *Chlamydomonas* biology and experimental methodology, see Harris [35]. The purpose of this review is to examine how *Chlamydomonas* receives and interprets the environmental cues that stimulate progression through the sexual life cycle.

Chlamydomonas offers several advantages for the study of signal transduction pathways. Studies of signalling in the tissues of multicellular organisms can be confounded by the presence of many different types of cell. Because, in the laboratory, we can grow large populations of *Chlamydomonas* in a uniform physiological state, application of an external stimulus can elicit a response that is synchronized in the entire population of cells. This, and the relative ease of assaying flagellar-mediated behaviours, has fostered rapid progress in our understanding of signal transduction in this organism. It is now clear that the second messengers, cAMP, IP_3, and Ca^{2+}, play important roles in mediating the responses of *Chlamydomonas*, as do several calcium-binding proteins, kinases and ion channels. Genetic studies are providing important clues about the interrelationships of these signalling molecules.

Excluded from this chapter are the advances in our understanding of the signalling pathways that mediate the motile responses of *Chlamydomonas* to light (recently reviewed by [37, 106]). In addition to sexual signalling and the photoresponses, *Chlamydomonas* cells respond to certain external stimuli by shedding their flagella [36, 67, 68, 70].

An overview of the role of environmental signals in the sexual life cycle of *Chlamydomonas* is shown in Fig. 1. In the sections that follow, I will deal in turn with each of these signals and the signal transduction pathways that are elicited.

Gametogenesis

The morphological distinctions between vegetative and gametic *Chlamydomonas* cells are not easily seen by light microscopy. By electron microscopy, the most apparent differences are the appearance of golgi-derived vesicles and the appearance of mating-type-specific mating structures ([54]; see Fig. 2). Gametes can also be distinguished from vegetative cells biochemically, physiologically and behaviourally (e.g. [2, 32, 74, 75]). Changes in gene expression occur during

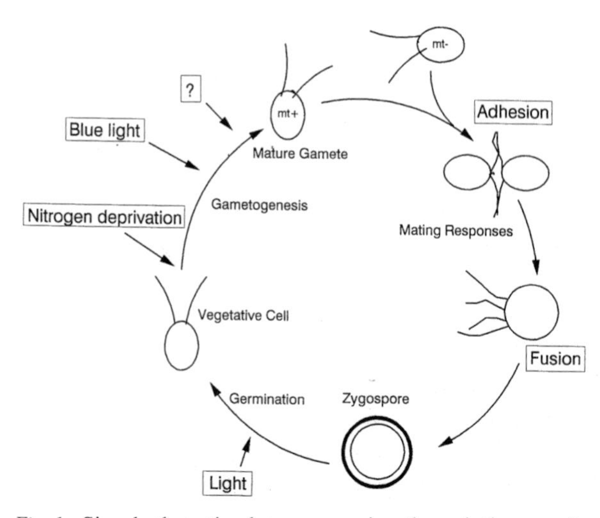

Fig. 1. Signals that stimulate progression through the sexual life cycle of *C. reinhardtii*. Nitrogen deprivation and blue-light each activate signalling pathways which culminate in changes in gene expression responsible for the production of sexually competent gametes. Another signalling cascade is triggered by the flagellar adhesion of complementary mature gametes. This signalling cascade(s) triggers the mating responses, culminating in cell fusion. Fusion then provides the stimulus for a pathway that triggers production of the third cell type, the thick-walled zygospore within which meiosis occurs. Environmental cues (probably light) stimulate completion of the cycle by initiating the pathway of germination, which results in the hatching of vegetative zoospores. Each of these signalling pathways will be discussed in this review.

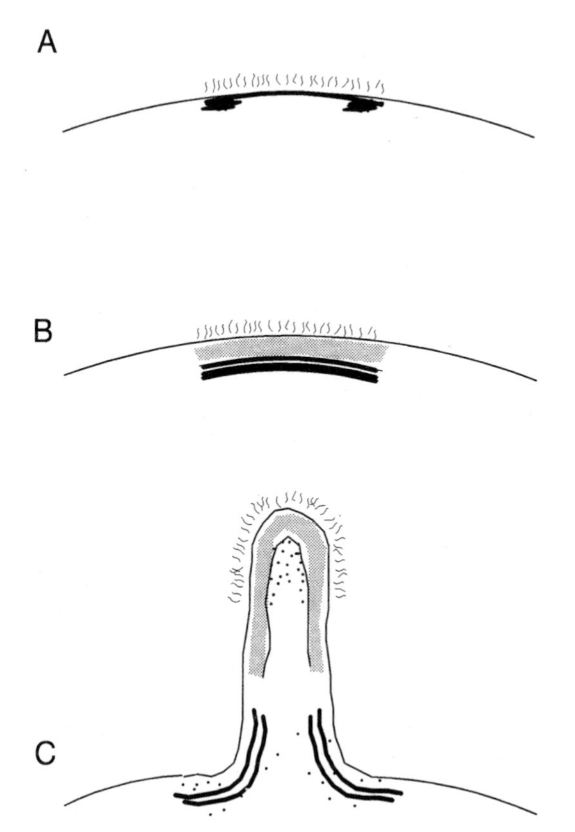

Fig. 2. The mating structures of *C. reinhardtii*. Panels A and B illustrate the morphology of the mating structures of mature mt⁻ and mt⁺ gametes, respectively. Panel C illustrates an activated mt⁺ mating structure as it would appear after flagellar adhesion between the mt⁺ and an mt⁻ gamete. These sketches are derived from the electron micrographs of Goodenough *et al.* [31] and Detmers *et al.* [14].

gametogenesis and a number of proteins are only expressed in gametes [95, 96]. Mature gametes can return to the vegetative state by addition of nitrogen or incubation in the dark [75]. But what are the signals that trigger the program of development that produces gametes?

At least two environmental signals (nitrogen deprivation and blue light) are necessary for gametogenesis of *C. reinhardtii*. The blue light signal has been well characterized [5, 104], but little is known about the signal generated by nitrogen deprivation. New screens designed to isolate mutants in the gametogenic pathways have yielded five *C. reinhardtii* strains specifically defective in gametogenesis: *dif-1*, *dif-2*, *lrg-1*, *lrg-2* and *C4* [9, 10, 73] and more are likely to be reported soon. The *imp-11* and *iso-1* mutants are also defective in gametogenesis, perhaps at an early point in the gametogenic pathway [11, 31]. As our understanding of the developmental pathways is refined, it should become possible to specifically target the signal transduction pathways for genetic and molecular analysis. Moreover, recent improvements in the technology of *Chlamydomonas* molecular biology [34, 47, 60, 79, 93, 100] should facilitate the cloning of genes involved in this developmental pathway.

Environmental signals for gametogenesis

Although 40 years have passed since Sager and Granick [71] reported that nitrogen deprivation can induce sexual differentiation in *C. reinhardtii*, we know little about the mechanism involved. In the laboratory, *Chlamydomonas* cells grow well on a variety of media, both liquid and solidified with agar (see [35]). Depletion of nitrogen from any of these sources will yield functional gametes, but the subcellular morphology of gametes formed under different growth conditions may vary [54]. It is clear that nitrogen deprivation is an important stimulus for gametogenesis, but it is also clear that nitrogen deprivation induces metabolic and morphological changes unrelated to the developmental programme. Presumably, one or more of the many changes that facilitate survival during nitrogen starvation has been co-opted as a signal for the programme of gametic differentiation. Matsuda *et al.* [57] propose that the signal in *C. reinhardtii* is a drop in intracellular $[NH_4^+]$.

Sexually mature gametes of *C. reinhardtii* are not produced in the dark. Nitrogen deprivation in the dark produces pregametes. Pregametes are cells which are not competent to mate, but are competent to receive the blue-light signal necessary to achieve mating competence [96]. Trier *et al.* [96] gave pulses of light to dark-grown cells and demonstrated that a period of nitrogen starvation (> 5 h) is required before the cells are competent to respond to a pulse of light and develop into mature gametes. Because an organic source of carbon (acetate) will sustain the cells in the dark, but will not facilitate differentiation, it was concluded that the need for light was not photosynthetic. A few earlier studies (e.g. [54]) reported that gametic differentiation could proceed to completion in the dark, but it now seems likely that these cells were exposed to light (N. Martin, personal communication). Beck and Acker [5] report that, like the blue-light responses of higher plants, a low fluence rate is sufficient to trigger the gametogenic pathway. Weissig and Beck [104] report the action spectrum for the light-dependent step in gametogenesis to have two maxima at 370 and 450 nm, similar to the spectrum of blue-light responses in higher plants.

There may be a third environmental stimulus, in addition to nitrogen deprivation and blue light, that is required for completion of the gametogenic program. The state of differentiation of cells grown to nitrogen starvation on agar plates is not clear. After 7 days of constant illumination on TAP plates (see Harris [35] for the formulation of this and other growth media), the cells have depleted the nitrogen [54]. Because these cells are often flagella-less, they cannot be immediately assayed for the ability to mate. However, within 2 h in liquid media, the cells have grown flagella and are mating-competent [54]. Is the acquisition of mating competence a simple consequence of the acquisition of flagella by cells that are already gametes, or do the cells differentiate as the flagella grow? Because 2 h is a much shorter period of time than is required for gametogenesis when liquid-grown cells are transferred to nitrogen-free media (usually >10 h is required), it has been suggested that the cells have differentiated while on the plates [54]. In support of this idea, mt^+ cells, harvested directly from the plates, express mating structures (U. Goodenough, personal communication). In contrast, Saito *et al.* [75] state that cells grown on plates contain neither cell body agglutinin activity, nor lytic enzyme activity in the soluble fraction of their cell homogenates, suggesting to these workers that the cells 'can be classified as vegetative'. While these cells are not phenotypically mature gametes, it seems unlikely that they are truly vegetative either. Certainly we know that flagella-less cells on plates are stimulated to grow flagella when the plates are flooded. What environmental cue is triggering flagellar growth? Is the same environmental cue required to stimulate the final stages of gametic differentiation?

The environmental signals for gametogenesis in *C. eugametos* differ somewhat from those in *C. reinhardtii*. While depletion of nitrogen is specifically required to induce gamete formation in *C. reinhardtii* other nutrient stresses may initiate differentiation in *C. eugametos* [94], but this issue has not been thoroughly explored. It is possible that a common metabolic signal triggers the developmental programme in both, or even all, species of *Chlamydomonas*. [NH_4^+] may be the intracellular signal for *C. reinhardtii*. In *C. monoica* there is an optimal amount of NO_3^- required for the formation of mature gametes [99]. This result suggests that a minimal amount of nitrogen is required for the cells to progress through the cell cycle, but then the cells must be nitrogen-depleted (i.e. perhaps it is necessary that intracellular [NH_4^+] is low) during G_1 (G_0?) in order to undergo gametogenesis ([108], reviewed by van den Ende [99]).

Although the nutrient deprivation signal may be common, the requirement for a light signal appears to be different in the different species: *C. reinhardtii* requires light for differentiation, whereas *C. eugametos* does not. It is interesting that some strain of *C. eugametos* require a short pulse of light to activate their agglutinins. This trait is controlled by a single gene that is not linked to mating type. The molecular basis of this activation is not understood, but it is thought that a light-induced chemical change in the proteins is occurring on the flagellar surface, i.e. the light is not thought to be activating a signalling cascade ([49, 50], reviewed by van den Ende [99]).

Signal transduction pathways of gametogenesis

After nitrogen depletion, *Chlamydomonas* cells no longer grow and divide [44]. However, new proteins continue to be synthesized [43, 44, 95] with a pattern of gene expression distinct from cells grown in the presence of nitrogen [95, 96]. Some of these 'gamete-specific' proteins are not specifically involved in sexual differentiation. For example, protein M, a periplasmically localized l-amino-acid oxidase, is probably important for the scavenging of ammonium from extracellular amino acids [99]. In contrast, other proteins specifically synthesized under conditions of nitrogen starvation play important roles in the production of mature gametes. Two of these have been identified by their function in the mating reactions of mature gametes: the agglutinins and gamete lytic enzyme (GLE; discussed below). But it has been difficult to identify other transcripts or proteins whose expression is both induced by nitrogen deprivation and required for gametogenesis. Von Gromoff and Beck [101] have identified four genes (*gas3*, *gas18*, *gas28* and *gas96*), in addition to *gle* and the agglutinins, that are expressed specifically in gametes, but there is no evidence that any of these are specifically involved in gametogenesis. However, it is clear that the nitrogen deprivation signal is perceived and transduced into a change in the pattern of gene expression which is ultimately responsible for the conversion of vegetative cells into gametes. We do not yet know enough to distinguish the molecular machinery of the signal transduction pathway that receives and interprets the nitrogen deprivation signal from the molecular machinery of the developmental program that produces gametes.

One productive approach will continue to be the dissection of gametogenesis by the identification of non-mating mutants. In order to distinguish mutations that disrupt the signalling and/or developmental pathway(s) of gametogenesis from mutations that produce defective gametes, it is necessary to supplement the functional definition of gametes (cells with the ability to mate) with a biochemical definition. Fully differentiated gametes display functional mating-type-specific agglutinin molecules on their flagellar surface, have stores of agglutinins in association with the cell body (see below for a discussion of the agglutinins), carry stores of inactive GLE, and possess rudimentary mating structures [25, 61, 86, 99]. Saito *et al.* [75] reasoned that although mutants defective in any one of these would be unable to mate successfully, the developmental pathway for gametic differentiation would likely be intact, whereas a mutant with a defect in differentiation is predicted to be deficient in multiple aspects of the gametic phenotype. By this criterion, most of the non-mating mutants isolated to date are defective in expression of only one of either functional agglutinins, mating structures, GLE, or flagella, and thus are not likely to be defective in the pathway of gametic differentiation [18, 19, 29–31, 41, 55, 56]. Saito and Matsuda [73] have identified two temperature-sensitive mutants, *dif-1* and *dif-2*, that are defective in the signalling and/or developmental pathway and do not produce any of the measured gametic qualities at 35 °C. Because *dif-1* and *dif-2* gametes produced at 25 °C dedifferentiate into vegetative cells when transferred to 35 °C (even in the absence of nitrogen), Saito and Matsuda [73] postulate that the products of these two genes (they are unlinked to each other and to the mating type locus, see below) encode positive regulators of gametogenesis.

In *C. reinhardtii*, RNA and protein synthesis are required in order for blue light to induce the completion of gametogenesis [96]. Thus it appears that the blue-light signal is inducing the expression of developmentally important genes. Blue light is an important determinant of growth and development in higher plants where it also exerts many of its effects via changes in gene expression (reviewed by J. Chory, this volume, and [12, 86, 52]). Intensive biochemical and genetic analysis of higher plants, particularly *Arabidopsis*, has lead to the identification of a putative blue-light receptor [3] and implicated a heterotrimeric G protein and a protein kinase activity in the blue-light signalling pathway (see reviews by T.D. Elich and J. Chory, H. Ma and C. Walker, this volume). *Chlamydomonas* is poised to contribute to the rapid progress in this field. Buerkle *et al.* [9] have isolated two *C. reinhardtii* mutants, *lrg-1* and *lrg-2*, with defects specific to the blue-light pathway of gametogenesis. *Lrg-1* cells have lost the requirement for a light signal during gametogenesis, whereas *lrg-2* cells require an increased fluence rate of blue light in order to complete differentiation. These workers hypothesize that both *lrg-1* and *lrg-2* participate in the blue-light-activated signal transduction cascade. The molecular characterization of these mutants will provide interesting clues to this pathway.

Beck and Acker [5] suggest that early changes in gene expression are triggered by nitrogen deprivation and later changes in gene expression are responses to light. Because the early genes of nitrogen deprivation (*gas3* and *gas96*) are turned on with normal kinetics in *C4*, another mutant that does not produce mature gametes [10, 101] but the late genes (*gas18*, *gas28* and *gle*) are not expressed, it has been concluded that the *C4* defect may be required for the progression of differentiation beyond an early initial phase [101]. Alternatively, *gas3* and *gas96* may play no role in gametogenesis and the *C4* gene product might be pivotal in the initiation of the gametogenesis program. Because Saito and Matsuda ([73]; discussed above) used continuous light during gametogenesis, we do not yet know whether *dif-1* and/or *dif-2* are defective in the light-activated or in the nitrogen-deprivation-activated pathway.

Another approach to dissecting the gametogenic pathway is to study mutants that are confused about their sexual identity. Differences in the phenotypes of mt^+ and mt^- gametes are clearly the consequence of differences in the gametogenic programs of these two cell types (see Fig. 2). Normally, the specific program to be executed in an individual cell is controlled primarily by the mt^+ or mt^- allele of the complex mating type locus [15, 85]. Exciting work in the Goodenough laboratory has led to the identification of two mutants, *imp-11* and *iso-1*, whose genetic control of mating type is disrupted [11, 31]. From a signalling perspective, *iso-1* is particularly interesting: within a clonal mt^- population, individual cells are either mt^- or pseudo-mt^+, expressing mt^+ agglutinins and mating structures, but lacking traits encoded by the absent plus allele of the mating type locus [11, 28]. Given an identical genetic make-up and environment, what determines whether a cell will develop as a mt^- or pseudo-mt^+ gamete? Presumably there is a stochastic event early in the developmental pathway, perhaps as early as the signalling cascade triggered by the environmental factors that initiate gametogenesis. The wild-type *iso-1* gene product presumably regulated this event. Therefore, the *iso-1* gene, which is unlinked to mating type but is only expressed in mt^- cells, may encode an important regulatory element, possibly a transcription factor, in the environmentally triggered signalling cascade. Because *iso-1* was generated by insertional mutagenesis, facilitating cloning by the plasmid rescue technique (e.g. [93]) we expect that molecular information will soon supplement the biological characterization of this interesting mutant.

The mating responses

Initial contact between *Chlamydomonas* gametes is via their flagella. Flagellar interactions initiate a signalling cascade that triggers a series of mating responses, the preludes to fusion. In *C. reinhardtii*, flagellar adhesion leads to activation of the gamete lytic enzyme (GLE, a metalloprotease) which has accumulated in the periplasm during gametogenesis [48, 91]. The proteolytic activity of GLE results in the shedding of gametic cell walls within seconds of flagellar adhesion (reviewed by Adair and Snell [1]). Adhesion also triggers actin polymerization and the consequent activation of the fertilization tubule [14] tipping (migration of membrane protein to the tips of the flagella [26, 28]) and flagellar tip activation (a change in the morphology of the flagellar tips [60]). Wall-less, and with activated mating structures held juxtaposed by flagellar adhesion, the cells fuse. The mating responses of *C. eugametos* are similar, but the gametes do not shed their walls until after fusion of the fertilization tubules [61, 99]. In both species, the adhesiveness of the flagella is lost upon fusion.

Agglutination is the initial signal

Initial contact between gametes of opposite mating type is mediated by adhesion of complementary agglutinins, gender- and species-specific proteins expressed on the flagellar surface of mature gametes. The agglutinins are hydroxyproline-rich glycoproteins about which we know a great deal. (For references and detailed discussions of these interesting molecules, see reviews by Goodenough [25] and Musgrave [61]). Relevant to the present discussion is the analogy between the agglutinins and extracellular matrix (ECM) proteins [25]. First, the agglutinins are not integral membrane proteins; they are fibrous extracellular proteins held on the flagellar surface by their association with the as yet unidentified agglutinin-binding protein(s) [2, 45, 72, 81]. Second, the agglutinins are homologous to proteins of the *C. reinhardtii* and *C. eugametos* cell wall, both structurally and biochemically (see Goodenough [25] for references). Hydoxyproline-rich glycoproteins are also important components of the cell wall of higher plants [84] and the cell wall is, of course, an ECM. In animal systems, ECM is emerging as much more than a simple structural entity. It is now clear that ECM plays a dynamic role in determining the form and function of multicellular assembleges [42, 53]. For example, by binding to specific receptors expressed on the plasma membranes of cells, specific ECM proteins can trigger a signalling cascade (perhaps both biochemical and cytoskeletal) that results in developmentally significant changes in gene expression. In *Chlamydomonas*, flagellar adhesion, mediated by the agglutinins, is the signal that initiates the mating responses. The mt^+ and mt^- agglutinins are distinct proteins in both *C. reinhardtii* and *C. eugametos*, but there are strong structural similarities: they are very large proteins (ca. 2×10^6 kDa), each with a globular head, a rod-like domain, a flexible region, and a terminal hook (see [25]). The receptors for the agglutinins have not been identified (complementary agglutinins?), but the signal generated on the flagellar surface is transmitted to the cell body where many of the mating responses occur.

How does agglutination generate a signal?

Flagellar agglutinins display high affinity for the flagella of complementary gametes, but complex changes in flagellar adhesiveness occur during mating. First, adhesion causes a deactivation of the agglutinins [74, 88] by an unknown mechanism [40, 87, 90]. Recruitment of new agglutinins is required to maintain flagellar adhesiveness [40]. The agglutinins, either individually or as multimeric units, appear to be multivalent because their interactions result in the formation of patches of membrane proteins. The crosslinking or 'patching' (analogous to the patching of membrane proteins observed in the activation of lymphocytes and mast cells; see [25]), initiates a signal cascade which triggers the mating responses.

Patching of integral membrane proteins, via agglutinin crosslinking, appears to be the stimulus for signal generation. Treatment of a single mating type with an antibody directed against flagellar proteins causes iso-agglutination of the cells, whereas treatment with a monovalent form of the antibody does not [28]. Furthermore, *C. reinhardtii* mutants that lack agglutinins can be induced to undergo the mating responses when their flagellar proteins are cross-linked with antisera directed against the flagellar surface [28]. In *C. eugametos*, flagella can be stripped of extrinsic proteins and the integral flagellar membrane proteins can be cross-linked by wheat germ agglutinin (reviewed by Musgrave [61]). Again, the artificially induced patching of membrane proteins stimulates the effects of agglutinin interactions, suggesting that the clustering of integral membrane proteins probably generates the intracellular signal. Although we do not yet understand how the cross-linking of membrane proteins is transduced into a signal, we do know that both cAMP and calcium play important roles in eliciting the mating responses (Fig. 3).

cAMP is an important second messenger for the mating responses

It is well established that cAMP is a key second messenger in the pathways initiated by agglutination. Pijst *et al.* [66] working with *C. eugametos* and Pasquale and Goodenough [65] working with *C. reinhardtii*, noted large (ca. 10-fold) accumula-

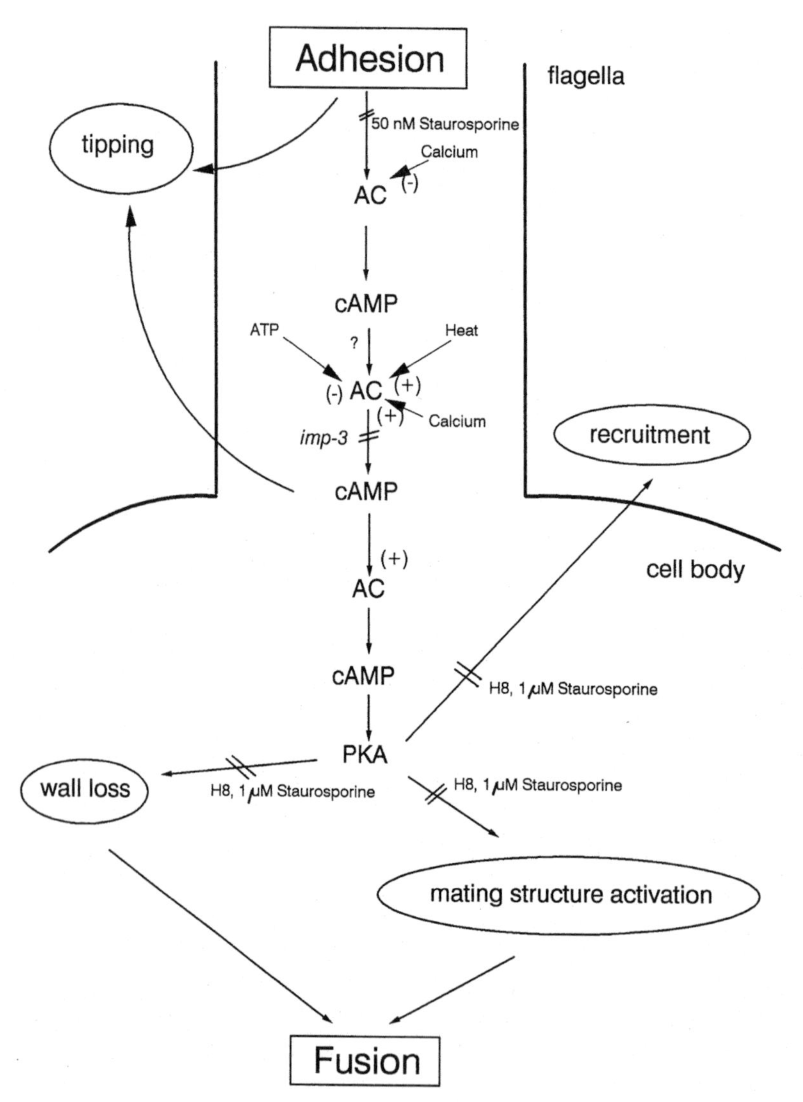

Fig. 3. Model of the signalling pathways induced by sexual adhesion. AC, adenylyl cyclase; PKA, cAMP-dependent protein kinase. See text for discussion.

tions of cAMP in gametes undergoing agglutination. Because, in *C. reinhardtii*, the exogenous application of a permeant analogue of cAMP (dibutryl-cAMP) causes recruitment of agglutinins, cell wall loss, mating structure activation and flagellar tip activation in gametes of a single mating type (i.e. in the absence of flagellar agglutination), it appears that a cAMP signal couples agglutination to these responses [24, 51, 65]. Moreover, treatment with dibutryl-cAMP can rescue mating in non-agglutinating mutants [65], suggesting that cAMP activates not only the known mating responses, but all necessary responses. The simplest hypothesis is that agglutination-induced patching directly activates adenylyl cyclase and that the cAMP produced then signals each of the subsequent mating responses through cAMP-dependent kinase cascades and/or the gating of cAMP-gated ion channels. However, the experimental data suggest that the real story is more complex (Fig. 3).

Adenylyl cyclases in Chlamydomonas

One approach to understanding how agglutination might activate adenylyl cyclase has been to study the modes of regulation of this enzyme in *Chlamydomonas*. Flagellar membranes from *C. reinhardtii* gametes express a Mn^{2+}-stimulated adenylyl cyclase activity on the order of 0.5 nmol min^{-1} mg^{-1}. The flagellar adenylyl cyclase is insensitive to GTP, GTPγS, and AlF_4^-, suggesting that it is not regulated by G proteins, nor does it respond to forskolin [65, 111]. Surprisingly, adenylyl cyclase activity in preparations of flagella from mature gametes can be stimulated 10-fold by a brief incubation at elevated temperatures [76, 111]. Zhang *et al.* [111] mimicked the 45 °C stimulation of cyclase activity by incubating the flagellar preparations in the protein kinase inhibitor staurosporin (1 μM) but not by the non-hydolysable ATP analogue, App(NH)p. Because the cyclase activity was inhibited by ATP, these authors hypothesized that the gametic flagellar adenylyl cyclase of *C. reinhardtii* is inhibited by a heat-sensitive protein kinase. The Goodenough laboratory has also observed the activation of this activity by heat and its inhibition by ATP, but in their hands stauroporin does not block the inhibitory effects of ATP and the non-hydrolysable analogue, App(NH)p is as inhibitory as ATP, suggesting that the inhibitory effects of the nucleotides is not due to phosphorylation [76]. It is important to note that flagellar preparations from vegetative cells express a cyclase activity that is not activated by heat or by Mn^{2+}, suggesting that a specific adenylyl cyclase activity is expressed in mature gametes [76, 109]. Gametic flagellar adenylyl cyclase activity is stimulated two-fold if, prior to isolation of the flagella, the gametes are allowed to undergo normal agglutination with cells of the opposite mating type or if a single mating type is induced to undergo iso-agglutination by antibody treatment [76]. This exciting result would neatly tie together the stimulation of adenylyl cyclase activity and sexual adhesion, except that cross-linking of the flagellar proteins of vegetative cells with anti-flagellar antibody produces the same result [76]. It appears that the 'patching'-induced 2-fold stimulation of adenylyl cyclase is common to gametic and vegetative flagella but, as discussed above, only the gametic flagellar activity can be stimulated by heat and Mn^{2+}. Zhang and Snell [110] have recently reported a 3-fold stimulation of gametic adenylyl cyclase by *in vitro* adhesion of isolated flagella of mt^+ and mt^- cells. In control experiments, neither mixing of vegetative flagella of the two mating types nor gametic flagella of a single mating type produced the stimulation of adenylyl cyclase activity. This result, taken together with the results of Saito *et al.* [76], suggests that, while patching of membrane proteins may stimulate the same adenylyl cyclase activity in both vegetative and gametic cells, only gametes possess the means of inducing patching (i.e. they carry complementary agglutinins and agglutinin receptors).

But how is the patching of membrane proteins coupled to the stimulation of adenylyl cyclase? Zhang and Snell [110] report that 50 nM staurosporin, a much lower concentration than used in the earlier studies [76, 100, 111], inhibits adhesion-induced activation of adenylyl cyclase, suggesting that the adhesion-stimulated cyclase activity is up-regulated by phosphorylation. At first this result appears at odds with the earlier report of an inhibitory effect of phosphorylation [109]. However, the results are theoretically reconciled by the proposal that two different steps in the activation of adenylyl cyclase are differently regulated by phosphorylation (see Fig. 3). The first step is inhibited by 50 nM staurosporin in a dibutryl-cAMP-rescuable manner and is not inhibited by a different protein kinase inhibitor, H-8, whereas the second step is inhibited by H-8 and higher concentrations of staurosporin (1 μM) and the inhibition cannot be rescued by dibutryl-cAMP. So far these experiments have only been done with adenylyl cyclase stimulated by flagellar adhesion. The model depicted in Fig. 3 predicts that 50 nM staurosporin will block the 2-fold stimulation induced by the artificial patching of membrane proteins [76].

Although both vegetative and gametic flagellar adenylyl cyclases can respond to the patching of as yet undefined membrane proteins, gametes are endowed for a unique response to this stimulation. As discussed by Saito *et al.* [76], the gametic flagella could express either two different isozymes of adenylyl cyclase, or two different mechanisms for regulation of a single isozyme. The two flagellar adenylyl cyclase activities have been distinguished pharmacologically and genetically.

The second, and putatively gamete-specific, adenylyl cyclase activity is defective in a non-mating mutant of *C. reinhardtii*. Gametes of the mutant strain, *imp-3*, agglutinate normally but agglutination does not elicit any of the mating responses and the cells do not fuse [29]. The agglutination-induced accumulation of cAMP that occurs in a wild-type mating is not observed in a cross of *imp-3* mt^+ and mt^- gametes [76]. Moreover, *imp-3* matings can be rescued if the gametes are treated with dibutryl-cAMP [76]. Therefore, *imp-3* is defective in transduction of the agglutination signal into a cAMP signal. Because the patching-induced, 2-fold stimulation of adenylyl cyclase is observed in both vegetative and gametic flagella, this activity is unlikely to directly produce the accumulation of cAMP necessary for sexual signalling. Consistent with this idea, *imp-3* flagella show the normal patching-induced stimulation of adenylyl cyclase activity [76]. However, flagellar membranes of *imp-3* gametes do not express the 10-fold heat-induced stimulation of adenylyl cyclase observed in wild-type gametes [76]. These data provide strong support for the idea that the heat-induced activation is related to production of the cAMP signal that elicits the mating responses (Fig. 4). As discussed above, Zhang *et al.* [109, 111] suggest that heat treatment is mimicking the inhibition of a kinase (or the activation of a phosphatase). If this is true, we need to learn how agglutination leads to a net dephosphorylation of the adenylyl cyclase or a regulator of adenylyl cyclase.

The 10-fold accumulation of cAMP observed during a normal mating reaction may not be a direct consequence of the hypothesized 10-fold activation of the gamete-specific flagellar adenylyl cyclase (Fig. 3). Instead, the cAMP produced in the flagella may serve to activate an isozyme of adenylyl cyclase located in the cell body. Gametic cell bodies express 90% of total adenylyl cyclase activity on a per cell basis, and the activity of the cell body enzyme is stimulated ca. 1.5-fold during mating [76]. Saito *et al.* have proposed a sequential model of cyclase activation: the adhesion-induced two-fold activation of flagellar cyclase stimulates the 10-fold activation of flagellar cyclase which is followed temporally and, they suggest, causally by the 1.5-fold stimulation of cell body cyclase. Because Cd^{2+} inhibits the 2-fold stimulation and all mating responses, but not the *in vitro* heat-induced 10-fold stimulation, they hypothesize that the 2-fold stimulation is the initial step in the signalling cascade. The caveat is that Cd^{2+} is the only inhibitor with this property and it may be inhibiting the patching-induced activation of cyclase and the mating responses via independent mechanisms. Alternatively, Cd^{2+} could inhibit an important response to patching, thereby preventing both patching-induced activation of cyclase and patching-induced activation of the mating responses. Nevertheless, the model of Saito *et al.* [76] hypothesizes that the cell body cyclase is directly activated by cAMP that diffuses from the flagella. This hypothesis is tentatively supported by an extremely modest 1.2-fold stimulation of cell body cyclase by cAMP *in vitro* [76].

Calcium plays a role in sexual signalling

An increase of cytosolic free $[Ca^{2+}]$ is a key feature of many signal transduction pathways. The influx of extracellular Ca^{2+} and the release of Ca^{2+} from different intracellular Ca^{2+} pools play distinct roles in Ca^{2+} signalling [6]. There is substantial evidence that Ca^{2+} plays an important role in the mating reactions of *Chlamydomonas*, but what that role is, and where the Ca^{2+} comes from, are both complex issues.

Many pharmacological agents, including La^{2+}, Cd^{2+}, lidocaine, trifluoroperizine (TFP), W7, colchicine and amiprophos-methyl (APM) block the adhesion-induced accumulation of cAMP (and therefore the mating responses) without blocking adhesion [13, 33, 65, 90]. The Goodenough lab [33, 76] refers to these agents as 'upstream inhibitors' because they prevent the coupling of adhesion to cAMP accumulation but do not interfere with cellular responses if dibutyryl-cAMP is supplied. Such a diversity of inhibitors suggests that there is more than one step coupling adhesion to cAMP accumulation. Nevertheless, Goodenough *et al.* [33] have pointed out that among their other activities, these agents are all known to interfere with Ca^{2+} signalling (La^{3+} and Cd^{2+} block Ca^{2+} channels; TFP and W7 are calmodulin antagonists; APM and lidocaine can affect Ca^{2+} fluxes; see Goodenough *et al.* [33] for references). Their effects on mating can be attenuated by the addition of excess Ca^{2+} [33, 89]. Furthermore, several of these agents are known to inhibit other Ca^{2+}-mediated behaviours in *Chlamydomonas*, including phototaxis, deflagellation, and flagellar motility [7, 63]. Clearly, it is not necessary to suggest that each and every upstream inhibitor is affecting the same, Ca^{2+}-dependent signalling step, but the pharmacological data strongly implicates a role for Ca^{2+} in coupling sexual adhesion to the accumulation of cAMP. If Ca^{2+} is involved, where does it come from, and what is its role?

Several lines of evidence suggest that sexual adhesion is not stimulating a general increase in cytosolic-free [Ca^{2+}]. First, attempts to induce mating responses with Ca^{2+} ionophores have failed [8, 65]. Second, large increases in intracellular [Ca^{2+}] causes the cells to deflagellate [78]. Third, using the length of the nucleus-basal body connector (a Ca^{2+}-dependent contractile apparatus [77]) as an indirect method of assessing intracellular [Ca^{2+}] suggests that Ca^{2+} levels temporarily *decrease* during mating [33]. However, it is important to note that *Chlamydomonas* flagella are rich in Ca^{2+}-binding proteins such as calmodulin and centrin [23, 38, 77, 80] and it is likely that Ca^{2+} responses are mediated rapidly and locally.

Whether extracellular Ca^{2+} is required for mating has been difficult to discern. Bloodgood and Levin [8] reported that *C. reinhardtii* mates with high efficiency in the absence of extracellular Ca^{2+}, but Goodenough *et al.* [33] report a reduction in mating efficiency in Ca^{2+}-depleted media. Because EDTA inhibits cell wall lysis (GLE is a metalloprotease) and strips agglutinins from the flagellar surface [25], Goodenough *et al.* [33] used repeated washes in media with no added Ca^{2+} in order to deplete extracellular Ca^{2+}. In so doing, they learned that over a 30 min incubation, [Ca^{2+}] in the buffer increased from 0.04 to ca. 25 μM [33]. Apparently the gametes are releasing Ca^{2+} into the medium. Consequently, direct documentation of whether extracellular Ca^{2+} is required at the time of mating is problematic.

Goodenough *et al.* [33] argue that extracellular Ca^{2+} is necessary for mating, but whether the requirement is for a specific influx of Ca^{2+} or is necessary to maintain the membrane in an unperturbed state has not yet been determined. The Cd^{2+} block of the mating responses and of the patching-induced activation of adenylyl cyclase (described above) may be mediated by the blockade of a plasma membrane Ca^{2+} channel [33, 76]. As described above, the Cd^{2+} block of mating responses can be rescued by dibutyryl-cAMP, suggesting that Cd^{2+} is preventing the activation of adenylyl cyclase. Consequently, one of the hypotheses presented by these authors is that the patching-induced activation of adenylyl cyclase is mediated by an influx of Ca^{2+}. This possibility would appear to be ruled out by the observation that La^{3+}, a particularly potent blocker of Ca^{2+} channels, does not block the patching-induced activation of adenylyl cyclase, but does block mating in a dibutyryl-cAMP-rescuable manner ([76]; see also Quarmby and Hartzell [67] and references therein).

Several laboratories have suggested that the release of Ca^{2+} from intracellular stores plays an important role in sexual signalling. Bloodgood and Levin [8] observed a short-lived stimulation of ^{45}Ca efflux within the first 6 min of mating, suggesting that the cells were clearing the cytosol of the sudden excess of Ca^{2+} by extruding it into the medium. (Only a small component of this efflux can be attributed to the shedding of the gametic cell walls [8, 33]). Spectrographic measurements using the Ca^{2+}-indicator dye Arsenazo III, also indicate a mating-stimulated efflux of Ca^{2+} [33]. However, the mating-stimulated Ca^{2+} efflux appears to be a response rather than a trigger of cAMP accumulation. First, dibutryl-cAMP stimulates a comparable Ca^{2+} efflux, and second, the upstream inhibitors, Cd^{2+} and TFP, which block accumulation of cAMP, also block mating-induced efflux [34]. But the fact that Ca^{2+} efflux is stimulated suggests that a local and/or transient increase in intracellular free [Ca^{2+}] does occur. The Ca^{2+} efflux might reflect an intracellular signalling event, perhaps even a Ca^{2+}-mediated activation of adenylyl cyclase. Heat stimulation of the gametic flagellar adenylyl cyclase is augmented by Ca^{2+} in a dose-dependent manner from 10–500 nM free [Ca^{2+}] [76]. Basal activity of the gametic flagellar adenylyl cyclase is inhibited by Ca^{2+} when assayed *in vitro* [76, 109].

The release of Ca^{2+} from intracellular stores in other cell types is specifically regulated by the opening of Ca^{2+} channels by at least three different signals: IP_3, Ca^{2+}, and cADP-ribose. Cytosolic IP_3 gates the IP_3-receptor, one class of intracellular Ca^{2+} channel [6]. Ca^{2+} gates this and another class of intracellular Ca^{2+} channel, the ryanodine receptor [6]. There is now compelling evidence that the third activator of intracellular Ca^{2+} channels, cyclic ADP-ribose, probably in synergy with Ca^{2+}, also releases Ca^{2+} from ryanodine-sensitive pools [20, 21, 92, 105]. These two types of channel can occur in the same cell and can control release of Ca^{2+} from apparently distinct pools [16, 102]. To my knowledge, no one has yet examined *Chlamydomonas* for the presence of cADPR or cADPR-sensitive Ca^{2+} channels, but a signal-responsive production of IP_3 is well-documented in both *C. reinhardtii* and *C. eugametos* [62, 68].

In *C. reinhardtii*, IP_3 does not accumulate during mating [33], although the possibility of a transient or localized accumulation cannot be ruled out. The situation appears to be different in *C. eugametos*. In this species, activation of the mating responses by dibutryl-cAMP is less complete than it is for *C. reinhardtii*, and IP_3 may play an important role as a second messenger in sexual signalling (reviewed by Musgrave [61]). Although cAMP accumulates rapidly after gametes are mixed and dibutryl-cAMP triggers many of the mating responses, dibutryl-cAMP does not activate the mating structures of *C. eugametos* [83]. Musgrave *et al.* [62] demonstrated than ethanol stimulates both IP_3 accumulation and the pseudo-activation of mating structures. Because earlier evidence [82] suggested that increases in intracellular Ca^{+2} trigger mating structure activation in this species, Musgrave *et al.* [62] hypothesize that IP_3 used to produce the Ca^{2+} signal during mating. Consistent with this idea, IP_3 begins to accumulate after 5 min of mating, correlating with the timing of mating structure activation in this species [61]. We are therefore left with the question, if IP_3 is the second messenger for mating structure activation in *C. eugametos*, and cAMP does not stimulate IP_3 production, then what signal stimulates the accumulation of IP_3?

Activation of the mating responses

Flagellar agglutination is the external signal that stimulates the mating responses in a *Chlamydomonas* gamete. In the sections above I have reviewed the state of our understanding of agglutination and the signals it generates. In this section I will briefly focus on how the signals are translated into action. The reader is referred to Goodenough [25, 27], Musgrave [61] and Snell [86] for more thorough discussions of this topic.

In response to agglutination of *C. reinhardtii* gametes, pro-GLE is proteolytically cleaved to GLE, the active form of lysin, the inner cell wall is digested and the walls are shed like eggshells. One study strongly suggests that agglutination induces secretion of a serine protease (p-lysinase) which is the activator of GLE [91] although it remains possible that GLE is activated autocatalytically [48]. Whatever the details of the pathway leading to GLE activation turn out to be, the pathway appears to be triggered by cAMP – either an accumulation of endogenous cAMP in response to the patching of membrane proteins, or the exogenous application of dibutryl-cAMP [65]. Cell wall loss is inhibited by the protein kinase inhibitors staurosporin or H-8, strongly suggesting that cAMP is exerting its effects via the activation of a cAMP-dependent protein kinase [33, 65]. Cell-cell fusion is similarly induced by cAMP and inhibited by the protein kinase inhibitors [33].

Transport of agglutinins from the cell body to the flagellar surface is necessary in order for flagellar adhesion to be prolonged until fusion can occur [87]. It is not clear whether the cell body pool of agglutinins is stored cytoplasmically or on the cell body plasma membrane [40, 76, 86]. Like cell wall lysis and cell fusion, recruitment of agglutinins can be stimulated by cAMP [24, 26, 49, 51]. However, the mechanism must differ in that recruitment of agglutinins can occur in the presence of staurosporin and H-8 [33]. The agglutinins, pro-GLE and *p*-lysinase must all be secreted either during gametogenesis or during the activation of the mating responses. It has been hypothesized that both the agglutinins and *p*-lysinase are secreted in response to sexual agglutination. If either of these secretion events is regulated, it will be interesting to learn how a cAMP signal is coupled to secretion.

The mating structures of mt^+ and mt^- gametes are distinct (see Fig. 2) and the mechanisms of activation may also differ. As discussed above, a cAMP-dependent protein kinase appears to mediate the activation of the *C. reinhardtii* mating structures. For mt^+, this activation involves the assembly of actin filaments and the movement of surface coat material, termed the fringe, to the tip of the fertilization tubule [14, 31]. *Imp-1* and *imp-II* mutants do not form the mt^+ fringe and are unable to fuse [31]. For mt^-, the mating structure approximately doubles in size, with the fringe at its tip [31]. In *C. eugametos*, mating structure activation is not stimulated by cAMP, but may be produced in response to an IP_3-generated Ca^{2+} signal (see [61]).

Flagellar tip activation refers to a morphological change that occurs at the tip of the flagella (from slender to bulbous), as a result of elongation of the A microtubes and the deposition of electron-dense material [59]. Little is known about the role flagellar tip activation plays in the mating process. It is of interest here to note that lidocaine inhibits all of the mating responses except flagellar tip activation [89].

The fifth mating response, known as tipping, involves the transport of patches of cross-linked agglutinins to the tips of the flagella, a process analogous to 'capping' in lymphocyte activation (see [25]). Unlike all of the other mating responses discussed so far, tipping requires flagellar adhesion. Application of dibutryl-cAMP to unmated gametes does not activate tipping [24, 51]. But adhesion alone is not sufficient to induce tipping. When treatment with either Cd^{2+} or La^{3+} is used to inhibit cAMP accumulation in agglutinating cells (see above), flagellar tipping does not occur [26]. The protein kinase inhibitors, staurosporin and H-8 also block tipping [33]. Thus tipping requires both flagellar adhesion and, probably, a cAMP-dependent protein kinase phosphorylation event.

Fusion and formation of the zygospore

When the mating responses, described above, successfully culminate in cell fusion, a new series of events is set in motion. Within minutes, the flagella lose their adhesiveness, cAMP levels drop, and there is a change in gene expression [25, 27]. Several cDNA clones for genes that are transcribed within minutes of fertilization have been isolated [17, 58, 97, 107]. Some of these gene products participate in the formation of the dessication-resistant zygotic cell wall [107]. Another intriguing aspect of early zygotic development is the destruction of the mt^- chloroplast DNA (see review by Gillham *et al.* [22]). Uniparental inheritance of chloroplast genomes is common to all photosynthetic eukaryotes. In *Chlamydomonas*, the selective degradation of mt^- chloroplast DNA is controlled, at least in part, by products of the *Ezy-1* mating type-linked gene cluster [4, 27]. Thus, a crow-bar has been inserted into a crevice of this long-standing problem and perhaps we will soon see the inner workings of uniparental inheritance. Meanwhile, the signals and signal transduction pathways that communicate the event of cell fusion to the changes in cAMP levels, flagellar adhesion, flagellar resorption and early zygotic gene expression, remain largely unknown. Musgrave and colleagues [113] have demonstrated that activation of phospholipase C correlates with cell fusion in *C. eugametos*. Because fusion precedes flagellar deagglutination, the specific inhibition of deagglutination (but not fusion) by 0.1 μM staurosporine suggests to these authors that IP_3 and DAG may be the second messengers produced as a consequence of fusion. If so, how does fusion activate phospholipase C? How is phospholipase C activation coupled to the changes which produce a zygote?

Germination

Zygotes must mature (undergo meiosis) before they are competent to germinate. Mature zygotes can be induced to germinate by environmental signals, particularly light. Wegener and Beck [104] have identified three different genes that are expressed >24 h after mating. The messages for these genes disappear upon light-induced germination, but the temporal patterns are gene-specific. Based solely on temporal correlations, these authors hypothesize that the *zymC* gene product may participate in the attainment of competence for germination or in the initiation of germination itself. However, we know virtually nothing about the nature of the light signal that triggers germination, how it is received or how it is transduced into the responses that lead to the hatching of vegetative cells.

Acknowledgements

Drs H.C. Hartzell, U.W. Goodenough and W.S. Sale are warmly thanked for their constructive critiques and insightful discussions. I also appreciate the comments of my colleagues from Amsterdam, especially Teun Munnik. I thank the NSF for financial support.

References

1. Adair WS, Snell WJ: Organization and *in vitro* assembly of the Chlamydomonas reinhardtii cell wall. In: Varner JE (ed) Self Assembling Architecture, pp. 25–41. A.R. Liss, New York (1990).
2. Adair W, Monk B, Cohen R, Wang C, Goodenough U: Sexual agglutinins from the Chlamydomonas flagellar membrane. J Biol Chem 257: 4593–4602 (1982).
3. Ahmad M, Cashmore AR: HY4 gene of *A. thaliana* encodes a protein with characteristics of a blue-light photoreceptor. Nature 366: 162–166 (1993).
4. Armbrust EV, Ferris PJ, Goodenough UW: A mating type-linked gene cluster expressed in *Chlamydomonas* zygotes participates in the uniparental inheritance of the chloroplast genome. Cell 74: 801–811 (1993).
5. Beck CF, Acker A: Gametic differentiation of *Chlamydomonas reinhardtii*. Plant Physiol 98: 822–826 (1991).
6. Berridge MJ: Inositol trisphosphate and calcium signalling. Nature 361: 315–325 (1993).
7. Bloodgood RA: Directed movement of ciliary and flagellar membrane components: a review. J Biochem Cell Biol 63: 608–620 (1992).
8. Bloodgood RA, Levin EN: Transient increase in calcium efflux accompanies fertilization in *Chlamydomonas*. J Cell Biol 97: 397–404 (1983).
9. Buerkle S, Gloeckner G, Beck C: *Chlamydomonas* mutants affected in the light-dependent step of sexual differentiation. Proc Natl Acad Sci USA 90: 6981–6985 (1993).
10. Bulte L, Bennoun P: Translational accuracy and sexual differentiation in *Chlamydomonas reinhardtii*. Curr Genet 18: 155–160 (1990).
11. Campbell AM, Rayala HJ, Goodenough UW: Mating-type minus gametic mutant *iso-1* converts *Chlamydomonas reinhardtii* from heterothallism to incomplete homothallism. Mol Biol Cell 4: 147a (1993).
12. Deng XW: Fresh view of light signal transduction in plants. Cell 76: 423–426 (1994).
13. Detmers PA, Condeelis J: Trifluoperazine and W-7 inhibit mating in *Chlamydomonas* at an early stage of gametic interaction. Exp Cell Res 163: 317–326 (1986).
14. Detmers PA, Goodenough UW, Condeelis J: Elongation of the fertilization tubule in *Chlamydomonas*: new observations on the core microfilaments and the effect of transient intracellular signals on their structural integrity. J Cell Biol 97: 522–532 (1983).
15. Ebersold WT, Levine RP, Levine EE, Olmsted A: Linkage maps in *Chlamydomonas reinhardii*. Genetics 47: 531–543 (1962).
16. Ellisman MH, Deerinck TJ, Ouyang Y, Beck CF, Tanksley SJ, Walton PD, Airey JA, Sutko JL: Identification and localization of ryanodine binding proteins in the avian central nervous system. Neuron 5: 135–146 (1990).

17. Ferris PJ, Goodenough UW: Transcription of novel genes, including a gene linked to the mating-type locus, induced by *Chlamydomonas* fertilization. Mol Cell Biol 7: 2360–2366 (1987).
18. Forest CL: Specific contact between mating structure membranes observed in conditional fusion-defective Chlamydomonas mutants. Exp Cell Res 148: 143–154 (1983).
19. Forest CL, Togasaki RK: Proc Natl Acad Sci USA 72: 3652–3655 (1975).
20. Galione A: Ca^{2+}-induced Ca^{2+} release and its modulation by cyclic ADP-ribose. Trends Pharmacol Sci 13: 304–306 (1992).
21. Galione A, Lee HC, Busa WB: Ca^{2+} induced Ca^{2+} release in sea urchin egg homogenates: modulation by cyclic ADP-ribose. Science 253: 1143–1146 (1991).
22. Gillham NW, Boynton JE, Harris EH: Transmission of plastid genes. In: Bogorad L, Vasil IK (eds) Cell Culture and Somatic Cell Genetics, vol. 7A, pp. 55–92. Academic Press, New York (1991).
23. Gitelman SE, Witman GB: Purification of calmodulin from *Chlamydomonas*: calmodulin occurs in cell bodies and flagella. J Cell Biol 87: 764–770 (1980).
24. Goodenough UW: Cyclic AMP enhances the sexual agglutinability of *Chlamydomonas* flagella. J Cell Biol 109: 247–252 (1989).
25. Goodenough UW: *Chlamydomonas* mating interactions. Microbial Cell-Cell Interactions: 71–111 (1991).
26. Goodenough U: Tipping flagellar agglutinins by gametes on *Chlamydomonas reinhardtii*. Cell Motility Cytoskeleton 25: 179–189 (1993).
27. Goodenough UW: Development in *Chlamydomonas* and related organisms. Annu Rev Plant Physiol Plant Mol Biol, in press (1994).
28. Goodenough UW, Jurivich D: Tipping and mating-structure activation induced in Chlamydomonas gametes by flagellar membrane antisera. J Cell Biol 79: 680–693 (1978).
29. Goodenough UW, Hwang C, Martin H: Isolation and genetic analysis of mutant strains of *Chlamydomonas reinhardi* defective in gametic differentiation. Genetics 82: 169–186 (1976).
30. Goodenough UW, Hwang C, Warren J: Sex-limited expression of gene loci controlling flagellar membrane agglutination in the *Chlamydomonas* mating reaction. Genetics 89: 235–243 (1978).
31. Goodenough UW, Detmers PA, Hwang C: Activation for cell fusion in *Chlamydomonas*. Analysis of wild-type gametes and non-fusing mutants. J Cell Biol 92: 378–386 (1982).
32. Goodenough UW, Adair WS, Collin-Osdoby P, Heuser JE: Structure of the *Chlamydomonas* agglutinin and related flagellar surface proteins *in vitro* and *in situ*. J Cell Biol 101: 924–941 (1985).
33. Goodenough UW, Shames B, Small L, Saito T, Crain RC, Sanders MA, Salisbury JL: The role of calcium in the *Chlamydomonas reinhardtii* mating reaction. J Cell Biol 121: 365–374 (1993).
34. Hall LM, Taylor KB, Jones DD: Expression of a foreign gene in *Chlamydomonas reinhardtii*. Gene 124: 75–81 (1993).
35. Harris EH: The *Chlamydomonas* Sourcebook, 1st ed., vol. 1. Academic Press, Berkeley (1989).
36. Hartzell LB, Hartzell HC, Quarmby LM: Mechanisms of flagellar excision. I. The role of intracellular acidification. Exp Cell Res 208: 148–153 (1993).
37. Harz H, Nonnengasser C, Hegemann P: The photoreceptor current of the green alga *Chlamydomonas*. Phil Trans R Soc Lond B 338: 39–52 (1992).
38. Huang B, Mengersen A, Lee VD: Molecular cloning of cDNA for caltractin, a basal body-associated calcium-binding protein: homology in its protein sequence with calmodulin and the yeast *cdc-31* gene product. J Cell Biol 197: 133–140 (1988).
39. Hunnicutt G, Snell WJ: Rapid and slow mechanisms for loss of cell adhesiveness during fertilization in *Chlamydomonas*. Devel Biol 147: 216–224 (1991).
40. Hunnicutt GR, Kosfiszer MG, Snell WJ: Cell body and flagellar agglutinins in *Chlamydomonas reinhardtii*: the cell body plasma membrane is a reservoir for agglutinins whose migration to the flagella is regulated by a functional barrier. J Cell Biol 111: 1605–1616 (1990).
41. Hwang C, Monk BC, Goodenough UW: Linkage of mutations affecting minus flagellar membrane agglutinability to the mt^- mating-type locus of *Chlamydomonas*. Genetics 99: 41–47 (1981).
42. Jones PL, Schmidhauser C, Bissell MJ: Regulation of gene expression and cell function by extracellular matrix. Crit Rev Eukaryot Gene Expr 3: 137–154 (1993).
43. Jones R: Physiological and biochemical aspects of growth and gametogenesis in *Chlamydomonas reinhardtii*. Ann NY Acad Sci 175: 649–659 (1970).
44. Jones RF, Kates JR, Keller SJ: Protein turnover and macromolecular synthesis during growth and gametic differentiation in *Chlamydomonas reinhardtii*. Biochim Biophys Acta 157: 589–598 (1968).
45. Kalshoven HW, Musgrave A, van den Ende H: Mating receptor complex in the flagellar membrane of *Chlamydomonas eugametos* gametes. Sex Plant Reprod 3: 77–87 (1990).
46. Kaufman L: Transduction of blue-light signals. Plant Physiol 102: 333–337 (1993).
47. Kindle KL: High-frequency nuclear transformation of *Chlamydomonas reinhardtii*. Proc Natl Acad Sci USA 87: 1228–1232 (1990).
48. Kinoshita T, Fukuzawa H, Shimada T, Saito T, Matsuda Y: Primary structure and expression of a gamete lytic enzyme in *Chlamydomonas reinhardtii*: similarity of functional domains to matrix metalloproteases. Proc Natl Acad Sci USA 89: 4693–4697 (1992).
49. Kooijman R, Elzenga TJM, de Wildt P, Musgrave A, Schuring F, van den Ende H: Light dependence of sexual

agglutinability in *chlamydomonas eugametos*. Planta 169: 370–378 (1986).

50. Kooijman R, de Wildt P, Homan WL, Musgrave A, van den Ende H: Light affects flagellar agglutinability in *Chlamydomonas eugametos* by modification of the agglutinin molecules. Plant Physiol 86: 216–223 (1988).
51. Kooijman R, de Wildt P, van den Briel W, Tan S, van den Ende H: Cyclic AMP is one of the intracellular signals during the mating of *Chlamydomonas eugametos*. Planta 181: 529–537 (1990).
52. Li HM, Washburn T, Chory J: Regulation of gene expression by light. Curr Opin Cell Biol 5: 455–460 (1993).
53. Lin CQ, Bissel MJ: Multi-faceted regulation of cell differentiation by extracellular matrix. FASEB J 7: 737–743 (1993).
54. Martin NC, Goodenough UW: Gametic differentiation in *Chlamydomonas reinhardtii*. J Cell Biol 67: 587–605 (1975).
55. Matsuda Y, Tamaki S, Tsubo Y: Mating type specific induction of cell wall lytic factor by agglutination of gametes in *Chlamydomonas reinhardtii*. Plant Cell Physiol 19: 1253–1261 (1978).
56. Matsuda Y, Saito T, Umemoto T, Tsubo Y: Transmission patterns of chloroplast genes after polyethylene glycol-induced fusion of gametes in non-mating mutants of *Chlamydomonas reinhardtii*. Curr Genet 14: 53–58 (1988).
57. Matsuda Y, Shimada T, Sakamoto Y: Ammonium ions control gametic differentiation and dedifferentiation in *Chlamydomonas reinhardtii*. Plant Cell Physiol 33: 909–914 (1992).
58. Matters GL, Goodenough UW: A gene/pseudogene tandem duplication encodes a cysteine-rich protein expressed during zygote development in *Chlamydomonas reinhardtii*. Mol Gen Genet 232: 81–88 (1992).
59. Mesland DAM, Hoffman JL, Caligor E, Goodenough UW: Flagellar tip activation stimulated by membrane adhesions in *Chlamydomonas* gametes. J Cell Biol 84: 599–617 (1980).
60. Mitchell DR, Kang Y: Identification of oda6 as a *Chlamydomonas* dynein mutant by rescue with the wild-type gene. J Cell Biol 113: 835–842 (1991).
61. Musgrave A: Mating in *Chlamydomonas*. In: Round FE, Chapman DJ (eds), Progress in Phycological Research, vol. 9, pp. 193–237. Biopress, Bristol (1993).
62. Musgrave A, Kuin H, Jongen M, de Wildt P, Schuring F, Klerk H, van den Ende H: Ethanol stimulates phospholipid turnover and inositol 1,4,5-trisphosphate production in *Chlamydomonas eugametos* gametes. Planta 186: 442–449 (1992).
63. Musgrave A, Schuring F, Munnik T, Visser K: Inositol 1,4,5-trisphosphate as fertilization signal in plants: testcase *Chlamydomonas eugametos*. Planta 171: 280–284.
64. Nultsch W, Pfau J, Dolle R: Effects of calcium channel blockers on phototaxis and motility of *Chlamydomonas reinhardtii*. Arch Microbiol 144: 393–397 (1986).
65. Pasquale SM, Goodenough UW: Cyclic AMP functions as a primary sexual signal in gametes of *Chlamydomonas reinhardtii*. J Cell Biol 105: 2279–2292 (1987).
66. Pijst HLA, van Driel R, Janssens PMW, Musgrave A, van den Ende H: Cyclic AMP is involved in sexual reproduction of *Chlamydomonas eugametos*. FEBS Lett 174: 132–136 (1984).
67. Quarmby LM, Hartzell HC: Two distinct, calcium-mediated, signal transduction pathways can trigger deflagellation in *Chlamydomonas reinhardtii*. J Cell Biol 124: 807–815 (1994).
68. Quarmby LM, Yueh YG, Cheshire JL, Keller LR, Snell WJ, Crain RC: Inositol phospholipid metabolism may trigger flagellar excision in *Chlamydomonas reinhardtii*. J Cell Biol 116: 737–744 (1992).
69. Ranum LPW, Thompson MD, Schloss JA, Lefebvre PA, Silflow CD: Mapping flagellar genes in *Chlamydomonas* using restriction fragment length polymorphisms. Genetics 120: 109–122 (1988).
70. Rosenbaum JL, Carlson K: Cilia regeneration in *Tetrahymena* and its inhibition by colchicine. J Cell Biol 40: 415–425 (1969).
71. Sager R, Granick S: Nutritional control of sexuality in *Chlamydomonas reinhardi*. J Gen Physiol 37: 729–742 (1954).
72. Saito T, Matsuda Y: Sexual agglutinin of mating-type minus gametes in *Chlamydomonas reinhardii*. Exp Cell Res 152: 322–330 (1984).
73. Saito T, Matsuda Y: Isolation and characterization of *Chlamydomonas* temperature-sensitive mutants affecting gametic differentiation under nitrogen-starved conditions. Curr Genet 19: 65–71 (1991).
74. Saito T, Tsubo Y, Matsuda Y: Synthesis and turnover of cell body-agglutinin as a pool for flagellar surface-agglutinin in *Chlamydomonas reinhardii*. Arch Microbiol 142: 207–210 (1985).
75. Saito T, Tsubo Y, Matsuda Y: A new assay system to classify non-mating mutants and to distinguish between vegetative cell and gamete in *chlamydomonas reinhardtii*. Curr Genet 14: 59–63 (1988).
76. Saito T, Small L, Goodenough UW: Activation of adenylyl cyclase in *Chlamydomonas reinhardtii* by adhesion and by heat. J Cell Biol 122: 137–147 (1993).
77. Salisbury JL, Sanders MA, Harpst L: Flagellar root contraction and nuclear movement during flagellar regeneration in Chlamydomonas reinhardtii. J Cell Biol 105: 1799–1805 (1987).
78. Sanders MA, Salisbury JL: Centrin plays an essential role in microtubule severing during flagellar excision in *Chlamydomonas reinhardtii*. J Cell Biol 79: 795–806 (1994).
79. Schnell RA, Lefebvre PA: Isolation of the *Chlamydomonas* Regulatory Gene NIT2 by Transposon Tagging. Genetics society of America, vol. 124, pp. 737–747 (1993).

80. Schulze D, Robenek H, McFadden GI, Melkonian M: Immunolocalization of a calcium-modulated contractile protein in the flagellar apparatus of green algae: the nucleus-basal body connector. Eur J Cell Biol 45: 51–61 (1987).
81. Schuring F, Smeenk JW, Homan WL, Musgrave A, van den Ende H: Occurrence of *O*-methylated sugars in surface glycoconjugates in *chlamydomonas eugametos*. Planta 170: 322–327 (1987).
82. Schuring F, Brederoo J, Musgrave A, van den Ende H: Increase in calcium triggers mating structure activations in *Chlamydomonas eugametos*. Federation of Microbiological Societies, vol. 71, pp. 237–240 (1990).
83. Schuring F, Musgrave A, Elders MCC, Teunissen Y, Homan WL, van den Ende H: Fusion-defective mutants of *Chlamydomonas eugametos*. Protoplasma 162: 108–119 (1991).
84. Showalter AM, Rumeau D: Molecular biology of plant cell wall hydroxyproline-rich glycoproteins. In: Adair WS, Mecham RP (eds) Organization and Assembly of Plant and Animal Extracellular Matrix, pp. 247–281. Academic Press, San Diego (1990).
85. Smith GM, Regnery DC: Inheritance of sexuality in *Chlamydomonas reinhardi*. Proc Natl Acad Sci USA 36: 246–248 (1950).
86. Snell WJ: Signal transduction during fertilization in *Chlamydomonas*. In: Kurjan J (ed) Signal Transduction: Prokaryotic and Simple Eukaryotic Systems, pp. 255–277. Academic Press, New York (1993).
87. Snell WJ, Moore WS: Aggregation-dependent turnover of flagellar adhesion molecules in *Chlamydomonas* gametes. J Cell Biol 84: 203–210 (1980).
88. Snell WJ, Roseman S: Kinetics of adhesion and deadhesion of *chlamydomonas* gametes. J Biol Chem 254: 10820–10829 (1979).
89. Snell WJ, Buchanan M, Clausell A: Lidocaine reversibly inhibits fertilization in *Chlamydomonas*: a possible role for calcium in sexual signalling. Cell Biol 94: 607–612 (1982).
90. Snell WJ, Kosfizer MG, Clausell A, Perillo N, Imam S, Hunnicutt G: A monoclonal antibody that blocks adhesion of *Chlamydomonas* mt^{+} gametes. J Cell Biol 103: 2449–2456 (1986).
91. Snell WJ, Eskue WA, Buchanan MJ: Regulated secretion of a serine protease that activates an extracellular matrix-degrading metalloprotease during fertilization in *Chlamydomonas*. J Cell Biol 109: 1689–1694 (1989).
92. Takasawa S, Nata K, Yonekura H, Okamoto H: Cyclic ADP-ribose in insulin secretion from pancreatic beta cells. Science 259: 370–373 (1993).
93. Tam LW, Lefebvre PA: Cloning of flagellar genes in *Chlamydomonas reinhardtii* by DNA insertional mutagenesis. Genetics 135: 375–384 (1993).
94. Tomson AM, Demets R, Bakker NPM, Stegwee D, van den Ende H: Gametogenesis in liquid cultures of *Chlamydomonas eugametos*. J Gen Microbiol 131: 1553–1560 (1985).
95. Treier U, Beck CF: Changes in gene expression patterns during the sexual life cycle of Chlamydomonas reinhardtii. Physiol Plant 83: 633–639 (1991).
96. Treier U, Fuchs S, Weber M, Wakarchuk WW, Beck CF: Gametic differentiation in *Chlamydomonas reinhardtii*: light dependence and gene expression patterns. Arch Microbiol 152: 572–577 (1989).
97. Uchida H, Kawano S, Sato N, Kuroiwa T: Isolation and characterization of novel genes which are expressed during the very early stage of zygote formation in *Chlamydomonas reinhardtii*. Curr Genet 24: 296–300 (1993).
98. Vallon O, Bulte L, Kuras R, Olive J, Wollman F: Extensive accumulation of an extracellular *L*-amino-acid oxidase during gametogenesis of *Chlamydomonas reinhardtii*. Eur J Biochem 215: 351–360 (1993).
99. Van den Ende H: Vegetative and gametic development in the green alga *Chlamydomonas*. Adv Bot Res, in press (1994).
100. Van Winkle-Swift KP: *Chlamydomonas* surrenders. Nature 358: 106–107 (1992).
101. Von Gromoff ED, Beck CF: Genes expressed during differentiation of *Chlamydomonas reinhardtii*. Mol Gen Genet 241: 415–421 (1993).
102. Walton PD, Airey JA, Sutko JL, Beck CF, Mignery GA, Sudhof TC, Deerinck TJ, Ellisman MH: Ryanodine and inositol trisphosphate receptors coexist in avian cerebellar purkinje neurons. J Cell Biol 113: 1145–1157 (1991).
103. Wegener D, Beck CF: Identification if novel genes specifically expressed in *Chlamydomonas reinhardtii* zygotes. Plant Mol Biol 16: 937–946 (1991).
104. Weissig H, Beck CF: Action spectrum for the light-dependent step in gametic differentiation of *Chlamydomonas reinhardtii*. Plant Physiol 97: 118–121 (1990).
105. White AM, Watson SP, Galione A: Cyclic ADP-ribose-induced Ca^{2+} release from rat brain microsomes. FEBS Lett 318: 259–263 (1993).
106. Witman GB: *Chlamydomonas* phototaxis. Trends Cell Biol 3: 403–408 (1993).
107. Woessner JP, Goodenough UW: Molecular characterization of a zygote wall protein: an extensin-like molecule in *Chlamydomonas reinhardtii*. Plant Cell 1: 901–911 (1989).
108. Zachleder V, Jakobs M, van den Ende H: Relationship between gametic differentiation and the cell cycle in the green alga *Chlamydomonas eugametos*. J Gen Microbiol 137: 1333–1339 (1991).
109. Zhang Y, Snell WJ: Differential regulation of adenylylcyclases in vegetative and gametic flagella of *Chlamydomonas*. J Biol Chem 268: 1786–1791 (1993).
110. Zhang Y, Snell WJ: Flagellar adhesion-dependent regulation of *Chlamydomonas* adenylyl cyclase *in vitro*: a possible role for protein kinases in sexual signalling. J Cell Biol 617–624 (1994).
111. Zhang Y, Ross EM, Snell WJ: ATP-dependent regulation of flagellar adenylylcyclase in gametes of *Chlamydomonas reinhardtii*. J Biol Chem 266: 22954–22959 (1991).

Plant Molecular Biology **26**: 1289–1303, 1994.

Control of cell proliferation during plant development

P. Ferreira [1], A. Hemerly [1], M. Van Montagu [1,*] and Dirk Inzé [2]
[1] *Laboratorium voor Genetica (*author for correspondence);* [2] *Laboratoire associé de l'Institut National de la Recherche Agronomique (France), Universiteit Gent, B-9000 Gent, Belgium*

Received 31 March 1994; accepted 26 April 1994

Key words: cell division, G_1 to S transition, G_2 to m transition, plant development

Abstract

Knowledge of the control of cell division in eukaryotes has increased tremendously in recent years. The isolation and characterization of the major players from a number of systems and the study of their interactions have led to a comprehensive understanding of how the different components of the cell cycle apparatus are brought together and assembled in a fine-tuned machinery. Many parts of this machine are highly conserved in organisms as evolutionary distant as yeast and animals. Some key regulators of cell division have also been identified in higher plants and have been shown to be functional homologues of the yeast or animal proteins. Although still in its early days, investigations into the regulation of these molecules have provided some clues on how cell division is coupled to plant development.

Introduction

The first appreciation of the cell cycle as consisting of four distinct phases was made in root tips of *Vicia faba* [57]. Despite this initial finding, the lack of powerful genetic models such as yeast or of a well characterized biochemical system such as the *Xenopus* oocytes, has hindered progress and, consequently, the understanding of cell division in plants. Nevertheless, the plant cell cycle is mechanistically very similar to that of other eukaryotes. This observation leads to an almost instinctive prediction that many of the molecules involved in controlling cell division described in other systems should, first, be present in plants, and second, have similar functions as their yeast and animals counterparts. The first part of this prediction was confirmed by the isolation of several plant homologues of known cell cycle genes [5, 12, 27–29, 47–49, 52, 54, 55, 60, 93, 109]. The second part still awaits confirmation, although the functional complementation of yeast temperature-sensitive mutants by plant *cdc2* and phosphatase 1 homologues and the triggering of germinal vesicle breakdown in *Xenopus* oocytes by soybean, *Arabidopsis* and maize cyclins [49, 52, 124] demonstrated that the prediction is at least in some cases true. On the other hand, the machinery itself that drives cell cycle progression is only one of the levels of cell cycle control [114]. In addition, the cell cycle must work as an orderly number of steps and, reasonably, the cell must have mechanisms to sense if one event was properly completed before going to the subsequent one. These mechanisms are known collectively as checkpoints [46]. Finally, a third level which controls proliferation acts on the critical decisions on whether or when a cell must divide. In this review, we survey what is known about the control of proliferation in yeast and animals, and compare that to the current state of knowledge in plant systems. Other particular aspects of plant cell division such as the control of cell division in plant cultures [39] and cytokinesis [148] or during plant development [63, 137] have been extensively addressed. Specific and/or general facets of the eukaryotic cell cycle have been reviewed recently [96, 105, 108, 110, 130, 132].

The cell cycle

In a simplified scheme, a typical cell cycle is composed of four successive phases (Fig. 1). The M phase, which consists of mitosis (nuclear division) and cytokinesis (cytoplasmic division), is followed by an interval known as the G_1 phase (G = gap). Then, the cell replicates its DNA during the S phase. Another interval, the G_2 phase, separates the end of DNA synthesis from the next M phase. Cell cycle progression is mainly controlled at two crucial transition points. First, there is a critical point late in G_1, known in mammalian cells as the restriction point [118] and START in yeast [26, 123]. At this point it is decided whether to stay in the cycle or to go out of the cycle. A second crucial point occurs at G_2. Before going into mitosis, a cell senses if certain conditions are met. For instance, replication of the DNA must have been completed, so that it can be equally divided between the daughter cells. Events at this stage are known collectively as the G_2 to M transition. To gain insight into the control of proliferation, one must first take a look at the events during these two critical points. Entry into mitosis is best understood in fission yeast and marine invertebrates. On the other hand, the budding yeast and mammalian cell cultures have extensively been used in the elucidation of the regulatory mechanisms that operate during the G_1 to S transition. For the sake of brevity, we will restrict most of the discussion in each case to the two systems.

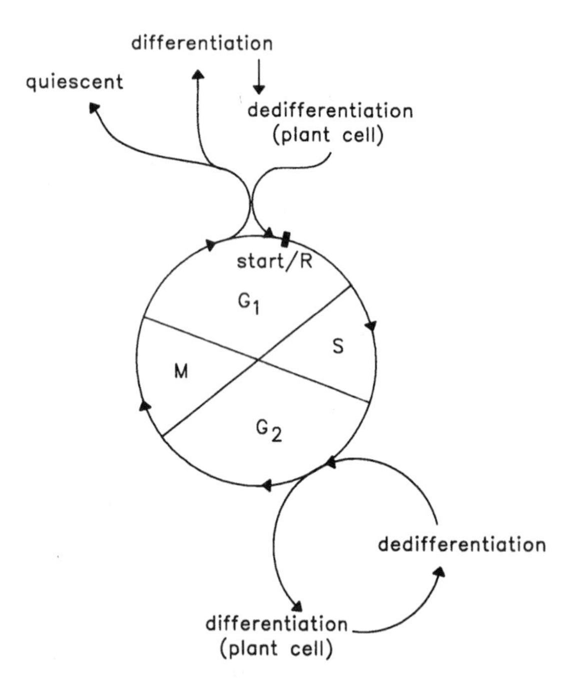

Fig. 1. Schematic representation of the eukaryotic cell cycle. Plant cells are capable of differentiate and dedifferentiate both in G_1 and G_2.

The G_2 to M transition

Research on two very different model systems – the genetic analysis of the yeast cell cycle, and the biochemical characterization of the maturation-promoting factor (MPF) of *Xenopus* oocytes – has established the universality of the mechanisms that regulate entry into mitosis [112]. A scheme that summarizes our current knowledge of the G_2/M transition is presented in Fig. 2.

A protein kinase composed of a catalytic subunit, cdc2, and a regulatory subunit, cyclin, is rate-limiting for triggering mitosis. Cdc2 is a 34 kDa serine/threonine kinase also known as $p34^{cdc2}$ which takes its name from the fission yeast *cdc2* gene (*CDC28* in the budding yeast) [80, 113]. cdc2-homologous kinases are now named cdk (for cyclin-dependent kinase). Whereas cdc2 is a cdk (cdk1), its initial name is maintained for historical reasons. Genetic complementation of a temperature-sensitive *cdc2* strain has shown that mammalian cells have a functional homologue [75]. The biochemical purification of *Xenopus* MPF and the subsequent demonstration that

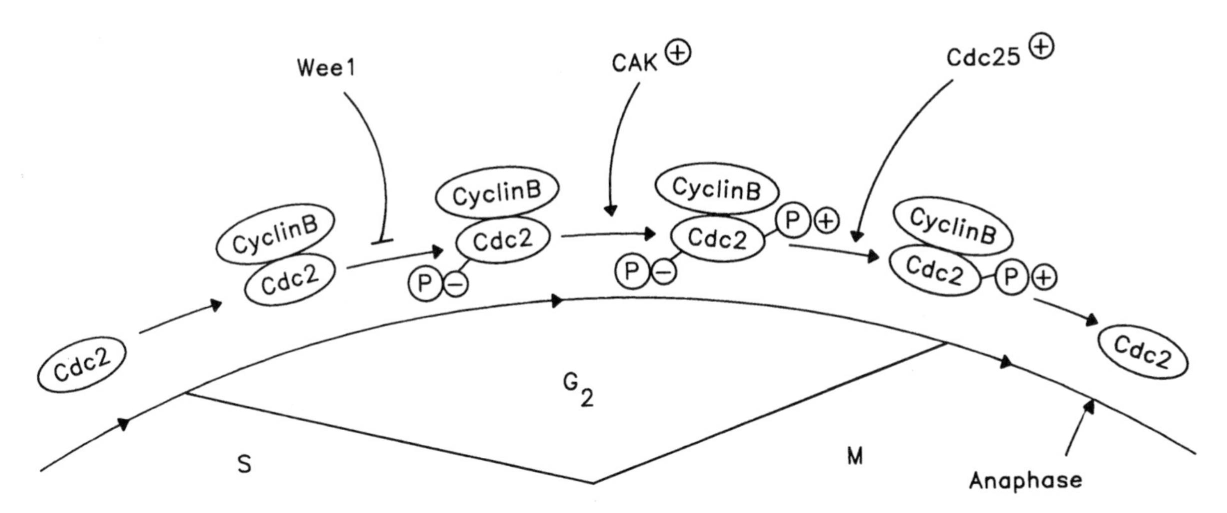

Fig. 2. Schematic illustration of the events leading to the activation of $p34^{cdc2}$ at the G_2 to M boundary. For sake of simplicity, phosphorylation/dephosphorylation steps are represented as one directional event. Until the onset of mitosis, these reactions should be seen as in equilibrium.

$p34^{cdc2}$ was one of the two components of kinase activity merged the genetic and the biochemical approaches [34].

Cyclins were first described as proteins periodically synthesized before each mitosis in marine invertebrate oocytes [24]. Based on differences in steady state, and amino acid sequence homology, cyclins were initially classified into an A-type and a B-type. Subsequent work has shown that microinjection of cyclin A and B could activate MPF and trigger mitosis [120, 139, 147]. Furthermore, in cell-free *Xenopus* egg extracts depleted from all its endogenous mRNAs, MPF activation, chromosome condensation and nuclear envelop breakdown can be induced through addition of cyclin B RNA alone [102]. Next, cyclin B was identified as the second component of MPF [35, 74]. Interestingly, at the time of purification of MPF, genetic analysis in fission yeast had already shown that the *cdc13* gene interacts with *cdc2* during the G_2/M transition [6]. Cloning of *cdc13* revealed that it encodes a B-type cyclin [7, 133]. Although there is good evidence that cyclin A plays a role in initiation of DNA replication [11, 37, 117], the role of cyclin A in mitosis is not definitely understood.

The presence of $p34^{cdc2}$ and accumulation of cyclin B is necessary but not sufficient to activate MPF. Phosphorylation at key residues of *cdc2* regulates, both positively and negatively, the activity of the kinase (reviewed in [2, 17]). In fission yeast, the kinase is kept inactive by phosphorylation at the amino acid Tyr-15. Before mitosis, the phosphate is removed and the kinase activated. Furthermore, a mutated *cdc2*Phe15 enters mitosis prematurely, confirming that phosphorylation at Tyr-15 prevents *cdc2* activation [40]. In animals, a similar mechanism is present, except that besides Tyr-15, Thr-14 is phosphorylated as well [72, 111]. A universal mechanism to control the timing of mitosis by phosphorylation and dephosphorylation of Tyr-15 has been proposed [112]. Surprisingly, substitution of the corresponding tyrosine residue in the budding yeast cdc28 protein does not accelerate mitosis, notwithstanding that the wild-type cdc28 protein is phosphorylated in Tyr-15 [1, 135]. The enzymology responsible for phosphorylation/dephosphorylation of Tyr-15 has been described in some detail. The phosphatase that removes the phosphate from Tyr-15, and possibly Thr-14 as well [77], takes its name from the fission yeast *cdc25* gene [20, 36, 94, 127]. Interestingly, activation of the cdc25 phosphatase seems to be mediated by interactions with cyclin B [33, 64, 153]. In *Schizosaccharomyces pombe*, another phosphatase, Pyp3$^+$, acts cooperatively with cdc25 to dephosphorylate Cdc2 on Tyr-15 [95]. In fission yeast, two kinase products of the *wee1*$^+$ and *mik1* genes, are responsible for Tyr-15 phosphorylation [83, 128, 129]. Recently, a human functional homologue of *wee1* (WEE1HU) was isolated by complementation of a temperature-sensitive fission yeast strain [59]. The kinase that acts on Thr-14 is not known yet.

In fission yeast, $p34^{cdc2}$ is positively regulated by phosphorylation at Thr-167 (Thr-161 in the human cdc2) [41]. The phosphate seems to be necessary for cyclin binding and kinase activity (reviewed in [17]). Although it is still a matter of debate (see in [119]) if Thr-161 phosphorylation is an absolute requirement for cyclin binding, there is compelling evidence that it is important for stabilization of the cyclin/cdc2 heterodimer [85]. While in yeast the protein responsible for this event has yet to be identified, such a protein, called CAK (for cdc2-activating kinase) has been described in starfish and *Xenopus* oocytes. CAK is a cdc2-related kinase and likewise needs other regulatory subunit(s) [31, 122, 134]. The isolation of a CAK homologous kinase from rice suggests that plants may have a similar mechanism of cdc2 activation [48].

Inactivation of the kinase complex is achieved at least in part by destruction of the cyclin moiety. A stretch of amino acids in the N-terminal region known as the destruction box, followed by a number of positively charged residues, is a target for proteolytic degradation through a ubiquitin-mediated pathway. Eggs injected with a truncated cyclin B lacking the N-terminal part are blocked in M phase with high levels of kinase [103].

The G_1 to S transition

The machinery that controls the G_1 to S transition is less conserved between yeasts and animals. In spite of that, there are some similarities and a few generalizations can be made. In both systems, S phase is initiated by activating a cdk kinase. In addition, like in G_2, the kinase is a heterodimer composed of one cdk subunit, and one cyclin regulatory subunit (though G_1 cyclins are much more divergent than their mitotic counterparts). While in both animal and yeast cells cdks are phosphorylated, the role of these post-translational modifications is yet poorly understood and the activity of the kinase may be mainly regulated by transcriptional control of the cyclin partner (although the recent discovery of CKIs shows that the situation is probably more complex, see below). Notwithstanding the similarities, we feel that a separate discussion of the budding yeast and animal G_1 to S transition will allow a more in-depth description of the molecular mechanisms that controls the timing of initiation of DNA replication.

The budding yeast G_1 to S transition

In budding yeast, START is defined as the transition point in the cell cycle, late in G_1, after which the cell becomes committed to a new mitotic cycle [123]. There are two key requirements that should be met to specify START. First, after the transition, haploid cells are no longer sensitive to cell cycle arrest caused by pheromones (in haploid cells, conjugation is launched by the secretion of hormones of the opposite mating type). Second, before START, if not enough nutrients are available in the medium, cells will enter a quiescent state, and as a result will not go into the S phase [46]. A model that summarizes the present knowledge of the G_1 to S transition in budding yeast is shown in Fig. 3.

The catalytic subunit of the protein kinase that operates at START is the same as that responsible for the G_2/M transition, cdc28 (the same holds true for cdc2 in fission yeast). Yet, cdc28 associates with a distinct group of cyclins: CLN1, CLN2 and CLN3. CLN1 and CLN2 are closely related proteins and were isolated in genetic screens for suppressors of cdc28 mutations [43]. CLN3 is a more distant relative and was identified as a dominant mutation that causes cells to divide at smaller sizes or in the presence of the mating pheromone [14, 104]. Initial work indicated that all three G_1 cyclins were genetically redundant: while deletion of all three cyclins is

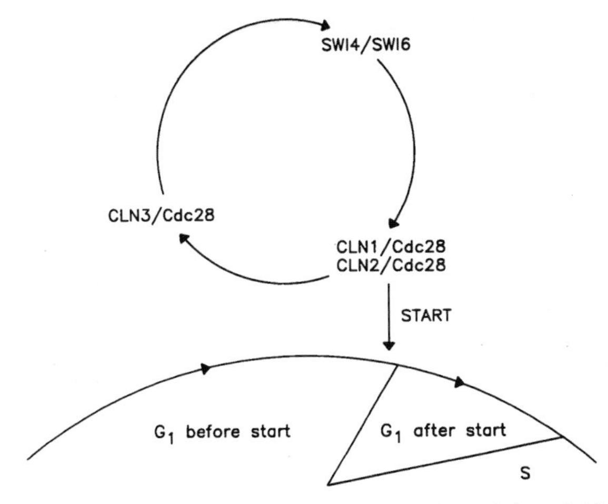

Fig. 3. A simplified diagram of the activation of the cdc28 kinase at START.

lethal, the progression to the S phase is achieved in a cell expressing only one of the CLNs [125]. Yet, several independent lines of evidence indicate that they execute different tasks. CLN1 and CLN2 transcripts and protein levels are cell cycle-regulated (see below), and histone H1 kinase associated with either Cln1 or Cln2 parallels the protein abundance, with a peak in late G_1, just before START [149]. On the other hand, Cln3 is a rare protein with a weak H1-associated kinase that does not oscillate through the cell cycle [143]. An artificial burst of CLN3 accelerates START and induces five other cyclins (CLN1, CLN2, HCS26, ORFD and CLB5) and the transcription factor SWI4 [143] (see below). In similar experiments, overexpression of CLN1 is much less efficient to trigger START. In addition, in a strain were CLN1 and CLN2 were deleted, transcription of CLB5, HCS26 and ORFD depends on CLN3. In spite of its weak associated kinase activity, CLN3 might be an upstream activator of G_1 cyclins which directly activates START [143].

As discussed above, the abundance of CLN1 and CLN2 mRNAs oscillate during the cell cycle. There is strong evidence that the abundance of the proteins might be regulated by transcriptional control (reviewed in [66]). Transcription of both CLN1 and CLN2 genes depends on a transcription factor called SBF (Swi4/6-binding factor). Swi4 and Swi6 were first described as involved in transcription of the mating-type switch endonuclease HO [9]. Swi6 is also part of a transcription factor called MBF (for MluI-binding factor) which is important for transcription of DNA synthesis genes. Binding sites for SBF, called SCB (Swi4/6-dependent cell cycle box), present in both CLN1 and CLN2 promoter regions, are sufficient to confer late G_1 transcription on reporter genes [10, 91]. In swi4 mutants, the level of Cln1 and Cln2 proteins is drastically reduced and causes the cells' arrest just before START [106, 115]. Moreover, overexpression of CLN2 can overcome the lethality caused by the absence of SBF [115].

START is dependent on both cdc28 and G_1 cyclins. Yet, CLN1 and CLN2 transcription requires an active cdc28 kinase. To explain this apparent contradiction, a positive feedback involved in the activation of cdc28 has been proposed [15, 16]. In this model, low levels of cdc28 kinase activate SBF. SBF would then induce CLN1 and CLN2 transcription. G_1 cyclin accumulation would close the loop through further activation of cdc28 (Fig. 4).

In the fission yeast, the *cdc10* gene encodes a transcription factor essential for START [81]. The predicted sequence has homology in distinct domains with Swi4 and Swi6. In mammalian cells, the E2F transcription factor is necessary for periodic transcription of genes needed for cell cycle progression [107]. A structural similarity between the DNA-binding domains of SWI4, cdc10 and E2F-1 factor has been noticed [73]. In addition, two almost perfect SCB-binding sites are found in the promoter region of an *Arabidopsis* cyclin which is expressed in late G_1 [30]. It sug-

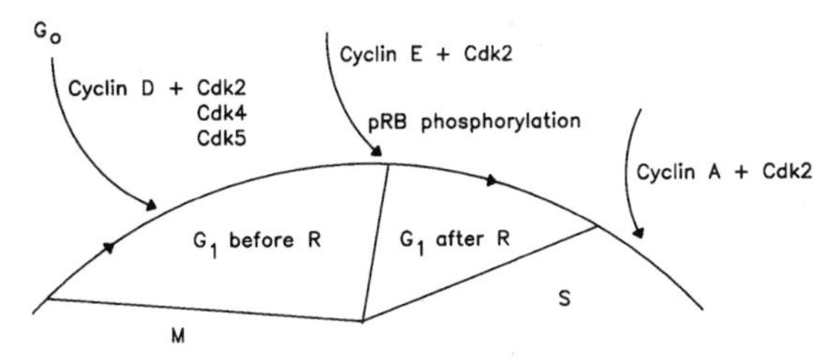

Fig. 4. A schematic outline of the cyclin-cdk interactions during G_1 to S transition in mammalian cells.

gests that some aspects of the transcriptional regulation of cell cycle progression have been conserved throughout evolution.

The animal G_1 to S transition

The mammalian G_1 cyclins, known as cyclin C, D and E were isolated through their capacity to rescue a triple CLN-defective yeast strain [79]. Subsequently, it was shown that cyclin D consists of a multigene family, D1, D2 and D3, which most likely are not functionally redundant [61, 88, 151]. Cyclin D1 was also identified independently as a putative oncogene (PRAD1) [100]. A simplified model of the G_1 to S transition in mammalian cells is shown in Fig. 4.

Whereas truncated forms of cyclin A and B are also capable of rescuing the yeast triple CLN mutant, only cyclin C, D and E are expressed during G_1. While the cyclin C levels oscillate only very little with a small peak in early G_1, cyclin E is periodically expressed with maximum levels at the G_1 to S transition [69, 79]. Cyclin D rises before cyclin E when quiescent cells are stimulated with serum, but in cycling cells, the cyclin D level fluctuates minimally. In proliferating cells, the level of cyclin D declines drastically if growth factors are restricted, indicating that cyclin D might be more related to growth than to the cycling apparatus [89].

Several distinct cdks have been isolated from mammalian cells. Of these, three can complement START in budding yeast: cdc2 (cdk1), cdk2 and cdk3 [92]. However, most of the data indicate that cdc2 only function at the G_2 to M transition [44, 141]. On the other hand, cdk2 is necessary for DNA synthesis and might be responsible for the G_1 to S transition [25]. A function for cdk3 has not been conclusively established, although there are indications that it may function at the G_1 to S transition [144]. Cyclin E associates preferentially with cdk2 [19, 70], while D-type cyclins form complexes also with cdk4 and cdk5 [68, 90, 150].

Besides associating to each other, cyclins and cdks bind to a number of other proteins (reviewed in [130]). These associations are used as a means to sequester basic regulatory molecules needed later in the cell cycle, such as the transcription factor E2F or the product of the tumour suppressor retinoblastoma gene (pRB). Although in some cell types pRB function might not be essential [76], the loss of RB is frequently associated with tumour development [8]. A cycle of phosphorylation/dephosphorylation regulates pRB activity during the cell cycle. In early G_1, hypophosphorylated pRB is found associated with transcription factors, some of them related to S-phase events. Thus, cell cycle progression might be restricted by sequestering of essential factors by pRB. Late in the cycle, pRB phosphorylation provokes dissociation of the bound regulatory proteins. The release of the transcription factors may be important in the control of G_1 to S transition [56]. Cyclins D form complexes with pRB, albeit D1 less efficiently than D2 and D3 [68]. Although *in vitro* pRB can be phosphorylated by different cyclin-cdk complexes, cdk4-cyclin D and/or cdk2-cyclin E are most likely responsible for pRB inactivation [131].

Recent data suggest that cell cycle progression can be prevented by proteins that bind and inactivate the kinase (CDK inhibitors or CKIs; reviewed in [58]). One such CKI, a protein of 21 kDa (Pic1), associates and prevents the activation of several cyclin-dependent kinases [21, 42, 45, 152]. Pic1 expression is correlated with suppression of growth. In addition, the promoter of Pic1 has a binding site for the p53 tumour suppressor protein. A reporter gene under the control of the Pic1 promoter can be induced by p53. Together, these data suggest that p53 may suppress cell proliferation by controlling the transcription of Pic1, which in turn would maintain inactive cyclin-cdk complexes.

Cyclin D acts earlier than cyclin E, and the latter might be the rate-limiting regulatory subunit of the cdk for G_1 to S transition. Microinjection of cyclin D into serum-stimulated cells is inhibitory when executed during the middle of G_1, but not close to the G_1 to S boundary [3]. On the other hand, overexpression of cyclin E shortens the G_1 phase and decreases the cell size required to enter the S phase [116].

Control of proliferation in plants

It has become apparent that the general features of the cell cycle machinery are well conserved in all eukaryotes. Nonetheless, the signalling pathways that integrate cell division and development have probably evolved in connection with the unique modes of development of each organism. Postembryonic development is quite distinct in plants and animals (reviewed in [63]). If the comparisons are restricted to the control of proliferation, a few specific dissimilarities can be listed. First, plant cells can quit the cell cycle and differentiate either in G_1 or in G_2 [145]. Second, and probably most relevant to this discussion, upon stimulation, most mature plant cells can dedifferentiate and re-enter the cell cycle, regardless of being in G_1 or G_2 [4, 145]. Furthermore, a clear state of quiescence (comparable to G_o in animals and quiescence in yeasts) has yet to be demonstrated for plants.

In higher animals there is a stringent positive correlation between levels of $p34^{cdc2}$ and the proliferative state of the cell [71, 78]. Initial attempts to define a correlation between mitotic activity and levels of $p34^{cdc2}$ and/or cdc2 mRNA in plants were not conclusive. A positive association between meristematic activity and cdc2 mRNA levels was observed for *Arabidopsis*, maize and alfalfa, although transcripts were also detected in non-dividing tissues [13, 28, 54]. On the other hand, in wheat leaves and in segments of pea roots, constant amounts of $p34^{cdc2}$ or mRNA levels on a *per cell* basis have been measured, independently of the mitotic activity of the tissue [63, 65]. Careful examination of the temporal and spatial expression patterns of the *cdc2a* gene in *Arabidopsis* unravelled a more complex scenario than a yes or no situation [53, 87]. *In situ* hybridization studies by Martinez and colleagues showed that the cdc2a mRNA is present in all plant meristems. In addition, signals were detected in parenchyma and pericycle cells from radish roots, which are non-dividing cells. Hemerly and co-workers, in a study based on promoter-β-glucuronidase (GUS) analyses, confirmed and extended these observations. *Cdc2a* expression is induced 30 min after wounding, whereas concomitant induction of cell division was not observed. Likewise, apical meristem of dark-grown plants, which is not actively dividing, also exhibits *cdc2* expression. Assuming that *cdc2a* expression reflects $p34^{cdc2}$ levels (but not necessarily cdc2 kinase activity), the interpretation is that there is a positive correlation between *cdc2a* expression and a physiological competence for cell division. Contrary to animals, mature plant cells retain some $p34^{cdc2}$ that can be activated to reinitiate cell division. The presence of mRNAs of the two soybean *cdc2* genes in differentiated tissues was similarly interpreted as a reflection of differences in the status of cell differentiation between plant and animal cells [93].

Thus, re-entering the cell cycle is not limited by $p34^{cdc2}$ accumulation, at least not in the initial steps. A close examination of processes in other systems in which cell cycle progression is triggered by the activation of a limiting factor yields three immediate candidates: cyclins, cdc25 and MAP kinase. In the initial cleavage cycles of frog oocytes, periodic cyclin synthesis is the only requirement to activate mitosis [24, 96, 102]. These cycles lack a G_1 and G_2 phase, and accumulation of mitotic cyclins to a threshold level starts mitosis and division. Mitotic-like cyclins have been isolated from various plants including carrot [49], soybean [49], *Arabidopsis* [30, 52], pea [63], alfalfa [55], *Antirrhinum majus* [32], maize [124], tobacco (C. Bergounioux, personal communication), and *Sesbania rostrata* (S. Goormachtig, personal communication). By sequence homology, the plant cyclins can be grouped into three classes, and are closer to the A- and B-type cyclins than to any other animal or yeast cyclin (Fig. 5) [124].

In animal and yeast somatic cells, the oscillations of cyclin transcripts during the cell cycle are parallelled by fluctuations in the accumulation of the corresponding gene product. It is not unreasonable to suppose that the same is true for plant cyclins. Indeed, in partially synchronized alfalfa cell cultures, an antibody against human cyclin A immunopreciptated a histone H1-associated kinase activity in early S phase [84]. Preliminary

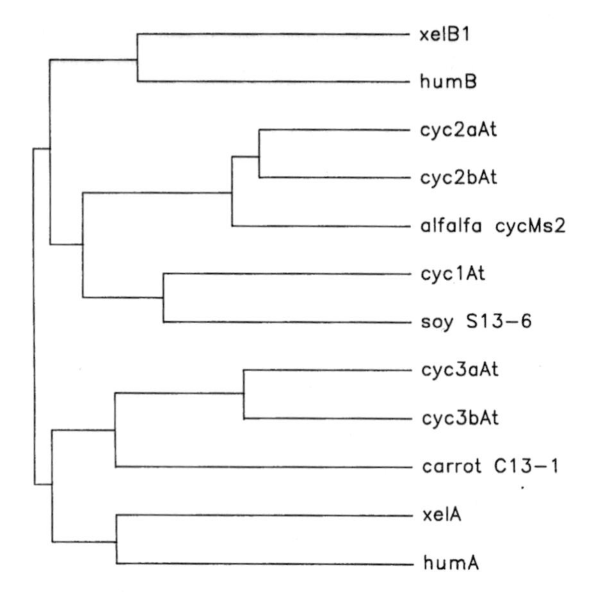

Fig. 5. A phylogenetic tree of cyclin sequences from plants, *Xenopus laevis* cyclins A and B1, and mammalian cyclins A and B. The *Xenopus* A and B1 cyclins are from Minshull *et al.* [97, 98], the human A and B cyclins are from Wang *et al.* [146], and Pines and Hunter [121], respectively, the soybean S13–6 and the carrot C13–1 cyclins are from Hata *et al.* [49], the *Arabidopsis* cyc1At cyclin is from Hemerly *et al.* [52], the alfalfa cycMs2 cyclin is from Hirt *et al.* [55], and the *Arabidopsis* cyc2aAt, cyc2bAt, cyc3aAt, and cyc3bAt are from Ferreira *et al.* [30].

results on the examination of cyclin expression indicate that, different from the situation with cdc2, steady-state levels of cyclin mRNAs are more strictly correlated with mitotic activity. Much larger amounts of cyclin mRNA were detected in carrot somatic embryos than in relatively slow-growing pre-embryogenic masses [49]. Likewise, more cyclin transcripts were present in the rapidly growing symbiotic nodules, than in uninoculated roots [49]. In *Arabidopsis* cell cultures, cyclin transcripts are far more abundant than in organized tissues [52]. Starvation of alfalfa cell cultures led to a decrease of mRNA levels in two different cyclins, although apparently the kinetics of disappearance of the two messengers is different [53]. In maize, a tight association between accumulation of cyclin transcripts and mitotic activity of the different tissues was observed [124].

A detailed examination of the spatial and temporal patterns of expression of one of the *Arabidopsis* cyclins, the mitotic cyc1At, allowed three observations. First, expression is restricted to zones of mitotic activity, and is not induced or present in situations that reflect a competence for cell division, such as after wounding or in the apical meristem of dark-grown seedlings [53]. This correlation is maintained throughout plant development. Second, in the root apical meristem, where the rate of proliferation is not very high, the pattern of cyclin expression is cell autonomous. This pattern most likely reflects the distribution of dividing cells in G_2 in the meristems. Third, cyclin expression is an early event during *de novo* induction of cell division, such as in lateral root formation from pericycle cells [30].

The cyclin 1 gene from *Antirrhinum majus*, which is highly homologous to the *Arabidopsis cyc1At* gene, also exhibits a cell-specific pattern of expression in inflorescence meristems [32]. Furthermore, Fobert and co-workers demonstrated that the dispersed expression reflects cell cycle transcription of the cyclin 1 gene. The particular pattern of expression of cyclin genes is consistent with the hypothesis that the control of cyclin expression is, at least in part, responsible for the activation of the cdc2 kinase and, consequently, the induction of cell division.

This working hypothesis may, however, not account for the whole story. It is likely that, depending on the developmental pathway, plants might have available distinct mechanisms to turn on proliferation. For instance, in dormant buds of poplar trees, there is a gradient of cyclin and *cdc2* expression levels, with the highest levels in the upper buds. Moreover, the gradient is somewhat sharper for cyclin 1 than for cdc2 (A. Rhode, personal communication). Dormant and axillary buds are possibly in a further stage of competence for cell division (than pericycle cells, for instance), with a cdc2-cyclin complex already formed waiting to be activated.

One candidate to activate the kinase could be a yet unidentified plant *cdc25* homologue. In *Drosophila*, the first 13 embryonic cell division cycles are driven by stores of maternal transcripts of cell cycle control genes. However, in the 14th cycle, the maternal transcripts are destroyed and transcription from the larva starts. It has been demonstrated that transcription of one of the *Drosophila* cdc25 homologous genes, *string*, is the factor that limits cell cycle progression in the 14th, 15th and 16th cycles (reviewed in [114]). Nevertheless, to our knowledge, a plant *cdc25* has not yet been cloned. Furthermore, tyrosine phosphorylation in plants has still to be demonstrated and it remains possible that the cdc2 kinase is activated only via removal of the phosphate from the plant equivalent of Thr-14.

The recently isolated MAP kinase homologous cDNAs from alfalfa [18, 67], pea [136] and *Arabidopsis* [99] are other candidates to play a role in activating the kinase. MAP kinase seems to be involved both in the activation of quiescent cells arrested in G_0 and of MPF in *Xenopus* oocytes arrested at the G_2 to M boundary (reviewed in [126]). Mizoguchi and co-workers showed that mitogenic stimulation, by treatment of auxin-starved tobacco cells with 2,4-dichlorophenoxyacetic acid, led to a rapid and transient activation of a protein kinase activity typical of MAP kinase. Furthermore, auxin treatment also caused an augmentation in protein kinase activity towards recombinant *Arabidopsis* MAP kinase [99]. Therefore, like in other systems, mitogenic stimulation in plants may involve a MAP kinase phosphorylation cascade. Here, the complication is that MAP kinase is part of signal transduction cascade that is triggered by *ras* and a true *ras* homologue has not yet been identified in plants [140].

No doubt, there should be several other contenders for the position of key regulator(s) of proliferation and the above discussion has to be regarded as an estimation. As pointed out before, many plant cells differentiate and are capable of re-entering the cell cycle both in G_1 and G_2. Plant homologues of G_1 regulators could be limiting factors for cells that differentiate in G_1. Because the players involved in G_1 to S transition are less conserved, they are harder to isolate by sequence homology. Nevertheless, PCNA, which was recently shown to be part of cyclin D quaternary complexes, has been isolated from various plants [50, 86, 138]. Alternatively, one cannot exclude that plants may have their own unique set of regulators acting on the cell cycle apparatus. Nevertheless, efforts to isolate plant genes involved in cell cycle control by other means than screening by homology resulted in most cases in the isolation of genes that turn out to be involved in cell cycle regulation in other systems [62].

What are the signals that mediate stimulation of cell division in plants?

Plant hormones have long been acknowledged as cell division regulators (reviewed in [23]). Yet, they have many more effects besides promoting cell division. The pleiotropic consequences of hormone action can best be appreciated in the complex phenotypes displayed by mutants with an altered hormone response [22]. Nevertheless, the availability of plant cell cycle control genes provides tools for examining the influence of growth regulators on mitotic activity. Freshly isolated protoplasts can be stimulated to divide if provided with adequate concentrations of auxins and cytokinins. Under these circumstances, a burst of *cdc2* expression follows the initiation of cell division [5, 53, 84]. Transcription of the *cdc2a Arabidopsis* gene is similarly induced by either an auxin or a cytokinin alone [53]. Dedifferentiation of carrot cotyledons can also be induced by auxin, and is paralleled by an increase in $p34^{cdc2}$-like protein levels [38]. Oscillations in the steady state of *cdc2* mRNA followed auxin-induced embryogenesis of alfalfa cell cultures [54].

In intact *Arabidopsis* plants, transcription of *cdc2* can be induced or inhibited depending on the phytohormone applied. Moreover, there is a correlation between *cdc2* expression and mitotic activity in the provoked morphogenetic programme [53]. Thus, *cdc2a* expression is induced during auxin-induced lateral root formation [53, 87]. Likewise, cytokinin-induced *cdc2a* expression can be associated with primary thickening of the roots [53]. In soybean roots, differential accumulation of *cdc2* transcripts occurs after auxin treatment in the elongation zone [93].

Presuming that phytohormones do not specify developmental programmes, but simply ensure their expression [142], what could be the signals that regulate proliferation in intact plants? One hypothesis is that oscillations in cytoplasmic calcium could control cell cycle progression [142]. Indeed, in other eukaryotic systems calcium and calmodulin are required for the activation of cell cycle control kinases (reviewed in [82]). Calcium however is a second messenger in many plant metabolic pathways [51, 101]. Therefore, another hierarchic layer, which integrates developmental programmes and cell division, has to be superimposed. The isolation of less pleiotropic mitogens like the nod factors may allow a thorough examination of the signalling cascade controlling proliferation.

Work in animal and yeast systems has shown that the activity of the cell cycle control genes is mostly modulated at the post-translational level. In contrast, a good part of the plant work carried out so far has been on transcription analysis. Consequently, the interpretations are largely inferential. The recent localization of $p34^{cdc2}$ in the preprophase band [13] underscores the need for a biochemical approach for the study of cell division.

The patterns of cell division in apical meristems are very intricate. To account for the different rates of cell division in the files or groups of cells, the activity of the cell cycle regulators must be regulated temporally and spatially. It is conceivable that even with all the tools available it will take some time before a clear picture of cell cycle regulation in meristems emerges. Nevertheless, a thorough investigation of cell cycle regulation in mutants with affected meristematic activity could help to understand the mechanisms acting upon cell division in normal plants.

On the other hand, the availability of more tractable systems such as nodule formation in legume plants, the inception of lateral roots, somatic embryogenesis, or the activation of dormant or axillary buds may provide cues on how proliferation is overseen by plant developmental pathways.

Acknowledgements

The authors thank V. Sundaresan and S. de Vries for critical reading of the manuscript. This work was supported by grants from the Belgian Programme on Interuniversity Poles of Attraction (Prime Minister's Office, Science Policy Programming, No 38), and 'Vlaams Actieprogramma Biotechnologie' (ETC 002). P.F. and A.S.H. are indebted to the Conselho Nacional de Desenvolvimento Científico e Tecnológico (CNPq 204081/82-2) and Coordenação de Aperfeiçoamento de Pessoal de Nível Superior (CAPES 1387/89-12) respectively, for predoctoral fellowships. D.I. is research engineer and research director of the Institut National de la Recherche Agronomique (France).

References

1. Amon A, Surana U, Muroff I, Nasmyth K: Regulation of $p34^{CDC28}$ tyrosine phosphorylation is not required for entry into mitosis in *S. cerevisiae*. Nature 355: 368–371 (1992).
2. Atherton-Fessler S, Hannig G, Piwnica-Worms H: Reversible tyrosine phosphorylation and cell cycle control. Sem Cell Biol 4: 433–442: (1993).
3. Baldin V, Lukas J, Marcote MJ, Pagano M, Draetta G: Cyclin D1 is a nuclear protein required for cell cycle progression in G_1. Genes Devel 7: 812–821 (1993).
4. Bergounioux C, Perennes C, Brown SC, Sarda C, Gadal P: Relation between protoplast division, cell-cycle stage and nuclear chromatin structure. Protoplasma 142: 127–136 (1988).
5. Bergounioux C, Perennes C, Hemerly AS, Qin LX, Sarda C, Inzé D, Gadal P: A *cdc2* gene of *Petunia hybrida* is differentially expressed in leaves, protoplasts and during various cell cycle phases. Plant Mol Biol 20: 1121–1130 (1992).
6. Booher R, Beach D: Interaction between *cdc13*$^+$ and *cdc2*$^+$ in the control of mitosis in fission yeast; dissociation of the G_1 and G_2 roles of the *cdc2*$^+$ protein kinase. EMBO J 6: 3441–3447 (1987).
7. Booher R, Beach D: Involvement of *cdc13*$^+$ in mitotic control in *Schizosaccharomyces pombe*: possible interaction of the gene product with microtubules. EMBO J 7: 2321–2327 (1988).
8. Bookstein R, Lee W-H: Molecular genetics of the retinoblastoma suppressor gene. CRC Crit Rev Oncogen 2: 211–227 (1991).
9. Breeden L, Mikesell GE: Cell cycle-specific expression

of the *SWI4* transcription factor is required for the cell cycle regulation of *HO* transcription. Genes Devel 5: 1183–1190 (1991).

10. Breeden L, Nasmyth K: Cell cycle control of the yeast *HO* gene: *cis*- and *trans*-acting regulators. Cell 48: 389–397 (1987).
11. Cardoso MC, Leonhardt H, Nadal-Ginard B: Reversal of terminal differentiation and control of DNA replication: cyclin A and cdk2 specifically localize at subnuclear sites of DNA replication. Cell 74: 979–992 (1993).
12. Colasanti J, Tyers M, Sundaresan V: Isolation and characterization of cDNA clones encoding a functional $p34^{cdc2}$ homologue from *Zea mays*. Proc Natl Acad Sci USA 88: 3377–3381 (1991).
13. Colasanti J, Cho S-O, Wick S, Sundaresan V: Localization of the functional $p34^{cdc2}$ homolog of maize in root tip and stomatal complex cells: association with predicted division sites. Plant Cell 5: 1101–1111 (1993).
14. Cross FR: *DAF1*, a mutant gene affecting size control, pheromone arrest, and cell cycle kinetics of *Saccharomyces cerevisiae*. Mol Cell Biol 8: 4675–4684 (1988).
15. Cross FR, Tinkelenberg AH: A potential positive feedback loop controlling *CLN1* and *CLN2* gene expression at the start of the yeast cell cycle. Cell 65: 875–883 (1991).
16. Dirick L, Nasmyth K: Positive feedback in the activation of G1 cyclins in yeast. Nature 351: 754–757 (1991).
17. Draetta G: cdc2 activation: the interplay of cyclin binding and Thr-161 phosphorylation. Trends Cell Biol 3: 287–289 (1993).
18. Duerr B, Gawienowski M, Ropp T, Jacobs T: MsERK1: a mitogen-activated protein kinase from a flowering plant. Plant Cell 5: 87–96 (1993).
19. Dulić V, Lees E, Reed S.I: Association of human cyclin E with periodic G^1-S phase protein kinase. Science 257: 1958–1961 (1992).
20. Dunphy WG, Kumagai A: The cdc25 protein contains an intrinsic phosphatase activity. Cell 67: 189–196 (1991).
21. El-Deiry WS, Tokino T, Velculescu VE, Levy DB, Parsons R, Trent JM, Lin D, Mercer WE, Kinzler KW, Vogelstein B: *WAF1*, a potential mediator of p53 tumor suppression. Cell 75: 817–825 (1993).
22. Estelle M: The plant hormone auxin: insight in sight. BioEssays 14: 439–444 (1992).
23. Evans ML: Functions of hormones at the cellular level of organization. In: Scott TK (ed) Hormonal Regulation of Development. Encyclopedia of Plant Physiology, New Series, vol. 10, pp. 22–79. Springer-Verlag, Berlin (1984).
24. Evans T, Rosenthal ET, Youngblom J, Distel D, Hunt T: Cyclin: a protein specified by maternal mRNA in sea urchin eggs that is destroyed at each cleavage division. Cell 33: 389–396 (1983).
25. Fang F, Newport JW: Evidence that the G1-S and G2-M transitions are controlled by different cdc2 proteins in higher eukaryotes. Cell 66: 731–742 (1991).
26. Fantes P: Cell cycle control in *Schizosaccharomyces pombe*. In: Nurse P, Streiblova E (eds) The Microbial Cell Cycle, pp. 109–125. CRC Press, Boca Raton, FL (1984).
27. Feiler HS, Jacobs TW: Cloning of the pea *cdc2* homologue by efficient immunological screening of PCR products. Plant Mol Biol 17: 321–333 (1991).
28. Ferreira PCG, Hemerly AS, Villarroel R, Van Montagu M, Inzé D: The *Arabidopsis* functional homolog of the $p34^{cdc2}$ protein kinase. Plant Cell 3: 531–540 (1991).
29. Ferreira CG, Hemerly AS, Van Montagu M, Inzé D: A protein phosphatase 1 from *Arabidopsis thaliana* restores temperature sensitivity of a *Schizosaccharomyces pombe* $cdc25^{ts}/wee1^{-}$ double mutant. Plant J 4: 81–87 (1993).
30. Ferreira P, Hemerly A, de Almeida Engler J, Bergounioux C, Burssens S, Van Montagu M, Engler G, Inzé D: Three discrete classes of *Arabidopsis* cyclins are expressed during different intervals of the cell cycle. Proc Natl Acad Sci USA, submitted (1994).
31. Fesquet D, Labbé J-C, Derancourt J, Capony J-P, Galas S, Girard F, Lorca T, Shuttleworth J, Dorée M, Cavadore J-C: The *MO15* gene encodes the catalytic subunit of a protein kinase that activates cdc2 and other cyclin-dependent kinases (CDKs) through phosphorylation of Thr-161 and its homologues. EMBO J 12: 3111–3121 (1993).
32. Fobert PR, Coen ES, Murphy GJP, Doonan JH: Patterns of cell division revealed by transcriptional regulation of genes during the cell cycle in plants. EMBO J 13: 616–624 (1994).
33. Galaktionov K, Beach D: Specific activation of cdc25 tyrosine phosphatases by B-type cyclins: evidence for multiple roles of mitotic cyclins. Cell 67: 1181–1194 (1991).
34. Gautier J, Norbury C, Lohka M, Nurse P, Maller J: Purified maturation-promoting factor contains the product of a Xenopus homolog of the fission yeast cell cycle control gene *cdc2*$^{+}$. Cell 54: 433–439 (1988).
35. Gautier J, Minshull J, Lohka M, Glotzer M, Hunt T, Maller JL: Cyclin is a component of maturation-promoting factor from *Xenopus*. Cell 60: 487–494 (1990).
36. Gautier J, Solomon MJ, Booher RN, Bazan JF, Kirschner MW: cdc25 is a specific tyrosine phosphatase that directly activates $p34^{cdc2}$. Cell 67: 197–211 (1991).
37. Girard F, Strausfeld U, Fernandez A, Lamb NJC: Cyclin A is required for the onset of DNA replication in mammalian fibroblasts. Cell 67: 1169–1179 (1991).
38. Gorst JR, John PCL, Sek FJ: Levels of $p34^{cdc2}$-like protein in dividing, differentiating and dedifferentiating cells of carrot. Planta 185: 304–310 (1991).
39. Gould AR: Control of the cell cycle in cultured plant cells. CRC Crit Rev Plant Sci 1: 315–344 (1983).

40. Gould KL, Nurse P: Tyrosine phosphorylation of the fission yeast *cdc2*$^+$ protein kinase regulates entry into mitosis. Nature 342: 39–45 (1989).
41. Gould KL, Moreno S, Owen DJ, Sazer S, Nurse P: Phosphorylation at Thr-167 is required for *Schizosaccharomyces pombe* p34^{cdc2} function. EMBO J 10: 3297–3309 (1991).
42. Gu Y, Turck CW, Morgan DO: Inhibition of CDK2 activity *in vivo* by an associated 20K regulatory subunit. Nature 366: 707–710 (1993).
43. Hadwiger JA, Wittenberg C, Richardson HE, de Barros Lopes M, Reed SI: A family of cyclin homologues that control the G_1 phase in yeast. Proc Natl Acad Sci USA 86: 6255–6259 (1989).
44. Hamaguchi JR, Tobey RA, Pines J, Crissman HA, Hunter T, Bradbury EM: Requirement for p34^{cdc2} kinase is restricted to mitosis in the mammalian cdc2 mutant FT210. J Cell Biol 117: 1041–1053 (1992).
45. Harper JW, Adami GR, Wei N, Keyomarsi K, Elledge SJ: The p21 Cdk-interacting protein Cip1 is a potent inhibitor of G1 cyclin-dependent kinases. Cell 75: 805–816 (1993).
46. Hartwell LH, Weinert TA: Checkpoints: controls that ensure the order of cell cycle events. Science 246: 629–634 (1989).
47. Hashimoto J, Hirabayashi T, Hayano Y, Hata S, Ohashi Y, Suzuka I, Utsugi T, Toh-E A, Kikuchi Y: Isolation and characterization of cDNA clones encoding *cdc2* homologues from *Oryza sativa*: a functional homologue and cognate variants. Mol Gen Genet 233: 10–16 (1992).
48. Hata S: cDNA cloning of a novel cdc2$^+$/CDC28-related protein kinase from rice. FEBS Lett 279: 149–152 (1991).
49. Hata S, Kouchi H, Suzuka I, Ishii T: Isolation and characterization of cDNA clones for plant cyclins. EMBO J 10: 2681–2688 (1991).
50. Hata S, Kouchi H, Tanaka Y, Minami E, Matsumoto T, Suzuka I, Hashimoto J: Identification of carrot cDNA clones encoding a second putative proliferating cell-nuclear antigen, DNA polymerase delta auxiliary protein. Eur J Biochem 203: 367–371 (1992).
51. Hedrich R, Busch H, Raschke K: Ca^{2+} and nucleotide dependent regulation of voltage dependent anion channels in the plasma membrane of guard cells. EMBO J 9: 3889–3892 (1990).
52. Hemerly A, Bergounioux C, Van Montagu M, Inzé D, Ferreira P: Genes regulating the plant cell cycle: isolation of a mitotic-like cyclin from *Arabidopsis thaliana*. Proc Natl Acad Sci USA 89: 3295–3299 (1992).
53. Hemerly AS, Ferreira PCG, de Almeida Engler J, Van Montagu M, Engler G, Inzé D: *cdc2a* expression in *Arabidopsis thaliana* is linked with competence for cell division. Plant Cell 5: 1711–1723 (1993).
54. Hirt H, Páy A, Györgyey J, Bakó L, Németh K, Bögre L, Schweyen RJ, Heberle-Bors E, Dudits D: Complementation of a yeast cell cycle mutant by an alfalfa cDNA encoding a protein kinase homologous to p34^{cdc2}. Proc Natl Acad Sci USA 88: 1636–1640 (1991).
55. Hirt H, Mink M, Pfosser M, Bögre L, Györgyey J, Jonak C, Gartner A, Dudits D, Heberle-Bors E: Alfalfa cyclins: differential expression during the cell cycle and in plant organs. Plant Cell 4: 1531–1538 (1992).
56. Hollingsworth RE Jr, Chen P-L, Lee W-H: Integration of cell cycle control with transcriptional regulation by the retinoblastoma protein. Curr Opin Cell Biol 5: 194–200 (1993).
57. Howard A, Pelc SR: Synthesis of deoxyribonucleic acid in normal and irradiated cells and its relation to chromosome breakage. Heredity (Suppl) 6: 216–273 (1953).
58. Hunter T: Braking the cycle. Cell 75: 839–841 (1993).
59. Igarashi M, Nagata A, Jinno S, Suto K, Okayama H: *Wee1*$^+$-like gene in human cells. Nature 353: 80–83 (1991).
60. Imajuku Y, Hirayama T, Endoh H, Oka A: Exon-intron organization of the *Arabidopsis thaliana* protein kinase genes *CDC2a* and *CDC2b*. FEBS Lett 304: 73–77 (1992).
61. Inaba T, Matsushime H, Valentine M, Roussel MF, Sherr CJ, Look AT: Genomic organization, chromosomal localization, independent expression of human cyclin D genes. Genomics 13: 565–574 (1992).
62. Ito M, Komamine A: Molecular mechanisms of the cell cycle in synchronous cultures of plant cells. J Plant Res (Special issue 3): 17–28 (1993).
63. Jacobs T: Control of the cell cycle. Devel Biol 153: 1–15 (1992).
64. Jessus C, Beach D: Oscillation of MPF is accompanied by periodic association between cdc25 and cdc2-cyclin B. Cell 68: 323–332 (1992).
65. John PCL, Sek FJ, Carmichael, JP, McCurdy DW: p34^{cdc2} homologue level, cell division, phytohormone responsiveness and cell differentiation in wheat leaves. J Cell Sci 97: 627–630 (1990).
66. Johnston LH: Cell cycle control of gene expression in yeast. Trends Cell Biol 2: 353–357 (1992).
67. Jonak C, Páy A, Bögre L, Hirt H, Heberle-Bors E: The plant homologue of MAP kinase is expressed in a cell cycle-dependent and organ-specific manner. Plant J 3: 611–617 (1993).
68. Kato J-y, Matsushime H, Hiebert SW, Ewen ME, Sherr CJ: Direct binding of cyclin D to the retinoblastoma gene product (pRb) and pRb phosphorylation by the cyclin D-dependent kinase CDK4. Genes Devel 7: 331–342 (1993).
69. Koff A, Cross F, Fisher A, Schumacher J, Leguellec K, Philippe M, Roberts, J.M: Human cyclin E, a new cyclin that interacts with two members of the *CDC2* gene family. Cell 66: 1217–1228 (1991).
70. Koff A, Giordano A, Desai D, Yamashita K, Harper JW, Elledge S, Nishimoto T, Morgan DO, Franza BR, Roberts JM: Formation and activation of a cyclin

E-cdk2 complex during the G_1 phase of the human cell cycle. Science 257: 1689–1694 (1992).
71. Krek W, Nigg EA: Structure and developmental expression of the chicken *CDC2* kinase. EMBO J 8: 3071–3078 (1989).
72. Krek W, Nigg EA: Differential phosphorylation of vertebrate $p34^{cdc2}$ kinase at the G1/S and G2/M transitions of the cell cycle: identification of major phosphorylation sites. EMBO J 10: 305–316 (1991).
73. La Thangue NB, Taylor WR: A structural similarity between mammalian and yeast transcription factors for cell-cycle-regulated genes. Trends Cell Biol 3: 75–76 (1993).
74. Labbé J-C, Capony J-P, Caput D, Cavadore J-C, Derancourt J, Kaghad M, Lelias J-M, Picard A, Dorée M: MPF from starfish oocytes at first meiotic metaphase is a heterodimer containing one molecule of cdc2 and one molecule of cyclin B. EMBO J 8: 3053–3058 (1989).
75. Lee MG, Nurse P: Complementation used to clone a human homologue of the fission yeast cell control gene *cdc2*. Nature 327: 31–35 (1987).
76. Lee EY-H, Chang C-Y, Hu N, Wang Y-CJ, Lai C-C, Herrup K, Lee W-H, Bradley A: Mice deficient for Rb are nonviable and show defects in neurogenesis and haematopoiesis. Nature 359: 288–294 (1992).
77. Lee MS, Ogg S, Wu M, Parker LL, Donoghue DJ, Maller JL, Piwnica-Worms H: cdc25+ encodes a protein phosphatase that dephosphorylates p34cdc2. Mol Biol Cell 3: 73–84 (1992).
78. Lehner CF, O'Farrell PH: *Drosophila cdc2* homologues: a functional homolog is coexpressed with a cognate variant. EMBO J 9: 3573–3581 (1990).
79. Lew DJ, Dulić V, Reed SI: Isolation of three novel human cyclins by rescue of G1 cyclin (Cln) function in yeast. Cell 66: 1197–1206 (1991).
80. Lörincz AT, Reed SI: Primary structure homology between the product of yeast cell division control gene *CDC28* and vertebrate oncogenes. Nature 307: 183–185 (1984).
81. Lowndes NF, McInerny CJ, Johnson AL, Fantes PA, Johnston LH: Control of DNA synthesis genes in fission yeast by the cell-cycle gene $cdc10^+$. Nature 355: 449–453 (1992).
82. Lu KP, Means AR: Regulation of the cell cycle by calcium and calmodulin. Endocrine Rev 14: 40–58 (1993).
83. Lundgren K, Walworth N, Booher R, Dembski M, Kirschner M, Beach D: mik1 and wee1 cooperate in the inhibitory tyrosine phosphorylation of cdc2. Cell 64: 1111–1122 (1991).
84. Magyar Z, Bakó L, Bögre L, Dedeoğlu D, Kapros T, Dudits D: Active *cdc2* genes and cell cycle phase-specific cdc2-related kinase complexes in hormone-stimulated alfalfa cells. Plant J 4: 151–161 (1993).
85. Marcote MJ, Knighton DR, Basi G, Sowadski JM, Brambilla P, Draetta G, Taylor SS: A three-dimensional model of the Cdc2 protein kinase: localization of cyclin- and Suc1-binding proteins and phosphorylation sites. Mol Cell Biol 13: 5122–5131 (1993).
86. Markley N-A, Bonham-Smith PC, Moloney MM: Molecular cloning and expression of a cDNA encoding the proliferating cell nuclear antigen from *Brassica napus* (oilseed rape). Genome 36: 459–466 (1993).
87. Martinez MC, Jørgensen J-E, Lawton MA, Lamb CJ, Doerner PW: Spatial pattern of *cdc2* expression in relation to meristem activity and cell proliferation during plant development. Proc Natl Acad Sci USA 89: 7360–7364 (1992).
88. Matsushime H, Roussel MF, Sherr CJ:. Novel mammalian cyclin (CYL) genes expressed during G1. Cold Spring Harbor Symp Quant Biol 56: 69–74 (1991).
89. Matsushime H, Roussel MF, Ashmun RA, Sherr CJ: Colony-stimulating factor 1 regulates novel cyclins during the G1 phase of the cell cycle. Cell 65: 701–713 (1991).
90. Matsushime H, Ewen ME, Strom DK, Kato J-Y, Hanks SK, Roussel MF, Sherr CJ: Identification and properties of an atypical catalytic subunit ($p34^{PSK-J3}$/cdk4) for mammalian D type G1 cyclins. Cell 71: 323–334 (1992).
91. McIntosh EM, Atkinson T, Storms RK, Smith M: Characterization of a short, *cis*-acting DNA sequence which conveys cell cycle stage-dependent transcription in *Saccharomyces cerevisiae*. Mol Cell Biol 11: 329–337 (1991).
92. Meyerson M, Enders GH, Wu C-L, Su L-K, Gorka C, Nelson C, Harlow E, Tsai L-H: A family of human cdc2-related protein kinases. EMBO J 11: 2909–2917 (1992).
93. Miao G-H, Hong Z, Verma DPS: Two functional soybean genes encoding $p34^{cdc2}$ protein kinases are regulated by different plant developmental pathways. Proc Natl Acad Sci USA 90: 943–947 (1993).
94. Millar JBA, McGowan CH, Lenaers G, Jones R, Russell P: p80cdc25 mitotic inducer is the tyrosine phosphatase that activates $p34^{cdc2}$ kinase in fission yeast. EMBO J 10: 4301–4309 (1991).
95. Millar JBA, Lenaers G, Russell P: *Pyp3* PTPase acts as a mitotic inducer in fission yeast. EMBO J 11: 4933–4941 (1992).
96. Minshull J: Cyclin synthesis: who needs it? BioEssays 15: 149–155 (1993).
97. Minshull J, Blow JJ, Hunt T: Translation of cyclin mRNA is necessary for extracts of activated *Xenopus* eggs to enter mitosis. Cell 56: 947–956 (1989).
98. Minshull J, Golsteyn R, Hill CS, Hunt T: The A- and B-type cyclin associated cdc2 kinases in *Xenopus* turn on and off at different times in the cell cycle. EMBO J 9: 2865–2875 (1990).
99. Mizoguchi T, Gotoh Y, Nishida E, Yamaguchi-Shinozaki K, Hayashida N, Iwasaki T, Kamada H, Shinozaki K: Characterization of two cDNAs that encode MAP kinase homologues in *Arabidopsis thaliana* and analysis of the possible role of auxin in inactivating such

kinase activities in cultured cells. Plant J 5: 111–122 (1994).
100. Motokura T, Bloom T, Kim HG, Jüppner H, Ruderman JV, Kronenberg HM, Arnold A: A novel cyclin encoded by a *bcl1*-linked candidate oncogene. Nature 350: 512–515 (1991).
101. Moysset L, Simon E: Role of calcium in phytochrome-controlled nyctinastic movements of *Albizzia lophantha* leaflets. Plant Physiol 90: 1108–1114 (1989).
102. Murray AW, Kirschner MW: Cyclin synthesis drives the early embryonic cell cycle. Nature 339: 275–280 (1989).
103. Murray AW, Solomon MJ, Kirschner MW: The role of cyclin synthesis and degradation in the control of maturation promoting factor activity. Nature 339: 280–286 (1989).
104. Nash R, Tokiwa G, Anand S, Erickson K, Futcher AB: The *WHI1*$^{+}$ gene of *Saccharomyces cerevisiae* tethers cell division to cell size and is a cyclin homolog. EMBO J 7: 4335–4346 (1988).
105. Nasmyth K: Control of the yeast cell cycle by the Cdc28 protein kinase. Curr Opin Cell Biol 5: 166–179 (1993).
106. Nasmyth K, Dirick L: The role of *SWI4* and *SWI6* in the activity of G1 cyclins in yeast. Cell 66: 995–1013 (1991).
107. Nevins JR: E2F: a link between the Rb tumor suppressor protein and viral oncoproteins. Science 258: 424–429 (1992).
108. Nigg EA: Targets of cyclin-dependent protein kinases. Curr Opin Cell Biol 5: 187–193 (1993).
109. Nitschke K, Fleig U, Schell J, Palme K: Complementation of the cs *dis2*-11 cell cycle mutant of *Schizosaccharomyces pombe* by a protein phosphatase from *Arabidopsis thaliana*. EMBO J 11: 1327–1333 (1992).
110. Norbury C, Nurse P: Animal cell cycles and their control. Annu Rev Biochem 61: 441–470 (1992).
111. Norbury C, Blow J, Nurse P: Regulatory phosphorylation of the p34^{cdc2} protein kinase in vertebrates. EMBO J 10: 3321–3329 (1991).
112. Nurse P: Universal control mechanism regulating the onset of M-phase. Nature 344: 503–507 (1990).
113. Nurse P, Bissett Y: Gene required in G_1 for commitment to cell cycle and in G_2 for control of mitosis in fission yeast. Nature 292: 558–560 (1981).
114. O'Farrell PH: Cell cycle control: many ways to skin a cat. Trends Cell Biol 2: 159–163 (1992).
115. Ogas J, Andrews BJ, Herskowitz I: Transcriptional activation of CLN1, CLN2, and a putative new G1 cyclin (*HCS26*) by SWI4, a positive regulator of G1-specific transcription. Cell 66: 1015–1026 (1991).
116. Ohtsubo M, Roberts JM: Cyclin-dependent regulation of G_1 in mammalian fibroblasts. Science 259: 1908–1912 (1993).
117. Pagano M, Pepperkok R, Verde F, Ansorge W, Draetta G: Cyclin A is required at two points in the human cell cycle. EMBO J 11: 961–971 (1992).
118. Pardee AB: G_1 events and regulation of cell proliferation. Science 246: 603–608 (1989).
119. Pines J: Clear as crystal? Curr Biol 3: 544–547 (1993).
120. Pines J, Hunt T: Molecular cloning and characterization of the mRNA for cyclin from sea urchin eggs. EMBO J 6: 2987–2995 (1987).
121. Pines J, Hunter T: Isolation of a human cyclin cDNA: evidence for cyclin mRNA and protein regulation in the cell cycle and for interaction with p34^{cdc2}. Cell 58: 833–846 (1989).
122. Poon RYC, Yamashita K, Adamczewski JP, Hunt T, Shuttleworth J: The cdc2-related protein p40^{MO15} is the catalytic subunit of a protein kinase that can activate p33^{cdk2} and p34^{cdc2}. EMBO J 12: 3123–3132 (1993).
123. Pringle J, Hartwell L: The *Saccharomyces cerevisiae* cell cycle. In: Strathern J, Jones E, Broach J (eds) The Molecular Biology of the Yeast Saccharomyces, pp. 97–142. Cold Spring Harbor Laboratory Press, Cold Spring Harbor, NY (1981).
124. Renaudin J-P, Colasanti J, Rime H, Yuan Z, Sundaresan V: Cloning of four cyclins from maize indicates that higher plants have three structurally distinct groups of mitotic cyclins. Proc Natl Acad Sci USA, in press (1994).
125. Richardson HE, Wittenberg C, Cross F, Reed SI: An essential G1 function for cyclin-like proteins in yeast. Cell 59: 1127–1133 (1989).
126. Ruderman JV: MAP kinase and the activation of quiescent cells. Curr Opin Cell Biol 5: 207–213 (1993).
127. Russell P, Nurse P: *cdc25*$^{+}$ functions as an inducer in the mitotic control of fission yeast. Cell 45: 145–153 (1986).
128. Russell P, Nurse P: The mitotic inducer *nim1*$^{+}$ functions in a regulatory network of protein kinase homologues controlling the initiation of mitosis. Cell 49: 569–576 (1987).
129. Russell P, Nurse P: Negative regulation of mitosis by *wee1*$^{+}$, a gene encoding a protein kinase homolog. Cell 49: 559–567 (1987).
130. Sherr CJ: Mammalian G_1 cyclins. Cell 73: 1059–1065 (1993).
131. Sherr CJ: The ins and outs of *RB*: coupling gene expression to the cell cycle clock. Trends Cell Biol 4: 15–18 (1994).
132. Solomon MJ: Activation of the various cyclin/cdc2 protein kinases. Curr Opin Cell Biol 5: 180–186 (1993).
133. Solomon M, Booher R, Kirschner M, Beach D: Cyclin in fission yeast. Cell 54: 738–740 (1988).
134. Solomon MJ, Harper JW, Shuttleworth J: CAK, the P34^{cdc2} activating kinase, contains a protein identical or closely related to p40^{MO15}. EMBO J 12: 3133–3142 (1993).
135. Sorger PK, Murray AW: S-phase feedback control in budding yeast independent of tyrosine phosphorylation of p34^{cdc28}. Nature 355: 365–368 (1992).
136. Stafstrom JP, Altschuler M, Anderson DH: Molecular

cloning and expression of a MAP kinase homologue from pea. Plant Mol Biol 22: 83–90 (1993).
137. Staiger C, Doonan J: Cell division in plants. Curr Opin Cell Biol 5: 226–231 (1993).
138. Suzuka I, Hata S, Matsuoka M, Kosugi S, Hashimoto J: Highly conserved structure of proliferating cell nuclear antigen (DNA polymerase delta auxiliary protein) gene in plants. Eur J Biochem 195: 571–575 (1991).
139. Swenson KI, Farrell KM, Ruderman JV: The clam embryo protein cyclin A induces entry into M phase and the resumption of meiosis in *Xenopus* oocytes. Cell 47: 861–870 (1986).
140. Terryn N, Van Montagu M, Inzé D: GTP-binding proteins in plants. Plant Mol Biol 22: 143–152 (1993).
141. Th'ng JPH, Wright PS, Hamaguchi J, Lee MG, Norbury CJ, Nurse P, Bradbury EM: The FT210 cell line is a mouse G2 phase mutant with a temperature-sensitive *CDC2* gene product. Cell 63: 313–324 (1990).
142. Trewavas AJ: Growth substances, calcium and the regulation of cell division. In: Bryant JA, Francis D (eds) The Cell Division Cycle in Plants. Society for Experimental Biology Seminar Series, vol. 26, pp. 133–156. Cambridge University Press, Cambridge (1985).
143. Tyers M, Tokiwa G, Futcher B: Comparison of the *Saccharomyces cerevisiae* G_1 cyclins: Cln3 may be an upstream activator of Cln1, Cln2 and other cyclins. EMBO J 12: 1955–1968 (1993).
144. Van den Heuvel S, Harlow E: Distinct roles for cyclin-dependent kinases in cell cycle control. Science 262: 2050–2054 (1993).
145. Van 't Hof J: Control points within the cell cycle. In: Bryant JA, Francis D (eds) The Cell Division Cycle in Plants. Society for Experimental Biology Seminar Series, vol. 26, pp. 1–13. Cambridge University Press, Cambridge (1985).
146. Wang J, Chenivesse X, Henglein B, Bréchot C: Hepatitis B virus integration in a cyclin A gene in a hepatocellular carcinoma. Nature 343: 555–557 (1990).
147. Westendorf JM, Swenson KI, Ruderman JV: The role of cyclin B in meiosis I. J Cell Biol 108: 1431–1444 (1989).
148. Wick SM: Spatial aspects of cytokinesis in plant cells. Curr Opin Cell Biol 3: 253–560 (1991).
149. Wittenberg C, Sugimoto K, Reed SI: G1-specific cyclins of S. cerevisiae cell cycle periodicity regulation by mating pheromone, and association with the $p34^{CDC28}$ protein kinase. Cell 62: 225–237 (1990).
150. Xiong Y, Zhang H, Beach D: D type cyclins associate with multiple protein kinases and the DNA replication and repair factor PCNA. Cell 71: 505–514 (1992).
151. Xiong Y, Menninger J, Beach D, Ward DC: Molecular cloning and chromosomal mapping of *CCND* genes encoding human D-type cyclins. Genomics 13: 575–584 (1992).
152. Xiong Y, Hannon GJ, Zhang H, Casso D, Kobayashi R, Beach D: p21 is a universal inhibitor of cyclin kinases. Nature 366: 701–704 (1993).
153. Zheng W-F, Ruderman JV: Functional analysis of the P box, a domain in cyclin B required for the activation of Cdc25. Cell 75: 155–164 (1993).

Plant Molecular Biology **26**: 1305–1313, 1994.

Signal molecules involved in plant embryogenesis

Ed D.L. Schmidt, Anke J. de Jong and Sacco C. de Vries*
Department of Molecular Biology, Wageningen Agricultural University, Dreijenlaan 3, 6703 HA Wageningen, The Netherlands (author for correspondence)*

Received 10 May 1994; accepted 11 May 1994

Key words: zygotic embryogenesis, somatic embryogenesis, chitinase, nod factor, signal molecules

Abstract

In plant embryogenesis, inductive interactions mediated by diffusable signal molecules are most likely of great importance. Evidence has been presented that at late globular stages in plant embryogenesis, perturbation of the polar auxin transport results in abberrant embryo morphology. *Rhizobium* lipo-oligosaccharides or Nod factors are a newly discovered class of bacterial molecules that are able to trigger initial steps in root nodule development in legumes. Part of the activity of Nod factors may be directed towards alteration of endogenous plant growth regulator balance. The same bacterial Nod factors promoted the formation of globular embryos in the carrot cell line ts11. Whether there exist plant analogues of the Nod factors and whether these molecules are active as a more universal control system perhaps designed to initiate and or mediate gradients in auxin and cytokinin remains to be determined.

Introduction

Currently there is a wide and increasing interest in the molecular-genetic analysis of plant embryogenesis. Detailed descriptions of gametogenesis and embryogenesis have been the subject of many recent studies [1–8] as well as reviews [9–15] and therefore will only be recapitulated briefly.

The male gametophytes or pollen grains are formed in the anther, while the female gametophyte or embryo sac is formed in the pistil. In angiosperms that exhibit the polygonum type [16] it consists of seven cells: the egg cell, two synergids, the central cell and three antipodal cells. The polarized egg cell and synergids are positioned at the micropylar pole of the embryo sac. The polarity of the egg cell is reflected in the position of the nucleus and most of the cytoplasm at the chalazal side of the cell, while the micropylar part is highly vacuolated. Dual fertilization of the diploid central cell and the haploid egg cell results in the endosperm and the zygote respectively. The first zygotic division is asymmetrical and yields a small apical cell and a large basal cell. The basal cell remains positioned at the micropylar pole of the embryo sac, and undergoes a series of transversal divisions to form the suspensor. From the uppermost cell of the suspensor, the hypophysial cell, the centre of the future root meristem is formed [17]. The apical cell undergoes three divisions, resulting in the octant stage embryo proper. Tangential divisions then set apart the protoderm cells, from which the epidermis is derived. Development of the *Arabidopsis* embryo from fertilization, through the octant, globular, triangular, heart, torpedo and bent-cotyledon stages, to the mature desiccated embryo, has been subdivided into a sequence of 20 different stages [3].

While the above-described developmental sequence that gives rise to the *Arabidopsis* embryo appears to involve a highly predictable series of divisions, it is important to note that this is but one of the types of embryo development in plants [16]. Many variations in the plants and positions of early cell divisions and in suspensor morphology have been described, with apparently little or no consequence for the eventual seedling morphology. Recently the systematic genetic dissection of the formation of the zygotic embryo has been initiated [18–21]. Based on the mutant phenotypes obtained, a division of the young embryo along the longitudinal axis into an apical, central and basal region was proposed [20, 22]. A second, radial pattern, superimposed on the apical-basal pattern and consisting of the vascular, ground and epidermal tissues, appears to be established independently.

Somatic or asexual embryogenesis is the process by which somatic cells develop into plants through the same characteristic morphological stages as their zygotic counterparts. For dicots these are the globular, heart and torpedo stages. The ability to form embryos that do not orignate from a fertilized egg cell is quite widespread among plants. It may occur naturally as in *Malaxis*, where somatic embryos form spontanously on the leaf tips [23], or in the form of apomictic processes [24]. Under *in vitro* conditions somatic embryos can either form directly on the surface of an organized tissue such as a leaf or stem segment, from protoplasts or from microspores, or indirectly via an intermediary step of callus or suspension culture [25]. By virtue of their excellent experimental accessibility, somatic embryogenesis is exploited to isolate plant-produced molecules that have promotive effects on the formation of somatic embryos [26]. Both somatic embryogenesis as well as *in vitro* cultured zygotic embryos are being employed to try and answer long-standing questions concerning the role of 'classical' plant growth regulators such as auxin in embryogenesis [27, 28].

Signal molecules in zygotic embryogenesis

Two mechanisms appear to be universally used in animal embryogenesis to initiate cell differentiation. The first of these is a polarization of cellular determinants, sometimes, but not always, followed by asymmetric cell division [29]. In plants, fertilization is followed by an extensive redistribution of organelles resulting in polarization of the zygote. Asymmetric cell division occurs frequently in plants, and the analysis of the *Arabidopsis gnom* mutant [30], in which the mutant phenotype appears correlated with the inability to perform a normal unequal division of the zygote, clearly shows that this mechanism is indeed of crucial importance in plant embryogenesis. The second mechanism consists of the interaction between an inducing cell or tissue and a responding cell or tissue brought about by specific signal molecules. Signal molecules that are produced outside of a group of equivalent cells are defined as inducers [31]. Cellular communication between adjacent cells can occur by cell-surface-located signal molecules able to act as inducers. Cells that are not in direct contact with an inducer-producing cell can be influenced if the signal molecule is diffusable, usually resulting in a concentration gradient. Signal molecules are defined as morphogens when the slope of their concentration gradient provides reference points for the formation of a pattern. The local concentration of the morphogen then determines the response of the cells.

Caenorhabditis elegans has served as an example for the existence of a series of segregating cytoplasmic determinants that result in a rigid cell lineage as a means to generate patterns in the embryo and differentiated cells later on in development [32]. Recently described mutants provide evidence that a number of maternally expressed genes that encode nuclear proteins and cell surface proteins similar to the *lin-12* family are involved in cell-inductive processes [33].

In plant development cell position rather than developmental history is considered to be essential for the formation of the somatic tissues [34–37]. This implies that cell-inductive processes and the use of signal molecules that act at short range and over longer distances might be important in the organization of the plant embryo. However, no direct evidence is available that cell-inductive processes are indeed occurring in plants. A sequential and transient expression of a plasma membrane arabinogalactan protein (AGP) epitope, recognized by the monoclonal antibody JIM8, was observed in reproductive tissues and the suspensor of an early globular embryo of *Brassica*. Pennell *et al.* [38] speculated that the JIM8 epitope may be a marker for a cell-inductive process in plants. Whether AGPs themselves act as signal molecules in zygotic embryogenesis is not known. There is however some evidence from somatic embryogenesis (see next section) that particular AGPs may indeed have a direct biological function in embryogenesis. If so, they will be likely to act as short-range inducer molecules over a distance of a few cells only. A fascinating result was recently reported for *Fucus* [39] where, upon laser microsurgery of two-celled embryos, it was established that prolonged contact of a thallus cell protoplast with the wall of the ablated rhizoid cell resulted in the formation of cells with the characteristics of the rhizoid cell. This experiment provides evidence for the presence of stable non-diffusible wall components able to change the fate of algal cells, and it will be of great importance to establish whether similar experimental systems are feasible in higher plants. The ability to perform *in vitro* fertilization [40] makes this a realistic option.

A glimpse of the role that plant growth regulators may play as signal molecules during embryogenesis has been provided by the characterization of *Arabidopsis* mutants, in which the normal balance of auxin and cytokinins was disturbed. The *pin-1* mutant was isolated as one of the *Arabidopsis* flower mutants, and found to have a severely reduced ability for auxin transport. The phenotype of *pin-1* appears to be a gross abnormality of the inflorescence axis, as well as formation of abnormal flowers and leaves. This phenotype could be reproduced by germinating wild-type seeds in the presence of polar auxin transport inhibitors [41]. The embryos of this mutant exhibit a fused collar-like arrangement of their cotyledon primordia, as opposed to the normal bilateral arrangement [27]. Application of the polar auxin transport inhibitor to excised and *in vitro* cultured zygotic embryos of *Brassica* resulted in a similar fused-cotyledon phenotype as observed for the *pin-1* mutant [27], and in the Arabidopsis *gnom* mutant [30]. These results are interpreted to suggest that polar auxin transport determines the transition from radial to bilateral symmetry in the globular dicot embryo. The *amp-1* mutant [42], exhibiting an elevated level of cytokinins, has a complex and pleiotropic phenotype affecting photomorphogenesis and flowering time, as well as an increase in the number of tricot and tetracot seedlings. Multiplication of cotelydons is also observed in embryo mutants of the *hauptling* type [15, 20]. It appears therefore that interference with the local balance between auxin and cytokinin primarily influences the formation of the proper number and orientation of cotyledon primordia in the late globular embryo. Because chemical inhibitors such as NPA and TIBA act as inhibitors of auxin efflux carriers [43], application will result in elevated intracellular concentrations in all cells subjected to the inhibitor. Without the possibility to monitor the concentration of (active) growth regulators at the individual cell level, it is not easy to draw up conclusive models of the way polar transport of auxin is involved in the formation of cotyledon primordia [44]. However, all models assume that during the formation of the globular embryo gradients in these growth regulators are established by an unequal distribution of auxin-synthesizing cells. Whether this is the result of unequal distribution of cytoplasmic determinants in the zygote or the effect of an earlier acting signal molecule remains to be determined. In the next part of this review we will describe some recent findings regarding somatic embryogenesis, which may provide clues to the identity of additional signal molecules involved in plant embryogenesis.

Signal molecules in somatic embryogenesis

The possibility of somatic plant cells in culture to acquire embryogenic potential has been exploited to establish experimental systems that allow the identification of molecules able to promote or influence the formation of somatic embryos. An important and as yet unsolved question in these studies is whether the findings are only relevant to the somatic embryo system, or whether they are also applicable to zygotic embryogenesis.

In *Daucus*, the usual strategy to start an embryogenic suspension culture is to expose explants to a high concentration of auxin. After reinitiation of cell division and a period of proliferation of the released explant cells in the presence of auxin, a small subset of the cell population becomes embryogenic [45]. These embryogenic cells are usually in the form of clusters of small cytoplasmic cells, referred to as proembryogenic masses [46]. In contrast to the non-embryogenic cells, these proembryogenic masses become insensitive to auxin [47]. By the time the embryos developed from the proembryogenic masses reach the plantlet stage, auxin sensitivity is regained. An important question concerning the formation of embryogenic cells in suspension cultures is whether they are formed continuously from non-embryogenic cells or are derived from a subset of explant cells, which is propagated independently [48]. Application of techniques to follow the developmental fates of individual single cells present in embryogenic suspension cultures has revealed a striking heterogeneity in embryogenic single cell types and early cell division patterns that was nonetheless not readily apparent in the morphology of the resulting somatic embryos [49].

Recent evidence suggests that particular purified AGPs isolated from the culture medium of embryogenic *Daucus* lines and from dry *Daucus* seeds were able to promote the formation of proembryogenic masses, even in previously non-embryogenic *Daucus* cell lines, when added in nanomolar concentrations. Other AGPs, isolated from the medium of a non-embryogenic line, acted negatively on the formation of proembryogenic masses [50]. These results show that specific members of the family of AGPs are involved in the formation of embryogenic clusters. Although the underlying mechanisms are unclear, these observations, together with earlier ones employing unfractionated conditioned medium [51], suggest that molecules totally different from conventional plant growth regulators are able to direct the transition of somatic cells into embryogenic cells. Since cell-surface AGPs turn over very rapidly [52], and their expression is clearly developmentally regulated [53], they are likely candidates for molecules able to mediate developmental processes in plants, perhaps by a cell-inductive mechanism [38].

Chitinases and chitin-containing signal molecules

Recently de Jong *et al.* [26] have identified a 32 kDa endochitinase able to lift the arrest in embryo development of the temperature-sensitive carrot variant ts11. Later, de Jong *et al.* [54] have shown that chitin-containing bacterial signal molecules, Nod factors, are able to mimic the effect of the 32 kDa endochitinase. These results suggest that the 32 kDa endochitinase is involved in the generation of plant analogues of these compounds, but this hypothesis assumes that a plant substrate is indeed available for plant chitinases, and that the phenotypical result of both the chitinase and the Nod factor act via the same mechanism.

Enzymes that catalyse hydrolysis of *N*-acetyl-β-D-glucosaminide β-1,4 linkages in chitin and in chito-oligomers are classified as either exo- or endochitinases. In 1965, Powning and Irzykiewicz [55] proposed a role for chitinases in a defence mechanism against chitin-containing parasites from the soil. As has been demonstrated repeatedly, plant chitinase activity is increased by fungal elicitors [56–58], and is part of the hypersensitive response to pathogen attack [59, 60]. For some plant-produced chitinases the ability to inhibit fungal growth *in vitro* has been shown [61–63]. Some of the isolated plant chitinases only possess antifungal activity in combination with β-1,3-glucanase activity [64], while several others do not have antifungal activity *in vitro* at all [65].

Transgenic tobacco seedlings, expressing a bean chitinase gene, show an increased ability to survive in soil infested with the fungal pathogen *Rhizoctonia solani* [66]. Chitinase genes are also expressed in the absence of pathogens [67–70], a finding that is usually explained by assuming that they are part of a continuously present defense mechanism.

The results presented by de Jong *et al.* [26] represent the first direct evidence for a role of at least one plant chitinase in somatic embryogenesis. At the same time these results raise the question what the natural substrate of these enzymes might be, assuming that it is the catalytic property of the 32 kDa endochitinase which is required for the rescue of ts11 embryo development. Employing an indirect approach by testing putative products of chitinase activity for their effect on ts11 embryo development, the *Rhizobium* Nod factor NodRlv-V(Ac, C18:4) proved to be able to stimulate ts11 embryo formation with a similar efficiency as the 32 kDa endochitinase [54]. This result was interpreted to suggest that the 32 kDa endochitinase acts by releasing Nod factor-like signal molecules from plant-produced precursors. However, it must be kept in mind that presently it is not known whether the 32 kDa endochitinase and NodRlv-V(Ac, C18:4) act via the same or via different mechanisms. Therefore, it remains possible that there is no direct relationship between both observations.

When boiled homogenates of bean plants are treated with a bean endochitinase, no *N*-acetylglucosamine-containing fragments can be detected, indicating that chitin-like substrates are not present in the plant [71]. Employing cytochemical labelling, Benhamou and Asselin [72] have shown binding of chitinases and wheat germ agglutinin, a plant lectin that is specific for oligomers of *N*-acetylglucosamine, to secondary cell walls of a variety of Solanaceae. This binding can be abolished by prior treatment with lipase. Although lipid-linked *N*-acetylglucosamine residues have not been identified as structural components of plant cell walls [73], the results of Benhamou and Asselin [72] suggest that the *N*-acetylglucosamine residues detected may be present in the form of glycolipids. Incubation of membrane fractions from bean or pea stems with *N*-acetyl[^{14}C]glucosamine and analysis of the lipophilic fraction show incorporation of *N*-acetyl[^{14}C]glucosamine only in the highly charged dolichol pyrophosphate oligosaccharides, the intermediates in the biosynthesis of glycoproteins [74, 75]. These intermediates contain two *N*-acetylglucosamines at most per lipid molecule. The minimal length of an *N*-acetylglucosamine chain required to serve as substrate for endochitinases is three [76], indicating that the dolichol pyrophosphate oligosaccharides are unable to serve as substrate for chitinases. Another class of glycolipids in plants that contain *N*-acetylglucosamine are of the sphingolipid type. They are present in seeds and commonly contain complex oligosaccharides rather than a single saccharide and a variety of sugar components including *N*-acetylglucosamine, glucuronic acid, inositol, galactose, arabinose and fucose [77]. However, a stretch of three *N*-acetylglucosamines has not been detected in such glycolipids.

To test the hypothesis whether lipo-oligosaccharides analogous to the *Rhizobium* Nod factors occur in plants, *Lathyrus* flowers have been labelled with *N*-acetyl[^{14}C]glucosamine [78]. Some of the extracted lipophilic compounds migrate similarly to the rhizobial lipo-oligosaccharides on TLC plates. To determine the nature of the oligosaccharide chain, the lipophilic compounds have been treated with commercial chitinase and analysed on TLC plates. The appearance of at least three new spots after chitinase treatment suggests that lipophilic compounds can indeed be hydrolysed by chitinases. Indirect evidence for the presence of chitin-like molecules in plants has been provided by the introduction of *Rhizobium* nodulation genes *nodA* and *nodB*, involved in the biosynthesis of the Nod factors, into tobacco plants [79, 80]. Schmidt *et al.* [81] have shown that NodB together with the NodA protein are sufficient to produce small, heat-stable compounds that stimulate mitosis in legume and non-legume protoplasts. The NodB gene product is an oligosaccharide-modifying enzyme that deacetylates the non-reducing *N*-acetylglucosamine residues of chito-oligosaccharides [82], while the NodA protein is involved in the *N*-acylation of the chito-oligosaccharide, essential for the coupling of the sugar and fatty acid parts of the active Nod factor [83]. Expression of *nodA* and *nodB* in tobacco results in plants with morphological abnormalities such as wrinkled leaves, reduced growth and compact inflorescences. These experiments suggest that tobacco is able to produce substrate molecules that can be used by the *nodA*- and *nodB*-encoded proteins to synthesize growth-controlling factors. Cultured tomato cells also respond to both chitin fragments as well as Nod factors by a rapid and transient alkalinization of the culture medium [84].

Taken together, although no substrate molecules for chitinases have been unequivocally identified in plants, several lines of circumstantial evidence indicate that substrates for chitinases may indeed be present. It will now be of great interest to identify plant substrates for chitinases, because they may act as inactive precursors of lipo-oligosaccharides, with chitinases in the role of enzymes releasing these molecules. If plant chitinases indeed are able to release signal molecules from plant-produced precursor molecules, they may not only be able to release but also to inactivate lipo-oligosaccharide signal molecules, like they may do with bacterial Nod factors [85, 86].

Do lipo-oligosaccharides represent a novel class of plant growth regulators?

Besides the *Rhizobium* lipo-oligosaccharides, other oligosaccharides can also influence plant growth and development at concentrations several orders of magnitude below those of more conventional plant growth regulators. These oligosaccharides are termed oligosaccharins [87, 88]. One of these, XXFG, inhibits 2,4-D-stimulated elongation of pea stem segments in a dose-dependent fashion [89, 90] Hence, oligosaccharins can act either as intermediates in the growth hormone-regulating mechanism or interact with a consecutive site somewhere in the cascade of events normally triggered by auxin.

Purified Nod factors, when applied to legume seedlings at concentrations as low as 10^{-12} M, stimulate differentiation of epidermal cells into root hairs, deformation of root hairs, induction of early nodulin genes related to the infection process in root epidermal cells, and induction of cell divisions in the inner cortex of the roots [see for reviews 91–93]. These events normally occur during the early nodulation process. Induction of early nodulin gene expression, cortical cell division and subsequent meristem formation can also be triggered by auxin transport inhibitors [94, 95], suggesting that a change in the endogenous balance of growth regulators plays a role in the initiation of nodule formation. A cytokinin-producing gene, cloned into *Escherichia coli*, is sufficient to provide this bacterium with the ability to induce cortical cell divisions in alfalfa roots [96]. Although there are still several possibilities to explain these findings, one attractive hypothesis is that bacterial Nod factors are able to modulate the endogenous auxin-cytokinin balance in the root cortex. A plausible but unproven explanation for the effect of Nod factors on carrot cells may be that they alter the endogenous growth regulator balance of the susceptible cells.

Acknowledgements

We thank Ton Bisseling and Gerd Jürgens for valuable comments on the manuscript. We are grateful to our colleagues for communicating unpublished results. Primary research in our laboratory is supported by the Foundation for Biological Research, subsidized by the Netherlands Organization for Scientific Research (A.J.d.J.), the Technology Foundation, subsidized by the Netherlands Organization for Scientific Research (E.D.L.S.) and the Biotech-PTP programme of the European Community.

References

1. Castle LA, Meinke DW: Embryo-defective mutants as tools to study essential functions and regulatory processes in plant embryo development. Semin Devel Biol 4: 31–39 (1993).

2. Cresti M, Blackmore S, van Went JL: Atlas of Sexual Reproduction in Flowering Plants. Springer-Verlag, Berlin (1992).
3. Jürgens G, Mayer U: Arabidopsis. In: Bard J (ed) A Colour Atlas of Developing Embryos. Wolfe, London (1994).
4. Mansfield SG, Briarty LG: Development of the free-nuclear endosperm in *Arabidopsis thaliana* L. Arabidopsis Inf Serv 27: 53–64 (1990).
5. Mansfield SG, Briarty LG: Endosperm cellularization in *Arabidopsis thaliana* L. Arabidopsis Inf Serv 27: 65–72 (1990).
6. Mansfield SG, Briarty LG: Early embryogenesis in *Arabidopsis thaliana*. II. The developing embryo. Can J Bot 69: 461–476 (1991).
7. Mansfield SG, Briarty LG, Erni S: Early embryogenesis in *Arabidopsis thaliana*. I. The mature embryo sac. Can J Bot 69: 447–460 (1991).
8. Webb MC, Gunning BES: Embryo sac development in *Arabidopsis thaliana*. I. Megasporogenesis, including the microtubular cytoskeleton. Sex Plant Reprod 3: 244–256 (1990).
9. de Jong AJ, Schmidt EDL, de Vries SC: Early events in higher-plant embryogenesis. Plant Mol Biol 22: 367–377 (1993).
10. Thomas TL: Gene expression during plant embryogenesis and germination: an overview. Plant Cell 5: 1401–1410 (1993).
11. Zimmerman JL: Somatic embryogenesis: a model for early development in higher plants. Plant Cell 5: 1411–1423 (1993).
12. Yeung EC, Meinke DW: Embryogenesis in angiosperms: development of the suspensor. Plant Cell 5: 1371–1381 (1993).
13. Lindsey K, Topping JF: Embryogenesis: a question of pattern. J Exp Bot 44: 359–374 (1993).
14. West MAL, Harada JJ: Embryogenesis in higher plants: an overview. Plant Cell 5: 1361–1369 (1993).
15. Jürgens G, Torres Ruiz RA, Berleth T: Embryonic pattern formation in flowering plants. Annu Rev Genet 28, in press (1994).
16. Johri BM: Embryology of Angiosperms. Springer-Verlag, Berlin (1984).
17. Dolan L, Janmaat K, Willemsen V, Linstead P, Poethig S, Roberts K, Scheres B: Cellular organization of the *Arabidopsis thaliana* root. Development 119: 71–84 (1993).
18. Patton DA, Meinke DW: Ultrastructure of arrested embryos from lethal mutants of *Arabidopsis thaliana*. Am J Bot 55: 807–819 (1990).
19. Errempalli D, Patton D, Castle L, Mickelson L, Hansen K, Schnall J, Feldmann K, Meinke D: Embryonic lethals and T-DNA insertional mutagenesis in *Arabidopsis*. Plant Cell 3: 149–157 (1991).
20. Jürgens G, Mayer U, Torres Ruiz RA, Berleth T, Miséra S: Genetic analysis of pattern formation in the *Arabidopsis* embryo. Development, Suppl 91.1: 27–38 (1991).
21. Meinke DW: Embryonic mutants of *Arabidopsis thaliana*. Devel Genet 12: 382–392 (1991).
22. Mayer U, Torres Ruiz RA, Berleth T, Miséra S, Jürgens G: Mutations affecting body organization in the *Arabidopsis* embryo. Nature 353: 402–407 (1991).
23. Taylor RL: The foliar embryos of *Malaxis paludosa*. Can J Bot 45: 1553–1556 (1967).
24. Koltunow AM: Apomixis: embryo sacs and embryos formed without meiosis or fertilization in ovules. Plant Cell 5: 1425–1437 (1993).
25. Williams EG, Maheswaran G: Somatic embryogenesis: factors influencing coordinate behavior of cells as an embryogenetic group. Ann Bot 57: 443–462 (1986).
26. de Jong AJ, Cordewener J, Lo Schiavo F, Terzi M, Vandekerckhove J, van Kammen A, de Vries SC: A carrot somatic embryo mutant is rescued by chitinase. Plant Cell 4: 425–433 (1992).
27. Liu C-m, Xu Z-h, Chua N-h: Auxin polar transport is essential for the establishment of bilateral symmetry during early plant embryogenesis. Plant Cell 5: 621–630 (1993).
28. Schiavone FM, Cooke TJ: Unusual patterns of somatic embryogenesis in domesticated carrot: developmental effects of exogenous auxins and auxin transport inhibitors. Cell Diff 21: 53–62 (1987).
29. Gurdon JB: The generation of diversity and pattern in animal cells. Cell 68: 185–199 (1992).
30. Mayer U, Büttner G, Jürgens G: Apical-basal pattern fromation in the *Arabidopsis* embryo: studies on the role of the *gnom* gene. Development 117: 149–162 (1993).
31. Greenwald I, Rubin GM: Making a difference: the role of cell-cell interactions in establishing separate identities for equivalent cells. Cell 68: 271–281 (1992).
32. Sulton JE, Schierenberg E, White JG, Thomson JN: The embryonic development of *Caenorhabditis elegans*. Devel Biol 100: 64–119 (1983).
33. Wood WB, Edgar LG: Patterning in the *C. elegans* embryo. Trends Genet 10: 49–54 (1994).
34. Dawe RK, Freeling M: Cell lineage and its consequences in higher plants. Plant J 1: 3–8 (1991).
35. Furner IJ, Pumfrey JE: Cell fate in the shoot apical meristem of *Arabidopsis thaliana*. Development 115: 755–764 (1992).
36. Irish VF, Sussex IM: A fate map of the *Arabidopsis* embryonic shoot apical meristem. Development 115: 745–753 (1992).
37. Poethig RS: Genetic mosaic and cell lineage analysis in plants. Trends Genet 5: 273–277 (1989).
38. Pennell RI, Janniche L, Kjellbom P, Scofield GN, Peart JM, Roberts K: Developmental regulation of a plasma membrane arabinogalactan protein epitope in oilseed rape flowers. Plant Cell 3: 1317–1326 (1991).
39. Berger F, Taylor A, Brownlee C: Cell fate determination by the cell wall in early *Fucus* development. Science 263: 1421–1423 (1994).

40. Dumas C, Mogensen HL: Gametes and fertilization: maize as a model for experimental embryogenesis in flowering plants. Plant Cell 5: 1337–1348 (1993).
41. Okada K, Ueda J, Komaki MK, Bell CJ, Shimura Y: Requirement of the auxin polar transport system in early stages of *Arabidopsis* floral bud formation. Plant Cell 3: 677–684 (1991).
42. Chaudhury AM, Letham S, Craig S, Dennis ES: *Amp-1*: a mutant with high cytokinin levels and altered embryonic pattern, faster vegetative growth, constitutive photomorphogenesis, and precocious flowering. Plant J 4: 907–916 (1993).
43. Rubery PH: Auxin transport. In: Davies PJ (ed) Plant Hormones and their Role in Plant Growth and Development, pp. 341–362. Martinus Nijhoff, Hingham, MA (1987).
44. Cooke TJ, Racusen RH, Cohen JD: The role of auxin in plant embryogenesis. Plant Cell 5: 1494–1495 (1993)
45. de Vries SC, Booij H, Meyerink P, Huisman G, Wilde DH, Thomas TL, van Kammen A: Acquisition of embryogenic potential in carrot cell-suspension cultures. Planta 176: 196–204 (1988).
46. Halperin W: Alternative morphogenetic events in cell suspensions. Am J Bot 53: 443–453 (1966).
47. Lo Schiavo F, Filippini F, Cozzani F, Vallone D, Terzi M: Modulation of auxin-binding proteins in cell suspensions. I. Differential responses of carrot embryo cultures. Plant Physiol 97: 60–64 (1991).
48. van Engelen FA, de Vries SC: Extracellular proteins in plant embryogenesis. Trends Genet 8: 66–70 (1992).
49. Toonen MAJ, Hendriks T, Schmidt EDL, Verhoeven HA, van Kammen A, de Vries SC: Description of somatic embryo-forming single cells in carrot suspension cultures employing video cell tracking. Planta 194: 565–572 (1994).
50. Kreuger M, van Holst GJ: Arabinogalactan proteins are essential in somatic embryogenesis of *Daucus carota* L. Planta 189: 243–248 (1993).
51. de Vries SC, Booij H, Janssens R, Vogels R, Saris L, Lo Schiavo F, Terzi M, van Kammen A: Carrot somatic embryogenesis depends on the phytohormone-controlled presence of correctly glycosylated extracellular proteins. Genes Devel 2: 462–476 (1988).
52. van Holst GJ , Klis FM, de Wildt PJM, Hazenberg CAM, Buijs J, Stegwee D: Arabinogalactan protein from a crude cell organelle fraction of *Phaseolus vulgaris* L. Plant Physiol 68: 910–913 (1981).
53. Knox JP, Day S, Roberts K: A set of cell surface glycoproteins forms an early marker of cell position, but not cell type in the root apical meristem of *Daucus carota* L. Development 106: 47–56 (1989).
54. de Jong AJ, Heidstra R, Spaink HP, Hartog MV, Meijer EA, Hendriks T, Lo Schiavo F, Terzi M, Bisseling T, van Kammen A, de Vries, SC: Rhizobium lipooligosaccharides rescue a carrot somatic embryo mutant. Plant Cell 5: 615–620 (1993).
55. Powning RF, Irzykiewicz H: Studies on the chitinase systems in bean and other seeds. Comp Biochem Physiol 14: 127–133 (1965).
56. Ishige F, Mori H, Yamazaki K, Imaseki H: Cloning of a complementary DNA that encodes an acidic chitinase which is induced by ethylene and expression of the corresponding gene. Plant Cell Physiol 34: 103–111 (1993).
57. Kurosaki F, Tashiro N, Nishi A: Chitinase induction in carrot cell cultures treated with various fungal compounds. Biochem Internatl 20: 99–106 (1990).
58. Kirsch C, Hahlbrock K, Kombrink E: Purification and characterization of extracellular, acidic chitinase isoenzymes from elicitor-stimulated parsley cells. Eur J Biochem 213: 419–425 (1993).
59. Metraux JP, Boller T: Local and systemic induction of chitinase in cucumber plants in response to viral, bacterial and fungal infections. Physiol Mol Plant Path 28: 161–169 (1986).
60. Margis-Pinheiro M, Metz-Boutigue MH, Awade A, De Tapia M, Le Ret M, Burkard G: Isolation of a complementary DNA encoding the bean PR4 chitinase: an acidic enzyme with an amino-terminus cysteine-rich domain. Plant Mol Biol 17: 243–253 (1991).
61. Schlumbaum A, Mauch F, Vogeli U, Boller T: Plant chitinases are potent inhibitors of fungal growth. Nature 324: 365–367 (1986).
62. Broekaert WF, van Parijs J, Allen AK, Peumans WJ: Comparison of some molecular, enzymatic and antifungal properties of chitinases from thorn-apple, tobacco and wheat. Physiol Mol Plant Path 33: 319–331 (1988).
63. Huynh QK, Hironaka CM, Levine EB, Smith CE, Borgmeyer JR, Shah DM: Antifungal proteins from plants. Purification, molecular cloning, and antifungal properties of chitinases from maize seed. J Biol Chem 267: 6635–6640 (1992).
64. Sela-Buurlage MB, Ponstein AS, Bres-Vloemans SA, Melchers LS, van den Elzen PJM, Cornelissen BJC: Only specific tobacco (*Nicotiana tabacum*) chitinases and β-1,3-glucanases exhibit antifungal activity. Plant Physiol 101: 857–863 (1993).
65. Woloshuk CP, Meulenhoff JS, Sela-Buurlage M, van den Elzen PJM, Cornelissen BJC: Pathogen-induced proteins with inhibitory activity toward *Phytophthora infestans*. Plant Cell 3: 619–628 (1991).
66. Broglie K, Chet I, Holliday M, Cressman R, Biddle P, Knowlton, S, Mauvais CJ, Broglie R: Transgenic plants with enhanced resistance to the fungal pathogen *Rhizoctonia solani*. Science 254: 1194–1197 (1991).
67. Shinshi H, Mohnen D, Meins F Jr: Regulation of a plant pathogenesis-related enzyme: inhibition of chitinase and chitinase mRNA accumulation in cultured tobacco tissue by auxin and cytokinin. Proc Natl Acad Sci USA 84: 89–93 (1987).
68. Lotan T, Ori N, Fluhr R: Pathogenesis-related proteins are developmentally regulated in tobacco flowers. Plant Cell 1: 881–887 (1989).

69. Herget T, Schell J, Schreier PH: Elicitor-specific induction of one member of the chitinase gene family in *Arachis hypogaea*. Mol Gen Genet 224: 469–476 (1990).
70. Kaufmann H, Kirch H, Wemmer T, Peil, A, Lottspeich F, Uhrig H, Salamini F, Thompson R: Sporophytic and gametophytic self-incompatibility. In: Cresti M, Tiezzi A (eds) Sexual Plant Reproduction, pp. 115–125. Springer-Verlag, Berlin (1992).
71. Boller T, Gehri A, Mauch F, Vogeli U: Chitinase in bean leaves: induction by ethylene, purification, properties and possible function. Planta 157: 22–31 (1983).
72. Benhamou N, Asselin A: Attempted localization of a substrate for chitinases in plant cells reveals abundant *N*-acetyl-*D*-glucosamine residues in secondary walls. Biol Cell 67: 341–350 (1989).
73. Bolwell GP: Synthesis of cell wall components: aspects of control. Phytochemistry 27: 1235–1253 (1988).
74. Lehle L, Fartaczek F, Tanner W, Kauss H: Formation of polyprenol-linked mono- and oligosaccharides in *Phaseolus aureus*. Arch Biochem Biophys 175: 419–426 (1976).
75. Durr M, Bailey DS, Maclachlan G: Subcellular distribution of membrane-bound glycosyltransferases from pea stems. Eur J Biochem 97: 445–453 (1979).
76. Usui T, Matsui H, Isobe K: Enzymic synthesis of useful chito-oligosaccarides utilizing transglycosylation by chitinolytic enzymes in a buffer containing ammonium sulfate. Carbohydr Res 203: 65–77 (1990).
77. Carter HE, Betts BE, Strobach DR: Biochemistry of the sphingolipids. XVII. The nature of the oligosaccharide component of phytoglycolipid. Biochemistry 3: 1103–1107 (1964).
78. Spaink HP, Aarts A, Bloemberg GV, Folch J, Geiger O, Schlaman HRM, Oates JE, van de Snade K, van Spronsen P, van Brussel AAN, Wijfjes AHM, Lugtenberg BJJ: Rhizobial lipo-oligosaccharides: their biosynthesis and their role in the plant. In: Nester E (ed) Advances in Molecular Genetics of Plant-Microbe Interactions, vol 2, pp. 151–162. Kluwer Academic Publishers, Dordrecht (1993).
79. Schmidt J, Röhrig H, John M, Wieneke U, Stacey G, Koncz C, Schell J: Alteration of plant growth and development by *Rhizobium nodA* and *nodB* genes involved in the synthesis of oligosaccharide signal molecules. Plant J 4: 651–658 (1993).
80. Spaink HP, Sheeley DM, van Brussel AAN, Glushka J, York WS, Tak T, Geiger O, Kennedy EP, Reinhold, VN, Lugtenberg BJJ: A novel highly unsaturated fatty acid moiety of lipo-oligosaccharide signals determines host specificity of *Rhizobium*. Nature 354: 125–130 (1991).
81. Schmidt J, Wingender R, John M, Wieneke U, Schell J: *Rhizobium meliloti nodA* and *nodB* genes are involved in generating compounds that stimulate mitosis of plant cells. Proc Natl Acad Sci USA 85: 8578–8582 (1988).
82. John M, Röhrig H, Schmidt J, Wieneke U, Schell J: *Rhizobium* NodB protein involved in nodulation signal synthesis is a chitooligosaccharide deacetylase. Proc Natl Acad Sci USA 90: 625–629 (1993).
83. Röhrig H, Schmidt J, Wieneke U, Kondorosi E, Barlier I, Schell J, John M: Biosynthesis of lipooligosaccharide nodulation factors: *Rhizobium* NodA protein is involved in *N*-acylation of the chitooligosaccharide backbone. Proc Natl Acad Sci USA, in press (1994).
84. Staehelin C, Granado J, Müller J, Wiemken A, Mellor, RB, Felix G, Regenass M, Broughton WJ, Boller T: Perception of *Rhizobium* nodulation factors by tomato cells and inactivation by root chitinases. Proc Natl Acad Sci USA 91: 2196–2200 (1994).
85. Vasse J, De Billy F, Truchet G: Abortion of infection during the *Rhizobium meliloti*-alfalfa symbiotic interaction is accompanied by a hypersensitive reaction. Plant J 4: 555–566 (1993).
86. Staehelin C, Schultze M, Kondoros E, Mellor RB, Boller T, Kondorosi A: Structural modifications in *Rhizobium meliloti* Nod factors influence their stability against hydrolysis by root chitinases. Plant J 5: 319–330 (1994).
87. Darvill AG, Augur C, Bergmann C, Carlson RW, Cheong J-j, Eberhard S, Hahn MG, Ló V-m, Marfà V, Meyer B, Mohnen D, O'Neill MA, Spiro MD, van Halbeek H, York WS, Albersheim P: Oligosaccharins-oligosaccharides that regulate growth, development and defence responses in plants. Glycobiology 2: 181–198 (1992).
88. Aldington S, Fry SC: Oligosaccharins. Adv Bot Res 19: 1–101 (1993).
89. York WS, Darvill AG, Albersheim P: Inhibition of 2,4-dichlorophenoxyacetic acid-stimulated elongation of pea stem segments by a xyloglucan oligosaccharide. Plant Physiol 75: 295–297 (1984).
90. McDougall GJ, Fry SC: Structure-activity relationships for xyloglucan oligosaccharides with antiauxin activity. Plant Physiol 89: 883–887 (1989).
91. Fisher RF, Long SR: *Rhizobium*-plant signal exchange. Nature 357: 655–660 (1992).
92. Spaink HP: Rhizobial lipo-oligosaccharides: answers and questions. Plant Mol Biol 20: 977–986 (1992).
93. Vijn I, Das Neves L, van Kammen A, Franssen H, Bisseling T: Nod factors and nodulation in plants. Science 260: 1764–1765(1993).
94. Hirsch AM, Bhuvaneswari TV, Torrey JG, Bisseling T: Early nodulin genes are induced in alfalfa root outgrowths elicited by auxin transport inhibitors. Proc Natl Acad Sci USA 86: 1244–1248 (1989).
95. Van de Wiel C, Norris JH, Bochenek B, Dickstein R, Bisseling T, Hirsch AM: Nodulin expression and ENOD2 localization in effective, nitrogenfixing and ineffective, bacteria-free nodules of alfalfa (*Medicago sativa*). Plant Cell 2: 1009–1017 (1990).
96. Cooper JB, Long SR: Morphogenetic rescue of *Rhizobium meliloti* nodulation mutants by trans-zeatin secretion. Plant Cell 6: 215–225 (1994).

Plant Molecular Biology **26**: 1315–1327, 1994.

Initial events in phytochrome signalling: still in the dark

Tedd D. Elich and Joanne Chory*
Plant Biology Laboratory, The Salk Institute for Biological Studies, P.O. Box 85800, San Diego, CA 92186–5800, USA (author for correspondence)*

Received 6 April 1994; accepted 26 April 1994

Key words: phytochrome, signal transduction, photomorphogenesis, plant development, light regulation

Introduction

Photosynthetic organisms require light to provide the energy necessary to convert CO_2 into sugars. Given the non-motile nature of most plants, it is not surprising then that plants have developed systems to alter their growth and developmental patterns in response to the light environment. Indeed, plants exhibit such photomorphogenic responses throughout all stages of the life cycle, from seed germination to floral induction. By studying the wavelength dependence of different responses, two major classes of photoreceptors have been implicated in mediating photomorphogenesis in higher plants: blue/UV photoreceptors and red/far-red photoreceptors. It should be noted at the outset that extensive co-action between these pigment systems has been observed in various species. Until recently, however, only the red/far-red light photoreceptors, called phytochromes, have been identified and characterized at the molecular level. Because they remain the photoreceptors about which most is known, phytochromes will be the sole focus of this review. We note, however, the exciting prospects opened up by the recent cloning of a probable blue-light receptor encoded by the *Arabidopsis hy*4 locus [1]. For a recent overview of blue/UV receptors and responses, the reader is referred to the chapters by Senger and Schmidt, and Horwitz in reference [34].

When setting out to write a review concerning phytochrome and its involvement in plant photomorphogenesis, one is confronted with a problem: what is left to review or say? In this regard, we refer the readers to a number of excellent reviews which have all appeared since 1993 [10, 18, 40, 61, 63, 82, 86, 91]. In addition, a number of chapters in the new edition of 'Photomorphogenesis in plants' [34] are also relevant, most notably those by Cherry and Vierstra; Furuya and Song; Mancinelli; Pratt; Quail; Roux; Rudiger and Thummler; Smith; and Vierstra. To focus our review, we propose that the major goal facing phytochrome researchers today is to elucidate what happens immediately subsequent to phytochrome phototransformation, i.e., what does phytochrome interact with and how? With this in mind, our review will focus on recent advancements that may contribute to this goal.

Phytochrome molecular properties

The hallmark of a classic phytochrome response is its red/far-red light reversibility. This characteristic is a result of phytochrome's ability to exist in two spectrally distinct forms: Pr, the red-light-absorbing form (λ_{max} 660 nm), and Pfr, the far-red-light-absorbing form (λ_{max} 730 nm). These two forms are photointerconvertible so that irradiation with red light converts Pr to Pfr, and irradiation with far-red light converts Pfr back to Pr. The Pfr form traditionally has been thought to be the physiologically active form since the induction of most responses can be correlated with the formation of Pfr. Likewise, far-red irradiation given subsequent to Pfr formation can negate many responses. Recently, however, this dogma has been challenged by studies that have implicated a role for Pr in regulating gravitropism [43] and seed germination [64, 76]. In the few cases where molecular information is available, it is apparent that a primary means by which phytochrome exerts its effects is through regulating the transcription of nuclear genes [62].

Depending on the plant species, phytochrome polypeptides are approximately 120–130 kDa and contain a single, covalently attached, linear tetrapyrrole chromophore which is responsible for visible light absorption [34]. The phytochrome polypeptide folds into two major domains separated by a protease-susceptible hinge region: a ca. 70 kDa N-terminal chromophore-bearing domain and a ca. 55 kDa C-terminal domain [39]. Native phytochrome exists as a dimer in solution [31, 39]. The dimerization site resides on the C-terminal domain [31] and thus the molecule is visualized as an apparent tripartite structure by electron microscopy [30, 53].

Photoconversion of Pr to Pfr is likely to involve a Z to E isomerization about the C15 double bond between the C- and D-tetrapyrrole rings [69, 82]. In addition, the phytochrome polypeptide undergoes conformational changes throughout its length as detected by numerous techniques including protease susceptibility [39], phosphorylation by exogenous kinases [48, 92], and immunoreactivity [11, 29, 72]. In general, these studies indicate that the immediate N-terminus of phytochrome is more exposed in the Pr form, while regions of the C-terminus are more exposed in the Pfr form. An obvious and long held hypothesis is that these conformational changes result in differential interactions with subsequent component(s) of the signal transduction chain(s) linking phytochrome photoconversion to physiological responses.

Over the past several years, evidence from physiological, spectrophotometric, and immunochemical studies has demonstrated that there are at least two pools of phytochrome in plants [25, 62]. One pool, called type I phytochrome, predominates in etiolated tissue and is the phytochrome that has been purified and extensively characterized. Type I phytochrome is light-labile and present at substantially lower levels in green tissue than in etiolated tissue. Conversely, type II phytochrome is light-stable and present at approximately the same relatively low levels in light grown and etiolated tissue. A molecular basis for this heterogeneity became clear with the finding that phytochrome is encoded by a multigene family in all higher plants examined in detail. For example, in *Arabidopsis thaliana*, phytochrome is encoded by 5 genes named *PHYA-E* [75]. Preliminary characterization of the protein products of the first three demonstrated that *PHYA* encodes a light-labile phytochrome while *PHYB* and *PHYC* encode light-stable phytochromes [79]. The finding of multiple phytochromes immediately suggested that individual phytochromes might have specific physiological roles. Recently, this hypothesis could be directly tested through the use of mutants lacking specific phytochromes.

Phytochrome mutants

The application of genetic approaches to study light-mediated plant development has proven fruitful. A variety of phytochrome response mutants, both in the photoreceptors and downstream components, have been isolated in *Arabidopsis* [10, 61, 63]. Of considerable excitement is the recent cloning of the COP1 [19] and DET1 loci (Pepper, Delaney, Washburn, Poole and Chory, submitted) which were defined by mutations that resulted in light-grown phenotypes in dark-grown plants. These are the first phytochrome signal transduction pathway components that have been identified at the molecular level; however, both proteins probably act in the nucleus ([19]; Pepper *et al.*, submitted) and are therefore likely to be at least one step removed from phytochrome photoconversion. For the purposes of this review, the most interesting of the characterized mutants are those deficient in individual phytochromes since they can be used to address questions about the number and specificity of phytochrome signalling pathways, and can provide information about regions of the molecule important for function.

A class of *Arabidopsis* mutants, called *hy* mutants, were isolated based on a long hypocotyl phenotype when grown in the light and thus were candidate phytochrome mutants [10, 61]. Of the six loci, *hy*1, *hy*2, and *hy*6 were found to be defective in chromophore biosynthesis [59] while *hy*3 was found to encode PHYB [65]. In addition to elongated hypocotyls, *phyB* mutants also exhibit elongated petioles, stems, and root hairs. Furthermore, *phyB* mutants flower earlier, contain less chlorophyll, and are defective in end-of-day far-red-light responses and shade avoidance responses as compared to wild-type plants [50, 65, 66].

More recently, *phyA* mutants were isolated by virtue of exhibiting a long hypocotyl phenotype when grown under continuous far-red light [17, 52, 60, 64, 90]. This screen was based on the hypothesis, now proven correct, that far-red high-irradiance responses were mediated by light-labile phytochrome [46, 78]. Interestingly, an elongated hypocotyl under far-red light is the only obvious phenotype *phyA* mutants exhibit [17, 52, 60, 64, 90].

With these tools in hand, a careful analysis was performed comparing different light responses of wild-type plants to those of *phyA* null mutants, *phyB* null mutants, and *phyA/phyB* double mutants [64]. Effects that were specific to a single mutation as well as effects that only appeared in the double mutants were found. For example, double mutants, but neither single mutant, showed significant reduction in *CAB* expression induced by red light pulses. Additionally, only double mutants showed reduced cotyledon development when grown in red light; however, defects in cotyledon development under far-red light were specific to *phyA* mutants. Other phytochrome-specific responses deduced from the mutants include inhibition of hypocotyl elongation which appears to be a PHYA-specific function under far-red light, whereas PHYB appears to be the predominant regulator of this response under red and white light. Likewise, germination defects in far-red light and darkness were specific to *phyA* and *phyB* mutants, respectively. Interestingly, the *phyB* mutation could suppress the far-red light germination deficiency in *phyA* mutants leading to the conclusion that the Pr form of PHYB actively inhibits this response. This same conclusion was reached independently in studies using single mutants only [76]. Similarly, it was also recently concluded that the Pr form of PHYB actively promotes shoot gravitropism [43]. Thus, contrary to accepted dogma, it appears that both the Pr and Pfr forms of this phytochrome can activate signal transduction pathways.

These studies have clearly demonstrated that PHYA and PHYB have both overlapping and distinct roles in regulating *Arabidopsis* development. The general emerging view is that PHYA has a limited role in normal *Arabidopsis* photomorphogenesis, primarily restricted to de-etiolation and far-red responses. Conversely, PHYB seems to be an important mediator of *Arabidopsis* photomorphogenesis throughout the life cycle.

The demonstration that some phenotypes can only be observed in the double mutant indicates that both PHYA and PHYB can activate a shared signal transduction pathway with the simplest explanation being that they interact with a common component. The finding of PHYB Pr- and Pfr-specific effects suggests that either there are PHYB form-specific interacting components, or component(s) capable of interacting with both forms of PHYB in a qualitatively different manner. Likewise, the presence of responses specific to PHYA and PHYB suggest that there may also be components that interact specifically with either PHYA or PHYB. This conclusion is only tentative, however, since phytochrome-specific responses could also be explained by cell- and tissue-specific expression, though there is currently no evidence of this. Furthermore, PHYA-specific responses in particular may simply be a result of the biochemical properties of this phytochrome that make it the only species capable of forming an inductive level of Pfr under certain conditions (e.g. far-red light) [46, 78].

In addition to information about signalling pathways, *phy* mutants can be useful in localizing regions of the protein important for function. With both *phyA* and *phyB* mutants, weak alleles were identified which had wild-type protein levels but contained missense mutations in coding regions [17, 65]. Such mutations in general can fall into two classes: those that affect spectral activity and those that do not. The first class would include mutations that interfered with chromophore attachment or photoreversibility and would not be relevant for the purposes at hand. The second class is of more interest since it could include mutations in interaction domains.

There are two known *phyA* alleles which contain missense mutations, both in conserved residues. The *phyA*-103 allele (formerly called *hy*8–3) results in a Gly-to-Glu conversion at position 727 [17] while the *phyA*-205 allele results in a Val-to-Met change at position 631 [64]. Since apparently normal phytochrome difference spectra could be detected in etiolated tissue in both cases [17, 60], these missense mutations must fall into the second class defined above. We note that both mutated residues reside in the C-terminal domain, in proximity to the hinge region [39]. This region has been shown to be more exposed in the Pfr form by proteolysis [39] and phosphorylation by exogenous kinases [48, 92]. Thus, this area of PHYA is a particularly good candidate for an interaction domain; however, the possibility that these mutations have indirect effects (e.g. involvement in transducing protein conformational changes to the actual interaction domain) cannot be ruled out.

At present, only one *phy*B missense mutation is known. The 4–117 allele contains a mutation that converts a conserved His-to-Tyr at position 283 [65]. The proximity of this mutation to the chromophore attachment site at Cys-357 suggests that it may be of the class that affects spectral activity. Because of the low levels of PHYB, however, it is very difficult to determine whether endogenous protein has spectral activity. Fortunately, this can now be directly tested using recombinant phytochrome expression systems that have been developed (discussed below).

Phytochrome overexpression studies

Site-directed mutagenesis and deletion analysis of phytochrome expressed in transgenic plants provides a complementary approach to identify functionally important regions of this photoreceptor. The most desirable system would involve transforming *phy* null mutants with corresponding *phy* constructs behind the endogenous promoter. Since *phy* mutants have become available only recently, however, ectopic overexpression of heterologous phytochromes in wild-type plants has been the method of choice. The first such studies involved the expression of monocot PHYA cDNAs behind the 35S promoter in tobacco and tomato [5, 32, 33, 51]. Subsequently, overexpression in *Arabidopsis* of an oat PHYA cDNA [6], and of rice and *Arabidopsis* PHYB cDNAs [88], has also been reported. In all cases, the overexpressed phytochromes conferred light-dependent phenotypes demonstrating biological activity. In general, phytochrome-overexpressing transgenic plants are dwarfed and often are greener and show reduced apical dominance as compared to wild-type controls [5, 6, 32, 33, 51, 88]. In this regard, we note a caveat that extends to all overexpression studies. The fact that the dwarf phenotype can be induced by overexpressed PHYA, whereas *phyA* mutants appear morphologically indistinguishable from wild-type plants except under far-red light [17, 52, 60, 64, 90], demonstrates that ectopically overexpressed PHYAs can activate responses that are not normally attributable to PHYA in wild-type plants.

Nonetheless, these systems have allowed the *in vivo* analysis of functional domains of phytochrome important to the biological activity exhibited in the heterologous systems. For example, expression of oat phytochrome lacking amino acids 7 to 69 had little or no effect in tobacco whereas expression of the full-length protein results in up to a four-fold reduction in height of adult plants [9]. The N-terminally deleted phytochrome was photoreversible indicating that this region is not required for chromophore attachment or photoconversion; however, blue-shifted absorbance spectra and increased dark-reversion were observed [9]. These findings are consistent with the involvement of the N-terminus in chromophore-protein interactions [26], and additionally indicate an importance in biological activity. This latter conclusion was supported by a subsequent study wherein the first ten serine residues of rice PHYA (within 20 amino acids from the N-terminus) were changed to alanines and the resulting phytochrome expressed in tobacco [80]. Compared to the full-length rice PHYA control, the serine-less phytochrome exhibited greater biological activity with respect to inhibiting hypocotyl elongation. The rationale for these experiments was the hypothesis that serines in this regions may be a target site for regulatory phosphorylation [80]. In this regard, the analogous serine-rich region of oat PHYA is known to be a substrate for phosphorylation by phytochrome-associated kinase activity [48, 92]. While the possible functional significance of phytochrome phosphorylation awaits further proof, these studies do support a role of the N-terminus in biological activity.

C-terminal deletion analysis of oat PHYA expressed in tobacco has also been performed leading to a number of interesting results [8]. Most notably was the finding that deletion of as few as 35 amino acids from the C-terminus completely eliminated the dwarf phenotype conferred by the full-length molecule, but had no effect on dimerization or spectral properties [8]. These results clearly demonstrate a functional importance of the C-terminus in the heterologous system. Of additional interest was the finding that deletions of 210 or more amino acids resulted in monomeric phytochrome, directly demonstrating that the region between amino acids 919 and 1093 is necessary for dimerization *in vivo* [8]. Finally, further C-terminal deletions demonstrated that residues 1 to 398 were sufficient for chromophore attachment and photoreversibility; however, the resulting truncated phytochrome exhibited blue-shifted absorbance properties. Normal spectral properties were observed in truncated phytochrome containing residues 1 to 652 demonstrating that residues 399 to 652, in addition to the N-terminus, are involved in chromophore-protein interactions [8].

A recent analysis of oat PHYA expressed in *Arabidopsis* [4] provides somewhat of a different picture. Hypocotyl lengths of wild-type seedlings grown under different light conditions were compared to those of transgenic seedlings expressing either the full-length oat protein or oat PHYA lacking either the first 52 N-terminal residues, the last 512 C-terminal residues, or the C-terminal internal residues 617–686 [4]. Unlike the case in tobacco [9], in this system oat phytochrome lacking the N-terminus exhibited as much biological activity as the full-length protein in red or white light. The differences between the tobacco (see above) and *Arabidopsis* studies were not due to residues 52–69 [4]. Of more interest was the finding that under far-red light, expression of the N-terminal truncated phytochrome resulted in a dominant negative effect (i.e. the seedlings were taller than non-transgenic controls) [4]. In addition, both C-terminal deletions conferred a dominant negative phenotype under all light conditions [4]. The similar effects of both C-terminal deletions may indicate a specific importance of residues 617 to 686. In this regard, we note that this region contains the analogous conserved Val that is mutated in the *Arabidopsis phyA*-205 allele [64]. Alternatively, the internal deletion may simply interfere with proper C-terminal folding or conformational changes.

The dominant negative effects uncovered in this study indicate that the chromophore-bearing domain between residues 52 and 617 is involved in interactions with a transduction chain component. This interaction would appear to be Pfr-specific since no dominant negative effects were observed upon expression of a non-photoconvertible phytochrome containing a Cys-to-Ser change at residue 322 (the chromophore attachment site). Furthermore, this interaction is not dependent upon dimerization since the C-terminal-truncated phytochrome is a monomer [4]. Presumably, further interactions with either the bound component or additional components are then required to activate the signal transduction cascade leading to inhibition of hypocotyl elongation. These further interactions appear to always require the C-terminus, while an additional N-terminal interaction is needed to elicit the response under far-red light. The requirement of the N-terminus for the far-red response, which is specific to PHYA in wild-type plants, would suggest that there are transduction components which exhibit PHYA-specific interactions.

The lack of dominant negative effects in studies of similar oat PHYA deletion products expressed in tobacco may simply be due to the different systems and responses being assayed: hypocotyl length in young *Arabidopsis* seedlings grown under continuous artificial light [4] versus height of adult tobacco plants grown under natural diurnal light cycles [8, 9]. For example, the required phytochrome interacting components may be more limiting in young *Arabidopsis* seedlings under these conditions and thus more easily titrated out by the overexpressed phytochrome. Despite the differences, however, both studies have implicated both the N- and C-terminus in being important for biological activity. The major contradictory result between the two studies involves the different effects of N-terminal deletions under white light: eliminating biological activity in tobacco [9] while having no effect on biological activity in *Arabidopsis* [4]. If one assumes that the measured responses are phenomenologically the same and involve conserved mechanisms in *Arabidopsis* and tobacco, this discrepancy between the two studies has no easy explanation.

Phytochrome signalling mechanisms

Biochemical and pharmacological studies have implicated the involvement of numerous signalling mechanisms in phytochrome responses including phosphorylation, G-proteins and Ca^{2+}. While not leading to the identification of any transduction chain components, these studies do provide information with respect to the types of proteins that may be involved in phytochrome signalling pathways. We will attempt here to provide a brief overview of some of the more interesting and recent findings in this area with no intention of providing an exhaustive review.

Numerous reports have suggested an involvement of phosphorylation in phytochrome responses. These include reports of red-light-regulated protein phosphorylation [14, 24, 56, 68], phosphorylation-regulated binding of factors to the promoters of phytochrome-regulated genes [13, 36, 70], and light-regulated expression of genes encoding putative protein kinases [42]. In addition, the C-terminus of higher-plant phytochromes show sequence similarities to bacterial histidine kinase catalytic domains (discussed in detail below) [71]. Furthermore, preparations of highly purified oat phytochrome have been shown to contain polycation-stimulated serine/threonine kinase activity [92, 94] capable of phosphorylating the serine-rich N-terminus of the photoreceptor [48]. These reports suggest the possibility that phytochrome itself is a protein kinase. In support of this hypothesis was the demonstration that phytochrome has an ATP-binding site [93]. While subsequent reports have provided evidence suggesting that the co-purifying kinase activity may not be intrinsic to the photoreceptor [28, 35], the question of whether higher-plant phytochrome is a kinase must still be considered unresolved.

It does appear, however, that phytochrome from the moss *Ceratodon purpureus* possesses kinase activity [2, 83]. The phytochrome gene from this organism contains three exons. The first two exons encode a phytochrome chromophore domain most similar to that of fern phytochrome [83]. In contrast, the third exon encodes protein sequences that are not related to phytochrome, but rather show significant similarities to the conserved regions of protein kinase catalytic domains [83]. Preliminary biochemical evidence has been presented supporting the notion that this phytochrome possesses protein kinase activity [2]. The structure of this gene has obvious functional implications. With regard to higher-plant phytochromes, a natural hypothesis is that the two functional domains were separated during evolution. Therefore, it will be of interest to see whether higher plants contain homologues of the *Ceratodon* phytochrome kinase domain.

There is extensive evidence that Ca^{2+} is involved in phytochrome responses. In this regard, we refer the reader to an exhaustive review of the literature [85]. Of recent note was a study providing convincing evidence that transient increases in cytosolic-free Ca^{2+} levels are involved in phytochrome-regulated swelling of wheat protoplasts [73]. It was demonstrated that protoplast swelling in response to red-light pulses was preceded by Ca^{2+} transients. Both protoplast swelling and the Ca^{2+} transients were reversible by a far-red pulse given subsequent to the red pulse. Finally, Ca^{2+} transients induced by photolytic release of caged probes also resulted in protoplast swelling demonstrating a cause-effect relationship [73].

Heterotrimeric G-proteins have also been reported to be involved in phytochrome signal transduction based on inhibitor studies of *CAB* gene expression in soybean cell cultures [67]. Recently, a model combining G-proteins, Ca^{2+}, and cGMP in a phytochrome signal transduction cascade has emerged from a series of novel single-cell micro-injection studies [3, 54]. Chloroplast development, anthocyanin accumulation, and expression of a photoregulated *CAB*-GUS reporter gene were monitored in single sub-epidermal cells of the tomato PHYA-deficient *aurea* mutant micro-injected with various compounds. As a foundation for this work, it was shown that cells micro-injected with purified oat PHYA, but not un-injected cells, exhibited anthocyanin accumulation, chloroplast development, and reporter gene expression [54]. Subsequent extensive studies involving micro-injection of inhibitors, activators, and putative signalling compounds led to a model in which PHYA photoconversion leads to the activation of a heterotrimeric G-protein(s) which in turn activates three subsequent pathways: one involving cGMP and leading to anthocyanin accumulation; one involving Ca^{2+}/calmodulin which results in expression of PSII-associated proteins, ATP synthase, and Rubisco; and one requiring involvement of both cGMP and Ca^{2+}/calmodulin which results in expression of Cyt b_6f and PSI-associated proteins [3, 54]. Since G-protein activation is the earliest event in the proposed pathway, heterotrimeric G-proteins remain the only possible phytochrome interacting component identified by these studies. In this regard, at least one $G\alpha$ subunit has been cloned from plants [45]. We note, however, that G-proteins are membrane-associated and normally coupled to transmembrane receptors [7]. This might suggest the requirement for additional components to couple phytochrome photoconversion to G-protein-regulated pathways.

While clearly an important contribution in the elucidation of phytochrome signalling pathways, there are several caveats to these experiments. First, *aurea* is not simply a *phyA* mutant [74]. This is probably reflected by the fact that the *aurea* phenotype, including the deficiencies in the responses being dissected, is significantly different than that of *Arabidopsis phyA* mutants [17, 52, 60, 64, 90]. Therefore, the micro-injected PHYA is likely leading to responses that are not attributable to endogenous PHYA in normal plants. This is also apparent from the fact that the micro-injected PHYA could elicit these same responses in epidermal cells [54] which do not normally exhibit these responses. Therefore, as with ectopic overexpression studies (see above), one must be careful about extending conclusions from non-natural responses to the mechanisms of endogenous phytochrome signalling pathways. In addition, it is clear that many plant growth and developmental responses that are regulated by phytochrome are also influenced by other factors such as environmental conditions, stress, hormones, and other photoreceptors. It seems likely that signalling components like Ca^{2+}, cGMP, and G-proteins will be involved in numerous pathways in plants as they are in animals. Thus, analyzing responses by pharmacological approaches has the inherent danger of short-circuiting pathways leading to the misinterpretation of specific pathway intermediates.

Are bacterial signalling mechanisms involved in phytochrome pathways?

Bacteria can alter their behavior and metabolism in response to environmental signals such as nitrogen availability, osmolarity, and the presence of nutrients or toxins. The signal transduction pathways that activate these responses involve functionally and structurally conserved protein modules arranged in simple circuits (for reviews see [57, 58]). In the simplest scenario, two components are involved: a sensor consisting of an input domain coupled to a transmitter module, and a response regulator consisting of a receiver module coupled to an output domain. These modular functions can be present on the same protein, different proteins, or any combination therein. Signal transduction between transmitter and receiver modules is via phosphorylation/dephosphorylation reactions. Typically, the sensor is a transmembrane protein with its input domain residing in the periplasm and its transmitter module in the cytoplasm. In response to signals perceived by the input domain, transmitter modules undergo autophosphorylation on a unique histidine residue. This phosphohistidine then serves as the phosphoryl donor in the phosphorylation of an aspartyl residue on the receiver module. The phosphorylation state of the receiver, which can also be regulated by transmitter phosphatase activity in some cases, influences the activity of the output domain leading to responses such as alterations in flagellar rotation or gene expression [57, 58].

Transmitter modules involved in different signalling pathways exhibit significant similarities to one another at the amino acid level. In particular, there are five especially conserved regions, each of which contains several nearly invariant residues [58]. Four of these regions are thought to be important for catalytic activity while the other one contains the histidine which is autophosphorylated [58]. Surprisingly, the C-terminus of phytochrome exhibits considerable homology to these canonical sequences of bacterial transmitters [58, 71]. While different phytochromes lack a number of the most conserved residues, most notably the conserved histidine, it should be noted that cases of bacterial transmitters lacking one of the conserved regions are known. For example, CheA, which is involved in chemotaxis, lacks the histidine-containing region but does have a histidine in a non-conserved region which serves as the site of phosphorylation [58].

In addition to amino acid homology, other functional and structural similarities between phytochrome and bacterial transmitters are apparent. As indicated previously, transgenic plant studies have implicated phytochrome's C-terminus, containing the putative transmitter domain, as being important for biological function [4, 8]. Like bacterial transmitter modules, this region of phytochrome is coupled to a signal perception domain: the N-terminal chromophore domain. In bacterial sensor proteins, signal perception is thought to be relayed to the transmitter module through protein conformational changes [57, 58]. It is known that photoconversion of phytochrome, which requires only the N-terminal domain, results in C-terminal conformational changes. Finally, transmitters generally function as homodimers, and autophosphorylation is thought to be intermolecular [57, 58] indicating close contact between the canonical regions of the respective monomers. Phytochrome is a homodimer and regions necessary for dimerization [8, 31] overlap with the regions that exhibit transmitter homology [58, 71].

Phytochrome was the first reported example of a eukaryotic protein with significant sequence similarities to bacterial transmitters. Recently, even more extensive transmitter homologies have been found in the ETR1 protein of plants which is involved in ethylene signal transduction [7], and in the SLN1 protein of yeast which may be involved in regulating growth [55]. These findings provide strong evidence that bacterial two-component signal transduction mechanisms occur in eukaryotes, including plants, and suggest that such mechanisms may in fact be widespread. Unfortunately, with respect to phytochrome, no further evidence supporting or refuting the biological significance of the transmitter homology has appeared in print. These regions could serve as mutagenesis target sites for overexpression studies which are prevalent in phytochrome research. Additionally, an obvious experiment would be to look for histidine-kinase activity in phytochrome. We note, however, that even if phytochrome lacked this activity, the transmitter sequence similarities may still be meaningful. It has recently been reported that the *Bacillus subtilis* protein SpoIIAB, which regulates σ^F-dependent transcription, shows amino acid similarities to transmitters but exhibits serine- rather than histidine-kinase activity [47]. As noted previously, serine/threonine kinase activity is known to co-purify with phytochrome and may be an intrinsic property of the photoreceptor itself [92, 93, 94].

Recombinant phytochrome expression systems

Ultimately, a mechanistic understanding of how phytochrome phototransformation activates signal transduction pathways will require knowledge of phytochrome's molecular structure. The ability to purify to homogeneity full-length native phytochrome from etiolated plant sources has existed for about 11 years [44, 87]. While purified phytochrome has been subjected to detailed biochemical and photochemical characterization, it apparently has been refractile to crystallization studies as judged by the lack of reports to the contrary. Possible reasons for this include that: (1) the purification protocols utilize etiolated tissue and thus the products are predominantly PHYA which may be more recalcitrant to crystallization than one of the other phytochromes; (2) the purified products were heterogeneous due to the presence of multiple phytochromes and heterogeneity is detrimental to crystal formation; and (3) the exposure to any light may disrupt a phytochrome crystal.

The expression of recombinant native phytochrome in other organisms may circumvent these problems. Since the only known activity of phytochrome is its photochemical activity, this remains the only test to assess the structural fidelity of recombinant phytochrome. The ability to reconstitute holophytochrome from recombinant apoprotein is a prerequisite for spectral assays. The feasibility of reconstitution became apparent from investigations into chromophore biosynthesis and holoprotein assembly *in vivo* [21, 23]. These studies indicated that chromophore attachment could occur post-translationally, and demonstrated that phycocyanobilin, the readily available cleaved prosthetic group from the phycobiliprotein phycocyanin, could serve as a substrate for chromophore attachment. The resulting apophytochrome-phycocyanobilin adduct is fully photoreversible but exhibits blue-shifted absorbance properties due to the presence of an ethyl group rather than the naturally occurring vinyl group, on the tetrapyrrole D-ring [23]. This work culminated with the demonstration of *in vitro* assembly of spectrally active holophytochrome from phycocyanobilin and phytochrome apoprotein either purified from plants [22] or produced by *in vitro* translation [38]. Attachment does not require additional enzymes or cofactors indicating that the chromophore C-S lyase activity is a property of the phytochrome apoprotein. Production of *bona fide* holophytochrome has since been realized using the natural chromophore precursor, phytochromobilin, produced from isolated plastids [81, 89] or as a by-product of phycobiliprotein methanolysis [12].

These advancements have led to the realization of holophytochrome assembly from recombinant apoprotein produced in *Escherichia coli* [89] and yeast [15, 37, 89]. To date, reconstitution of oat PHYA [12, 41, 89], pea PHYA [15, 16], tobacco PHYB [37], and *Arabidopsis* PHYB (Elich and Chory, unpublished) has been achieved. Thus, as long as one has a cDNA, the ability to produce a homogenous phytochrome species is now at hand. This ability will facilitate the biochemical and photochemical characterization of the low-abundance light-stable phytochromes, and could provide a source of protein for crystallization studies. The major obstacle to these goals at present remains the relatively low yields of native phytochrome that have been obtained. While *E. coli* systems are capable of expressing high levels of phytochrome, the majority of the protein in high-expression systems is denatured with much of it in inclusion bodies [4]. The large size of phytochrome makes renaturation procedures difficult, though the use of chaperonins [20, 27, 49] provides some hope. At present, expression in yeast appears to be the best system for producing native apoprotein; however, the reported yields in crude soluble extracts have only been around 10–80 μg/g fresh weight cells [37, 89]. Such yields make the conventional purification of milligram quantities of phytochrome a daunting prospect. Alternatively, other systems like insect cells may provide higher yields than those obtainable with either yeast or *E. coli.*

Finally, while the ability to produce spectrally active holoprotein is a measure of structural fidelity, this may not be the best form of phytochrome for crystallization studies. Conversely, the presence of the chromophore greatly assists purification assays and may provide structural rigidity that could be an asset to crystallization. This suggests that the optimum form of phytochrome for crystallization would contain a non-photoconvertible chromophore analogue. In this regard, it has been demonstrated that phycoerythrobilin, the cleaved prosthetic group of the phycobiliprotein phycoerythrin, satisfies this criterion by virtue of containing a reduced C15–16 bond between the C- and D-tetrapyrrole rings [41].

Concluding remarks

While considerable progress is being made in defining events in phytochrome signal transduction pathways, a glaring gap in our knowledge remains the lack of identification of phytochrome interacting components. The ability to address this question biochemically has existed for quite some time yet no putative interacting components have been reported. Studies with mutants deficient in particular phytochromes suggest that there may be multiple interacting components of varying specificity. Pharmacological and biochemical studies of phytochrome responses have implicated the involvement of universal signalling components and mechanisms but have not led to the identification of responsible proteins. Genetic approaches have identified mutations in genes whose products are formally in the phytochrome signal transduction pathway; however, the majority of these appear to be in the photoreceptor itself or in downstream components. The involvement of universal signalling components in the early part of the pathway may preclude their identification by genetic approaches since such components undoubtedly would be involved in the integration of numerous signal transduction pathways; therefore, mutations in them may not give readily anticipated phenotypes or may be lethal.

Clearly other approaches are needed in the quest for phytochrome interacting proteins. These might include screening for *phy* suppressors (particularly missense allele-specific ones), as well as techniques like the two-hybrid system or expression screening. The latter studies will be facilitated by the advancements that have been made in identifying regions of the phytochrome protein that are likely to be involved in activating signal transduction pathways. Missense mutations, suggestive homologies, conformational changes which are non-essential to photoconversion, and a requirement for biological activity all indicate a particular importance of the C-terminus in this regard.

Acknowledgement

T.D.E. is supported by an Amgen Fellowship.

References

1. Ahmad M, Cashmore AR: *HY4* gene of *A. thaliana* encodes a protein with characteristics of a blue-light photoreceptor. Nature 366: 162–166 (1993).
2. Algarra P, Linder S, Thummler F: Biochemical evidence that phytochrome of the moss *Ceratodon purpureus* is a light-regulated protein kinase. FEBS Lett 315: 69–73 (1993).
3. Bowler C, Neuhaus G, Yamagata H, Chua N-H: Cyclic GMP and calcium mediate phytochrome phototransduction. Cell 77: 73–81 (1994).
4. Boylan M, Douglas N, Quail PH: Dominant negative suppression of *Arabidopsis* photoresponses by mutant

Phytochrome A sequences identifies spatially discrete regulatory domains in the photoreceptor. Plant Cell 6: 449–460 (1994).
5. Boylan MT, Quail PH: Oat phytochrome is biologically active in transgenic tomatoes. Plant Cell 1: 765–773 (1989).
6. Boylan MT, Quail PH: Phytochrome A overexpression inhibits hypocotyl elongation in transgenic *Arabidopsis*. Proc Natl Acad Sci USA 88: 10806–10810 (1991).
7. Chang C, Kwok SF, Bleecker AB, Meyerowitz EM: *Arabidopsis* ethylene-response gene ETR1: similarity of product to two-component regulators. Science 262: 539–544 (1993).
8. Cherry JR, Hondred D, Walker JM, Keller JM, Hershey HP, Vierstra RD: Carboxy-terminal deletion analysis of oat phytochrome A reveals the presence of separate domains required for structure and biological activity. Plant Cell 5: 565–575 (1993).
9. Cherry JR, Hondred D, Walker JM, Vierstra RD: Phytochrome requires the 6-kDa N-terminal domain for full biological activity. Proc Natl Acad Sci USA 89: 5039–5043 (1992).
10. Chory J: Out of darkness: mutants reveal pathways controlling light-regulated development in plants. Trends Genet 9: 167–172 (1993).
11. Cordonnier M-M: Monoclonal antibodies: molecular probes for the study of phytochrome. Photochem Photobiol 49: 821–831 (1989).
12. Cornejo J, Beale SI, Terry MJ, Lagarias JC: Phytochrome assembly. The structure and biological activity of 2(R),3(E)-phytochromobilin derived from phycobiliproteins. J Biol Chem 267: 14790–14798 (1992).
13. Datta N, Cashmore AR: Binding of pea nuclear protein to promoters of certain photoregulated genes is modulated by phosphorylation. Plant Cell 1: 1069–1077 (1989).
14. Datta N, Chen Y-R, Roux SJ: Phytochrome and calcium stimulation of protein phosphorylation in isolated pea nuclei. Biochem Biophys Res Comm 128: 1403–1408 (1985).
15. Deforce L, Furuya M, Song P-S: Mutational analysis of the pea Phytochrome A chromophore pocket: chromophore assembly with apophytochrome A and photoreversibility. Biochemistry 32: 14165–14172 (1994).
16. Deforce L, Tomizawa K-I, Ito N, Farrens D, Song P-S, Furuya M: *In vitro* assembly of apophytochrome and apophytochrome deletion mutants expressed in yeast with phycocyanobilin. Proc Natl Acad Sci USA 88: 10392–10396 (1991).
17. Dehesh K, Franci C, Parks BM, Seeley KA, Short TW, Tepperman JM, Quail PH: *Arabidopsis HY8* locus encodes Phytochrome A. Plant Cell 5: 1081–1088 (1993).
18. Deng X-W: Fresh view of light signal transduction in plants. Cell 76: 423–426 (1994).
19. Deng X-W, Matsui M, Wei N, Wagner D, Chu AM, Feldmann KA, Quail PH: COP1, an *Arabidopsis* regulatory gene, encodes a protein with both a zinc-binding motif and a Gβ homologous domain. Cell 71: 791–801 (1992).
20. Edgerton MD, Santos MO, Jones AM: *In vivo* suppression of phytochrome aggregation by the GroE chaperonins in *Escherichia coli*. Plant Mol Biol 21: 1191–1194 (1993).
21. Elich TD, Lagarias JC: Phytochrome chromophore biosynthesis. Both 5-aminolevulinic acid and biliverdin overcome inhibition by gabaculine in etiolated *Avena sativa* seedlings. Plant Physiol 84: 304–310 (1987).
22. Elich TD, Lagarias JC: Formation of a photoreversible phycocyanobilin-apophytochrome adduct *in vitro*. J Biol Chem 264: 12902–12908 (1989).
23. Elich TD, McDonagh AF, Palma LA, Lagarias JC: Phytochrome chromophore biosynthesis. Treatment of tetrapyrrole-deficient *Avena* explants with natural and non-natural bilatrienes leads to formation of spectrally active holoproteins. J Biol Chem 264: 183–189 (1989).
24. Fallon KM, Shacklock PS, Trewavas AJ: Detection of very rapid red light-induced calcium-sensitive protein phosphorylation in etiolated wheat (*Triticum aestivum*) leaf protoplasts. Plant Physiol 101: 1039–1045 (1993).
25. Furuya M: Molecular properties and biogenesis of Phytochrome I and II. Adv Biophys 25: 133–167 (1989).
26. Furuya M, Song P-S: Assembly and properties of holophytochrome. In: Kendrick RE, Kronenberg GHM (eds) Photomorphogenesis in Plants, 2nd ed., pp. 51–70. Kluwer Academic Publishers, Dordrecht (1994).
27. Grimm R, Donaldson GK, van der Vies SM, Schafer E, Gatenby AA: Chaperonin-mediated reconstitution of the phytochrome photoreceptor. J Biol Chem 268: 5220–5226 (1993).
28. Grimm R, Gast D, Rudiger W: Characterization of a protein-kinase activity associated with phytochrome from etiolated oat (*Avena sativa* L) seedlings. Planta 178: 199–206 (1989).
29. Holdsworth ML, Whitelam GC: A monoclonal antibody specific for the red-light-absorbing form of phytochrome. Planta 172: 539–547 (1987).
30. Jones AM, Erickson HP: Domain structure of phytochrome from *Avena sativa* visualized by electron microscopy. Photochem Photobiol 49: 479–483 (1989)
31. Jones AM, Quail PH: Quaternary structure of 124-kilodalton phytochrome from *Avena sativa* L. Biochemistry 25: 2987–2995 (1986).
32. Kay SA, Nagatani A, Keith B, Deak M, Furuya M, Chua N-H: Rice phytochrome is biologically active in transgenic tobacco. Plant Cell 1: 775–782 (1989).
33. Keller JM, Shanklin J, Vierstra RD, Hershey HP: Expression of a functional monocotyledonous phytochrome in transgenic tobacco. EMBO J 8: 1005–1012 (1989).
34. Kendrick RE, Kronenberg GHM (eds): Photomorphogenesis in Plants, 2nd ed. Kluwer Academic Publishers, Dordrecht (1994).

35. Kim I-S, Bai U, Song P-S: A purified 124-kDa oat phytochrome does not possess a protein kinase activity. Photochem Photobiol 49: 319–323 (1989).
36. Klimczak LJ, Schindler U, Cashmore AR: DNA binding activity of the *Arabidopsis* G-box binding factor GBF1 is stimulated by phosphorylation by casein kinase II from broccoli. Plant Cell 4: 87–98 (1992).
37. Kunkel T, Tomizawa K-I, Kern R, Furuya M, Chua N-H, Schafer E: *In vitro* formation of a photoreversible adduct of phycocyanobilin and tobacco apophytochrome B. Eur J Biochem 215: 587–594 (1993).
38. Lagarias JC, Lagarias DM: Self-assembly of synthetic phytochrome holoprotein *in vitro*. Proc Natl Acad Sci USA 86: 5778–5780.
39. Lagarias JC, Mercurio FM: Structure function studies on phytochrome. Identification of light-induced conformational changes in 124-kDa *Avena* phytochrome *in vitro*. J Biol Chem 260: 2415–2423 (1985).
40. Li H-M, Washburn T, Chory J: Regulation of gene expression by light. Curr Opin Cell Biol 5: 455–460 (1993).
41. Li L, Lagarias JC: Phytochrome assembly. Defining chromophore structural requirements for covalent attachment and photoreversinbility. J Biol Chem 267: 19204–19210 (1992).
42. Lin X, Feng X-H, Watson JC: Differential accumulation of transcripts encoding protein kinase homologs in greening pea seedlings. Proc Natl Acad Sci USA 88: 6951–6855 (1991).
43. Liscum E, Hangarter RP: Genetic evidence that the red-absorbing form of phytochrome B mediates gravitropism in *Arabodopsis thaliana*. Plant Physiol 103: 15–19 (1993).
44. Litts JC, Kelly JM, Lagarias JC: Structure-function studies on phytochrome. Preliminary characterization of highly purified phytochrome from *Avena sativa* enriched in the 124-kilodalton species. J Biol Chem 258: 11025–11031 (1983).
45. Ma H, Yanofsky MF, Meyerowitz EM: Molecular cloning and characterization of *GPA*1, a G protein α subunit gene from *Arabidopsis thaliana*. Proc Natl Acad Sci USA 87: 3821–3825 (1990).
46. Mancinelli AL: The physiology of phytochrome action. In: Kendrick RE, Kronenberg GHM (eds) Photomorphogenesis in Plants, 2nd ed., pp. 51–70. Kluwer Academic Publishers, Dordrecht (1994).
47. Min K-T, Hilditch CM, Diederich B, Errington J, Yudkin MD: σ^F, the first compartment-specific transcription factor of *B subtilis*, is regulated by an anti-σ factor that is also a protein kinase. Cell 74: 735–742 (1993).
48. McMichael RW, Lagarias JC: Phosphopeptide mapping of *Avena* phytochrome phosphorylated by protein kinases *in vitro*. Biochemistry 29: 3872–3878 (1990).
49. Mummert E, Grimm R, Speth V, Eckerskorn C, Schiltz E, Gatenby AA, Schafer E: A TCP1-related molecular chaperone from plants refolds phytochrome to its photoreversible form. Nature 363: 644–648 (1993).
50. Nagatani A, Chory J, Furuya M: Phytochrome B is not detectable in the *hy3* mutant of *Arabidopsis*, which is deficient in responding to end-of-day far-red light treatments. Plant Cell Physiol 32: 1119–1122 (1991).
51. Nagatani A, Kay SA, Deak M, Chua N-H, Furuya M: Rice type I phytochrome regulates hypocotyl elongation in transgenic tobacco seedlings. Proc Natl Acad Sci USA 88: 5207–5211 (1991).
52. Nagatani A, Reed RW, Chory J: Isolation and initial characterization of *Arabidopsis* mutants that are deficient in Phytochrome A. Plant Physiol 102: 269–277 (1993).
53. Nakasako M, Wada M, Tokutomi S, Yamamoto KT, Sakai J, Kataoka M, Tokunaga F, Furuya M: Quarternary structure of pea Phytochrome I dimer studied with small-angle X-ray scattering and rotary-shadowing electron microscopy. Photochem Photobiol 52: 3–12 (1990).
54. Neuhaus G, Bowler C, Kern R, Chua N-H: Calcium/calmodulin-dependent and -independent phytochrome signal transduction pathways. Cell 73: 937–952 (1993).
55. Ota I, Varshavsky A: A yeast protein similar to bacterial two-component regulators. Science 262: 566–569 (1993).
56. Otto V, Schafer E: Rapid phytochrome-controlled protein phosphorylation and dephosphorylation in *Avena sativa*. Plant Cell Physiol 29: 1115–1121 (1988).
57. Parkinson JS: Signal transduction schemes of bacteria. Cell 73: 857–871 (1993).
58. Parkinson JS, Kofoid EC: Communication modules in bacterial signalling proteins. Annu Rev Genet 26: 71–112 (1992).
59. Parks BM, Quail PH: Phytochrome-deficient *hy1* and *hy2* long hypocotyl mutants of *Arabidopsis* are defective in phytochrome chromophore biosynthesis. Plant Cell 3: 1177–1186 (1991).
60. Parks BM, Quail PH: *hy8*, a new class of *Arabidopsis* long hypocotyl mutants deficient in functional phytochrome A. Plant Cell 5: 39–48 (1993).
61. Pepper A, Delaney T, Chory J: Genetic interactions in plant photomorphogenesis. Sem Devel Biol 4: 15–22 (1993).
62. Quail PH: Phytochrome: a light-activated molecular switch that regulates plant gene expression. Annu Rev Genet 25: 389–409 (1991).
63. Reed JW, Chory J: Mutational analyses of light-controlled seedling development in *Arabidopsis*. Sem Cell Biol, in press (1994).
64. Reed JW, Nagatani A, Elich TD, Fagan M, Chory J: Phytochrome A and Phytochrome B have overlapping but distinct functions in *Arabidopsis* development. Plant Physiol, in press (1994).
65. Reed JW, Nagpal P, Poole DS, Furuya M, Chory J: Mutations in the gene for the red/far red light receptor phytochrome B alter cell elongation and physiological responses throughout *Arabidopsis* development. Plant Cell 5: 147–157 (1993).
66. Robson PRH, Whitelam GC, Smith H: Selected components of the shade-avoidance syndrome are displayed in a normal manner in mutants of Arabidopsis thaliana and

Brassica rapa deficient in Phytochrome B. Plant Physiol 102: 1179–1184 (1993).

67. Romero LC, Lam E: Guanine nucleotide binding protein involvement in early steps of phytochrome-regulated gene expression. Proc Natl Acad Sci USA 90: 1465–1469 (1993).
68. Romero LC, Biswal B, Song P-S: Protein phosphorylation in isolated nuclei from etiolated *Avena* seedlings. Effects of red/far-red light and cholera toxin. FEBS Lett 282: 347–350 (1991).
69. Rudiger W, Thummler F: The phytochrome chromophore. In: Kendrick RE, Kronenberg GHM (eds) Photomorphogenesis in Plants, 2nd ed., pp. 51–70. Kluwer Academic Publishers, Dordrecht (1994).
70. Sarokin L, Chua N-H: Binding sites for the two novel phosphoproteins, 3AF5 and 3AF3, are required for *rbc* S-3A expression. Plant Cell 4: 473–483 (1992).
71. Schneider-Poetsch HAW: Signal transduction by phytochrome: phytochromes have a module related to the transmitter modules of bacterial sensor proteins. Photochem Photobiol 56: 839–846 (1992).
72. Schneider-Poetsch HAW, Braun B, Rudiger W: Phytochrome – all regions marked by a set of monoclonal antibodies reflect conformational changes. Planta 177: 511–514 (1989).
73. Shacklock PS, Read ND, Trewavas AJ: Cytosolic free calcium mediates red light-induced photomorphogenesis. Nature 358: 753–755 (1992).
74. Sharma R, Lopez-Juez E, Nagatani A, Furuya M: Identification of photo-inactive phytochrome A in etiolated seedlings and photo-active phytochrome B in green leaves of the *aurea* mutant of tomato. Plant J 4: 1035–1042 (1993).
75. Sharrock RA, Quail PH: Novel phytochrome sequences in *Arabidopsis thaliana*: structure, evolution, and differential expression of a plant regulatory photoreceptor family. Gene Devel 3: 1745–1757 (1989).
76. Shinomura T, Nagatani A, Chory J, Furuya M: The induction of seed germination in *Arabidopsis thaliana* is regulated principally by Phytochrome B and secondarily by Phytochrome A. Plant Physiol 104: 363–371 (1994).
77. Simon MI, Strathmann MP, Gautam N: Diversity of G proteins in signal transduction. Science 252: 802–808 (1991).
78. Smith H, Whitelam GC: Phytochrome, a family of photoreceptors with multiple physiological roles. Plant Cell Environ 13: 695–707 (1990).
79. Somers DE, Sharrock RA, Tepperman JM, Quail PH: The *hy3* long hypocotyl mutant of *Arabidopsis* is deficient in phytochrome B. Plant Cell 3: 1263–1274 (1991).
80. Stockhaus J, Nagatani A, Halfter U, Kay S, Furuya M, Chua N-H: Serine-to-alanine substitutions at the amino-terminal region of phytochrome A result in an increase in biological activity. Genes Devel 6: 2364–2372 (1992).
81. Terry MJ, Lagarias JC: Holophytochrome assembly. Coupled assay for phytochromobilin synthase *in organello*. J Biol Chem 266: 22215–22221 (1991).
82. Terry MJ, Wahleithner JA, Lagarias JC: Biosynthesis of the plant photoreceptor phytochrome. Arch Biochem Biophys 306: 1–15 (1993).
83. Thummler F, Dufner M, Kreisl P, Dittrich P: Molecular cloning of a novel phytochrome gene of the moss *Ceratodon purpureus* which encodes a putative light-regulated protein kinase. Plant Mol Biol 20: 1003–1017 (1992).
84. Tomizawa K-I, Ito N, Komeda Y, Uyeda TQP, Takio K, Furuya M: Characterization and intracellular distribution of pea phytochrome I polypeptides expressed in *E. coli*. Plant Cell Environ 32: 95–102 (1991).
85. Tretyn A, Kendrick RE, Wagner G: The role(s) of calcium ions in phytochrome action. Photochem Photobiol 54: 1135–1155 (1991).
86. Vierstra RD: Illuminating phytochrome functions. Plant Physiol 103: 679–684 (1993).
87. Vierstra RD, Quail PH: Purification and initial characterization of 123-kilodalton phytochrome from *Avena*. Biochemistry 22: 2498–2505 (1983).
88. Wagner D, Tepperman JM, Quail PH: Overexpression of phytochrome B induces a short hypocotyl phenotype in transgenic *Arabidopsis*. Plant Cell 3: 1275–1288 (1991).
89. Wahleithner JA, Li L, Lagarias JC: Expression and assembly of spectrally active recombinant holophytochrome. Proc Natl Acad Sci USA 88: 10387–10391 (1991).
90. Whitelam GC, Johnson E, Peng J, Carol P, Anderson ML, Cowl JS, Harberd NP: Phytochrome A null mutants of *Arabidopsis* display a wild-type phenotype in white light. Plant Cell 5: 757–768 (1993).
91. Whitelam GC, Harberd NP: Action and function of phytochrome family members revealed through the study of mutant and transgenic plants. Plant Cell Environ 17: 615–625 (1994).
92. Wong Y-S, Cheng H-C, Walsh DA, Lagarias JC: Phosphorylation of *Avena* phytochrome *in vitro* as a probe of light-induced conformational changes. J Biol Chem 261: 12089–12097 (1986).
93. Wong Y-S, Lagarias JC: Affinity labeling of *Avena* phytochrome with ATP analogs. Proc Natl Acad Sci USA 86: 3469–3473 (1989).
94. Wong Y-S, McMichael RW Jr, Lagarias JC: Properties of a polycation-stimulated protein kinase activity associated with purified *Avena* phytochrome. Plant Physiol 91: 709–718 (1989).

Plant Molecular Biology **26**: 1329–1341, 1994.

Mechanical signalling, calcium and plant form

Anthony Trewavas [1,*] and Marc Knight [2]
[1] *Molecular Signalling Group, Institute of Cell and Molecular Biology, University of Edinburgh, Edinburgh EH9 3JH, UK (* author for correspondence);* [2] *Plant Sciences, South Parks Road, University of Oxford, Oxford OX 1 3RB, UK*

Received 2 March 1994; accepted 26 April 1994

Key words: aequorin, cytosol calcium, mechanosensing, morphogenesis, touch, wind

Abstract

Calcium is a dynamic signalling molecule which acts to transduce numerous signals in plant tissues. The basis of calcium signalling is outlined and the necessity for measuring and imaging of calcium indicated. Using plants genetically transformed with a cDNA for the calcium-sensitive luminescent protein, aequorin, we have shown touch and wind signals to immediately increase cytosol calcium. Touch and wind signal plant cells mechanically, through tension and compression of appropiate cells. Many plant tissues and cells are very sensitive to mechanical stimulation and the obvious examples of climbing plants, insectivorous species as well as other less well-known examples are described. Touch sensing in these plants may be a simple evolutionary modification of sensitive mechanosensing system present in every plant. The possibility that gravitropism may be a specific adaptation of touch sensing is discussed. There is a growing appreciation that plant form may have a mechanical basis. A simple mechanical mechanism specifying spherical, cylindrical and flat-bladed structures is suggested. The limited morphological variety of plant tissues may also reflect mechanical specification. The article concludes with a discussion of the mechanisms of mechanical sensing, identifying integrin-like molecules as one important component, and considers the specific role of calcium.

Calcium acts as a universal signalling molecule

Calcium is the most dynamic of known signalling molecules. No other signalling molecule receives such attention or is the subject of such intense interest. The use of technologies which image the distribution of calcium in single cells have uncovered an enormous variety of signalling mechanisms. These discoveries have illustrated the diverse way in which calcium acts to fundamentally regulate metabolism and development and must stand as one of the prime achievements of biological research in the past decade.

Figure 1 is a simple cartoon which outlines events in calcium signalling. Such diagrams are not intended to be realistic descriptions of calcium signalling but do describe calcium signalling at one level of understanding. Calcium enters the cytosol through specific calcium channels – protein pores whose existence in plant cells is now clearly established by patch clamp studies (see references and discussion in [18, 55]). The resting level of the Ca^{2+} in the cytoplasm is about 100 nM and is maintained at this very low level by plasma membrane and other Ca^{2+} ATPases [16]. Upon signalling, intracellular calcium, $[Ca^{2+}]_i$, in the cytosol is increased and combines with calmodulin, the primary plant calcium

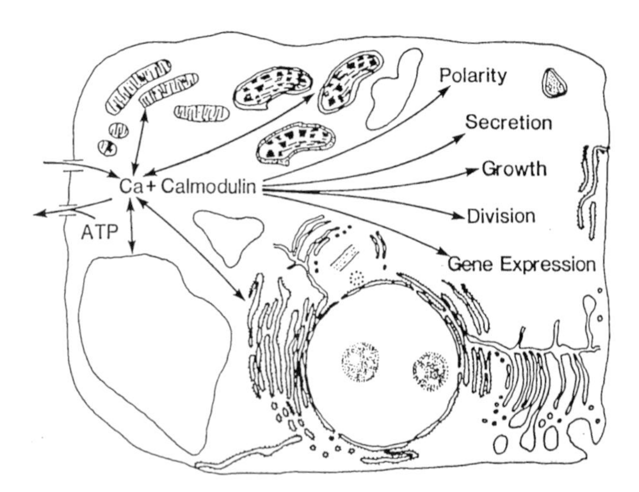

Fig. 1. A cartoon summarising the basic elements of calcium signalling. The cartoon illustrates the entry of extracellular calcium through plasma membrane bound channels in to the cytoplasm, expulsion via plasma membrane-located Ca^{2+} ATPases and interrelations with calcium in other organelles. Combination with calmodulin the primary calcium receptor influences numerous cellular responses.

receptor, or other calcium-binding proteins. Subsequent cellular processes are then initiated usually via activation of appropiate protein kinases.

The cytosolic Ca^{2+} pool is in communication with storage pools in the vacuole and the rough endoplasmic reticulum (ER) which may contain up to 10–1000 mM Ca^{2+}. Both the vacuole and the rough ER membranes contain Ca^{2+} channels and ATPases and Ca^{2+} mobilisation may also be regulated via Ca^{2+}/H^+ exchange.

The situation with mitochondria, chloroplasts and nuclei is less clear. Originally these organelles were regarded as merely helping to regulate cytosolic $[Ca^{2+}]_i$ and to act as Ca^{2+} stores. Increasingly, the perception is that intraorganelle $[Ca^{2+}]_i$ may itself manipulate oxidative respiration, photosynthetic CO_2 fixation and, most notably, gene expression [2, 44, 45, 61]. These three organelles have all been shown to contain calcium-binding proteins and calmodulin. This would suggest that $[Ca^{2+}]_i$ acts as a co-ordinating molecule able to mobilise both cytoplasmic and organelle activities after signalling. Of these three organelles there is currently most interest in the nucleus. This organelle can maintain a free Ca^{2+} concentration different to the cytosol, it can signal directly through its own calmodulin and possesses the enzymology to hydrolyse phosphatidylinositol 4,5-bisphosphate (PIP_2), synthesising the calcium mobilising agent, inositol 1,4,5-trisphosphate (IP_3) [2, 61].

Although we are still unsure of many of the details of calcium signalling we are sure of its importance in the transduction of a variety of signals in plant cells. The Molecular Signalling Group at Edinburgh has employed fluorescence ratio imaging and photometry of $[Ca^{2+}]_i$ coupled with UV photolysis of caged Ca^{2+} and caged IP_3 to mimic observed $[Ca^{2+}]_i$ transients. In one case caged abscisic acid has also been used [1]. These results establish clearly an involvement of $[Ca^{2+}]_i$ in stomatal aperture control, red light transduction through phytochrome, pollen/stigma incompatibility and the direction of pollen tube growth [1, 17, 19, 43, 60]. Other data [46, 50, 56] clearly establish the presence of a calcium gradient in pollen tubes which is required for the maintenance of growth. We have described numerous other signals which modify $[Ca^{2+}]_i$ but these observations do not yet have the necessary confirmation of significance using caged calcium release [36–38]. Although such methods establish an involvement of $[Ca^{2+}]_i$ in transduction they do not debar the involvement of other signalling processes. Signal transduction pathways are complicated networks with numerous parallel and interlinked events; the proportion of flux through each of the main pathways shifts as circumstances change [1].

Much early scepticism of the significance of $[Ca^{2+}]_i$ signalling concerned the large variety of signals $[Ca^{2+}]_i$ was believed to transduce. There is not much variety available in the kinetics of concentration; it can increase, decrease or stay the same. Some additional variety can be introduced through signalling-induced relocation of calmodulin or release of calmodulin from bound forms by phosphorylation [42]. Furthermore, a variety of calmodulins with different sequences and different cell locations increase possible complexity in transduction [5]. A novel signalling mechanism, the induction of waves of $[Ca^{2+}]_i$

and frequency modulation of waves by different signals, represents a further rich source of signalling and transduction information [31]. Further complexity follows from the spatial constraints exerted on the movement of $[Ca^{2+}]_i$ in the cytoplasm. Estimates of the $[Ca^{2+}]_i$ diffusion constant suggest it to be 100-fold lower than in free solution [3]. This diffusional constraint is believed to result from the presence of numerous calcium-binding proteins attached to the cytoskeleton or located on membrane structures. As a consequence, transient signalling, which is localised to regions of the plasma membrane (for example by receptor clustering), initiates an equally localised response to regions of adjacent cytoplasm. Although we in the Molecular Signalling Group have made many hundreds of observations of the distribution of $[Ca^{2+}]_i$ in signalled plant cells, we have yet to observe cells in which $[Ca^{2+}]_i$ is uniformly distributed, except in the resting state. Even ionophores do not produce even distributions of $[Ca^{2+}]_i$. Finally, cell age determines the cellular metabolic apparatus capable of responding to a change in $[Ca^{2+}]_i$, increasing the variety of specific signalling mechanisms available.

Figure 1 emphasises the necessity of measurement and the imaging of $[Ca^{2+}]_i$ in order to understand calcium signalling. That emphasis may be seen to be misplaced as further detailed knowledge emerges, but it provides a route towards an improved understanding.

Touch and wind signals initiate immediate increases in $[Ca^{2+}]_i$

The Edinburgh Molecular Signalling Group has developed a novel method for measuring and imaging $[Ca^{2+}]_i$. Aequorin is a calcium-sensitive luminescent protein from the jellyfish *Aequorea victoria*. In the yellyfish high levels of aequorin are found in specific cells. When attacked by a predator, $[Ca^{2+}]_i$ rapidly elevates in these cells causing the jelly fish to luminesce, thus inhibiting predation.

Aequorin is a 21 kDa protein composed of an apoprotein, apoaequorin, and an imidazolopyrazine luminophore, coelenterazine. On interaction with Ca^{2+}, coelenterazine is oxidized to coelenteramide and luminescent light is emitted. Calibration of emitted light against free Ca^{2+} is relatively straightforward in free solution and in single cells [38]. There are numerous types of coelenterazine available which form aequorins with very different properties (for references and use, see [38]). Calibration of $[Ca^{2+}]_i$ transients can be conducted using e-coelenterazine which forms an aequorin with a bimodal spectrum. Luminescence ratio measurements at two different wavelengths can be linearly related to Ca^{2+} concentration freeing $[Ca^{2+}]_i$ measurements from variations in aequorin concentration.

We have genetically transformed *Nicotiania plumbaginifolia* to express apoaequorin and showed that incubation of the transformed seedlings in coelenterazine led to reconstitution of actively reporting aequorin [36]. Since aequorin is a soluble protein we have thus constituted luminous plants whose luminosity directly reports cytosolic $[Ca^{2+}]_i$. The aequorin was expressed in all three primary regions of the seedling, the cotyledons, stem and root. Targeting of aequorin to different cell compartments or different cell types, different tissues or different developmental ages provides a $[Ca^{2+}]_i$ measuring method of extraordinary range and power.

These transformed seedlings were found to be touch-sensitive [36]. By inserting a fine wire through the luminometer port and arranging a transformed seedling so that a kink in the wire touched it on one rotation, we were able to observe (transient) bursts of luminescent light each time the seedling was touched (Fig. 2). Touch therefore immediately increased $[Ca^{2+}]_i$.

We were able to observe that the seedlings we touched by this method showed slight movement. Was the increase in $[Ca^{2+}]_i$ the result of movement or did it result from touch damage of hairs and epidermal cells? To clarify this situation we sought to induce movement by means which did not involve touch. Seedlings were fixed in the luminometer by their roots and subjected to small blasts of air from a syringe mimicking a wind stimulation [37]. These stimuli caused the plant

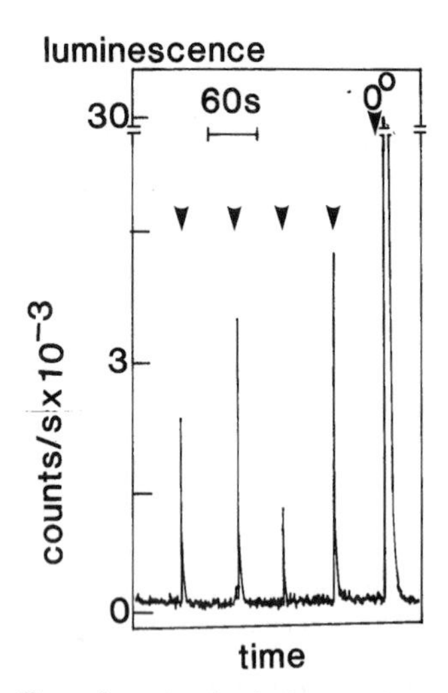

Fig. 2. The effect of touch stimulation on the luminescenc of tobacco seedlings transformed with the calcium-sensitive luminescent protein aequorin. One week old tobacco seedlings transformed with aequorin were placed in a luminometer and gently touched with a fine wire once every minute for four minutes (indicated by arrows). After this stimulation the seedlings were irrigated with ice-cold water (0 °C). Adapted from Knight *et al.* [36].

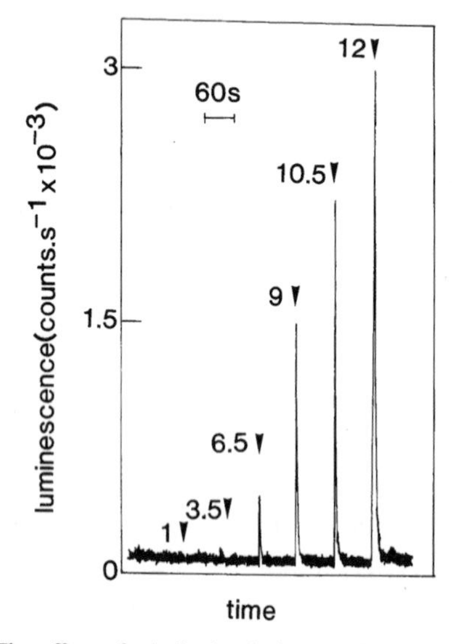

Fig. 3. The effect of wind stimulation on the luminescence of tobacco seedlings transformed with the calcium-sensitive luminescent protein, aequorin. A single tobacco seedling was fixed in a luminometer by the roots and stimulated by gusts of wind from a syringe. Wind forces are in newtons (N) and are indicated as numbers beside the trace. An increasing signal strength elicits an increasing luminescence. Adapted from Knight *et al.* [37].

to rock slightly (ca. 10° from vertical) about the hypocotyl/root junction causing alternate compression and tension in cells in this region. Providing the wind force exceeded a threshold value, an immediate burst of luminescent light was again observed. As the wind force increased, the seedling was observed to rock for longer periods and the detected luminescent light increased in parallel (Fig. 3). From this we concluded that while changes in compression and tension of cells continued (resulting from movement), $[Ca^{2+}]_i$ continued to rise. When the seedling stopped moving, $[Ca^{2+}]_i$ returned to the resting level.

The source of movement-induced $[Ca^{2+}]_i$ may be intracellular

We investigated a range of calcium channel blockers, calmodulin-binding inhibitors and other inhibitors of calcium metabolism on wind-induced $[Ca^{2+}]_i$ changes [37]. Inhibitory effects were only observed with ruthenium red at low concentrations. Ruthenium red is believed to inhibit mitochondrial calcium uptake or to inhibit release of Ca^{2+} from the rough ER [14]. The site of action of ruthenium red is not understood in plant cells but it is generally assumed to modify $[Ca^{2+}]_i$ release from internal stores.

Touch and wind mechanically signal plant cells. Further evidence for the role of $[Ca^{2+}]_i$

The touch sensitivity of plants is well characterised. Early compendia by Darwin [12, 13] and Pfeffer [53] characterising and describing climbing and other touch-sensitive plants illustrate the enormous variety which climb via the agency of touch-sensitive stems, petioles, tendrils, flower peduncles and roots. While the usual response to touch is differential growth, in other cases adhesive pads or even roots may form. In a separate set of experiments, Darwin [13] described the responsiveness of roots from numerous species which bend in response to touch. Numerous insectivorous species, *Mimosa* and *Berberis* stamens round up the list of obviously touch-sensitive plants and tissues.

Touch sensitivity is found amongst many flowering families. Thus touch sensitivity is unlikely to be a unique evolutionary adaptation arising completely *de novo* on numerous occasions but an evolutionary amplification or modification of a facility present in most or all higher plants. Climbing plants and others have simply coupled touch to an easily visible growth response. There are other numerous examples of less obvious touch-sensitive responses. There are reports in the literature that: (1) coleoptiles stroked with a cork rod bend to the stroked side [63]; (2) etiolated plants stroked with sheets of paper respond as if exposed to white light (the stems are shortened and thickened [6, 7]; (3) plant stems which are rubbed respond by synthesising ethylene and exhibit subsequent stem shortening and thickening, a phenomenon termed thigmomorphogenesis [32]; (4) stroking roots interrupts their growth for several hours [26]. Fungi are also thigmomorphogenic. *Uromyces* hyphae respond to the lip of a guard cell by forming an appressorium [28]. Pollen is also touch-sensitive and grows in defined directions on material with repetitively spaced ridges [27].

Studies in the past century with a variety of agents to touch plants led Pfeffer (quoted in [4]) to conclude that touch stimulation relies upon a localised shearing or differential deformation by the applied mechanical force. Tension and compression induced by touch and wind stimuli [36, 37] also induce localised shearing and differential deformation. *Taraxacum* peduncles held at their base were subjected to a 2 g bending stress causing about 10° bowing for five minutes [10]. Cells in the peduncle were thus subjected to tension and compression. After release from the bending stress, the peduncle continued to grow but bent away from the compressed side after a slight delay. This bowing treatment clearly mimics the compression and tension stimulus we imposed by wind stimulation on tobacco seedlings [37]. We would suggest that many touched, rubbed or stroked tissues similarly undergo slight movement and thus tension and compression during the stimulation. Transformed tobacco seedlings have been permanently bent in a luminometer and a $[Ca^{2+}]_i$ spike has been observed [36, 37].

Wind is a well recognised morphogenic stimulus [25]. In woody plants the wind-induced rocking of the stem about the roots leads to a diversion of carbohydrate resources to induce stem thickening and lignification (references in [33]). The stem becomes more resistant to sway which helps reduce further shoot damage on further episodes of wind stimulation. However the extent of thickening depends on the extent of stimulation thus rendering the character a good example of plasticity [68]. It is thought that the primary yield differences between glasshouse-grown crops and their equivalents in the field reflect the absence of wind stimulation in the greenhouse. The diversion of carbohydrate to increase stem thickening otherwise reduces the food reverse available to provision the seeds.

The effects of wind can be mimicked by intermittent shaking to sway the plant, a phenomenon called seismomorphogenesis [33]. There is a consequent reduction in height and an increase in stem thickening in both etiolated and light-grown plants [7]. Neels and Harris [47–49] reported that shaking induced premature dormancy in young trees and substantial height and yield reductions in maize.

The mechanism whereby touch and wind stimuli modify plant growth and development is not known but changes in $[Ca^{2+}]_i$ may transduce both stimuli. Aside from the measurements described above, Jones and Mitchell [33] reported that EGTA and calmodulin-binding inhibitors negated rub-induced growth reductions in soybean. Much more dramatically, Braam and Davies [5] showed that touch stimulation (mainly rubbing or stroking) of *Arabidopsis* massively induced the expression of five touch genes, three of which were identified as calmodulin or calmodulin-related proteins. These important data demonstrate that mechanical stimulation can specifically modify gene expression and again indicate a substantive role for $[Ca^{2+}]_i$.

Plant tissues are very sensitive to mechanical stimulation

Actual figures for plant sensitivity to touch stimulation are hard to come by. Darwin reported that tendrils were sensitive to the weight of a cotton thread which he measured to be 4 mg [13]. Pfeffer reported that some tendrils were sensitive to weights of 25 μg thus making the tendrils more sensitive than human touch [53]. Only 30 s of shaking per day is sufficient to mimic wind sway and reduce height 6-fold in young trees. Clifford *et al.* showed that bowing of only 10° was sufficient to induce counteractive growth [10].

These data suggest that plant cells are extraordinarily sensitive to slight mechanical stress. It is thought that a touch stimulation is only exercised when unequal pressure is exerted at different points producing deformation of the epidermis [53]. If tissue deformation is the normal means of sensing touch the degree of deformation by a few milligrams weight on a cell wall maintained in shape by 8 atm (800 kPa) turgor must be very tiny. Deformation will be experienced first by the cell wall but information has to be conveyed to the interior of the cell if a response is to be generated. The cell wall should then be adhesively connected to both the plasma membrane and at least to the cortical cytoskeleton which is located in the non-streaming part of the cytoplasm. If these three constituents were not connected then mechanical stress would cause slippage of the cell wall over the plasma membrane and in turn over the cytoskeleton. As a consequence, cells would not maintain a specifically positioned cytoskeleton with respect to the plasma membrane. Polarized secretion leading to growth of defined regions of the wall (side walls compared to end walls for example) could not be maintained. With slippage, mechano-transduction mechanisms become more difficult to visualise.

Direct evidence supports the presence of an inter-linked cytoskeleton/plasma membrane/cell wall continuum in plant cells [72] and in animal cells [11, 29]. Studies on the fixation of polarity and localised growth in the developing *Fucus* zygote, the regeneration of differentiation initiated by protoplasting, studies on plasmolysed cells and the spatial localisation of secretion of many plant cells are some of the evidence which implies the existence of this continuum.

Some sensitive plants have been reported to contain specific thigmosensitive cells. However the simplest way to make selected tissues more sensitive is probably to increase the surface area density of mechanoreceptive constituents.

Gravitropic sensing may be a special form of touch sensing

The commonest view of gravisensing assigns statoliths an important role. Statoliths are large, subcellular, sedimentable bodies. In higher plants statoliths are usually identified with amyloplasts and gravisensing is believed to be restricted to statocytes, specific cells which contain very prominent amyloplast structures. Sedimentation of the amyloplast onto cellular membranes is thought to initiate gravisensing. The isolation of *Arabidopsis* mutants with little or no detectable starch which are still gravitropically sensitive [9] and the rediscovery of gravity-regulated streaming rates in giant algal cells [71] has led to a reappraisal of ideas on gravisensing. These suggest that while statoliths help refine and improve gravisensing [35, 71], they may not be the absolute necessity once thought [9].

An intriguing alternative to statoliths, first proposed around the turn of the century, has recently been resuscitated [54, 71]. This suggests that gravisensing may result from detection by individual cells of the weight of the cytoplasmic bulk on the cortical cytoplasm, the plasma membrane and the appropriate cell wall – a sort of inside-out deformation or touch sensing. Calculations by Wayne *et al.* [71] show this to be a tenable hypothesis for giant algal cells but a weaker possibility for the much smaller higher-plant cell. It has also been suggested that statoliths simply increase the weight of the cytoplasm thus making the statocyte a more reliable gravitropic detector [71]. Deformation or mechanical stress is increased and this improved sensitivity explains the widespread distribution of statocytes in graviperceptive tissues.

However sensing of cytoplasmic weight by individual higher-plant cells may not be necessary. Branches of trees which grow horizontally experience tension and compression stresses between the upper and lower parts. While the maximum stress will be experienced by the outermost cells there will be effective gradients of tension/compression throughout the tissue. Compression induces the formation of reaction wood to counterbalance the stress. Weights as little as 50 mg can induce the formation of reaction wood [20]. A horizontally placed but unsupported coleoptile or root should also experience tension and compression, with the maximum stress exerted on the upper and lower epidermal cells although of a lower magnitude than in a branch. Epidermal cells are believed to exert primary control on extension growth [39] and differential mechanical stress on these may be sufficient to initiate growth rate modification. In horizontally supported tissues, the bottom epidermal cells will still experience compression from the weight of cells above (in addition to any touch response). Based on the calculations provided by Wayne *et al.* [71] this mechanism could contribute to gravitropic sensing.

Pickard and Ding [54] have suggested that gravitropic sensing may be an evolutionary refinement of a general growth co-ordinating mechanism. Local mechanical stresses and strains occur as cells expand unevenly in growing tissues. These mechanical stresses are sensed and the subsequent fine turning of expansion rates ensures co-ordinated growth. It has also been hypothesised [66, 67] that this fine-tuning may be a function of auxin, smoothing out uncoordinated patterns of growth. The resulting auxin-induced cell synchronisation may then result in increased growth rates. We have recently showed that gravi-stimulating *Arabidopsis* roots leads to a 3–4-fold accumulation of calmodulin mRNA [62]. This observation again suggests a metabolic similarity between touch-induced responses [5] and gravity responses. Numerous observations relate gravitropic phenomena to calcium (e.g. [64] and references therein) deepening the possible relation to touch.

Is there a mechanical basis to plant form?

Plant form has a holistic basis. 'Die Pflanze bildet Zellen, nicht Zellen bilden Pflanzen' (quoted in [65]) is an important statement which summarises traditional recognition of the essential holistic character to morphogenesis. The evidence which so impressed previous generations of plant morphologists is derived from numerous experimental observations. Chief among them are the regeneration of cell differentiation and tissue patterns and finally plant form from wounded plant tissues [40]. In seeking to identify cellular characteristics from which holistic tissue properties might emerge, the most obvious is the contiguous character of tissue cell walls interacting with turgor pressure. A change in the tensile strength of the wall (or the turgor) of any one cell will inevitably have consequent effects on the shapes of surrounding cells. If larger numbers of cells change wall strength or change turgor pressure the effects will be increasingly experienced throughout the tissue-influencing form. The shape of any organ will reflect the balance of mechanical forces experienced inside it. Any of the primary tissue shapes should possess unique mechanical stress patterns resulting from inequalities of tension between the constituent cells. As growth continues these stress patterns will change. *If the mechanical stress can be sensed and transduced to specify the further directions of growth a mechanical basis to the generation of form emerges.* There is potential to specify both the primary shapes of plant organs and the positional distribution of many inside-tissues based on sensing of lines of stress through the tissue.

D'Arcy Thompson [65], like Hooke and Grew before him, was fascinated by the similarity of the shapes of parenchymatous cells to the shapes adopted by soap bubbles in a froth. Single bubbles are spherical because they assume a shape in which the surface tensions (the mechanical stresses) of the film are at a minimum. Similar minimal mechanical stress requirements account for the shape of froth-enclosed bubbles; they would also help ensure that the surface area of interaction of the bubbles be at a minimum too.

D'Arcy Thompson [65] hypothesised that newly formed cell walls act like a fluid film and that where they join older, more rigid walls, minimum energy (stress) considerations place this new wall at an angle of 90° – a suggestion in accord with observation. As this new cell wall acquires a rigidity similar to those to which it is joined the angle of juncture assumes 120° equalising intercellular tension – an hypothesis supported by many observations.

These important observations can be used to generate simple models of form based on three premises.

1. The plane of cell division can be aligned by mechanical stress. Observations of cell division alignment induced in callus by externally applied pressure support this premise as does the alignment of cell plates induced during induction of reaction wood by compression [20, 41]. These observations suggest that mechanical stress can be sensed and transduced into rearrangements of a cytoskeletal structure specifying the plane of division.

2. The mechanical strength of individual cells walls can be different amongst the constituent cells of growing tissues. Bisection of growing sections causes the tissue to bow outwards, an observation first made in the last century. The accepted explanation for this observation is that the peripheral layers are less easily deformable than inner tissues. Peripheral cell walls are 6–20 times thicker than inner walls and evidence suggests that growing tissues are constrained by an epidermal 'straitjacket' [38]. Growing tissues contain a number of different cell types. Each cell type should have a different wall structure with different tensile strengths thus providing permanent internal tension within growing and developing tissues. The outermost epidermal wall may be the most resistant to deformation and even in 1906 was drawn as thicker than other epidermal cell walls [53].

3. The directions of growth and division be set to minimise the mechanical stress throughout the growing tissue and that a mechanism for sensing and acting upon mechanical stress be available to plant cells. This is discussed in the last section of this article.

Kutschera [38] has suggested that growing stems with their epidermal 'straitjacket' can be regarded for some purposes as like a giant turgid cell. Analogously, the spherical form of many fruits and berries can be understood as resulting from less deformable, more rigid peripheral layers surrounding a mass of inner dividing cells with softer cell walls. Conceptually this is equivalent to a bubble or a balloon providing the deformability of all epidermal cells is uniform. The spherical shape simply minimises the mechanical stress characteristics built into the tissue. Organs such as shoots and roots have a circular cross section which is the structure of minimal mechanical stress which results from the greater tension exerted by the peripheral layers. The shape is tubular rather than spherical because division and growth are constrained to the tip where wall deformability is higher. Many apical regions possess a cone-like shape and this shape can be conceived as resulting from a gradient of decreasing peripheral cell deformability as cells age. In both these organs the direction of growth is clearly specified by the directionality of tissue tension and stresses.

With an initial asymmetry built into the mechanical tension within the meristem, aspects of phyllotaxis can be described [59]. 'The behaviour of cells in the meristem is detemined not by any character or properties of their own but by the positions and the forces to which they are subject. As soon as the tension of adjacent cell walls becomes unequal then the form alters' [65]. The shoot apex in *Vinca* is not a perfect hemisphere implying a non-uniformity of tension within the apex which slowly becomes elliptical as leaves grow out [59]. The new primordial leaves thus emerge at nodal points of tension/stress within the apex. The outgrowing leaf continually modifies torsional stress within the apex until a new nodal point in tension emerges and phyllotaxis recommences. The formation and maintenance of the tunica can be understood since new cross walls will be jointed to the thicker, more rigid outer epidermal cell wall in the growing tissue and thus remain at an angle of 90° [65]. Circummutation is perhaps a later indication of an earlier asymmetry of growth in the apex.

An initially tubular primordial leaf can be turned into a flat blade by limited spatial strengthening of the peripheral walls on top and bottom of the tissue; dividing cells are literally squeezed out at either side where there is less resistance to division and growth. In other primordial leaves which are effectively flat at birth, localised peripheral layer strengthening can again account for the form. The production of form becomes a kind of structural epigenesis in which the sensing of tension and compression followed by adjustments of growth direction become critical elements.

Mechanical hypotheses explaining the production of form are attractive because of the simplicity of mechanism they offer [51]. These hypotheses have gone largely unrecognised because of a lack of appreciation that plants can sensitively perceive mechanical stress and more importantly can act upon it. However mechanical bases to the production of form have been championed for many years by Paul Green (see [59]) and, more lately, by Brian Goodwin and coworkers [23, 24]. There has also been a dominance of this topic of investigation by notions of chemically specified positional information, a still unsubstantiated theory [51]. No doubt aspects of both mechanisms contribute.

How are mechanical signals sensed? The role of calcium

Predominant interest in animal cells centres upon mechanical sensing by integrins [11, 21, 29, 30]. These data will be discussed since it is most likely that equivalents exist in plant cells as considered later.

Integrins are plasma membrane-spanning proteins composed of two separate subunits. Regions of the integrin molecule penetrate the extracellular matrix where they bind to fibronectin, vitronectin and laminins and perhaps other extracellular adhesive molecules. The cytoplasmic face of the integrin molecule acts as a nucleation or anchorage site for the attachment of cytoskeletal structures through the linking proteins vinculin, talin and actinin. At least 19 different integrins have been characterised so far and it seems likely that each may perform slightly different functions in mechanical signalling or be located in different cell types; some act to nucleate microfilaments and some intermediate filaments for example.

Many motile animal cells possess focal adhesion sites through which they attach themselves to the substratum. Integrins are specifically concentrated in focal adhesion sites [8]. Here they act as a focus for the formation of a cytoskeletal network (so-called stress fibres) which spreads from these sites throughout the cell. The attached cytoskeletal network is under tension and this provides shape or form to the cell. If the focal adhesion sites are disrupted, cells will round up and become quiescent. Integrins therefore directly mediate mechanical stress between the extracellular matrix and the cytoskeleton.

A direct role for integrins in sensing mechanical stress was demonstrated by Wang *et al.* [70]. Vitronectins have a conserved peptide region, notably the sequence RGD (Arg-Gly-Asp), which binds to integrins and can be used to probe integrin function. Tiny ferromagnetic beads were coated with RGD peptide and allowed to adhere to endothelial cells where they attached themselves to the exposed face or integrin molecules. On attachment the beads initiate integrin clustering, the step believed to induce signalling. Talin, vinculin and actinin were observed to be recruited into the region of bead attachment indicating the formation of microfilaments attached to the clustered integrins. A magnetic force was then applied to twist the bead and mechanically stress the attached microfilaments. Surprisingly, integrin-attached microfilaments showed increasing stiffness (ratio of stress to strain) in direct proportion to the applied stress. In fact the bead could only be rotated by 25°. By using appropiate inhibitors the authors demonstrated that the microfilaments had become attached to, and become part of, a cytoskeletal network stretching throughout the cell. This network continually rearranges as stress is increased in an attempt to resist further torsional stress. Mechanotransduction may thus be mediated simultaneously at multiple locations inside the cell through force-induced rearrangements of an integrated cytoskeleton.

Integrin clustering and thus signalling can be induced by anti-integrins. Integrin signalling increases $[Ca^{2+}]_i$, changes phosphatidylinositol (PI) metabolism and membrane protein phosphorylation. There are specific alterations in gene expression. Since nuclei are sometimes held in a basket of mirofilaments, restructuring of the cytoskeleton after integrin signalling may directly modify chromatin structure and thus gene expression. However in one well-characterised case it is nuclear $[Ca^{2+}]_i$ which is specifically increased [61]. Altered gene expression may then directly result from combination with nuclear calmodulin and altered nuclear calmodulin dependant protein kinase activity. Increases in $[Ca^{2+}]_i$ may also act to promote microfilament reconstruction through activation of gelsolin issuing from force-induced cytoskeletal rearrangements. Since mechanical stress can modify the plane of division [41], integrins may be involved in directing the position of the preprophase band of microtubules.

Recent exploration of integrin signalling has indicated that intracellular events can also influence the conformation and ligand binding affinity of the extracellular domain of integrins. This 'inside-out' modification of signal transduction appears to be mediated through the cytoplasmic domains [21]. The integrin surface repertoire is altered, enormously increasing the range and affinity for different ligands. Thus internal changes can determine which components of the extracellular matrix are recognised and modify the sensitivity to mechanical stress. This 'inside out' signalling is reminiscent of the way in which gravitropic stimulation might be an inside-out version of the touch response.

Are there integrins in plant cells? Present indications are tantalising. Sequences of integrins are not highly conserved so direct sequence equivalents are unlikely [72]. However vitronectin-like molecules have been detected in numerous plant and even algal species [22, 57, 69]. Vitronectin along with fucoidan is concentrated in the growing rhizoid wall of the early *Fucus* zygote. If this vitronectin acts as attachment sites for integrin-like molecules and microfilaments inside the cell this would help explain the localised secretion of polysaccharide-containing vesicles which feed the growing wall. Vitronectins are part of the adhesion mechanism used by the early zygote to attach itself to a substratum. In the absence of adhesion the rhizoid fails to develop and instead the zygote becomes a ball of undifferentiated cells. Adhesion may be an unrecognised but important aspect of plant morphogenesis. Schindler *et al.* [58] and Wyatt and Karpita [72] have used the RGD peptide to inhibit vitronectin binding by integrins in plant cells. In both cases modified adhesion of the plasma membrane to the cell wall was observed.

There is no reason to suppose that mechanical stress sensing is confined to integrins in plant cells. The wall is compositionally complex and theoretically any of the primary wall constituents could bind to an appropiate and specific plasma membrane-spanning protein that could be used for mechanosensing. Indications that wall components are sensed may be deduced from the known effects of oligosaccharides in defense reactions and cell development [72]. Polysaccharides secreted by one cell could promote cytoskeletal rearrangements in adjacent cells thus leading to phenomena such as homeogenetic induction [40].

The role of calcium in mechanical sensing and transduction has yet to be well defined but a function in microfilament rearrangement via gelsolin is one good possibility. Kirchofer *et al.* [34] demonstrated a very direct requirement for calcium ions for integrin attachment to vitronectin and thus for adhesion. A role for plant cell wall calcium is thus suggested in addition to pectin binding. Goodwin and collaborators [23, 24] have suggested that the cortical cytoskeleton is an integrated visoelastic network under tension whose properties are directly modified by the free $[Ca^{2+}]_i$. They have suggested that nodal stress points appear in this network at critical Ca^{2+} concentrations which could then be used to specify local areas of secretion of wall-softening enzymes. They demonstrate that their mechanism could explain aspects of morphogenesis in *Acetabularia*. Their ideas are in part based on equivalent mechanical signalling mechanisms proposed to underlie animal morphogenesis [51, 52]. Ding and Pickard [15] have characterised stretch-activated calcium channels in onion epidermal cells which cluster in opening activity. Stretch channels are an alternative means for signalling mechanical stress but may require a significant change in membrane area before activation. Ding and Pickard [15] have suggested these stretch channels might be directly connected to integrins suggesting a very basic role for such channels in signalling.

Conclusions

This short article has illustrated aspects of mechanical sensing and transduction in plant cells. The importance of mechanical stress in the production of form has almost certainly been underestimated and hopefully a resurgence of interest will clarify the relevance to morphogenesis. There is increasing attention being paid to the plasma membrane/cytoskeleton/cell wall continuum and this should help clarify the role of mechanical stress. The significance of $[Ca^{2+}]_i$ to mechanical signalling is not yet clear but circumstantial evidence strongly suggests it is an important component of the transduction process. The next step is to identify the origins of mechanically induced changes in $[Ca^{2+}]_i$ and the mechanism whereby increases occur. These are in hand.

Acknowledgements

We are grateful to Dr R. Lyndon and Dr M. Fricker for critical reading and correction of this manuscript.

References

1. Allan AC, Fricker MD, Ward J, Beale M, Trewavas AJ: Caged ABA induced calcium transients in guard cells of *Commelina communis* are dependant on previous growth temperature. Plant Cell 6: 1319–1328.
2. Bachs O, Carafoli E: Calmodulin and calmodulin-binding proteins in liver cell nuclei. J Biol Chem 262: 10786–10790 (1987).
3. Barritt GJ: Communication Within Animal Cells. Oxford University Press, Oxford (1992).
4. Bentrup FW: Reception and transduction of electrical and mechanical stimuli. In: Haupt WB, Feinlieb ME (eds) Physiology of Movements. Encyclopedia of Plant Physiology, New Series, vol. 7, pp. 42–70 (1979).
5. Braam J, Davies RW: Rain-, wind- and touch-induced expression of calmodulin and calmodulin related genes in *Arabidopsis*. Cell 60: 357–364 (1990).
6. Bunning E: Uber die Verhinderung des Etiolements. Ber Deut Bot Ges 59: 2–9 (1941).
7. Bunning E, Lempnau C: Über die Wirkung mechnischer und photischer Reize auf die Gewebe und Organbildung von *Mimosa pudica*. Ber Deut Bot Ges 67: 10–18 (1954).
8. Burridge K, Fath K, Kelly T, Nuckolls G, Turner C: Focal adhesions. Annu Rev Cell Biol 4: 487–525 (1988).
9. Caspar T, Pickard BG: Gravitropism by a starchless mutant of *Arabidopsis*; implications for the starch statolith theory. Planta 177: 185–197 (1989).
10. Clifford PE, Fensom DS, Munt BJ, McDowell WE: Lateral stress initiates bending responses in dandelion peduncles; a clue to geotropism? Can J Bot 60: 2671–2673 (1982).
11. Damsky CH, Werb Z: Signal transduction by integrin receptors for extracellular matrix: cooperative processing of extracellular information. Curr Opin Cell Biol 4: 772–781 (1992).
12. Darwin C: The Power of Movement in Plants. John Murray, London (1880).
13. Darwin C: The Movements and Habits of Climbing Plants. John Murray, London (1891).
14. Denton RM, McCormack JG, Edyell NJ: Role of calcium ions in the regulation of intramitochondiral metabolism. Biochem J 190: 107–117 (1980).
15. Ding JP, Pickard BG: Mechanosensory calcium selective cation channels in epidermal cells. Plant J 3: 83–110 (1993).
16. Evans DE, Briars SA, Williams LE: Active calcium transport by plant cell membranes. J Exp Bot 42: 285–303 (1991).
17. Franklin-Tong VE, Ryde JP, Read ND, Trewavas AJ, Franklin C: The self incompatability response in *Papaver rhoeas* is mediated by free cytosolic calcium. Plant J 4: 163–177 (1993).
18. Gilroy S, Bethke PC, Jones RL: Calcium homeostasis in plants. J Cell Sci 106: 453–462 (1993).
19. Gilroy S, Read ND, Trewavas AJ: Elevation of cytosol calcium using caged calcium and caged inostiol phosphate initiates stomatal closure. Nature 346: 769–771 (1990).
20. Gilroy S, Trewavas AJ: Signal sensing and signal transduction across the plasma membrane. In: Larsson C,

Moller IM (eds) The Plant Plasma Membrane, pp. 203–233. Springer-Verlag, Heidelberg (1990).
21. Ginsburg MH, Xiaoping D, Plow EF: Inside-out integrin signalling. Curr Opin Cell Biol 4: 766–771 (1992).
22. Goodner R, Quatrano RS: *Fucus* embryogenesis: a model to study the establishment of polarity. Plant Cell 5: 1471–1481 (1993).
23. Goodwin BC, Briere C, O'Shea POP: Mechanisms underlying the formation of spatial structure in cells. Soc Gen Microbiol Symp 23: 1–9 (1987).
24. Goodwin BC, Trainor LEH: Tip and whorl morphogenesis in *Acetabularia* by calcium regulated strain fields. J Theor Biol 117: 79–106 (1985).
25. Grace J: Plant Response to Wind. Academic Press, London (1977).
26. Hanson JB, Trewavas AJ: Regulation of plant cell growth; the changing perspective. New Phytol 90: 1–18 (1982).
27. Hirouchi T, Suda S: Thigmotropism in the growth of pollen tubes of *Lilium longiflorum*. Plant Cell Physiol 16: 377–381 (1975).
28. Hoch HC, Staples RC, Whitehead B, Comeau J, Wolf ED: Signalling for growth orientation and cell differentiation by surface topography in *Uromyces*. Science 235: 1659–1662 (1987).
29. Hynes RO: Integrins: A family of cell surface receptors. Cell 48: 549–554 (1987).
30. Ingber D: Integrins as mechanochemical transducers. Curr Opin Cell Biol 3: 841–848 (1991).
31. Jaffe LJ: Classes and mechanisms of calcium waves. Cell Calcium 14: 736–745 (1993).
32. Jaffe MJ: Thigmomorphogenesis; the response of plant growth and development to mechanical perturbation. Planta 114: 143–157 (1973).
33. Jones RS, Mitchell CA: Calcium ion involvment in growth inhibition of mechanically-stressed soybean seedlings. Physiol Plant 76: 598–602 (1989).
34. Kirchhofer D, Grzesiak J, Pierschbascher MD: Calcium as a potential physiological regulator of integrin mediated cell adhesion. J Biol Chem 266: 4471–4477 (1991).
35. Kiss JZ, Sack FD: Severely-reduced gravitropism in dark grown hypocotyls of a starch deficient mutant of *Nicotiania sylvestris*. Planta 180: 123–130 (1989).
36. Knight MR, Campbell AK, Smith SM, Trewavas AJ: Transgenic plant aequorin reports the effects of touch and cold shock and fungal elicitors on cytosolic calcium. Nature 352: 524–526 (1991).
37. Knight MR, Smith SM, Trewavas AJ: Wind-induced plant motion immediately increases cytosolic calcium. Proc Natl Acad Sci USA 89: 4967–4972 (1992).
38. Knight MR, Read ND, Campbell AK, Trewavas AJ: Imaging calcium dynamics in living plants using semisynthetic recombinant aequorins. J Cell Biol 121: 83–90 (1993).
39. Kutschera U: The role of the epidermis in the control of elongation growth in stems and coleoptiles. Bot Acta 105: 246–253 (1992).
40. Lang A: Progressiveness and contagiousness in plant differentiation and development. Encyclopedia of Plant Physiology vol 15 (1), pp. 409–424 (1964).
41. Linthilac PM, Vesecky TB: Stress-induced alignment of division plane in plant tissues grown *in vitro*. Nature 307: 363–364 (1984).
42. Liu Y, Storm DR: Dephosphorylation of neuromodulin by calcineurin. J Biol Chem 264: 12800–12804 (1989). Trends Pharmacol 11: 107–111 (1990).
43. Malho R, Read ND, Pais MS, Trewavas AJ: Role of cytosolic free calcium in the reorientation of pollen tube growth. Plant J 5: 331–341 (1994).
44. McCormack JG, Halestrap AP, Denton RM: Role of calcium ions in regulation of mammalian intramitochondrial metabolism. Physiol Rev 70: 391–425 (1990).
45. Melkonian B, Burchet M, Kreimer G, Latzko E: Binding and possible function of calcium in the chloroplast. Curr Topics Plant Biochem Physiol 9: 38–46 (1990).
46. Miller DB, Callahan DA, Gross DJ, Hepler PK: Free Ca^{2+} gradient in growing pollen tubes of *Lilium*. J Cell Sci 101: 7–12 (1992).
47. Neel PL, Harris RW: Motion-induced inhibition of elongation and induction of dormancy in *Liquidamber*. Science 173: 58–59 (1971).
48. Neel PL, Harris RW: Growth inhibition by mechanical stress. Science 174: 961–962 (1972).
49. Neel PL, Harris RW: Tree seedling growth: effect of shaking. Science 175: 918–919 (1972).
50. Obermeyer G, Weisenseel MH: Calcium channel blocker and calmodulin antagonists affect the gradient of free calcium ions in lily pollen tubes. Eur J Cell Biol 56: 319–327 (1991).
51. Odell GM, Oster G, Alberch P, Burnside B: The mechanical basis of morphogenesis. Devel Biol 85: 446–462 (1981).
52. Oster GF, Murray JD, Harris AK: Mechanical aspects of mesenchymal morphogenesis. J Embryol Exp Morphol 78: 83–125 (1983).
53. Pfeffer W: The Physiology of Plants, vol 3 (translated by A.J. Ewart). Clarendon Press, Oxford (1906).
54. Pickard BD, Ding JP: Gravity sensing by higher plants. Adv Comp Envir Physiol 10: 81–110 (1992).
55. Poovaiah BW, Reddy ASN: Calcium and signal transduction in plants. Crit Rev Plant Sci 12: 185–211 (1993).
56. Rathore KS, Cork RJ, Robinson KR: A cytoplasmic gradient of Ca^{2+} is correlated with the growth of lily pollen tubes. Devel Biol 148: 612–619 (1991).
57. Sanders LC, Wang CS, Walling LL, Lord EM: A homolog of the substrate adhesion factor vitronectin occurs in four species of flowering plants. Plant Cell 3: 629–635 (1991).
58. Schindler M, Meiners S, Cheresh DA: RGD-dependent linkage between plant cell wall and plasma membrane; consequences for growth. J Cell Biol 108: 1955–1965 (1989).
59. Selker JML, Steucek GL, Green PB: Biophysical mecha-

nisms for morphogenetic progressions at the shoot apex. Devel Biol 153: 29–43 (1992).
60. Shacklock P, Read ND, Trewavas AJ: Cytosolic free calcium mediates red light induced photomorphogeneis. Nature 358: 753–755 (1992).
61. Shankar G, Davison I, Helfrich MP, Mason WT, Horton MA: Integrin receptor mediated mobilisation of intranuclear calcium in rat osteoclasts. J Cell Sci 105: 61–68 (1993).
62. Sinclair W, Oliver I, Maher P, Trewavas AJ: Effect of gravistimulation on calmodulin mRNA in wild type and mutant *Arabidopsis* plants. Plant Physiol, submitted (1994).
63. Stark P: Weitere Untersuchungen über das Restantengesetz beim Haptotropismus. Jahrb Wiss Bot 61: 126–167 (1921).
64. Stinemetz CL, Kuzmanoff KM, Evans ML, Jarret HW: Correlations between calmodulin activity and gravitropic sensitivity in primary roots of maize. Plant Physiol 84: 1337–1342 (1987).
65. Thompson DA: On Growth and Form. Cambridge University Press, Cambridge, UK (1942).
66. Trewavas AJ: How do plant growth substances work? Plant Cell Envir 4: 203–228 (1981).
67. Trewavas AJ: How do plant growth substances work? II. Plant Cell Envir 14: 1–12 (1991).
68. Trewavas AJ, Knight MR: The regulation of shape and form by cytosolic calcium. In: Ingram D, Hudson A (eds) Shape and Form in Plant and Fungal Cells, pp. 221–233. Academic Press, London (1992).
69. Wagner VT, Brian L, Quatrano RS: Role of a vitronectin-like molecule in embryo adhesion of the brown alga *Fucus*. Proc Natl Acad Sci USA 89: 3644–3648 (1992).
70. Wang N, Butler JP, Ingber D: Mechanotransduction across the cell surface and through the cytoskeleton. Science 260: 1124–1127 (1993).
71. Wayne R, Staves MP, Leopold AC: Gravity dependent polarity of cytoplasmic streaming in *Nitellopsis*. Protoplasma 155: 43–57 (1990).
72. Wyatt SE, Carpita NC: The plant cytoskelton-cell wall continuum. Trends Cell Biol 3: 413–417 (1993).

Plant Molecular Biology **26**: 1343–1356, 1994.

Plasmodesmata: composition, structure and trafficking

Bernard L. Epel
Botany Department, Tel Aviv University, George S. Wise Faculty of Life Sciences, Tel Aviv University, Tel Aviv 69978, Israel

Received and accepted 13 September 1994

Key words: conductivity, movement, plasmodesmata, size exclusion limit, proteins, viruses

Abstract

Plasmodesmata are highly specialized gatable trans-wall channels that interconnect contiguous cells and function in direct cytoplasm-to-cytoplasm intercellular transport. Computer-enhanced digital imaging analysis of electron micrographs of plasmodesmata has provided new information on plasmodesmatal fine structure. It is now becoming clear that plasmodesmata are dynamic quasi-organelles whose conductivity can be regulated by environmental and developmental signals. New findings suggest that signalling mechanisms exist which allow the plasmodesmatal pore to dilate to allow macromolecular transport. Plant viruses spread from cell to cell via plasmodesmata. Two distinct movement mechanisms have been elucidated. One movement mechanism involves the movement of the complete virus particle along virus-induced tubular structures within a modified plasmodesma. Apparently two virus-coded movement proteins are involved. A second movement mechanism involves the movement of a non-virion form through existing plasmodesmata. In this mechanism, the viral movement protein causes a rapid dilation of existing plasmodesmata to facilitate protein and nucleic acid movement. Techniques for the isolation of plasmodesmata have been developed and information on plasmodesma-associated proteins is now becoming available. New evidence is reviewed which suggests that plasmodesmatal composition and regulation may differ in different cells and tissues.

Introduction

In higher plants, cell-to-cell communication as well as nutrient transport may be symplastic, via junctional structures called plasmodesmata or apoplastic, via membrane-associated receptors, trans-membranous gatable channels and/or transporters. Plasmodesmata are highly specialized gatable cytoplasmic trans-wall channels that interconnect contiguous cells and function in the direct cytoplasm-to-cytoplasm intercellular movement of water, nutrients, small signalling molecules and, in certain cases, of macromolecules.

Plasmodesmata were thought for many years to be nonselective pores, passively allowing the bi-directional movement of molecules between adjacent cells. During the past few years this static concept of the plasmodesmata has been changing and it is now clear that plasmodesmata are dynamic selective entities with the capacity to 'gate'. Evidence is now emerging which suggests that plasmodesmatal composition and regulation may differ in different cells and tissues. In this review, we consider new studies on plasmodesmatal structure, composition, on the regulation of gating, and on virus-induced alterations in plasmodesmatal conductance.

Plasmodesmatal structure

Early electron micrography studies of plasmodesmata led to a simple general consensus model describing the plasmodesma as a wall-embedded plasmalemma-lined unbranched cylinder that contains a central axial component generally termed the desmotubule. For a review of early structural aspects of plasmodesmata, see the book by Gunning and Robards [29]. The desmotubule was considered to be derived from and continuous with the endoplasmic reticulum of adjoining cells. Recent studies suggest that the endoplasmic reticulum (ER) component of a plasmodesma is a derivative of and continuous with cortical ER [30, 31, 52]. In some cells, the outer regions of the plasmodesma are constricted and form what has been termed the neck region, a structure which is not a general feature of all plasmodesmata [56]. The cylindrical space between the desmotubule and the plasmalemma has been referred to as the cytoplasmic annulus or sleeve. Most models consider the cytoplasmic annulus as the pathway through which transport occurs. Electron micrographs show that in the orifice region, the ER was apparently constricted and no lumen was detected. Lucas *et al.* [38] have suggested that, since it is now clear that the desmotubule is not a tubule but a modified ER, its nomenclature should be changed to *appressed ER*. This new nomenclature may not be appropriate for a number of reasons. The term 'appressed' refers to laminar structures while the ER within the plasmodesma is more like a cylinder. A more appropriate descriptive term might be *constricted ER* rather than *appressed ER*. Furthermore, within the more interior regions of the plasmodesmata, the ER often balloons and a lumen is present, i.e. it is neither appressed or constricted. Most importantly, we feel that one must be cautious in interpreting static electron micrographs. The ER within the plasmodesmatal channel may be a dynamic structure, constricting and dilating as signalling and function dictate. We feel that the term 'desmotubule' could be retained. However, if one feels the need to use a functional neutral term, then one can refer to the 'plasmodesma ER component' (PERC). It is entirely conceivable that within the plasmodesmata the ER has undergone major modifications, and that different parts of the plasmodesmatal ER component have different compositions.

The substructural detail of plasmodesmata in higher plants has been examined by various workers and a number of models have been proposed (see review by Robards and Lucas [55]; [7, 14, 68]). Recently, two models of plasmodesmatal structure were proposed based on computer-enhanced digital imaging analysis of electron micrographs of plasmodesmata from a dicotyledonous plant [14] and from a C_4 grass [7].

Ding *et al.* [14] examined the substructure of the plasmodesmata of the C_3 dicot *Nicotiana tabacum* following cryofixation and freeze substitution. According to their model, the plasmodesma is depicted as a complex pore-containing proteinaceous particles embedded in the inner leaflet of the plasma membrane and in the outer leaf of the desmotubule. Spoke-like filamentous strands apparently connect the globular proteins of the outer leaf of the desmotubule to the proteins embedded in the encasing PD plasma membrane. The central region of the desmotubule, the so-called rod, an electron-dense region, is depicted as being composed of a series of particles, probably protein, that are embedded in the lipid of the fused inner leaflet of the ER membrane. In their model, the desmotubule is constricted and no lumen is present. It was suggested that the electron-dense particles embedded in the inner leaflet of the plasmalemma and the outer leaflet of the desmotubule form a convoluted channel that functions as a molecular sieve, determining the size exclusion properties. It was further hypothesized that the radial spokes may function in dynamically altering the effective radius of the cytoplasmic sleeve, thus regulating the size exclusion limit of the plasmodesma.

Botha *et al.* [7] employed similar computer enhancement techniques with plasmodesmata at the Kranz mesophyll bundle sheath interface of *Thermeda trianda*, a C_4 grass. Their model showed many similarities to that of Ding *et al.* [14] but with a number of significant differences. Associated with the inner plasmalemma leaflet and the outer desmotubule wall were particles ca. 2.5–3 nm in diameter, presumably proteins that abutted into the so-called cytoplasmic sleeve. In the central region of the desmotubule were electron-dense particles also about 2.5 nm in diameter. In contrast to Ding *et al.* [14], Bothe *et al.* [7] detected the presence of a lumen of about 2.5 nm in diameter between the central protein core and the desmotubule wall. Furthermore, filamentous connections were observed between the central rod particles of the desmotubule and the desmotubule wall. This was in contrast to the model of Ding *et al.* [14] for the plasmodesma for *N. tabacum*, in which the spoke-like filamentous strands connected proteins of the outer leaf of the desmotubule to the proteins embedded in the encasing PD plasma membrane. The data and model present by Bothe *et al.* [7] brings into question the general consensus that desmotubules are constricted and can not function in cell-to-cell transport. Additional studies are needed to address this question.

Other workers have reported on the presence of structures, termed 'sphincters', either external to [49] or within [56] the neck region of plasmodesmata. It was suggested that these 'sphincters' function in the regulation of plasmodesmatal conductance. It should be noted that these conclusions were based on structural evidence alone and since these structures are not universal, it is difficult to assess their involvement in the regulation of plasmodesmatal conductance.

Transport through plasmodesmata

Transport through plasmodesmata has generally been studied by measuring the cell-to-cell movement of plasma membrane-impermeable fluorescent dyes of different sizes and properties. The classical studies of Goodwin and co-workers [23, 27], of Tucker [71] and of Terry and Robards [67] led to a consensus that only molecules of less then about 800–1000 Da pass freely through plasmodesmata. Recent studies, however, are leading to a re-evaluation of the size exclusion limit (SEL) and selectivity of plasmodesmata. These new studies indicate not only that the plasmodesma can be gated, but that the size exclusion limit can be modulated by environmental and developmental signals.

The diffusion selectivity of plasmodesmata to a series of fluorescein labelled probes was re-examined quantitatively by Tucker and Tucker [73] employing fluorescein labelled (F-) mono-, di-, tri- and quatra-amino acids. Their data suggested that the plasmodesmata of the stamenal hairs of *Setcreasea purpurea* have a size exclusion limit of about 800 Da and select for small hydrophilic molecules with a charge from −2 to −4. Molecules, however, that have attached aromatic amino acid such as phenylalanine and tryptophan exhibited low mobility and their kinetic curves, as well as those of F-$(meth)_2$ and F-$(his)_2$ generally did not fit simple diffusion kinetics. When carboxyfluorescein (CF), a highly mobile probe was micro-injected into the cytoplasm of cells previously micro-injected with probes whose transport was apparently blocked, the CF passed readily from cell to cell. This implied that the aberrantly mobile probes had neither blocked nor closed the permeation pores but that the plasmodesmata exhibited specific selectivity against these molecules.

Azide treatment markedly altered the conductance properties of *Setcreasea purpurea* plasmodesmata [74]. In untreated cells, probes such as F-Phe, F-Try, F-$(meth)_2$, F-$(his)_2$ and the FITC-labelled octa-peptide angiotensin II did not pass through plasmodesmata or passed with non-diffusion kinetics. However, following azide treatment all probes passed with ease with their diffusion coefficients increasing by about 400%. The data indicated that azide treatment may have caused the plasmodesmatal pore to dilate.

Further support for the notion that the size exclusion limit of plasmodesmata is much more extended then previously believed and that size selectivity can be modulated comes from studies by Cleland *et al.* [10]. They found that under aerobic conditions or in the absence of inhibitors of respiration, the SEL of wheat root PD, as probed with LYCH or F-dextrans, was about 1 kDa. However, after either azide (1 or 10 mM) treatment or flushing with nitrogen, the SEL increased within 30–60 min to 3–4 kDa and sometimes to as high a mass as 7–10 kDa.

The effects of azide and anaerobiosis are pleiotropic, causing an inhibition in oxidative respiration, a decrease in ATP, an increase in Ca^{2+} and a lowering of cytoplasmic pH. Evidence has previously been presented which suggests that the gating of plasmodesmata [3, 42, 72] and of animal junctional pore structures, the gap junctions [12, 57, 58, 65, 66, 70] may be regulated by phosphorylation. Changes in Ca_2^+ and ATP levels possibly alter protein kinase and/or phosphatase activity resulting in a change in the phosphorylation level of plasmodesmatal proteins and thus altering the structure of the conducting channels within the plasmodesmata and resulting in a modulation of the SEL. I hypothesize that the phosphorylation of plasmodesmatal components results in the down-regulation of plasmodesmatal conductance while dephosphorylation results in a dilation of the conducting channel and up-regulates conductance.

Regulation of closure by differential pressure

The gating of plasmodesmata has been shown to be affected by various environmental and developmental signals [11, 18, 50, 51, 53, 60, 61]. In giant algae such as *Chara* and *Nitella*, pressure gradients between cells increased electrical resistance across the node and inhibited intercellular transport of radioactively labelled assimilates [11, 53]. In higher plants, direct evidence for pressure-generated closure was recently provided in a study by Oparka and Prior [50]. In this study, a pressure gradient was generated between adjacent cells of leaf tricomes of *Nicotiana clevelandii* with a micropressure probe and its effect on intercellular transport followed by measuring the cell-to-cell movement of the fluorescent probe Lucifer Yellow. A pressure differential in excess of 0.2 MPa was required to close plasmodesmata completely. The degree of plasmodesmatal closure was dependent on the magnitude of the pressure differential between the two cells. Just raising or lowering the internal pressure was not sufficient, a pressure differential between two adjacent cells was essential. If a cell was wounded by puncturing, cells acropetal to the wounded cell lost pressure but remained in communication with each other and with the wounded cell. However, cells basipetal to the punctured cell which were in direct contact with the plant body (leaf) retained turgor, and communication with the wounded cell was terminated.

The above findings have obvious implications for wounding. It was suggested that one of the first responses during cell wounding would be a pressure-generated closure of plasmodesmata in the cells adjacent to the wound, provided that the tissue had an adequate water supply and there was no general loss of turgor in the cell adjacent to the wound site. If there was a loss of turgor in tissues adjacent to the wound site then these cells would remain in symplastic communication with the wounded cell and leakage out into the wound region would occur. Based on these observations, we suggest that if excised tissue would be placed in an isotonic solution, no pressure differential would be established between the wounded cell and adjacent intact cells and the plasmodesmata in the wounded cells would not close. The plasmodesmata of the wounded cell would then continue to communicate with adjacent cells. This phenomenon could be exploited to introduce non-membrane permeable molecules into cells without the necessity of microinjection. This hypothesis could explain the uptake of polar tracers into intact cells of fresh cut sections as reported by Burnell [8] for bundle sheath strand from C_4 plants, by McCaskill *et al.* [40] for glandular trichomes from peppermint leaves, and by Tucker and Tucker [73] for severed stamenal hairs of *Setcreasea purpurea* and by Wang and Fisher for fresh crease tissue of developing wheat grains [77].

Regulation of gating by light

Plasmodesmatal gating can also be regulated by light. Epel and Erlanger [18] measured symplastic transport in the mesocotyl tissue of dark grown maize seedlings using the fluorescent symplastic probe carboxyfluorescein. They reported that lateral symplastic transport from the stele into the mesocotyl cortex was inhibited by a prior white-light irradiation. The inhibitory effect of a prior white-light irradiation was completely photo-modulatable by terminal far-red and far-red/red irradiations, suggesting the involvement of phytochrome. It was suggested that the modulation of plasmodesmatal conductance by light, and possibly by other environmental and/or developmental signals, might modulate growth and development, in part by establishing or altering symplastic domains and by channelling cell-to-cell transport of nutrients and growth regulations. As we have previously pointed out, large gradients of growth regulators and solutes would exist at interfaces between domains and sub-domains with resultant effects on growth and differentiation [17].

Cell- and tissue-specific differences in SEL

There is accumulating evidence that the size exclusion limit of plasmodesmata may differ in different cells and tissues. It was found that when fresh-cut sections of the crease tissue of developing wheat grains were incubated in fluorochrome solutions of normally impermeable apoplastic probes such as Lucifer Yellow, dye was rapidly absorbed via unsealed plasmodesmata [77]. Once absorbed, the dye moved symplastically. Using a size-graded series of probes, the size exclusion limit (SEL) for the post-phloem pathway was estimated. In most, perhaps all, cells of the crease tissue with the exception of the pericarp, the effective molecular diameter of the conducting channel was estimated to be about 6.2 nm vs. about 2.5–3 nm from other studies with other plants and tissues.

A number of recent studies suggest that plasmodesmata between sieve elements and their companion cells may traffic in proteins [6, 24]. In higher plants, mature sieve elements are generally enucleated and lack ribosomes. Since these cells can function for prolonged periods, either the proteins in these cells have a longer lifetime or proteins may be transported into these cells, most likely from adjacent companion cells via plasmodesmata. Aphids and other sap-sucking insects obtain food by inserting their stylet into a functional sieve element. The content of sieve elements can be analyzed by severing the stylet and collecting the pressure-mediated exudation of the phloem sap. Fisher *et al.* [24] investigated protein turnover in sieve tubes of wheat by ^{35}S-methionine labelling of leaves and by the use of aphid stylets to sample the sieve tube contents. About 200 different soluble proteins were present in the sieve tube, many of which underwent rapid labelling. Given the constant protein composition along the path, it was concluded that most proteins were loaded at the source (the leaf) and unloaded at the sink (the grain). Certain proteins showed very rapid turnover suggesting some selectivity of movement from companion cells into sieve tubes. If movement through these plasmodesmata is indeed specific for particular proteins, then it must be concluded that these plasmodesmata do not function as simple molecular sieves. It is unclear whether the potential to transport proteins between cells is a unique property of sieve tubes and their companion cells or whether protein transport can normally occur between all cells. Molecular studies must be designed to test whether the potential for protein transport requires special targeting sequences. In order to examine whether signal sequences are involved in selective plasmodesmatal transport of proteins, genes of proteins able to transverse plasmodesmata should be cloned and sequenced and putative consensus sequences for plasmodesmatal targeting identified. By employing chimeric reporter proteins it may be possible to identify small targeting sequences. Micro-injection experiments with fluorescently labelled phloem proteins could be employed to test whether plasmodesmata in other tissues are competent to transport these phloem-resident proteins.

Since membrane lipids and proteins within the sieve tubes probably also undergo turnover, there must be some mechanism for the transport of membranes or membrane components, both lipids and proteins, into sieve tubes. This could be accomplished through migration within the plasma membrane of the plasmodesmata or via budding-off of desmotubule membranes and incorporation into the plasma membrane of the sieve tube. A recent study using the technique of fluorescence redistribution after photobleaching (FRAP) has provided evidence that the ER membranes of plasmodesmata can serve as a dynamic diffusion pathway for the movement of lipids and lipid signalling molecules between contiguous cells [28]. In this study, either the plasma membranes or the ER membranes of cultured soybean suspension cells were labelled with a range of fluorescent lipids or phospholipid analogues. After photobleaching of the fluorescent probe in a target cell, the transport of the lipid from adjacent cells into the target cell was monitored under a confocal microscope. In the case of ER-located fluorescent probes, there was clear evidence for intercellular communication between contiguous cells. No detectable intercellular communication was observed for fluorescent probes residing exclusively in the plasma membrane despite the fact that plasma membrane-located probes showed considerable lateral mobility within the plasma membrane of a single cell.

Further evidence that plasmodesmata between companion cells and sieve elements can selectively transport proteins was provided by experiments with the phloem exudate from pumpkin. The phloem exudate from pumpkin contains two abundant basic proteins, termed PP1 and PP2 that are involved in slime plug formation. These two proteins are expressed exclusively in developing and mature sieve elements and their companion cells [62]. P-protein synthesis is thought to occur either in the immature sieve elements or in the companion cells prior to transport into sieve elements. This transport from companion cells into sieve elements has been postulated to occur via plasmodesmata. Thompson and coworkers [6] used a combined molecular and structural approach to investigate the temporal and spatial appearance of these two P-proteins. To obtain evidence for the site of P-protein synthesis, PP1 and PP2 mRNA were localized by *in situ* hybridization (PP1 [6], PP2, personal communication). PP1 and PP2 antisense transcripts hybridized to mRNA only in companion cells within the phloem of hypocotyl tissues in both the bundle and extrafascicular phloem tissue. Not withstanding the above data, the exact site of P-protein synthesis in the differentiating phloem tissue is still unclear. Thompson suggests three possible scenarios regarding the site of P-protein synthesis: (1) protein synthesis occurs exclusively in the companion cells and proteins are transported into sieve elements; (2) immature nucleated sieve elements synthesize stable proteins which are also synthesized in companion cells; and (3) a combination of 1 and 2 where protein synthesis occurs in both cell types and protein transport occurs from the companion cells to the enucleated mature sieve element. Developmental studies using *in situ* localization and/or using transgenic plants containing reporter constructs under the transcriptional regulation of the PP promoter should aid in determining the correct model. If indeed P-proteins are synthesized in companion cells and transported via plasmodesmata to contiguous sieve elements, this could indicate that the plasmodesmata between the companion cells and sieve element have different properties and possibly different composition then other plasmodesmata. Alternatively, it could be hypothesized that plasmodesmata connecting the companion cells with a sieve element have a unique regulatory mechanism for passage of large molecules. As indicated above, under conditions of low ATP, plasmodesmata can dilate, allowing passage of large molecules. If this regulation is due to phosphorylation/dephosphorylation of plasmodesmatal proteins, regulation of plasmodesmatal dilation may be by cell-specific protein kinases or phosphatases.

Kikuyama *et al.* [35] examined the SEL of plasmodesmata in the characean plant *Nitella* using a number of fluorescent probes included dextrans and proteins. In experiments measured in the time frame of minutes, they obtained results similar to those observed with higher plants, i.e. the SEL was about 1000. However, if transport was measured after 24 h, it was found that myoglobin (20 kDa) and egg albumin (45 kDa) were transported from cell to cell, but bovine serum albumin 70 kDa was not transported. It would be of importance to quantify the transport rates for these proteins and determine whether the energy charge of the cell has any effect on the transport capacity for these large molecules.

Virus plasmodesma interactions

Most, if not all, plant viruses spread from cell to cell via plasmodesmata [1, 32, 39]. Several studies over the past few years have focused on the process of viral cell-to-cell spread and have revealed the involvement of virus-encoded movement proteins (MP). Studies with different classes of viruses indicate at least two distinct movement mechanisms. One mechanism, exemplified by tobacco mosaic virus [16], red-clover necrotic mosaic virus [25] and bean dwarf mosaic geminivirus [48], involves a virus-encoded movement protein that facilitates the passage of a non-virion form of the virus. The second mechanism, exemplified by the comovirus cowpea mosaic virus (CPMV), involves the movement of the complete virus particles along tubular structures through plasmodesmata [75, 78]. In this type of mechanism the desmotubule is absent and specific virus-induced tubules are assembled in the plasmodesma, through which virus particles move from one cell to the other [76]. Two virus-encoded movement proteins are involved, one of which becomes incorporated into the virus-induced tubules; the function of the second protein which is not incorporated into the tubule is unclear. It is still an open question whether existing plasmodesmata are modified or whether new connections between cells are induced by the virus. Cell-to-cell movement through virus-induced tubules has also been reported for a number of other important viruses from different virus groups [76].

The mechanism for the movement of the non-virion form apparently does not entail a modification in the composition of existing plasmodesmata. Recent data indicate that the virus movement protein apparently interacts directly with an intact plasmodesma, causing it to dilate and to potentiate the passage of itself and other large biomolecules [25, 48]. Previous immunological studies suggested that MP may become incorporated into plasmodesmata [2, 15, 45, 69]. It was suggested that in transgenic tobacco plants transformed with MP30, the MP was incorporated only into secondary plasmodesmata during their formation and caused a significant increase in the SEL from 800 Da (control) to >10 kDa [15]. It was hypothesized that only these modified secondary plasmodesmata have altered SEL and function in viral transport [15].

Recent experimental evidence suggests that the hypothesis that *de novo* synthesis of modified secondary plasmodesmata containing MP as a prerequisite for virus spread is incorrect. In these experiments, various putative viral movement proteins were directly introduced in mesophyll cells and their cell-to-cell movement measured. Evidence from these microinjection studies indicated that the injected movement protein altered the state of existing plasmodesmata almost immediately, potentiating the trafficking of macromolecules [25, 48]. When FITC-labelled 35 kDa movement protein of the red-clover necrotic mosaic virus (RCNMV) was microinjected into cowpea mesophyll cells, it spread from cell to cell within seconds. By contrast, an alanine scanning mutant movement protein (mutant protein 278 which was unable to promote virus spread) did not spread out of the injected cell. The wild-type RCNMV movement protein also promoted cell-to-cell spread of fluorescently labelled RCNMV RNAs. Upon co-injection of a 10 kDa Fluorescently labelled dextran (F-dextran) with native MP, the F-dextran moved from cell to cell. In the absence of the MP only F-dextran molecules of less than 1 kDa moved. It was reported that only those alanine scanning mutant proteins that promote virus movement in infected plants increased the SEL with but one exception: mutant 280, which did not promote virus movement *in vivo* [26] but which did increase the SEL. The finding that this mutant protein up-regulated the SEL of plasmodesmata but did not facilitate viral movement suggested that there may be a distinct domain responsible for opening plasmodesmata that is different from that for mediating viral spread.

Similar results were obtained with the movement protein for the bean dwarf mosaic geminivirus [48]. This virus possesses a bipartite genome divided between two circular single-stranded (ss) DNA molecules, DNA-A and DNA-B. Encapsidation and replication are encoded by DNA-A while systemic spread functions are encoded by DNA-B. Frame-shift mutations or single-amino-acid substitutions in either the BL1 or BR1 open reading frames encoded on DNA-B-abolished systemic movement of the mutated form of the virus but did not affect DNA-B replication in protoplasts. These results led to the suggestion that these genes are essential for the systemic spread of the infectious form of BDMV. When FITC-labelled BL1 protein was microinjected into bean cells, the movement protein spread from cell to cell within seconds. Upon co-injection of BL1 with fluorescently labelled dextrans, it was found that 10 kDa but not 29 kDa F-dextran moved from cell to cell within 1–2 min. The BL1 movement protein could also facilitate cell-to-cell movement of double- but not single-stranded DNA. The BL1 movement protein potentiated both its own movement and that of ssDNA but did not potentiate the movement of BR1, a second viral protein that mediates the movement of ssDNA and dsDNA out of nuclei. It is unclear whether the BL1 movement protein could potentiate the movement of proteins other than itself.

The finding that MP up-regulates plasmodesmatal conductance within seconds is not in accord with the hypothesis that viral movement occurs only through *de novo* synthesized MP-modified secondary plasmodesmata. The large-scale accumulation of MP in secondary plasmodesmata is apparently unrelated to the movement mechanism. These results suggest that MP, in some unknown manner, almost immediately up-regulates the plasmodesmatal SEL, with no need for *de novo* secondary plasmodesmata formation. It is suggested that the MP interacts with regulatory factors associated with plasmodesmata that normally function in gating plasmodesmatal conductivity and/or selectivity. In order to understand how viruses alter plasmodesmata, basic information is required on the composition of plasmodesmata.

Plasmodesmatal composition

First biochemical and molecular information on plasmodesmata came from studies demonstrating the presence of animal connexin-homologous proteins in plants [33, 34, 36, 41, 42, 43, 44, 59, 81, 82]. Schindler and co-workers noted numerous functional homologies between plasmodesmata and between animal gap junctions, structures involved in cytoplasmic cell-to-cell communication in animals. These functional analogies included similar behavior with regard to SEL tracer dyes, similar electrical properties and similar modes of down-regulation by effectors of protein kinase C. It was suggested that there may be biochemical homologies between gap junction proteins and plasmodesmatal proteins. Immunological data suggested that dicotyledonous plants contain peptides immunologically related to the mammalian gap junction proteins, connexin32 [33, 41, 42, 44, 81], and connexin26 [33, 34, 59]. Indirect evidence suggested that these cross-reacting proteins [33, 34, 59, 81] may be plasmodesma-associated. More definitive evidence for the presence of a connexin homologous protein in plants was provided by immunocytochemical studies in maize by Yahalom *et al.* [82]. Using affinity-purified antibodies against two different gap junction proteins, connexin32 and connexin43, it was shown by indirect immunogold labelling of thin sections that maize mesocotyl plasmodesmata contain two different proteins that cross-react with connexin gap junction antibodies. A connexin32 antiserum cross-reacted with a 27 kDa maize plasmodesma-associated protein termed PAP27, while an affinity-purified antiserum against connexin43 labelled a 26 kDa protein termed PAP26. PAP26 immunolocalized along the entire length of the plasmodesma as well as to plasmalemma regions surrounding the plasmodesma orifice. PAP27 immunolocalized to outer regions of the plasmodesma and is apparently a peripheral membrane protein. After tissue homogenization and differential centrifugation it was found not only associated with the wall fraction but also with other membrane fractions as well as the soluble fraction [36, 82]. After repeated passages through a nitrogen pressure bomb and repeated washes in a buffer containing high chelator concentrations it dissociated from the wall fraction. In contrast, PAP26 was concluded to be an integral membrane protein; it was found associated only to the wall fraction and could not be extracted with chelators, high salt, high pH or Triton X-100 (Yahalom and Epel, unpublished results).

Meiners *et al.* [44] isolated and sequenced a cDNA clone, termed CX32, from an *Arabidopsis* expression library that encoded for a protein that cross-reacted with an animal connexin32 antibody. They attempted to align its deduced amino acid sequence with that of a rat connexin and concluded that the *Arabidopsis* CX32 protein showed significant homologies to that of the animal connexin. It was suggested that the plant protein was a connexin homologue. Mushegian and Koonin [47] re-analyzed the alignment employing a more statistically rigorous alignment program and suggested that the alleged plant connexin from *Arabidopsis* was unrelated to animal connexins but was more closely related to a protein kinase-like protein.

At present no sequence data are available either for the two confirmed plasmodesma-associated maize proteins PAP26 and PAP27 [82] or for the *Vicia faba* 21 kDa wall-associated protein that cross-reacted with antibodies to mouse liver connexin26 [33]. It would be of great interest to see if these proteins are truly connexin-homologous or only show very limit homologies at a small number of epitopes.

Further characterization of plasmodesma-associated proteins was made possible by the development of techniques for plasmodesma isolation. Initial work was performed with only semi-clean cell wall fractions that contained embedded plasmodesmata. [46, 82] but which contained contaminating subcellular organelles and membranes [36]. Yahalom *et al.* [82] detected well over 20 proteins associated with such a semi-clean maize mesocotyl wall fraction including PAP26 and PAP27. Monzer and Kloth [46] examined proteins associated with a wall fraction obtained from shoot tips of *Solanum nigrum*. A comparison of the proteins extracted from the various cell fraction with those associated with the wall revealed that 2 proteins with apparent molecular masses of 28 and 43 kDa were highly enriched in this wall fraction. Turner and Roberts (personal communication) have examined proteins associated with isolated walls from a 2 mm region of the root tip of maize seedlings. This meristematic region has very high concentrations of primary plasmodesmata. Turner found that there were three major wall associated proteins with apparent molecular weights of 100, 70 and 40 kDa that could be extracted with Triton X-100 or with CHAPS, and that the extraction with these detergents apparently removed plasmodesmata from the walls. Blackman and Overall (personal communication) have taken a unique approach to examining proteins associated with plasmodesmata. In studies with the higher alga *Chara*, they isolated the internode wall which contain no plasmodesmata and walls of nodal cells which contain very high concentrations of plasmodesmata. It should be noted that the plasmodesmata in these cells are structurally different from those of higher plants in that they do not contain a desmotubule. Associated with the walls containing plasmodesmata, they identified 4 unique proteins with apparent molecular masses of 53, 27, 26 and 20 kDa that were absent in the walls of the internode cells which are devoid of plasmodesmata. Monoclonal antibodies are being generated against these proteins and will be used in immunolocalization studies.

Kotlizky *et al.* [36], in order to enrich for plasmodesmatal proteins and to reduce or eliminate spurious non-plasmodesmatal proteins, developed a procedure for preparing a clean wall fraction that contained wall-embedded plasmodesmata and that was devoid of contaminating cytoplasm, organelles and non-relevant membranes. At least 10 proteins were associated with this clean wall fraction [21, 22, 36], including at least two calcium-dependent protein kinases [22, Yahalom and Epel, unpublished data].

With the advent of this wall isolation procedure, it became feasible to attempt to isolate plasmodesmata by enzymatic digestion of the encasing cell wall. Since most if not all commercial cellulase preparations contain proteases and lipases which cause a partial degradation of the plasmodesmata, it was necessary to include a cocktail of protease inhibitors in order to reduce the degree of proteolysis [22]. After overnight enzymatic hydrolysis of the encasing cell wall with a low-protease cellulase from *Trichoderma reesi* and after differential centrifugation of the hydrolysate, a fraction was obtained which contained free plasmodesmata, and plasmodesmata aggregates enmeshed in an apparently cellulase resistant network [22, Epel and Van Lent, unpublished data]. About 8 putative plasmodesmata-associated proteins (pPAP) with apparent molecular masses of 64, 51, 41, 32, 26, 21, 17 and 15 kDa as determined by SDS-PAGE were identified and isolated [20, 21].

Epel and co-workers generated polyclonal antibodies against the 17, 26, 32, 41, 51 and 64 kDa wall-associated polypeptides and showed by western blot analysis of different cell fractions that the 17, 26, 32, and 51 kDa proteins were detectable only in the wall fraction, suggesting that these proteins are probably unique to the plasmodesmata (Epel, Levi and Erlanger, unpublished results). An unequivocal and final confirmation that these proteins are truly plasmodesma-associated proteins requires that each protein be immunolocalized to the plasmodesma. Immunolocalization studies, in addition to those performed with PAP26 and PAP27, have now been performed with the 41 kDa protein [21]. Silver-enhanced immunogold light microscopy showed that the 41 kDa protein was associated with the walls of cells both in the stele and cortex. The immunolabelling pattern was trans-wall and punctate. Electron microscopic immunogold labelling localized the polypeptide to plasmodesmata and to electron-dense cytoplasmic structures which are apparently Golgi membranes. The observation of gold particles over all parts of the plasmodesmata suggested that the protein was spread over the whole length of the structure. Unfortunately, immunogold labelling does not have sufficient resolution to determine more precisely where in the fine structure of the plasmodesmata the protein is located.

PAP41 is apparently a peripheral membrane protein. Following tissue homogenation, the 41 kDa protein was found not only to be associated with the wall fraction but was also present in the soluble fraction and in heavy- and light-membrane fractions. Treatments that release peripheral-bound membrane protein such as 3 M NaCl or 100 mM Na_2CO_3 pH 11, released PAP41 as a soluble protein from the wall/plasmodesmatal fraction (Yahalom, Katz and Epel, unpublished results). Interestingly, overnight incubation of the wall fraction in the presence of 2% Triton X-100 did not solubilize this protein. The protein has been isolated and an internal peptide isolated and microsequenced. Katz, Kotlizky and Epel (unpublished results) screened a λZAP maize cDNA library with antibodies to PAP41 and with oligonucleotide probes synthesized according to a microsequenced internal peptide. One clone has been partially sequenced and shown to contain the sequence for the internal peptide. A full-length sequence should soon be available.

The finding that 41 kDa protein was also associated with Golgi-like structures suggested that this protein is probably transported to plasmodesmata via Golgi membranes. Primary plasmodesmata are apparently formed in part as the result of fusion of Golgi at the phragmoplast at the laying down of the primary cell wall [30]. The region of the mesocotyl sectioned and probed was from the 5 mm region below the coleoptile node, a region undergoing rapid elongation. In this region secondary plasmodesmata formation would be occurring [19].

In animals, the connexins, a family of related gap junction proteins [37], exhibit heterologous expression in various tissues [4, 5, 13, 79]. It was suggested that the diversity of connexins may be to serve the different functional needs of the cell type; differences in conductance and differences in regulatory mechanisms have been shown [9, 63, 64]. Kotlizky, in the laboratory of Epel, obtained data that suggest that plasmodesmata, much like gap junctions, may be composed of different molecular subunits, depending on the source tissue or plant organ [20]. They showed by western blot analysis that PAP26, PAP41 and pPAP17 varied in the level of expression in different organs. PAP26 was present in leaf and mesocotyl but was essentially absent in the root. pPAP17 was present in root, and mesocotyl but was undetectable in the leaf. In contrast to the other 2 proteins, PAP41 was found in all organs tested. In the cortex and stele of the mesocotyl, all three proteins exhibited a qualitatively similar pattern of expression, albeit the specific concentration of all three was higher in the cortex than in the stele. It should be noted that the quantitative differences in expression between the cortex and the stele were more pronounced in both pPAP17 and PAP26 than in PAP41. This variation in expression both in plant organs and in plant tissues hints at a possible specialization of plasmodesmata in different tissue and organs.

Accumulating evidence suggests that in animal cells the regulation of transport through gap junction channels is by a phosphorylation mechanism. Several gap junction proteins (connexins) are phosphorylated by protein kinase C, calmodulin-dependent protein kinase and c-AMP dependent protein kinase [57, 58, 65, 66, 70]. Correlations were found between the phosphorylation of gap junction proteins and the gating of the gap junction channel [12, 54, 57]. Available evidence suggests that the gating of plasmodesmatal conductance may also be via a phosphorylation/dephosphorylation of as yet unidentified regulatory elements [3, 10, 72, 74].

Support for this hypothesis comes from *in vitro* phosphorylation studies with isolated clean wall fraction and isolated plasmodesmata [83, Yahalom and Epel, unpublished results]. In the presence of labelled ATP, several proteins including PAP41 and pPAP17 were phosphorylated by one or several endogenous Ca^{2+}-dependent protein kinase(s). *In situ* phosphorylation on nitrocellulose paper revealed the presence of at least 2 polypeptides that undergo autophosphorylation with an apparent MW of 51–56 kDa. The cell wall/plasmodesma-associated protein kinases were activated by Ca^{2+}, but not by phospholipids or calmodulin. They are probably tightly bound to the cell wall/plasmodesma fraction since proteins were phosphorylated even after overnight extraction with either 3 M NaCl, 2% Triton X-100 or 100 mM Na_2CO_3 (pH 11). While LiCl concentrations of up to 4 M increased phosphorylation, overnight extraction with 8 M LiCl released the endogenous kinases from the cell wall/plasmodesmata and completely abolished phosphorylation. It should be pointed out that direct evidence that these protein kinases function *in vivo* in phosphorylating plasmodesma-associated proteins is still lacking. Furthermore, strategies must be designed which will allow for the testing of the hypothesis that these kinases function in the regulation of plasmodesmatal conductance.

Prospects

Plasmodesmata regulate the movement of nutrients, ion, signalling molecules and, in certain cases, macromolecules. This movement can be gated and is under developmental and environmental control. Indirect evidence indicates the involvement of protein phosphorylation mechanisms. The nature of the signalling mechanisms is unclear and needs to be explored. Biochemical and molecular studies must be extended to identify and characterize plasmodesmatal components and regulatory elements. Studies must also be extended to cell- and organ-specific differences in plasmodesmatal structure, composition and regulation. Most studies to date have been devoted to maize. Studies should be extended to other organisms. Once we have identified and characterized plasmodesmatal components and characterized their genes we will have molecular tools for exploring mechanisms by which plants coordinate intercellular and tissue-tissue interactions. These molecular tools will also allow us to explore how viruses interact with plasmodesmata and exploit them as conduits for virus spread.

Acknowledgements

I thank Avital Yahalom and Guy Kotlizky for critically reading this manuscript. Work from the author's laboratory was supported by the Israeli Science Foundation, by the Joint German-Israeli Research Program, MOST, NCRD (project DI-SNAT 1036), by the Dutch-Israeli Agricultural Research Program and by the Karse-Epel Fund for Botanical Research at Tel Aviv University.

References

1. Atabekov JG, Taliansky ME: Expression of a plant virus-coded transport function by different viral genomes. Adv Virus Res 38: 201–248 (1991).
2. Atkins D, Hull R, Wells B, Roberts K, Moore P, Beachy RN: The tobacco mosaic virus 30 K movement protein in transgenic tobacco plants is localized to plasmodesmata. J Gen Viorol 72: 209–211 (1991).
3. Baron-Epel O, Hernandez D, Jiang LW, Meiners S, Schindler M: Dynamic continuity of cytoplasmic and membrane compartments between plant cells. J Cell Biol 106: 715–721 (1988).
4. Beyer EC, Paul DL, Goodenough DA: Connexin43: A protein from rat heart homologous to a gap junction protein from liver. J Cell Biol 105: 2621-2629 (1987).
5. Beyer EC, Goodenough DA, Paul DL: The connexins, a family of releated gap junction proteins. In: Hertzberg EL, Johnson RG (eds) Gap Junctions, pp. 165–175. Alan Liss, New York (1988).
6. Bostwick DE, Dannenhoffer JM, Skaggs MI, Lister RM, Larkin BA, Thompson GA: Pumpkin phloem lectin genes are specifically expressed in companion cells. Plant Cell 4: 1539–1548 (1992).
7. Botha CEJ, Hartley BJ, Cross RHM: The ultrastructure and computer-enhanced digital image analysis of plasmodesmata at the Kranz mesophyll-bundle sheath interface of *Thermeda triandra* var *imberbis* (Retz) A. Camus in conventionally fixed leaf blades. Ann Bot 72: 255–261 (1993).
8. Burnell JN: An enzymatic method for measuring the molecular weight exclusion limit of plasmodesmata of bundle sheath cells of C_4 plants. J Exp Bot 39: 1575–1580 (1988).
9. Burt JM, Spray DC: Ionotropic agents modulate gap junctional conductance between cardiac myocytes. Am J Physiol 254: 1206–1210 (1988).
10. Cleland RE, Fujiwara T, Lucas WJ: Plasmodesmatal-mediated cell-to-cell transport in wheat roots is modulated by anaerobic stress. Protoplasma 178: 81–85 (1994).
11. Cote R, Thain JF, Fensom DS: Increase in electrical resistance of plasmodesmata of *Chara* induced by an applied pressure gradient across nodes. Can J Bot 40: 509–511 (1987).
12. Crow DS, Beyer EC, Paul DL, Kobe SS, Lau AF: Phosphorylation of connexin43 gap junction protein in uninfected and rous sarcoma virus-transformed mammalian fibroblasts. Mol Cell Biol 10: 1754–1763 (1990).
13. Dermietzel R, Hwang TK: Structural and molecular diversity of gap junctions. In: Robards AW, Lucas WJ, Pitts JD, Jongsma HJ, Spray DC (eds) Parallels in Cell to Cell Junctions in Plants and Animals, pp. 1–12. NATO ASI Series, Springer-Verlag, Berlin (1990).
14. Ding B, Turgeon R, Parthasarathy MV: Substructure of freeze-substituted plasmodesmata. Protoplasma 169: 28-41 (1992).
15. Ding B, Haudenshield JS, Hull RJ, Wolf S, Beachy RN, Lucas WJ: Secondary plasmodesmata are specific sites of localization of the tobacco mosaic virus movement protein in transgenic tobacco plants. Plant Cell 4: 915–928 (1992).
16. Doem RM, Oliver MJ, Beachy RN: The 30-kilodalton gene product of tobacco mosaic virus potentiates virus movement. Science 237: 389–394 (1987).
17. Epel BL, Bandurski RS: Tissue to tissue symplastic communication in the shoots of etiolated corn seedlings. Physiol Plant 79: 604–609 (1990).
18. Epel BL, Erlanger M: Light regulates symplastic communication in etiolated corn seedlings. Physiol Plant 83: 149–153 (1991).
19. Epel BL, Warmbrodt RD, Bandurski RS: Studies on the longitudinal and lateral transport of IAA in the shoots of etiolated corn seedlings. J Plant Physiol 140: 310–318 (1992).
20. Epel BL, Kotlizky G, Kuchuck B, Shurtz S, Yahalom A, Katz A: Characterization of plasmodesmatal-associated proteins. Abstracts, Second International Workshop on Basic and Applied Research in Plasmodesmatal Biology (Oosterbeek, Netherlands), pp. 3–5 (1992).
21. Epel B, Yahalom A, Katz A, Kotlizky G, Cohen L, Erlanger M, Levi R, Levi N, Van Lent J: Researches on plasmodesmatal composition and function. Biol Plant 36: S239 (1994).
22. Epel BL, Kuchuck B, Kotlizky G, Shurtz S, Erlanger M, Yahalom A: Isolation and characterization of plasmodesmata. In: Galbraith DW, Bourgue DP, Bohnert HJ (eds) Methods in Cell Biology: Plant Cell Biology. Academic Press, New York, in press.
23. Erwee MG, Goodwin PB: Characterization of *Egeria densa* Planch. leaf symplast. Inhibition of intercellular movement of fluorescent probes by group II ions. Planta 158: 320–328 (1983).
24. Fisher DB, Wu Y, Ku MSB: Turnover of soluble proteins

in the wheat sieve tube. Plant Physiol 100: 1433–1441 (1992).
25. Fujiwara T, Giesman-Cookmeyer D, Ding B, Lommel SA, Lucas WJ: Cell-to-cell trafficking of macromolecules through plasmodesmata potentiated by the red clover necrotic mosaic virus movement protein. Plant Cell 5: 1783–1794 (1993).
26. Giesman-Cookmeyer D, Lommel A: Alanine scanning mutagenesis of a plant virus movement protein identified three functional domains. Plant Cell 5: 973–982 (1993).
27. Goodwin PB: Molecular size limit for movement in the symplast of *Elodea* leaf. Planta 157: 124–130 (1983).
28. Graski S, de Feijter AW, Schindler M: Endoplasmic reticulum forms a dynamic continuum for lipid diffusion between contiguous soybean root cells. Plant Cell 5: 25–38 (1983).
29. Gunnings BES, Robards AW: Intercellular Communication in Plants: Studies on Plasmodesmata. Springer-Verlag, Berlin/Heidelberg (1976).
30. Hepler PK: Endoplasmic reticulum in the formation of the cell plate and plasmodesmata. Protoplasma 111: 121–133 (1982).
31. Hepler PK, Palevitz BA, Lancelle SA, McCauley MM, Lichtscheidl: Cortical endoplasmic reticulum in plants. J Cell Biol 96: 355–373 (1990).
32. Hull R: The movement of viruses in plants. Annu Rev Phytopath 27: 213–240 (1989).
33. Hunte C, Schnabl H, Traub O, Willecke K, Schultz M: Immunological evidence of connexin-like proteins in the plasma membrane of *Vicia faba* L. Bot Acta 105: 104–110 (1992).
34. Hunte C, Janssen M, Schulz M, Traub O, Willecke K, Schnabl H: Age-dependent modification and further localization of the cx 26-like protein form *Vicia faba* L. Bot Acta 106: 207–212 (1993).
35. Kikuyama M, Hara Y, Shimada K, Yamamoto K, Hiramoto Y: Intercellular transport of macromolecules in *Nitella*. Plant Cell Physiol 33: 413–417 (1992).
36. Kotlizky G, Shurtz S, Yahalom A, Malik Z, Traub O, Epel BL: Isolation and characterization of plasmodesmata embedded in clean maize cell walls. Plant J 2: 623–630 (1992).
37. Kumar NM, Gilula NB: Molecular biology and genetics of gap junction channels. Sem Cell Biol 3: 3–16 (1992).
38. Lucas WJ, Ding B, van der Schoot C: Plasmodesmata and the supracellular nature of plants. New Phytol 435–476 (1993).
39. Maule AJ: Virus movement in infected plants. Crit Rev Plant Sci 9: 457–473 (1991).
40. McCaskill D, Gershenzon J, Croteau R: Morphology and monoterpene biosynthetic capacities of secretory cell clusters isolated from grandular trichromes of peppermint (*Mentha piperita* L.). Planta 187: 445–454 (1992).
41. Meiners S, Schindler M: Immunological evidence for gap junction polypeptide in plant cells. J Biol Chem 262: 951–953 (1987).
42. Meiners S, Baron-Epel O, Schindler M: Intercellular communication-filling in the gaps. Plant Physiol 88: 791–793 (1988).
43. Meiners S, Schindler M: Characterization of a connexin homologue in cultured soybean cells and diverse plant organs. Planta 179: 148–155 (1989).
44. Meiners S, Xu A, Schindler M: Gap junction protein homologue from *Arabidopsis thaliana*: evidence for connexins in plants. Proc Natl Acad Sci USA 88: 4119-4122 (1991).
45. Moore PJ, Fenczik CA, Doem CM, Beachy RN: Developmental changes in plasmodesmata in transgenic tobacco expressing the movement protein of tobacco mosaic virus. Protoplasma 170: 115-127 (1992).
46. Monzer J, Kloth S: The preparation of plasmodesmata from plant tissue homogenates: access to the biochemical characterization of plasmodesmata-related polypeptides. Bot Acta 104: 82–84 (1991).
47. Mushegian AR, Koonin EV: The proposed plant connexin is a protein kinase-like protein. Plant Cell 5: 998–999 (1993).
48. Noueiry AO, Lucas WJ, Gilbertson RL: Two proteins of a plant DNA virus coordinate nuclear and plasmodesmal transport. Cell 76: 925–932 (1994).
49. Olensen P: The neck constriction in plasmodesmata: evidence for a peripheral sphincter-like structure revealed by fixation and tannic acid. Planta 144: 349–358 (1979).
50. Oparka KJ, Prior DAM: Direct evidence for pressure-generated closure of plasmodesmata. Plant J 2: 741–750 (1992).
51. Oparka KJ: Signalling via plasmodesmata – the neglected pathway. Sem Cell Biol 4: 131–138 (1993).
52. Oparka KJ, Prior DAM, Crawford JW: Behavior of plasma membrane, cortical ER and plasmodesmata during plasmolysis of onion epidermal cells. Plant Cell Environ 17: 163–171 (1994).
53. Reid RJ, Overall RL: Intercellular communication in *Chara*: factors affecting transnodal electrical resistance and solution fluxes. Plant Cell Environ 15: 507–517 (1992).
54. Reynhout JK, Lampe PD, Johnson RG: An activation of protein kinase C inhibits gap junction communication between culture bovin lens cells. Exp Cell Res 198: 337–342 (1992).
55. Robards AW, Lucas WJ: Plasmodesmata. Annu Rev Plant Physiol 41: 369–419 (1990).
56. Robinson-Beers K, Evert RF: Fine structure of plasmodesmata in mature leaves of sugarcane. Planta 184: 307–318 (1991).
57. Saez JC, Spray DC, Nairn AC, Hertzberg EL, Greengard P, Bennett MVL: cAMP increases junctional conductance and stimulates phosphorylation of the 27 kDa principal gap junction polypeptide. Proc Natl Acad Sci USA 83: 2473–2477 (1986).
58. Saez JC, Nairn AC, Czernik AJ, Spray DC, Hertzberg EL, Greengard P, Bennett MVL: Phosphorylation of con-

nexin32, a hepatocyte gap junction protein by cAMP-dependent protein kinase, protein kinase C, and Ca^{2+}/calmodulin-dependent protein kinase II. Eur J Biochem 192: 263–273 (1990).

59. Schulz M, Traub O, Knop M, Willecke K, Schnabl H: Immunofluorescent localization of connexin26-like protein at the surface of mesophyll protoplasts from *Vicia faba* L. and *Helianthus annuus* L. Bot Acta 105: 111–115 (1992).
60. Shepherd VA, Goodwin PB: Seasonal patterns of cell-to-cell communication in *Chara corallina* Klein ex Welld. I. Cell-to-cell communication in vegetative lateral branches during winter and spring. Plant Cell Environ 15: 137–150 (1992).
61. Shepherd VA, Goodwin PB: Seasonal patterns of cell-to-cell communication in *Chara corallina* Klein ex Welld. Cell-to-cell communication during the development of antheridia. Plant Cell Environ 15: 151–162 (1992).
62. Smith LM, Sabnis DD, Johnson RPC: Immunochemical localization of phloem lectin from *Cucurbita maxima* using peroxidase and colloidal-gold labels. Planta 170: 461–470 (1987).
63. Spray DC, Saez JC, Brosius D, Bennett MVL, Hertzberg EL: Isolated liver gap junction gating of transjunctional current is similar to that in intact pairs of rat hepatocytes. Proc Natl Acad Sci USA 83: 5494–5497 (1986).
64. Spray DC: Electrophysiological properties of gap junction channels. In: Robards AW, Lucas WJ, Pitts JD, Jongsma HJ, Spray DC (eds) Parallels in Cell to Cell Junctions in Plants and Animals, pp. 63–85. Nato ASI Series H 46. Springer-Verlag, Berlin/Heidelberg (1990).
65. Takeda A, Hashimoto E, Yamamura H, Shimazu T: Phosphorylation of liver gap junction protein by protein kinase C. FEBS Lett 210: 169–172 (1987).
66. Takeda A, Saheki S, Shimazu T, Takeuchi N: Phosphorylation of the 27 kDA gap junction protein by protein kinase C *in vitro* and in rat hepatocytes. J Biochem 106: 723–727 (1989).
67. Terry BR, Robards AW: Hydrodynamic radius alone governs the mobility of molecules through plasmodesmata. Planta 171: 145–157 (1987).
68. Tilney L, Cooke T, Connely P, Tilney M: The structure of plasmodesmata as revealed by plasmolysis, detergent extraction and protease digestion. J Cell Biol 112: 739–747 (1991).
69. Tomenius K, Clapham D, Meshi T: Localization by immunogold cytochemistry of the virus-coded 30 K protein in plasmodesmata of leaves infected with tobacco mosaic virus. Virology 160: 363-371 (1987).
70. Traub O, Look J, Paul D, Willecke K: Cyclic adenosine monophosphate stimulates biosynthesis and phosphorylation of the 26 kDa gap junction protein in cultured mouse hepatocytes. Eur J Cell Biol 43: 48–54 (1987).
71. Tucker EB: Translocation in the staminal hairs of *Setcreasea purpurea*. I. A study of cell ultrastructure and cell-to-cell passage of molecular probes. Protoplasma 113: 193–202 (1982).
72. Tucker EB: Inositol biphosphate and inositol triphosphate inhibit cell-to-cell passage of carboxyfluorescein in staminal hairs of *Setcreasea purpurea*. Planta 174: 358–363 (1988).
73. Tucker EB, Tucker JE: Cell-to-cell selectivity in staminal hairs of *Setcreasea purpurea*. Protoplasma 174: 36–44 (1993).
74. Tucker EB: Azide treatment enhances cell-to-cell diffusion in staminal hairs of *Setcreasea purpurea*. Protoplasma 174: 45–49 (1993).
75. Van Lent J, Wellink J, Goldbach R: Evidence for the involvement of the 58 K proteins in intercellular movement of cowpea mosaic virus. J Gen Virol 71: 219–223 (1990).
76. Van Lent J, Storms M, van de Meer F, Wellink J, Goldbach R: Tubular structures involved in movement of cowpea mosaic virus are also formed in infected cowpea protoplasts. J Gen Virol 72: 2615–2623 (1991).
77. Wang N, Fisher DB: The use of fluorescent tracers to characterize the post-phloem transport pathway in maternal tissues of developing wheat grains. Plant Physiol 104: 17–27 (1994).
78. Wellink J, van Lent JWM, Verver J, Sijen T, Goldbach RW, van Kammen AB: The cowpea mosaic virus M RNA-encoded 48-kilodalton protein is responsible for induction of tubular structures in protoplasts. J Virol 67: 3660–3664 (1993).
79. Willecke K, Traub O: Molecular biology of mammalian gap junctions. In: Demello WC (ed) How Cells Communicate. CRC Press Boca Raton, FL (1989).
80. Willecke K, Hennemann H, Herbers K, Heynkes R, Kozjek G, Look J, Stutenkemper R, Traub O, Winterhager E, Nicholson B: Molecular heterogeneity of gap junctions in different mammalian tissues. In: Robards AW, Lucas WJ, Pitts JD, Jongsma HJ, Spray DC (eds) Parallels in Cell to Cell Junctions in Plants and Animals, pp. 21–34. NATO ASI Series H 46. Springer-Verlag, Berlin/Heidelberg (1990).
81. Xu A, Meiners S, Schindler M: Immunological investigations of relatedness between plant and animal connexins. In: Robards AW, Lucas WJ, Pitts JD, Jongsma HJ, Spray DC (eds) Parallels in Cell to Cell Junctions in Plants and Animals, pp. 171–183. NATO ASI Series H 46. Springer-Verlag, Berlin/Heidelberg (1990).
82. Yahalom A, Warmbrodt RD, Laird DW, Traub O, Revel JP, Willecke K, Epel BL: Maize mesocotyl plasmodesmata proteins cross-react with connexin gap junction protein antibodies. Plant Cell 3: 407–417 (1991).
83. Yahalom A, Kutchuk, Katz A, Kotlizky G, Epel B: Studies on the phosphorylation of presumptive plasmodesmatal associated proteins in maize mesocotyl. Abstracts, Second International Workshop on Basic and Applied Research in Plasmodesmatal Biology (Oosterbeek, The Netherlands), pp. 6–9 (1992).

Plant Molecular Biology **26**: 1357–1377, 1994.

Genetic analyses of signalling in flower development using *Arabidopsis*

Kiyotaka Okada [1,*] and Yoshiro Shimura [1,2]
[1] *Division 1 of Gene Expression and Regulation, National Institute for Basic Biology, Okazaki 444, Japan (* author for correspondence);* [2] *Department of Biophysics, Faculty of Science, Kyoto University, Kyoto 606, Japan*

Received and accepted 18 June 1994

Key words: *Arabidopsis thaliana*, floral development, floral meristem, flower mutants, floral organ development, inflorescence meristem, signalling

Abstract

Flower development can be divided into four major steps: phase transition from vegetative to reproductive growth, formation of inflorescence meristem, formation and identity determination of floral organs, and growth and maturation of floral organs. Intercellular and intracellular signalling mechanisms must have important roles in each step of flower development, because it requires cell division, cell growth, and cell differentiation in a concerted fashion. Molecular genetic analysis of the process has started by isolation of a series of mutants with unusual flowering time, with aberrant structure in inflorescence and in flowers, and with no self-fertilization. At present more than 60 genes are identified from *Arabidopsis thaliana* and some of them have cloned. Although the information is still limited, several types of signalling systems are revealed. In this review, we summarize the present genetic aspects of the signalling network underlying the processes of flower development.

Introduction

Distinct from animals, the basic plan of the plant body is the continuous repetitive growth of a structural unit. The unit of aerial portion is a shoot which is composed of a stem, a meristem at the top of stem, and leaves continuously formed at the meristem. Flowers are considered to be specifically differentiated shoots, because they are composed of a short stem and four kinds of floral organs, sepals, petals, stamens and carpels. Floral organs are considered to be homologous organs of leaves, because their anatomy and developmental process are quite similar. We can easily notice that they are flat and laterally symmetric, and have the same branching pattern of major vascular bundles. The process of flower development can be divided into four major steps: phase transition from vegetative to reproductive growth, formation of inflorescence meristem, formation and identity determination of floral organs, and growth and maturation of floral organs. Several different types of signalling mechanism, between or within cells, must have important roles in each step of flower development, because each step requires cell division, cell growth and cell differentiation in a concerted fashion. Without any signalling system, it will be impossible to show the drastic changes of cells such as formation of organ primordia at the symmetrical positions. In order to unveil the signalling mechanism and its genetic regulatory systems, genetic, biochemical, anatomical and physiological analyses of the processes of flower development are being undertaken using *Arabidopsis thaliana* [52, 56, 63, 69], *Antirrhinum majus* [20, 21, 84], *Petunia hybrida* [103], *Zea mays* [45, 106] and other plant species. In this review, we summarize the present aspects of the genetic network underlying the flower development of *Arabidopsis*. Genes and mutants of *Arabidopsis* involved in the flower development are listed in Table 1. Results of studies on other plant species are almost consistent with those of *Arabidopsis*; therefore, it is strongly suggested that the signalling systems in flower development concluded from *Arabidopsis* studies are fundamental and common to other dicot plants, possibly to all angiosperms.

Transition from vegetative to reproductive growth

Floral initiation of *Arabidopsis* is known to be induced by several environmental factors: photoperiod, cold treatment (vernalization), gibberellin treatment, and nutrients. When wild-type plants are grown at 22 °C under continuous illumination or long-day conditions, floral buds appear after forming 6–8 rosette leaves. The shape of the apical meristem was observed with microscopes to examine the timing of transition from vegetative to reproductive growth. Swelling of the apical dome increases at about 10 days after imbibition, and the first flower primordium appears at about 15 days after imbibition [4, 58]. In the short-day condition, the timing of flowering delays to more than 40 days. When the seeds are sown in pots and incubated at 4 °C in low light, flowering time is reduced [55].

By characterizing the responses of the late-flowering mutants to photoperiod and vernalization, more than 40 mutants of 12 different loci were classified into 3 phenotypic classes [42, 46, 55]. The *fca*, *fpa*, *fve*, *fy* and *ld* mutants are sensitive to both photoperiod and vernalization, whereas the *co* (new name of *fg*) and *gi* (new name of *fb*) mutants are insensitive to both environmental conditions. The rest of the mutants, *fe*, *fd*, *fhc*, *ft* and *fwa*, are sensitive to daylength, but less sensitive or insensitive to cold treatment. In addition to the mutagen-induced mutants, additional late-flowering genes were identified by analyzing other ecotypes of *Arabidopsis*. Late-flowering ecotypes from Europe carry a late-flowering gene, named *FLA* [45]. Two late-flowering genes, *FRI* and *KRY*, are found in Scandinavian natural populations [18]. These mutants are responsive to vernalization. Based on the characterization of the mutants, a model of flowering pathways with genetic and environmental factors is proposed [42]. Attempts to clone the late-flowering genes are being made by chromosome walking or by tagging [4, 75]. Recently, the LUMINIDEPENDENS (LD) gene has been cloned [46]. The gene product is suggested to be a transcriptional factor because predicted amino acid sequence contains two nuclear localization signals and a glutamine-rich region which is known to work as transcription activation domain.

Another set of genes controlling the flowering time have been identified as mutants which initiate flowers earlier than wild type. *elf1* (for *early flowering 1*) and *elf2* mutants are sensitive to photoperiod, but *elf3* is not [112]. The *tfl* (for *terminal flower*) mutant shows early flowering and sensitivity to photoperiod as well as morphological deficiency in the inflorescence meristem identity (see next section) [88, 87, 112]. A mutant of a putative protein kinase gene, *tousled* (*tsl*), shows a late-flowering phenotype, as well as pleiotropic morphological defects (see next section). Another mutant, *embryonic flower* (*emf*), generates flowers immediately after germination bypassing vegetative growth [95]. It is suggested that the *EMF* gene activates vegetative growth and represses flower initiation.

Studies of *Arabidopsis* mutants have revealed the involvement of other factors in the process of flower initiation. First, a plant hormone, gibberellin, is crucial to flower initiation [110]. A gibberellin-insensitive mutant, *gai*, flowers early in a long-day condition, and shows a late-flowering phenotype in a short-day condition.

Table 1. Genes involved in the regulation of flower development in *Arabidopsis thaliana.*

Genes	Mutant phenotype	Proposed function of gene product	References
1. Genes regulating transition from vegetative to reproductive growth			
Late-flowering genes (*CO, FCA, FD, FE, FHA, FLA, FPA, FRI, FT, FVE, FY, GI*)	delayed flowering		4, 18, 42, 45, 55, 75
LUMINIDEPENDENS (*LD*)	delayed flowering	contains nuclear-localization signals and glutamine-rich region/transcription factor	46
EARLY FLOWERING1.2.3 (*ELF1,2,3*)	early flowering		112
TERMINAL FLOWER (*TFL*)	early flowering/conversion of inflorescence meristems to floral meristems		3, 86, 87, 113
TOUSLED (*TSL*)	delayed flowering/reduced number of floral organs/curling of cauline leaves	protein kinase	80
EMBRYONIC FLOWER (*EMF*)	generation of inflorescence meristem without vegetative growth		95
GIBBERELLIN DEFICIENT (*GA1*)	late flowering/defective in gibberellin biosynthesis	*ent*-kaurene synthase	94, 110
GIBBERELLIN INSENSITIVE (*GAI*)	late flowering/insensitive to gibberellin		110
PHYTOCHROME A (*FRE/HY8/PHY2/PHYA*)	long hypocotyl in continuous far-red light/less sensitive to night break	phytochrome A	64, 70, 76, 109
LONG HYPOCOTYL3 (*HY3/PHYB*)	early flowering/long hypocotyl	phytochrome B	29, 32, 76
LONG HYPOCOTYL2 (*HY2*)	early flowering/long hypocotyl	biosynthesis of chromophore	29, 32
PHOSPHOGLUCOMUTASE (*PGM*)	late flowering/deficient of starch synthesis	phosphoglucomutase	13
STARCH OVER-PRODUCTION (*SOP*)	late flowering/deficient of starch degradation		14
2. Genes involved in formation of inflorescence meristem			
TERMINAL FLOWER (*TFL*)	conversion of inflorescence meristems to floral meristems		see above
LEAFY (*LFY*)	partial conversion of floral meristems to inflorescence meristems	transcription factor	35, 81, 107
APETALA1 (*AP1*)	production of axillary flowers in flowers/homeotic conversion of sepals to leaves/absense of petals in strong mutants	MADS box protein transcription factor	10, 31, 38, 53, 54

Table 1. (Continued)

Genes	Mutant phenotype	Proposed function of gene product	References
APETALA2 (*AP2*)	homeotic conversion of sepals to leaves and petals to stamens in weak mutants/homeotic conversion of sepals to carpels/petals and stamens are mostly absent in strong mutants	negative regulator of AGAMOUS	6, 8, 41, 43, 66, 69
CAULIFLOWER (*CAL*)	single mutant is phenotypically wild/conversion of flower meristems to inflorescence meristems in combination with *ap1* mutations		10
CLAVATA1 (*CLV1*)	large meristem/more carpels		17, 23, 47, 66
PIN-FORMED (*PIN*)	forming no floral buds or deformed flowers	a component of auxin polar transport system?	28, 30, 67
FILAMENTOUS FLOWER (*FIL*, *Fl*-54)	after forming several flowers, development of floral meristem stops immaturely, but recovers later/few floral organs/lack of anthers/partial homeotic conversions of sepals to petals, stamens to petals, stamens to carpels		41
ACAULIS1 (*ACL1*)	few flowers/short inflorescence axis		12
AUXIN RESISTANCE1 (*AXR1*)	short inflorescence axis/bushy inflorescence/resistant to auxin/no root gravitropism		27, 48
AUXIN RESISTANCE2 (*AXR2*)	short inflorescence/resistant to auxin		99
3. Genes regulating formation and identity determination of floral organs			
APETALA1 (*AP1*)	see above		see above
APETALA2 (*AP2*)	see above		see above
APETALA3 (*AP3*)	homeotic conversion of petals to sepals and stamens to carpels	MADS box protein transcription factor	6, 8, 38, 39, 68
PISTILLATA (*PI*)	homeotic conversion of petals to sepals and stamens to carpels	MADS box protein transcription factor	6, 8, 28, 34, 66
AGAMOUS (*AG*)	homeotic conversion of stamens to petals/indeterminate floral meristem/no pistil	MADS box protein transcription factor	6–8, 26, 59, 111
SUPERMAN (*SUP*)	more stamens/small pistil	regulator of PI and AP3 expression	9, 82
FILAMENTOUS FLOWER (*FIL*, *Fl*-54)	see above		see above
TOUSLED (*TSL*)	see above		see above
FASCIATA1, 2 (*FAS1, 2*)	fewer petals & stamens/narrow sepals & petals/stem flattening		47
CLAVATA1 (*CLV1*)	see above		see above

Table 1. (Continued)

Genes	Mutant phenotype	Proposed function of gene product	References
4. Genes regulating floral organ growth and maturation			
BICAUDAL (*BIC, Fl-89*)	two clumps of stigmatic papillae and two horny structures at the top of pistils/narrow sepals & petals		41
FIDDLEHEAD (*FDH*)	fusion of buds, floral organs and leaves		50, 51
FILAMENTOUS FLOWER (*FIL, Fl-54*)	stamens lacking anthers		see above
ANTHERLESS (*AT*)	stamens lacking anthers or with anthers converted to sepals		15, 16
male-sterile mutants (*MS1, MS2, MS3, MS4, MS5, MS7, MS8, MS15, MSK, MSW, MSX, MSZ, BM3*)	deficient pollen development		1, 16, 25, 77
MALE-STERILE H (*MSH*)	no dehiscence of anthers		25
APT (*BM3*)	abortive pollen development after meiosis/lack of APRT activity	adenine phosphoribosyl-transferase (APRT)	61, 77
POP1 (POLLEN-PISTIL INTER-ACTIONS)	lack of extracellular pollen coat, tryphine		73
QUARTET 1, 2 (*QRT1, 2*)	pollen grains released in tetrads		74
BELL1 (*BEL1*)	abortive ovules/conversion of ovules to carpel-like structures		60, 79
SHORT INTEGUMENTS1 (*SIN1*)	abortive ovules		79
OVULE MUTANT-2,3 (*OVM2, 3*)	defective integument development		78

Flowering is promoted by cold treatment, whereas gibberellin-deficient mutants, *ga1–3*, flower late in a long-day condition, is unable to flower in a short-day condition, and is insensitive to vernalization. The *ga1* mutant dramatically responds to sprayed GA_3 and produces flowers as promptly as GA-treated wild-type plants. These results indicate that the process of flowering requires a certain level of GA and GA-mediated signal transduction systems. Second, phytochromes A and B are involved in the response to daylength. Phytochrome A-deficient mutants, *phyA*, are insensitive to photoperiod [76], whereas phytochrome B-deficient mutants, *phyB* and *hy3*, and a chromophore-defective mutant, *hy2*, are responsive to photoperiod and initiate flowers earlier than wild-type plants [29, 32, 76]. Interestingly, *hy3* and *hy2* mutations compensate for the late-flowering phenotype when combined with some of the late-flowering mutations, because flowering time becomes shorter in double mutants of *hy3* or *hy2* with one of the later-flowering mutants, *fca*, *fwa* or *co*, under continuous light [32]. Third, starch degradation and movement in plant body is considered to be responsible to flower initiation, because retardation of flowering is observed in starchless mutant, phosphoglucomutase-defective (*pgm*), and in starch degradation-deficient mutant, starch overproducer (*sop*) [13, 14].

Control of flowering initiation is a multifactorial process including many genes which may regulate a network of signalling pathways. The search of key factors such as florigen and anti-florigen has not been successful so far. Although more detailed physiological, genetic and molecular studies are required to clarify the nature of signalling pathways, a recent hypothesis postulates at least two promotive pathways and two inhibitory pathways in the process [5].

Formation of the inflorescence meristem

Immediately after reproductive growth has started, the vegetative shoot is converted to an inflorescence. Floral buds are formed one by one at the apex, and the axis elongates. The inflorescence meristem has three functions: (1) formation of floral meristem on its flanks; (2) formation of axis tissues at its bottom; and (3) maintenance of the inflorescence meristem at the top. Recent mutational studies have revealed that different sets of genes regulate the three functions. The early-flowering gene, *TFL* (for *terminal flower*), is known to govern the 2nd and the 3rd functions, because *tfl* mutant shows conversion of the indeterminate inflorescence meristem to a flower, and stop of the growth of inflorescence [3, 86]. In several cases, a large flower with an excess number of floral organs are formed at the terminal part. Lateral branches also terminate in a flower. The first function of the inflorescence meristem is regulated by several genes. In *leafy* mutants (*lfy*), floral meristem develops into an inflorescence carrying aberrant flowers without petals and stamens [35, 81, 107]. The cloned *LFY* gene shows extensive sequence homology with the *Antirrhinum majus* gene, *FLORICAULA* (*FLO*), which has the same mutant phenotype to *LFY* [19, 107]. LFY and FLO are suggested to be transcription factors, because they have a proline-rich domain and an acidic domain characteristic of transcriptional activating domains. It is suggested that the *LFY* gene is involved in the decision of identity and maintenance of the floral meristem, because *LFY* is expressed in young flower primordia, but not in inflorescence meristems [107]. Double mutant analysis with *lfy* and other flower mutants have revealed that (1) LFY interacts with APETALA1 (AP1) and APETALA2 (AP2); and (2) the *lfy* mutation is epistatic to *pistillata* (*pi*) and *agamous* (*ag*) mutations [35]. Further genetic analysis of multiple mutants has led to draw a simple model of inflorescence development [87]. According to this model, three genes, AP1, AP2 and LFY, promote growth of the floral meristem and repress development of the inflorescence meristem. Instead, TFL promotes inflorescence meristem and suppresses floral meristem. TFL may repress expression of *AP1*, *AP2* and *LFY* genes, whereas the three genes may negatively control *TFL* expression (Fig. 1). Mutual activation of *AP1*, *AP2* and *LFY* is also proposed by other groups [10, 83]. The model of the positive and negative interactions is supported by the observations of *in situ* hybridization. Expression of *LFY* and *AP1* genes normally observed in floral buds of early

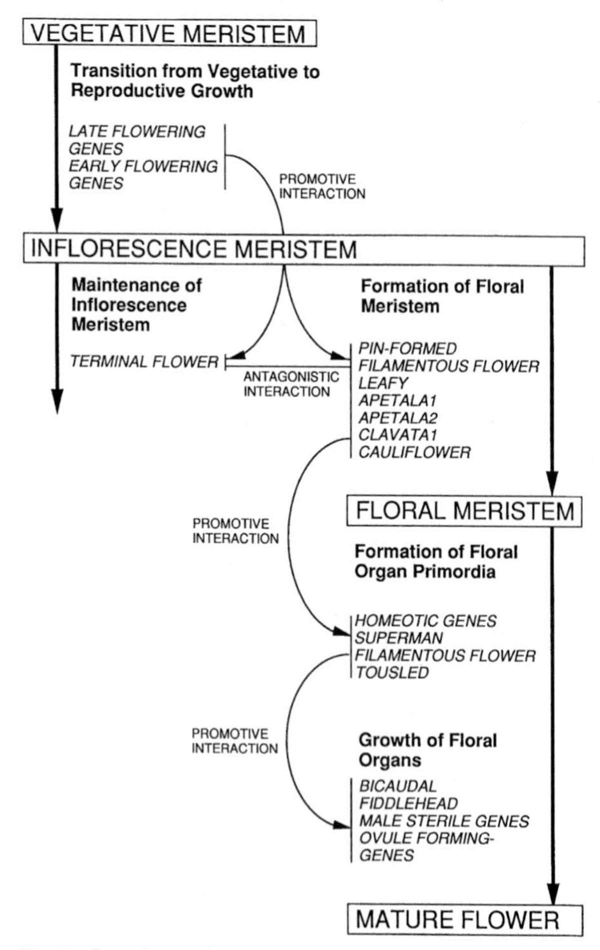

Fig. 1. Genetic regulatory network in the process of flower development. Genes involved in each step are listed in italics. Thick vertical arrows indicate temporal change of the fate of meristemic cells.

stages is also observed in the converted floral meristem at the top of inflorescence axis of the *tfl* mutant [10, 31, 107].

Two genes, *CAULIFLOWER* (*CAL*) and *CLAVATA1* (*CLV1*), are also involved in establishing and maintenance of floral meristem. Although strains carrying a single mutation in *CAL* gene have normal inflorescence, double mutants having mutations in both *CAL* and *AP1* in homozygous state convert a floral meristem to an inflorescence meristem with highly branched flowers, just like the famous vegetable cauliflower [10]. Both *AP1* and *LFY* are positively regulated by *CAL*. Interestingly, the flower morphology of the triple mutant, *ap1 cal tfl*, and that of another triple mutant, *ap1 cal lfy*, are indistinguishable from that of the *ap1 tfl* double mutant and that of the *ap1 lfy* double mutant, respectively [10]. The epistasis of *tfl* and *lfy* mutations to *cal* in an *ap1* mutant background indicates the close interaction of the gene products in the process of floral meristem establishment. *CLV1* is known to have a function that determines the size of vegetative, inflorescence and floral meristems. The large floral meristem of the *clv1* mutant generates a flower with a big pistil of 4 carpels [17, 23, 47, 66]. When *clv1* is combined with the *lfy*, *ap1* or *tfl* mutant, the *clv1* mutation enhances the phenotypes of single mutants, indicating that *CLV1* is involved in determining meristem identity by interacting with *LFY*, *AP1* and possibly with *TFL* [10, 87].

In addition to the five genes described above, we would like to emphasize that two more genes, *PIN-FORMED* (*PIN*) and *FILAMENTOUS FLOWER* (*FIL*), are playing important roles in establishing floral meristems. Phenotype of *pin* mutants are pleiotropic: they show structural abnormalities in embryo, phyllotaxis, leaves, inflorescence and flowers [30, 67]. The typical inflorescence of the *pin* mutant looks like the tip of a pin because it fails to generate floral buds (Fig. 2A). The naked apex often shows segments which may correspond to internodes between flowers. Because the inflorescence axes continuously grow at a normal rate, it is suggested that the *pin* mutant is missing the first of the three functions of the inflorescence meristem described above, formation of floral meristem, but is still keeping the other two functions, formation of axis tissues and maintenance of the meristem. In some cases, however, deformed flowers are formed at the top of axes (Fig. 2B). The flowers often lack stamens, but have a few wide petals or many narrow petals, and usually have a pistil at the center. In extreme cases, a pistil-like structure with no other organs is formed. The pistils are sterile, because ovules are not fully developed. In addition to the structural defects of the inflorescence, the *pin* mutants have aberrant phenotypes in other organs. The position of the two cotyledons are not symmetric like that of wild type. In some cases two cotyledons are fused to one. Rosette leaves are often wide with a vein branched at the base. Phyllotaxis on the inflorescence axes is also abnormal; two lateral shoots are often formed in 'opposite' positions in the mutant, whereas lateral shoots are formed in 'alternate' positions in wild type [30, 67].

We found wild-type plants grown on agar medium containing an auxin polar transport inhibitor, 9-hydroxyfluorene-9-carboxylic acid (HFCA) or N-(1-naphthyl)phthalamic acid (NPA), to be a phenocopy of the *pin* mutant [67]. Because almost all of the pleiotropic phenotypes of the mutant were observed in the drug-treated plants, we assumed that the activity of auxin polar transport is reduced in the *pin* mutant, and that reduction of the activity is the cause of structural abnormalities. This assumption was supported by the result of direct measurement of the auxin transport activity. The activity of the inflorescence of the mutant is about 10% of that of wild type [66]. Further supporting evidence for the assumption was obtained from *in vitro* embryo culture of Indian mustard, *Brassica juncea* [49]. Abnormal embryos with fused cotyledons similar to these of the *pin* mutant were induced by auxin polar transport inhibitors. These results indicate that the auxin polar transport system plays an important role in the process of floral meristem establishment. In order to examine at which stage(s) of flower bud development the normal auxin transport system is required, a series of shift experiments have been done (Junichi Ueda *et al.*,

Fig. 2. Phenotypes of wild type and mutant *Arabidopsis* flowers. A. Inflorescence of the wild type (right) and of the *pin-formed* (*pin*) mutant (left). B. A deformed flower of the *pin* mutant. C. Inflorescence of the *filamentous flower* (*fil*) mutant. Phases 1, 2, 3 represent the four fertilized siliques at the bottom, filaments and sepal-like structures in the middle region, and a cluster of flowers at the top, respectively. D. Inflorescence of the *fil tfr* (*terminal flower*) double mutant. Only the phase 1 flowers are formed. E.

A flower of the *fil ap1* (*apetala1*) double mutant. A flower is converted to an inflorescence. F. A flower of the weak allele of the *ap1* mutant, the *Fl-1* strain. Four normal petals are formed. G. A flower of the strong allele of the *ap1* mutant, *ap1–1*. No petals are formed. H. A flower of the weak allele of the *ap2* (*apetala2*) mutant, *ap2–1*. Four sepals and four petals are converted to leaf-like structures and staminoid petals, respectively. I. A flower of the strong allele of the *ap2* mutant, *ap2–3*. The flower has two sepals converted to carpels, lacks petals and stamens. J. A flower of the weak *ap1*/weak *ap2* double mutant. A flower is converted to an inflorescence with flowers lacking petals. K. A flower of the heavy *ap2 ag* double mutant. Staminoid petals are formed.

manuscript in preparation). Wild-type plants germinated and grown for several days under continuous light on agar medium containing HFCA were transferred onto agar medium without the drug. After incubation on the medium for 3 weeks, the structure of the first flower was examined. More than 80% of the first flower was normal when the plants were transferred at 4 days after germination. However, when transferred at 6 days after germination, nearly 80% of the plants developed an inflorescence of a complex form composed of two regions; the lower region shows *pin* mutant-specific segments with no floral buds, but the upper region bears normal floral buds. This structure indicates that development of the first flower was repressed, but the repression did not work for the development of later flowers. A series of experiments indicated that the presence of transport inhibitor before 4 days after germination do not influence floral bud formation, but that the presence of the drug at 6 days or more after germination represses the normal development of the first floral meristem. Shift experiments in the reverse direction help to determine the period that the drug is effective for development of the first flower. When wild-type plants germinated and grown on agar medium without the drug were transferred onto medium containing the drug at 10 days after germination, nearly 70% of plants showed pin-formed inflorescence; however, when transferred at 14 days after germination, all of the first flowers were normal. Therefore, by combining the results of shift experiments, it is strongly indicated that formation of the first floral meristem is inhibited by auxin polar transport inhibitors if the drug is supplied from 5 to 12 days after germination. This 'critical 7 days' includes the timing of transition from vegetative to reproductive growth, and precedes the timing of formation of the first floral meristem of visible size. Therefore, reduction of auxin polar transport activity may have crucial effects on floral meristem formation at very early stages, even if considering that the drug remains for several days in plant body after removal of the drug from the medium.

We have also noticed that *FILAMENTOUS FLOWER* (*FIL*) gene is regulating establishment and development of floral meristem. A *fil* mutant (former name is *Fl-54*) shows several structural defects in flowers and inflorescences [44]. Although wild-type plants bear flowers in a spiral arrangement at more or less constant intervals along the inflorescence axis, the mutant has three different phases of flower formation under growth conditions of constant temperature, illumination, nutrients and water supply (Fig. 2C). After flower initiation, the *fil* mutant elongates an inflorescence axis normally, and forms about 10 flowers (phase 1). The flowers show several structural abnormalities in shape, number and position of floral organs and have long peduncles. Numbers of sepals, petals and stamens are decreased. Most of the stamens lack anther sacs. In some flowers, the top of filaments is white and flat like a petal, or covered by stigmatic papillae. Homeotic conversion is also observed in petals and sepals. Some petals carry pollen sacs. Margins of sepals are sometimes white, showing partial conversion to petals. In flowers lacking some floral organs, the remaining organs are not positioned symmetrically. Pistils are less affected by the mutation and set seeds. After forming the phase 1 flowers, formation of floral buds is stopped at the inflorescence meristem, and instead, more than 10 filaments and more than 10 sepal-like structures are formed (phase 2). Because the sepal-like structure is facing the inflorescence axis as the abaxial sepal, and sits on a peduncle-like short stem, we suppose this structure corresponds to a floral bud whose development is stopped at very early stages after forming only the abaxial sepal. It would be worth noting that the abaxial sepal is located at the outermost on the floral meristem, and develops first in the 16 floral organs in a flower [63, 89]. The filament would be a peduncle without flowers, because in a rare case, a small floral bud is formed at the top of a filament. After forming the phase 2 cluster, a cluster of flowers is formed again (phase 3). Structural defects of the phase 3 flowers are heavier than those of the phase 1 flowers. A sepal-like structure is often attached to the long peduncle, suggesting that the phase 3 flowers originate from the sepal-like structure, but the developmental process is not stopped. Usually, petals are missing, and sepals and stamens are converted to filaments with no extra organs at the top. Pistils look normal and fertile. The phase shift of inflorescence is also observed in lateral shoots as well as in the main stem. Phases are changed in an order of 1–2–3, but phase 1 is omitted in some lateral shoots coming later. In summary, the *FIL* gene may have two different functions: one in early stages of floral bud development and the other in formation and growth of floral organ primordia. A lack of the first function results in the formation of filaments or sepal-like structures instead of flowers. A defect of the second function causes a decreased number of floral organs and misdifferentiation of the organs. The phase shift observed in the mutant inflorescence strongly suggests involvement of a cell-cell communication system in the early stage of floral meristem development. Although synchronization of the phases in main and side shoots of individual plant is not clear, one can postulate that some diffusible factor(s) are participating in the communication. One possible explanation of the phase shift is that the amount of the factor(s) or the amount of receptor molecules of the factor(s) changes to high-low-high as plants grow, and phase 2 results when the amount decreases to a level lower than a threshold level enough to support the development of floral meristem to a mature flower. In order to examine the interactions with products of other genes regulating establishment and growth of the floral meristem, a series of double mutants were constructed (Komaki *et al.* and Okada *et al.*, manuscripts in preparation). Changes of the three phases are observed when combined with *lfy*, *ap1*, *ap2* and *clv1* mutants. The *fil tfl* double mutant has only the phase 1 flowers, because growth of the inflorescence meristem terminates before shifting to phase 2 (Fig. 2D). In combination with *fil* and a strong allele of the *ap1* mutant, *ap1–1*, phase 1 flowers are converted to inflorescence meristem bearing many flowers (Fig. 2E). When combined with weak alleles of *ap1* or *ap2* mutant, the flower structure of double mutants resembles that of a heavy allele of the *ap1* or *ap2* mutant, respectively. *FIL* also interacts with *CLV1*, because inflorescences of the *fil clv1* double mutant are fasciated and form a clump of meristemic tissue covered by stigmatic papillae when aged more than 6 weeks. These results show that *FIL* controls the process of floral meristem formation in combination with *AP1*, *AP2* and *CLV1*.

Several other genes are reported to control growth of inflorescence axis. *ACAULIS1* (*ACL1*) has a very short inflorescence axis and bears less than 10 flowers. Because cells of the mutant at the internode are small but the number of cells is not changed, a short axis is considered to result from the defect in cell elongation process. The size of the inflorescence meristem of the *acl1* mutant is reduced to about half of that of wild type [102]. A defect in internode elongation is also observed in two auxin-resistant mutants, *axr1* and *axr2*. The short inflorescence of the *axr1* mutant is caused by a decreased number of cells [48]. However, in the case of the *axr2* mutant, cell elongation is repressed [99]. The heavy allele of the *axr1* mutant is bushy. It generates about 30 thin-stemmed primary inflorescences [27], indicating that auxin has a role in activation of branching and growth of lateral shoots.

Formation and identity determination of floral organs

A flower originates from a floral meristem which is composed of three layers of undifferentiated cells (L1, L2 and L3) [104, 105]. Analysis of chimeras between tomato strains differing in number of carpels has shown that number and size of carpels are not determined by the genotype of cells in the outer cell layers, L1 and L2, but by the genotype of L3 cells [96]. Because coordinated cell division and expansion between cells of the three layers are necessary to generate floral organs, this result indicates the importance of cell-cell communication in the vertical direction in the floral meristem. Intercellular communication in the horizontal direction within the meristem is also known to be indispensable for organizing the floral organs in a flower. If an immature floral meristem is cut along the median line and has been incubated for several days, partly or completely divided flowers are produced [24, 33, 90]. The structure of the resulting flower is dependent on the stage of flower development at which the microsurgery is done. When the floral meristem is bisected before forming sepals, two almost complete flowers are regenerated. However, if bisected when sepal primordia are already formed but primordia of other organs are not formed yet, two flowers with normal numbers of petals, stamens and carpels are formed, but the number of sepals is not compensated. A series of median bisection experiments suggests that the identity of the single flower is based on cell-cell communication between the center and peripheral regions of meristem, and that the position of floral organs is fixed on the basis of the cell-cell communication system. The molecular nature of the communication system in either vertical or horizontal direction in the floral meristem is not clarified yet. Although similar experiments have not been repeated in *Arabidopsis*, the same communication system is undoubtedly involved in the process of floral bud formation in this plant. Further investigations are necessary to identify genes controlling the system.

Recent genetic and molecular studies of homeotic mutants of *Arabidopsis* and *Antirrhinum* have provided a beautiful model explaining how the floral organ identity is determined [8, 10, 22, 31]. This model, designated the ABC model [22, 52], is based on three assumptions: (1) the floral meristem is simplified to be composed of four concentric regions, named whorl 1, 2, 3 and 4 from outside to inside; (2) three groups of homeotic genes are expressed in different whorls, namely group A genes in whorls 1 and 2, group B genes in whorls 2 and 3, and group C genes in whorls 3 and 4; (3) the fate of organ primordia is determined by the combination of homeotic genes expressed in the whorl where the primordia are located, namely expression of group A genes leads to sepals, expression of group A and B genes to petals, expression of group B and C genes to stamens, and expression of group C genes to carpels. *Arabidopsis* genes identified to correspond to

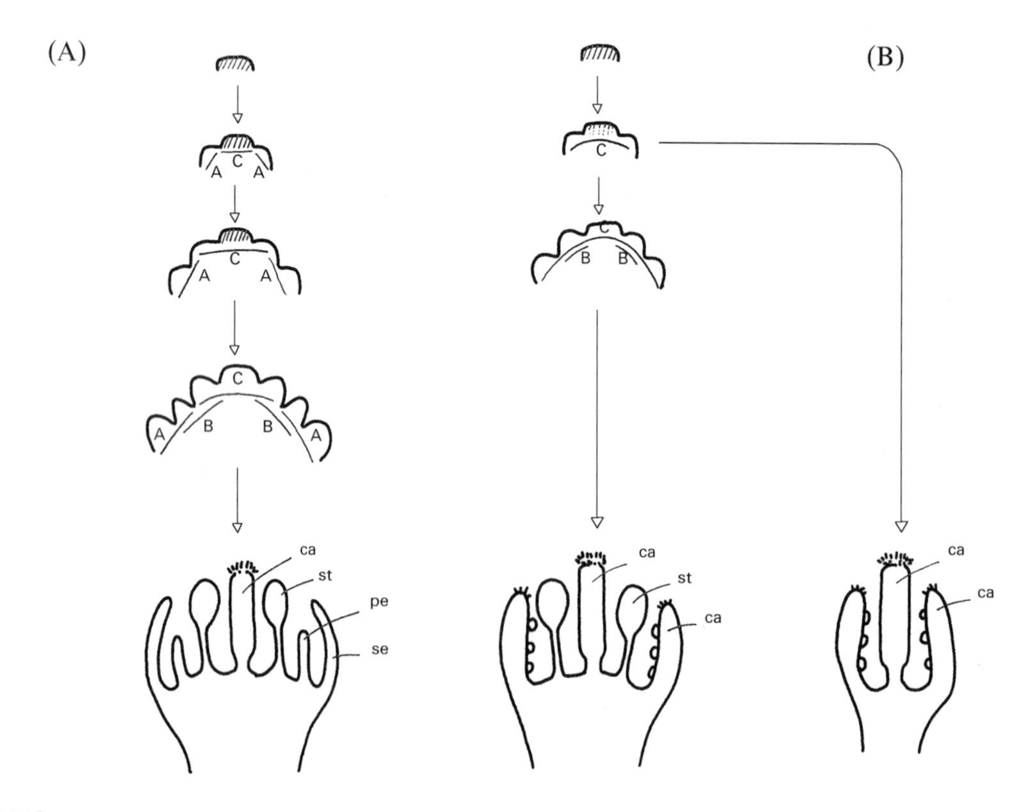

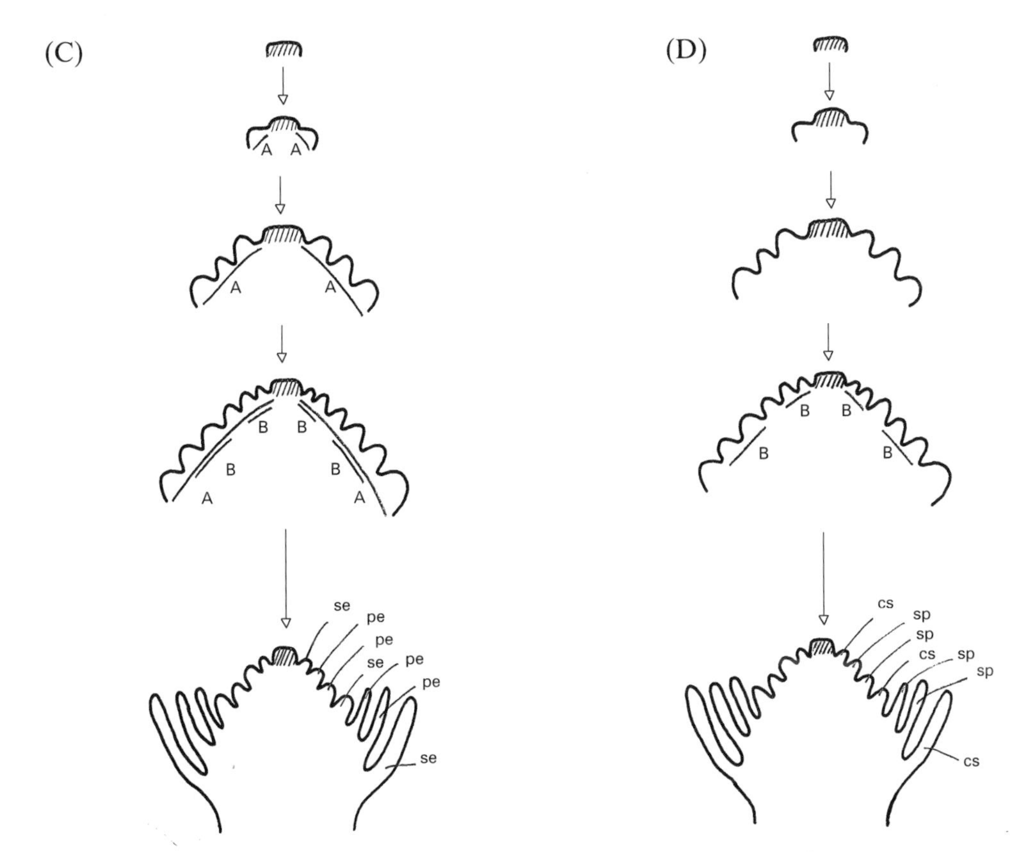

Fig. 3. Schematic representations of the modified model of whorl formation and floral organ identity. Bars A, B and C in the figures indicate the region where groups A, B and C of organ identity genes are expressed, respectively. Hatched regions indicate meristemic cells. Figures at the bottom represent the final structure of flowers. se, sepal; pe, petal; st, stamen; ca, carpel; sp, staminoid petal; cs, carpeloid sepal. (A). Development of wild-type flowers. At early stages, group A genes are expressed at the marginal region, and group C genes are expressed at the central region of the floral meristem. When the amount of group C gene products exceeds the threshold level, proliferation of meristemic cells is halted, and the carpel primordia are formed at the top of meristem. At a later stage, group B genes are expressed in the 2nd and 3rd whorls. The combination of group A, B and C genes determines the identity of organ primordia. (B). A flower of a group A gene mutant. In case of heavy mutants of group A genes, a heavy allele of *ap1* or a heavy allele of *ap2*, expression of group C genes is enhanced and distributed to the whole regions of meristem. Therefore, the amount of group C gene products increases faster and reaches the threshold level before the meristemic cells proliferate to form 4 whorls. If proliferation of the meristemic cells stops after 2 whorls have been formed, organ primordia develop to carpels. If 3 whorls are formed, group B genes are expressed at the intermediate whorl. Organ primordia on the intermediate whorl develop to stamens. The same situation is expected in the weak *ap1*/weak *qp2* double mutant. (C). A flower of a group C gene mutant. Lack of group C gene expression leads the meristemic cells to proliferate eternally, and to make many whorls indeterminately. Expression of group A genes also continues. If it is assumed that group B genes are expressed discontinuously, floral organs repeat the pattern of sepals-petals-petals. (D). A flower of a heavy allele of an *ap2-ag* double mutant. Many whorls are formed owing to the lack of group C genes. If group B genes are expressed in the same discontinuous pattern as that of C, organs repeat the pattern of carpeloid sepals-staminoid petals-staminoid petals.

the A, B, and C groups are *AP1* and *AP2* (group A), *APETALA3* (*AP3*) and *PISTILLATA* (*PI*) (group B), and *AGAMOUS* (*AG*) (group C), which are known to cause several different types of homeotic conversion of floral organs [7, 8, 22, 31]. A similar set of genes have also been studied in *Antirrhinum majus*, namely *SQUAMOSA* (*AP1* homologue), *DEFICIENS* (*AP3* homologue), *GLOBOSA* (*PI* homologue) and *PLENA* (*AG* homologue) [11, 12, 20, 22, 37, 84, 91, 101]. Interestingly, all of the homeotic genes listed above except *AP2* encode a protein containing a conserved motif, the MADS box [11, 37, 39, 40, 53, 54, 68, 84, 91, 92, 101]. The MADS-box-containing proteins are considered to be transcription factors, because well-characterized MADS-box proteins of human (SRF) and yeast (MCM1) promote expression of other genes as transcription factors [65, 71], and because some of the MADS-box genes, SRF, MCM1, *DEFICIENS* and *AG*, are known to bind to a DNA sequences containing a consensus sequence, 5′-$CC(A/T)_6GG$ – 3′ (termed CArG box) [36, 62, 65, 72, 85, 88]. *AP2* gene product, however, does not contain the MADS box, but encodes a putative transcription factor, because this protein contains a nuclear localization signal and a highly acidic serine-rich domain which is found in several DNA-binding proteins [69].

In situ hybridization experiments using probes of the homeotic genes, *AP1*, *AP3*, *PI* and *AG*, showed a unique pattern of localized expression as expected from the model [7, 28, 31, 39, 108, 111]. However, expression of *AP2* is not localized in whorls 1 and 2 as predicted from the model but in all regions of floral meristem, even in leaf, stem and root at low levels [68]. It is suggested that *AP2* behaves as group A genes only in combination with some other gene product(s); a possible candidate is *AP1*. The localized expression of the homeotic genes is explained by negative and positive regulatory interactions between the genes, namely (1) *AP3*, *PI* and *AG* are activated by *AP1* and *LFY*; (2) *AP1* is repressed by *AG*; (3) *AP3* and *PI* are repressed by another gene, *SUPERMAN* (*SUP*) [9, 82]. The regulatory circuit is supported by the ectopic expression experiments using transgenic plants [40, 54, 59].

These lines of results are consistent and strongly support the ABC model, at least assumptions (2) and (3). However, as for assumption (1), the model does not explain how the number of whorls is determined. In addition, flower structure of some homeotic mutants is not fully explained from the model. As mentioned in Bowman *et al.* [7, 8], it is not easy to understand why strong alleles of *ap2* mutants lack organs in whorls 2 and 3 (petals and stamens) (Fig. 2I) and nevertheless the missed whorls are recovered in the strong *ap2 ag* double mutant (Fig. 2K). The recovery of the second whorl organs in the *ap2 ag* double mutant is explained as a result of ectopic expression of AG, but its mechanism is not shown [7, 8, 30]. In the case of the strong *ap1 ag* double mutant, petals are recovered in flowers of aged plants; nevertheless, the strong *ap1* allele usually lacks flowers [30]. In addition, it is also difficult to explain why petals are missing in the weak *ap1*/weak *ap2* double mutant, although both weak *ap1* and *ap2* mutants have four petals (Fig. 2F, H, J) [10, 38, 41, 43].

Figure 3 shows a modified version of the organ identity model which can explain the difficulties described above. This model is based on three additional assumptions. First, full expression of group C genes stops proliferation of undifferentiated meristemic cells at the top of the floral meristem, and converts the meristemic cells to carpel primordia. Second, because expression of group A and C genes (or activation of the gene products) is mutually repressed, expression level of group C genes as well as distribution of cells expressing group C genes are controlled by a balance of expression between group A and group C genes. Third, whorls are formed from the meristemic cells from the marginal to the central region of floral meristem. Following to the model of wild-type flower development, group A genes are expressed at the marginal region, whereas group C genes are expressed at the central region of floral meristem at very early stages of flower development (Fig. 3A). As floral meristem grows, whorls are formed one by one from the marginal region of the meristem. The expression level of group C genes increases at the central region, and finally reaches a threshold level high enough to stop further proliferation of floral meristemic cells. At this stage, normal proliferation of meristemic cells results in the formation of four whorls. Group B genes express at later stages than group A and C genes. The identity of primordia formed on each whorl is determined by a combination of the expression of group A, B, and C genes. In case of a heavy allele of group A mutants, expression of group C genes is not restricted to the inner region of meristem. The enhanced expression of group C genes will stop the proliferation of meristemic cells at an earlier stage than that of wild-type flowers, so that the meristem stops its growth before it has formed four whorls. As a result, only two or three whorls are formed (Fig. 2I). In case of flowers of three whorls, expression of group B may be permitted at the intermediate whorl. Primordia formed on both the outermost and the innermost whorl differentiate to carpels, and primordia on the middle whorl to stamens (Fig. 3B). A similar mechanism will work in case of strong alleles of the *ap1* mutant and weak *ap1*/weak *ap2* double mutants (Fig. 2). This model indicates that whorls 2 and 3 present in wild-type flowers are not deleted in the mutant flowers, but only two or three whorls are formed, and the whorls differentiate as whorls 1, 3 (in case of flowers of three whorls) and 4 of wild type. On the contrary, in the *ag* mutant, a lack of group C gene function leads to continuous proliferation of the meristemic cells (Fig. 3C). Indeterminate growth of the meristem results in formation of many whorls. Assuming that group B genes will be expressed at every second and third whorls following to a mechanism somewhat similar to that of wild type, the reiterated pattern of sepals-petals-petals is formed. In strong *ap2 ag* double mutants, a lack of group C gene function may lead to the formation of many whorls as shown in the *ag* mutant. If it is assumed that group B genes are expressed at the same regions as those postulated for the *ag* mutant, floral organs are formed in the reiterated pattern of carpeloid sepals-staminoid petals-staminoid petals (Fig. 3D). In case of a strong *ap1* (*ap1-1*) *ag* double mutant, petals are recovered in flowers of aged plants, although flowers of young plants do not have petals or petal-like organs, but instead bracts with an axillary shoot are repeatedly formed [31, 38]. The failure of recovery of petal-related organs in the young plants of the double mutant can be explained as that the initial activation of *LFY* by the *AP1* gene product is abolished by the *ap1* mutation, and that floral meristem is partially converted to leafy inflorescence. The modified model (Fig. 3) shows that the number of whorls is determined by the balance of group A and C homeotic genes, and that mutations of the genes affect the number of whorls as well as the identity of organ primordia. We believe this model to explain the architecture of flowers more easily than the original ABC model. It would be worthy of further evaluation.

Several other genes are reported to control the number of floral organs. As described in the preceding section, flowers of the *fil* mutant have decreased numbers of sepals, petals and/or stamens [41]. The position of remaining organs deviates from the symmetric arrangement observed in wild-type flowers. FIL is possibly required for correct arrangement of floral organ primordia. Mutants of the putative protein kinase gene, *TOUSLED* (*TSL*), form flowers of decreased number of sepals, petals, and stamens [80]. Electron microscopic analysis revealed that the symmetric pattern of initiation of sepal primordia is altered. Pleiotropic phenotypes of *tsl* and *fil* are somewhat similar, but the two genes are not allelic, because *TSL* is mapped on chromosome 5 [80] whereas *FIL* is on chromosome 2 (Komaki *et al.*, unpublished results). Reduction of floral organs is also reported in stem-fasciation mutants, *fasciata1* (*fas1*), and *fasciata2* (*fas2*) [47]. Mutations of *CLV1* cause enlargement of apical and floral meristems, and form four carpel primordia and an extra meristemic tissue at the innermost whorl [17, 23, 47, 66].

Floral organ growth and differentiation

The organ identity genes activate new sets of genes that promote growth and differentiation of the floral organ primordia. This step is most likely mediated by cascades of transcriptional regulation, because the organ identity genes encode proteins having characteristics of transcriptional factors. Searches of the target genes are being done using several different procedures, such as screening of organ-specific cDNA clones, differential screening of cDNA libraries of wild-type and homeotic mutant flowers, and direct biochemical or immunochemical isolation of the DNA fragment-MADS box protein complex. It is still difficult to have a general scope of the regulatory cascade, because only a few putative target genes have been isolated from several plant species. An extracellular leucine-rich repeat protein gene, *FIL2*, is specifically expressed in stamens and carpels of *Antirrhinum majus* [93]. Because the *FIL2* gene is not expressed in a homeotic mutant, *deficiens*, and has a CArG box, the binding motif of MADS box proteins, in its promoter region, it is strongly suggested that *FIL2* is a target of the *DEFICIENS* gene product. The ascorbate oxidase gene of *Brassica napus* has a CArG box in its promoter [2]. This gene is specifically expressed in pollens at later stages of the maturation process. In *Petunia hybrida*, a family of DNA-binding protein containing two repeats of a zinc finger motif is expressed specifically in floral organs; namely, *EPF1* is mainly expressed in petals, and *EPF2–5* is transcribed in petals and stamens, whereas *EPF2–7* is expressed in sepals and petals at high level but in other floral organs at low level [97, 98]. The organ-specific pattern of gene expression and the presence of the CArG box in the promoter region of *EPF1* and *EPF2–5* genes strongly suggest that the *EPF* genes are positively regulated by the MADS-box-containing floral organ identity genes. Because EPF1 has a DNA-binding activity in the promoter region of *EPSPS*, the 5-enolpyruvylshikimate-3-phosphate synthase gene, which is a key enzyme in the pigment synthesis pathway in petals, it is postulated that a transcription regulatory cascade, MADS-box-containing petal identity genes (groups A and B) – *EPF1* – *EPSPS*, is working in petals of *Petunia* [97]. DNA-binding analyses of SRF and MCM1, MADS-box proteins of human and yeast, demonstrate that the proteins bind to the target genes as ternary complex with accessory proteins [62, 100]. The MADS-box-containing plant genes are also postulated to act as transcription regulators in the form of a multicomponent protein complex.

There are several other genes that control growth and maturation of floral organs. *BICAUDAL* (*BIC*, a new name of *Fl*-89) has a function in pistil development [41]. Flowers of the *bic* mutant have a pistil with two clumps of stigmatic papillae and two horn-shaped structures at the top. Sepals and petals of the mutant are narrower than those of the wild type. In the process of normal growth of pistils, two carpel primordia develop to a cylinder-formed pistil as a result of coordinate growth and fusion of the carpels. The opening at the top of young pistils is closed at the later stage [66, 89]. The *bic* mutant is defective in the fusion of the two carpels, and in the closing at the top of pistils. The loss of fusion process of carpels is drastically enhanced in the *bic ap1* double mutant, indicating an interaction of the two gene products in pistil development (Komaki *et al.*, unpublished results). Another interesting gene, *FIDDLEHEAD* (*FDH*), is reported to be possibly involved in the ontogenetic fusion process [50]. Floral organs and leaves of the *fdh* mutant fuse and form structures reminiscent of fern fiddleheads, by altering epidermal competence to adhere to cell walls. *FDH* appears to be a regulatory gene controlling the carpel-specific fusion program.

Another important process in floral organ maturation is development of gametes. Male-sterile mutants are classified into four groups: defective to form functional stamens, deficient in microsporogenesis, lacking tryphine in pollens, and deficient in dehiscence of anthers [16]. Two genes are reported to generate stamens with deformed anthers, *ANTHERLESS* [15, 16] and *FIL* [41], mutant flowers of these genes have stamens with no anthers or anthers converted to sepals. A number of male-sterile (MS) genes control microsporogenesis [16, 25]. One of the MS genes, *MS2*, has been shown to have nucleotide sequences homologous to an open reading frame found in the wheat mitochondrial genome, though the relationship between *MS2* and the cytoplasmic male sterility remains unknown [1]. Abortion of pollen development also results from a mutation of adenine phosphoribosyl transferase, a key enzyme in the salvage pathway of purine synthesis [61, 77]. Two genes are known to control formation of outer walls of pollen grains. One gene, *POP1*, is responsible for the synthesis of extracellular pollen coat, lipoidic tryphine [73]. Failing germination of the mutant pollens on the stigma demonstrates that tryphine is required for the cell signalling in the pollen-stigma interactions. Other genes, *QRT1 & 2*, cause cell wall fusion of four pollen grains originated from the single mother cell [74]. This mutation will permit tetrad analysis to be available in *Arabidopsis*. In addition, *MSH* is involved in the dehiscence of anthers [25]. Mutants defective in ovule development are also studied. Four genes are reported to be involved in the control of integument development [78, 79]. The *ovm3* mutant lacks both outer and inner integuments, whereas the *bell* mutant lacks inner integument. The structure of ovules is abnormal in the *sin1* mutant because of aberrant cell divisions in the integuments. Embryo sac of *ovm2* mutant looks to be replaced by nuclear cells. At the later stages of pistil development, cells of the outer integuments of *bell* mutant sometimes develop to a carpel-like structure [60]. Similar homeotic conversion of ovules to carpels is observed in an *ap2* mutant, *ap2–6* [60].

Summary and future perspectives

Although they are a scattered set of observations, the *Arabidopsis* mutants summarized in this review demonstrate a flow of genetic information leading the process of flower development. As shown in Fig. 1, the flow branches after inflorescence meristem has been formed. One branch aims to maintain the inflorescence meristem, and the other one supports formation and development of floral meristem. The floral meristem develops to a floral bud bearing a set of floral organs. Because the concerted patterns of cell division, cell elongation and cell differentiation are observed in every step of the process, intercellular and intracellular signalling systems are undoubtedly underlying the regulatory network of the genes controlling the process. Although the information is still limited, at least four different types of signalling systems are revealed. First, cascades of transcriptional regulation play major roles in the steps of formation of floral meristem and of formation and development of floral organ primordia, because many putative transcription factors are involved in the steps. Second, plant hormones may function as signals of intercellular communication in flower initiation, and in floral bud formation. Gibberellin has been shown to be involved in the control of flowering time. The auxin transport system is required in the step of formation of floral meristem. Third, as shown in a mutant of a protein kinase gene, *TOUSLED*, protein kinase is involved in the determination of floral organ primordia, as well as in the growth of leaves and roots. Fourth, the regulatory pathway of flower initiation may share some steps with the signalling pathways of photomorphogenesis, because several phytochrome mutants show a delayed-flowering phenotype.

As shown in Table 1, more than sixty genes have been identified to be involved in flower development. However, the molecular function of the genes mostly remain to be clarified. There would be four different approaches for the future study of signalling pathways. The first is cloning of the mutated genes using efficient procedures such as gene tagging, subtraction, or orthodox chromosome walking. Functions of the isolated genes will be examined *in vivo* by analyzing the phenotype of transgenic plants which overexpress or ectopically express the genes. The second is isolation of new types of mutants or new alleles of known genes. A wide spectrum of mutations is required to identify the genetic regulatory networks of signalling pathways. The third is the examination of the phenocopy of the mutants. Treating wild-type plants with drugs of known action or with artificial conditions will help us to postulate the molecular function of mutated genes. The fourth is integration of the accumulated data of genetic, physiological, anatomical, biochemical and molecular studies on flower development. Close exchange of information between researchers of different approaches is required.

Acknowledgements

We thank Dr Junichi Ueda of the University of Osaka Prefecture and Dr Hiroshi Takatsuji of the National Institute of Agrobiological Resources for communicating results prior to publication, and present and past members of our laboratory who worked with us in the flower project, in particular Drs Callum J. Bell, Sumie Ishiguro, Toshiro Ito, Masako K. Komaki, Yoichi Ono, Shinichiro Sawa, Hideaki Shiraishi and Azusa Yano. We also thank Ms Hideko Nonaka, Ms Akiko Kawai and Ms Yuko Takahashi for technical assistance. The research in our laboratory was supported by grants from the Japanese Ministry of Education, Science and Culture, by funds from the Human Frontier Science Program, and by funds from the Joint Studies Program for Advanced Studies from the Science and Technology Agency of Japan.

References

1. Aarts MGM, Dirkse WG, Stiekema WJ, Pereira A: Transposon tagging of a male sterility gene in *Arabidopsis*. Nature 363: 715–717 (1993).
2. Albani D, Sardana R, Robert LS, Altosaar I, Arnison PG, Fabijanski SF: A *Brassica napus* gene family which shows sequence similarity to ascorbate oxidase is expressed in developing pollen. Molecular characterization and analysis of promoter activity in transgenic tobacco plants. Plant J 2: 331–342 (1992).
3. Alvarez J, Guli CL, Yu X-H, Smyth DR: *Terminal flower*: A gene affecting inflorescence development in *Arabidopsis thaliana*. Plant J 2: 103–116 (1992).
4. Araki T, Komeda Y: Analysis of the role of the late-flowering locus, *GI*, in the flowering of *Arabidopsis thaliana*. Plant J 3: 231–239 (1993).
5. Bernier G, Havelange A, Houssa C, Petitjean A, Lejeune P: Physiological signals that induce flowering. Plant Cell 5: 1147–1155 (1993).
6. Bowman JL, Smyth DR, Meyerowitz EM: Genes directing flower development in *Arabidopsis*. Plant Cell 1: 37–52 (1989).
7. Bowman JL, Drews GN, Meyerowitz EM: Expression of the *Arabidopsis* floral homeotic gene *AGAMOUS* is restricted to specific cell types late in flower development. Plant Cell 3: 749–758 (1991).
8. Bowman JL, Smyth DR, Meyerowitz EM: Genetic interactions among floral homeotic genes of *Arabidopsis*. Development 112: 1–20 (1991).
9. Bowman JL, Sakai H, Jack T, Weigel D, Mayer U, Meyerowitz EM: *SUPERMAN*, a regulator of floral homeotic genes in *Arabidopsis*. Development 114: 599–615 (1992).
10. Bowman JL, Alvarez J, Weigel D, Meyerowitz EM, Smyth DR: Control of flower development in *Arabidopsis thaliana* by *APETALA1* and interacting genes. Development 119: 721–743 (1993).
11. Bradley D, Carpenter R, Sommer H, Hartley N, Coen E: Complementary floral homeotic phenotypes result from opposite orientations of a transposon at the *plana* locus of *Antirrhinum*. Cell 72: 85–95 (1993).
12. Carpenter R, Coen ES: Floral homeotic mutations produced by transposon-mutagenesis in *Antirrhinum majus*. Genes Devel 4: 1483–1493 (1990).
13. Casper T, Huber SC, Somerville C: alterations in growth, photsynthesis, and respiration in a starchless mutant of *Arabidopsis thaliana* (L.) deficient in chloroplast phosphoglucomutase activity. Plant Physiol 79: 11–17 (1985).
14. Casper T, Lin T-P, Kakefuda G, Benbow L, Preiss J, Somerville C: Mutants of *Arabidopsis* with altered regulation of starch degradation. Plant Physiol 95: 1181–1188 (1991).
15. Chaudhury AM, Craig S, Farrell L, Bloemer K, Dennis ES: Genetic control of male-sterility in higher plants. Aust J Plant Physiol 19: 419–425 (1992).
16. Chaudhury AM: Nuclear genes controlling male fertility. Plant Cell 5: 1277–1283 (1993).
17. Clark SE, Running MP, Meyerowitz EM: *CLAVATA1*, a regulator of meristem and floral development in *Arabidopsis*. Development 119: 397–418 (1993).
18. Clark JH, Dean C: Mapping FRI, a locus controlling flowering time and vernalization response in *Arabidopsis thaliana*. Mol Gen Genet 242: 81–89 (1994).
19. Coen ES, Romero JM, Doyle S, Elliott R, Murphy G, Carpenter R: *floricaula*: a homeotic gene required for flower development in *Antirrhinum majus*. Cell 63: 1311–1322 (1990).
20. Coen ES: The role of homeotic genes in flower development and evolution. Annu Rev Plant Physiol Plant Mol Biol 42: 241–279 (1991).
21. Coen ES, Carpenter R: The metamorphosis of flowers. Plant Cell 5: 1175–1181 (1993).
22. Coen ES, Meyerowitz EM: The war of whorls: genetic interactions controlling flower development. Nature 353: 31–37 (1991).
23. Crone W, Lord EM: Flower development in the organ number mutant, *clavata1*, of *Arabidopsis thaliana* (Brassicaseae). Am J Bot 80: 1419–1256 (1993).
24. Cusick F: Studies of floral morphogenesis. I. Median bisections of flower primordia in *Primula bulleyana* Forrest. Trans Roy Soc Edinb 63: 153–166 (1956).
25. Dawson J, Wilson ZA, Aarts MGM, Braithwaite AF, Briarty LG, Mulligan BJ: Microspore and pollen development in six male-sterile mutants of *Arabidopsis thaliana*. Can J Bol 71: 629–638 (1993).
26. Drews GN, Bowman JL, Meyerowitz EM: Negative regulation of the *Arabidopsis* homeotic gene *AGAMOUS* by *APETALA2* product. Cell 65: 991–1002 (1991).
27. Estelle MA, Somerville C: Auxin-resistant mutants of *Arabidopsis thaliana* with an altered morphology. Mol Gen Genet 206: 200–206 (1987).

28. Goto K, Meyerowitz EM: Function and regulation of the *Arabidopsis* floral homeotic gene *PISTILLATA*. Genes Devel 8: 1548–1560 (1994).
29. Goto N, Kumagai T, Koornneef M: Flowering reponses to light-breaks in photomorphogenic mutants of *Arabidopsis thaliana*, a long-day plant. Physiol Plant 83: 209–215 (1991).
30. Goto N, Katoh N, Kranz AR: Morphogenesis of floral organs in *Arabidopsis*: predominant carpel formation of the *pin-formed* mutant. Jap J Genet 66: 551–567 (1991).
31. Gustafson-Brown C, Savidge CB, Yanofsky MF: Regulation of the *Arabidopsis* floral homeotic gene *APETALA1*. Cell 76: 131–143 (1994).
32. Halliday KJ, Koornneef M, Whitelam GC: Phytochrome B and at least one other phytochrome mediate the asselerated flowering response of *Arabidopsis thaliana* L. to low red/far-red ratio. Plant Physiol 104: 1311–1315 (1994).
33. Hicks GS, Sussex IM: Organ regeneration in sterile culture after median bisection of the flower primordia of *Nicotiana tabacum*. Bot Gaz 132: 350–363 (1971).
34. Hill JP, Lord EM: Floral development in *Arabidopsis thaliana*: a comparison of the wild type and the homeotic pistillata mutant. Can J Bot 67: 2922–2936 (1989).
35. Huala E, Sussex IM: *LEAFY* interacts with floral homeotic genes to regulate *Arabidopsis* floral development. Plant Cell 4: 901–913 (1992).
36. Huang H, Mizukami Y, Hu Y, Ma H: Isolation and characterization of the binding sequences for the product of the *Arabidopsis* floral homeotic gene *AGAMOUS*. Nucl Acids Res 21: 4769–4776 (1993).
37. Huijser P, Klein J, Lönnig W-E, Meijer H, Saedler H, Sommer H: Bracteromania, an inflorescence anomaly, is caused by the loss of function of the MADS-box gene squamosa in *Antirrhinum majus*. EMBO J 11: 1239–1249 (1992).
38. Irish VF, Sussex IM: Function of the *apetala*-1 gene during *Arabidopsis* floral devleopment. Plant Cell 2: 741–753 (1990).
39. Jack T, Brockman LL, Meyerowitz EM: The homeotic gene *APETALA3* of *Arabidopsis thaliana* encodes a MADS box and is expressed in petals and stamens. Cell 68: 683–697 (1992).
40. Jack T, Fox GL, Meyerowitz EM: Arabidopsis homeotic gene *APETALA3* ectopic expression: transcriptional and posttranscriptional regulation determine floral organ identity. Cell 76: 703–716 (1994).
41. Komaki MK, Okada K, Nishino E, Shimura Y: Isolation and characterization of novel mutants of *Arabidopsis thaliana* defective in flower development. Development 104: 195–203 (1988).
42. Koornneef M, Hanhart CJ, van der Veen JH: A genetic and physiological analysis of late flowering mutants in *Arabidopsis thaliana*. Mol Gen Genet 229: 57–66 (1991).
43. Kunst L, Klenz JE, Martinez-Zapater J, Haughn GW: *AP2* gene determines the identity of perianth organs in flowers of *Arabidopsis thaliana*. Plant Cell 1: 1131–1135 (1989).
44. Langdale JA, Irish EE, Nelson TM: Action of the Tunicate locus on maize floral development. Devel Genet 15: 176–187 (1994).
45. Lee I, Bleecker A, Amasino R: Analysis of naturally occuring late flowering in *Arabidopsis thaliana*. Mol Gen Genet 237: 171–176 (1993).
46. Lee I, Aukerman MJ, Rore SL, Lohman KN, Michaelis,SD, Weaver LM, John MC, Feldmann KA, Amasino RM: Isolation of *LIMINIDEPENDENS*: a gene involved in the control of flowering time in *Arabidopsis*. Plant Cell 6: 75–83 (1994).
47. Leyser HMO, Furner IJ: Characterization of three shoot apical meristem mutants of *Arabidopsis thaliana*. Development 116: 397–403 (1992).
48. Lincoln C, Britton JH, Estelle M: Growth and development of the *axr1* mutants of *Arabidopsis*. Plant Cell 2: 1071–1080 (1990).
49. Liu C-L, Xu Z-h, Chua N-H: Auxin polar transport is essential for the establishment of bilateral symmetry during early plant embryogenesis. Plant Cell 5: 621–630 (1993).
50. Lolle SJ, Cheung AY, Sussex IM: *Fiddlehead*: an *Arabidopsis* mutant constitutively expressing an organ fusion program that involves interactions between epidermal cells. Devel Biol 152: 383–392 (1992).
51. Lolle SJ, Cheung AY: Promiscuous germination and growth of wild type pollen from Arbidopsis and related species on the shoot of the *Arabidopsis* mutant, *fiddlehead*. Devel Biol 155: 250–258 (1993).
52. Ma H: The unfolding drama of flower development: recent results from genetic and molecular analyses. Genes Devel 8: 745–756 (1994).
53. Mandel MA, Gustafson-Brown C, Savidge B, Yanofsky MF: Molecular characterization of the *Arabidopsis* floral homeotic gene *APETALA1*. Nature 360: 273–277 (1992).
54. Mandel MA, Bowman J, Kempin SA, Ma H, Meyerowitz EM, Yanofsky MF: Manipulation of flower structure in transgenic tobacco. Cell 71: 133–143 (1992).
55. Martinez-Zapater JM, Somerville CR: Effect of light quality and vernalization on late-flowering mutants of *Arabidopsis thaliana*. Plant Physiol 92: 770–776 (1990).
56. Meyerowitz EM: *Arabidopsis*, a useful weed. Cell 56: 263–269 (1989).
57. Meyerowitz EM, Smyth DR, Bowman JL: Abnormal flowers and pattern formation in floral development. Development 106: 209–217 (1989).
58. Miksche JP, Brown JAM: Development of vegetative and floral meristems of *Arabidopsis thaliana*. Am J Bot 52: 533–537 (1965).
59. Mizukami Y, Ma H: Ectopic expression of the floral homeotic gene *AGAMOUS* in transgenic *Arabidopsis* plants alters floral organ identity. Cell 71: 119–131 (1992).

60. Modrusan Z, Reiser L, Feldmann KA, Fischer RL, Haughn GW: Homeotic transformation of ovules into carpel-like structures in *Arabidopsis*. Plant Cell 6: 333–349 (1994).
61. Moffatt B, Somerville C: Positive selection for male-sterile mutants of *Arabidopsis* lacking adenine phosphoribosyl transferase activity. Plant Physiol 86: 1150–1154 (1988).
62. Mueller CGF, Nordheim A: A protein domain conserved between yeast MCM1 and human SRF directs ternary complex formation. EMBO J 10: 4219–4229 (1991).
63. Müller A: Zur Charakterisierung der Blüten und Infloreszenzen von *Arabidopsis thaliana* (L.) Heynh. Kulturpflanze 9: 364–393 (1961).
64. Nagatani A, Reed JW, Chory J: Isolation and initial characterization of *Arabidopsis* mutants that are deficient in phytochrome A. Plant Physiol 102: 269–277 (1993).
65. Norman C, Runswick M, Pollock R, Treisman R: Isolation and properties of cDNA clones encoding SRF, a transcription factor that binds to the *c*-fos serum response element. Cell 55: 989–1003 (1988).
66. Okada K, Komaki MK, Shimura Y: Mutational analysis of pistil structure and development of *Arabidopsis thaliana*. Cell Diff Devel 28: 27–38 (1989).
67. Okada K, Ueda J, Komaki MK, Bell CJ, Shimura Y: Requirement of the auxin polar transport system in early stages of *Arabidopsis* floral bud formation. Plant Cell 3: 677–684 (1991).
68. Okamoto H, Yano A, Shiraishi H, Okada K, Shimura Y: Genetic complementation of a floral homeotic mutation, *apetala3*, with an *Arabidopsis thaliana* gene homologous to *DEFICIENS* of *Antirrhinum majus*. Plant Mol Biol 26: 465–472 (1994).
69. Okamuro JK, den Boer BGW, Jofuku KD: Regulation of *Arabidopsis* flower development. Plant Cell 5: 1183–1193 (1993).
70. Parks BM, Quail PH: *hy8*, a new class of *Arabidopsis* long hypocotyl mutants deficient in functional phytochrome A. Plant Cell 5: 39–48 (1993).
71. Passmore S, Maine GT, Elble R, Christ C, Tye B-K: *Saccharomyces cerevisiae* protein involved in plasmid maintenance is necessary for mating of *MAT* a cells. J Mol Biol 204: 593–606 (1988).
72. Passmore S, Elble R, Tye B-K: A protein involved in minichromosome maintenance in yeast binds a transcriptional enhancer conserved in eukaryote. Genes Devel 3: 921–935 (1989).
73. Preuss D, Lemieux B, Yen G, Davis RW: A conditional sterile mutation eliminates surface components from *Arabidopsis* pollen and disrupts cell signalling during fertilization. Genes Devel 7: 974–985 (1993).
74. Preuss D, Rhee SY, Davis RW: Tetrad analysis possible in *Arabidopsis* with mutation of the *QUARTET* (*QRT*) genes. Science 264: 1458–1460 (1994).
75. Putterill J, Robson F, Lee K, Coupland G: Chromosome walking with YAC clones in *Arabidopsis*: isolation of 1700kb of contiguous DNA on chromosome 5, including a 300kb region containing the flowering-time gene CO. Mol Gen Genet 239: 145–157 (1993).
76. Reed JW, Nagatini A, Elich TD, Fagan M, Chory J: Phytochrome A and phytochrome B have overlapping but distinct functions in *Arabidopsis* development. Plant Physiol 104: 1139–1149 (1994).
77. Regan SM, Moffatt BA: Cytochemical analysis of pollen development in wild-type *Arabidopsis* and a male-sterile mutant. Plant Cell 2: 877–889 (1990).
78. Reiser L, Fischer RL: The ovule and the embryo sac. Plant Cell 5: 1291–1301 (1993).
79. Robinson-Beers K, Pruitt RE, Gasser CS: Ovule development in wild-type *Arabidopsis* and two female-sterile mutants. Plant Cell 4: 1237–1249 (1992).
80. Roe JL, Rivin CJ, Sessions RA, Feldmann KA, Zambryski PC: The tousled gene in *A. thaliana* encodes a protein kinase homolog that is required for leaf and flower development. Cell 75: 939–950 (1993).
81. Schultz EA, Haughn GW: LEAFY: a homeotic gene that regulates inflorescence development in *Arabidopsis*. Plant Cell 3: 771–781 (1991).
82. Schultz EA, Pickett FB, Haughn GW: The *FLO10* gene product regulates the expression domain of homeotic genes *AP3* and *PI* in *Arabidopsis* flowers. Plant Cell 3: 1221–1237 (1991).
83. Schultz EA, Haughn GW: Genetic analysis of the floral initiation process (FLIP) in *Arabidopsis*. Development 119: 745–765 (1993).
84. Schwarz-Sommer Zs, Huijser P, Nacken W, Saedler H, Sommer H: Genetic control of flower development by homeotic genes in *Antirrhinum majus*. Science 250: 931–936 (1990).
85. Schwarz-Sommer Zs, Hue I, Huijser P, Flor P, Hansen R, Tetens F, Lönnig W-E, Saedler H, Sommer H: Characterization of the Antirrhinum floral homeotic MADS-box gene *deficiens*: evidence for DNA binding and autoregulation of its persistent expression throughout flower development. EMBO J 11: 251–263 (1992).
86. Shannon S, Meeks-Wagner DR: A mutation in the *Arabidopsis TFL1* gene affects inflorescence meristem development. Plant Cell 3: 877–892 (1991).
87. Shannon S, Meeks-Wagner DR: Genetic interactions that regulate inflorescence development in *Arabidopsis*. Plant Cell 5: 639–655 (1993).
88. Shiraishi H, Okada K, Shimura Y: Nucleotide sequences recognized by the AGAMOUS MADS domain of *Arabidopsis thaliana in vitro*. Plant J 4: 385–398 (1993).
89. Smyth DR, Bowman JL, Meyerowitz EM: Early flower development in *Arabidopsis*. Plant Cell 2: 755–767 (1990).
90. Soetiarto SR, Ball E: Ontogenetical and experimental studies of the floral apex of *Portulaca grandiflora*. 2. Bisection of the meristem in successive stages. Can J Bot 47: 1067–1081 (1969).

91. Sommer H, Beltrán J-P, Huijser P, Pape H, Lönnig W-E, Saedler H, Schwarz-Sommer Zs: *Deficiens*, a homeotic gene involved in the control of flower morphogenesis in *Antirrhinum majus*: the protein shows homology to transcription factors. EMBO J 9: 605–613 (1990).
92. Sommer H, Nacken W, Beltran P, Juijser P, Pape H, Hansen R, Flor P, Saedler H, Schwarz-Sommer Zs: Properties of *deficiens*, a homeotic gene involved in the control of flower morphogenesis in *Antirrhinum majus*. Development (Suppl) 1: 169–175 (1991).
93. Steinmayr M, Motte P, Sommer H, Saedler H, Schwarz-Sommer Zs: FIL2, an extracellular leucine-rich repeat protein, is specifically expressed in *Antirrhinum* flowers. Plant J 5: 459–467 (1994).
94. Sun TP, Goodman HM, Ausubel FM: Cloning the *Arabidopsis* GA1 locus by genomic subtraction. Plant Cell 4: 119–128 (1992).
95. Sung ZR, Belachew A, Shunong B, Bertrand-Garcia R: EMF, an *Arabidopsis* gene required for vegetative shoot development. Science 258: 1645–1647 (1992).
96. Szymkowiak EJ, Sussex IM: The internal meristem layer (L3) determines floral meristem size and carpel number in tomato periclinal chimeras. Plant Cell 4: 1089–1100 (1992).
97. Takatsuji H, Mori M, Benfey FN, Ren L, Chua N-H: Characterization of a zinc finger DNA-binding protein expressed specifically in Petunia petals and seedlings. EMBO J 11: 241–249 (1992).
98. Takatsuji H, Nakamura N, katsumoto Y: A new family of zinc finger proteins in petunia: structure, DNA sequence recognition, and floral organ-specific expression. Plant Cell 6: 947–958 (1994).
99. Timpte CS, Wilson AK, Estelle M: Effects of the *axr2* mutation of *Arabidopsis* on cell shape in hypocotyl and inflorescence. Planta 188: 271–278 (1992).
100. Treisman R, Marais R, Wynne J: Spacial flexibility in ternary complexes between SRF and its accessory proteins. EMBO J 11: 4631–4640 (1992).
101. Tröbner W, Ramirez L, Motte P, Hue I, Juijser P, Lönnig W-E, Saedler H, Sommer H, Schwarz-Sommer Zs: *GLOBOSA*: a homeotic gene which interacts with *DEFICIENS* in the control of *Antirrhinum* floral organogenesis. EMBO J 11: 4693–4704 (1992).
102. Tsukaya H, Naito S, Rédei GP, Komeda Y: A new class of mutations in *Arabidopsis thaliana*, *acaulis1*, affecting the development of both inflorescences and leaves. Development 118: 751–764 (1993).
103. Van der Krol AR, Chua N-H: Flower development in petunia. Plant Cell 5: 1195–1203 (1993).
104. Vaughan JG, Jones FR: Structure of the angiosperm inflorescence apex. Nature 171: 751–752 (1953).
105. Vaughan JG: The morphology and growth of the vegetative and reproductive apices of *Arabidopsis thaliana* (L.) Heynh., *Capsella bursa-pastoris* (L.) Medic, and *Anagallis arvensis* L. J Linn Soc Lond Bot 55: 279–301 (1955).
106. Veit B, Schmidt RJ, Hake S, Yanofsky MF: Maize floral development: new genes and old mutants. Plant Cell 5: 1205–1215 (1993).
107. Weigel D, Alvarez J, Smyth DR, Yanofsky MF, Meyerowitz EM: *LEAFY* controls floral meristem identity in *Arabidopsis*. Cell 69: 843–859 (1992).
108. Weigel D, Meyerowitz EM: Activation of floral homeotic genes in *Arabidopsis*. Science 261: 1723–1726 (1993).
109. Whitelam GC, Smith H: Retention of phytochrome-mediated shade avoidance responses in phytochrome-deficient mutants of *Arabidopsis*, cucumber and tomato. J Plant Physiol 139: 119–125 (1991).
110. Wilson RN, Heckman JW, Somerville CR: Gibberellin is required for flowering in *Arabidopsis thaliana* under short days. Plant Physiol 100: 403–408 (1992).
111. Yanofsky MF, Ma H, Bowman JL, Drews GN, Feldmann KA, Meyerowitz EM: The protein encoded by the *Arabidopsis* homeotic gene *agamous* resembles transcription factors. Nature 346: 35–39 (1990).
112. Zagotta MT, Shannon S, Jacobs C, Meeks-Wagner DR: Early-flowering mutants of *Arabidopsis thaliana*. Aust J Plant Physiol 19: 411–418 (1992).

Plant Molecular Biology **26**: 1379–1411, 1994.

Oligosaccharins: structures and signal transduction

François Côté and Michael G. Hahn*
*Complex Carbohydrate Research Center and Department of Botany, University of Georgia, 220 Riverbend Road, Athens, GA 30602–4712, USA (*author for correspondence)*

Received and accepted 13 September 1994

Key words: elicitors, oligosaccharins, phytoalexin synthesis, phytoalexin accumulation, signal transduction

Introduction

Oligosaccharins are complex carbohydrates that can function in plants as molecular signals that regulate growth, development, and survival in the environment [3]. Studies of plant-microorganism interactions yielded the first evidence that oligosaccharins could serve as biological signals. Much of this research focused on the synthesis and accumulation of antimicrobial phytoalexins in response to microbial attack. Phytoalexin synthesis and accumulation are observed not only after microbial infection, but also after treatment of plant tissue with cell-free extracts of microbial origin. The active components in these extracts are commonly referred to as 'elicitors'. The term 'elicitor' was originally used to refer to molecules and other stimuli that induce the synthesis and accumulation of phytoalexins in plant cells [130], but is now commonly used for molecules that stimulate any plant defense mechanism [68, 70, 71, 104]. Recent reviews [56, 72, 107, 204] provide an overview of the structures of diverse elicitors, several of which are oligosaccharins, and their activities. In addition to those oligosaccharins that function as elicitors, more recent research has suggested the involvement of oligosaccharins in normal plant growth and development [5, 56a, 206].

This review will provide a brief overview of the structures and activities of six different oligosaccharins (see Fig. 1). Three originate from polysaccharides present in the fungal wall, two others, from plant primary cell wall polysaccharides, while the last group originates from glycoproteins of plant and fungal origin. The lipo-oligosaccharidic oligosaccharins [148a] synthesized by bacterial symbionts of plants will not be considered here, since they are discussed in another contribution to this volume. We will focus on those oligosaccharins that have been purified to apparent homogeneity and whose structures have been determined. There are a number of studies, which will not be discussed here, that suggest the involvement of additional oligosaccharides as signals in plants, although the oligosaccharin has not been purified and characterized (see review [5]).

The purification of a complex carbohydrate to homogeneity often requires elaborate methodology due to the potential structural complexity of these molecules. Therefore, some attention will be given to the purification and characterization procedures developed for each of the six oligosaccharins discussed in this review. We will also provide an overview of the current understanding of signal transduction pathways activated by oligosaccharins, highlighting our studies on the recognition of a hepta-β-glucoside elicitor by specific binding protein(s) in soybean plasma membranes. We will argue that the use of homogeneous preparations of these signal molecules is essential in order to interpret unambiguously the results of signal transduction studies. We will finally discuss the biological significance of the oligosaccharins in the context of the perception and

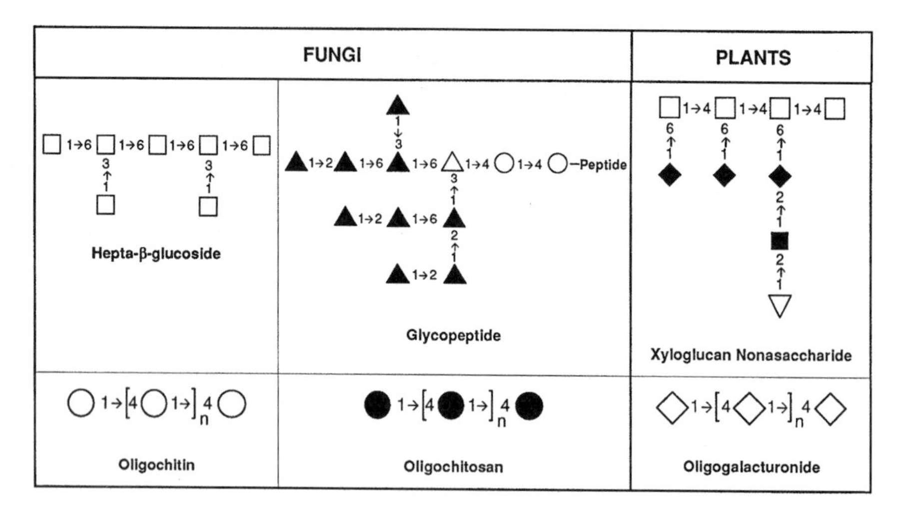

Fig. 1. Structures of oligosaccharins. The oligogalacturonide elicitor and the xyloglucan nonasaccharide were isolated from plant cell walls, the hepta-β-glucoside, oligochitin and oligochitosan elicitors were isolated from fungal cell walls, and the glycopeptide fragments were isolated from yeast invertase. The most active oligogalacturonides have a degree of polymerization between 11 and 14 ($n = 9$–12). A minimum size of at least a tetramer ($n = 2$) or a heptamer ($n = 5$) is required, respectively, for activity of the oligochitin and oligochitosan elicitors. The symbols used represents the following glycosyl residues: □, β-glucose; ▲, α-mannose; △, β-mannose; ○, β-*N*-acetyl-glucosamine; ◆, α-xylose; ■, β-galactose; ▽, α-fucose; ●, β-glucosamine; ◇, α-galacturonic acid.

response of plant cells to signals originating not only from within the organism but also from the external environment, pointing out questions that still require further research.

Purification and biological activities

Oligosaccharins derived from fungal wall polysaccharides

Glucan-derived oligosaccharins

Elicitor-active glucans were first detected in the culture filtrates of *Phytophthora sojae*, a phytopathogenic oomycete [10], and were later also purified from a commercially available yeast extract [106]. Glucan elicitors have been shown to induce phytoalexins in several legumes including soybean [11, 214], chickpea [101], bean [48], alfalfa [136], and pea [26], and also in the solanaceous species, potato [48] and sweet pepper [26]. These elicitors were shown to be composed of 3-, 6-, and 3,6-linked β-glucosyl residues [10, 106], a composition very similar to β-glucans that are important structural components of various mycelial walls [16]. Thus, subsequent research focused on elicitors released from mycelial walls.

Partial hydrolysis of the mycelial walls of *P. sojae* using either hot water [11] or 2 N trifluoroacetic acid [214] releases a complex mixture of oligo- and polysaccharides. This mixture contains β-glucan fragments ranging in size from short oligomers up to polysaccharides ($M_r > 100\,000$) [12, 214]. Oligoglucans in this mixture have been shown to elicit phytoalexin accumulation [214] and production of hydroxyproline-rich glycoproteins [202] in soybean, and to induce resistance to viruses [141] and activate glycine-rich protein gene expression [33] in tobacco. There may be other, as yet unidentified, activities in the oligoglucan mixture as well. The mycelial wall hydrolysate also contains a large number of biologically inactive molecules and non-carbohydrate elicitors. For example, a proteinaceous component of the crude *P. sojae* elicitor preparation elicits phytoalexin synthesis in suspension-cultured parsley cells [182]. This protein elicitor is not active in soybean [182]. The following sections focus on the glucan elicitors of phytoalexin accumulation, and describe the purification of a hepta-β-glucoside elicitor and structure-activity studies that define essential structural elements of this elicitor.

Oligoglucoside elicitor purification. Fractionation of the mixture of oligoglucosides generated by partial acid hydrolysis of mycelial walls of *P. sojae* on a size-exclusion column revealed that elicitor-active oligosaccharides were present in all fractions containing oligomers of degree of polymerization (DP)$\geq$6 [214]. The heptamer-enriched fraction was further fractionated on a series of normal- and reversed-phase high-performance liquid chromatography (HPLC) columns [214]. It was estimated that the original heptaglucoside mixture contained >100 structurally distinct heptaglucosides, based on the number of peaks observed in the various chromatographic steps. Homogeneous preparations of the aldehyde-reduced forms (i.e. the hexa-β-glucosyl glucitols) of one elicitor-active hepta-β-glucoside (compound **1**, Fig. 2) and of seven other elicitor-inactive hepta-β-glucosides were obtained in amounts sufficient to determine their structures [213]. The structure of the elicitor-active hepta-β-glucoside [212] was subsequently confirmed by its chemical synthesis [92, 153, 180, 227].

The ability of the chemically synthesized, unreduced hepta-β-glucoside elicitor to induce phytoalexin accumulation in soybean cotyledons is identical to that of the corresponding hexa-β-glucosyl glucitol purified from fungal wall hydrolysates [212]. Both are active at concentrations of ca. 10 nM, making them some of the most active elicitors of phytoalexin accumulation yet observed. The seven other hexa-β-glucosyl glucitols that were purified from the partial hydrolysates of fungal cell walls had no elicitor activity over the concentration range ($\leq$400 μM) tested [214]. These results provided the first evidence that specific structural features are required for an oligo-β-glucoside to be an effective elicitor of phytoalexin accumulation. Structure-activity studies carried out in our laboratory have confirmed and expanded upon these initial findings.

Structure-activity studies of oligoglucoside elicitors. Sixteen oligo-β-glucosides (**2–14** and **16–18**, see Fig. 2), structurally related to the elicitor-active hepta-β-glucoside **1** (Fig. 2), were chemically synthesized ([27, 92, 121, 180], and unpublished results of N. Hong, T. Ogawa, R. Verduyn and J. van Boom). This group of oligoglucosides was instrumental in allowing the identification of structural features essential for effective elicitation of phytoalexin accumulation in soybean cotyledon tissue [45, 46]. Three oligo-β-glucosides (**4**, **7**, and **8**) have the same branching pattern as that of elicitor-active hepta-β-glucoside **1** [213] and have an EC_{50} (concentration required to give a half-maximum biological response) of ca. 10 nM. Hexa-β-glucoside **4** is the minimum fully elicitor-active structure [46]. Increasing the length of this hexaglucoside by the addition of glucosyl residues at the reducing end of the molecule (compounds **1**, **7**, and **8**) has no significant effect on its elicitor activity. In contrast, removing glucosyl residues from the active hexa- or heptaglucoside (compounds **2**, **3**, **16**, **17**, and **18**) or rearranging their side-chains (compound **5**) results in molecules with significantly lower elicitor activity. For example, removal of the non-reducing terminal backbone glucosyl residue from hexaglucoside **4** (compound **3**) results in a 4000-fold reduction in elicitor activity, suggesting that this glucosyl residue has a particularly important function [45].

Several recently synthesized oligoglucosides (R. Verduyn and J. van Boom, unpublished results) have been used to assess the importance of the side-chain glucosyl residue closest to the reducing end of the elicitor-active oligoglucosides. These oligoglucosides, a tetramer, a pentamer and a hexamer (compounds **16–18**), are significantly less active (1000- to 10000-fold) than the elicitor-active hepta-β-glucoside **1** (J.-J. Cheong and M. G. Hahn, unpublished results). The demonstration that a linear, 6-linked hepta-β-glucoside is inactive provides further evidence that the side-chain glucosyl residues are required for elicitor activity of oligoglucosides [46].

An additional set of hexasaccharides (compounds **9–14**, Fig. 2) has been synthesized (N. Hong and T. Ogawa, unpublished results) in which one or the other terminal glucosyl residue at the non-reducing end of hexa-β-glucoside **4** was modified. Thus, replacement of the side-chain glucosyl residue of the terminal trisaccharide with

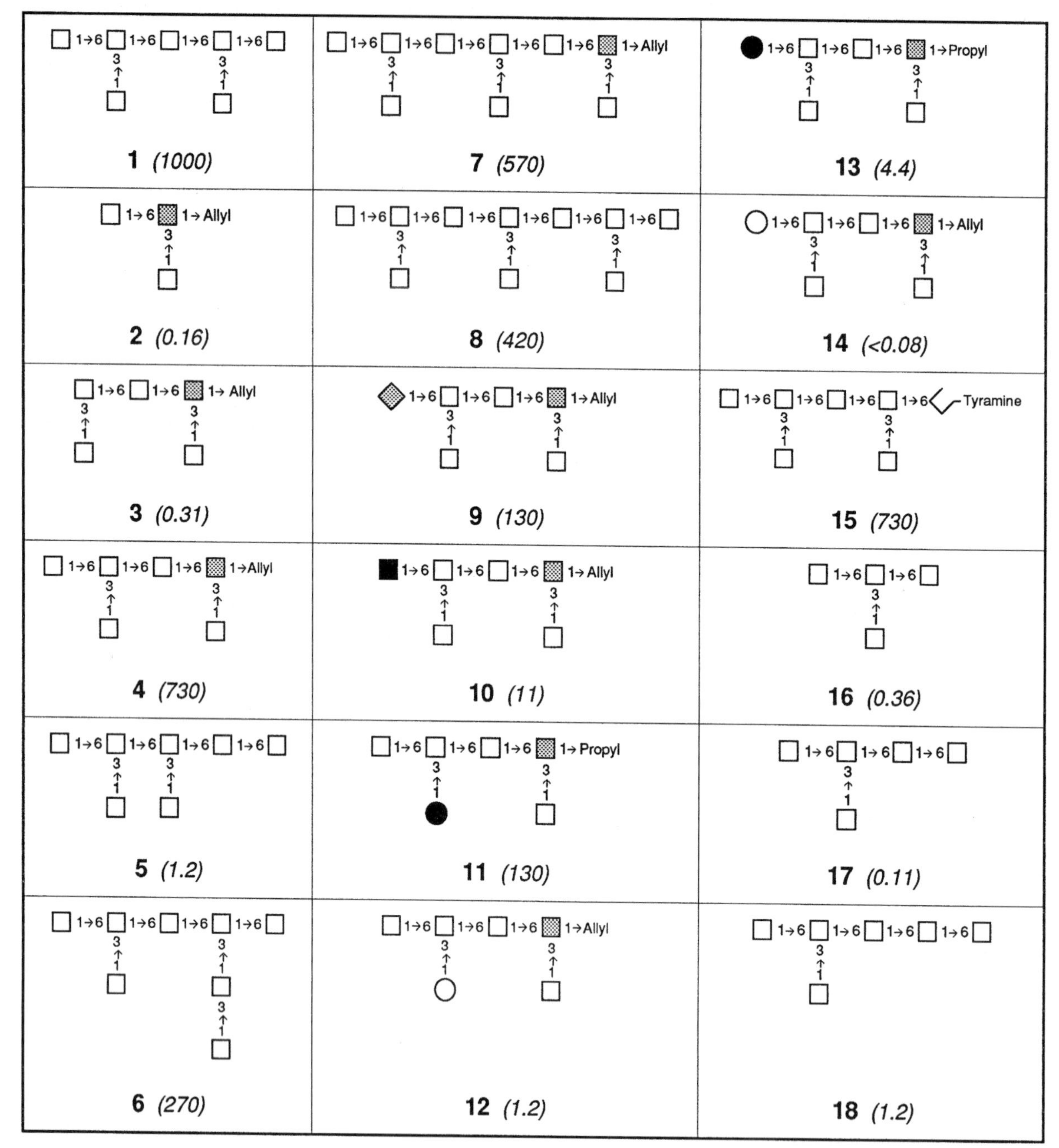

Fig. 2. Structures of synthetic oligoglucosides used in biological activity and ligand-binding studies in soybean [45–47]. The number in parentheses following the compound number represents its relative elicitor activity which is defined as the concentration of an oligosaccharide required to give half-maximum induction of phytoalexin accumulation ($A/A_{std} = 0{,}5$) in the soybean cotyledon bioassay corrected to the standard curve for hepta-β-glucoside **1** (arbitrarily = 1000; [46]). The symbols used represents the following glycosyl residues: □, β-glucose; ▩, α-glucose; ◇, β-xylose; ■, β-galactose; ●, β-glucosamine; ○, β-*N*-acetyl-glucosamine; ‹ in **15**, amino-glucitol.

a β-glucosaminyl (compound **11**) or a *N*-acetyl-β-glucosaminyl residue (compound **12**) reduces the elicitor activity ca. 10- and ca. 1000-fold, respectively. The corresponding modifications of the non-reducing terminal backbone glucosyl residue (compounds **13** and **14**, respectively) result in even greater decreases in elicitor activity (ca. 100- and ca. 10000-fold, respectively). Substitution of the same glucosyl residue with a xylosyl residue (compound **9**) or a galactosyl residue (compound **10**) reduces the activity about 10- and 100-fold, respectively [45]. These data, together with those described earlier, prove that all three non-reducing terminal glucosyl residues present in the hepta-β-glucoside elicitor are essential for the ability of the molecule to effectively elicit phytoalexin accumulation in soybean.

The reducing terminal glucosyl residue of the hepta-β-glucoside elicitor, on the other hand, is not essential for biological activity. Phytoalexin elicitor assays of reducing-end derivatives of hepta-β-glucoside **1** demonstrated that attachment of an alkyl or aromatic group to the oligosaccharide (e.g. compound **15**) does not have a significant effect on its EC_{50} [46]. A tyramine-coupled derivative of hepta-β-glucoside **5** is slightly more active (ca. 2.5-fold) than underivatized hepta-β-glucoside **5**. Coupling tyramine or benzylhydroxylamine to maltoheptaose, a structurally unrelated 4-linked hepta-α-glucoside, yields derivatives with no detectable elicitor activity. These results establish that coupling of aromatic groups to biologically inactive oligoglucosides does not endow those oligosaccharides with phytoalexin elicitor activity. Thus, it is possible to prepare a fully active, radio-iodinated form of compound **15** for use as a labeled ligand to search for the presence of elicitor-specific, high-affinity binding sites in soybean membranes (see Signal transduction section below).

Chitin- and chitosan-derived oligosaccharins

Oligosaccharide fragments derived from chitin, a linear polysaccharide composed of 4-linked β-*N*-acetylglucosaminyl residues, and chitosan, its de-*N*-acetylated derivative (see Fig. 1), have been shown to elicit defense responses in various plants. Chitin and chitosan polysaccharides are structural components of the cell walls of many fungi [16].

Purification. Chitosan and chitin-derived oligomers can be generated from commercially available chitosan derived from the shells of crustaceans. Depending on the source, between 80–90% of the glucosaminyl residues are de-*N*-acetylated [131]; thus, chitin is generally obtained by re-*N*-acetylation of chitosan [169]. Oligomers of either chitin or chitosan are generated by partial acid hydrolysis of the corresponding polymer, and the products fractionated on high resolution size-exclusion columns. Pure chitin oligomers are separated from partially *N*-acetylated fragments by cation-exchange chromatography. The fragments are then further purified by HPLC on amine-modified silica columns [85]. Alternatively, *N*-acetyl-chitooligosaccharides can be prepared by re-*N*-acetylation of the corresponding chitosanoligosaccharides [236]. Chitin- and chitosan-derived oligosaccharides generated from fungal spores and mycelial walls have also been purified by liquid chromatography [131]. Chitin and chitosan fragments of DP 3 to 6 are commercially available [21, 144].

Induction of plant defense responses by oligochitins and -chitosans. Heterogeneous preparations of chitosan and chitosan-derived fragments elicit the accumulation of phytoalexins in pea pods [102, 103a, 232], and in suspension-cultured soybean [143] and parsley [49] cells. Other plant defenses induced by chitosan-derived oligosaccharides include the accumulation of defense-related proteinase inhibitors in tomato and potato leaves [185, 232–234], the synthesis of callose, a β-1,3-glucan, in suspension-cultured parsley [49], tomato [100] and *Catharanthus roseus* [130] cells, and the inhibition of fungal growth [103a].

Oligosaccharide fragments of chitosan and chitin have both been shown to induce defense-related lignification of the walls of suspension-cultured slash pine (*Pinus elliottii*) cells [149, 150] The deposition of both callose and lignin are thought to function in plant defense by strengthening the plant cell wall. Chitosan has also been reported to induce, in mechanically wounded leaves, protection from subsequent infection with viruses [187, 188], though the mechanisms underlying this induced resistance are unknown. Chitosan elicits free radical formation that is thought to lead to cell death, cell wall thickening, and chitinase secretion in suspension-cultured rice cell [160].

The structural requirements for bioactive chitosan- and chitin-derived oligosaccharides have not, until recently, been investigated in detail. In most studies, the fragments used have been heterogeneous in size, and often, in structure (for example, in the extent and pattern of *N*-acetylation). The oligosaccharides generally must have a DP > 4 to induce a biological response, but beyond that requirement, it is not possible to generalize about structural features essential for functionality. For example, de-*N*-acetylated and partially *N*-acetylated oligoglucosamines with DP > 4 are both effective elicitors of proteinase inhibitors in tomato and potato leaves [232, 233]. In contrast, chitosan-derived, but not chitin-derived, oligosaccharides with DP > 7 elicit pisatin accumulation in pea pods [102, 103a, 131, 132]. Interestingly, the nature of the substituent at the reducing end of the oligochitosans has a major impact on the activity of these oligosaccharides [103a]. Oligochitosans of DP = 8 carrying an *O*-methyl group at the reducing end are active, while those carrying an *O*-methoxyphenyl group are inactive. These results stand in notable contrast to experiments with chitin- and glucan-oligomers, where the nature of the substituent at the reducing end of the oligosaccharide has no significant impact on the abilities of the oligosaccharides to induce plant responses [21, 46]. In suspension-cultured rice cell, the specificity is again different. Chitosan and its oligomers have little or no activity, while homogenous preparations of chitin oligomer with a minimum DP of 4 induce diterpene phytoalexin biosynthesis and accumulation. In this system, oligomers of DP ≥ 6 are the most active [196, 236]. Other bioactivities also have diverse structural requirement. For example, partial *N*-acetylation or chemical fragmentation reduces the ability of chitosan to elicit callose formation in *C. roseus* cells [129]. Finally, chitosan-derived oligosaccharides do not induce lignification in wounded wheat leaves, although the chitosan and chitin polysaccharides and chitin-derived oligosaccharides with DP > 4 are effective elicitors of lignification in the leaves [15].

Oligosaccharins derived from plant cell wall polysaccharides

Pectin-derived oligosaccharins

Homogalacturonan, the pectic polysaccharide composed of 1,4-linked α-D-galactosyluronic acid residues (Fig. 1) and its partially methyl-esterified derivative, is an important structural component of the primary cell walls of higher plants [14, 177]. Partial depolymerization of homogalacturonan generates oligogalacturonides that exhibit various regulatory effects in plants (Table 1) including the elicitation of defense responses, the regulation of growth and development, and the induction of rapid responses at the cell surface [2, 198, 206].

Purification of oligogalacturonides. The biological activities of oligogalacturonides have often been determined using heterogeneous mixtures of oligosaccharides containing both active and inactive molecules (Table 1). The bioactive molecules were not, in all cases, purified to homogeneity and shown to be oligogalacturonides. However, the active molecules must contain several contiguous 1,4-linked α-D-galactosyluronic acid since they are inactivated by treatment with homogeneous endo-polygalacturonases (EPGase). Chemically synthesized homogalacturonides [173] are required to obtain unequivocal evidence for bioactivity, although homogeneous and well-characterized oligogalacturonides generated from cell walls can be used in their absence since they are more readily available [217]. Homogeneous oligogalacturonides are essential for the identification of oligogalacturonide-specific receptors and for determining the effects of oligogalacturonides on the physico-chemical properties of membranes.

The preparations of oligogalacturonides used for biological studies have been generated by various methods from different source materials and purified to varying degrees. The heterogeneous nature of many of the utilized oligogalacturonide

Table 1. Biological activities of α-1,4-oligogalacturonides

Activity	Oligogalacturonides				Plant	Reference
	DP range[1]	Single DP[2]	Concentration[3]	EPG[4]		
Plant defense responses						
Induction of phytoalexins	8–13 (12)[5]	+	10^{-5}	+	soybean	[60, 110, 175]
	3–12 (6 & 10)	+	ND[6]	ND	soybean	[139]
	9–15 (13)	+	10^{-4}	ND	castor bean	[125, 233]
	≥ 3	–	ND	ND	parsley	[61]
	9–15	–	ND	ND	bean	[69, 224]
	ca. 20	–	ND	ND	pea	[233]
Induction of shikonin	12–15 (14)	+	$<10^{-6}$	+	*Lithospernum*	[223]
Induction of PAL	>9	–	ND	ND	carrot	[167]
	9–15	–	ND	ND	bean	[69, 224]
Induction of chalcone synthase	9–15	–	ND	ND	bean	[224]
Induction of β-1,3-glucanase	≥ 3	–	ND	ND	parsley	[61]
Induction of chitinase	ND	–	ND	ND	tobacco	[36]
Induction of lignin	8–11 (11)	+	10^{-7}	ND	cucumber	[199]
	8–15	–	ND	ND	castor bean	[37]
	9–15	–	ND	ND	bean	[224]
Induction of proteinase inhibitors	2–6 (6) & up to 20	+	$\leq 10^{-4}$	NA[7]	tomato	[28, 80, 233]
	2–3 (2)	+	$<10^{-6}$	NA	tomato	[170]
	1–7 & 10–20	–	ND	ND	tomato	[225]
Induction of isoperoxidases	ND	–	ND	ND	castor bean	[37]
Inhibition of hypersensitive response	ND	–	ND	ND	tobacco	[13]
Elicitation of necrosis	ND	–	ND	ND	cowpea	[40]
Development and growth						
Regulation of TCL morphogenesis:						
flower formation	10–14	+	10^{-7}	+	tobacco	[74, 159]
root formation	(12–14)	+	10^{-7}	ND	tobacco	[24, 74]
Inhibition of auxin-induced elongation	>8	–	$<10^{-4}$	ND	pea stem	[34]
Induction of ethylene	>8 & 5–19	–	ND	ND	tomato	[35, 38]
	5–19	–	ND	ND	pear	[39]
Rapid responses at the cell surface						
Induction of H_2O_2 & oxidative burst	ND	–	ND	ND	soybean	[7, 147]
Rapid depolarization of plasma membrane	1–7 & 10–20	–	ND	ND	tomato	[225]
Efflux of K^+, influx of Ca^{2+}	10–16 (12)	+	10^{-7}	ND	tobacco	[161]
	>9	–	ND	ND	carrot	[167]
Enhanced *in vitro* phosphorylation of ca. 34 kDa protein	14–20	+	ca. 10^{-7}	ND	tomato	[81]

[1] DP (degree of polymerization) range of oligogalacturonides that show the designated biological activity.

[2] Single purified oligomers have been tested (+). No data on single oligomers (–).

[3] Estimate of the concentration of the most active oligogalacturonides needed to give a half-maximum biological response. A concentration is given only when purified oligogalacturonides were assayed.

[4] Sensitivity of the biological activity of the oligogalacturonides to digestion with endopolygalacturonase. (+) indicates sensitivity to the enzymes.

[5] Numbers in parentheses are the DP of the most active oligogalacturonide.

[6] ND: DP of active oligogalacturonides not determined; therefore, molar concentration not determined. The sensitivity of the biological activity of the oligogalacturonides to digestion with endopolygalacturonase has not been determined.

[7] NA: not applicable because short oligomers (dimer and trimer) that are not substrates for endopolygalacturonase were used.

preparations complicates the interpretation of the biological activity studies. Oligogalacturonides can be released from homogalacturonans present in plant primary cell walls either by partial acid hydrolysis of the walls [110, 175], or by treating the walls with pectic-degrading enzymes such as EPGase [14, 43, 62, 64, 82, 125, 171, 200] or endo-pectate lyase [43, 59, 60, 86, 171, 201]. The released oligomers are sometimes de-esterified using a mild alkali treatment in order to reduce the complexity of the mixture and facilitate purification. De-esterification is also required in order for EPGase digestion to be used to establish the oligogalacturonide nature of the active molecules; EPGase will only cleave a de-esterified substrate. A more readily available starting material for the preparation of bioactive oligogalacturonides is polygalacturonic acid (PGA) obtainable from commercial sources. While cell wall homogalacturonans are methyl-esterified to varying degrees, PGA is a methyl de-esterified pectin. Similar, if not identical, active oligogalacturonides are released from cell walls or PGA by both chemical or enzymatic fragmentation [60, 217].

The mixtures of oligogalacturonides generated by fragmentation of either plant cell walls or polygalacturonic acid are heterogeneous in size, including galacturonic acid and oligomers with DP between 2 and 20. The released oligogalacturonides in such mixtures are usually separated, by low-pressure anion-exchange chromatography [110, 125, 175, 200], into fractions enriched for specific sizes of oligogalacturonide. Such size-fractionated oligomers have been used in many of the bioactivity studies. However, recent evidence has shown that oligogalacturonides, particularly those of DP > 10, isolated by low pressure anion-exchange chromatography of polygalacturonic acid digests, contain significant amounts (ca. 30%) of modified oligogalacturonides [59, 217]. For example, a fraction containing predominantly a tridecagalacturonide was further analyzed by high-performance anion-exchange chromatography in combination with pulsed amperometric detection (HPAEC-PAD) and resolved into seven peaks, three of which had galactaric acid-containing oligogalacturonides [217]. Galactaric acid, the C-1 oxidized derivative of galacturonic acid, may be formed during commercial processing of PGA or it may be a natural product formed by the action of a plant galacturonic acid oxidase [191]. The HPAEC-purified, apparently homogeneous, tridecagalacturonide is an active elicitor [217]. None of the galactaric acid-containing oligogalacturonides elicit phytoalexin accumulation in soybean tissue (A. Koller, J.-J. Cheong and M. O'Neill, unpublished results), although relatively high concentrations activate the *in vitro* anionic peroxidase-catalyzed oxidation of indole-3-acetic acid [190]. These results emphasize the importance of obtaining homogenous oligogalacturonides in order to avoid uncertainty in the attribution of biological activity to homo-oligogalacturonides. The various activities of oligogalacturonides and an evaluation of the purity of the material used are summarized in Table 1.

Elicitation of plant defense responses by oligogalacturonides. The presence of endogenous molecules (elicitors) capable of inducing defense responses in plants was first observed in experiments in which the placement of frozen-thawed bean stem segments in contact with healthy stem segments resulted in the accumulation of phytoalexins in the healthy stem segments [114]. Subsequently, elicitor-active material was found to be present in extracts of autoclaved bean (*Phaseolus vulgaris*) hypocotyls [115] and isolated soybean cell walls [110]. Elicitor-active molecules present in these extracts were later identified as linear oligomers of 1,4-linked α-D-galactosyluronic acid residues [110, 175].

The nature of the plant defense response(s) induced by oligogalacturonides depends on the plant being studied. Thus, oligogalacturonides have been shown to induce the accumulation of phytoalexins in soybean [59, 60, 110, 175], castor bean [125], bean [69, 224], pea [233] and parsley [61]. Oligogalacturonides also induce the accumulation of anti-microbial shikonins in suspension-cultured *Lithospermum erythrorhizon* cells [223]. In soybean and parsley, oligogalacturonide and oligo-β-glucoside elicitors (see above) act synergistically, i.e. the concentrations required to elicit phytoalexins when both elicitors are present are less than the concentration required for each elicitor to elicit phytoalexins individually [58, 61]. This synergistic effect has been observed with a synthetic dodecagalacturonide, but not with the decagalacturonide [109, 172, 173]. Other oligogalacturonide-induced defense responses include the induction of glycosylhydrolases (β-glucanase, chitinase, lysozyme) in parsley [61] and tobacco [36], the increased deposition of lignin in cucumber [199] and castor bean [37], and the accumulation of proteinase inhibitors in tomato [28, 80].

The size range of oligogalacturonides that activate defense responses is usually quite narrow. For example, a size range between 10 and 15 galactosyluronic acid residues is generally required to elicit most of the plant defense responses listed above. Although the reason for the frequently observed size dependence of the response to oligogalacturonides is not known, this requirement suggests that oligogalacturonides need ten or more galactosyluronic acid residues in order to assume a solution conformation or oligomeric structure that is biologically active. Evidence that oligogalacturonides undergo a conformational transition at DP>10 has been obtained from binding studies with a monoclonal antibody that preferentially binds oligogalacturonides of DP≥10 [152]. However, di- and trigalacturonides induce proteinase inhibitor production in tomato [170], suggesting that in some tissues there may be a different size requirement for biological activity.

Whether or not esterification is important for the biological activity of oligogalacturonides has been partially assessed with somewhat contradictory results. A crude mixture of oligosaccharides released from the cell wall by EPGase is active in a tobacco thin-cell layer flower induction assay both before and after de-esterification [159]. Re-treatment of the de-esterified cell wall-derived oligosaccharides with EPGase destroys the flower-inducing activity, showing that the active component contains 1,4-linked α-D-galactosyluronic acid residues. A crude preparation of rhamnogalacturonan II, containing oligogalacturonides, elicits soybean phytoalexin accumulation, whether or not it had been de-esterified [110]. However, casbene synthetase induction in castor bean by oligomers of galacturonic acid of DP 12–15 was significantly decreased by *in vitro* methyl esterification of the carboxylate groups [125]. The elicitor activity was restored by re-de-esterification.

Induction of plant growth and development by oligogalacturonides. One major effect of oligogalacturonides in tobacco is the regulation of morphogenesis. The ability of undifferentiated totipotent plant cells to differentiate and develop into specialized organs was believed to be mediated mainly by factors such as hormones and light [208]. Thin-cell-layer (TCL) explant bioassays [168] have been used to study the effects of phytohormones and plant cell wall-derived fragments on plant morphogenesis [24, 74, 155, 159]. Tobacco TCL explants containing five to ten cell layers of floral stem tissue, when incubated for 24 days on culture medium containing the phytohormones auxin and cytokinin, form flowers, vegetative shoots, or roots [168]. The particular organ that forms depends on the concentrations of auxin and cytokinin in the culture medium. Oligogalacturonides with DP 10 to 14, generated by EPGase treatment of sycamore cell walls or polygalacturonic acid, induce flower formation and inhibit root formation in tobacco TCLs or leaf explants with a half-maximal response at 400 nM (1 μg/ml) [24, 74, 159]. Oligogalacturonides with DP between 12 and 14 are the most active oligomers [159]. It is noteworthy that the concentrations of oligogalacturonides required to affect morphogenesis are 10- to 100-fold lower than those required for induction of plant defense responses (Table 1).

Plant cell wall-derived fragments can also regulate diverse developmental processes such as cell elongation and fruit ripening. A crude mixture of oligogalacturonides inhibits auxin-induced elongation of pea stem segments at concentrations of 300 μg/ml [34]. Individual oligomers of DP 10 to 15 are active at lower concentrations (40 μg/ml; F. Cervone, personal communication). The effect of oligogalacturonides on fruit ripening is believed to be mediated through enhanced production of ethylene. Indeed, treatment with oligogalacturonides of DP 5 to 19 induces an increase in ethylene production in ripening tomato fruits [35, 38, 166a] and suspension-cultured pear cells [39]. However, increased ethylene production is also observed under stress conditions, some of which are elicited by oligogalacturonides (see section above). Therefore, the effect of oligogalacturonides on ripening might be a consequence of a more general stress response. Additional studies are required to determine if oligogalacturonides induce ethylene biosynthesis directly by increasing the synthesis of 1-amino-cyclopropane-1-carboxylic acid (ACC) synthase [39] and/or other enzymes of the biosynthetic pathway, or if oligogalacturonide-induced ethylene synthesis is a secondary effect of the plant's response to these bioactive fragments.

Xyloglucan-derived oligosaccharins

Xyloglucans were first detected in seeds [140], but were later found in the primary cell wall of higher plants [20]. They are part of the hemicelluloses which are functionally defined as those branched cell wall polysaccharides that form strong noncovalent associations with cellulose microfibrils. The dynamic nature of the cellulose-xyloglucan cross-linking is hypothesized to be a major factor in controlling the rate of cell wall expansion, thereby regulating cell growth [20, 116]. Xyloglucan consists of a backbone of 1,4-linked β-D-glucosyl residues, ca. 75% of which are substituted at carbon 6 (C-6) with α-D-xylosyl residues. One-third to one-half of the α-D-xylosyl residues in the xyloglucans isolated from most dicotyledonous plants are substituted at C-2 with a β-D-galactosyl residue, and the galactosyl residue is itself sometimes substituted with an α-L-fucosyl residue. The galactosyl residue can also be substituted with either one or two *O*-acetyl groups [54, 237a]. Xyloglucans from graminaceous plants appear to be significantly less frequently substituted with side chains than the xyloglucans found in dicotyledonous plants. The structure, function, and biological activities of xyloglucans, have been the subject of extensive reviews [5, 116].

Purification. Xyloglucan can be fragmented by treatment with a purified endo-β-1,4-glucanase isolated from *Trichoderma reesei* and the released oligosaccharides separated into size classes by high-resolution size-exclusion chromatography [20, 238]. Further purification of these oligosaccharides is accomplished by normal-phase HPLC on amino-bonded silica [119, 166], reversed-phase HPLC on octadecyl silica [75, 119, 135, 238], and HPAEC-PAD [135, 166]. Reversed-phase HPLC is a powerful procedure for the purification of xyloglucan fragments. However, because the separation of the α- and β-anomers of reducing oligosaccharides leads to complicated chromatograms, it is usually necessary to reduce the oligosaccharides with $NaBH_4$ to form their corresponding oligoglycosyl alditols in order to simplify the chromatography. Xyloglucan oligoglycosyl alditols are also easier to purify than their corresponding reducing oligosaccharides, and their structures are more readily determined by ^1H-nuclear magnetic resonance (NMR) spectroscopy and fast atom bombardment-mass spectroscopy (FAB-MS) [119, 238].

Endo-β-1,4-glucanase attacks the cellulosic backbone of xyloglucan where it is not branched, typically at every fourth glucosyl residue. Thus, the most commonly observed xyloglucan fragments released by the fungal endo-β-1,4-glucanase are oligosaccharides XXXG, XXFG, and XLFG (see Fig. 3). Difucosylated undecasaccharide XFFG (Fig. 3) is present at very low levels in the digests of xyloglucans isolated from the walls of sycamore cells and rapeseed hulls [118]. Oligosaccharides XXXG, XXFG, XLFG and XXLG (Fig. 3) represent structural motifs that are present in most of the xyloglucans found in dicotyledonous plants. However, xyloglucans isolated from the Solanaceae have xylosyl residues substituted with 1,2-linked α-L-arabinosyl residues.

Xyloglucan fragments inhibit auxin-stimulated growth. There is evidence that a small portion of

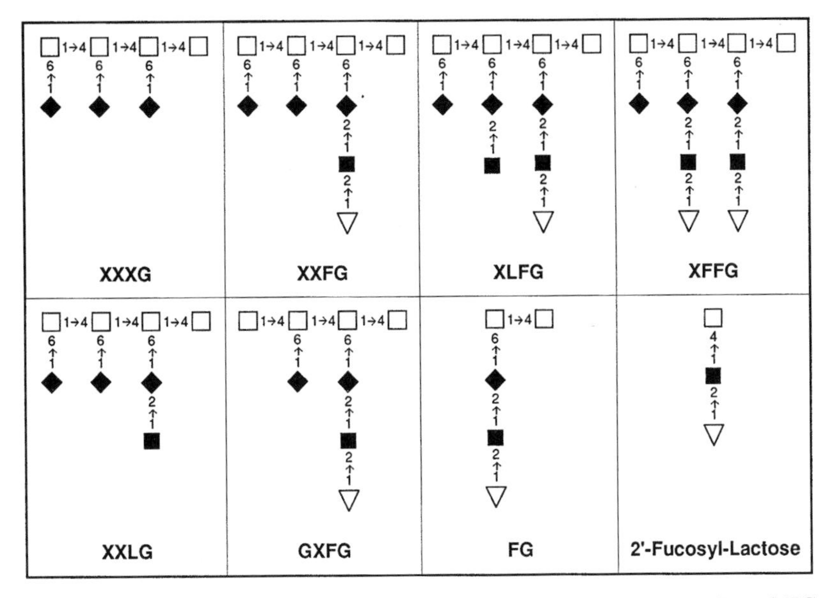

Fig. 3. Structure of xyloglucan oligosaccharides. Compounds XXXG, XXFG, XLFG, XFFG, and FG are released by *endo*-glucanase digestion of the xyloglucan polysaccharide. Compounds XXLG and GXFG can be generated by enzymatic treatment of XXFG. The nomenclature used for the xyloglucan-derived oligosaccharide fragments is based on recommendations adopted by a panel of researchers working with these molecules; the letters identify the terminal glycosyl residue of each side chain of the fragments [91]. The symbols used represents the following glycosyl residues: open squares, β-glucose (**G**); diamonds, α-xylose (**X**); solid squares, β-galactose (**L**); triangles, α-fucose (**F**).

plant cell wall xyloglucan is subject to turnover during auxin-promoted growth, resulting in the production and accumulation of xyloglucan oligosaccharides [117]. Some of these oligosaccharides may participate in regulating auxin-induced growth. The xyloglucan nonasaccharide, XXFG, is able, at ca. 10 nM, to inhibit 2,4-dichlorophenoxyacetic acid (2,4-D)- as well as gibberellic acid-induced elongation of pea stem segments [123, 235, 237]. The heptasaccharide XXXG has no growth-inhibiting effect. The observed inhibiting activity of the nonamer was shown not to be a phytotoxic effect, as concentrations higher and lower than 10 nM have less growth-inhibiting effect. These observations were confirmed by the demonstration that the chemically synthesized XXFG [207] inhibits 2,4-D-stimulated growth of pea stems to the same extent and at the same concentrations as the nonasaccharide isolated from xyloglucan [9]. Chemically synthesized XXXG, like the cell wall-derived molecule, does not significantly inhibit auxin-stimulated growth.

Other effects have been reported for the nonasaccharide XXFG. For example, at nanomolar concentrations, XXFG stimulates glucan synthases I and II activities in carrot protoplasts cultured in regeneration medium containing auxin [76]. The viability of the protoplasts was also increased. Furthermore, XXFG, when present in the culture medium of *Rubus fruticosus*, induces an increase in the activities of cell wall-bound carboxymethylcellulase and chitinase [128].

It has been of interest to determine which glycosyl residues of nonasaccharide XXFG are required for inhibition of auxin-stimulated growth. The reduced XXFG, as well as an octasaccharide GXFG (Fig. 3) resulting from α-xylosidase-treatment of XXFG, were found to be as active as native XXFG in inhibiting auxin-induced pea stem growth [9]. The closely related octasaccharide, XXLG, which lacks only the terminal fucosyl residue of the nonamer, does not inhibit 2,4-D-stimulated growth of pea stems. Together with the data obtained using the heptasaccharide XXXG, these results suggest that the terminal fucosyl residue of nonasaccharide XXFG is essential for the growth-inhibiting activity, while the terminal xylosyl residue farthest from the reducing terminus and the reducing glucosyl residue are not required for activity. An undecasaccharide (Fig. 3, XFFG) containing two fucosyl-galactosyl side chains, is more effective than the nonasaccharide in inhibiting 2,4-D-stimulated growth. Surprisingly, decasaccharide XLFG, which is identical to the undecasaccharide except that it lacks the fucosyl residue farthest from the reducing end, was reported not to be active despite the presence of a fucosyl-galactosyl side-chain [164]. The fucose-containing pentasaccharide FG and 2′-fucosyl-lactose are less effective than the nonamer in inhibiting 2,4-D-stimulated growth of pea stems [89]. The chemically synthesized pentasaccharide, FG, has been shown to be active in another system, suppressing fusicoccin-stimulated elongation of pumpkin cotyledons at concentrations of 10 nM; the same molecule did not affect cytokinin (6-BAP)-induced growth of the cotyledons [184].

Oligosaccharins derived from plant and fungal glycoproteins

Glycopeptide oligosaccharins

A number of glycoproteins have been identified that induce responses in plants, particularly defenses responses (reviewed in [6]). In some cases, the intact native glycoprotein is required (e.g. for enzyme activity), and for others the activity resides in a non-glycosylated peptide (e.g. the peptide elicitor from *Phytophthora sojae* [176, 183]). Recent studies of some of these glycoproteins have demonstrated that their carbohydrate portion is essential for the bioactivity, and resulted in the identification of glycopeptides as a new class of oligosaccharins, which will be the focus of our discussion.

Biological activities

A partially purified extract from yeast was shown to induce ethylene biosynthesis and phenylalanine-ammonia-lyase (PAL) activity, but not callose deposition, in suspension-cultured tomato cells [84, 100]. The activity of the yeast extract differs from that of chitosan oligomers, which induce all of those responses [100]. The molecules in the yeast extract that are responsible for these activities were subsequently shown to be glycopeptides [18]. Similar glycopeptide fragments could be generated from yeast invertase by chymotrypsin treatment, and were shown to be elicitors of ethylene biosynthesis [17]. Structure-activity studies using different glycan side-chains and peptide sequences have shown that the elicitor activity depends more on the glycan than on the peptide structure [17, 18]. The most active glycopeptide elicitors, at concentration of 5–10 nM, induce ethylene biosynthesis and PAL activity. The peptides are glycosylated with $Man_{10}GlcNAc_2$ and $Man_{11}GlcNAc_2$ glycan side chains (see Fig. 1) [17]. Release of the oligosaccharides from the glycopeptides by treatment with endo-β-*N*-acetylglucosaminidase H (endo-H) results in the loss of elicitor activity. However, the free *N*-glycans suppress the activity of the glycopeptides, suggesting that the oligosaccharides interact with the same binding site as the glycopeptide [17, 18]. Such high-mannose oligosaccharides have not been found in plant glycoproteins.

Two biologically active glycopeptides have been purified from the spore germination fluid of the pea pathogen, *Mycosphaerella pinodes*. These two molecules, suppressins A (GalNAc-O-Ser-Ser-Gly) and B (Gal-GalNAc-O-Ser-Ser-Gly-Asp-Glu-Thr), suppress the elicitation of the accumulation of the pea phytoalexin, pisatin, by crude fungal elicitor preparations [216, 243], although the concentrations required for suppression are quite high (≥ 220 and 80 μM respectively). In addition to its suppressor activity, suppressin B at a concentration of 320 μM inhibits plasma membrane ATPase activity by 80%. The relative importance of the peptide and the carbohydrate portions of the molecules have not been established, and no structure-activity studies have been performed. The authors speculated that the suppressins bind to a putative pea membrane receptor and thereby disrupts the function of essential membrane components such as ATPase [216].

Biologically active mannose- and xylose-containing *N*-glycans have been isolated from suspension-cultured cells of *Silene alba*, and are thought to be generated by glycoprotein proteolysis [194]. The free oligosaccharides $Man_5GlcNAc$, $Man_3(Xyl)GlcNAc(Fuc)GlcNAc$, and $Man_5(Xyl)GlcNAc(Fuc)GlcNAc$ apparently act as growth factors during the early development of flax [193], and stimulate ripening of tomato fruit as measured by red color development and ethylene production in the intact fruit [192]. Here again, detailed structure-activity studies have not been carried out.

Signal transduction

Cells that make up an organism have complex and diverse mechanisms for perceiving and responding to stimuli originating not only from within the organism but also from the external environment. The process of cellular signalling can be divided into three steps: signal perception, usually by cellular receptors that specifically recognize the signal (e.g. [105, 120]); signal transduction and amplification, that is, transmission of the signal to its site(s) of action within the cell, either directly or indirectly; and signal translation, that is, conversion of the signal into a specific cellular response such as activation of specific genes. Most of our understanding of the biochemical basis of signal transduction mechanisms has come from studies in bacterial and mammalian systems. In contrast to these systems, little is known about cellular signalling mechanisms in plants, but research is rapidly expanding using both biochemical and molecular approaches [93, 126]. Much of this effort was initially based on ideas and paradigms from animal systems [181]. However, it has become clear that, while plant analogues for molecules involved in animal signal cascades exist, these analogues have structural and functional properties that are unique. Detailed biochemical understanding of cellular signalling mechanisms in plants will require the isolation and characterization of the molecules involved in these processes. The investigation of the structures and activities of oligosaccharins have yielded important insights into how plant cells respond to external signals and provided useful model systems for molecular studies on signal perception, signal transduction, and gene regulation in plants (for recent reviews, see [65, 68, 145, 209]). Thus, binding sites for oligoglucoside, oligochitin and glycopeptide oligosaccharins have recently been identified and are discussed in detail below. The study of perception of the other oligosaccharins is more problematic, mainly due to the difficulty of generating a biologically active derivative of the oligosaccharin that can be used for binding studies. As a consequence, the translation of oligosaccharide signals such as oligogalacturonides have been more studied than their perception.

Oligoglucoside signal transduction

The first step in the signal transduction pathway induced by the hepta-β-glucoside elicitor is likely to be its recognition by a specific receptor. Indeed, the specificity of the response of soybean tissue to oligoglucoside elicitors of phytoalexin accumulation [46, 214] discussed above, suggests that a specific receptor for the hepta-β-glucoside elicitor exists in soybean cells. Several studies utilizing heterogeneous mixtures of mycelial glucan fragments indicated that binding sites for glucan fragments exist in soybean membranes [53, 186, 211, 239, 240]. In particular, binding studies carried out with partially purified elicitor-active glucans from *P. sojae* mycelial walls demonstrated the presence of high-affinity glucan-binding sites on soybean root plasma membranes [211] and protoplasts from suspension-cultured soybean cells [53]. Lower-affinity binding sites for a radiolabeled preparation of glucanase-released glucan fragments [240] and radiolabeled laminarin, a β-1,3-glucan [239], have also been identified in soybean membranes. Since the glucan preparations utilized in these binding studies were not homogenous, it is difficult to determine whether the observed binding was specific for the biologically active oligoglucosides.

Subsequent research utilizing hepta-β-glucoside elicitor coupled either to radio-iodinated aminophenethylamine [52] or to tyramine [47] as the labeled ligand has provided further evidence that specific membrane-localized glucan elicitor-binding sites exist in soybean cells. The hepta-β-glucoside elicitor binding sites are present in membranes prepared from every major organ of young soybean plants [47]. These elicitor-binding sites co-migrate with an enzyme marker (vanadate-sensitive H^+-ATPase) for plasma membranes in isopycnic sucrose density gradients [45], confirming earlier results obtained with partially purified labeled glucan fragments [211]. Binding of the radiolabeled hepta-β-glucoside elicitor to the root membranes is saturable over a concentration range of 0.1 to 5 nM, which is somewhat lower than the range of concentrations (6 to 200 nM) required to saturate the bioassay for phytoalexin accumulation [46, 214]. Data analysis (Scatchard, Hill and Woolf plots) of the ligand saturation experiments indicates that the root membranes possess only a single class of high-affinity hepta-β-glucoside binding sites (apparent $K_d \approx 1$ nM) of relatively low abundance ($B_{max} \approx 1$ pmol/mg protein) [47, 52]. These binding sites are inactivated by heat or pronase treatment [47], suggesting that the molecule(s) responsible for the binding are proteinaceous. Binding of the active hepta-β-glucoside to the membrane preparation is reversible, indicating that the elicitor does not become covalently attached to the binding protein(s) [47, 87].

The membrane-localized, elicitor-binding proteins exhibit a high degree of specificity with respect to the oligoglucosides that they bind. More importantly, the ability of an oligoglucoside to bind to soybean root membranes correlates with its ability to induce phytoalexin accumulation (Fig. 4) [45, 47]. Oligo-β-glucosides with high elicitor activity are efficient competitors of the radiolabeled elicitor, while biologically less active oligo-β-glucosides are less efficient. Thus, four oligo-β-glucosides ranging in size from hexamer to decamer (compounds **1**, **4**, **7**, and **8**, Fig. 2), that are indistinguishable in their abilities to induce phytoalexin accumulation [46], are equally

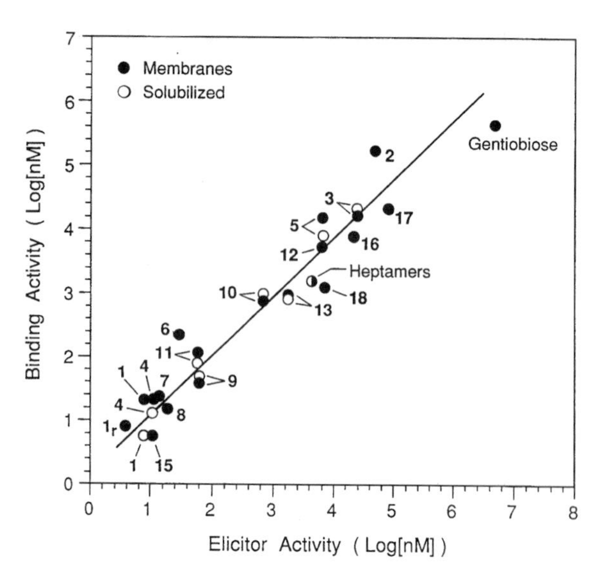

Fig. 4. Correlation of the elicitor activities of oligoglucosides with their affinities for the membrane-localized (●) or detergent-solubilized (○) hepta-β-glucoside-binding protein(s). The relative elicitor activity is defined as the concentration of an oligosaccharide required to give half-maximum induction of phytoalexin accumulation ($A/A_{std} = 0.5$) in the soybean cotyledon bioassay compared to the standard curve for hepta-β-glucoside **1** [46]. The binding activity is defined as the concentration of oligosaccharide required to give 50% inhibition of the binding of radiolabeled hepta-β-glucoside **15** to its binding protein(s). Data points are identified with numbers identifying the oligoglucosides (see Fig. 2); 1_r = reduced hepta-β-glucoside **1**; Heptamers = mixture of heptaglucosides prepared from mycelial wall hydrolysates of *P. sojae* [214]. Reprinted from [108] with permission of the American Society of Plant Physiologists.

effective competitive inhibitors of binding of radiolabeled **15** to soybean root membranes. The abilities of structurally modified oligo-β-glucosides to compete with radiolabeled hepta-β-glucoside **15** are reduced in proportion to the reduction in the biological activities of these oligo-β-glucosides (Fig. 4).

The results of the structure activity [46] and ligand-binding [45, 47] studies demonstrate that those structural elements of the hepta-β-glucoside required to elicit phytoalexin synthesis are also essential for efficient binding of the elicitor to its putative receptor. These essential structural features include the non-reducing terminal backbone glucosyl residue and the two side-chain terminal glucosyl residues of the hepta-β-glucoside elicitor.

The distribution of side-chain glucosyl residues along the backbone of the molecule is also important for recognition of the elicitor by the binding proteins. The combined results of the biological assays [46] and the binding studies [45, 47] provide strong evidence that the binding proteins are physiological receptors for the hepta-β-glucoside elicitor.

Characterization of the hepta-β-glucoside elicitor-binding proteins requires their solubilization from the membranes and subsequent purification. The low abundance of the elicitor-binding proteins in the soybean root membranes [47, 52] suggests that about a 10000-fold purification will be required to attain binding protein homogeneity, assuming a receptor molecular mass of ca. 100 kDa. Thus, the development of a purification strategy that includes ligand-affinity chromatography is essential. The following paragraphs summarize the results from solubilization studies and describe recent progress toward purification of glucan elicitor-binding proteins.

Fully functional glucan elicitor-binding proteins can be solubilized from soybean root microsomal membranes by several detergents [45, 50]. The non-ionic detergents *n*-dodecylmaltoside, *n*-dodecylsucrose and Triton X-114 each solubilize between 40 and 60% of the elicitor-binding activity from pelleted, frozen membranes in a single extraction. At least 75% of the elicitor-binding activity can be solubilized by *n*-dodecylsucrose treatment of freshly prepared membranes (unpublished results of the authors). The *n*-dodecylsucrose-solubilized binding protein preparation loses about half of its binding activity after 3 weeks at 4 °C. This loss of binding activity can be prevented in large measure by inclusion of the serine and thiol protease inhibitor, leupeptin, in the buffer during preparation of the membranes. The zwitterionic detergent *n*-dodecyl-*N*,*N*-dimethyl-3-ammonio-1-propane-sulfonate (ZW 3-12) is almost as effective at low detergent concentrations (ca. 0.3%) as the non-ionic detergents, but higher concentrations of the zwitterionic detergent yield reduced recoveries of the elicitor-binding protein(s) [45, 50].

The solubilized elicitor-binding proteins retain their high affinity for the hepta-β-glucoside elicitor. For example, the *n*-dodecylsucrose-solubilized binding proteins have an apparent K_d of 1.8 nM [45]. A similar apparent dissociation constant has been obtained for ZW 3-12-solubilized binding proteins [45, 51, 87]. More importantly, the solubilized hepta-β-glucoside binding proteins and the membrane-bound proteins have the same specificity for elicitor-active oligoglucosides (Fig. 4), irrespective of which detergent is used for solubilization [45]. The successful solubilization of the elicitor-binding proteins with retention of their elicitor-binding properties is a crucial first step toward the purification of these proteins.

Recently, progress has been made towards identification and purification of hepta-β-glucoside elicitor-binding proteins. Size-exclusion chromatography of detergent-solubilized hepta-β-glucoside elicitor-binding proteins indicates that elicitor-binding activity is associated primarily with large detergent-protein micelles (M_r > 200000 [45, 50]). Indeed, the highest specific elicitor-binding activity is present in detergent-protein micelles having a $M_r > 669000$. Moreover, more than 95% of the elicitor-binding activity is retained after concentration of the solubilized membrane proteins using membrane filters with molecular mass cut-offs of 100 and 300 kDa (unpublished results). Little or no elicitor-binding activity is associated with the smallest detergent-protein micelles [45]. The large elicitor-binding protein-detergent complexes can be disrupted only with a combination of high detergent concentrations and sonication; a treatment that results in the loss of 70% of the binding activity (unpublished results). These data suggest that the elicitor-binding proteins exist in multimeric protein complexes. Indeed, the fraction of solubilized membrane proteins retained on affinity columns carrying either an immobilized mixture of elicitor-active fungal glucans [51, 87] or immobilized hepta-β-glucoside elicitor (unpublished results) and subsequently eluted with free ligand contains several polypeptides as visualized by SDS-PAGE. Photo-affinity labeling experiments carried out on solubilized soybean membrane preparations suggests that at least three of these polypeptides (ca. 70, 100, and 170 kDa) carry elicitor-binding domains [51]. Preliminary results indicate that the affinity-purified hepta-β-glucoside binding proteins retain the same specificity for elicitor-active oligoglucosides that was observed for the membrane-localized and crude detergent-solubilized proteins (unpublished results). Further purification of these proteins and assays of the polypeptides, either alone or in combination, for elicitor-binding activity will be required to determine whether all of the proteins found in the affinity-purified fraction are essential for elicitor-binding activity. The specific activity of the affinity-purified elicitor-binding protein(s) is still at least an order of magnitude lower than theoretically predicted for purification to homogeneity ([87]; unpublished results). This could be a reflection of either inactivation of the proteins following elution from the affinity column or incomplete removal of the eluting ligand from the binding proteins.

The possible mechanisms for the transmission of the glucan elicitor signal within plant cell have been examined in several studies. For example, neither cAMP [111] nor phosphoinositides [221], two important second messengers in animal signal transduction, appear to be involved in glucan elicitor signal transduction. In contrast, glucan-induced phytoalexin accumulation in suspension-cultured soybean cells is inhibited when the extracellular calcium concentration is reduced [219]. Furthermore, a calcium ionophore was shown to induce phytoalexin accumulation [219], while selected calcium channel blockers inhibit elicitor-induced increases in chalcone synthase activity in soybean cells [73]. Thus, calcium ion fluxes appear to have a role in the glucan elicitor signalling pathway. Vanadate, an inhibitor of plasma membrane H^+-ATPase, mimics the effect of a crude water-insoluble cell wall elicitor from *Phytophthora cambivora* on suspension-cultured peanut cells at concentrations matching those required for inhibition of H^+-ATPase isolated from plasma membranes. Such results led those authors to propose that changes in the H^+-ATPase-associated membrane potentials are intermediates in the signal cascade leading to phytoalexin accumulation [220]. Finally, receptor-mediated endocytosis has been observed, using confocal microscopy, by following the internalization of a β-glucan elicitor-gold conjugate bound to the plasma membrane of soybean protoplasts (L. Griffing, personal communication). It is not known whether the endocytosis is part of the signal transduction pathway or a desensitization method to remove both signal and receptors from the cell surface.

The role of protein phosphorylation in the glucan elicitor-stimulated signalling pathway has been investigated in several studies. Rapid changes in the pattern of protein phosphorylation were observed after treating suspension-cultured soybean cells with an heterogeneous glucan elicitor preparation [94], but the roles, if any, of the phosphoproteins in the signalling pathway leading to phytoalexin accumulation remain unknown. A protein kinase inhibitor, K-252a, when in the presence of pure hepta-β-glucoside or partially purified β-glucan elicitors, synergistically induces chalcone synthase activity in soybean cell cultures [73]. In apparent contrast with these results, several protein phosphatase inhibitors, including okadaic acid, were recently found to induce the production of daidzein, an isoflavone precursor of the soybean phytoalexins, in soybean cotyledons [157]. Moreover, okadaic acid stimulates the expression of phenylalanine-ammonialyase in suspension-cultured soybean cells, but this effect is blocked by the addition of K-252a [157], contrasting with the synergistic stimulation of soybean chalcone synthase activity by this protein kinase inhibitor and β-glucans [73].

The effect of the various signal translation inhibitors on phytoalexin accumulation has not been assessed directly; instead their effect on the induction of early enzymes in the biosynthetic pathway and on the accumulation of precursors was measured. The inducibility of the early enzymes in the phenylpropanoid pathway in response to various stimuli is a matter of debate [31, 66]. Precursors of glyceollin, such as the isoflavone daidzein, are present in soybean tissue as glycosyl conjugates [97]; soybean cotyledons contain particularly high and constant levels of these conjugates [95]. After an incompatible pathogen infection, the daidzein conjugates are rapidly hydrolyzed; subsequently, high levels of glyceollin accumulate [97]. In a compatible interaction, conjugate hydrolysis is retarded, and only low levels of glyceollin accumulate [97]. A spatial study of the elicitation of phytoalexin accumulation in soybean cotyledons has shown that glyceollin and conjugates of the isoflavonoid precursors accumulate to high levels in cells adjacent to the elicitor application point; in more distant cell layers, only the precursor conjugates accumulate [96]. Abiotic elicitors, unlike biotic elicitors such as oligoglucosides, induce glyceollin accumulation and a decrease in the amounts of the daidzein conjugates. Such abiotic elicitors may act by inducing the release of daidzein from a preexisting pool [96]. Thus, okadaic acid-induced production of daidzein [157] may be interpreted as showing that the phosphatase inhibitor is itself a abiotic elicitor in soybean cells. In conclusion, although evidence has been presented suggesting that protein phosphorylation-dephosphorylation plays a role in elicitor signal transduction, it is not clear at which step of the signalling cascade, or the biosynthetic pathway, the phosphorylated proteins act.

The results of most studies to date on glucan elicitor signal transmission within plant cells must be interpreted cautiously since they have been carried out with heterogeneous glucan preparations, thereby making it difficult to relate the observed effects unambiguously to the glucan elicitors themselves. Clearly, investigations of the signalling pathway downstream of the signal perception merit re-examination utilizing the homogeneous hepta-β-glucoside elicitor.

Oligochitin and -chitosan signal transduction

The cellular signalling pathway triggered by oligochitosan and oligochitin elicitors is not known. Treatment of suspension-cultured soybean cells with chitosan induces electrolyte fluxes across the plasma membrane [229, 244, 245] that were proposed to result in the activation of the Ca^{2+}-dependent callose synthase [142]. In contrast, chitosan-induced defense responses in pea pods are not correlated with membrane leakage and Ca^{2+} flux [133, 134]. Based on studies of the correlations between the DP of the chitosan oligomers, the degree of *N*-acetylation of these oligosaccharides, and the abilities of oligochitosans to elicit callose synthesis, it has been suggested that the oligochitosan elicitors interact primarily with regularly spaced negative charges on the plant plasma membrane rather than with a discrete receptor [129]. Evidence has also been presented suggesting that protein phosphorylation is not involved in the chitosan-induced synthesis of callose in tomato cells [100].

Chitin oligosaccharides have been shown to induce rapid responses in plant cells that may be part of the signal cascade leading to the induction of plant defense responses. For example, the elicitation of phytoalexin production in rice cell cultures by chitin oligomers of DP$\geq$6 at sub-nanomolar concentrations [196] is accompanied by a rapid and transient membrane depolarization [144], while shorter oligomers induced neither response. The oligochitin-induced membrane depolarization is rapid (within 1 minute), suggesting that a change in membrane potential is one of the initial events of the signal transduction pathway leading to the synthesis of phytoalexins. Chitin oligomers of DP$\geq$4 also stimulate alkalinisation of the medium of suspension-cultured tomato cells at sub-nanomolar concentrations [85]. More than a 1000-fold higher concentration of the chitin trimer was required to induce an equivalent increase in the extracellular pH [85]. The reducing end of the chitin fragments could be modified without any loss of activity as long as four *N*-acetylglucosaminyl residues in their pyranose form remained intact [21]. The chitosan pentamer showed no activity in tomato, even in the micromolar range. The extracellular alkalinization of tomato cell cultures by chitin oligomers is also accompanied by changes in the pattern of protein phosphorylation [85]. The activity of oligochitins on tomato cells is only transient, and it is followed by a refractory state where the cells are no longer responsive to subsequent signal [21]. Moreover, oligochitins and a lipochitooligosaccharride nodulation factor from *Rhizobium leguminosarum* desensitize tomato cells to each other's action [218]. These results suggest the operation of an endocytic mechanism to remove both signal and receptor from the cell surface. However, these rapid responses of cells to oligochitins have yet to be directly tied to a specific signalling pathway triggered by these oligosaccharide elicitors.

A binding site for *N*-acetylchitooctaose (β-1,4-*N*-acetylglucosaminooctaose) with high affinity (apparent K_d ca. 5.4 nM) and low abundance (B_{max} ca. 0.3 pmol/mg protein) is present in membrane preparations of suspension-cultured rice cells. The binding specificity has not been investigated in detail, but the de-*N*-acetylated oligomer (DP = 7) did not compete for the binding site even at concentrations four orders of magnitude higher than were required for the *N*-acetylchitooligosaccharide elicitor [215]. More recently, the presence of a high-affinity binding site in tomato cell membranes and on the surface of whole cells was demonstrated using a radioactive derivative of the chitin pentamer active in the tomato cell-culture alkalinization assay [21]. The membrane-localized binding site has an apparent K_d of 32 nM and is of low abundance (B_{max} ca. 2.65 pmol/mg protein) [21]. Ligand specificity studies demonstrated that oligomers of DP 4 and 5 are effective competitors, while smaller oligomers are significantly less effective. Thus, there is a direct correlation between the ability of a chitin oligomer to bind to tomato membranes and its ability to induce alkalinisation of the medium of suspension-cultured tomato cells [21, 85]. A Nod factor isolated from *Rhizobium leguminosarum* is able to compete for the oligochitin binding site in tomato membrane preparations at concentrations similar to those required for the chitin pentamer; such results are consistent with the observation that some Nod factors stimulate alkalinisation of the medium of suspension-cultured tomato cells [21, 218]. The binding site has been solubilized from tomato membranes using the same non-ionic detergents that have proven useful in solubilization of the hepta-β-glucoside elicitor-binding protein(s) (T. Boller, personal communication). Successful solubilization of the oligochitin-binding activity will greatly facilitate the purification and characterization of the macromolecules responsible for the observed binding.

Glycopeptide signal transduction

Detailed studies of the signal transduction pathway(s) induced by glycopeptides oligosaccharins are in their infancy. A radiolabeled derivative of the yeast glycopeptide elicitor has been prepared with apparent full retention of biological activity. Using this ligand, a high-affinity binding site for the yeast glycopeptide elicitor has been found to be present in tomato cell membrane preparations [19]. Ligand saturation experiments have shown that the apparent K_d is 3 nM for the radiolabeled glycopeptide elicitor (carrying $Man_{10}GlcNAc_2$). The concentration of binding sites is very low (19 fmol/mg membrane protein), which is likely to hamper efforts to purify the binding protein(s) in amounts sufficient for their characterization. Free glycan side chains generated by endo-H treatment of the glycopeptides, competed with the glycopeptides for the binding *in vitro* [19]. These free glycans also act as suppressors of the biological activity of the glycopeptides. This is the first example in plants where carbohydrate-containing agonists and antagonists have been identified that appear to interact with the same binding site.

Oligogalacturonide signal transduction

The mechanism(s) by which oligogalacturonides induce responses in plant cells remain unknown, although recent research has demonstrated that oligogalacturonides induce several rapid responses at the plant cell surface. The effects of oligogalacturonides on ion fluxes across the plasma-membrane have been examined in several studies. For example, the membranes of tomato leaf mesophyll cells [225] are depolarized within 5 min of exposure to relatively high concentrations (1 mg/ml) of oligogalacturonide mixtures. Lower concentrations (10 μg/ml) of size-specific oligogalacturonides (DP 12 to 15) induce, within 5 min, a transient stimulation of K^+ efflux, alkalinisation of the medium, depolarization of the plasma membrane, and a decrease in the external Ca^{2+} concentration in suspension-cultured tobacco cells [161]. This last effect of oligogalacturonides has also been observed in carrot protoplast culture medium [167]. Prior depletion of calcium in the growth medium of tobacco cells abolished the oligogalacturonide-stimulated K^+ efflux. Size-heterogeneous mixtures of oligomannuronides and oligoguluronides are ca. 400-fold less effective in inducing K^+ efflux, suggesting that the biological response to oligogalacturonides is structurally specific. Furthermore, treatment of tobacco cells with an oligoguluronide preparation did not result in a measurable decrease in the Ca^{2+} concentration of the incubation medium. The decrease of Ca^{2+} concentration in the incubation medium can be explained by either an increase in the Ca^{2+} influx, or by the formation of intermolecular complexes (dimers) between oligogalacturonides with a DP > 10 in the presence of Ca^{2+} [137, 138, 189], thereby immobilizing the divalent cation in the medium or in the cell wall. The latter hypothesis has led to the suggestion that an interaction between oligogalacturonide-Ca^{2+} complexes and plant cells induces the localized physiological and morphological responses [79, 206], although no evidence for a specific binding site (receptor) for such complexes has been presented. Altogether, these observations provide evidence that size-specific oligogalacturonides, at low concentrations, can specifically induce rapid responses at the plant cell surface.

There is evidence that at least some plant cells become rapidly desensitized to oligogalacturonides. For example, tobacco cells treated with bioactive oligogalacturonides to induce K^+ efflux respond less intensively to a second treatment with the same fragments [161]. Moreover, oligogalacturonide-induced ethylene biosynthesis in cultured pear cells results in a time-dependent decrease in the ability of the cells to produce ethylene in response to a second elicitor treatment [39]. These desensitization effects may be explained by internalization of oligogalacturonide receptors from the plasma membrane [122].

Size-specific oligogalacturonides (DP 14 and 15) have been shown to enhance the *in vitro* phosphorylation of a 34 kDa protein associated with purified plasma membranes isolated from potato and tomato leaves [80, 81]. The phosphorylation of the 34 kDa protein does not appear to be directly related to the induction of proteinase inhibitor accumulation in tomato, since oligogalacturonides with DP < 14 induce the production of these inhibitors but do not enhance *in vitro* phosphorylation of the 34 kDa protein [80]. Oligoguluronide preparations, although less effective than oligogalacturonides, induce proteinase inhibitor activity and *in vitro* phosphorylation of the 34 kDa protein in tomato [80]. However, the DP and purity of the active oligoguluronides used in the bioassay were not reported.

Suspension-cultured soybean cells release H_2O_2 within 5 min of exposure to heterogeneous oligogalacturonide mixtures [7, 147], and evidence suggesting the involvement of GTP-binding proteins [146] and phospholipase C [148] in this signalling pathway has been presented. Recent data suggest that the oxidative burst induced by oligogalacturonides results from a separate signalling pathway from that leading to phytoalexin accumulation [57].

The rapid effects produced by oligogalacturonides at the plant cell surface have not yet been correlated with any of the known plant responses induced by oligogalacturonides. Thus, further investigation of the oligogalacturonide-induced membrane responses will be required in order to couple these responses to specific signalling pathways. One critical gap in our knowledge is information about the molecules responsible for initial recognition of the active oligogalacturonides. Derivatives of biologically active oligogalacturonides have been made that would be suitable for the preparation of radiolabeled ligands for use in ligand-binding studies. However, their synthesis is more complicated than for the other oligosaccharins described in previous sections due to the occurrence of side reactions involving the carboxylic acid groups in the oligogalacturonides. Homogeneous tyraminylated and biotinylated derivatives of the tridecagalacturonide have been prepared (M. Spiro and B. Ridley, personal communication). The preparation of a tyrosine hydrazone of the dodecagalacturonide has also been described [122], although the purity and structure of the derivative were not documented. Saturable binding of the radio-iodinated form of the latter derivative to intact soybean cells has been reported [154] and evidence suggesting uptake of the bound oligogalacturonide via receptor-mediated endocytosis has been presented [122]. Most recently, a fluorescein derivative of a tetradecagalacturonide-enriched fraction was prepared and binding of this derivative to the surface of soybean protoplasts was visualized using a novel silver-enhanced immunogold technique [67]. The observed binding is competable by a large excess of unlabeled heterogeneous oligogalacturonides, although only 50% competition was achieved and high standard deviations for the data were observed [67]. However, in no case has the specificity of the observed binding of the oligogalacturonides been clearly demonstrated, raising questions about the significance of the findings with respect to oligogalacturonide signal transduction. These preliminary studies clearly indicate that the molecular tools are now in hand to pursue the identification and characterization of oligogalacturonide-binding proteins.

Biological relevance

Almost all of the investigations of oligosaccharins described above have utilized *in vitro* preparations, often made using harsh chemical treatments that are unlikely to be encountered *in planta*. Legitimate questions can thus be raised as to whether oligosaccharins actually are present *in vivo*, how they are released from larger complex carbohydrates, how that release is regulated, and whether the oligosaccharins can interact with their apparent target tissue to effect a biological response. Several recent studies have begun to address these questions and are summarized below.

There is evidence suggesting that bioactive fragments can be generated *in vivo* for three oligosaccharins: oligogalacturonides, xyloglucan fragments, and oligoglucosides. First, the results of preliminary studies have shown that oligogalacturonides are present in the growth medium of suspension-cultured sycamore maple cells (A. Darvill, personal communication); their presence in intact plants remains to be investigated. The presence of biologically active oligo-galacturonides in extracts of ripening tomato pericarp-tissue has also been documented [164a]. Secondly, hypotheses were formulated proposing that bioactive xyloglucan oligosaccharides could be generated by auxin-inducible endo-β-(1,4)-glucanases; such oligosaccharides could, in turn, participate in the regulation of plant growth [237]. Indeed, it has been demonstrated that the nonasaccharide XXFG is produced *in vivo* by the partial hydrolysis of xyloglucan rather than by *de novo* synthesis [88, 165]. Finally, elicitor-active glucans are released into the extracellular medium by germinating spores of *Phytophthora sojae* cultured in the absence of any plant tissue [230]. These glucans share common structural features with those released by chemical hydrolysis. These results suggest that elicitor-active glucans are released by the oomycete as a result of normal growth and development.

Enzymatic hydrolysis offers the most likely biological mechanism for the generation of oligosaccharins. Plants and pathogens are known to produce a number of hydrolytic enzymes that could be involved in this process. For example, plant β-1,3-glucanases and chitinases exist as families of isoforms whose members are expressed either constitutively or induced following pathogen infection or elicitor treatment [29]. These enzymes are mostly localized in either the cell wall, the extracellular space, or the vacuole [30, 163]. The induction of glucanase and chitinase is often observed in coordination [30]. Inducible glucanases, chitinases and chitosanases are thought to function in plant-pathogen interactions directly by inhibiting fungal growth [99, 103, 131, 162, 210]. Alternatively, any of these enzymes may generate biologically active glucans, oligochitins and -chitosans from the mycelial walls of an invading pathogen during the plant-pathogen interaction [15, 131, 196], thereby triggering plant defenses. The demonstration that bioactive glucan-, chitin- or chitosan-derived oligosaccharides are generated at the plant-pathogen interface by hydrolytic enzymes would provide evidence that these fragments do have a role *in vivo*. However, the release of bio-active fragments by purified enzymes has been demonstrated only in the case of glucanases. Several proteins with β-1,3-glucanase activity have been purified from soybean and shown to release elicitor-active fragments from the mycelial walls of *P. sojae* [113, 241]. The expression of these glucanases appear to be differentially regulated. One glucanase has been cloned and its RNA is induced by exogenous application of ethylene [222]. Two other β-1,3-glucanases are induced by both pathogen infection and the abiotic elicitor, mercuric chloride [113]. The complete structure of a glucanase-released elicitor-active oligoglucoside has, as yet, not been determined, precluding a comparison with the previously characterized chemically generated hepta-β-glucoside elicitor.

Other examples of hydrolytic enzymes possibly involved in the generation of oligosaccharins during a plant-pathogen interaction, are the microbial pectic-degrading enzymes, in particular endo-polygalacturonase and endo-pectate lyase. These enzymes are among the first to be secreted by pathogens when cultured on plant cell walls [127, 158]. On the other hand, a number of different pectic-modifying enzymes exist in uninfected plant tissues; such enzymes have the potential to generate bioactive oligogalacturonides *in planta* by de-esterification of pectic methyl esters and cleavage of homogalacturonans (e.g. [164a]). In contrast, the *in vivo* generation of glycopeptide elicitors by degradative enzymes such as proteases has not yet been addressed experimentally. The localization of such enzymes within plant tissues and the presence of bioactive glycopeptide structures as components of the glycoproteins produced by phytopathogenic microbes also await investigation.

The action of the glycosylhydrolases discussed above ultimately results *in vitro* in the rapid and total fragmentation of the plant and fungal cell wall polysaccharides into short, biologically inactive oligomers. Therefore, regulatory mechanisms must exist *in planta* in order for the action of wall polysaccharide-degrading enzymes to result in the production of longer, biologically active oligosaccharides. The involvement of specific inhibitors of the glycosylhydrolases has been presented as one possible regulatory mechanism. For example, the identification of pathogen-produced glucanase inhibitors has been reported [4, 112]. However, the role of such inhibitors in regulating the enzymatic activity, and hence elicitor-releasing activity, of plant glucanases remains unknown. Indeed, a protein secreted by *P. sojae* can specifically inhibit one β-1,3-glucanase, while having no activity on a second. Evidence for the involvement of enzyme inhibitors of pectic-degrading enzymes is more substantial. Thus, a polygalacturonase-inhibiting protein (PGIP), which has broad specificity against fungal endo-polygalacturonases [1, 42], can reduce the activity of these enzymes *in vitro* by 99.7% [41]; plant endopolygalacturonases are not affected by PGIP [42]. The reduced activity of the EPGase results in the accumulation of biologically active oligogalacturonides with DP > 10 that have half-lives of hours rather than minutes [44]. One important consequence of the action of PGIP at the site of infection might be to raise the local concentration of oligogalacturonides sufficiently high to trigger plant defense responses, thereby shifting the action of oligogalacturonides away from their regulatory role in growth and development in the healthy plant [23]. The gene encoding bean PGIP has been cloned and its expression has been studied. Interestingly, PGIP mRNA levels increase after treatment with elicitors such as oligogalacturonides, oligoglucosides and salicylic acid, and after wounding [25]. Western blot analysis using an antibody specific to PGIP, confirmed the increased expression in salicylic acid-treated and wounded tissues. Moreover, an immunocytological study using the same antibody in bean infected with *Colletotrichum lindemuthianum* has shown that PGIP accumulates preferentially in cells adjacent to the infection site [25]. The activity of endo-pectate lyases is not affected by PGIP [42]. However, evidence has been presented that, at the physiological pH of the plant cell wall (pH 5–6), the activity of endo-pectate lyase is reduced to such an extent that the formation of biologically active oligogalacturonides of DP > 10 is again favored [63].

The ability of the released oligosaccharins to move freely in the apoplast and the vascular tissues to their target site is also an important consideration with regard to the biological relevance of such signals. In systemic responses, the target site can be localized in tissues or organs at a distance from the sites where the signal was generated. Experiments addressing this question have been described using oligogalacturonides. Studies of the movement of radiolabeled oligogalacturonides (DP 6) *in planta* have shown that these molecules move freely in the vascular system, although modifications such as the shortening of the chain and esterification occurred [156]. In contrast, radiolabeled pectic fragments applied to a wound were shown not to move to adjacent tissues [22]. Whether or not the movement of oligogalacturonides in plants is size-dependent is not known.

At the cellular level, the ability of oligosaccharins to pass through the plant cell wall in order to interact with plasma membrane-localized receptors is of concern. For example, non-esterified homogalacturonans can aggregate via a calcium-mediated cooperative mechanism [98, 332]. This property of homogalacturonans has been used to localize pectin components in the plant cell wall using labeled oligogalacturonides of DP ca. 30 [228]. Furthermore, the binding of a monoclonal antibody which recognizes a calcium-induced conformation of pectin [151], is inhibited by oligogalacturonides of DP 9 and 10, but only in the presence of Ca^{2+} [152]. These results suggest that bioactive oligogalacturonides in the presence of Ca^{2+} could be bound to partially de-esterified stretches of pectin, thereby immobilizing the fragments in the wall matrix. The immobilization of oligochitosan in the cell wall and at the cell surface might occur through the interaction of regularly spaced negative charges on galacturonans and on the plasma membrane with the positive charges on this oligosaccharin. The differences between the activities of chitosan and chitin fragments might be attributable to the ability of oligochitosan, but not oligochitin, to participate in such ionic interactions with the cell wall. Until evidence is provided that specific plasma membrane-localized receptors exist for oligogalacturonides and oligochitosans, one cannot rule out that these oligosaccharins exert their biological effects by interactions within the cell wall.

Another mechanism that could prevent biologically active oligosaccharides from reaching their target site at the plasma membrane is their use as substrates by cell wall-localized enzymes. Such enzymes could integrate the oligosaccharin into the wall (e.g. glycosyltransferase), or modify or remove glycosyl residues, thereby destroying the activity of the oligosaccharin. For example, xyloglucan endo-transglycosylase cuts xyloglucan and transfers the remaining attached portion of the polysaccharide to an acceptor xyloglucan, usually another polysaccharide. Small xyloglucan fragments can act as competing acceptors for this enzyme, thus enhancing cell expansion and promoting elongation of tissues [90], a role in apparent contrast with their observed biological activity. Furthermore, a xyloglucan oligosaccharide-specific α-xylosidase has been identified and extracted from pea stems [178] and germinating *Nasturtium* seeds [77]. The purified enzyme only releases the xylosyl residue attached to the terminal non-reducing end glucosyl residue of the nonamer XXFG, to yield the octamer GXFG. α-Fucosidases that cleave the fucosyl residue from the bioactive nonamer have been detected in *Nasturtium* cotyledons [78] and purified from almond emulsin [242] and etiolated pea stems [8, 179]. Such enzymes may play a role in the regulation of the activity of XXFG, albeit further studies of their expression and regulation will be necessary to substantiate this hypothesis.

Conclusions

The study of any physiological event mediated by receptor-effector interactions goes through several experimental stages. One could define these stages as being the identification of the biological response and the development of an effective and reproducible biological assay for the response, the description and the characterization of signal molecules that trigger this response, and finally, the understanding of the underlying mechanisms involved in the perception and the translation of the signals leading to the observed biological events. The results of extensive research in a number of laboratories over the past 15 years has clearly established that oligosaccharins are important signal molecules that play major roles in biological phenomena involved in plant-pathogen interactions and plant developmental processes. The structural characterization of these molecules has been in some cases, a major experimental challenge. However, knowledge of the structures of oligosaccharins and the biological responses to these signals has now progressed to the point where detailed studies of the modes of action of oligosaccharins can be undertaken.

Oligosaccharides were first described as signals in plants. Due to the complexity of the plant cell wall, it is probable that more plant-derived oligosaccharins will be identified, possibly with novel activities. Equally complex, the cell surfaces of the many microorganisms that interact with plants could also provide a pool of oligosaccharide signals in addition to those reviewed here. It is becoming equally clear that oligosaccharides can act as signals in organisms other than plants. For example, oligosaccharides derived from plant cell walls have been recently found to affect the expression of specific genes in bacteria [203], although the active component has not been purified and characterized. Bacterial oligosaccharide-containing molecules such as LPS and K-antigens are also known to trigger, or prevent, specific and general defense responses in mammals [124, 195]; the same bacterial components are known to play a role in plant-bacterial interactions, although the specific functions have not yet been determined [174, 197]. Fungal-derived oligosaccharides can act as regulators of the immune responses in animals [226]. Endogenous mammalian oligosaccharides such as fragments of heparin and heparan sulphate have drawn attention because of their involvement in the regulation of blood coagulation [32] and fibroblast growth factor activity [231]. Finally, human receptors for fungal β-glucans have been identified [55]. In addition, oligosaccharide serve as ligands for the cell surface selectins, which have important roles in cell-cell interactions [54, 83], although a regulatory activity of the oligosaccharide ligands has yet to be demonstrated. The identification of new activities of oligosaccharides in a broader range of organisms, together with extensive structure activity and signal transduction studies, will undoubtedly broaden our understanding of the crucial role of these molecules as biological signals and yield valuable insights into how cells perceive and respond to stimuli from their environment [205].

Progress in elucidating the mechanisms by which oligosaccharins exert their effects on plant cells has often been hampered by the heterogeneity of the oligosaccharide preparations used in many studies. Detailed biochemical investigations of the cellular signalling pathways triggered by oligosaccharins require the isolation and/or chemical synthesis of homogeneous oligosaccharides in order to unambiguously assign the observed effects to single inducer-stimulated pathways. Recent improvements in the techniques for the purification of oligosaccharides and contributions from synthetic organic chemists, have made available homogeneous preparations of oligosaccharides in milligram quantities. For example, the availability of chemically synthesized hepta-β-glucoside elicitor and structurally related, less active agonists, has lead to the tentative identification of a binding protein (putative receptor) for this elicitor. Similarly, research on the other oligosaccharides, particularly on oligochitin and glycopeptide fragments is also beginning to shed light on their modes of action.

We have summarized the binding characteristics of putative receptors for three oligosaccharins, namely the chitooligosaccharide, the yeast glycopeptide and the hepta-β-glucoside elicitor. It is somewhat surprising that no binding studies have been performed with xyloglucan oligosaccharins, given the availability of pure and homogenous preparations of the different oligomers. It would be interesting to know whether the effect of xyloglucan fragments on auxin-stimulated growth in pea and *Nasturtium* is mediated via a specific membrane receptor, or by interaction with one of the auxin-binding proteins. The high affinity, saturability, and ligand specificity of the oligosaccharin-binding proteins identified thus far, strongly suggests that these are physiological receptors for their respective oligosaccharin. Moreover, the low apparent K_d values for these binding proteins indicates maximal occupancy of the putative receptor, thereby fully triggering the cellular transduction pathways at low signal concentrations. However, the purification and characterization of oligosaccharin-binding proteins and/or the isolation of the gene(s) encoding these proteins will be required to prove that the binding proteins function as physiological receptors. Evidence supporting a role for these protein(s) in the cellular signalling pathway leading to the biological response could be obtained by reconstitution of an oligosaccharin responsive system using purified binding protein(s) or by functional expression of the genes encoding the binding protein(s). It will be of great interest to determine the specific functions of oligosaccharin receptors in the plant signal transduction pathways, and to see if new paradigms for signal perception and signal translation emerge from these studies.

Acknowledgements

Our work on glucan elicitor-binding proteins is supported by a grant from the National Science Foundation (MCB-9206882), and in part by the Department of Energy-funded Center for Plant and Microbial Complex Carbohydrates (DE-FG09-93ER20097). F.C. has been supported in part by a post-doctoral fellowship from the Natural Sciences and Engineering Research Council of Canada. We are grateful to P. Garegg (University of Stockholm, Sweden), T. Ogawa (RIKEN, Japan), J. van Boom (University of Leiden, Netherlands) and their colleagues for their generous gifts of synthetic oligoglucosides. We also thank Rob Alba, Alan Darvill, Debra Mohnen, Malcolm O'Neill, Brad Reuhs, Mark Spiro and Myron Williams for helpful discussions. The technical assistance of Anna-Maria Sult, Jürg Enkerli and Paul Garlich, and the drawing skills of Carol L. Gubbins are also gratefully acknowledged.

References

1. Albersheim P, Anderson AJ: Proteins from plant cell walls inhibit polygalacturonases secreted by plant pathogens. Proc Natl Acad Sci USA 68: 1815–1819 (1971).
2. Albersheim P, Darvill AG: Oligosaccharins. Sci Am 253(3): 58–64 (1985).
3. Albersheim P, Darvill AG, McNeil M, Valent BS, Sharp JK, Nothnagel EA, Davis KR, Yamazaki N, Gollin DJ, York WS, Dudman WF, Darvill JE, Dell A: Oligosaccharins: Naturally occurring carbohydrates with biological regulatory functions. In: Ciferri O, Dure L, III (eds) Structure and Function of Plant Genomes, pp. 293–312. Plenum, New York, (1983).
4. Albersheim P, Valent BS: Host-pathogen interactions. VII. Plant pathogens secrete proteins which inhibit enzymes of the host capable of attacking the pathogen. Plant Physiol 53: 684–687 (1974).
5. Aldington S, Fry SC: Oligosaccharins. Adv Bot Res 19: 1–101 (1993).
6. Anderson AJ: The biology of glycoproteins as elicitors. In: Kosuge T, Nester E (eds) Plant-Microbe Interactions: Molecular and Genetic Perspectives, vol. 3, pp. 87–130. McGraw-Hill, New York (1989).
7. Apostol I, Heinstein PF, Low PS: Rapid stimulation of an oxidative burst during elicitation of cultured plant cells. Role in defense and signal transduction. Plant Physiol 90: 109–116 (1989).
8. Augur C, Benhamou N, Darvill A, Albersheim P: Purification, characterization, and cell wall localization of an α-fucosidase that inactivates a xyloglucan oligosaccharin. Plant J 3: 415–426 (1993).
9. Augur C, Yu L, Sakai K, Ogawa T, Sinaÿ P, Darvill AG, Albersheim P: Further studies of the ability of xyloglucan oligosaccharides to inhibit auxin-stimulated growth. Plant Physiol 99: 180–185 (1992).
10. Ayers AR, Ebel J, Finelli F, Berger N, Albersheim P: Host-pathogen interactions. IX. Quantitative assays of elicitor activity and characterization of the elicitor

present in the extracellular medium of cultures of *Phytophthora megasperma* var. *sojae*. Plant Physiol 57: 751–759 (1976).
11. Ayers AR, Ebel J, Valent B, Albersheim P: Host-pathogen interactions. X. Fractionation and biological activity of an elicitor isolated from the mycelial walls of *Phytophthora megasperma* var. *sojae*. Plant Physiol 57: 760–765 (1976).
12. Ayers AR, Valent B, Ebel J, Albersheim P: Host-pathogen interactions. XI. Composition and structure of wall-released elicitor fractions. Plant Physiol 57: 766–774 (1976).
13. Baker CJ, Mock N, Atkinson MM, Hutcheson S: Inhibition of the hypersensitive response in tobacco by pectate lyase digests of cell wall and of polygalacturonic acid. Physiol Mol Plant Path 37: 155–167 (1990).
14. Baldwin EA, Biggs RH: Cell-wall lysing enzymes and products of cell-wall digestion elicit ethylene in citrus. Physiol Plant 73: 58–64 (1988).
15. Barber MS, Bertram RE, Ride JP: Chitin oligosaccharides elicit lignification in wounded wheat leaves. Physiol Mol Plant Path 34: 3–12 (1989).
16. Bartnicki-Garcia S: Cell wall chemistry, morphogenesis, and taxonomy of fungi. Annu Rev Microbiol 22: 87–108 (1968).
17. Basse CW, Bock K, Boller T: Elicitors and suppressors of the defense response in tomato cells. Purification and characterization of glycopeptide elicitors and glycan suppressors generated by enzymatic cleavage of yeast invertase. J Biol Chem 267: 10 258–10 265 (1992).
18. Basse CW, Boller T: Glycopeptide elicitors of stress responses in tomato cells. *N*-linked glycans are essential for activity but act as suppressors of the same activity when released from the glycopeptides. Plant Physiol 98: 1239–1247 (1992).
19. Basse CW, Fath A, Boller T: High affinity binding of a glycopeptide elicitor to tomato cells and microsomal membranes and displacement by specific glycan suppressors. J Biol Chem 268: 14 724–14 731 (1993).
20. Bauer WD, Talmadge KW, Keegstra K, Albersheim P: The structure of plant cell walls. II. The hemicellulose of the walls of suspension-cultured sycamore cells. Plant Physiol 51: 174–187 (1973).
21. Baureithel K, Felix G, Boller T: Specific, high affinity binding of chitin fragments to tomato cells and membranes. Competitive inhibition of binding by derivatives of chitin fragments and a nod factor of *Rhizobium*. J Biol Chem 269: 17 931–17 938 (1994).
22. Baydoun EA-H, Fry SC: The immobility of pectic substances in injured tomato leaves and its bearing on the identity of the wound hormone. Planta 165: 269–276 (1985).
23. Bellincampi D, De Lorenzo G, Cervone F: Oligogalacturonides as signal molecules in plant pathogen interactions and in plant growth and development. IAPTC Newsl, in press (1994).
24. Bellincampi D, Salvi G, De Lorenzo G, Cervone F, Marfà V, Eberhard S, Darvill A, Albersheim P: Oligogalacturonides inhibit the formation of roots on tobacco explants. Plant J 4: 207–213 (1993).
25. Bergmann CW, Ito Y, Singer D, Albersheim P, Darvill AG, Benhamou N, Nuss L, Salvi G, Cervone F, De Lorenzo G: Polygalacturonase-inhibiting protein accumulates in *Phaseolus vulgaris* L. in response to wounding, elicitors and fungal infection. Plant J 5: 625–634 (1994).
26. Bhandal IS, Paxton JD: Phytoalexin biosynthesis induced by the fungal glucan polytran L in soybean, pea, and sweet pepper tissues. J Agric Food Chem 39: 2156–2157 (1991).
27. Birberg W, Fügedi P, Garegg PJ, Pilotti Å: Syntheses of a heptasaccharide β-linked to an 8-methoxycarbonyl-oct-1-yl linking arm and of a decasaccharide with structures corresponding to the phytoelicitor active glucan of *Phytophthora megasperma* f.sp. *glycinea*. J Carbohydr Chem 8: 47–57 (1989).
28. Bishop PD, Pearce G, Bryant JE, Ryan CA: Isolation and characterization of the proteinase inhibitor-inducing factor from tomato leaves. Identity and activity of poly- and oligogalacturonide fragments. J Biol Chem 259: 13 172–13 177 (1984).
29. Bol JF, Linthorst HJM, Cornelissen BJC: Plant pathogenesis-related proteins induced by virus infection. Annu Rev Phytopath 28: 113–138 (1990).
30. Boller T: Hydrolytic enzymes in plant disease resistance. In: Kosuge T, Nester EW (eds) Plant-Microbe Interactions. Molecular and Genetic Perspectives, vol. 2, pp. 385–413. Macmillan, New York (1987).
31. Bonhoff A, Loyal R, Ebel J, Grisebach H: Race:cultivar-specific induction of enzymes related to phytoalexin biosynthesis in soybean roots following infection with *Phytophthora megasperma* f. sp. *glycinea*. Arch Biochem Biophys 246: 149–154 (1986).
32. Bourin M-C, Lindahl U: Glycosaminoglycans and the regulation of blood coagulation. Biochem J 289: 313–330 (1993).
33. Brady KP, Darvill AG, Albersheim P: Activation of a tobacco glycine-rich protein gene by a fungal glucan preparation. Plant J 4: 517–524 (1993).
34. Branca C, De Lorenzo G, Cervone F: Competitive inhibition of the auxin-induced elongation by α-D-oligogalacturonides in pea stem segments. Physiol Plant 72: 499–504 (1988).
35. Brecht JK, Huber DJ: Products released from enzymically active cell wall stimulate ethylene production and ripening in preclimacteric tomato (*Lycopersicon esculentum* Mill.) fruit. Plant Physiol 88: 1037–1041 (1988).
36. Broekaert WF, Peumans WJ: Pectic polysaccharides elicit chitinase accumulation in tobacco. Physiol Plant 74: 740–744 (1988).
37. Bruce RJ, West CA: Elicitation of lignin biosynthesis and isoperoxidase activity by pectic fragments in

suspension-cultures of castor bean. Plant Physiol 91: 889–897 (1989).
38. Campbell AD, Labavitch JM: Induction and regulation of ethylene biosynthesis and ripening by pectic oligomers in tomato pericarp discs. Plant Physiol 97: 706–713 (1991).
39. Campbell AD, Labavitch JM: Induction and regulation of ethylene biosynthesis by pectic oligomers in cultured pear cells. Plant Physiol 97: 699–705 (1991).
40. Cervone F, De Lorenzo G, Degrà L, Salvi G: Elicitation of necrosis in *Vigna unguiculata* Walp. by homogeneous *Aspergillus niger* endo-polygalacturonase and by α-D-galacturonate oligomers. Plant Physiol 85: 626–630 (1987).
41. Cervone F, De Lorenzo G, Degrà L, Salvi G, Bergami M: Purification and characterization of a polygalacturonase-inhibiting protein from *Phaseolus vulgaris* L.. Plant Physiol 85: 631–637 (1987).
42. Cervone F, De Lorenzo G, Pressey R, Darvill AG, Albersheim P: Can *Phaseolus* PGIP inhibit pectic enzymes from microbes and plants? Phytochemistry 29: 447–449 (1990).
43. Cervone F, De Lorenzo G, Salvi G, Bergmann C, Hahn MG, Ito Y, Darvill A, Albersheim P: Release of phytoalexin elicitor-active oligogalacturonides by microbial pectic enzymes. In: Lugtenberg BJJ (ed) Signal Molecules in Plants and Plant-Microbe Interactions. NATO ASI Series, vol. H36, pp. 85–89. Springer-Verlag, Heidelberg, FRG (1989).
44. Cervone F, Hahn MG, De Lorenzo G, Darvill A, Albersheim P: Host-pathogen interactions. XXXIII. A plant protein converts a fungal pathogenesis factor into an elicitor of plant defense responses. Plant Physiol 90: 542–548 (1989).
45. Cheong J-J, Alba R, Côté F, Enkerli J, Hahn MG: Solubilization of functional plasma membrane-localized hepta-β-glucoside elicitor binding proteins from soybean. Plant Physiol 103: 1173–1182 (1993).
46. Cheong J-J, Birberg W, Fügedi P, Pilotti Å, Garegg PJ, Hong N, Ogawa T, Hahn MG: Structure-activity relationships of oligo-β-glucoside elicitors of phytoalexin accumulation in soybean. Plant Cell 3: 127–136 (1991).
47. Cheong J-J, Hahn MG: A specific, high-affinity binding site for the hepta-β-glucoside elicitor exists in soybean membranes. Plant Cell 3: 137–147 (1991).
48. Cline K, Wade W, Albersheim P: Host-pathogen interactions. XV. Fungal glucans which elicit phytoalexin accumulation in soybean also elicit the accumulation of phytoalexins in other plants. Plant Physiol 62: 918–921(1978).
49. Conrath U, Domard A, Kauss H: Chitosan-elicited synthesis of callose and of coumarin derivatives in parsley cell suspension cultures. Plant Cell Rep 8: 152–155 (1989).
50. Cosio EG, Frey T, Ebel J: Solubilization of soybean membrane binding sites for fungal β-glucans that elicit phytoalexin accumulation. FEBS Lett 264: 235–238 (1990).
51. Cosio EG, Frey T, Ebel J: Identification of a high-affinity binding protein for a hepta-β-glucoside phytoalexin elicitor in soybean. Eur J Biochem 204: 1115–1123 (1992).
52. Cosio EG, Frey T, Verduyn R, Van Boom J, Ebel J: High-affinity binding of a synthetic heptaglucoside and fungal glucan phytoalexin elicitors to soybean membranes. FEBS Lett 271: 223–226 (1990).
53. Cosio EG, Pöpperl H, Schmidt WE, Ebel J: High-affinity binding of fungal β-glucan fragments to soybean (*Glycine max* L.) microsomal fractions and protoplasts. Eur J Biochem 175: 309–315 (1988).
54. Cummings RD, Smith DF: The selectin family of carbohydrate-binding proteins: structure and importance of carbohydrate ligands for cell adhesion. BioEssays 14: 849–856 (1992).
55. Czop JK, Austen KF: A β-glucan inhibitable receptor on human monocytes: Its identity with the phagocytic receptor for particulate activators of the alternative complement pathway. J Immunol 134: 2588–2593 (1985).
56. Darvill AG, Albersheim P: Phytoalexins and their elicitors: a defense against microbial infection in plants. Annu Rev Plant Physiol 35: 243–275 (1984).
56a. Darvill A, Augur C, Bergmann C, Carlson RW, Cheong J-J, Eberhard S, Hahn MG, Lo V-M, Marfà V, Meyer B, Mohnen D, O'Neill MA, Spiro MD, van Halbeek H, York WS, Albersheim P: Oligosaccharins – oligosaccharides that regulate growth, development and defense responses in plants. Glycobiology 2: 181–198 (1992).
57. Davis D, Merida J, Legendre L, Low PS, Heinstein P: Independent elicitation of the oxidative burst and phytoalexin formation in cultured plant cells. Phytochemistry 32: 607–611 (1993).
58. Davis KR, Darvill AG, Albersheim P: Host-pathogen interactions. XXXI. Several biotic and abiotic elicitors act synergistically in the induction of phytoalexin accumulation in soybean. Plant Mol Biol 6: 23–32 (1986).
59. Davis KR, Darvill AG, Albersheim P, Dell A: Host-pathogen interactions. XXIX. Oligogalacturonides released from sodium polypectate by endopolygalacturonic acid lyase are elicitors of phytoalexins in soybean. Plant Physiol 80: 568–577 (1986).
60. Davis KR, Darvill AG, Albersheim P, Dell A: Host-pathogen interactions. XXX. Characterization of elicitors of phytoalexin accumulation in soybean released from soybean cell walls by endopolygalacturonic acid lyase. Z Naturforsch 41c: 39–48 (1986).
61. Davis KR, Hahlbrock K: Induction of defense responses in cultured parsley cells by plant cell wall fragments. Plant Physiol 85: 1286–1290 (1987).
62. Davis KR, Lyon GD, Darvill AG, Albersheim P: Host-pathogen interactions. XXV. Endopolygalacturonic acid lyase from *Erwinia carotovora* elicits phytoalexin accu-

mulation by releasing plant cell wall fragments. Plant Physiol 74: 52–60 (1984).
63. De Lorenzo G, Cervone F, Hahn MG, Darvill A, Albersheim P: Bacterial endopectate lyase: evidence that plant cell wall pH prevents tissue maceration and increases the half-life of elicitor-active oligogalacturonides. Physiol Mol Plant Path 39: 335–344 (1991).
64. De Lorenzo G, Ito Y, D'Ovidio R, Cervone F, Albersheim P, Darvill AG: Host-pathogen interactions. XXXVII. Abilities of the polygalacturonase-inhibiting proteins from four cultivars of *Phaseolus vulgaris* to inhibit the *endo*polygalacturonases from three races of *Colletotrichum lindemuthianum*. Physiol Mol Plant Path 36: 421–435 (1990).
65. de Wit PJGM: Molecular characterization of gene-for-gene systems in plant-fungus interactions and the application of avirulence genes in control of plant pathogens. Annu Rev Phytopath 30: 391–418 (1992).
66. Dhawale S, Souciet G, Kuhn DN: Increase of chalcone synthase mRNA in pathogen-inoculated soybeans with race-specific resistance is different in leaves and roots. Plant Physiol 91: 911–916 (1989).
67. Diekmann W, Herkt B, Low PS, Nürnberger T, Scheel D, Terschüren C, Robinson DG: Visualization of elicitor binding loci at the plant cell surface. Planta 195: 126–137 (1994).
68. Dixon RA: The phytoalexin response: elicitation, signalling and control of host gene expression. Biol Rev 61: 239–291 (1986).
69. Dixon RA, Jennings AC, Davies LA, Gerrish C, Murphy DL: Elicitor active components from French bean hypocotyls. Physiol Mol Plant Path 34: 99–115 (1989).
70. Dixon RA, Lamb CJ: Molecular communication in interactions between plants and microbial pathogens. Annu Rev Plant Physiol Plant Mol Biol 41: 339–367 (1990).
71. Ebel J: Phytoalexin synthesis: The biochemical analysis of the induction process. Annu Rev Phytopath 24: 235–264 (1986).
72. Ebel J, Cosio EG: Elicitors of plant defense responses. Int Rev Cytol 148: 1–36 (1994).
73. Ebel J, Cosio EG, Feger M, Frey T, Kissel U, Reinold S, Waldmüller T: Glucan elicitor-binding proteins and signal transduction in the activation of plant defence. In: Nester EW, Verma DPS (eds) Advances in Moleclar Genetics of Plant-Microbe Interactions, vol. 2, pp. 477–484. Kluwer Academic Publishers, Dordrecht, Netherlands (1993).
74. Eberhard S, Doubrava N, Marfà V, Mohnen D, Southwick A, Darvill A, Albersheim P: Pectic cell wall fragments regulate tobacco thin-cell-layer explant morphogenesis. Plant Cell 1: 747–755 (1989).
75. El Rassi Z, Tedford D, An J, Mort A: High-performance reversed-phase chromatographic mapping of 2-pyridylamino derivatives of xyloglucan oligosaccharides. Carbohydr Res 215: 25–38 (1991).
76. Emmerling M, Seitz HU: Influence of a specific xyloglucan-nonasaccharide derived from cell walls of suspension-cultured cells of *Daucus carota* L. on regenerating carrot protoplasts. Planta 182: 174–180 (1990).
77. Fanutti C, Gidley MJ, Reid JSG: A xyloglucan-oligosaccharide-specific α-D-xylosidase or *exo*-oligoxyloglucan-α-xylohydrolase from germinated nasturtium (*Tropaeolum majus* L.) seeds. Purification, properties and its interaction with a xyloglucan-specific *endo*-(1→4)-β-D-glucanase and other hydrolases during storage-xyloglucan mobilisation. Planta 184: 137–147 (1991).
78. Farkas V, Hanna R, Maclachlan G: Xyloglucan oligosaccharide α-L-fucosidase activity from growing pea stems and germinating nasturtium seeds. Phytochemistry 30: 3203–3207 (1991).
79. Farmer EE, Moloshok TD, Ryan CA: *In vitro* phosphorylation in response to oligouronide elicitors: structural and biological relationships. Curr Top Plant Biochem Physiol 9: 249–258 (1990).
80. Farmer EE, Moloshok TD, Saxton MJ, Ryan CA: Oligosaccharide signalling in plants: Specificity of oligouronide-enhanced plasma membrane protein phosphorylation. J Biol Chem 266: 3140–3145 (1991).
81. Farmer EE, Pearce G, Ryan CA: *In vitro* phosphorylation of plant plasma membrane proteins in response to the proteinase inhibitor inducing factor. Proc Natl Acad Sci USA 86: 1539–1542 (1989).
82. Favaron F, Alghisi P, Marciano P: Characterization of two *Sclerotinia sclerotiorum* polygalacturonases with different abilities to elicit glyceollin in soybean. Plant Sci 83: 7–13 (1992).
83. Feizi T: Oligosaccharides that mediate mammalian cell-cell adhesion. Curr Opin Struct Biol 3: 701–710 (1993).
84. Felix G, Grosskopf DG, Regenass M, Basse CW, Boller T: Elicitor-induced ethylene biosynthesis in tomato cells. Characterization and use as a bioassay for elicitor action. Plant Physiol 97: 19–25 (1991).
85. Felix G, Regenass M, Boller T: Specific perception of subnanomolar concentrations of chitin fragments by tomato cells: Induction of extracellular alkalinization, changes in protein phosphorylation, and establishment of a refractory state. Plant J 4: 307–316 (1993).
86. Forrest RS, Lyon GD: Substrate degradation patterns of polygalacturonic acid lyase from *Erwinia carotovora* and *Bacillus polymyxa* and release of phytoalexin-eliciting oligosaccharides from potato cell walls. J Exp Bot 41: 481–488 (1990).
87. Frey T, Cosio EG, Ebel J: Affinity purification and characterization of a binding protein for a hepta-β-glucoside phytoalexin elicitor in soybean. Phytochemistry 32: 543–550 (1993).
88. Fry SC: *In vivo* formation of xyloglucan nonasaccharide: a possible biologically active cell-wall fragment. Planta 169: 443–453 (1986).
89. Fry SC: The structure and functions of xyloglucan. J Exp Bot 40: 1–11 (1989).

90. Fry SC, Aldington S, Hetherington PR, Aitken J: Oligosaccharides as signals and substrates in the plant cell wall. Plant Physiol 103: 1–5 (1993).
91. Fry SC, York WS, Albersheim P, Darvill A, Hayashi T, Joseleau J-P, Kato Y, Lorences EP, Maclachlan GA, McNeil M, Mort AJ, Reid JSG, Seitz HU, Selvendran RR, Voragen AGJ, White AR: An unambiguous nomenclature for xyloglucan-derived oligosaccharides. Physiol Plant 89: 1–3 (1993).
92. Fügedi P, Birberg W, Garegg PJ, Pilotti Å: Syntheses of a branched heptasaccharide having phytoalexin-elicitor activity. Carbohydr Res 164: 297–312 (1987).
93. Gilroy S, Trewavas A: Signal sensing and signal transduction across the plasma membrane. In: Larsson C, Möller IM (eds) The Plant Plasma Membrane, pp. 203–232. Springer-Verlag, Berlin (1990).
94. Grab D, Feger M, Ebel J: An endogenous factor from soybean (*Glycine max* L.) cell cultures activates phosphorylation of a protein which is dephosphorylated *in vivo* in elicitor-challenged cells. Planta 179: 340–348 (1989).
95. Graham TL: Flavonoid and isoflavonoid distribution in developing soybean seedling tissues and in seed and root exudates. Plant Physiol 95: 594–603 (1991).
96. Graham TL, Graham MY: Glyceollin elicitors induce major but distinctly different shifts in isoflavonoid metabolism in proximal and distal soybean cell populations. Mol Plant-Microbe Interact 4: 60–68 (1991).
97. Graham TL, Kim JE, Graham MY: Role of constitutive isoflavone conjugates in the accumulation of glyceollin in soybean infected with *Phytophthora megasperma*. Mol Plant-Microbe Interact 3: 157–166 (1990).
98. Grant GT, Morris ER, Rees DA, Smith PJC, Thom D: Biological interactions between polysaccharides and divalent cations: the egg-box model. FEBS Lett 32: 195–198 (1973).
99. Grenier J, Asselin A: Some pathogenesis-related proteins are chitosanases with lytic activity against fungal spores. Mol Plant-Microbe Interact 3: 401–407 (1990).
100. Grosskopf DG, Felix G, Boller T: A yeast-derived glycopeptide elicitor and chitosan or digitonin differentially induce ethylene biosynthesis, phenylalanine ammonia-lyase and callose formation in suspension-cultured tomato cells. J Plant Physiol 138: 741–746 (1991).
101. Gunia W, Hinderer W, Wittkampf U, Barz W: Elicitor induction of cytochrome P-450 monooxygenases in cell suspension cultures of chickpea (*Cicer arietinum* L.) and their involvement in pterocarpan phytoalexin biosynthesis. Z Naturforsch 46c: 58–66 (1991).
102. Hadwiger LA, Beckman JM: Chitosan as a component of pea-*Fusarium solani* interactions. Plant Physiol 66: 205–211 (1980).
103. Hadwiger LA, Line RF: Hexosamine accumulations are associated with the terminated growth of *Puccinia striiformis* on wheat isolines. Physiol Mol Plant Path 19: 249–255 (1981).
103a. Hadwiger LA, Ogawa T, Kuyama H: Chitosan polymer sizes effective in inducing phytoalexin accumulation and fungal suppression are verified with synthesized oligomers. Mol Plant-Microbe Interact 7: 531–553 (1994).
104. Hahlbrock K, Scheel D: Biochemical responses of plants to pathogens. In: Chet I (ed) Innovative Approaches to Plant Disease Control, pp. 229–254. John Wiley, New York (1987).
105. Hahn MG: Animal receptors: examples of cellular signal perception molecules. In: Lugtenberg BJJ (ed) Signal Molecules in Plants and Plant-Microbe Interactions. NATO ASI Series, vol. H36, pp. 1–26. Springer-Verlag, Heidelberg (1989).
106. Hahn MG, Albersheim P: Host-pathogen interactions. XIV. Isolation and partial characterization of an elicitor from yeast extract. Plant Physiol 62: 107–111 (1978).
107. Hahn MG, Bucheli P, Cervone F, Doares SH, O'Neill RA, Darvill A, Albersheim P: Roles of cell wall constituents in plant-pathogen interactions. In: Kosuge T, Nester EW (eds) Plant-Microbe Interactions: Molecular and Genetic Perspectives, vol. 3, pp. 131–181. McGraw-Hill, New York (1989).
108. Hahn MG, Cheong J-J, Alba R, Côté F: Oligosaccharide elicitors: Structures and signal transduction. In: Schultz J, Raskin I (eds) Plant Signals in Interactions with other Organisms, pp. 24–46. American Society of Plant Physiologists, Rockville, MD (1993).
109. Hahn MG, Cheong J-J, Birberg W, Fügedi P, Pilotti Å, Garegg P, Hong N, Nakahara Y, Ogawa T: Elicitation of phytoalexins by synthetic oligoglucosides, synthetic oligogalacturonides, and their derivatives. In: Lugtenberg BJJ (ed) Signal Molecules in Plants and Plant-Microbe Interactions. NATO ASI Series, vol. H36, pp. 91–17. Springer-Verlag, Heidelberg, FRG (1989).
110. Hahn MG, Darvill AG, Albersheim P: Host-pathogen interactions. XIX. The endogenous elicitor, a fragment of a plant cell wall polysaccharide that elicits phytoalexin accumulation in soybeans. Plant Physiol 68: 1161–1169 (1981).
111. Hahn MG, Grisebach H: Cyclic AMP is not involved as a second messenger in the response of soybean to infection by *Phytophthora megasperma* f.sp. *glycinea*. Z Naturforsch 38c: 578–582 (1983).
112. Ham K-S, Darvill AG, Albersheim P: A fungal pathogen secretes a protein that specifically inhibits a β-1,3-glucanase pathogenesis-related protein of its host. Unpublished work (1993).
113. Ham K-S, Kauffmann S, Albersheim P, Darvill AG: Host-pathogen interactions. XXXIX. A soybean pathogensis-related protein with β-1,3-glucanase activity releases phytoalexin elicitor-active heat-stable fragments from fungal walls. Mol Plant-Microbe Interact 4: 545–552 (1991).
114. Hargreaves JA, Bailey JA: Phytoalexin production by hypocotyls of *Phaseolus vulgaris* in response to consti-

tutive metabolites released by damaged bean cells. Physiol Plant Path 13: 89–100 (1978).

115. Hargreaves JA, Selby C: Phytoalexin formation in cell suspensions of *Phaseolus vulgaris* in response to an extract of bean hypocotyls. Phytochemistry 17: 1099–1102 (1978).
116. Hayashi T: Xyloglucans in the primary cell wall. Annu Rev Plant Physiol Plant Mol Biol 40: 139–168 (1989).
117. Hensel A, Brummell DA, Hanna R, Maclachlan G: Auxin-dependent breakdown of xyloglucan in cotyledons of germinating nasturtium seeds. Planta 183: 321–326 (1991).
118. Hisamatsu M, Impallomeni G, York WS, Albersheim P, Darvill AG: A new undecasaccharide subunit of xyloglucans with two α-L-fucosyl residues. Carbohydr Res 211: 117–129 (1991).
119. Hisamatsu M, York WS, Darvill AG, Albersheim P: Characterization of seven xyloglucan oligosaccharides containing from seventeen to twenty glycosyl residues. Carbohydr Res 227: 45–71 (1992).
120. Hollenberg MD: Structure-activity relationships for transmembrane signalling: The receptor's turn. FASEB J 5: 178–186 (1991).
121. Hong N, Ogawa T: Stereocontrolled syntheses of phytoalexin elicitor-active β-D-glucohexaoside and β-D-gluconononaoside. Tetrahedron Lett 31: 3179–3182 (1990).
122. Horn MA, Heinstein PF, Low PS: Receptor-mediated endocytosis in plant cells. Plant Cell 1: 1003–1009 (1989).
123. Hoson T, Masuda Y: Effect of xyloglucan nonasaccharide on cell elongation induced by 2,4-dichlorophenoxyacetic acid and indole-3-acetic acid. Plant Cell Physiol 32: 777–782 (1991).
124. Jann B, Jann K: Structure and biosynthesis of the capsular antigens of *Escherichia coli*. Curr Top Microbiol Immunol 150: 19–42 (1990).
125. Jin DF, West CA: Characteristics of galacturonic acid oligomers as elicitors of casbene synthetase activity in castor bean seedlings. Plant Physiol 74: 989–992 (1984).
126. Jones AM: Surprising signals in plant cells. Science 263: 183–184 (1994).
127. Jones TM, Anderson AJ, Albersheim P: Host-pathogen interactions. IV. Studies on the polysaccharide-degrading enzymes secreted by *Fusarium oxysporum* f.sp. *lycopersici*. Physiol Plant Path 2: 153–166 (1972).
128. Joseleau JP, Cartier N, Chambat G, Faik A, Ruel K: Structural features and biological activity of xyloglucans from suspension-cultured plant cells. Biochimie 74: 81–88 (1992).
129. Kauss H, Jeblick W, Domard A: The degree of polymerization and *N*-acetylation of chitosan determine its ability to elicit callose formation in suspension cells and protoplasts of *Catharanthus roseus*. Planta 178: 385–392 (1989).
130. Keen NT: Specific elicitors of plant phytoalexin production: determinants of race specificity in pathogens? Science 187: 74–75 (1975).
131. Kendra DF, Christian D, Hadwiger LA: Chitosan oligomers from *Fusarium solani* pea interactions, chitinase/β-glucanase digestion of sporelings and from fungal wall chitin actively inhibit fungal growth and enhance disease resistance. Physiol Mol Plant Path 35: 215–230 (1989).
132. Kendra DF, Hadwiger LA: Characterization of the smallest chitosan oligomer that is maximally antifungal to *Fusarium solani* and elicits pisatin formation in *Pisum sativum*. Exp Mycol 8: 276–281 (1984).
133. Kendra DF, Hadwiger LA: Calcium and calmodulin may not regulate the disease resistance and pisatin formation responses of *Pisum sativum* to chitosan or *Fusarium solani*. Physiol Mol Plant Path 31: 337–348 (1987).
134. Kendra DF, Hadwiger LA: Cell death and membrane leakage not associated with the induction of disease resistance in peas by chitosan or *Fusarium solani* f. sp. *phaseoli*. Phytopathology 77: 100–106 (1987).
135. Kiefer LL, York WS, Albersheim P, Darvill AG: Structural characterization of an arabinose-containing heptadecasaccharide enzymically isolated from sycamore extracellular xyloglucan. Carbohydr Res 197: 139–158 (1990).
136. Kobayashi A, Tai A, Kanzaki H, Kawazu K: Elicitor-active oligosaccharides from algal laminaran stimulate the production of antifungal compounds in alfalfa. Z Naturforsch 48c: 575–579 (1993).
137. Kohn R: Ion binding on polyuronates-alginate and pectin. Pure Appl Chem 42: 371–397 (1975).
138. Kohn R: Binding of divalent cations to oligomeric fragments of pectin. Carbohydr Res 160: 343–353 (1987).
139. Komae K, Komae A, Misaki A: A 4,5-unsaturated low molecular oligogalacturonide as a potent phytoalexin-elicitor isolated from polygalacturonide of *Ficus awkeotsang*. Agric Biol Chem 54: 1477–1484 (1990).
140. Kooiman P: The constitution of *Tamarindus*-amyloid. Rec Trav Chim 80: 849–865 (1961).
141. Kopp M, Rouster J, Fritig B, Darvill A, Albersheim P: Host-pathogen interactions. XXXII. A fungal glucan preparation protects Nicotianae against infection by viruses. Plant Physiol 90: 208–216 (1989).
142. Köhle H, Jeblick W, Poten F, Blashek W, Kauss H: Chitosan-elicited callose synthesis in soybean cells as a Ca^{2+}-dependent process. Plant Physiol 77: 544–551 (1985).
143. Köhle H, Young DH, Kauss H: Physiological changes in suspension-cultured soybean cells elicited by treatment with chitosan. Plant Sci Lett 33: 221–230 (1984).
144. Kuchitsu K, Kikuyama M, Shibuya N: *N*-acetylchitooligosaccharides, biotic elictor for phytoalexin production, induce transient membrane depolarization in suspension-cultured rice cells. Protoplasma 174: 79–81 (1993).
145. Lamb CJ, Lawton MA, Dron M, Dixon RA: Signals

and transduction mechanisms for activation of plant defenses against microbial attack. Cell 56: 215–224 (1989).

146. Legendre L, Heinstein PF, Low PS: Evidence for participation of GTP-binding proteins in elicitation of the rapid oxidative burst in cultured soybean cells. J Biol Chem 267: 20 140–20 147 (1992).

147. Legendre L, Rueter S, Heinstein PF, Low PS: Characterization of the oligogalacturonide-induced oxidative burst in cultured soybean (*Glycine max*) cells. Plant Physiol 102: 233–240 (1993).

148. Legendre L, Yueh YG, Crain R, Haddock N, Heinstein PF, Low PS: Phospholipase C activation during elicitation of the oxidative burst in cultured plant cells. J Biol Chem 268: 24 559–24 563 (1993).

148a. Lerouge P: Symbiotic host specificity between leguminous plants and rhizobia is determined by substituted and acylated glucosamine oligosaccharide signals. Glycobiology 4: 127–134 (1994).

149. Lesney MS: Growth responses and lignin production in cell suspensions of *Pinus elliottii* 'elicited' by chitin, chitosan or mycelium of *Cronartium quercum* f.sp. *fusiforme*. Plant Cell Tiss Organ Cult 19: 23–31 (1989).

150. Lesney MS: Effect of 'elicitors' on extracellular peroxidase activity in suspension-cultured slash pine (*Pinus elliottii* Engelm.). Plant Cell Tiss Organ Cult 20: 173–175 (1990).

151. Liners F, Letesson J-J, Didembourg C, Van Cutsem P: Monoclonal antibodies against pectin. Recognition of a conformation induced by calcium. Plant Physiol 91: 1419–1424 (1989).

152. Liners F, Thibault J-F, Van Cutsem P: Influence of the degree of polymerization of oligogalacturonates and of esterification pattern of pectin on their recognition by monoclonal antibodies. Plant Physiol 99: 1099–1104 (1992).

153. Lorentzen JP, Helpap B, Lockhoff O: Synthese eines elicitoraktiven Heptaglucansaccharides zur Untersuchung pflanzlicher Abwehrmechanismen. Angew Chem 103: 1731–1732 (1991).

154. Low PS, Legendre L, Heinstein PF, Horn MA: Comparison of elicitor and vitamin receptor-mediated endocytosis in cultured soybean cells. J Exp Bot 44 (Suppl): 269–274 (1993).

155. Lozovaya VV, Zabotina OA, Rumyantseva NI, Malihov RG, Zihareva MV: Stimulation of root development on buckwheat thin cell-layer explants by pectic fragments from pea stem cell walls. Plant Cell Rep 12: 530–533 (1993).

156. MacDougall AJ, Rigby NM, Needs PW, Selvendran RR: Movement and metabolism of oligogalacturonide elicitors in tomato shoots. Planta 188: 566–574 (1992).

157. MacKintosh C, Lyon GD, Mackintosh RW: Protein phosphatase inhibitors activate anti-fungal defence responses of soybean cotyledons and cell cultures. Plant J 5: 137–147 (1994).

158. Mankarios AT, Friend J: Polysaccharide degrading enzymes of *Botrytis allii* and *Sclerotium cepivorum*. Enzyme production in culture and the effect of the enzymes on isolated onion cell walls. Physiol Plant Path 17: 93–104 (1980).

159. Marfà V, Gollin DJ, Eberhard S, Mohnen D, Darvill A, Albersheim P: Oligogalacturonides are able to induce flowers to form on tobacco explants. Plant J 1: 217–225 (1991).

160. Masuta C, Van Den Bulcke M, Bauw G, Van Montagu M, Caplan AB: Differential effects of elicitors on the viability of rice suspension cells. Plant Physiol 97: 619–629 (1991).

161. Mathieu Y, Kurkdijan A, Xia H, Guern J, Koller A, Spiro M, O'Neill M, Albersheim P, Darvill A: Membrane responses induced by oligogalacturonides in suspension-cultured tobacco cells. Plant J 1: 333–343 (1991).

162. Mauch F, Mauch-Mani B, Boller T: Antifungal hydrolases in pea tissue. II. Inhibition of fungal growth by combinations of chitinase and β-1,3-glucanase. Plant Physiol 88: 936–942 (1988).

163. Mauch F, Staehelin LA: Functional implications of the subcellular localization of ethylene-induced chitinase and β-1,3-glucanase in bean leaves. Plant Cell 1: 447–457 (1989).

164. McDougall GJ, Fry SC: Structure-activity relationships for xyloglucan oligosaccharides with antiauxin activity. Plant Physiol 89: 883–887 (1989).

165. McDougall GJ, Fry SC: Xyloglucan nonasaccharide, a naturally-occurring oligosaccharin, arises *in vivo* by polysaccharide breakdown. J Plant Physiol 137: 332–336 (1991).

166. McDougall GJ, Fry SC: Purification and analysis of growth-regulating xyloglucan-derived oligosaccharides by high-pressure liquid chromatography. Carbohydr Res 219: 123–132 (1991).

166a. Melotto E, Greve LC, Labavitch JM: Cell wall metabolism. VII. Biologically active pectin oligomers in ripening tomato (*Lycopersicon esculentum* Mill.) fruits. Plant Physiol 106: 575–581 (1994).

167. Messiaen J, Read ND, Van Cutsem P, Trewavas AJ: Cell wall oligogalacturonides increase cytosolic free calcium in carrot protoplasts. J Cell Sci 104: 365–371 (1993).

168. Mohnen D, Eberhard S, Marfà V, Doubrava N, Toubart P, Gollin DJ, Gruber TA, Nuri W, Albersheim P, Darvill A: The control of root, vegetative shoot and flower morphogenesis in tobacco thin cell-layer explants (TCLs). Development 108: 191–201(1990).

169. Molano J, Durán A, Cabib E: A rapid and sensitive assay for chitinase using tritiated chitin. Anal Biochem 83: 648–656 (1977).

170. Moloshok T, Pearce G, Ryan CA: Oligouronide signalling of proteinase inhibitor genes in plants: structure-activity relationships of di- and trigalacturonic acids and

their derivatives. Arch Biochem Biophys 294: 731–734 (1992).
171. Moloshok T, Ryan CA: Di- and trigalacturonic acid and Delta^{4,5}-di- and Delta^{4,5}-trigalacturonic acids: Inducers of proteinase inhibitor genes in plants. Meth Enzymol 179: 566–569 (1989).
172. Nakahara Y, Ogawa T: Stereocontrolled, total synthesis of α-D-GalA-[(1→4)-α-D-GalA]$_8$-(1→4)-β-D-GalA-1→OPr, a synthetic model for phytoalexin elicitor-active oligogalacturonic acids. Carbohydr Res 167: c1–c7 (1987).
173. Nakahara Y, Ogawa T: Total synthesis of galactododecaosiduronic acid, an endogenous phytoalexin elicitor isolated from soybean cell wall. Tetrahedron Lett 30: 87–90 (1989).
174. Noel KD: Rhizobial polysaccharides required in symbioses with legumes. In: Verma DPS (ed) Molecular Signals in Plant-Microbe Communications, pp. 341–357. CRC Press, Boca Raton, FL (1992).
175. Nothnagel EA, McNeil M, Albersheim P, Dell A: Host-pathogen interactions. XXII. A galacturonic acid oligosaccharide from plant cell walls elicits phytoalexins. Plant Physiol 71: 916–926 (1983).
176. Nürnberger T, Nennstiel D, Jabs T, Sacks WR, Hahlbrock K, Scheel D: High-affinity binding of a fungal oligopeptide elicitor to parsley plasma membranes triggers multiple defense responses. Cell 78: 449–460 (1994).
177. O'Neil M, Albersheim P, Darvill A: The pectic polysaccharides of primary cell walls. In: Dey PM (ed) Methods in Plant Biochemistry, vol. 2, pp. 415–441. Academic Press, London (1990).
178. O'Neil RA, Albersheim P, Darvill AG: Purification and characterization of a xyloglucan oligosaccharide-specific xylosidase from pea-seelings. J Biol Chem 264: 20430–20437 (1989).
179. O'Neil RA, White AR, York WS, Darvill AG, Albersheim P: A gas chromatographic-mass spectrometric assay for glycosylases. Phytochemistry 27: 329–333 (1988).
180. Ossowski P, Pilotti Å, Garegg PJ, Lindberg B: Synthesis of a glucoheptaose and a glucooctaose that elicit phytoalexin accumulation in soybean. J Biol Chem 259: 11337–11340 (1984).
181. Palme K: Molecular analysis of plant signalling elements: relevance of eukaryotic signal transduction models. Int Rev Cytol 132: 223–283 (1992).
182. Parker JE, Hahlbrock K, Scheel D: Different cell-wall components from *Phytophthora megasperma* f.sp. *glycinea* elicit phytoalexin production in soybean and parsley. Planta 176: 75–82 (1988).
183. Parker JE, Schulte W, Hahlbrock K, Scheel D: An extracellular glycoprotein from *Phytophthora megasperma* f. sp. *glycinea* elicits phytoalexin synthesis in cultured parsley cells and protoplasts. Mol Plant-Microbe Interact 4: 19–27 (1991).
184. Pavlova ZN, Ash OA, Vnuchkova VA, Babakov AV, Torgov VI, Nechaev OA, Usov AI, Shibaev VN: Biological activity of a synthetic pentasaccharide fragment of xyloglucan. Plant Sci 85: 131–134 (1992).
185. Peña-Cortes H, Sanchez-Serrano J, Rocha-Sosa M, Willmitzer L: Systemic induction of proteinase-inhibitor-II gene expression in potato plants by wounding. Planta 174: 84–89 (1988).
186. Peters BM, Cribbs DH, Stelzig DA: Agglutination of plant protoplasts by fungal cell wall glucans. Science 201: 364–365 (1978).
187. Pospieszny H, Atabekov JG: Effect of chitosan on the hypersensitive reaction of bean to alfalfa mosaic virus. Plant Sci 62: 29–31 (1989).
188. Pospieszny H, Chirkov S, Atabekov J: Induction of antiviral resistance in plants by chitosan. Plant Sci 79: 63–68 (1991).
189. Powell DA, Morris ER, Gidley MJ, Rees DA: Conformations and interactions of pectins II. Influence of residue sequence on chain association in calcium pectate gels. J Mol Biol 155: 517–531 (1982).
190. Pressey R: Oxidized oligogalacturonides activate the oxidation of indoleacetic acid by peroxidase. Plant Physiol 96: 1167–1170 (1991).
191. Pressey R: Uronic acid oxidase in orange fruit and other plant tissues. Phytochemistry 32: 1375–1379 (1993).
192. Priem B, Gross KC: Mannosyl- and xylosyl-containing glycans promote tomato (*Lycopersicon esculentum* Mill.) fruit ripening. Plant Physiol 98: 399–401 (1992).
193. Priem B, Morvan H, Hafez AMA, Morvan C: Influence of a plant glycan of the oligomannoside type on the growth of flax plantlets. C R Acad Sci Paris III 311: 411–416 (1990).
194. Priem B, Solokwan J, Wieruszeski J-M, Strecker G, Nazih H, Morvan H: Isolation and characterization of free glycans of the oligomannoside type from the extracellular medium of a plant cell suspension. Glycoconjugate J 7: 121–132 (1990).
195. Raetz CRH: Biochemistry of endotoxins. Annu Rev Biochem 59: 129–170 (1990).
196. Ren Y-Y, West CA: Elicitation of diterpene biosynthesis in rice (*Oryza sativa* L.) by chitin. Plant Physiol 99: 1169–1178 (1992).
197. Reuhs BL, Carlson RW, Kim JS: *Rhizobium fredii* and *Rhizobium meliloti* produce 3-deoxy-D-manno-2-octulosonic acid-containing polysaccharides that are structurally analogous to group II K antigens (capsular polysaccharides) found in *Escherichia coli*. J Bact 175: 3570–3580 (1993).
198. Roberts K: Structures at the plant cell surface. Curr Opin Cell Biol 2: 920–928 (1990).
199. Robertsen B: Elicitors of the production of lignin-like compounds in cucumber hypocotyls. Physiol Mol Plant Path 28: 137–148 (1986).
200. Robertsen B: Endo-polygalacturonase from *Cladosporium cucumerinum* elicits lignification in cucumber

hypocotyls. Physiol Mol Plant Path 31: 361–374 (1987).
201. Robertsen B: Pectate lyase from *Cladosporium cucumerinum*, purification, biochemical properties and ability to induce lignification in cucumber hypocotyls. Mycol Res 94: 595–602 (1989).
202. Roby D, Toppan A, Esquerré-Tugayé M-T: Cell surfaces in plant-microorganism interactions V. Elicitors of fungal and of plant origin trigger the synthesis of ethylene and of cell wall hydroxyproline-rich glycoprotein in plants. Plant Physiol 77: 700–704 (1985).
203. Rong L, Carpita NC, Mort A, Gelvin SB: Soluble cell wall compounds from carrot roots induce the *picA* and *pgl* loci of *Agrobacterium tumefaciens*. Mol Plant-Microbe Interact 7: 6–14 (1994).
204. Ryan CA: Oligosaccharides as recognition signals for the expression of defensive genes in plants. Biochemistry 27: 8879–8883 (1988).
205. Ryan CA: Oligosaccharide signals: from plant defense to parasite offense. Proc Natl Acad Sci USA 91: 1–2 (1994).
206. Ryan CA, Farmer EE: Oligosaccharide signals in plants: A current assessment. Annu Rev Plant Physiol Plant Mol Biol 42: 651–674 (1991).
207. Sakai K, Nakahara Y, Ogawa T: Total synthesis of nonasaccharide repeating unit of plant cell wall xyloglucan: an endogenous hormone which regulates cell growth. Tetrahedron Lett 31: 3035–3038 (1990).
208. Salisbury FB, Ross CW: Plant Physiology. 3rd ed. Wadsworth Publishing Co., Belmont, CA (1985).
209. Scheel D, Parker JE: Elicitor recognition and signal transduction in plant defense gene activation. Z Naturforsch 45c: 569–575 (1990).
210. Schlumbaum A, Mauch F, Vögeli U, Boller T: Plant chitinases are potent inhibitors of fungal growth. Nature 324: 365–367 (1986).
211. Schmidt WE, Ebel J: Specific binding of a fungal glucan phytoalexin elicitor to membrane fractions from soybean *Glycine max*. Proc Natl Acad Sci USA 84: 4117–4121 (1987).
212. Sharp JK, Albersheim P, Ossowski P, Pilotti Å, Garegg PJ, Lindberg B: Comparison of the structures and elicitor activities of a synthetic and a mycelial-wall-derived hexa(β-D-glucopyranosyl)-D-glucitol. J Biol Chem 259: 11341–11345 (1984).
213. Sharp JK, McNeil M, Albersheim P: The primary structures of one elicitor-active and seven elicitor-inactive hexa(β-D-glucopyranosyl)-D-glucitols isolated from the mycelial walls of *Phytophthora megasperma* f.sp. *glycinea*. J Biol Chem 259: 11321–11336 (1984).
214. Sharp JK, Valent B, Albersheim P: Purification and partial characterization of a β-glucan fragment that elicits phytoalexin accumulation in soybean. J Biol Chem 259: 11312–11320 (1984).
215. Shibuya N, Kaku H, Kuchitsu K, Maliarik MJ: Identification of a novel high-affinity binding site for *N*-acetylchitooligosaccharide elicitor in the membrane fraction from suspension-cultured rice cells. FEBS Lett 329: 75–78 (1993).
216. Shiraishi T, Saitoh K, Kim HM, Kato T, Tahara M, Oku H, Yamada T, Ichinose Y: Two suppressors, supprescins A and B, secreted by a pea pathogen, *Mycosphaerella pinodes*. Plant Cell Physiol 33: 663–667 (1992).
217. Spiro MD, Kates KA, Koller AL, O'Neill MA, Albersheim P, Darvill AG: Purification and characterization of biologically active 1,4-linked α-D-oligogalacturonides after partial digestion of polygalacturonic acid with endopolygalacturonase. Carbohydr Res 247: 9–20 (1993).
218. Staehelin C, Granado J, Müller J, Wiemken A, Mellor RB, Felix G, Regenass M, Broughton WJ, Boller T: Perception of *Rhizobium* nodulation factors by tomato cells and inactivation by root chitinases. Proc Natl Acad Sci USA 91: 2196–2200 (1994).
219. Stäb MR, Ebel J: Effects of Ca^{2+} on phytoalexin induction by fungal elicitor in soybean cells. Arch Biochem Biophys 257: 416–423 (1987).
220. Steffens M, Ettl F, Kranz D, Kindl H: Vanadate mimics effects of fungal cell wall in eliciting gene activation in plant cell cultures. Planta 177: 160–168 (1989).
221. Strasser H, Hoffmann C, Grisebach H, Matern U: Are polyphosphoinositides involved in signal transduction of elicitor-induced phytoalexin synthesis in cultured plant cells? Z Naturforsch 41c: 717–724 (1986).
222. Takeuchi Y, Yoshikawa M, Takeba G, Tanaka K, Shibata D, Horino O: Molecular cloning and ethylene induction of mRNA encoding a phytoalexin elicitor-releasing factor, β-1,3-endoglucanase, in soybean. Plant Physiol 93: 673–682 (1990).
223. Tani M, Fukui H, Shimomura M, Tabata M: Structure of endogenous oligogalacturonides inducing shikonin biosynthesis in *Lithospermum* cell cultures. Phytochemistry 31: 2719–2723 (1992).
224. Tepper CS, Anderson AJ: Interactions between pectic fragments and extracellular components from the fungal pathogen *Colletotrichum lindemuthianum*. Physiol Mol Plant Path 36: 147–158 (1990).
225. Thain JF, Doherty HM, Bowles DJ, Wildon DC: Oligosaccharides that induce proteinase inhibitor activity in tomato plants cause depolarization of tomato leaf cells. Plant Cell Environ 13: 569–574 (1990).
226. Velupillai P, Harn DA: Oligosaccharide-specific induction of interleukin 10 production by $B220^+$ cells from schistosome-infected mice: A mechanism for regulation of $CD4^+$ T-cell subsets. Proc Natl Acad Sci USA 91: 18–22 (1994).
227. Verduyn R, Douwes M, van der Klein PAM, Mösinger EM, van der Marel GA, van Boom JH: Synthesis of a methyl heptaglucoside: analogue of the phytoalexin elicitor from *Phytophthora megasperma*. Tetrahedron 49: 7301–7316 (1993).

228. Vreeland V, Morse SR, Robichaux RH, Miller KL, Hua S-ST, Laetsch WM: Pectate distribution and esterification in *Dubautia* leaves and soybean nodules, studied with a fluorescent hybridization probe. Planta 177: 435–446 (1989).
229. Waldmann T, Jeblick W, Kauss H: Induced net Ca^{2+} uptake and callose biosynthesis in suspension-cultured plant cells. Planta 173: 88–95 (1988).
230. Waldmüller T, Cosio EG, Grisebach H, Ebel J: Release of highly elicitor-active glucans by germinating zoospores of *Phytophthora megasperma* f.sp. *glycinea*. Planta 188: 498–505 (1992).
231. Walker A, Turnbull JE, Gallagher JT: Specific heparan sulfate saccharides mediate the activity of basic fibroblast growth factor. J Biol Chem 269: 931–935 (1994).
232. Walker-Simmons M, Hadwiger L, Ryan CA: Chitosans and pectic polysaccharides both induce the accumulation of the antifungal phytoalexin pisatin in pea pods and antinutrient proteinase inhibitors in tomato leaves. Biochem Biophys Res Commun 110: 194–199 (1983).
233. Walker-Simmons M, Jin D, West CA, Hadwiger L, Ryan CA: Comparison of proteinase inhibitor-inducing activities and phytoalexin elicitor activities of a pure fungal endopolygalacturonase, pectic fragments, and chitosan. Plant Physiol 76: 833–836 (1984).
234. Walker-Simmons M, Ryan CA: Proteinase inhibitor synthesis in tomato leaves. Induction by chitosan oligomers and chemically modified chitosan and chitin. Plant Physiol 76: 787–790 (1984).
235. Warneck H, Seitz HU: Inhibition of gibberellic acid-induced elongation-growth of pea epicotyls by xyloglucan oligosaccharides. J Exp Bot 44: 1105–1109 (1993).
236. Yamada A, Shibuya N, Kodama O, Akatsuka T: Induction of phytoalexin formation in suspension-cultured rice cells by *N*-acetylchitooligosaccharides. Biosci Biotech Biochem 57: 405–409 (1993).
237. York WS, Darvill AG, Albersheim P: Inhibition of 2,4-dichlorophenoxyacetic acid-stimulated elongation of pea stem segments by a xyloglucan oligosaccharide. Plant Physiol 75: 295–297 (1984).
237a. York WS, Oates JE, van Halbeek H, Darvill AG, Albersheim P, Tiller PR, Dell A: Location of the *O*-acetyl substituents on a nonasaccharide repeating unit of a sycamore extracellular xyloglucan. Carbohydr Res 173: 113–132 (1988).
238. York WS, van Halbeek H, Darvill AG, Albersheim P: Structural analysis of xyloglucan oligosaccharides by ^{1}H-n.m.r. spectroscopy and fast-atom-bombardment mass spectrometry. Carbohydr Res 200: 9–31 (1990).
239. Yoshikawa M, Keen NT, Wang M-C: A receptor on soybean membranes for a fungal elicitor of phytoalexin accumulation. Plant Physiol 73: 497–506 (1983).
240. Yoshikawa M, Sugimoto K: A specific binding site on soybean membranes for a phytoalexin elicitor released from fungal cell walls by β-1,3-endoglucanase. Plant Cell Physiol 34: 1229–1237 (1993).
241. Yoshikawa M, Takeuchi Y, Horino O: A mechanism for ethylene-induced disease resistance in soybean: enhanced synthesis of an elicitor-releasing factor, β-1,3-endoglucanase. Physiol Mol Plant Path 37: 367–376 (1990).
242. Yoshima H, Takasaki S, Ito-Mega S, Kobata A: Purification of almond emulsin α-L-fucosidase I by affinity chromatography. Arch Biochem Biophys 194: 394–398 (1979).
243. Yoshioka H, Shiraishi T, Yamada T, Ichinose Y, Oku H: Suppression of pisatin production and ATPase activity in pea plasma membranes by orthovanadate, verapamil and a suppressor from *Mycosphaerella pinodes*. Plant Cell Physiol 31: 1139–1146 (1990).
244. Young DH, Kauss H: Release of calcium from suspension-cultured *Glycine max* cells by chitosan, other polycations, and polyamines in relation to effects on membrane permeability. Plant Physiol 73: 698–702 (1983).
245. Young DH, Köhle H, Kauss H: Effect of chitosan on membrane permeability of suspension-cultured *Glycine max* and *Phaseolus vulgaris* cells. Plant Physiol 70: 1449–1454 (1982).

Plant Molecular Biology **26**: 1413–1422, 1994.

Role of rhizobial lipo-chitin oligosaccharide signal molecules in root nodule organogenesis

Herman P. Spaink* and Ben J.J. Lugtenberg
Institute of Molecular Plant Sciences, Leiden University, Wassenaarseweg 64, 2333 AL Leiden, The Netherlands (author for correspondence)*

Received 1 April 1994; accepted in revised form 26 April 1994

Key words: plant-microbe interactions, nod metabolites, *nod* genes, oligosaccharins

Abstract

The role of oligosaccharide molecules in plant development is discussed. In particular the role of the rhizobial lipo-chitin oligosaccharide (LCO) signal molecules in the development of the root nodule indicates that oligosaccharides play an important role in organogenesis in plants. Recent results of the analyses of structures and of the biosynthesis of the LCO molecules are summarized in this paper. The knowledge and technologies that resulted from these studies will be important tools for further studying the function of LCO signals in the plant and in the search for analogous signal molecules produced by plants.

Introduction

The mechanisms underlying the formation of organs are, although one of the most intriguing problems in biology, still poorly understood. In animals, various genes and signal molecules involved in the local differentiation and dedifferentiation processes leading to organogenesis have been identified. However, in plants far less is known about organogenesis. Although several plant hormones, such as auxin and cytokinin, have been known for decades and have been studied intensively, their role in plant differentiation is still very poorly understood at the molecular level. The fact that they play a general role in many – if not all – morphogenic processes of the plant, as well as their apparent lack of specificity at a molecular level, strongly suggests that, like in animals, other, more specific (as yet undiscovered) signal molecules also have to play a role in plant development. During the east decade evidence has accumulated that several classes of oligosaccharides, called oligosaccharins, have strong effects on plant development (see [1, 17]). Recently, a novel class of oligosaccharin signal molecules has been discovered which plays a role in the host-specific interaction between rhizobial bacteria and leguminous plants leading to the nitrogen-fixing root nodules (see [22, 26, 61, 62, 71]). These signal molecules, which were shown to be lipo-chitin oligosaccharides (LCOs), are the first plant organogenesis-inducing factors discovered. Several results indicate that plants, and perhaps even animals, also use LCO analogues as signal molecules (see [62]). In this paper the role of the LCOs in plant development is discussed in the context of the oligosaccharin concept. Furthermore, the knowledge of the chemical structures and biosynthesis of LCOs is summarized since this will be an important tool in the future search for novel plant signal molecules.

The oligosaccharin concept

Oligosaccharins are defined as particular oligosaccharides which, at low concentrations, exert biological effects on plant tissue other than as carbon or energy sources [17]. The oligosaccharin concept emanates from the original discovery that a certain class of oligosaccharides acts as a potent elicitor of the plant defense response [4]. This concept was shown to be of a more general nature by the discoveries of various other classes of oligosaccharide elicitor molecules. Recently, it has been shown that even in animals, oligosaccharides can have a very strong signalling function (see [48]). This was demonstrated by Velupillai and Harn [70], who showed that the pentasaccharide LNFP-III produced by schistosome parasites is able to specifically trigger the production of cytokines by spleen B cells. For plants, the oligosaccharin concept is built around the assumption that hydrolytic enzymes of plant or parasitic origin are involved in the release of oligosaccharides from cell wall polysaccharides [17]. Indeed, some of the molecules derived by enzymatic treatment of cell wall material of the pathogenic fungus *Phytophthora megasperma* have been shown to be active elicitors of the hypersensitive response of the host plants. In this case the smallest active component appeared to be a branched heptasaccharide consisting of *D*-glucose [57]. Also in other cases oligosaccharins which elicit a defense response have been shown to be released by enzymatic treatment of cell wall material of the parasite or host plant [7, 11, 17, 23].

The function of oligosaccharins is not limited to that of signal molecules with a role in disease resistance. Some oligosaccharins have effects on plant development which are not obviously related to elicitor activity [39, 75]. Good examples are the xyloglucan-derived oligosaccharins which antagonize the growth promotion of pea stem segments by auxin at nanomolar concentrations [27, 75]. Although the effects of xyloglucan oligosaccharides are well documented in *in vitro* systems it has not been shown whether such molecules indeed play a role in the growth of intact pea plants [1].

The recently discovered LCOs produced by rhizobial bacteria are by definition oligosaccharins since they elicit various discernable effects on plants at low concentrations (see below). One of the effects that LCOs have in common with other oligosaccharins is that they can induce various effects on suspension-cultured plant cells in a species-non-specific way which, for instance, can be measured using electrophysiological techniques ([1], B. van Duijn and H.P. Spaink, unpublished results). A published example for such an effect of LCOs is the transient alkalinization of suspension cultures of tomato cells which occurs within 5 min after addition of LCOs to the culture medium [64]. However, LCOs differ from other classes of oligosaccharins discovered until now in the following respects: (1) LCOs are apparently not derived from a larger precursor by proteolytic cleavage; (2) the oligosaccharide is linked to a fatty acyl group; (3) LCOs are very plant species-specific in their activity on the (intact) host plant.

Structures and biosynthesis of rhizobial LCO

As indicated in Fig. 1, the LCOs produced by *Rhizobium*, *Azorhizobium* and *Bradyrhizobium* bacteria, collectively called rhizobia, uniformly consist of an oligosaccharide backbone of β-1,4-linked *N*-acetyl-*D*-glucosamine, varying in length between three and five sugar units. To the nitrogen of the non-reducing sugar moiety a fatty acid group is attached, the structure of which is variable (see [22, 26, 62]). In the cases of the LCOs produced by *R. meliloti* [35] and *R. leguminosarum* biovar. *viciae* [60] a special α,β-unsaturated fatty acid moiety can be present (for an example see Fig. 1). In the LCOs of other rhizobial species such a polyunsaturated fatty acyl group is not present but instead fatty acyl moieties are found of the classes which also commonly occur in the phospholipids of the cell membrane. The presence of other substitutions on the chitin backbone is dependent on the rhizobial strain (Table 1). Substitutions which have been found are: sulphate, acetyl, carbamoyl, glycerol and sugar moieties such as arabinose, 2-*O*-methylfucose or fucose. The latter two moieties can also contain additional acetyl or sulphate modifications. In the LCOs of some species an *N*-linked methyl group can also be present.

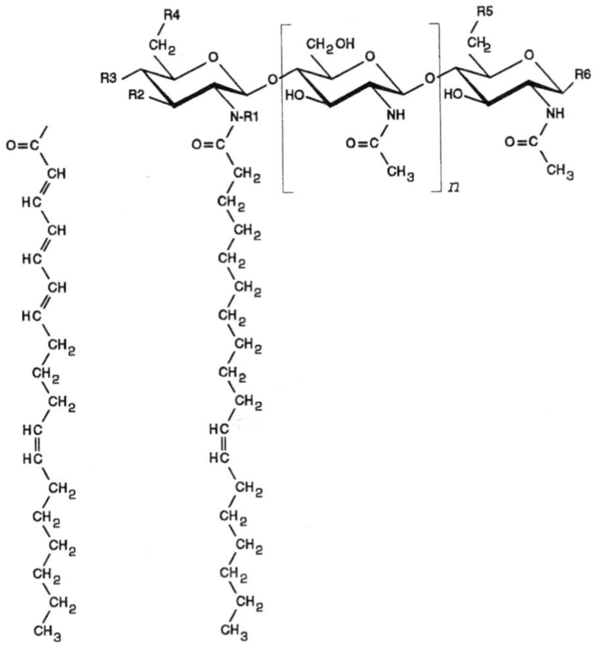

Most of the proteins encoded by the rhizobial *nod* genes play a crucial role in the biosynthesis of the LCOs [61]. The NodA, NodB and NodC proteins, which are called common Nod proteins because they are present in all rhizobia and are not involved in the determination of host specificity, are sufficient for the production of a basic LCO structure [60]. Recent results have given strong indications that the NodC and NodB proteins function as a chitin synthase and a chitin deacetylase, respectively [3, 13, 14, 18, 32, 33, 63] (Fig. 2). Since the NodA protein is essential for the production of LCOs, this protein has been

Fig. 1. Chemical structures of the rhizobial LCOs. The length of the chitin oligosaccharide backbone varies between 3 and 5 sugar units. The attached fatty acid shown is *cis*-vaccenic acid. At the left is indicated the highly unsaturated ($C_{18:4}$) moiety produced by *R. leguminosarum* biovar. *viciae* as an example of a special α,β-unsaturated fatty acyl moiety. The nature of other fatty acid moieties and of the substituents indicated by R1 to R5 is indicated in Table 1.

Table 1. Comparison of LCO structures produced by various rhizobia [1].

Producing strain	Specific lipid	Other substituents	Reference
R.l. bv. *viciae* RBL5560	$C_{18:4}$	R4, *O*-acetyl	[60]
R.l. bv. *viciae* TOM	$C_{18:4}$	R4, *O*-acetyl; R5, *O*-acetyl	[25]
R. meliloti 2011	$C_{16:2}$	R4(+/−), *O*-acetyl; R5, sulphate	[35, 67]
R. meliloti AK41	$C_{16:2}$ or $C_{16:3}$	R5, sulphate	[54]
Rhizobium NGR234	–	R1, *N*-methyl, R2 and R3(+/−), *O*-carbamoyl R5, 2-*O*-methylfucose or 2-*O*-methyl-3-*O*-sulphofucose or 2-*O*-methyl-4-*O*-acetylfucose	[42]
R. tropicii CFN299	–	R1, *N*-methyl; R5 (+/−), sulphate	[41]
R. fredii USDA257	–	R5, 2-*O*-methylfucose or fucose	[8]
B. japonicum USDA110	–	R5, 2-*O*-methylfucose	[50]
B. japonicum USDA135	–	R4(+/−), *O*-acetyl; R5, 2-*O*-methylfucose	[15]
B. japonicum USDA61	–	R1(+/−), *N*-methyl; R2 or R3 or R4 (+/−), carbamoyl; R4(+/−), *O*-acetyl; R5, 2-*O*-methylfucose or fucose; R6 (+/−), glycerol	[15]
A. caulinodans ORS571	–	R1, *N*-methyl; R4, *O*-carbamoyl; R5, *D*-arabinose	[38]

[1] Reference is made to the groups indicated in Fig. 1. A minus indicates that no α,β-unsaturated fatty acyl group is present but a common fatty acyl group like the *cis*-vaccenic acid moiety indicated in Fig. 1. If not indicated otherwise, R1 stands for hydrogen and R2, R3, R4 and R5 stand for hydroxyl groups. (+/−) indicates that such a group is not always present. Abbreviation: *R.l*, *R. leguminosarum*.

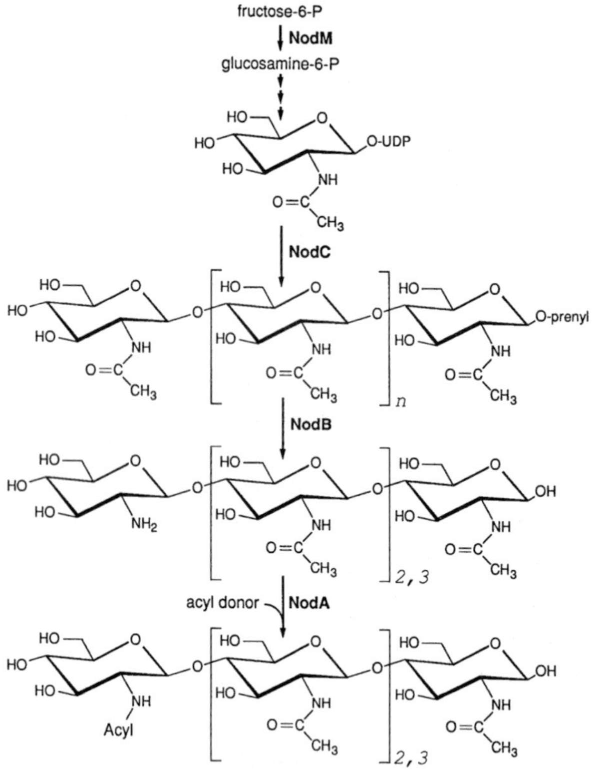

Fig. 2. Model for the functions for the NodA, NodB, and NodC proteins in the synthesis of rhizobial LCOs based upon data described in references [3, 13, 14, 18, 32, 33, 63]. The function of the NodM protein is postulated on basis of the results of Baev *et al.* [5] and Marie *et al.* [37]. The occurence of oligosaccharide metabolites attached to a prenyl carrier postulated to be produced by NodC protein has not been confirmed by structural analysis. NodC protein has also been indicated to produce free chitin oligomers ($n = 1, 2, 3$) in low quantities [63]. In addition to its deacetylating activity, NodB plays a major role in determining the quantity of produced oligosaccharides [63].

postulated to be involved in the addition of the fatty acyl moiety [10] (Fig. 2).

Other Nod proteins are involved in the synthesis or addition of various structural modifications as indicated in Table 2. These functions are in good agreement with their important roles in the determination of host-specificity. For some of these gene products their enzymatic function has also been shown by using *in vitro* test systems. In *R. meliloti* the NodP and NodQ proteins were shown to function together as ATP sulphurylase and adenosine 5′-phosphosulphate (APS) kinase, leading to the production of the sulphate donor 3′-phosphoadenosine 5′-phosphosulphate (PAPS) [55, 56]. The NodH protein acts as a sulphotransferase involved in the transfer of the sulphate moiety of PAPS to the reducing terminal sugar of the LCO acceptor [2, 35, 46]. The NodL protein, which is produced by various rhizobial species, is an acetyl transferase which is involved in the addition of the *O*-acetyl moiety to the non-reducing terminal sugar [10]. In addition to showing the biochemical functions of these Nod proteins, these results from *in vitro* analyses have also yielded valuable systems for obtaining radiolabelled derivatives of the LCO molecules which can be used in future studies devoted to their function in the plant.

Effects of LCOs on the host plant

At micromolar concentrations, externally applied purified LCO molecules can elicit in the inner

Table 2. *nod* or *nol* genes which have been shown to be involved in the addition of LCO substituents.

LCO substituent [1]	Gene involved	Reference
α,β-unsaturated fatty acid	*nodF* and *nodE*	[21, 29, 58, 60]
Sulphate (R5)	*nodP*, *nodQ* and *nodH*	[2, 46, 55, 56]
O-acetyl (R4)	*nodL*	[10, 60]
N-methyl (R1)	*nodS*	[28]
O-acetyl (R5)	*nodX*	[25]
2-*O*-methylfucose (R5)	*nodZ*	[65]

[1] Reference is made to the R groups indicated in Fig. 1.

cortex the formation of nodule primordia which are indistinguishable from the nodule primordia in the first stage of normal nodule organogenesis [60, 67]. Furthermore, as in plants which are infected by rhizobia, the primordia are only induced at certain positions in the plant root, namely the position where young root hairs emerge, opposite (or almost opposite) the protoxylem poles of the central cylinder [60, 67]. In the case of *Medicago* the nodule primordia are capable of further developing into full-grown nodules which have the anatomical and histological features of genuine rhizobium-induced nodules, such as apical meristems and peripheral vascular bundles and endodermis [67]. In *Vicia* this was never observed but instead the development of the nodules stops at a stage at which small outgrowths are externally visible on the roots [62]. Besides their role in the formation of the root nodule primordia, LCOs also seem to be involved in the bacterial infection process, as suggested by the induction of pre-infection thread structures in the outer cortex of *Vicia* roots by mitogenic LCOs in the absence of bacteria [69]. These pre-infection thread structures are characterized by the formation of so-called cytoplasmic bridges in the outer cortex which are radially aligned, giving the impression of cytoplasmic threads which cross the outer cortex. The formation of these structures, which are indistinguishable from those observed after infection with *R. leguminosarum* biovar. *viciae* bacteria, always precedes the formation of infection threads, and therefore they were named pre-infection thread structures. The formation of cytoplasmic bridges in vacuolated cells is preceded by polarization of the cell in which the nucleus moves to the centre of the cell just as in cells which are about to divide [6]. The process of pre-infection thread formation can therefore be interpreted as being the result of activation of the cell cycle as is the case of the formation of the nodule primordium in the inner cortex. The final result apparently is determined by the position of the cells in the cortex. An explanation for the local reaction of particular cortical cells to the rhizobial signals is given by the gradient hypothesis which postulates that a variation in concentration of a plant factor determines that only particular cortical cells respond towards the rhizobial signals [36, 69]. A factor from the central stele, which stimulated cell division in pea root explants at nanomolar concentrations has now been purified in our institute and was shown to be uridine ([59], G. Smit and J. Kijne, personal communication).

External application of LCOs, in concentrations varying between 10^{-8} and 10^{-12} M, can also elicit effects on root hairs of the respective host plants (see [22, 26, 61]). These effects, such as depolarization of membrane potential [24], curling, branching and swelling of the root hairs are probably related to the process of root hair curling which is also observed very early during the rhizobial infection process. Although the biological relevance of these phenotypes is not yet clear they have been shown to be very useful as semi-quantitative bioassays [30, 45].

At the molecular genetic level, several effects of LCO signals are observed which also occur during the rhizobial infection process. These effects include the induction of nodulin gene expression, for instance of the early nodulins ENOD12, ENOD5 and ENOD40, of which the expression in time and place is strongly correlated with the early steps in the symbiosis [31, 34, 52, 71, 74]. Transgenic plants which contain ENOD12-GUS reporter gene fusions have been constructed, providing a valuable molecular marker for studying LCO signal transduction in the plant [40]. Another effect of LCOs is the induction of flavonoid synthesis genes such as those encoding phenylalanine ammonia-lyase (PAL) and chalcone synthase (CHS). Flavonoid synthesis is condition-dependent since it is only detectable in roots not shielded from light [43, 44, 60, 68]. This induction process is correlated with the production of various new flavonoids which are capable of inducing the transcription of the *nod* genes [43, 44]. Savouré *et al.* [51] have shown that the cognate (*R. meliloti*) LCO signals also have various host-specific effects on gene expression in *Medicago* microcallus suspension cultures. At nanomolar concentrations a host-specific effect on the cell cycle was observed as was demonstrated by an increased expression of histone *H*3–1, *cdc2Ms* and the cyclin encoding gene *cycMs2*. A stimulation of the cell cycle was also indicated by enhanced thymidine incorporation, elevated number of S-phase cells and an increase in kinase activity of p34^{cdc2}-related complexes [51]. At higher concentrations (10^{-6} M) LCOs also induced the expression of the flavonoid synthesis gene encoding isoflavone reductase (IFR) [51].

There are several structural requirements that an LCO molecule has to fullfil in order to elicit biological effects on plant roots. The necessity of substituents, such as *O*-acetyl [60], sulphate [46] or 2-*O*-methylfucose [65], is dependent on the type of bioassay and the plant species tested. The presence of the fatty acyl substituent seems always to be required since chitin oligomers are inactive in several of the above-mentioned bioassays (see [61]). The host-specific unsaturated fatty acyl substituents are required in order to obtain nodule primordia on the roots of *Medicago* and *Vicia* plants [60, 67]. However, a special fatty acyl moiety is not required for other effects such as root hair deformation. Surprisingly, the presence of a fatty acyl substituent seems neither to be required to obtain nodule primordia when chitin oligosaccharides are delivered by ballistic microtargeting into the plant tissue (C. Sautter and H.P. Spaink, unpublished results). These results suggest that the fatty acyl group is involved in the delivery of the signal molecules inside the plant tissue.

Do plants produce chitin-derived signal molecules?

There are several indications that plants contain signal molecules that structurally resemble the rhizobial lipo-oligosaccharides.

1. In alfalfa a certain proportion of wild-type plants can spontaneously develop genuine root nodule structures in the absence of *Rhizobium* bacteria [66]. Since the number of root nodules as well as their position on the root are indistinguishable from those observed in the infected situation, this indicates that the plant is able to trigger the genes involved in the nodule formation process in the same way as *Rhizobium* bacteria do. Therefore it is possible that similar signal molecules are involved in the induction process in both cases.
2. Schmidt *et al.* [53] have shown that the *Rhizobium nodA* and *nodB* genes, when introduced singly or in combination into *Nicotiana* plants, have severe effects on plant development. One of the effects which was observed is that *nodB*-containing transgenic plants have abnormally formed leaves and flowers. Since these *nod* genes have an essential function in the biosynthesis of LCOs (Fig. 2), these results indicate that these *nod* genes interfere with the biosynthesis or structure of plant molecules which are involved in plant morphogenesis. They also suggest that such plant molecule(s) have structural homology with the bacterial LCOs.
3. De Jong *et al.* [20] have shown that the *Rhizobium* LCOs are able to rescue a temperature-sensitive somatic embryogenic mutant of *Daucus*. After addition of the LCOs in nanomolar concentrations, the ability of the mutant to form embryos was restored. In this heterologous test system the fatty acyl moiety of the LCOs was essential for activity. However, the presence of other structural modifications, like the *O*-acetyl moiety, did not influence activity [20]. Complementation of the embryogenic mutant could also be achieved by the addition of a 32 kDa endochitinase purified from wild-type *Daucus* [19]. Since chitin and its derivatives are currently the only possible known candidate substrates for this enzyme, it is tempting to speculate that the function of this chitinase is to release LCO-like molecules from larger polymers produced by *Daucus* cells. The observation that the expression of several other plant chitinases is correlated with plant development also indicates that chitin-like molecules occur in uninfected plants and could play a role in plant development (see [16, 62]).

Preservation of chitin-synthesis ability in various organisms

Since rhizobial bacteria and perhaps also plants are able to synthesize chitin oligosaccharides it is tempting to speculate about a general occurence of chitin in nature. Hardly anything is known about the occurrence of chitin derivatives in plants. In immunogold-labelling studies, using chitinase or wheat germ agglutinin as probes, Benhamou and Asselin [9] have obtained results which suggest that lipophilic chitin derivatives also occur in secondary plant cell walls of various plant species. Furthermore, using radioactive labelling studies we have recently obtained evidence that lipophilic molecules, which are susceptible to chitinase degradation, also occur in flowering *Lathyrus* plants ([62], Spaink *et al.*, unpublished results). In animals, improved methodologies for detecting chitin and chitin synthase genes has yielded results which also show that the classical notion that chitin only occurs in fungi and non-deuterostome animal taxa should be revised [73]. This is clearly indicated by the finding of chitin in the pectoral fins of the fish *Paralipophrys* [72]. The significant similarity of NodC protein, responsible for the oligomerization of the sugar backbone of the LCO (Fig. 2), with the DG42 protein, which is transiently expressed during embryogenesis of the frog [12, 47, 49, 62] suggests that chitin-like molecules might even play a role during embryogenesis in vertebrates.

Future prospects

The rhizobial LCO molecules are the first discovered examples of a novel class of signal molecules involved in plant organogenesis. Several lines of evidence indicate that plants and animals also produce chitin-derived oligosaccharide molecules. There are even indications that these molecules might play a role in the embryogenesis of plants as well as animals. Since plants are different from animals in that plants are continuously able to form new organs, it is not too far-fetched to speculate upon a generally conserved role of LCOs in the establishment of cell polarity and cell division, leading to the formation of new organs. The knowledge and the tools which have resulted from the study of the signal exchange in the nodulation process will be useful in the future search for such putative novel plant and animal signal molecules.

Acknowledgements

We thank Drs Jan Kijne (Leiden University), Gerrit Smit (Leiden University), Bert van Duijn (Leiden University) and Christof Sautter (ETH, Zürich) for communicating unpublished results. We are grateful to Drs Jane Thomas-Oates (Utrecht University) and Helmi Schlaman (Leiden University) for critically reading the manuscript. Herman Spaink was supported by the Royal Netherlands Academy of Arts and Sciences.

References

1. Aldington S, Fry SC: Oligosaccharins. Adv Bot Res 19: 1–101 (1993).
2. Atkinson EM, Ehrhardt DW, Long SR: *In vitro* activity of the *nodH* gene product from *Rhizobium meliloti*. In: Sixth International Symposium on Molecular Plant-Microbe Interactions, Program and Abstracts, abstract 45. University of Washington, Seattle, WA (1992).
3. Atkinson EM, Long SR: Homology of *Rhizobium meliloti* NodC to polysaccharide polymerizing enzymes. Mol Plant-Microbe Interact 5: 439–442 (1992).
4. Ayers AR, Ebel J, Finelli F, Berger N, Albersheim P: Host-pathogen interactions. IX. Quantitative assays of elicitor activity and characterization of the elicitor present in the extracellular medium of cultures of *Phytophthora megasperma* var. *sojae*. Plant Physiol 57: 751–759 (1976).
5. Baev N, Endre G, Petrovics G, Banfalvi Z, Kondorosi A: Six nodulation genes of nod box locus 4 in *Rhizobium meliloti* are involved in nodulation signal production: nodM codes for *D*-glucosamine synthetase. Mol Gen Genet 228: 113–124 (1991).
6. Bakhuizen R: The plant cytoskeleton in the *Rhizobium*-legume symbiosis. Ph.D. thesis, Leiden University, Netherlands (1988).
7. Barber MS, Bertram RE, Ride JP: Chitin oligosaccharides elicit lignification in wounded wheat leaves. Physiol Mol Plant Path 34: 3–12 (1989).
8. Bec-Ferté MP, Savagnac A, Pueppke SG, Promé JC: Nod factors from *Rhizobium fredii* USDA257. In: Palacios R, Mora J, Newton WE (eds) New Horizons in Nitrogen Fixation, pp. 157–158. Kluwer Academic Publishers, Dordrecht (1993).
9. Benhamou N, Asselin A: Attempted localization of a sub-

strate for chitinases in plant cells reveals abundant *N*-acetyl-D-glucosamine residues in secondary walls. Biol Cell 67: 341–350 (1989).

10. Bloemberg GV, Thomas-Oates JE, Lugtenberg BJJ, Spaink HP: Nodulation protein NodL of *Rhizobium leguminosarum O*-acetylates lipo-oligosaccharides, chitin fragments and *N*-acetylglucosamine *in vitro*. Mol Microbiol 11: 793–804 (1994).
11. Bowles DJ: Defense-related proteins in higher plants. Ann Rev Biochem 59:873–907 (1990).
12. Bulawa CE, Wasco W: Chitin and nodulation. Nature 353: 710 (1991).
13. Bulawa CE: CSD2,CSD3 and CSD4, genes required for chitin synthesis in *Saccharomyces cerevisiae*: the CSD2 gene product is related to chitin synthases and to developmentally regulated proteins in *Rhizobium* species and *Xenopus laevis*. Mol Cell Biol 12: 1764–1776 (1992).
14. Bulawa CE: Genetics and molecular biology of chitin synthesis in fungi. Annu Rev Microbiol 47: 505–534 (1993).
15. Carlson RW, Sanjuan J, Bhat R, Glushka J, Spaink HP, Wijfjes HM, van Brussel AN, Stokkermans TJW, Peters K, Stacey G: The structures and biological activities of the lipo-oligosaccharide nodulation signals produced by type I and type II strains of *Bradyrhizobium japonicum*. J Biol Chem 268: 18372–18381 (1993).
16. Collinge DB, Kragh KM, Mikkelsen JD, Nielsen KK, Rasmussen U, Vad K: Plant chitinases. Plant J 3: 31–40 (1993).
17. Davis KR, Darvill AG, Albersheim P, Dell A: Host-pathogen interactions. XXIX. Oligogalacturonides released from sodium polypectate by endopolygalacturonic acid lyase are elicitors of phytoalexins in soybean. Plant Physiol 80: 568–577 (1986).
18. Debellé F, Rosenberg C, Dénarié J: The *Rhizobium, Bradyrhizobium* and *Azorhizobium* NodC proteins are homologous to yeast chitin synthases. Mol Plant-Microbe Interact 3: 317–326 (1992).
19. de Jong AJ, Cordewener J, Lo Schiavo F, Terzi M, Vandekerckhove J, Van Kammen A, de Vries S: A carrot somatic embryo mutant is rescued by chitinase. Plant Cell 4: 425–433 (1992).
20. de Jong AJ, Heidstra R, Spaink HP, Hartog MV, Hendriks T, Lo Schiavo F, Terzi M, Bisseling T, Van Kammen A, de Vries S: A plant somatic embryo mutant is rescued by rhizobial lipo-oligosaccharides. Plant Cell 5: 615–620 (1993).
21. Demont N, Debellé F, Aurelle H, Dénarié J, Promé JC: Role of the *Rhizobium meliloti* nodF and nodE genes in the biosynthesis of lipo-oligosaccharidic nodulation factors. J Biol Chem 268: 20134–20142 (1993).
22. Dénarié J, Cullimore J: Lipo-oligosaccharide nodulation factors: a new class of signalling molecules mediating recognition and morphogenesis. Cell 74: 951–954 (1993).
23. Deverall BJ, Deakin AL: Genetic tests of the basis of wheat cultivar selectivity in symptom elicitation by preparations from rust pathogens. Physiol Mol Plant Path 30: 225–232 (1987).
24. Ehrhardt DW, Atkinson EM, Long SR: Depolarization of alfalfa root hair membrane potential by *Rhizobium meliloti* Nod factors. Science 256: 998–1000 (1992).
25. Firmin JL, Wilson KE, Carlson RW, Davies AE, Downie JA: Resistance to nodulation of cv. Afghanistan peas is overcome by *nodX*, which mediates an *O*-acetylation of the *Rhizobium leguminosarum* lipo-oligosaccharide nodulation factor. Mol Microbiol 10: 351–360 (1993).
26. Fisher RF, Long SR: *Rhizobium*-plant signal exchange. Nature 357: 655–660 (1992).
27. Fry SC, Aldington S, Hetherington PR, Aitken J: Oligosaccharides as signals and substrates in the plant cell wall. Plant Physiol 103: 1–5 (1993).
28. Geelen D, Mergaert P, Geremia RA, Goormachtig S, Van Montagu M, Holsters M: Identification of *nodSUIJ* genes in locus *1* of *Azorhizobium caulinodans*: evidence that *nodS* encodes a methyltransferase involved in Nod factor modification. Mol Microbiol 9: 145–154 (1993).
29. Geiger O, Spaink HP, Kennedy EP: Isolation of *Rhizobium leguminosarum* NodF nodulation protein: NodF carries a 4′-phosphopantetheine prosthetic group. J Bact 173: 2872–2878 (1991).
30. Heidstra R, Geurts R, Franssen H, Spaink HP, Van Kammen A, Bisseling T: A semi-quantitative root hair deformation assay to study the activity and fate of Nod factors. Plant Physiol 105: 787–797 (1994).
31. Horvath B, Heidstra R, Lados M, Moerman M, Spaink HP, Promé J-C, Van Kammen A, Bisseling T: Induction of pea early nodulin expression by Nod factors of *Rhizobium*. Plant J 4: 727–733 (1993).
32. John M, Röhrig H, Schmidt J, Wieneke U, Schell J: *Rhizobium* NodB protein involved in nodulation signal synthesis is a chitooligosaccharide deacetylase. Proc Natl Acad Sci USA 90: 625–629 (1993).
33. Kafetzopoulos D, Thireos G, Vournakis JN, Bouriotis V: The primary structure of a fungal chitin deacetylase reveals the function for two bacterial gene products Proc Natl Acad Sci USA 90: 8005–8008 (1993).
34. Kouchi H, Hata S: Isolation and characterization of novel nodulin cDNAs representing genes expressed at early stages of soybean nodule development. Mol Gen Genet 238: 106–119 (1993).
35. Lerouge P, Roche P, Faucher C, Maillet F, Truchet G, Promé JC, Dénarié J: Symbiotic host-specificity of *Rhizobium meliloti* is determined by a sulphated and acylated glucosamine oligosaccharide signal. Nature 344: 781–784 (1990).
36. Libbenga KR, Van Iren F, Bogers RJ, Schraag-Lamers MF: The role of hormones and gradients in the initiation of cortex proliferation and nodule formation in *Pisum sativum* L. Planta 114: 19–39 (1973).
37. Marie C, Barny M-A, Downie JA: *Rhizobium leguminosarum* has two glucosamine synthases, GlmS and NodM,

required for nodulation and development of nitrogen fixing nodules. Mol Microbiol 6: 843–851 (1992).

38. Mergaert P, Van Montagu M, Promé J-C, Holsters M: Three unusual modifications, a *D*-arabinosyl, a *N*-methyl, and a carbamoyl group, are present on the Nod factors of *Azorhizobium caulinodans* strain ORS571. Proc Natl Acad Sci USA 90: 1551–1555 (1993).
39. Mohnen D, Eberhard S, Marfà V, Doubrava N, Toubart P, Gollin DJ, Gruber TA, Nuri W, Albersheim P, Darvill AG: The control of root, vegetative shoot and flower morphogenesis in tobacco thin cell-layer explants (TCLs). Development 108: 191–201 (1990).
40. Pichon M, Journet EP, Dedieu A, de Billy F, Truchet G, Barker DG: *Rhizobium meliloti* elicits transient expression of the early nodulin gene *ENOD*12 in the differentiating root epidermis of transgenic alfalfa. Plant Cell 40: 1199–1211 (1992).
41. Poupot R, Martinez-Romero E, Promé J-C: Nodulation factors from *Rhizobium tropici* are sulphated or non-sulphated chitopentasaccharides containing an *N*-methyl-*N*-acylglucosaminyl terminus. Biochemistry 32: 10430–10435 (1993).
42. Price NPJ, Relić B, Talmont F, Lewin A, Promé D, Pueppke SG, Maillet F, Dénarié J, Promé J-C, Broughton WJ: Broad-host-range *Rhizobium* species strain NGR234 secretes a family of carbamoylated, and fucosylated, nodulation signals that are *O*-acetylated or sulphated Mol Microbiol 6: 3575–3584 (1992).
43. Recourt K: Flavonoids in the early *Rhizobium*-legume interaction. Ph.D. thesis, Leiden University, Netherlands (1991).
44. Recourt K, Schripsema J, Kijne JW, Van Brussel AAN, Lugtenberg BJJ: Inoculation of *Vicia sativa* subsp. *nigra* roots with *R. leguminosarum* biovar. *viciae* results in release of *nod* gene activating flavanones and chalcones. Plant Mol Biol 16: 841–852 (1991).
45. Relic B, Talmont F, Kopcinska J, Golinowski W, Promé J-C, Broughton WJ: Biological activity of *Rhizobium* sp. NGR234 Nod-factors on *Macroptilium atropurpureum*. Mol Plant-Microbe Interact 6: 764–774 (1993).
46. Roche P, Debellé F, Maillet F, Lerouge P, Faucher C, Truchet G, Dénarié J, Promé JC: Molecular basis of symbiotic host specificity in *Rhizobium meliloti*: *nodH* and *nodPQ* genes encode the sulphation of lipo-oligosaccharide signals. Cell 67: 1131–1143 (1991).
47. Rosa F, Sargent TD, Rebbert ML, Michaels GS, Jamrich M, Grunz H, Jonas E, Winkles JA, Dawid IB: Accumulation and decay of DG42 gene products follow a gradient pattern during *Xenopus* embryogenesis. Devel Biol 129: 114–123 (1988).
48. Ryan CA: Oligosaccharide signals: From plant defense to parasite offense. Proc Natl Acad Sci USA 91: 1–2 (1994).
49. Sandal NN, Marcker KA: Some nodulin and Nod proteins show similarity to specific animal proteins. In: Gresshoff PM, Roth LE, Stacey G, Newton WE (eds) Nitrogen Fixation: Achievements and Objectives, pp. 687–692. Chapman and Hall, New York (1990).
50. Sanjuan J, Carlson RW, Spaink HP, Bhat UR, Barbour WM, Glushka J, Stacey G: A 2-*O*-methylfucose moiety is present in the lipo-oligosaccharide nodulation signal of *Bradyrhizobium japonicum*. Proc Natl Acad Sci USA 89: 8789–8793 (1992).
51. Savouré A, Magyar Z, Pierre M, Brown S, Schultze M, Dudits D, Kondorosi A, Kondorosi E: Activation of the cell cycle machinery and the isoflavonoid biosynthesis pathway by active *Rhizobium meliloti* Nod signal molecules in *Medicago* microcallus suspensions. EMBO J 13: 1093–1102 (1994).
52. Scheres B, Van de Wiel C, Zalensky A, Horvath B, Spaink HP, Van Eck H, Zwartkruis F, Wolters A-M, Gloudemans T, Van Kammen A, Bisseling T: The ENOD12 gene product is involved in the infection process during the pea-*Rhizobium* interaction. Cell 60: 281–294 (1990).
53. Schmidt J, Röhrig H, John M, Wieneke U, Stacey G, Koncz C, Schell J: Alteration of plant growth and development by *Rhizobium nodA* and *nodB* genes involved in the synthesis of oligosaccharide signal molecules. Plant J 4: 651–658 (1993).
54. Schultze M, Quiclet-Sire B, Kondorosi E, Virelizier H, Glushka JN, Endre G, Géro SD, Kondorosi A: *Rhizobium meliloti* produces a family of sulphated lipo-oligosaccharides exhibiting different degrees of plant host specificity. Proc Natl Acad Sci USA 89: 192–196 (1992).
55. Schwedock J, Long SR: ATP sulphurylase activity of the *nodP* and *nodQ* gene products of *Rhizobium meliloti*. Nature 348: 644–647 (1990).
56. Schwedock J, Long SR: *Rhizobium meliloti* genes involved in sulphate activation: The two copies of *nodPQ* and a new locus *saa*. Genetics 132: 899–909 (1992).
57. Sharp JK, McNeil M, Albersheim P: The primary structures of one elicitor-active and seven elicitor-inactive hexa(β-*D*-glucopyranosyl)-*D*-glucitols isolated from the mycelian walls of *Phytophthora megasperma* f.sp. *glycinea*. J Biol Chem 259: 11321–11336 (1984).
58. Shearman CA, Rossen L, Johnston AWB, Downie JA: The *Rhizobium leguminosarum* nodulation gene *nodF* encodes a polypeptide similar to acyl-carrier protein and is regulated by *nodD* plus a factor in pea root exudate. EMBO J 5: 647–652 (1986).
59. Smit G, Lugtenberg BJJ, Kijne JW: Isolation of nodulation factor from the stele of the roots of *Pisum sativum*. In: Halick RB (ed) Molecular Biology of Plant Cell Growth and Development, Program and Abstracts, abstract 1427. Department of Biochemistry, Tucson, AR (1991).
60. Spaink HP, Sheeley DM, Van Brussel AAN, Glushka J, York WS, Tak T, Geiger O, Kennedy EP, Reinhold VN, Lugtenberg BJJ: A novel highly unsaturated fatty acid moiety of lipo-oligosaccharide signals determines host specificity of *Rhizobium*. Nature 354: 125–130 (1991).

61. Spaink HP: Rhizobial lipo-oligosaccharides: answers and questions. Plant Mol Biol 20: 977–986 (1992).
62. Spaink HP, Wijfjes AHM, Van Vliet, TB, Kijne JW, Lugtenberg BJJ: Rhizobial lipo-oligosaccharide signals and their role in plant morphogenesis: are analogous lipophilic chitin derivatives produced by the plant? Aust J Plant Physiol 20: 381–392 (1993).
63. Spaink HP, Wijfjes AHM, Van der Drift KMGM, Haverkamp J, Thomas-Oates JE, Lugtenberg BJJ: Structural identification of metabolites produced by the NodB and NodC proteins of *Rhizobium leguminosarum*. Mol Microbiol 13: 821–831 (1994).
64. Staehelin C, Granado J, Müller J, Wiemken A, Mellor RB, Felix G, Regenass M, Broughton WJ, Boller T: Perception of *Rhizobium* nodulation factors by tomato cells and inactivation by root chitinases. Proc Natl Acad Sci USA 91: 2196–2200 (1994).
65. Stacey G, Luka S, Sanjuan J, Banfalvi Z, Nieuwkoop AJ, Chun JY, Forsberg LS, Carlson R: *nodZ*, a unique host-specific nodulation gene, is involved in the fucosylation of the lipooligosaccharide nodulation signal of *Bradyrhizobgium japonicum*. J Bact 176: 620–633 (1994).
66. 66. Truchet G, Barker DG, Camut S, de Billy F, Vasse J, Huguet T: Alfalfa nodulation in the absence of *Rhizobium*. Mol Gen Genet 219: 65–68 (1989).
67. Truchet G, Roche P, Lerouge P, Vasse J, Camut S, De Billy F, Promé J-C, Dénarié J: Sulphated lipo-oligosaccharide signals of *Rhizobium meliloti* elicit root nodule organogenesis in alfalfa. Nature 351: 670–673 (1991).
68. Van Brussel AAN, Recourt K, Pees E, Spaink HP, Tak T, Wijffelman CA, Kijne JW, Lugtenberg BJJ: A biovar-specific signal of *Rhizobium leguminosarum* bv. viciae induces increased nodulation gene-inducing activity in root exudate of *Vicia sativa* subsp. *nigra*. J Bact 172: 5394–5401 (1990).
69. Van Brussel AAN, Bakhuizen R, Van Spronsen P, Spaink HP, Tak T, Lugtenberg BJJ, Kijne J: Induction of pre-infection thread structures in the host plant by lipo-oligosaccharides of *Rhizobium*. Science 257: 70–72 (1992).
70. Velupillai P, Harn DA: Oligosaccharide-specific induction of interleukin 10 production by B220$^+$ cells from schistosome-infected mice: a mechanism for regulation of CD4$^+$ T-cell subsets. Proc Natl Acad Sci USA 91: 18–22 (1994).
71. Vijn I, das Neves L, Van Kammen A, Franssen H, Bisseling T: Nod factors and nodulation in plants. Science 260: 1764–1765 (1993).
72. Wagner GP, Lo J, Laine R, Almeder M: Chitin in the epidermal cuticle of a vertebrate (*Paralipophrys trigloides*, Blenniidae, Teleostei). Experientia 49: 317–319 (1993).
73. Wagner GP: Evolution and multi-functionality of the chitin system. In: Molecular Approaches to Ecology and Evolution. Birkhäuser Verlag, Germany, in press.
74. Yang W-C, Katinakis P, Hendriks P, Smolders A, de Vries F, Spee J, Van Kammen A, Bisseling T, Franssen H: Characterization of *GmENOD40*, a gene showing novel patterns of cell-specific expression during soybean nodule development. Plant J 3: 573–585 (1993).
75. York WS, Darvill AG, Albersheim P: Inhibition of 2,4-dichlorophenoxyacetic acid-stimulated elongation of pea stem segments by a xyloglucan oligosaccharide. Plant Physiol 75: 295–297 (1984).

Plant Molecular Biology **26**: 1423–1437, 1994.

Fatty acid signalling in plants and their associated microorganisms

Edward E. Farmer
Institut de Biologie et de Physiologie Végétales, Bâtiment de Biologie, 1015 Lausanne, Switzerland

Received 10 March 1994; accepted in revised form 26 April 1994

Key words: fatty acids, signal transduction, wounding, jasmonic acid, traumatin, plant-microorganism interaction

Introduction

Fatty acid signals in plants have a long history. A fatty acid derivative, traumatic acid, was identified as a cell division promoter in wounded bean mesocarp and was among the first ever biologically active molecules isolated directly from a plant tissue. A structure for this molecule was published in 1939 [20]. Indolyl-3-acetic acid (IAA), an auxin isolated from urine in 1934, was definitively isolated from plant tissues in 1942 by Haagen-Smit *et al.* [32]. Like the complex story of the discovery of the structures of auxins, the true nature of traumatic acid has been confusing due to the more recent discovery that this compound may have become modified during the extraction leading to its discovery [77]. With renewed interest in fatty acid signalling, some of the old literature is being revised and many new facets of fatty acid signalling are emerging. One compound in particular, jasmonic acid, has been the focus of much attention as a signal that stimulates the expression of an array of wound-inducible and defence-related genes as well as playing roles in various developmental responses. While much of the current interest in plant-derived fatty acid signals centres on jasmonic acid, it is apparent that other biologically active fatty acids exist in plants.

Known fatty acid-derived signals in plants can be divided into several categories. These include the jasmonate family (octadecanoid-derived cyclopentanones and cyclopentanols; e.g. A and B in Fig. 1), the traumatin family and related alkenals resulting from the action of hydroperoxide lyases on fatty acid hydroperoxides (Fig. 1D, E, F), and other classes such as highly oxygenated fatty acid derivatives reminiscent of animal lipoxins (e.g. phaseolic acid, Fig. 1C). Within these presently known families are examples of volatile molecules (Fig. 1, compound E and the methyl ester of A), glycosides (Fig. 1B), conjugates with other, unrelated molecules, and fatty acids that are potentially unstable and reactive (especially the alkenals, Fig. 1D, E, F). Some of these properties have caused difficulties in experimentation, but on the whole add greatly to the interest of these compounds.

Of the presently known plant fatty acid signals jasmonic acid has been the most studied. Numerous reviews concerning jasmonates have appeared in the past five years [e.g. 3, 24, 33, 41, 49, 50, 62, 66, 72]. This review aims at summarizing some of the very recent information concentrating on jasmonate as a mediator of cellular responses. Progress in the quantitation of jasmonate and inhibition of its synthesis is discussed. Other fatty acid signals of interest in plants are highlighted, and finally some examples from the rapidly growing area of fatty acid signalling between plants and microorganisms are summarized.

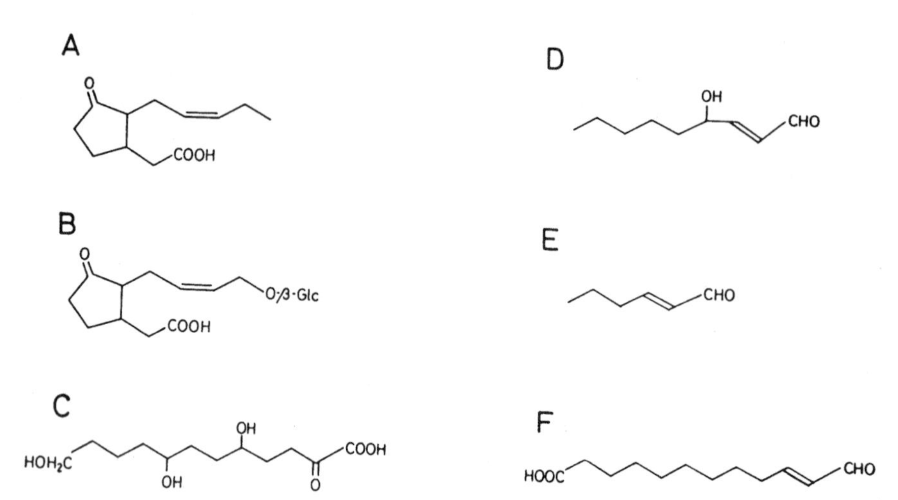

Fig. 1. Diverse signal functions for plant fatty acids. A. Jasmonic acid, involved in wound signalling, induces the synthesis of a variety of low- and high-molecular-weight defence-related compounds. Jasmonic acid is involved in several developmental responses including the stimulation of tendril coiling. Other closely related cyclopentanone fatty acids and their derivatives appear to play diverse roles in development. As a methyl ester jasmonate is a volatile floral scent. Methyl jasmonate might also act as an air-borne defence signal. B. Tuberonic acid-β-glucoside, a powerful natural inducer of tuberization in potato. The aglycone is less potent. C. Phaseolic acid (2-oxo-5,8,12-trihydroxydodecanoic acid) stimulates elongation in pea stem segments and induces α-amylase activity in barley endosperm. D. (2E)-4-Hydroxy-2-nonenal, a potent cytotoxin long known in animals; its synthesis by bean extracts was recently discovered. E. (2E)-hexenal, a toxin highly active against pests and pathogens; highly volatile, also a floral scent. F. Traumatin, 12-oxo-(10E)-dodecenoic acid, a cell division promoter in wounded bean parenchyma.

Jasmonates as wound signals

The jasmonate family of C12 fatty acids includes members which play roles in defence and development. Amongst the best studied jasmonate-responsive genes are many that are wound-inducible ones. Two classes of wound-inducible genes that have been studied in most detail are wound-inducible proteinase inhibitors and vegetative storage proteins. The potential roles of jasmonate in regulating the expression of these classes of genes has been the subject of much investigation which has been reviewed [24, 66].

A useful preliminary classification of wound-inducible genes and their response to jasmonate has been developed by Creelman *et al.* [14]. Wound-inducible proteins (in dicots) include representatives of the following classes: (1) phenylalanine ammonialyase and chalcone isomerase, (2) extensins, glycine-rich proteins and hydroxyproline-rich glycoproteins, (3) proteinase inhibitors and vegetative storage proteins (including some lipoxygenases), and (4) chitinase and β-glucanase. Jasmonate-inducible members of the first three classes have all been characterized. The final class of wound-inducible genes contains members that are ethylene-inducible although some representatives that are jasmonate-inducible are likely to emerge.

The list of genes which are both wound- and jasmonate-inducible is growing but there is still a lack of detailed information concerning monocots. A monocot protein which is clearly both wound- and jasmonate-inducible is a 61 kDa protein from barley leaves [4]. This protein is homologous to ribosome-inactivating proteins (W. Becker and K. Apel, personal communication) and may thus play a fundamental role in translational regulation. A better understanding of wound signalling in monocots is likely to emerge from work on wound-inducible Bowman-Birk trypsin inhibitor in maize seedlings [19]. In this system a wound signal is transmitted unidirectionally (upwards) to tissues distal to the wound site making this an attractive model for systemic signalling.

In summary, jasmonate does not stimulate the expression of *all* wound-inducible genes but may be an important and wide-spread regulator of many such genes throughout the plant kingdom. It will be interesting to look at the possible involvement of jasmonic acid (or other fatty acids) in the regulation of other classes of plant genes that could be classed as wound-inducible, for example the cell division control protein kinase $p34^{cdc2}$. While many wound-inducible genes are clearly jasmonate-inducible, the converse is not always true. Not all jasmonate-inducible genes are wound-inducible [5]. Further critical investigations of the relationship of jasmonate-inducibility to wound-inducibility are needed. In the meantime wound-inducible proteinase inhibitors provide an excellent system with which to study the role of jasmonate as a wound signal.

Systemically wound-inducible proteinase inhibitors are defence proteins which accumulate in the aerial parts of tomato plants upon wounding, for example during insect attack [30]. The systemic signal stimulating proteinase inhibitor gene expression at sites distal to the wound is an octadecapeptide, systemin [51]. A working model for the wound induction of proteinase inhibitor gene expression in tomato leaves has been published [25]. The model proposes that jasmonate is synthesized in response to systemin and possibly to oligogalacturonides, which may be locally released during pathogen attack [25]. The model suggests that jasmonate arises from the action of a lipase on membrane-associated linolenic acid, i.e. jasmonate is synthesized *de novo*. Jasmonate itself would then act to regulate the expression of proteinase inhibitor genes. In order to test this model several experimental approaches are being used and some of the chief questions being addressed are: are intracellular levels of jasmonic acid elevated on wounding leaves and do these levels correlate with gene expression? Is jasmonate synthesized *de novo* or released from cellular conjugates? Does jasmonate play a role in signalling downstream from other signals such as systemin or oligouronides, or , what is the signal order? Is a lipase or a lipase-like enzyme(s) responsible for linolenic acid release from cell membranes and how does jasmonate regulate gene expression? Answers to some of these questions are presented in the sections concerning cellular levels of jasmonate and inhibition of the octadecanoid pathway. In general, much of the work to date is consistent with an *in vivo* role for jasmonate as a regulator of wound-inducible genes but many details are lacking, especially considering the latter question: How does jasmonate regulate gene expression? The whole area concerning jasmonate receptors is unexplored. These is no evidence for the location of the receptor(s); they may reside near or in the nucleus or at the cell surface.

Jasmonates as developmental signals

Jasmonates are well known inducers of a wide variety of physiological responses in plants (reviewed in [41, 62, 74] and are likely to play roles in plant development. A number of interesting developmental phenomena that are jasmonate-inducible are currently under investigation. A concrete developmental role for a jasmonate family member is emerging from the search for a tuber-inducing factor in potatoes [41]. The discovery of the tuber-inducing stimulus as tuberonic acid β-glucoside [76] was a major breakthrough involving a great deal of demanding analytical work. The low levels of this substance and its high activity suggest that it may be a major regulator of tuberisation in potato, where it is synthesized in the leaves and transported downwards to the stolons. The fact that the β-glucoside of tuberonic acid is more active than the free acid indicates the potential importance of conjugated fatty acid metabolites as natural regulators.

The induction of low-molecular-weight metabolites by exogenous jasmonic acid has been investigated in a large number of plant suspension cultures [31]. In nearly every culture investigated (representing over 36 species) jasmonate has been found to induce the production of low-molecular-weight metabolites many of which have proposed roles in plant defence. The same compounds were in many cases induced by a yeast-derived elicitor which was shown to stimulate the accumulation of endogenous 3R,7S-jasmonic acid in suspension-cultured cells of a variety of dicot, monocot and gymnosperm species [47]. In all cases examined the increased synthesis of low-molecular-weight compounds correlated with the accumulation of endogenous 3R,7S-jasmonic acid. These quantitative data (some of which are presented in Table 1) have added to our knowledge of jasmonate as a signal, particularly since the kinetics of formation of a characterized stereoisomer was followed. By taking advantage of a suspension culture system, Mueller *et al.* [47] were also able to show that linolenic acid was released in response to elicitor treatment of cells, providing further evidence consistent with *de novo* synthesis of jasmonate prior to the induction of low-molecular-weight compounds. A recent parallel study of alkaloid synthesis in *Catharanthus* (periwinkle, Apocynaceae) and *Cinchona* (quinine, Rubiaceae) extends these studies to the whole plant level [1]. The seedlings of these two plants synthesize alkaloids at a precise developmental period. The plants are responsive to methyl jasmonate vapour, which increases the synthesis of alkaloids, only within this time frame. This work suggests that not all plant tissues appear to be jasmonate-responsive at all times.

One of the experimental systems of great interest in jasmonate research is the touch-induced coiling of the *Bryonia* tendril. *Bryonia dioica* is a dioecious climbing plant in the Cucurbitaceae. Both sexes produce tendrils which are mechanosensory organs specialized to localize solid objects, such as the stems of other plants, and then coil around them to provide support. Touching the ventral side of the tendril promotes a coiling reaction whereas touching the dorsal side of the tendril inhibits initiation of the coiling reaction. A series of developmental changes take place during touch-induced coiling. These include the lignification of a field of sclerenchyma cells (known as the Bianconi plate) as well as the proliferation of a ventral layer of collenchyma extending from the subepidermal cell layer to the lignified Bianconi plate [36]. Linolenic acid, 12-oxo-phytodienoic acid and jasmonic acid can all induce the coiling response as well as some of the developmental changes associated with coiling [22]. Endogenous levels of jasmonic acid increase dramatically on coiling ([2]; see Table 1) strongly suggesting an *in planta* role for this molecule. It has been suggested that jasmonate, or one of its precursors, may couple the primary response, mechanostimulation, to the coiling response [75]. The fact that jasmonate stimulates tendril coiling as well as cell-type-specific developmental changes makes this a system of great interest since the bryony tendril is a relatively simple organ with a well defined cellular architecture. Another feature of interest in the *Bryonia* tendril is the presence of numerous longitudinally extended air

Table 1. Stimulus-induced changes in jasmonic acid levels in plants.

Plant	Tissue	Resting JA level	Stimulus	Maximum JA level	Profile	Reference
Arabidopsis thaliana	L	40 ± 1.1 ng per g fwt	Wound	1360 ± 708 ng per g fwt	step	W.E. Weiler, pers. comm.
Lycopersicon esculentum	L	14 ± 3.8 ng per g fwt	Wound	52 ± 13.8 ng per g fwt	intermed.	W.E. Weiler, pers. comm. Albrecht *et al.* [2]
Glycine max	H	ca. 80 ng per g fwt	Wound	ca. 500 ng per g fwt	step	Creelman *et al.* [14]
Phaseolus vulgaris	CC	2 ng per g dwt	Elicitor	190 ng per g dwt	intermed.	Mueller *et al.* [47]
Rauvolfia canescens	CC	25 ng per g dwt	Elicitor	1370 ng per g dwt	spike	Gundlach *et al.* 1992
Eschscholtzia californica	CC	1.9 ng per g dwt	Elicitor	74 ng per g dwt	intermed.	Mueller *et al.* [47]
Taxus baccata	CC	3 ng per g dwt	Elicitor	40 ng per g dwt	step	Mueller *et al.* [47]
Agrostis tenuis	CC	10 ng per g dwt	Elicitor	1000 ng per g dwt	spike	Mueller *et al.* [47]
Avena sativa	L	14 ng per g fwt	Wound	49 ng per g fwt	intermed.	Albrecht *et al.* [2]
Bryonia dioica	L	15 ± 1.1 ng per g fwt	Wound	54 ± 5 ng per g fwt	intermed.	W.E. Weiler, pers. comm. Albrecht *et al.* [2]
Bryonia dioica	T	10 ± 3 ng per g fwt	Touch	52 ± 18.7 ng per g fwt	NYA	Weiler *et al.* [75]

[1] Abbreviations: L, leaf; H, hypocotyl; CC, cell culture; T, tendril; fwt, fresh weight; dwt, dry weight; NYA, not yet available.

spaces, particularly those in two bands of mesophyll-like cells which subtend the edges of the dorsal epidermis. It has been suggested that these air spaces play a role in the coiling response by allowing the air-borne intercellular transport of methyl jasmonate [22].

High levels of exogenous jasmonates frequently cause dramatic responses such as senescence or growth inhibition in plant tissues [62, 70]. Thus special care must be taken to provide physiologically relevant jasmonate concentrations in the study of the developmental role of jasmonate in plants. Perhaps related to the effects of jasmonates on growth inhibition and senescence in plants are the recent studies of Reinbothe *et al.* [56, 57, 58] who explored the effects of exogenous jasmonate on gene expression in barley leaf segments. In addition to the stimulation of expression of a number of genes involved in defence and development exogenous jasmonates have interesting inhibitory effects on the translation of a number of nuclear and plastid-encoded RNAs. The translation of a specific subset of cytoplasmic mRNAs (amongst which is the nuclear-encoded ribulose-1,5-bisphosphate carboxylase/oxygenase small subunit, SSU) is inhibited in jasmonate-treated barley leaf segments [56]. This raises a question about the functions these mRNAs have in common, and about the special features of jasmonate-inducible mRNAs that may enable their sustained high translation in jasmonate-treated leaf segments. A detailed study of the translation of SSU mRNA, normally associated with large polysomes in control leaf segments, showed that after exposure of leaf segments to methyl jasmonate the SSU transcripts became associated with smaller polysomes, probably due to decreased chain elongation during translation [57].

An *in vivo* and *in vitro* inhibition of translation of the plastid-encoded mRNA for ribulose-1,5-bisphosphate carboxylase/oxygenase large subunit (LSU) was also observed in jasmonate-treated leaf segments. In this case primer extension analysis showed that transcripts synthesized in the presence of jasmonate had undergone alternative processing of the 5′ end resulting in larger transcripts [58]. These alternative transcripts contain a 35 base sequence between −94 and −59 which is highly complementary to the extreme 3′ end of 16S rRNA. Formation of an intermolecular complex between the LSU RNA and ribosomal RNA would be a possible hindrance to translation of this message [58]. While the physiological importance of these effects is still unclear, these studies have the potential to yield much new information on translational control in a eukaryotic system. They have also emphasized the potential diversity of the effects of jasmonate on gene expression in plants.

Experiments to investigate the early stages of jasmonate interaction with the plant cell are needed to answer questions concerning the nature of jasmonate receptors. Of the many approaches available, mutagenesis to produce jasmonate-insensitive plants holds promise. Staswick *et al.* [67] have described an *Arabidopsis* mutant which fails to respond to jasmonate-inhibition of root growth or jasmonate-induction of proteins similar to vegetative storage proteins. It is hoped that the generation of mutants such as these will provide the necessary backgrounds from which to isolate genes involved in jasmonate signal transduction.

At this stage it should be noted that jasmonates are not the only plant-derived fatty acid signals known to regulate developmental phenomena in plants. Traumatins will be discussed below, although it is far from clear how widespread their roles are. Another fatty acid derivative, phaseolic acid (Fig. 1C) was characterized in the 1960s [55]. This trihydroxydodecanoic acid, purified from bean seeds, was shown to stimulate the growth of pea stem segments as well as induce α-amylase synthesis in barley endosperm [55]. Exogenous phaseolic acid also retarded senescence in barley leaf segments. This compound thus deserves further investigation, particularly in light of the fact that its regulation of α-amylase genes provides a sensitive assay for the barley compound. Several other reports of the synthesis of prostaglandin-like compounds in plants have appeared [e.g. 48] although their activity in plants, if any, was not reported.

Cellular levels of jasmonate in relation to gene expression

Jasmonates are known to be extremely widespread in the plant kingdom [44] but only recently have rapid, stimulus-induced changes in the cellular levels of jasmonic acid been investigated. Jasmonate accumulation as a consequence of wounding, elicitation or mechanical stimulation has now been monitored in a wide variety of plant species and tissues. Many of the data have come from gas chromatography-mass spectroscopy (GC-MS) studies which have the advantage of being able to quantitate individual stereoisomers of jasmonic acid. More recently monoclonal antibodies to jasmonic acid have provided a powerful analytical tool. Albrecht *et al.* [2] used 3R,7R-jasmonic acid coupled to haemocyanin or bovine serum albumin as immunogens. One of the resulting monoclonal antibodies, MAB JAH-1 8B4, has been characterized in detail and is employed in a competitive enzyme-linked immunosorbent assay (ELISA). This nonradioactive method is sensitive down to less than 0.1 pmol 3R,7R-jasmonic acid. The sensitivity for the other natural diastereomer 3R,7S-jasmonic acid (Fig. 3) is also very high, nearly 90% of that for 3R,7R jasmonic acid [2]. Thus these monoclonal antibodies are valuable analytical tools. Representative values for stimulus-induced changes in endogenous jasmonate levels in a variety of plant organs and cultures vary widely as shown in Table 1. For example, in elicitor-treated suspension cultured cells the maximum 3R,7S-jasmonic acid level was 74 ng per g dry weight in *Eschscholtzia* and 1000 ng/g in *Agrostis* [47]. Because of experimental design, stimulus-induced jasmonate values that have been published probably overestimate the levels of this substance actually necessary to induce physiological levels of gene expression under natural conditions. In the case of systemic wound-signalling it will be important to measure jasmonate levels in leaves distal to the wound site in addition to tissues adjacent to the wound site.

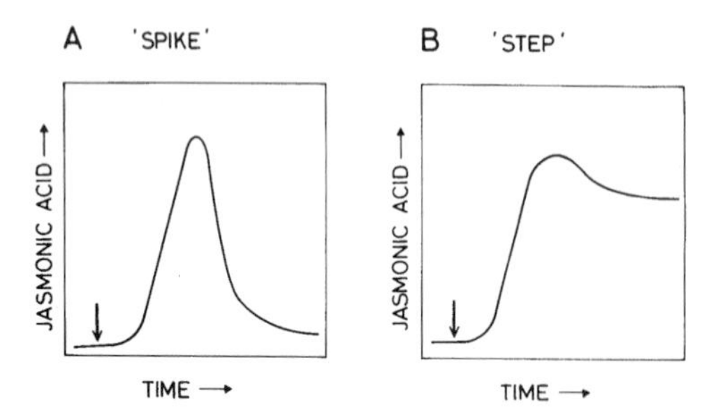

Fig. 2. Two common profiles, 'spike' and 'step', for the accumulation of jasmonic acid in plant tissues in response to various stimuli. Other known profiles fall between these two extremes. The arrows indicate timing of the stimulus leading to jasmonate induction. In some cases this is known to occur almost instantly after a wound stimulus.

Jasmonate peak profiles thus far appear to fall between two extremes, 'spikes' and 'steps', which are illustrated in Fig. 2. A spike profile was first reported by Gundlach *et al.* [31] in snakeroot (*Rauvolfia*) suspension cultures treated with a yeast cell wall elicitor. The step profile was first seen in wounded soybean hypocotyl tissue [14]. Since then a number of other measurements have revealed recurring examples of these two profiles, as well as intermediate type profiles. The shape of the jasmonate peak and its duration in the cell will depend to a large extent on how quickly and uniformly the stimulus reaches those cells able to respond. Fortunately data are available for both suspension cultures (where the stimulus should reach cells quickly) and intact plant tissues. The peak of endogenous jasmonate is in all cases so far reported of relatively long endurance. For example the wound induced accumulation of jasmonic acid in soybean hypocotyls lasted at least 24 h [14]. The elicitor-induced peaks of endogenous jasmonic acid in suspension cultures of *Phaseolus*, *Taxus*, *Eschscholtzia* and *Agrostis* lasted between ca. 3 and 6 h [47]. Wound-induced endogenous jasmonate peaks in the leaves of *Bryonia* lasted about 2.5 h [2]. In contrast to some second messenger-like compounds in animal cells, stimulus-induced jasmonic acid in plant cells can have a remarkably extended half-life. For example, after the induction of peak levels of diacylglycerol in platelets the level of this compound subsides rapidly towards baseline in about one minute [69]. Due to the observed long half-life of induced jasmonic acid pools in the plant cell, as well as other features of jasmonate signalling, this molecule should *not* be regarded as a classical second messenger. Jasmonates seem to have more in common with other forms of biological mediators such as prostaglandins, and may share features in common with paracrine signals [25]. Bearing in mind that prolonged exposure to elevated jasmonate levels can cause growth inhibition, it would be interesting to know if the jasmonate step profiles are associated with decreased or inhibited cell growth. In other words, are particular jasmonate profile shapes associated with particular responses?

A similar question concerning the biological activities of various jasmonic acid stereoisomers has been addressed. Newly synthesized jasmonic acid epimerises to form two diastereomers at or near thermodynamic equilibrium (Fig. 3). This epimerization takes place *in vivo* converting 3R,7S-jasmonic acid to the 3R,7R-diastereomer [47]. Different physiological responses are induced to different levels by the naturally occurring jasmonate isomers. Koda *et al.* [40] examined the effects of the four jasmonate stereoisomers (see review by Sembdner and Parthier [62]) on four different physiological responses and found that each response differed in its sensitivity to a given stereoisomer. An extrapolation of these results might mean that individual genes are differentially sensitive to jasmonate stereoisomers. Epimerisation of jasmonic acid within the cell (Fig. 3) may thus alter its biological activity or affects its turnover or conjugation within the cell.

Other ways that the cell could remove free jasmonate from the cellular pool include the formation of conjugates either with the carboxyl group

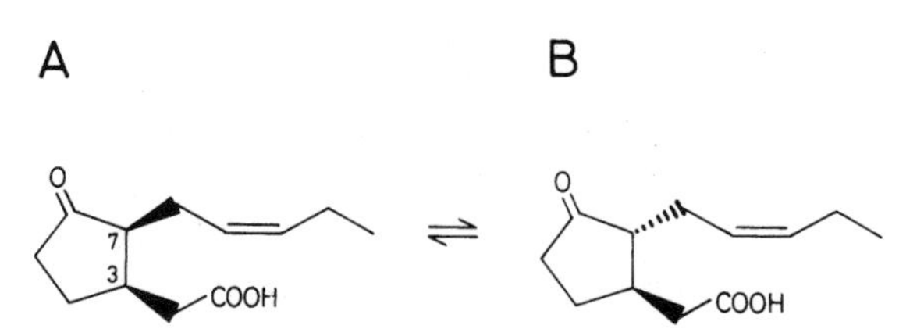

Fig. 3. The epimerization of biosynthetic 3R,7S-jasmonic acid (A) to 3R,7R-jasmonic acid (B) which occurs within plant cells and which could, in part, regulate the life-time of the active signal within the cell.

by esterification to amino acids or by the addition of glucose to hydroxyl groups introduced at carbons 11 or 12 [45, 46]. Conjugates of dihydrojasmonate with a variety of other molecules, especially the hydrophobic amino acids leucine, isoleucine and valine, have been characterized [45]. The possibility that simular conjugates arise from the proteolysis of polypeptides linked to jasmonic acid should be considered, and could perhaps be addressed using the monoclonal antibodies to jasmonic acid [2] which would be expected to have a high affinity for jasmonate-protein conjugates.

Inhibition of the octadecanoid signal pathway

Biochemical experiments to investigate the need for jasmonate biosynthesis in signal transduction have been conducted and cumulative evidence suggests that induced jasmonate is, at least in part, synthesized *de novo* in response to a stimulus and not simply released from 'storage' conjugates or pools. Ideally, disruption of the jasmonate biosynthetic pathway could be achieved with a genetic method, for example by expressing the mRNA for a jasmonate synthesizing enzyme in antisense. Until this is reported the available tools include a variety of inhibitors of fatty acid metabolism in plants, some of which have been borrowed from animal research. Experiments involving the use of lipoxygenase inhibitors are often difficult and it is possible that plant lipoxygenases involved in signal generation in plants differ somewhat from the animal enzymes for which the inhibitors were originally developed. Several such compounds have been employed to block the synthesis of wound-inducible proteinase inhibitors and vegetative storage proteins in leaves. These inhibitors include aspirin [18], and ibuprofen, *N*-propyl gallate, salicylhydroxamic acid and phenylbutazone [65]. A recently introduced inhibitor, ursolic acid, was shown to reduce endogenous levels of jasmonic acid in barley leaf segments as well as to inhibit the expression of a number of genes encoding jasmonate-inducible proteins [73]. This compound may prove to be a broadly useful inhibitor. Results strongly supporting the role of jasmonic acid in the wound induction of proteinase inhibitor gene expression were recently published. Peña-Cortes *et al.* [52] showed that aspirin blocked the accumulation of endogenous jasmonic acid in wounded leaves of tomato. Concomitant with reduced levels of jasmonate was the reduced accumulation of the messenger RNA for proteinase inhibitor II as well as the mRNAs for threonine dehydrase and cathepsin D-inhibitor. The proposed site of action of aspirin was on hydroperoxide dehydrase.

Recently a different type of inhibitor has emerged: sodium diethyldithiocarbamic acid (DIECA) was found to partially block the wound and systemin induction of proteinase inhibitors I and II in tomato leaves. A series of attempts to identify the site of action of DIECA led to the observations that this compound is a weak inhibitor of tomato leaf lipoxygenase activity but can react directly with 13(S)-hydroperoxylinolenic acid *in vitro* to form a compound with properties expected of 13-hydroxylinolenic acid (Farmer, Caldelari, Pearce, Walker-Simmons and Ryan, Plant Physiol, in press). Thus this inhibitor has at least two potential sites of action and this compound, or a similar less toxic derative, might prove useful since its action at the level of the lipoxygenase products (fatty acid hydroperoxides) would bypass the need to inhibit lipoxygenase directly. It must be noted that the apparent toxicity of DIECA might limit its utility as an inhibitor but some tentative conclusions have been drawn from its use in the tomato/proteinase inhibitor system. The most important conclusion from this study was that, since DIECA blocked the systemin-induced expression of proteinase inhibitor genes by interruption of fatty acid metabolism, the signal order systemin → jasmonate → gene is likely. Further experiments are necessary to test this proposed signal order.

Another way to inhibit jasmonate synthesis and thus to test various models for the regulation of wound-inducible gene expression would be to generate mutants incapable of producing wild-type levels of the jasmonate. These tests will be possible using *Arabidopsis* mutants deficient in the jasmonate precursor linolenic acid (M. McConn and J. Browse, 1993, Abstract B3, National Plant Lipid Cooperative Symposium, Minneapolis; J. Browse, personal communication) and it will be interesting to study jasmonate levels as well as the expression of wound-inducible genes in these plants.

The traumatin family

Fatty acid hydroperoxides (including the 13-(S)-hydroperoxide of linolenic acid, which is a precursor to jasmonic acid) can be metabolized by lytic reactions to yield two aldehydes (reviewed in [28]) which are candidates for further enzymic or nonenzymic oxidation to carboxylic acids. One of these molecules, dicarboxylic acid called traumatic acid, was isolated directly from the mesocarp of wounded beans and was shown to enhance cell proliferation at the wound site [20]. Traumatic acid has not received the attention it deserves because its effects appear to be limited to one or a few bean cultivars, and because it is likely that the traumatic acid originally identified by English *et al.* [20] arose through the oxidation of an unstable alkenal, traumatin [77]. It is this dicarboxylic acid that is commercially available and is often used in place of the less stable oxoacid which may have a higher biological activity [77]. For example, traumatic acid causes abscission of cotyledon petioles in cotton explants, and this effect is reduced by auxin treatment [68]. Several other studies reporting the effects of traumatin on plants or pathogens have employed traumatic acid and in some cases it is not clear if the authors were aware of the difference between these molecules. To further complicate the matter, traumatin itself arises through the rearrangements of an isomeric compound (9Z)12-oxo-dodecenoic acid (△9-ODA) [34]. This is often the primary metabolite produced from 18:2 or 18:3 fatty acid hydroperoxides by plant leaf extracts. For clarification the probable relationship between members of the traumatin family is outlined in Fig. 4. Further studies are needed to clarify the biosynthesis and biological activities of traumatin, △9-

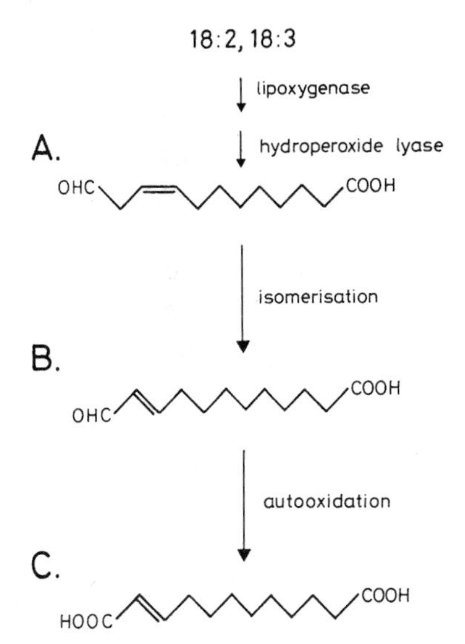

Fig. 4. Traumatin and its dodecanoid relatives. A. 12-oxo-9(Z)-dodecenoic acid (△9-ODA). B. Traumatin, 12-oxo-10(E)-dodecenoic acid (△10-ODA). C. Traumatic acid, 10(E) dodeca-1,12-dicarboxylic acid. Product A originates from the action of hydroperoxide lyase on 13-hydroperoxyoctadecanoids. This compound isomerizes in the presence of some plant extracts to B, 'traumatin'. Compound C, traumatic acid, is an oxidized derivative of B and was the structure originally characterized by Bonner's group in the 1930s as a wound hormone. The possibility that product B 'traumatin' is produced directly in some tissues is not excluded.

ODA and traumatic acid. It is already known that the levels of traumatin increase in wounded bean pod tissue [77] and it is interesting to note that the common precursor for jasmonate and traumatin, 13(S)-hydroperoxylinolenic acid, gives rise to two wound-inducible signals, one of which (jasmonate) is a defence signal and the other (traumatin) a wound-healing signal.

Traumatin is only one of a potentially larger class of alkenals that can be generated by the lytic action of hydroperoxide lyase on various fatty acid hydroperoxides. Since a potentially larger number of fatty acid hydroperoxides could be produced by the action of lipoxygenase on 1,4-Z,Z-pentadiene fatty acids, an even larger family of aldehydes could be generated *in planta*. The aldehyde function is labile and the double bond proximal to the aldehyde group is subject to *cis-trans* isomerization (e.g. Fig. 4). Many of these molecules are volatile and might rapidly diffuse through tissues and into the air. Examples of volatile and toxic alkenals are discussed in the next section.

Volatile and cytotoxic fatty acid oxidation products

Amongst the most toxic natural products of fatty acid oxidation are the unsaturated aldehydes (alkenals) and their more highly oxidized derivatives. Plants have long been known to produce large quantities of alkenals. One of the earliest papers on alkenal production highlighted both the wide-spread occurrence of this phenomenon in the plant kingdom and the potential toxicity of these molecules [60]. The antimicrobial properties of several unsaturated aldehydes, in particular (2E)-hexenal, are likely to play important roles in plant defence against fungi [42, 71], bacteria [15, 17] and arthropods [35, 42]. This volatile compound is transiently released into the air upon wounding leaves [60] and during the hypersensitive response of *Phaseolus* leaves to infection with *Pseudomonas syringae* pv. *phaseolicola* [15]. The volatility of (2E)-hexenal may contribute to its rapid diffusion through damaged or infected tissues. In addition to their toxicity, volatile alkenals such as (2E)-hexenal are often floral scent components [39].

A recent discovery of interest in plants is the production of (2E)-4-hydroxynonenal from (3Z)-nonenal by broad bean *Vicia faba* seed extracts [29]. (2E)-4-hydroxynonenal is a highly cytotoxic and genotoxic fatty acid derivative that can be produced in animal cells [21]. The possibility that this compound could act in defence against fungi has been investigated. (2E)-4-Hydroxynonenal was found to be an inhibitor of the growth of *Colletotrichun truncatum* and *Scherotium rolfsii* when added to the fungal growth medium at levels as low as 0.3 μmol per ml medium [71]. The effects of (2E)-4-hydroxynonenal on plant cells were not examined but judging by its toxicity to animal cells [21], this compound could potentially destroy plant cells at very low concentrations. Could unsaturated aldehydes or their derivatives be the cytotoxins or signals that kill plant cells during the hypersensitive response? If so, how is the generation of these toxic molecules regulated?

Methyl jasmonate vapour is a powerful signal for the induction of proteinase inhibitor gene expression in the leaves of several species of plants in the Solanaceae (e.g. tomato, potato, tobacco [23, 26]) and Fabaceae (e.g. alfalfa [26]). The low levels of this vapour necessary to trigger proteinase inhibitor gene expression, as well as to cause certain physiological responses like tendril coiling in *Bryonia* [22], suggest that volatile methyl jasmonate may play physiological roles. This compound was shown to be present on the leaf surfaces of sagebrush (*Artemisia tridentata*) in sufficient quantity to induce proteinase inhibitor accumulation in the leaves of neighbouring tomato plants under laboratory conditions [23]. The possible consequences of plants releasing this highly active signal into the air have not been investigated adequately, and the case of *A. tridentata* remains intriguing. This plant, North America's most abundant shrub (National Geographic, January 1989), could potentially affect neighbouring plants in a variety of ways. For example, it could induce defence responses or affect other aspects of physiology including growth inhibition or senescence. One possibility is that methyl jasmonate could inhibit the growth of competing plants, particularly climbing plants which are rarely observed on Great Basin *A. tridentata* populations (E. Farmer, unpublished; E.D. McArthur, personal communication).

Although a number of volatile fatty acid derivatives are known to be released by leaves there are still few examples of their role or utilization in nature. In contrast, a great many fatty acid-derived floral scents are known [39]. It is conceivable that the great number of fatty acid-derived floral scents may be paralleled in the production of volatile fatty acid derivatives by healthy or wounded leaves. Bearing in mind the volatile nature of many fatty acid derivatives, it will be interesting to reappraise the distribution and function of intercellular air spaces in plant tissues in connection with plant development and pathogenesis.

Fatty acid signalling between plants and microorganisms

The use of jasmonate as a spray for the protection of plants from pests and pathogens has been investigated in a number of systems. The idea behind these experiments was that jasmonate would induce native plant defences thus protecting the crop. Cohen *et al.* [13] observed that the application of jasmonic acid to potato plants caused local and systemic protection from the late blight fungus *Phytophthora infestans*. Although jasmonic acid was found to have a direct effect on the mode of germination of *P. infestans* sporangia, it is likely that the protective effect of jasmonic acid was due to the induction of antifungal defences since leaves distal to those sprayed with jasmonic acid were protected from late blight. The nature of the induced systemic defences was not established, but the authors speculated that an oxidative defence involving lipoxygenase activation may have been responsible for the protective effect of jasmonate.

A recent surprise was the observation that jasmonic acid, apart from its signalling effect in plants, may also act as a signal in fungi. Barley (*Hordeum*) plants sprayed with jasmonic acid solutions were found to be protected from infection by the pathogenic fungus *Erysiphe graminis* f.sp. *hordei* [61]. Several jasmonate-inducible proteins accumulated when barley was sprayed with jasmonic acid, however these proteins were found to have no antifungal activity against *Erysiphe*. Furthermore, pretreating plants with cordycepin (a transcription inhibitor) had no effect on the ability of jasmonic acid to protect barley against *E. graminis*. These first results suggested an unexpected and interesting result, that inhibition of pathogenesis was due, at least in part, to the direct effect of jasmonate on the life cycle of *Erysiphe*. Jasmonic acid was found to block the differentiation of appressoria (thus inhibiting penetration of the fungus) at concentrations which were not inhibitory to mycelial growth. These results highlight the potential complexity of the role of jasmonate in the plant-fungal interaction. However Schweizer *et al.* [61] caution that the levels of jasmonic acid which accumulate *in planta* are probably too low to have a direct effect on fungal differentiation.

The mycelial membranes of *Phytophthora infestans* contain C20 fatty acids (including arachidonic acid; Fig. 5B) which are powerful inducers of phytoalexin accumulation and browning in potato tuber tissue [6, 7]. Experiments suggest that lipoxygenase is necessary for defence gene induction by arachidonic acid [54]. However, the structures of the biologically active arachidonate metabolites are not yet known and the possibility remains that arachidonate can affect defence

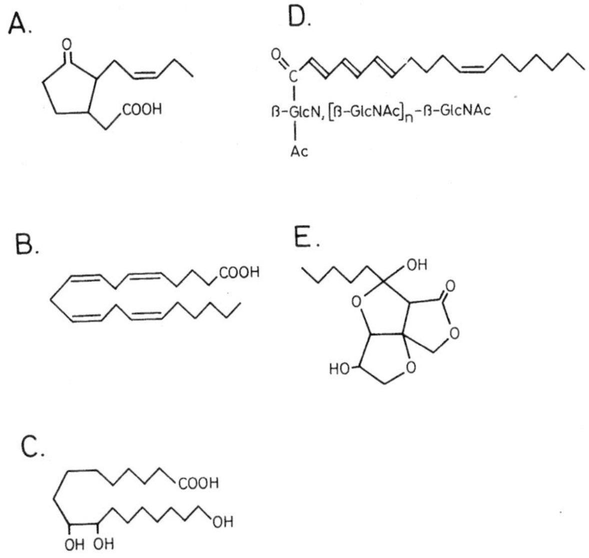

A. COOH

D. ß-GlcN, [ß-GlcNAc]$_n$-ß-GlcNAc Ac

B. COOH

E. OH O O O HO O

C. COOH OH OH OH

Fig. 5. Fatty acid derivatives involved in signalling between microorganisms and plants. A. Jasmonic acid: not only a signal molecule in plants, this fatty acid can inhibit appressorium formation in the fungus *Erysiphe* without inhibiting mycelial growth. B. Arachidonic acid, a component of the mycelial membranes of *Phytophthora infestans*, regulates a number of defence-related genes in potato tuber. C. 9,10,18-trihydroxyoctadecanoic acid, a component of plant cutin, is a powerful inducer of cutinase in the germinating spores of several pathogenic fungi including *Fusarium solani pisi* (*Nectria hematococca*). D. Nod Rlv-V(Ac,18:4) a decorated fatty acid produced by *Rhizobium leguminosarum* bv. *viciae*. This molecule causes root hair curling and cortical root cell division in pea. Modification of the fatty acid changes the ability of the signal to cause these effects in a given plant species. E. Syringolide 1, produced by bacteria (e.g. *Pseudomonas syringae* pv. *tomato*) expressing the avirulence gene *avrD*. Syringolide causes cell death only in plants expressing a complementary resistance gene, Rpg4, and is thus a specificity determinant in pathogenesis.

signal transduction by other mechanisms. Ricker and Bostock [59] identified several hydroperoxyeicosanoic acids generated by the application of arachidonic acid to potato tubers. None of these compounds including 5-hydroperoxyeicosatetraenoic acid (5-HPETE) had the ability to induce sesquiterpenoid phytoalexins. In contrast, Castoria *et al.* [10] reported that 5-HPETE could induce sesquiterpenoid phytoalexin accumulation in potato tuber tissue. If arachidonic acid hydroperoxides are able to signal phytoalexin accumulation, it will be interesting to see if the biologically active compound(s) is the fatty acid hydroperoxide or a downstream metabolite(s).

Recent studies on the regulation of a small family of HMG-CoA reductase (*hmg*) genes in potato tissues have revealed interesting patterns of differential gene regulation by fatty acids [11, 12]. The gene *hmg1* is wound-inducible and implicated in the synthesis of steroid glycoalkaloids which may be necessary for the restoration of membrane function during wound-healing. The wound-inducible *hmg1* gene is powerfully upregulated by jasmonic acid, but expression of this gene is suppressed by arachidonic acid [12]. This provides an example of a gene that appears to be differentially regulated by two different fatty acid signals of plant and fungal origin respectively. The genes *hmg2* and *hmg3* are involved in the biosynthesis of sesquiterpenoid phytoalexins and are, in contrast to *hmg1*, powerfully induced by exogenous arachidonic acid.

Fatty acid-derived signals involved in the interaction between plants and pathogens do not always derive from the latter. Plant cutin components including 10,16-dihydroxyhexadecanoic acid and 9,10,18-trihydroxyoctadecanoic acid are powerful upregulators of cutinase transcription in nuclei from the phytopathogenic fungus *Fusarium solani pisi* (*Nectria hematococca* [53]). Further evidence for a biological role for fatty acid polyhydroxides came from the observation that the mycelia of several phytopathogenic fungi lyse in the presence of 8-hydroxylinoleic acid. This compound is made by the soil fungus *Laetisaria arvalis*, which is a disease suppresser in some crops [9]. More recently a detailed study of linoleic acid metabolism by the wheat take-all fungus *Gaeumannomyces graminis* has revealed the production of an interesting array of oxygenated fatty acids including (8R)-hydroxylinoleic acid and 7,8-dihydroxylinoleic acid [8]. (8R)-Hydroxylinoleic acid was shown to be a sporulation regulator in *Aspergillus nidulans* [43]. Unsaturated monohydroxy fatty acids are also known to be antifungal [37]. It may only be a matter of time before these interesting molecules are found to have signalling roles in the plant.

The recent discoveries that several fatty acid-containing signal molecules can play governing roles in specificity determination and recognition phenomena has considerably broadened our understanding of plant-microorganism interactions. Nodulation (Nod) factors secreted by *Rhizobium* species can be regarded as fatty acids which are decorated through N-acetyl linkage to glucosamine oligomers. These complex signal molecules induce organogenesis leading to nodule formation in compatible legume species. The fatty acids found in Nod factors often contain trans (E) double bonds (e.g. Fig. 5D), and when conjugated could be potentially reactive. The fatty acid substituents vary between species and strains of *Rhizobium* and help to determine host specificity [27, 64]. This was probably the first indication that fatty acids could help control specificity in a recognition response. Increasing evidence suggests that Nod-like signals are made by plants themselves and may play a natural role in plant development [16].

Other fatty acid derivatives involved in recognition phenomena are syringolides, a small family of compounds thought to result from the conjugation of β-keto fatty acid and xylulose [38]. Syringolides are produced by gram negative bacteria carrying the avirulence gene *avrD* originally isolated from *Pseudomonas syringae* pv. *tomato*. Syringolides (Fig. 5E) cause cell death only in plants containing the cognate resistance gene Rpg4 [38]. It is not yet known why some bacteria make syringolides, and the possibility that they are signals for the bacterium itself cannot be ignored.

Conclusion

The number of biologically active fatty acids known in plants is increasing, although detailed information exists only for some members of the jasmonate family, chiefly jasmonic acid. This compound is a regulator of wound-inducible genes and its intracellular levels are transiently elevated in response to wounding. It can also regulate other defence-related genes. Members of the jasmonate family appear to play roles in a number of developmental responses including tuberization in potato, and tendril coiling, and in inducing the synthesis of secondary plant compounds. Jasmonate is known to occur in plants, fungi and insects; perhaps it is only a matter of time until it is found in higher animals. Increasingly the literature suggests that new fatty acid signals in plants will be discovered, although much remains to be learnt of biologically active fatty acids discovered many years ago, for example, traumatin and phaseolic acid. In animal cells fatty acid-derived signals are numerous with tens of biologically active fatty acid-derived molecules having been characterized and more awaiting discovery. Even considering the greater cellular complexity of higher animals, it is likely that more fatty acid signals exist in plants. The possibilities become more exciting when one considers the relationship between peptides and fatty acids. A large number of peptide signals in animals influence cellular lipid metabolism. If there are more peptide hormones in plants, are there more fatty acid signals? The recent discoveries of interkingdom fatty acid signalling have illuminated a wide variety of roles for these molecules, including roles in recognition phenomena. In the next few years more fatty acid derivatives involved in the interaction of plants and other organisms will probably be found and attention will turn increasingly towards receptors for these molecules.

Acknowledgements

The following people made data available prior to publication or otherwise assisted in providing useful information: K. Apel, R. Bostock, I.D. Brodowsky, D. Caldelari, D. Hildebrand, Y. Koda, J. Mullet, B. Parthier, U. Rüegg, P. Schweizer, W. Smith, B. Vick, C. Wasternack, E. Weiler, M.H. Zenk. I thank H. Berttoud for preparation of figures.

References

1. Aerts RJ, Gisi D, De Carolis E, De Luca V, Baumann TW: Methyl jasmonate vapor increases the developmentally controlled synthesis of alkaloids in *Catharanthus* and *Cinchona* seedlings. Plant J 5: 635–645 (1994).
2. Albrecht T, Kehlen A, Stahl K, Knöfel H-D, Sembdner G, Weiler EW: Quantification of rapid, transient increases in jasmonic acid in wounded plants using a monoclonal antibody. Planta 191: 86–94 (1993).
3. Anderson JM: Membrane-derived fatty acids as precursors to second messengers. In: Boss WF, Morre DJ (eds) Second Messengers in Plant Growth and Development, pp. 181–182. Liss, New York (1989).
4. Becker W, Apel K: Differences in gene expression between natural and artificially induced leaf senescence. Planta 189: 74–79 (1993).
5. Bolter CJ: Methyl jasmonate induces papain inhibitor(s) in tomato leaves. Plant Physiol 103: 1347–1352 (1993).
6. Bostock RM, Kuc JA, Laine RA: Eicosapentaenoic and arachidonic acids from *Phytophthora infestans* elicit fungitoxic sesquiterpenes in the potato. Science 212: 67–69 (1981).
7. Bostock RM, Yamamoto H, Choi D, Ricker KE, Ward BL: Rapid stimulation of 5-lipoxygenase activity in potato by the fungal elicitor arachidonic acid. Plant Physiol 100: 1448–1456 (1992).
8. Brodowsky IR, Hamberg M, Oliw EH: A linoleic acid 8(R)-dioxygenase and hydroperoxide isomerase of the fungus *Gaeumannomyces graminis*: Biosynthesis of 8(R)-hydroxylinoleic acid and 7(S),8(S)-dihydroxylinoleic acid from 8(R)-hydroperoxylinoleic acid. J Biol Chem 267: 14738–14745 (1992).
9. Bowers WS, Hoch HC, Evans PH, Katayama M: Thallophytic allelopathy: isolation and identification of laetisaric acid. Science 232: 105–106 (1986).
10. Castoria R, Fanelli C, Fabbri AA, Pasi S: Metabolism of arachidonic acid involved in its eliciting activity in potato tuber. Physiol Mol Plant Path 41: 127–137 (1992).
11. Choi D, Ward BL, Bostock RM: Differential induction and suppression of potato 3-hydroxy-3-methylglutaryl coenzyme A reductase genes in response to *Phytophthora infestans* and to its elicitor arachidonic acid. Plant Cell 4: 1333–1344 (1992).
12. Choi D, Bostock RM, Avdiushko S, Hildebrand DF: Lipid-derived signals that discriminate wound- and pathogen-responsive isoprenoid pathways in plants: methyl-jasmonate and the fungal elicitor arachidonic acid induce different HMG-CoA reductase genes and antimicrobial isoprenoids in *Solanum tuberosum* L. Proc Natl Acad Sci USA 91: 2329–2333 (1994).
13. Cohen Y, Gisi U, Mosinger E: Local and systemic protection against *Phytophthora infestans* induced in potato and tomato plants by jasmonic acid and jasmonic-methylester. Phytopath 83: 1054–1062 (1993).
14. Creelman RA, Tierney ML, Mullet JE: Jasmonic acid/methyl jasmonate accumulate in wounded soybean hypocotyls and modulate wound gene expression. Proc Natl Acad Sci USA 89: 4938–4941 (1992).
15. Croft KPC, Jüttner F, Slusarenko AJ: Volatile products of the lipoxygenase pathway evolved from *Phaseolus vulgaris* (L.) leaves inoculated with *Pseudomonas syringae* pv. *phaseolicola*. Plant Physiol 101: 13–24 (1993).
16. de Jong AJ, Heidstra R, Spaink HP, Hartog MV, Meijer EA, Hendriks T, Lo Schiavo F, Terzi M, Bisseling T, van Kammen A, de Vries SC: *Rhizobium* lipooligosaccharides rescue a carrot somatic embryo mutant. Plant Cell 6: 615–620 (1993)
17. Deng W, Hamilton-Kemp TR, Nelson MT, Andersen RA, Collins GB, Hildebrand DF: Effects of six-carbon aldehydes and alcohols on bacterial proliferation. J Agric Food Chem 41: 506–510 (1993).
18. Doherty HM, Selvendran RR, Bowles DJ: The wound response of tomato plants can be inhibited by aspirin and related hydroxybenzoic acids. Physiol Mol Plant Path 32: 377–384 (1988).
19. Eckelkamp K, Ehmann B, Schopfer P: Wound-induced systemic accumulation of a transcript coding for a Bowman-Birk trypsin inhibitor-related protein in maize (*Zea mays* L.) seedlings. FEBS Lettt 323: 73–76 (1993).
20. English J, Bonner J, Haagen-Smit AJ: Structure and synthesis of a plant wound hormone. Science 90: 329 (1939).
21. Esterbauer H, Schaur RJ, Zollner H: Chemistry and biochemistry of 4-hydroxynonenal, malonaldehyde and related aldehydes. Free Rad Biol Med 11: 81–128 (1988).
22. Falkenstein E, Groth B, Mithöfer A, Weiler EW: Methyl jasmonate and α-linolenic acid are potent inducers of tendril coiling. Planta 185: 316–322 (1991).
23. Farmer EE, Ryan CA: Interplant communication: airborne methyl jasmonate induces synthesis of proteinase inhibitors in plant leaves. Proc Natl Acad Sci USA 87: 7713–7716 (1990).
24. Farmer EE, Ryan CA: Octadecanoid-derived signals in plants. Trends Cell Biol 2: 236–241 (1992).
25. Farmer EE, Ryan CA: Octadecanoid precursors of jasmonic acid activate the synthesis of wound-inducible proteinase inhibitors. Plant Cell 4: 129–134 (1992).
26. Farmer EE, Johnson RR, Ryan CA: Regulation of expression of proteinase inhibitor genes by methyl jasmonate and jasmonic acid. Plant Physiol 98: 995–1002 (1992).

27. Fisher RF, Long SR: *Rhizobium*-plant signal exchange. Nature 357: 655–659 (1992).
28. Gardner HW: Recent investigations into the lipoxygenase pathway of plants. Biochim Biophys Acta 1084: 221–239 (1991).
29. Gardner HW, Hamberg M: Oxygenation of (3Z)-nonenal to (2E)-4-hydroxy-2-nonenal in the broad bean (*Vicia faba* L.). J Biol Chem 268: 6971–6977 (1993).
30. Green TR, Ryan CA: Wound-induced proteinase inhibitor in plant leaves: a possible defence mechanism against insects. Science 175: 776–77 (1972).
31. Gundlach H, Müller MJ, Kutchan TM, Zenk MH: Jasmonic acid is a signal transducer in elicitor-induced plant cell cultures. Proc Natl Acad Sci USA 89: 2389–2393 (1992).
32. Haagen-Smit AJ, Leach WD, Bergren WR: The estimation, isolation and identification of auxins in plant materials. Am J Bot 29: 500–506 (1942).
33. Hamburg M, Gardner HW: Oxylipin pathway to jasmonates: biochemistry and biological significance. Biochim Biophys Acta 1165: 1–18 (1992).
34. Hatanaka A, Kajiwara T, Sekija J: Biosynthetic pathway of C_6-aldehyde formation from linolenic acid in green leaves. Chem Phys Lipids 44: 341–361 (1987).
35. Hildebrand DF, Brown GC, Jackson DM, Hamilton-Kemp TR: Effects of some leaf-emitted volatile compounds on aphid population increase. J Chem Ecol 19: 1875–1887 (1993).
36. Kaiser I, Engelberth J, Groth B, Weiler EW: Touch and methyl jasmonate induced lignification in tendrils of *Bryonia dioica* Jacy. Bot Acta 107: 24–29 (1994).
37. Kato T, Yamaguchi Y, Hirano T, Yokohama T, Uyehara T, Yamanaka S, Harada T: Unsaturated hydroxyfatty acids, the self defence substances in rice plant against rice blast disease. Chem Lett: 409–412 (1984).
38. Keen NT, Sims JJ, Midland S, Yoder M, Jurnak F, Shen H, Boyd C, Yucul I, Lorang J, Murillo J: Determinants of specificity in the interaction of plants with bacterial pathogens. In: Nester EW, Verma PS (eds) Advances in Molecular Genetics and Plant-Microbe Interactions, pp. 211–220. Kluwer Academic Publishers, Dordrecht (1993).
39. Knudsen JT, Tollsten L, Bergstrom G: Floral scents – a checklist of volatile compounds isolated by headspace techniques. Phytochemistry 33: 253–280 (1993).
40. Koda Y, Kikuta Y, Kitahara T, Nishi T, Mori K: Comparisons of various biological activities of stereoisomers of methyl jasmonate. Phytochemistry 31: 1111–1114 (1992).
41. Koda Y: The role of jasmonic acid and related compounds in the regulation of plant development. Int Rev Cytol 135: 155–199 (1992).
42. Lyr H, Banasiak L: Alkenals, volatile defence substances in plants, their properties and activities. Acta Phytopath Acad Sci Hung 18: 3–12 (1983).
43. Mazur P, Nakanishi K, El-Zayat AAE, Champe SP: Structure and synthesis of sporogenic psi factors from *Aspergillus nidulans*. J Chem Soc Chem Commun 20: 1486–1487 (1991).
44. Meyer A, Miersch O, Büttner C, Dathe W, Sembdner G: Occurrence of the plant growth regulator jasmonic acid in plants. J Growth Regul 3: 1–8 (1984).
45. Meyer A, Schmidt J, Gross D, Jensen E, Rudolph A, Vorkefeld S, Sembdner G: Amino acid conjugates as metabolites of the plant growth regulator dihydrojasmonic acid in barley (*Hordeum vulgare*). J Plant Growth Regul 10: 17–25 (1991).
46. Meyer A, Gross D, Vorkfeld S, Kummer M, Schmidt J, Sembdner G, Schreiber K: Metabolism of the plant growth regulator dihydrojasmonic acid in barley shoots. Phytochemistry 28: 1007–1011 (1989).
47. Mueller MJ, Brodschelm W, Spannagl E, Zenk MH: Signalling in the elicitation process is mediated through the octadecanoid pathway leading to jasmonic acid. Proc Natl Acad Sci USA 90: 7490–7494 (1993).
48. Panossian AG, Avetissian GM, Mantsakanian VA, Batrakov SG, Vartanian SA, Gabrielian ES, Amroyan EA: Unsaturated polyhydroxy acids having prostagladin-like activity from *Bryonia alba*. II. Major components. Planta Med 47: 17–25 (1983).
49. Parthier B: Jasmonates: hormonal regulators or stress factors in leaf senescence? J Plant Growth Regul 9: 1–7 (1990).
50. Parthier B: Jasmonates, new regulators of plant growth and development: many facts and few hypotheses on their actions. Bot Acta 104: 446–454 (1991).
51. Pearce G, Strydom D, Johnson S, Ryan CA: A polypeptide from tomato leaves activates the expression of proteinase inhibitor genes. Science 253: 895–897 (1991).
52. Peña-Cortés H, Albrecht T, Prat S, Weiler EW, Willmitzer L: Aspirin prevents wound-induced gene expression in tomato leaves by blocking jasmonic acid biosynthesis. Planta 191: 123–128 (1993).
53. Podila GK, Dickman MB, Kolattukudy PE: Transcriptional activation of a cutinase gene in isolated fungal nuclei by plant cutin monomers. Science 242: 922–925 (1988).
54. Preisig CL, Kuć JA: Inhibition by salicylhydroxamic acid, BW755C, eicosatetraenoic acid, and disulfiram of hypersensitive resistance elicited by arachidonic acid or poly-*L*-lysine in potato tuber. Plant Physiol 84: 891–894 (1987).
55. Redemann CT, Rappaport L, Thompson RH: Phaseolic acid: a new plant growth regulator from bean seeds. In: Wightman F, Setterfield G (eds) Biochemistry and Physiology of Plant Growth Regulator Substances, pp. 109–124. Runge Press, Ottawa (1968).
56. Reinbothe S, Reinbothe C, Parthier B: Methyl jasmonate represses translation initiation of a specific set of mRNAs in barley. Plant J 4: 459–467 (1993).
57. Reinbothe S, Reinbothe C, Parthier B: Methyl jasmonate-regulated translation of nuclear-encoded chloroplast pro-

teins in barley (*Hordeum vulgare* L. cv. Salome). J Biol Chem 268: 10606–10611 (1993).
58. Reinbothe S, Reinbothe C, Heintzen C, Seidenbecher C, Parthier B: A methyl jasmonate-induced shift in the length of the 5′ untranslated region impairs translation of the plastid *rbcL* transcript in barley. EMBO J 12: 1505– 1512 (1993).
59. Ricker KE, Bostock RM: Eicosanoids in the *Phytophthora infestans*-potato interaction: lipoxygenase metabolism of arachidonic acid and biological activities of selected lipoxygenase products. Physiol Mol Plant Path 44: 65–80 (1994).
60. Schildknechkt VH, Rauch G: Die chemische Natur der Luftphytoncide von Blattpflanzen insbesondere von *Robinia pseudoacacia* Z Naturforsch 166: 422–429 (1961).
61. Schweizer P, Gees R, Mösinger E: Effect of jasmonic acid on the interaction of barley (*Hordeum vulgare* L.) with powdery mildew *Erysiphe graminis* f.sp. *hordei*. Plant Physiol 102: 503–511 (1993).
62. Sembdner G, Parthier B: The biochemistry and the physiological and molecular actions of jasmonates. Annu Rev Plant Physiol Plant Mol Biol 44: 569–589 (1993).
63. Spaink HP: Rhizobial lipo-oligosaccharides: answers and questions. Plant Mol Biol 20: 977–986 (1992).
64. Spaink HP, Sheeley DM, van Brussel AAN, Glushka J, York WS, Tak T, Geiger O, Kennedy EP, Reinhold VN, Lugtenberg BJJ: A novel highly unsaturated fatty acid moiety of lipo-oligosaccharide signals determines host specificity of *Rhizobium*. Nature 354: 125–130 (1991).
65. Staswick PE, Huang J-F, Yoon R: Nitrogen and methyl jasmonate induction of soybean vegetative storage protein genes. Plant Physiol 96: 130–136 (1991).
66. Staswick PE: Jasmonate, genes, and frageant signals. Plant Physiol 99: 804–807 (1992).
67. Staswick PE, Su W, Howell SH: Methyl jasmonate inhibition of root growth and induction of a leaf protein are decreased in an *Arabidopsis thaliana* mutant. Proc Natl Acad Sci USA 89: 6837–6840 (1992).
68. Strong FE, Kruitwagen E: Traumatic acid: an accelerator of abscission in cotton explants. Nature 215: 1380–1381 (1967).
69. Takai Y, Kikkawa U, Kaibuchi K, Nishizuka Y: Membrane phospholipid metabolism and signal transduction for protein phosphorylation. Adv Cyclic Nucl Prot Phosph Res 18: 119–149 (1984).
70. Ueda J, Kato J: Isolation and identification of a senescence-promoting substance from wormwood (*Artemisia absinthium* L.) Plant Physiol 66: 246–249 (1980).
71. Vaughn SF, Gardner HW: Lipoxygenase-derived aldehydes inhibit fungi pathogenic on soybean. J Chem Ecol 19: 2337–2345 (1993).
72. Vick BA: Oxygenated fatty acids of the lipoxygenase pathway. In: Moore TS (ed) Lipid Metabolism in Plants, pp. 167–191. CRC Press, London (1993).
73. Wasternack C, Atzorn R, Blume B, Leopold J, Parthier B: Ursolic acid inhibits synthesis of jasmonate-induced proteins in barley leaves. Phytochemistry 35: 49–54 (1994).
74. Weiler WE: Octadecanoid-derived signalling molecules involved in touch perception in a higher plant. Bot Acta 106: 2–4 (1993).
75. Weiler WE, Albrecht T, Groth B, Xia Z-Q, Luxem M, Liss H, Andert L, Spengler P: Evidence for the involvement of jasmonates and their octadecanoid precursors in the tendril coiling response of *Bryonia dioica*. Phytochemistry 32: 591–600 (1993).
76. Yoshikara T, Omer E-SA, Koshino H, Sakamura S, Kikuta Y, Koda Y: Structure of a tuber-inducing stimulus from potato leaves. Agric Biol Chem 53: 2835–2837 (1989).
77. Zimmerman DC, Coudron CA: Identification of traumatin, a wound hormone, as 12-oxo-trans-10-dodecenoic acid. Plant Physiol 63: 536–541 (1979).

Plant Molecular Biology **26**: 1439–1458, 1994.

The salicylic acid signal in plants

Daniel F. Klessig* and Jocelyn Malamy [1]
*Waksman Institute and Department of Molecular Biology and Biochemistry, Rutgers-The State University of New Jersey, P. O. Box 759, Piscataway, NJ 08855, USA (*author for correspondence);* [1] *Present address: Biology Department, 1009 Main Building, New York University, Washington Square, New York, NY 10003, USA*

Received 3 March 1994; accepted 26 April 1994

Key words: acquired resistance, active oxygen species, defense response, hypersensitive response, pathogenesis-related proteins, salicylic acid, signal transduction

Introduction

History

Plants are one of the world's richest sources of natural medicines. The use of plants and plant extracts for healing dates back to earliest recorded history. Today, such plant-derived medicines as quinine, digitalis, opiates and morphine are widely used, while new natural chemicals such as the putative anti-cancer drug taxol from yew tree bark are being characterized and developed.

The use of willow tree bark to relieve pain is believed to be as old as the 4th century B.C., when Hippocrates purportedly prescribed it for women during child birth [94, 145]. The active principle of willow remained a mystery until the 19th century when the salicylates, including salicylic acid (SA), methyl salicylate, saligenin (the alcohol of SA) and their glycosides, were isolated from extracts of different plants including willow. Soon thereafter SA was chemically synthesized, eventually leading to its widespread use. SA was subsequently replaced by the synthetic derivative acetylsalicylic acid (aspirin) which produces less gastrointestinal irritation yet has similar medicinal properties. Despite its long history the mode of action of SA is not fully understood. The finding that it plays a role in disease resistance responses in plants raises the possibility of some fascinating parallels between SA action in plants and animals.

Salicylic acid in plants

SA is one of numerous phenolic compounds, defined as compounds containing an aromatic ring with a hydroxyl group or its derivative, found in plants. There has been considerable speculation that phenolics in general function as plant growth regulators [1]. Exogenously supplied SA was shown to affect a large variety of processes in plants, including stomatal closure, seed germination, fruit yield and glycolysis (for review see [29]). However, some of these effects were also produced by other phenolic compounds. In addition, some effects of SA may have been caused by the general chemical properties of SA (as an iron chelator or acid) [95]. For these reasons, the significance of SA was not realized from these early studies. Only recently has there been evidence that SA has unique and specific regulatory roles.

Flowering and thermogenesis

The role of SA as an endogenous signalling molecule was first suggested in connection with flowering. Cleland and coworkers [23, 24] found that honeydew from aphids feeding on *Xanthium strumarium* contained an activity that induced flowering in duckweed (*Lemna gibba*) grown under a non-photoinductive light cycle. The flower-inducing factor could be extracted directly from the *Xanthium* phloem and was identified as SA. This was consistent with reports that exogenously applied SA was active in inducing flowering in both organogenic tobacco (*Nicotiana tabacum*) tissue culture [65] and whole plants of various species (for review see [95]). However, the possibility that SA is an endogenous signal for flowering remains in question since SA did not induce flowering when exogenously applied to *Xanthium*. Moreover, the endogenous levels of SA were the same in the phloem of vegetative and flowering *Xanthium*, and other substances were as effective as SA in flower induction (for review see [95]).

The first conclusive evidence implicating endogenous SA as a regulatory molecule resulted from studies of voodoo lilies (*Sauromatum guttatum*) [97, 99]. The spadix of the voodoo lily is thermogenic and exhibits dramatic increases in temperature during flowering. There are two periods of temperature increases in the spadix; a large, transient rise in endogenous SA levels was found to precede both periods. Furthermore, thermogenesis and the production of aromatic compounds associated with thermogenesis could be induced by treatment of spadix explants with SA, acetylsalicylic acid, or 2,6-dihydroxybenzoic acid but not with 31 structurally similar compounds.

The mechanism by which SA regulates heat production is beginning to emerge. During thermogenesis much of the electron flow in mitochondria is diverted from the cytochrome respiratory pathway to the alternative respiratory pathway [81]. The energy of electron flow through the alternative respiratory pathway is not conserved as chemical energy, but released as heat. The alternative respiratory pathway utilizes an alternative oxidase as the terminal electron acceptor. Rhoads and McIntosh found that expression of the alternative oxidase gene is induced by SA in voodoo lilies [103]. Interestingly, in nonthermogenic tobacco, SA treatment also caused a significant increase in alternative respiratory pathway capacity and a dramatic accumulation of alternative oxidase [105].

Disease resistance

The second process for which there is strong evidence that SA acts as a signal molecule is disease resistance. This has been an extremely active area of investigation during the past several years and is the focus of this review. Particular emphasis will be placed on progress made since the publication of several reviews in 1992 [29, 40, 73, 95, 96].

When a plant is infected with a pathogen to which it is resistant, a wide variety of biochemical and physiological responses are induced. Many of these responses are believed to protect the plant by restricting, or even eliminating, the pathogen and by limiting the damage it causes. In contrast, when a plant is infected with a pathogen to which it is susceptible, the pathogen replicates and frequently spreads throughout the plant, often causing considerable damage and even death of the host. Lack of resistance can be caused by an inability of the host plant to recognize or effectively respond to infection. Alternatively, the pathogen may have evolved strategies to overcome the plant's defense arsenal [59].

Although some plant-pathogen interactions lead to disease, most do not [72]. Plant disease resistance is often manifested as a restriction of pathogen growth and spread to a small zone around the site of infection. In many cases, this restriction is accompanied by localized death (necrosis) of host tissue. Together pathogen restriction and tissue necrosis characterize the hypersensitive response (HR) [78]. Often associated with this local response is the development, over a period of several days to a week, of enhanced resistance to a secondary infection by the same or even unrelated pathogens. This enhanced level of resistance can be manifested throughout the plant and is generally termed systemic acquired resistance (SAR) [22, 109, 110].

The fundamental processes involved in the HR

and SAR are not yet well understood but a large number of physiological, biochemical, and molecular changes have been noted that correlate with one or both of these responses. These include: (1) the synthesis and incorporation of hydroxyproline-rich glycoproteins (HRGPs), cellulose, callose, and phenolic polymers such as lignin into the cell wall to fortify this physical barrier; (2) the production of low molecular weight, antimicrobial compounds called phytoalexins; (3) the enhanced expression of genes encoding enzymes in the phenylpropanoid pathway, such as phenylalanine ammonia lyase (PAL), which often lead to the production of phytoalexins and other phenolic compounds; (4) the production of antiviral activities, some of which appear to be due to novel proteins; (5) the synthesis of proteinase inhibitors that block the activity of microbial and insect proteinases; (6) enhanced peroxidase activity, which is necessary for lignification and may be involved in crosslinking of cell wall protein; (7) the expression of genes encoding hydrolytic enzymes such as chitinases and β-1,3-glucanases that degrade the cell walls of microbes and may be involved in release of elicitor molecules; and (8) the synthesis of pathogenesis-related (PR) proteins.

Particular attention has been paid to the abundant PR proteins, a large group of proteins whose synthesis is induced by pathogen infection. These proteins have been divided into five or more unrelated families. Two of these families encode the hydrolytic β-1,3-glucanases (PR-2) and chitinases (PR-3), while the functions of the other families are poorly understood. The expression of many of the well characterized PR genes (e.g. PR-1 through PR-5) in tobacco has been correlated with resistance to a large variety of ([143]; for review see [17]) but not all [159] viral, bacterial and fungal pathogens. As a result, expression of PR genes is often used as a marker for induction of disease resistance. Moreover, studies with transgenic tobacco plants that overexpress some of these PR genes (PRs 1, 2, 3 and 5) have demonstrated that they enhance resistance to several fungal pathogens [4, 160, 15, 71a]. Antifungal activity *in vitro* has also been demonstrated for PR-2 through PR-5 proteins [71a, 79, 92, 142, 152]. For a more detailed discussion of the PR proteins and plant defense responses, the reader is referred to reviews by Carr and Klessig [17], Bowles [13], Bol *et al.* [10], Dixon and Harrison [34], Ryan [113], Linthorst [71], Madamanchi and Kuć [72], White and Antoniw [149], Ryals *et al.* [112], and Cutt and Klessig [30].

The diversity of the defense responses induced by pathogen attack suggests that they may be controlled by multiple signals acting through several pathways. Some of these signals need only act over short distances to induce defenses at the site of infection. However, for the development of SAR a signal must pass from the infection site to distal tissues. Grafting studies have demonstrated that a translocatable factor or signal can move from an infected leaf through the graft to the uninfected rootstock and induce SAR ([48]; for review see [72]). Considerable progress has been made in the past few years in the identification and characterization of several possible long-distance signals. These include systemin [80], jasmonates [44], electrical potentials [151], ethylene [91, 157] and SA. For a summary of the studies on systemic signals in plants, the reader is referred to two recent reviews [40, 73]. SA's putative role as a local and systemic signal is the subject of the next two sections of this review.

SA – a signal for defense

The first hint that SA might be involved in plant defense was provided by White [147] who found that injection of aspirin or SA into tobacco leaves enhanced resistance to subsequent infection by tobacco mosaic virus (TMV). This treatment also induced PR protein accumulation [5]. In addition to enhancing resistance to TMV in tobacco, SA also induced acquired resistance against many other necrotizing or systemic viral, bacterial, and fungal pathogens in a variety of plants ([144]; for review see [73]). (However, not all plant-pathogen systems respond to SA [106, 107, 158].) SA was also found to induce PR proteins in a wide range of both dicotyledonous and monocotyledonous plants including tomato [150], potato [148], bean [54], cucumber [84], cowpea [54, 148], rice [77, 121], garlic [134], soybean [28], azuki bean [55], sugar beet [45], *Arabidopsis thaliana* [131], and *Gomphrena globosa* [150].

Around the time that White [147] first demonstrated the effects of exogenous SA on defense responses, several reports showed a correlation between resistance and the levels of endogenous SA. For example, SA levels in extracts from the bark of different poplar species were shown to correlate with the resistance of the species to the fungal pathogen *Dothiciza populae* [93]. SA was also found to be toxic to several pathogens including *Colletotrichum falcatum*, *Fusarium oxysporum* [124] and *Agrobacterium tumefaciens* [114]. These latter results suggest that in some situations endogenous SA might inhibit pathogen growth due to its toxicity. However, SA was shown not to be directly toxic to cultures of *Colletotrichum lagenarium* [88] or preparations of TMV (Malamy and Klessig, unpublished results), even though it induces resistance to both of these pathogens in plants.

Since SA treatment of tobacco induces several of the same responses as TMV infection (i.e. acquired resistance and PR gene expression), it was postulated that SA acts by mimicking an endogenous phenolic signal that triggers PR gene expression and resistance [135]. However, by monitoring endogenous levels of SA in TMV-infected tobacco, Malamy and co-workers [74] provided evidence that strongly suggested a role for SA itself as a signal molecule. Infection of a TMV-resistant cultivar resulted in a dramatic increase (20–50-fold) in levels of endogenous SA in the TMV-inoculated leaves and a substantial rise (5–10-fold) in uninoculated leaves of the same plant. These increases were not seen in a nearly isogenic susceptible cultivar. The concentration of endogenous SA detected after infection (1–50 μM) was much lower than the concentration of SA used to exogenously induce PR gene expression and resistance (300–1000 μM). However, Raskin and colleagues [41, 156] demonstrated that when the concentration of endogenous SA in detached leaves or in plants grown hydroponically in different concentrations of SA reached levels similar to those detected in TMV-inoculated leaves, PR-1 gene expression and resistance were induced. Since the rise in endogenous SA paralleled or preceded the induction of PR-1 gene expression in both inoculated and uninoculated leaves of TMV-infected resistant plant [74], SA appears to play a role in the pathway leading to resistance responses.

Parallel studies in cucumber (*Cucumis sativus*) indicated that SA signalling is not unique to tobacco. A dramatic rise in SA levels (10–100-fold) was detected in the phloem exudates from cucumber leaves inoculated with tobacco necrosis virus, *C. lagenarium* [85] or *Pseudomonas syringae* pv. *syringae* [100, 124a]. These increases of SA in the phloem preceded both the appearance of SAR and the induction of peroxidase activity associated with resistance. Moreover, treatment of cucumber with SA also induced peroxidase activity and resistance to *C. lagenarium* [100, 124a]. Increases in SA levels recently have also been documented in *Arabidopsis thaliana* after infection with turnip crinkle virus ([132]; Dempsey, Wobbe and Klessig, unpublished results) or *P. syringae* [126] and in tobacco infected by tobacco necrosis virus, *P. syringae*, and *Peronospora tabacina* [120] and *Erwinia carotovora* [90a].

The work of Ward *et al.* [143] provided additional support for SA's involvement in disease resistance in tobacco. The expression of thirteen families of genes encoding peroxidase, acidic PR-1 through PR-5 proteins and their basic counterparts as well as several previously uncharacterized proteins were investigated. Expression of all thirteen genes were induced in TMV-inoculated leaves of resistant tobacco, while nine showed enhanced expression in uninoculated leaves of TMV-infected plants. These same nine genes, which included the acidic PR-1 through PR-5, basic PR-1, basic and acid class III chitinase, and PR-Q′ (a β-1,3-glucanase) genes, were induced by SA treatment. Thus, SA induced the same spectrum of genes activated during development of SAR upon TMV infection.

Further support for a signalling role for SA was provided by temperature shift experiments in the tobacco-TMV system. When TMV-resistant cultivars were inoculated and maintained at elevated temperatures (> 28 °C), they failed to synthesize PR proteins and the infection becomes systemic [47, 57]. However, when these infected plants were then transferred to lower temperatures (22–25 °C), PR gene expression was induced and resistance (HR) was restored. It was shown that temperatures which block the ability of tobacco to resist viral infection also inhibited increases in SA levels [75, 156]. When the resistance response was restored by shifting plants to lower temperatures, endogenous SA levels increased dramatically and preceded both PR-1 gene expression and necrotic lesion formation associated with resistance [75].

The strongest evidence for SA's involvement in plant defense comes from the elegant experiments of Gaffney *et al.* [46]. They constructed transgenic tobacco plants that constitutively express the *nahG* gene from *Pseudomonas putida* under the regulation of the 35S promoter of cauliflower mosaic virus (CaMV). *nahG* encodes salicylate hydroxylase, an enzyme which converts SA to catechol, a compound unable to induce SAR. SA levels in transgenic tobacco plants that accumulated substantial amounts of salicylate hydroxylase mRNA rose only 2- to 3-fold in the TMV-inoculated leaves compared to ca. 180-fold increase in untransformed control plants after TMV infection. These transgenic plants were subjected to a secondary inoculation of their upper leaves following a primary inoculation of their lower leaves to test the effect of reduced SA levels on the ability of the plant to establish SAR. Transgenic plants produced larger lesions in response to the secondary infection as compared to untransformed control plants, indicating a reduced ability to establish SAR. Furthermore, induction of genes associated with SAR such as the PR-1 genes was inhibited in the upper uninoculated leaves of TMV-infected *nahG* transgenic plants [141]. Surprisingly, the PR-1 genes were expressed in the TMV-inoculated tissue of *nahG* plants. Either the modest increase in SA in the inoculated *nahG* tissue is sufficient to induce these genes or another signal is involved.

In addition to its role in SAR development, SA may be involved in restricting the replication and spread of the pathogen from the initial sites of infection. It was observed that the increases in endogenous levels of SA were considerably greater in TMV-inoculated leaves than in uninoculated leaves [74], with the highest levels appearing in and around the infection sites [41]. In addition, larger primary necrotic lesions were formed on *nahG* transgenic tobacco than on non-transgenic controls after the initial infection by TMV, presumably due to the destruction of the SA signal by salicylate hydroxylase [46].

Taken together, these studies provide very strong support for SA's involvement in disease resistance. Levels of endogenous SA correlate with expression of defense-related genes and development of SAR while addition of exogenous SA induces defense responses and elimination of endogenous SA represses these responses. Further experiments are needed, however, to determine which of the several processes comprising disease resistance (e.g. lesion formation, restriction of pathogen growth or movement) are affected directly by SA. SA synthesis and response mutants, as well as the *nahG* plants, will help provide the necessary insights.

Is SA the translocated signal for SAR?

The experiments described above establish that SA plays a critical role in resistance and the development of SAR. However, it is unclear whether SA is the primary signal that travels from the inoculation site to distal tissues. Initial experiments suggested that SA might fulfill this function. First, the rise in SA levels preceded PR gene induction in uninoculated leaves of TMV-infected resistant tobacco [74]. Second, the large increase in SA in the phloem exudates from infected cucumber leaves preceded the development of SAR and induction of peroxidase activity in uninoculated leaves [85, 100, 124a]. Third, the appearance of chitinase (PR-3) in *P. lachrymans*-infected cucumber was preceded by an increase in SA levels in the upper, uninoculated leaves as well as in inoculated leaves [86]. These observations, together with the report that SA was in the phloem sap of TMV-infected tobacco [156], suggested that SA might be a primary mobile signal.

In contrast, two sets of experiments now strongly argue that SA is not the translocated signal for SAR. Since the systemic signal for SAR can pass through a graft junction [48], the requirement for SA in the induction of SAR and PR gene expression can be determined by grafting together *nahG* transgenic tobacco and untransformed tobacco (*Xanthi* nc) (141). When the scion (upper, grafted portion of the plant) was derived from the transgenic *nahG* plants, neither SAR nor PR-1 expression were induced in the scion after inoculation of the rootstock leaves with TMV, regardless of the origin of the rootstock. In contrast, an untransformed Xanthi nc scion grafted onto a *nahG* rootstock expressed PR-1 genes and showed SAR to a secondary infection by TMV or the fungal pathogen *Cercospora nicotianae* upon primary inoculation of the rootstock with TMV. These results indicate that (1) a signal other than SA can move from the rootstock to the scion after infection and (2) SA was required in the uninoculated leaves in the scion to mediate the translocated systemic signal.

The above results are consistent with and extend the observations in the cucumber-*P. syringae* system made by Hammerschmidt and co-workers [100, 124a]. When only one leaf on a cucumber plant was inoculated with *P. syringae*, increases in SA levels, peroxidase gene expression, and resistance were detected in the uninoculated leaves even if the inoculated leaf was removed as early as 4–6 h after infection. In contrast, SA was not detected in the phloem sap from the inoculated leaf until 8 h after infection. Thus, all the results to date are consistent with a model in which local infection leads to the production of an unidentified mobile factor or signal, which in turn requires SA in distal tissues for the establishment of SAR.

The model that SA is not the normal translocated signal is also consistent with the observation that when SA was applied locally via injection, it induced PR protein accumulation and enhanced resistance only in the treated tissue [136]. A likely explanation for this result comes from the studies of Métraux and coworkers [85] who showed that exogenous SA applied locally to leaves was not readily transported to other parts of the plant. However, detection of SA in the phloem sap of infected tobacco and cucumber leaves suggests that endogenous SA can be translocated. Perhaps endogenous SA and exogenously supplied SA are handled differently by the plants. This explanation, however, appears inconsistent with results from studies using chemical inducers of SA synthesis. Malamy *et al.* [76] found that local injection of polyacrylic acid or thiamine-HCl into tobacco leaves resulted in large, but transient increases in endogenous SA levels and stable accumulation of large amounts of a conjugated form of SA (see 'SA metabolism'). In contrast to TMV infection, the increase in SA levels and expression of PR-1 genes induced by these chemicals occurred only in the treated tissue. Thus, either endogenously produced SA is not readily translocated or transient elevation of SA is not sufficient for production (and translocation) of a systemic signal.

SA metabolism

In plants, SA is probably synthesized from phenylalanine [18], which is converted to *trans*-cinnamic acid by PAL. PAL is a key enzyme in the phenylpropanoid pathway that yields phytoalexins, lignins and hydroxybenzoic acids. There are two proposed pathways for the conversion of *trans*-cinnamic acid to SA; they differ in the order of β-oxidation and *ortho*-hydroxylation reaction. β-oxidation of *trans*-cinnamic acid produces benzoic acid, which can be hydroxylated to form SA. Alternatively, *ortho*-hydroxylation of *trans*-cinnamic acid forms *o*-coumaric acid, which can be converted to SA via β-oxidation.

Labelling studies by Yalpani and coworkers [153] indicate that in TMV-infected tobacco, SA is predominantly synthesized from benzoic acid (BA). The enzymatic activity responsible for converting BA to SA, BA 2-hydroxylase, was induced four- to five-fold by TMV infection [68].

BA treatment of tobacco plants also induced BA 2-hydroxylase activity. This latter result, together with the magnitude and timing of BA increases in TMV-infected plants, suggest that an increase in the BA pool is the primary cause of increased BA 2-hydroxylase activity. Thus, the rate-limiting step in SA production may be the formation of BA from *trans*-cinnamic acid or a conjugated form of BA [153].

Most phenolic acids in plants exist in the form of sugar conjugates. Glucose esters (glucose attachment through the carboxyl group) and glucosides (glucose attachment through the hydroxyl group) are particularly common. These derivatives of SA have been reported in several plants [9, 27, 51, 128, 133]. Several reports have demonstrated that the SA produced after TMV infection of tobacco is rapidly conjugated to glucose to form SA β-glucoside (SAG) [41, 75]. In these studies a large pool of SA could be identified after acid hydrolysis of crude extracts of TMV-infected tissues, indicating the presence of SA conjugates. These SA conjugates could be cleaved by β-glucosidase, establishing that they were SAG. Both SA and SAG were present at very low levels in uninfected plants but rose more or less in parallel after infection, with SAG becoming the predominant form. The same glucoside was formed from exogenously supplied radiolabelled SA within hours of application.

A UDP-glucose: SA glucosyltransferase that forms SAG from SA has been characterized in several plants. This enzyme activity is induced by SA in *Mallotus japonicus* [127], oats [154] and tobacco [39] and has been partially purified from all three plant species. In tobacco the enzyme activity is enhanced about 7-fold above basal levels between two and three days after infection with TMV, consistent with the rise in endogenous levels of SA.

Whether SAG accumulates in uninoculated portions of a plant is unclear. Enyedi *et al.* [41] detected free SA, but no SAG, in lower halves of tobacco leaves inoculated with TMV at their tips. They also found only SA in phloem exudates of inoculated leaves and in upper uninoculated leaves of TMV-infected plants. These results suggest that SAG is neither transported to nor synthesized in uninoculated sites. The authors speculate that these tissues may lack sufficient glucosyltransferase activity to convert SA to SAG. In contrast, Guo, Malamy and Klessig (unpublished results) have found both SA and SAG in the inoculated and uninoculated halves of TMV-infected leaves. Similarly, conjugated SA was found in uninoculated leaves of *P. lachrymans*-infected cucumber [86].

The existence of SAG suggests additional complexity in the modulation of the SA signal during defense responses. To test the bioactivity of SAG in the absence of SA, this compound was synthesized and injected into the extracellular spaces of tobacco leaves and PR-1 gene expression was subsequently monitored [53]. In these experiments SAG proved to be as active as SA in inducing PR-1 genes. However, isolation of extracellular fluid from SAG-injected leaves demonstrated that SAG was hydrolyzed to release SA in the extracellular spaces. Apparently, this released SA entered the surrounding cells and was reconjugated to form SAG. In accordance with this observation, a cell wall-associated β-glucosidase activity that converts SAG to SA has been detected (Malamy, Conrath and Klessig, unpublished results). The transient presence of SA in the injected leaves made it impossible to determine if SA or SAG was the active form. However, in studies of both phytohormones and phenolics the unconjugated forms have been shown to be active, while the glucose-conjugated forms are inactive [25, 26, 70, 102]. In addition, the SA-binding protein that is believed to transduce the SA signal fails to bind SAG (see 'Mechanism of action of SA'; [20]). Therefore, it seems likely that SA is active only in the free form.

Even if SAG is inactive, it may still serve as a storage form of SA. The spatial separation of an inactive glucoside and its β-glucosidase has been demonstrated for several phenolics (for review see [26]) and provides a means of regulating release of bioactive compounds. This compartmentation is reminiscent of the finding that SAG is intracellular while its putative β-glucosidase is extracellular. In addition, recent studies suggest that deconjugation plays an important role in regulating the activities of auxin [16, 43], cytokinin [42] and giberellin [116]. A model for the role of SAG in development of acquired resistance based on this type of regulation is presented in Fig. 1. In the model illustrated here, a primary infection by a pathogen such as TMV results in a HR, including formation of necrotic lesions and production of SA throughout the plant. As discussed above, much of this SA is converted to SAG. Subsequent challenge by a pathogen leads to cell damage and changes in membrane permeability at the infection site, resulting in the release of the local cellular stores of SAG into the extracellular spaces. Since extracellular SAG is hydrolyzed (see above), this would provide a high concentration of SA directly at the infection site where it could enter neighboring cells and induce defense responses. Through this process, the defense mounted during a second infection would be more effective than the initial defense. While some defense-related proteins would already be present due to their accumulation during the initial infection, the rapid release of SA from stored SAG would quickly induce the defense responses anew, when and where they were needed. This rapid and effective induction at the infection site would result in the more rapid restriction of the pathogen, and hence a smaller secondary lesion.

Besides SAG, additional SA derivatives have been detected in various plant species [27, 118, 128]. Moreover, when radiolabelled SA was supplied to tobacco leaves, ca. 20% could not be recovered as either SA or SAG [75]. At present, it is unclear whether other forms of SA are produced during defense responses and, if so, what their roles might be.

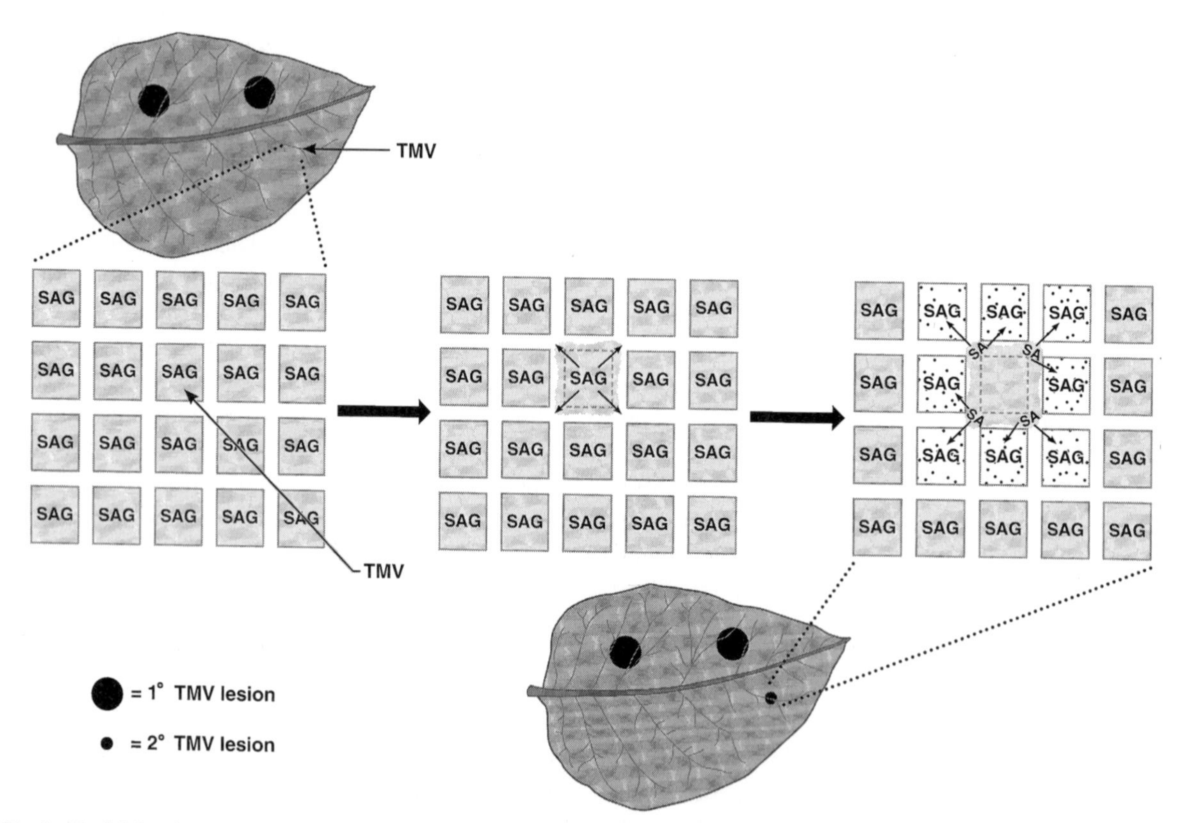

Fig. 1. Model for the role of SAG in acquired resistance. The primary infection of tobacco by TMV is shown by the large primary (1°) lesions on the top half of the leaf, while a secondary (2°) infection by TMV is shown on the lower half of the leaf. Development of acquired resistance is illustrated by reduced size of the secondary lesion. See text for details.

Mechanism of action of SA

As in animal and microbial systems, small biologically active compounds in plants such as phytohormones are believed to act through receptors. While it is still arguable whether or not SA is a hormone, particularly in light of the conflicting data concerning its translocation (see 'Is SA the translocated signal for SAR?'), it seems likely that identification of a cellular factor(s) which directly interacts with SA might shed light on SA's mode of action. This factor could be a receptor which perceives and transduces the SA signal or a SA-regulated cellular target such as an enzyme whose activity is altered by SA binding. Chen and Klessig [19, 20] have identified and characterized a soluble SA-binding protein (SABP) from tobacco which fits both descriptions of the factor.

The SABP is a 240–280 kDa complex which appears to be composed of four 57 kDa subunits. It has a binding affinity for SA (Kd 14 μM) which is consistent with the range of physiological concentrations of SA observed during the induction of defense responses. Its binding is also highly specific. Only those analogues of SA which are biologically active in the induction of PR genes and disease resistance (e.g. acetylsalicylic acid and 2,6-dihydroxybenzoic acid) effectively compete with SA for binding. Inactive analogues, though structurally very similar, are not bound by the SABP [19, 20].

Sequence analyses of the purified protein and a cDNA clone encoding the 57 kDa subunit indicated that SABP is highly homologus to catalase [21]. Indeed, SABP has catalase activity, as demonstrated by the ability of purified SABP to degrade H_2O_2 to H_2O and O_2. The size of the SABP complex and its subunits are also consistent with the structure of known catalases, which are composed of four identical or similar subunits of 50–60 kDa. SA was found to block SABP's catalase activity; it also inhibited the activity of catalases in crude tobacco leaf extracts. Furthermore, the effectiveness of different SA analogues to inhibit the catalase activity correlated with both their abilities to bind SABP and to induce defense responses. These results argue that the binding of SA is responsible for inhibition of catalase activity which may lead to the induction of defense responses.

Although surprising, the model that SA acts by inhibition of catalase activity is consistent with additional findings by Chen *et al.* [21]. Since the production of H_2O_2 is an ongoing process in plants (see legend to Fig. 2 for details), as expected, inhibition of catalase activity by treatment of tobacco plants with SA led to elevated H_2O_2 levels *in vivo*. Elevation of H_2O_2 levels by SA was as effective as that achieved by treatment with the plant and animal catalase inhibitor 3-amino-1,2,4-triazole (3AT). Furthermore, when endogenous H_2O_2 levels were raised by injecting leaves with H_2O_2 or 3AT, PR-1 gene expression was induced. PR-1 gene expression was also induced by injection of compounds that promote H_2O_2 generation *in vivo* such as glycolate and paraquat. (Glycolate is an intermediate in photorespiration which serves as a substrate for the generation of H_2O_2 by glycolate oxidase. Paraquat is a herbicide which is reduced *in vivo*. During its subsequent reoxidation superoxide anions are formed, which can be converted to H_2O_2 by superoxide dismutase [SOD].) In addition, the biologically inactive SA analogue 3-hydroxybenzoic acid, which differs from SA only in the placement of the hydroxyl group on the carbon ring, failed to induce either increases in H_2O_2 levels or PR-1 gene expression, consistent with its inability to bind SABP or inhibit its catalase activity. Moreover, preliminary experiments indicated that treatment of plants with 3AT or paraquat enhanced resistance to TMV infection (Chen and Klessig, unpublished results) as has been previously demonstrated for SA [147]. Taken together, these results strongly suggest that SA acts by blocking catalase activity, which results in elevated H_2O_2 levels. Hydrogen peroxide, or another active oxygen species (AOS) derived from it, then activates defense-related genes on the pathway to disease resistance, perhaps by acting as a second messenger. However, it should be noted that PR genes were activated considerably less effectively with H_2O_2 treatment than with SA treatment (Chen, Conrath and Klessig, unpu-

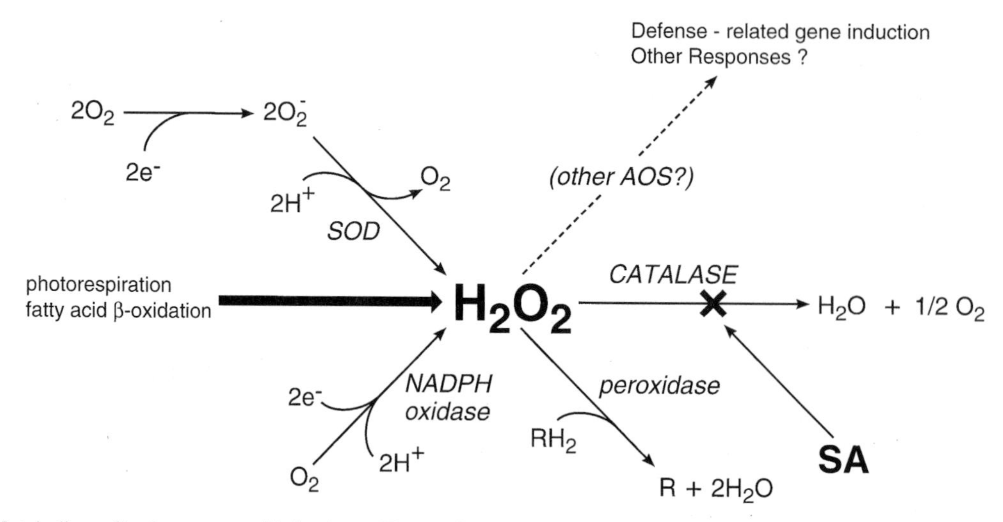

Fig. 2. Metabolism of hydrogen peroxide in plants. Two major sources of H_2O_2 are photorespiration and β-oxidation of fatty acids. H_2O_2 is also produced from superoxide anions (O_2^-) by SOD. The superoxide anion is formed by transfer of electrons (e^-) to molecular oxygen. These misdirected electrons often originate from electron transport during oxidative phosphorylation in the mitochondria and during photosynthesis in the chloroplasts. Membrane-bound, NADPH-dependent oxidases also generate H_2O_2. This is probably an important local source of H_2O_2 at the site of injury or infection. H_2O_2 can be converted to other AOS such as the highly reactive hydroxyl radical. H_2O_2, and other AOS derived from it, may be involved in a number of defense responses including the induction of defense-related genes such as the PR-1 genes. It may also lead to other responses yet to be defined. There are two major pathways involved in the breakdown of H_2O_2. As described in the text, H_2O_2 can be converted to H_2O and O_2 by catalase, a process that can be inhibited by SA. The other pathway involves breakdown of H_2O_2 by specific peroxidases either in the course of normal metabolism (e.g. in lignin formation) or via a specific pathway to eliminate toxic levels of H_2O_2 (e.g. Halliwell-Asada pathway). In addition, plant cells contain relatively high levels of glutathione, ascorbic acid and α-tocopherol which act as efficient AOS scavengers (for reviews see [12, 125]).

blished results; Ryals and Draper, personal communications). One explanation is that another signal in place of or in cooperation with H_2O_2 may be involved. Alternatively, the exogenously applied H_2O_2 may be too rapidly destroyed by catalases and oxidant scavengers to act as an effective inducer.

The participation of catalase and H_2O_2 in disease resistance is supported by several recent findings. Tobacco plants were constructed which constitutively express a copy of the catalase gene in an antisense orientation. These transgenic plants had reduced catalase levels and synthesized PR-1 proteins constitutively. In addition, there was a good correlation between the level of reduction in catalase and the amount of increased resistance to TMV in these transgenic lines (Chen and Klessig, unpublished results). Furthermore, several abiotic inducers of PR genes and disease resistance such as thiamine-HCl, polyacrylic acid, α-amino butyric acid, and $BaCl_2$ appear to act through SA since they induced the synthesis of SA and SAG in treated tobacco. One exception was 2,6-dichloroisonicotinic acid (INA) which did not stimulate production of SA and SAG [76]. In contrast, INA was found to bind and inhibit tobacco catalases, suggesting that its mode of action is similar to that of SA (Conrath, Chen, and Klessig, unpublished results). These results, in addition to the previous observations, strongly argue that SA and a broad range of defense-inducing compounds act by directly or indirectly modulating catalase activity.

Increases in the levels of AOS such as H_2O_2, superoxide anions or hydroxyl radicals have previously been linked to a number of processes associated with defense responses (for review see [125]). A large increase in the level of AOS, referred to as an oxidative or respiratory burst, occurs within minutes to hours after infection of plants by various pathogens which will initiate the HR (e.g. [36, 37]). The rapid kinetics of this oxidative burst diminishes the possibility that it results from SA inhibition of catalase during the initial infection, since elevated SA levels are not detected until 8–24 h or more after infection, depending on the system [74, 85, 100, 120, 126]. In contrast, during a secondary infection when free SA might be rapidly produced by hydrolysis of stored SAG, an oxidative burst might be mediated in part by SA.

Regardless, previous studies provide evidence that AOS can act as signals to elicit various defense responses. It has been suggested that the oxidative burst plays a role in killing the invading pathogen [60]. This has parallels to the immune response in mammals where phagocytic cells such as macrophages and neutrophils engulf their bacterial prey in a phagocytic vacuole where high concentrations of superoxide anion, H_2O_2, and other AOS can be generated. There is also evidence that AOS causes host plasma membrane damage [61] and cell death during the HR in plants [35–37]. In addition, increased lignification at the site of infection requires H_2O_2, as does the oxidative crosslinking of cell wall proteins in bean and soybean suspension culture cells after treatment with a fungal elicitor [14]. SA was found to induce this crosslinking, which can be explained by SA's ability to inhibit catalase's H_2O_2-scavenging activity. This could also explain SA's activation of superoxide dismutase (SOD) genes [11], as these genes are known to be induced by accumulation of AOS (for review see [12]). Finally, H_2O_2 may act as a second messenger for phytoalexin synthesis that accompanies the oxidative burst induced by fungal elicitor treatment of soybean suspension culture cells [6, 66].

Gene activation by H_2O_2 has been described in animal systems. Perhaps the best example involves the transcription factor NF-κB which mediates expression of genes associated with inflammatory, immune, and allergic responses. NF-κB exists in the cytoplasm in an inactive form, complexed with its inhibitory subunit IκB. H_2O_2-mediated activation results in dissociation of NF-κB from IκB. This allows NF-κB to enter the nucleus and activate genes by binding to specific motifs in promoter and enhancer elements (for review see [117]). Thus, there appear to be similar conditions (e.g. stress) in both animals and plants in which AOS are employed as second messengers. However, while activation of NF-κB-IκB complex does not require protein synthesis, the H_2O_2-mediated activation of the PR-1 genes by SA is inhibited by cycloheximide [130], suggesting that the mechanism of gene activation may be different.

SA signal transduction pathway(s)

Several other approaches are in progress to identify the various components of the SA signal transduction pathway(s). A popular approach is to first define the *cis*-acting elements in the target gene that are necessary for response to the signal and then to use these sequences to isolate the *trans*-acting factors. Currently, the *cis*-acting sequences of several SA-inducible genes have been partially analyzed by fusing various lengths of the promoter region to the *uidA* (GUS) reporter gene. Transgenic tobacco plants containing the chimeric genes were then constructed and analyzed. In some cases the presumptive SA-responsive region was fused to a minimal promoter, i.e. the core 35S promoter. The construct was then tested to determine if that region could confer SA inducibility to the heterologous promoter.

Of the SA-inducible promoters, the tobacco PR-1a promoter has been the most extensively characterized. 5′ end deletion analysis by Van de Rhee *et al.* [139] indicated that an important SA-responsive element was located between 643 and 689 bp upstream of the transcriptional start site. Moreover, fusion of the −625 to −902 region to the core 35S promoter imparted SA inducibility. Similarly, Uknes and co-workers [130] demonstrated that 661 bp 5′ of the transcription start site were sufficient to maintain SA inducibility while 318 bp were not. Beilmann *et al.* [7] also concluded that 335 bp of 5′-flanking DNA was insufficient to preserve SA responsiveness. In contrast, Ohashi and colleagues [90] observed a low level of induction of a chimeric gene containing only ca. 300 bp of 5′-flanking DNA and concluded that this region contained a SA-responsive element.

While the consensus from these studies is that one or more elements located between ca. −300 and −700 in the PR-1a promoter are required for high-level induction by SA, more recent studies indicated that the situation is more complex. Ohashi and coworkers [50] identified two independent binding sites at −37 to −61 and −168 to −179 for protein factors present in nuclear extracts from uninfected, and thus non-PR-expressing, *N. tabacum* cv. Samsun NN. In contrast, the factor was not detected with nuclear proteins from the interspecific hybrid of *N. glutinosa* × *N. debneyi*, which constitutively produces the PR-1 proteins [2, 89] and high levels of SA [155]. This result suggested that the PR-1a gene is negatively regulated through these two elements and their binding proteins. Although the two elements do not contain a common sequence, competition studies suggest that the same factor binds both. From an analysis of transgenic plants carrying fusion constructs of various segments of the PR-1a promoter with the core 35S promoter and GUS reporter gene, Van de Rhee and Bol [137] concluded that the PR-1a promoter contains a minimum of four regulatory elements located between nucleotides −902 to −691 (element 1), −689 to −643 (element 2), −643 to −287 (element 3) and −287 to +29 (element 4). In this study, these interacting elements appeared to function in a context-dependent manner, with elements 1, 2, and 3 functioning as positive regulators. Element 4 appeared to be important for maintaining the correct spacing between the more 5′ elements and the transcriptional start site. In all of the studies described in this section, each PR-1a promoter construct responded similarly to SA treatment and TMV infection, consistent with the hypothesis that SA is involved in the induction of PR-1 genes by infection.

The tobacco PR-2 genes, which encode the acidic β-1,3-glucanases, have also been characterized. By 5′ deletion analysis of the PR-2d promoter, Hennig *et al.* [52] mapped a *cis* element necessary for high-level SA inducibility between −321 and −607. Fusion of various segments of this promoter to the core 35S promoter suggested that there are at least two positive regulatory elements between nucleotides −318 and −607 that confer SA responsiveness (Shah and Klessig, unpublished results). 5′ deletion analysis of the PR-2b promoter demonstrated that as little as ca. 300 bp of 5′-flanking sequences were sufficient for low-level induction by SA (about 2-fold to 3-fold) while an additional 350 bp of 5′-flanking DNA were required for high-level induction [138]. Thus, in all three promoters (PR-1a, PR-2b and PR-2d) several regulatory elements appear to be involved in SA responsiveness and TMV inducibility.

The promoters of several other defense-related genes including those for the tobacco PR-5 protein [3] and glycine-rich protein [139] and the *Arabidopsis thaliana* acidic chitinase (PR-3 [115]) have been similarly, though less extensively, characterized. Sequence analysis of the 5′ untranscribed region of a barley β-1,3-glucanase identified a 10 bp motif (TCATCTTCTT) which is repeated several times [49]. This TCA motif is present in over 30 different plant genes which are known to be induced by various forms of stress including infection. A 40 kDa tobacco nuclear protein binds the TCA motif, and this binding activity is enhanced in nuclear extracts from SA-treated plants. Both the tobacco PR-1a and PR-2d genes contain multiple copies of the TCA motif in their 5′ ends. However, the minimum segments of these two promoters which confer SA inducibility to the heterologous core 35S promoter (i.e. −643 to −689 for PR-1a and −318 to −607 for PR-2d) do not carry the TCA motif. Moreover, there is as yet no evidence that the TCA motif, or concatemers of it, can confer SA responsiveness to a heterologous promoter. Thus, it is unclear what role this motif plays in SA inducibility.

Promoters of other genes such as alternative oxidase [103, 104], manganese SOD [11], and nopaline synthase [62] and the 35S promoter of CaMV [93a] are also stimulated by SA. In the

35S promoter SA responsiveness is mediated by the AS1 element. This element can also confer SA inducibility when fused to the ribulose bisphosphate carboxylase small subunit promoter. Qin and co-workers [93a] have reported modest activation of the core 35S promoter by SA in transgenic tobacco carrying a chimeric 35S:GUS reporter gene. At the RNA level induction was ca. 20-fold; however, when enzyme activity was assayed induction was only 3- to 5-fold. This inducing was very rapid, peaking by four hours after treatment, and was not sensitive to cycloheximide. However, other investigators have used the core 35S promoter, which contains one copy of the AS1 element, as a minimal heterologous promoter for fusion with potential SA-responsive regions from the PR-1 and PR-2 promoters. The core 35S promoter alone served as a negative control in these experiments, exhibiting only weak (about 2-fold), if any, stimulation by SA as measured by GUS activity [52, 90, 137, 139].

The genetic approach to defining components in a signalling pathway is very powerful and has been used extensively in bacteria, yeast, and *Drosophila*. This approach is also being applied to plants, particularly *Arabidopsis thaliana* which has a relatively rapid generation time, a small well-mapped genome, and the ability to be transformed (for review see [87]). Until recently, little was known about the pathology of this weed, but in the past few years a number of bacterial [31, 32, 123, 129, 146], fungal [63, 79a], viral [33, 56, 67, 82, 122, 126a], and nematode [119] pathogens of *Arabidopsis* have been identified. The cloning of several genes conferring resistance to some of these pathogens is in progress (e.g. [32, 64]).

As in tobacco and cucumber, SA appears to be involved in the defense responses of *Arabidopsis* to pathogen attack. PR genes were induced by SA treatment of *Arabidopsis* [131] and infection by turnip crinkle virus of the resistant ecotype Dijon lead to elevated levels of endogenous SA ([132]; Dempsey, Wobbe and Klessig, unpublished results). In addition, *Arabidopsis* contains a SA-binding activity and a SA-inhibitable catalase activity (Sánchez-Casas and Klessig, unpublished results). Finally, transgenic *Arabidopsis* plants which constitutively synthesize salicylate hydroxylase exhibited enhanced susceptibility to several pathogens (Uknes and Ryals, personal communication). Thus, *Arabidopsis* should provide an excellent opportunity to define components of the SA signal transduction pathway through genetic analysis. A number of mutants have already been isolated that may have lesions in this pathway [34a, 49a, 49b, 646].

Perspective

It is now clear that SA is an essential signal in development of SAR in several plant species and that this signal may be mediated, in part, by SA's ability to bind and inhibit the activity of catalases. However, many unanswered questions remain. Does SA play a role in the initial restriction of the pathogen? If so, is a similar mechanism involved as in the induction of SAR? Although there is good evidence for SA's involvement in defense responses in dicots (e.g. tobacco, cucumber, *Arabidopsis*), does it also participate in the defense of monocots, particularly the important grain crops?

It will be interesting to see whether all of the actions of SA in plant signal transduction are mediated by its inhibition of catalases or whether there are additional undiscovered modes of action. In animals salicylates appear to have multiple modes of action since these compounds exert a wide range of clinical effects including reduction of pain, fever, inflammation, blood clotting, and the risk of heart attacks and strokes. Aspirin inhibits the synthesis of prostaglandins from arachidonic acid, a fatty acid constituent of animal cell membranes [140]. Prostaglandins are potent compounds that can affect pain reception and can induce fever, swelling, platelet aggregation, and vasoconstriction. More recently, aspirin and aspirin-like compounds have also been found to perturb cell-cell communication, such as platelet aggregation and neutrophil activation. This perturbation may result from interference with G protein-mediated signal transduction and may account, in part, for aspirin's anti-clotting and anti-inflammatory activities (for review see [145]).

Recent studies suggest that SA may also have alternate mechanisms of action in plants. For example, although SA is present at high levels in rice [98], little, if any, SA-binding activity was detectable in this plant. In addition, rice catalase activity was not substantially inhibited by SA (Sánchez-Casas and Klessig, unpublished results). Thus, if SA is involved in pathways leading to disease resistance in rice, it probably is via an alternative mechanism. If SA is shown to have multiple modes of action in plants, it will be of interest to determine which processes (e.g. disease resistance and thermogenesis) are regulated by the different modes of action.

Another area that deserves further attention is the metabolism of SA. Although some of the steps in the pathway(s) are becoming clear and several of the enzymes are currently being characterized, much remains to be done. The enzymes have yet to be obtained in pure form and their respective genes have not been cloned. Relatively little is known concerning the regulation of these enzymes and their corresponding genes. Since synthesis of SA branches off the phenylpropanoid biosynthetic pathway, it is possible that SA increases are mediated by elevated levels of PAL [143]. PAL is a key enzyme in the phenylpropanoid pathway and is induced during defense responses in many plants [49c]. Ward and colleagues [143] have speculated on the possible connection between necrosis during HR and SA biosynthesis. Necrosis has been linked to induction of PAL [8, 38], and hence may actually precede SA production and subsequent development of SAR. Interestingly, an oxidative burst appears to be involved in formation of necrotic lesions. Doke and Ohashi [37] found that when TMV-infected tobacco were shifted from 30 °C (where viral replication and spread is not inhibited) to a lower temperature (20 °C) which allows the host to restrict viral replication and spread and form lesions, there was a rapid oxidative burst occurring from ten minutes to four hours after the shift. Infiltration of TMV-infected leaves with catalase, SOD or $NADP^+$ caused a reduction in lesion formation following the temperature shift. Thus, an early oxidative burst may be required for HR, which in turn may be necessary for production of SA, which then inhibits catalase activity leading to the second period of high AOS levels. Clearly more experiments are needed to address the relationships between these pathogen-induced processes.

In addition to the biosynthesis of SA, much remains to be learned about its catabolism. For example, the contribution of preformed stores of SAG to the SA pool during a secondary infection has not been established. In addition, the possible involvement of other, yet to be identified, forms of SA needs further investigation.

One of the most challenging problems will be defining the details of the SA signal transduction pathway(s). Some progress has been made with the identification and characterization of the SABP, the definition of several SA-responsive regulatory regions in the promoters of a number of defense-related genes, and the identification of potential *trans*-acting factors. However, our present picture represents only a rough outline of the pathway. In the future, the isolation and characterization of mutants in *Arabidopsis* will play an important role in deciphering this puzzle. Dissection of this complex pathway can also be facilitated by identifying multiple markers for different steps. Few markers now exist. In tobacco TMV infection leads to SA and ethylene production, but no intervening marker events are known except for the temperature sensitive step that precedes both ethylene and SA synthesis [75, 136, 156]. Dissection of downstream events are now possible due to the existence of both ethylene inhibitors and *nahG* transgenic plants, which selectively block responses to ethylene and SA, respectively. To provide further markers, Malamy *et al.* [76] tested chemical inducers of SAR and other defense responses and found that many, but not all, induce SA production. Three of the tested chemicals, polyacrylic acid, thiamine-HCl, and 2,6-dichloroisonicotinic acid (INA), enter the pathway at different points. Studies of the effects of these and other chemicals should allow their positioning in the pathway with respect to each other and the known marker events. Not only will chemical inducers aid in establishing the order of events, they will also provide a criteria for the identification of mutants defective at different steps in the signal transduction pathway. This combined biochemical and genetic approach should prove very powerful.

The complete signal transduction pathway(s) involving SA is likely to be highly complex. There may be multiple pathways leading to increased biosynthesis of SA. The isolation and characterization of plant disease resistance genes for several plant-pathogen systems that utilize SA (e.g. N locus in tobacco for TMV resistance; Baker, personal communication) should address this question and facilitate an understanding of how the SA signalling pathway is initiated. The SA signalling pathway itself may branch. There is likely to be extensive interplay between the SA signalling pathway and other pathways such as those for ethylene and jasmonic acid. For example, SA inhibits synthesis of ethylene in several suspension cell cultures including carrot [111], pear [69] and apple [108]. SA also blocks jasmonic acid biosynthesis [91a] and jasmonic acid induction of proteinase inhibitor synthesis in tomato (Ryan, personal communication). In contrast, pretreatment with SA potentiates or enhances phytoalexin production and incorporation of cell wall phenolics induced by suboptimal levels of a fungal elicitor in parsley suspension cells, although SA alone does not induce phytoalexin [58]. Signals from other pathways are also likely to both positively and negatively impact the SA pathway. For instance, ethylene appears to potentiate the SA-mediated induction of genes encoding acidic PR proteins in certain plants [64a]. Raz and Fluhr's [101] observation that inhibitors of ethylene action or biosynthesis blocked the induction of the acidic PR-3 chitinase by SA also suggest that ethylene and SA may act in concert.

A likely application of our emerging understanding of SA's role and mechanism of action in plant defense is crop protection. Enhanced disease resistance potentially can be achieved through the manipulation of SA metabolism. Alternatively, altering the reception of the SA signal or mimicking its effect through the use of other synthetic compounds may hold promise. The recent studies with INA suggest that the latter is a viable approach [83, 126, 131, 132, 143]. Finally, given the broad physiological effects of SA (and aspirin) in animal systems, insight into SA's mode of action(s) in plants may have implications beyond the plant world.

Acknowledgements

We would like to thank the investigators who contributed unpublished data to this review and members of the laboratory, particularly D'Maris Dempsey, for helpful comments and criticisms. Marline Boslet is gratefully acknowledged for assistance with the preparation of the manuscript. Our studies described in this review were partially supported by grants DCB-9003711 and MCB-9310371 from the National Science Foundation and 92-37301-7599 from the Department of Agriculture to D.F.K. and by a Benedict-Michael Predoctoral Fellowship to J.M.

References

1. Åberg B: Plant growth regulators. XLI. monosubstituted benzoic acids. Swedish J Agric Res 11: 93–105 (1981).
2. Ahl P, Gianinazzi S: b-protein as a constitutive component of highly (TMV) resistant interspecific hybrid of *Nicotiana glutinosa* × *Nicotiana debneyi*. Plant Sci Lett 26: 173–181 (1982).
3. Albrecht H, van de Rhee MD, Bol JF: Analysis of *cis*-regulatory elements involved in induction of a tobacco PR-5 gene by virus infection. Plant Mol Biol 18: 155–158 (1992).
4. Alexander D, Goodman RM, Gut-Rella M, Glascock C, Weymann K, Friedrich L, Maddox D, Ahl-Goy P, Luntz T, Ward E, Ryals J: Increased tolerance to two oomycete pathogens in transgenic tobacco expressing pathogenesis-related protein 1a. Proc Natl Acad Sci USA 90: 7327–7331 (1993).
5. Antoniw JF, White RF: The effects of aspirin and polyacrylic acid on soluble leaf proteins and resistance to virus infection in five cultivars of tobacco. Phytopath Z 98: 331–341 (1980).
6. Apostol I, Heinstein PF, Low PS: Rapid stimulation of an oxidative burst during elicitation of cultured plant cells. Plant Physiol 90: 109–116 (1989).
7. Beilmann A, Albrecht K, Schultze S, Wanner G, Pfitzner UM: Activation of a truncated PR-1 promoter by endogenous enhancers in transgenic plants. Plant Mol Biol 18: 65–78 (1992).
8. Bell JN, Ryder TB, Wingate VPM, Bailey JA, Lamb CJ:

Differential accumulation of plant defense gene transcripts in a compatible and an incompatible plant-pathogen interaction. Mol Cell Biol 6: 1615–1623 (1986).
9. Ben-Tal Y, Cleland CF: Uptake and metabolism of [^{14}C] salicylic acid in *Lemna gibba G3*. Plant Physiol 70: 291–296 (1982).
10. Bol JF, Linthorst HJM, Cornelissen BJC: Plant pathogenesis-related proteins induced by virus infection. Annu Rev Phytopath 28: 113–138 (1990).
11. Bowler C, Alliote T, De Loose M, Van Montagu M, Inzé D: The induction of manganese superoxide dismutase in response to stress in *Nicotiana plumbaginifolia*. EMBO J 8: 31–38 (1989).
12. Bowler C, Van Montagu M, Inzé D: Superoxide dismutase and stress tolerance. Annu Rev Plant Physiol Plant Mol Biol 43: 83–116 (1992).
13. Bowles D: Defense-related proteins in higher plants. Annu Rev Biochem 59: 873–907 (1990).
14. Bradley DJ, Kjelbom P, Lamb CJ: Elicitor- and wound-induced oxidative cross-linking of a proline-rich plant cell wall protein: a novel, rapid defense response. Cell 70: 21–30 (1992).
15. Broglie K, Chet I, Holliday M, Cressman R, Biddle P, Knowlton S, Mauvais CJ, Broglie R: Transgenic plants with enhanced resistance to the fungal pathogen *Rhizoctonia solani*. Science 254: 1194–1197 (1991).
16. Campos N, Bako L, Feldwisch J, Schell J, Palme K: A protein from maize labeled with azido-IAA has novel β-glucosidase activity. Plant J 2: 675–684 (1992).
17. Carr JP, Klessig DF: The pathogenesis-related proteins of plants. In: Setlow JK (ed) Genetic Engineering Principles and Methods, vol. 11, pp. 65–109. Plenum Press, New York/London (1989).
18. Chadha KC, Brown SA: Biosynthesis of phenolic acids in tomato plants infected with *Agrobacterium tumefaciens*. Can J Bot 52: 2041–2046 (1974).
19. Chen Z, Klessig DF: Identification of a soluble salicylic acid-binding protein that may function in signal transduction in the plant disease resistance response. Proc Natl Acad Sci USA 88: 8179–8183 (1991).
20. Chen Z, Ricigliano J, Klessig DF: Purification and characterization of a soluble salicylic acid-binding protein from tobacco. Proc Natl Acad Sci USA 90: 9533–9537 (1993).
21. Chen Z, Silva H, Klessig DF: Active oxygen species in the induction of plant systemic acquired resistance by salicylic acid. Science 262: 1883–1886 (1993).
22. Chester KS: The problem of acquired physiological immunity in plants. Quart Rev Biol 8: 275–324 (1933).
23. Cleland CF: Isolation of flower-inducing and flower-inhibiting factors from aphid honeydew. Plant Physiol 54: 899–903 (1974).
24. Cleland CF, Ajami A: Identification of the flower-inducing factor isolated from aphid honeydew as salicylic acid. Plant Physiol 54: 904–906 (1974).
25. Cohen JD, Bandurski RS: Chemistry and physiology of the bound auxins. Annu Rev Plant Physiol 33: 403–430 (1982).
26. Conn EE: Compartmentation of secondary compounds. In: Boudet AM, Alibert G, Marigo G, Lea PJ (eds) Annual Proceedings of the Phytochemical Society of Europe: Membranes and Compartmentation in the Regulation of Plant Functions, vol. 24, pp. 1–28. Clarendon Press, Oxford (1984).
27. Cooper-Driver G, Corner-Zamodits JJ, Swain T: The metabolic fate of hydroxybenzoic acids in plants. Z Naturforsch B 27: 943–946 (1972).
28. Crowell DN, John ME, Russell D, Amasino RM: Characterization of a stress-induced developmentally regulated gene family from soybean. Plant Mol Biol 18: 459–466 (1992).
29. Cutt JR, Klessig DF: Salicylic acid in plants: a changing perspective. Pharmaceut Technol 16: 26–34 (1992).
30. Cutt JR, Klessig DF: Pathogenesis-related proteins. In: Boller T, Meins Jr. F (eds) Plant Gene Research: Genes Involved in Plant Defense, pp. 209–243. Springer-Verlag, Wien/New York (1992).
31. Davis KR, Schott E, Ausubel FM: Virulence of selected phytopathogenic *Pseudomonas* in *Arabidopsis thaliana*. Mol Plant-Microbe Interact 4: 477–488 (1991).
32. Debener T, Lehnackers H, Arnold M, Dangl JL: Identification and molecular mapping of a single *Arabidopsis thaliana* locus determining resistance to a phytopathogenic *Pseudomonas syringae* isolate. Plant J 1: 289–302 (1991).
33. Dempsey DA, Wobbe KK, Klessig DF: Resistance and susceptible responses of *Arabidopsis thaliana* to turnip crinkle virus. Phytopathology 83: 1021–1029 (1993).
34. Dixon RA, Harrison MJ: Activation, structure and organization of genes involved in microbial defense in plants. Adv Genet 28: 165–234 (1990).
34a. Dietrich RA, Delaney TP, Uknes SJ, Ward ER, Ryals JA, Dangl JL: *Arabidopsis* mutants stimulating disease response. Cell 77: 565–577 (1994).
35. Doke N: Generation of superoxide anion by potato tuber protoplasts during the hypersensitive response to hyphal wall components of *Phytophthora infestans* and specific inhibition of the reaction by suppressors of hypersensitivity. Physiol Plant Path 23: 359–367 (1983).
36. Doke N: Involvement of superoxide anion generation in the hypersensitive response of potato tuber tissues to infection with an incompatible race of *Phytophthora infestans* and to the hyphal wall components. Physiol Plant Path 23: 345–357 (1983).
37. Doke N, Ohashi Y: Involvement of O_2^--generating systems in the induction of necrotic lesions on tobacco leaves infected with TMV. Physiol Mol Plant Path 32: 163–175 (1988).
38. Duchesne M, Fritig B, Hirth L: Phenylalanine ammonia-lyase in tobacco mosaic virus-infected hypersensitive tobacco; density-labelling evidence of de novo synthesis. Biochim Biophys Acta 485: 465–481 (1977).
39. Enyedi AJ, Raskin I: Induction of UDP-glucose: salicylic acid glucosyltransferase activity in tobacco mosaic virus-inoculated tobacco (*Nicotiana tabacum*) leaves. Plant Physiol 101: 1375–1380 (1993).
40. Enyedi AJ, Yalpani N, Silverman P, Raskin I: Signal molecules in systemic plant resistance to pathogens and pests. Cell 70: 879–886 (1992).
41. Enyedi AJ, Yalpani N, Silverman P, Raskin I: Localization, conjugation and function of salicylic acid in tobacco during the hypersensitive reaction to tobacco mosaic virus. Proc Natl Acad Sci USA 89: 2480–2484 (1992).

42. Estruch JJ, Chriqui D, Grossmann K, Schell J, Spena A: The plant oncogene *rolC* is responsible for the release of cytokinins from glucoside conjugates. EMBO J 10: 2889–2895 (1991).
43. Estruch JJ, Schell J, Spena A: The protein encoded by the *rolB* plant oncogene hydrolyzes indole glucoside. EMBO J 10: 3125–3128 (1991).
44. Farmer EE, Johnson RR, Ryan CA: Regulation of expression of proteinase inhibitor genes by methyl jasmonate and jasmonic acid. Plant Physiol 98: 995–1002 (1992).
45. Fleming TM, McCarthy DA, White RF, Antoniw JF, Mikkelsen JD: Induction and characterization of some of the pathogenesis-related proteins in sugar beet. Physiol Mol Plant Path 39: 147–160 (1991).
46. Gaffney T, Friedrich L, Vernooij B, Negrotto D, Nye G, Uknes S, Ward E, Kessmann H, Ryals J: Requirement of salicylic acid for the induction of systemic acquired resistance. Science 261: 754–756 (1993).
47. Gianinazzi S: Hypersensibilite aux virus, temperatures et proteines solubles chez le *Nicotiana tabacum* cv. Xanthi-nc. CR Acad Sci Paris D 270: 2382–2386 (1970).
48. Gianinazzi S, Ahl P: The genetic and molecular basis of b-proteins in the genus *Nicotiana*. Neth J Plant Path 89: 275–281 (1983).
49. Goldsbrough AP, Albrecht H, Stratford R: Salicylic acid-inducible binding of a tobacco nuclear protein to a 10 bp sequence which is highly conserved amongst stress-inducible genes. Plant J 3: 563–571 (1993).
49a. Greenberg JT, Ausubel FM: *Arabidopsis* mutants compromised for the control of cellular damage during pathogenesis and aging. Plant J 4: 327–341 (1994).
49b. Greenberg JT, Guo A, Klessig DF, Ausubel FM: Programmed cell death in plants: a pathogen-triggered response activated coordinately with multiple defense functions. Cell 77: 551–563 (1994).
49c. Hahlbrook K, Scheel D: Physiology and molecular biology of phenylpropanoid metabolism. Annu Rev Plant Physiol Plant Mol Biol 40: 347–369 (1989).
50. Hagiwara H, Matsuoka M, Ohshima M, Watanabe M, Hosokawa D, Ohashi Y: Sequence-specific binding of protein factors to two independent promoter regions of the acidic tobacco pathogenesis-related-1 protein gene (PR-1). Mol Gen Genet 240: 197–205 (1993).
51. Harborne JB: Phenolic glycosides and their natural distribution. In: Harborne JB (ed) Biochemistry of Phenolic Compounds, pp. 129–169. Academic Press, London (1964).
52. Hennig J, Dewey RE, Cutt JR, Klessig DF: Pathogen, salicylic acid and developmental dependent expression of a β-1,3-glucanase/GUS gene fusion in transgenic tobacco plants. Plant J 4: 481–493 (1993).
53. Hennig J, Malamy J, Grynkiewicz G, Indulski J, Klessig DF: Interconversion of the salicylic acid signal and its glucoside in tobacco. Plant J 4: 593–600 (1993).
54. Hooft van Huijsduijnen RAM, Alblas SW, de Rijk RH, Bol JF: Induction by SA of pathogenesis-related proteins and resistance to alfalfa mosaic virus infection in various plant species. J Gen Virol 67: 2143–2153 (1986).
55. Ishige F, Mori H, Yamazaki K, Imaseki H: Cloning of a complementary DNA that encodes an acidic chitinase which is induced by ethylene and expression of the corresponding gene. Plant Cell Physiol 34: 103–111 (1993).
56. Ishikawa M, Obata F, Kumagai T, Ohno T: Isolation of mutants of *Arabidopsis thaliana* in which accumulation of tobacco mosaic virus coat protein is reduced to low levels. Mol Gen Genet 230: 33–38 (1991).
57. Kassanis B: Some effects of high temperature on the susceptibility of plants to infection with viruses. Ann Appl Biol 39: 358–369 (1952).
58. Kauss H, Franke R, Krause K, Conrath U, Jeblick W, Grimmig B, Matern U: Conditioning of parsley (*Petroselinum crispum* L.) suspension cells increases elicitor-induced incorporation of cell wall phenolics. Plant Physiol 102: 459–466 (1933).
59. Keen NT: Pathogenic strategies for fungi. In: Lugtenberg B (ed) Recognition in Microbe-Plant Symbiotic and Pathogenic Interactions. NATO-ASI Series H, vol. 4, pp. 171–188. Springer-Verlag, Berlin/New York (1986).
60. Keppler LD, Baker CJ: O_2^--initiated lipid peroxidation in a bacteria-induced hypersensitive reaction in tobacco cell suspensions. Phytopathology 79: 555–562 (1989).
61. Keppler LD, Novacky A: Involvement of membrane lipid peroxidation in the development of a bacterially induced hypersensitive reaction. Phytopathology 76: 104–108 (1986).
62. Kim SR, Kim Y, An G: Identification of methyl jasmonate and salicylic acid response elements from the nopaline synthase (nos) promoter. Plant Physiol 103: 97–103 (1993).
63. Koch E, Slusarenko A: *Arabidopsis* is susceptible to infection by a downy mildew fungus. Plant Cell 2: 437–445 (1990).
64. Kunkel BN, Bent AF, Dahlbeck D, Innes RW, Staskawicz B: RPS2, an *Arabidopsis* disease resistant locus specifying recognition of *Pseudomonas syringae* expressing the avirulence gene avrRpt2. Plant Cell 5: 865–875 (1993).
64a. Lawton KA, Potter SL, Uknes S, Ryals J: Acquired resistance signal transduction in *Arabidopsis* is ethylene independent. Plant Cell 6: 581–588 (1994).
64b. Lawton KA, Uknes S, Friedrich L, Gaffney T, Alexander D, Goodman R, Métraux JP, Kessman H, Ahl-Goy P, Gut-Rella M, Ward D, Ryals J: The molecular biology of systemic acquired resistance. In: B. Fritig, M. Legrande (eds) Developments in Plant Biology, Mechanisms of Plant Defense Responses, pp. 422–432. Kluwer Academic Publishers, Dordrecht/Boston/London (1993).
65. Lee TT, Skoog F: Effects of substituted phenols on bud formation and growth of tobacco tissue culture. Physiol Plant 18: 386–402 (1965).
66. Legendre L, Rueter S, Heinstein PF, Low PS: Characterization of the oligogalacturonide-induced oxidative burst in cultured soybean (*Glycine max*) cells. Plant Physiol 102: 233–240 (1993).
67. Leisner SM, Howell SH: Symptom variation in different *Arabidopsis thaliana* ecotypes produced by cauliflower mosaic virus. Phytopathology 82: 1042–1046 (1992).
68. León J, Yalpani N, Raskin I, Lawton MA: Induction of benzoic acid 2-hydroxylase in virus-inoculated tobacco. Plant Physiol 103: 323–328 (1993).

69. Leslie CA, Romani RJ: Inhibition of ethylene biosynthesis by salicylic acid. Plant Physiol 88: 833–837 (1988).
70. Letham DS, Palni LMS: The biosynthesis and metabolism of cytokinins. Annu Rev Plant Physiol 34: 163–197 (1983).
71a. Liu D, Raghothama KG, Hasegawa PM, Bressan RA: Osmotin overexpression in potato delays development of disease symptoms. Proc Natl Acad Sci USA 91: 1888–1892 (1994).
71. Linthorst HJM: Pathogenesis-related proteins of plants. Crit Rev Plant Sci 10: 123–150 (1991).
72. Madamanchi NR, Kuć J: Induced systemic resistance in plants. In: Cole GT, Hoch HC (eds) The Fungal Spore and Disease Initiation in Plants and Animals, pp. 347–362. Plenum Press, New York (1991).
73. Malamy J, Klessig DF: Salicylic acid and plant disease resistance. Plant J 2: 643–654 (1992).
74. Malamy J, Carr JP, Klessig DF, Raskin I: Salicylic acid – a likely endogenous signal in the resistance response of tobacco to viral infection. Science 250: 1001–1004 (1990).
75. Malamy J, Hennig J, Klessig DF: Temperature-dependent induction of salicylic acid and its conjugates during the resistance response to tobacco mosaic virus infection. Plant Cell 4: 359–366 (1992).
76. Malamy J, Sánchez-Casas P, Hennig J, Guo A, Klessig DF: Dissection of the salicylic acid signalling pathway for defense responses in tobacco. Plant Physiol, submitted (1994).
77. Matsuta C, van den Bulcke M, Bauw G, van Montagu M, Caplan AG: Differential effects of elicitors on the viability of rice suspension cells. Plant Physiol 97: 619–629 (1991).
78. Matthews REF: Plant Virology, 3rd ed. Harcourt Brace Jovanovich, San Diego, CA (1991).
79. Mauch F, Mauch-Mani B, Boller T: Antifungal hydrolases in pea tissue. II. Inhibition of fungal growth by combinations of β-1,3-glucanase and chitinase. Plant Physiol 88: 936–942 (1988).
79a. Mauch-Mani B, Slusarenko A: Systematic acquired resistance in *Arabidopsis thaliana* induced by a predisposing infection with a pathogenic isolate of *Fusarium oxysporum*. Mol Plant-Microbe Interact 7: 378–383 (1994).
80. McGurl B, Pearce G, Orizco-Cardensa M, Ryan C: Structure, expression and antisense inhibition of the systemin precursor gene. Science 255: 1570–1573 (1992).
81. Meeuse BJD: Thermogenic respiration in aroids. Annu Rev Plant Physiol 26: 117–126 (1975).
82. Melcher U: Symptoms of cauliflower mosaic virus infection in *Arabidopsis thaliana* and turnip. Bot Gaz 150: 139–147 (1989).
83. Métraux JP, Ahl-Goy P, Staub T, Speich J, Steinemann A, Ryals J, Ward E: Induced resistance in cucumber in response to 2,6-dichloroisonicotinic acid and pathogens. In: Hennecke H, Verma DPS (eds) Advances in Molecular Genetics of Plant-Microbe Interactions, vol. 1, pp. 432–439. Kluwer Academic Publishers, Dordrecht (1991).
84. Métraux JP, Burkhart W, Moyer M, Dincher S, Middlesteadt W, Williams S, Payne G, Carnes M, Ryals J: Isolation of a complementary DNA encoding a chitinase with structural homology to a bifunctional lysozyme/chitinase. Proc Natl Acad Sci USA 86: 896–900 (1989).
85. Métraux JP, Signer H, Ryals J, Ward E, Wyss-Benz M, Gaudin J, Raschdorf K, Schmid E, Blum W, Inverardi B: Increase in salicylic acid at the onset of systemic acquired resistance in cucumber. Science 250: 1004–1006 (1990).
86. Meuwly Ph, Mölders W, Summermatter K, Sticher L, Métraux JP: Salicylic acid and chitinase in infected cucumber plants. Acta Hort, in press (1994).
87. Meyerowitz EM: *Arabidopsis*, a useful weed. Cell 56: 263–269 (1989).
88. Mills PR, Wood RKS: The effects of polyacrylic acid, aspirin and salicylic acid on resistance of cucumber to *Colletotrichum lagenarium*. Phytopath Z 111: 209–216 (1984).
89. Ohashi Y, Ohshima M, Itoh H, Matsuoka M, Watanabe S, Murakami T, Hosokawa D: Constitutive expression of stress-inducible genes, including pathogenesis-related 1 protein gene in a transgenic interspecific hybrid of *Nicotiana glutinosa* × *Nicotiana debneyi*. Plant Cell Physiol 33: 177–187 (1992).
90. Ohshima M, Itoh H, Matsuoka M, Murakami T, Ohashi Y: Analysis of stress-induced or salicylic acid-induced expression of the pathogenesis-related 1a protein gene in transgenic tobacco. Plant Cell 2: 95–106 (1990).
90a. Palva TK, Hurtig M, Saindrenan P, Palva ET: Salicylic acid-induced resistance to *Erwinia carotovora* subsp. *carotovora* in tobacco. Mol Plant-Microbe Interact 7: 356–363 (1994).
91. Pegg GF: The involvement of ethylene in plant pathogenesis. In: Heitefuss R, Williams PH (eds) Encyclopedia of Plant Physiology, New Series, vol. 4, pp. 582–591. Springer-Verlag, Heidelberg (1976).
91a. Peña-Cortés H, Albrecht T, Prat S, Water EW, Willmitzer L: Aspirin prevents wound-induced gene expression in tomato leaves by blocking jasmonic acid biosynthesis. Planta 191: 123–128 (1993).
92. Ponstein AS, Bres-Vloemans SA, Sela-Buurlage MB, van den Elzen PJM, Melchers LS, Cornelissen BJC: A novel pathogen- and wound-inducible tobacco (*Nicotiana tabacum*) protein with antifungal activity. Plant Physiol 104: 109–118 (1994).
93. Pucacka S: Role of phenolic compounds in the resistance of poplars to the fungus *Dothichiza populae*. Arbor Kornickie 25: 257–268 (1980).
93a. Qin XF, Holuigue L, Horvath DM, Chua N-H: Immediate early transcription activation by salicylic acid via the cauliflower mosaic virus *as*-1 element. Submitted (1994).
94. Rainsford KD: Aspirin and the Salicylates. Butterworth, London (1984).
95. Raskin I: Role of salicylic acid in plants. Annu Rev Plant Physiol Plant Mol Biol 43: 439–463 (1992).
96. Raskin I: Salicylate, a new plant hormone. Plant Physiol 99: 799–803 (1992).
97. Raskin I, Ehmann A, Melander WR, Meeuse BJD: Salicylic acid – a natural inducer of heat production in *Arum* lilies. Science 237: 1601–1602 (1987).
98. Raskin I, Skubatz H, Tang W, Meeuse BJD: Salicylic

acid levels in thermogenic and non-thermogenic plants. Ann Bot 66: 369–373 (1990).

99. Raskin I, Turner IM, Melander WR: Regulation of heat production in the inflorescences of an arum lily by endogenous salicylic acid. Proc Natl Acad Sci USA 86: 2214–2218 (1989).
100. Rasmussen JB, Hammerschmidt R, Zook M: Systemic induction of salicylic acid accumulation in cucumber after inoculation with *Pseudomonas syringae* pv. *syringae*. Plant Physiol 97: 1342–1347 (1991).
101. Raz V, Fluhr R: Calcium requirement for ethylene-dependent responses. Plant Cell 4: 1123–1130 (1992).
102. Reinecke DM, Bandurski RS: Auxin biosynthesis and metabolism. In: Davis PJ (ed) Plant Hormones and their Role in Plant Growth and Development, pp. 24–42. Martinus Nijhoff, Dordrecht (1988).
103. Rhoads DM, McIntosh L: Salicylic acid regulation of respiration in higher plants: alternative oxidase expression. Plant Cell 4: 1131–1139 (1992).
104. Rhoads DM, McIntosh L: The salicylic acid-inducible alternative oxidase gene *aox1* and genes encoding pathogenesis-related proteins share regions of sequence similarity in their promoters. Plant Mol Biol 21: 615–624 (1993).
105. Rhoads DM, McIntosh L: Cytochrome and alternative pathway respiration in tobacco; effects of salicylic acid. Plant Physiol 103: 877–883 (1993).
106. Roggero P, Pennazio S: Effects of salicylate on systemic invasion of tobacco plants by various viruses. J Phytopath 123: 207–216 (1988).
107. Roggero P, Pennazio S: Salicylate does not induce resistance to plant viruses, or stimulate pathogenesis-related protein production in soybean. Microbiologica 14: 65–69 (1991).
108. Romani RJ, Hess BM, Leslie CA: Salicylic acid inhibition of ethylene production by apple discs and other plant tissues. J Plant Growth Regul 8: 63–70 (1989).
109. Ross AF: Localized acquired resistance to plant virus infection in hypersensitive hosts. Virology 14: 329–339 (1961).
110. Ross AF: Systemic acquired resistance induced by localized virus infections in plants. Virology 14: 340–358 (1961).
111. Roustan JP, Latche A, Fallot J: Inhibition of ethylene production and stimulation of carrot somatic embryogenesis by salicylic acid. Biol Plant 32: 273–276 (1990).
112. Ryals J, Ward E, Ahl-Goy P, Métraux JP: Systemic acquired resistance: an inducible defence mechanism in plants. In: Wray JL (ed) Inducible Plant Proteins, pp. 205–229, Society for Experimental Biology, Seminar series 49 (1992).
113. Ryan CA: Proteinase inhibitors in plants: genes for improving defenses against insects and pathogens. Annu Rev Phytopath 28: 425–449 (1990).
114. Saint-Pierre B, Miville L, Dion P: The effects of salicylates on phenomena related to crown gall. Can J Bot 62: 729–734 (1984).
115. Samac DA, Shah DM: Developmental and pathogen-induced activation of the *Arabidopsis* acidic chitinase promoter. Plant Cell 3: 1063–1072 (1991).
116. Schneider G, Jensen E, Spray C, Phinney BO: Hydrolysis and reconjugation of gibberelin A20 glucosyl ester by seedlings of *Zea mays* L. Proc Natl Acad Sci USA 89: 8045–8048 (1992).
117. Schreck R, Baeuerle PA: A role for oxygen radicals as second messengers. Trends Cell Biol 1: 39–42 (1991).
118. Schultz M, Schnabl H, Manthe B, Schweihofen B, Casser I: Uptake and detoxification of salicylic acid by *Vicia faba* and *Fagopyrum esculentum*. Phytochemistry 33: 291–294 (1993).
119. Sijmons PC, Grundler FMW, von Mende N, Burrows PR, Wyss U: *Arabidopsis thaliana* as a new model host for plant-parasitic nematodes. Plant J 1: 245–254 (1991).
120. Silverman P, Nuckles E, Ye XS, Kuć J, Raskin I: Salicylic acid, ethylene, and pathogen resistance in tobacco. Mol Plant-Microbe Interact 6: 775–781 (1993).
121. Simmons CR, Litts JC, Huang N, Rodriguez RL: Structure of a rice β-glucanase gene regulated by ethylene, cytokinin, wounding, salicylic acid and fungal elicitors. Plant Mol Biol 18: 33–45 (1992).
122. Simon AE, Li XH, Lew JE, Stange R, Zhang C, Polacco M, Carpenter CD: Susceptibility and resistance of *Arabidopsis thaliana* to turnip crinkle virus. Mol Plant-Microbe Interact 5: 496–503 (1992).
123. Simpson RB, Johnson LJ: *Arabidopsis thaliana* as a host for *Xanthomonas campestris* pv. *campestris*. Mol Plant-Microbe Interact 3: 233–237 (1990).
124. Singh L: *In vitro* screening of some chemicals against three phytopathogenic fungi. J Indian Bot Soc 57: 191–195 (1978).
124a. Smith JA, Hammerschmidt R, Fulbright DW: Rapid induction of systemic induction of systemic resistance in cucumber by *Pseudomonas syringae* pv. *syringae*. Physiol Mol Plant Pathol 38: 223–235 (1991).
125. Sutherland MW: The generation of oxygen radicals during host plant responses to infection. Physiol Mol Plant Path 39: 79–93 (1991).
126. Summermatter K, Meuwly Ph, Mölders W, Métraux JP: Salicylic acid levels in *Arabidopsis thaliana* after treatments with *Pseudomonas syringae* or synthetic inducers. Acta Hort, in press (1994).
126a. Takahashi H, Goto N, Ehara Y: Hypersensitive response in cucumber mosaic virus-inoculated *Arabidopsis thaliana*. Plant J, in press (1994).
127. Tanaka S, Hayakawa K, Umetani Y, Tabata M: Glucosylation of isomeric hydroxybenzoic acids by cell suspension cultures of *Mallotus japonicus*. Phytochemistry 29: 1555–1558 (1990).
128. Towers GHN: Metabolism of phenolics in higher plants and microorganisms. In: Harborne JB (ed) Biochemistry of Phenolic Compounds, pp. 249–294. Academic Press, London (1964).
129. Tsuji J, Somerville SC, Hammerschmidt R: Identification of a gene in *Arabidopsis thaliana* that controls resistance to *Xanthomonas campestris* pv. *campestris*. Physiol Mol Plant Path 38: 57–65 (1991).
130. Uknes S, Dincher S, Friedrich L, Negrotto D, Williams S, Thompson-Taylor H, Potter S, Ward E, Ryals J: Regulation of pathogenesis-related protein-1a gene expression in tobacco. Plant Cell 5: 159–169 (1993).
131. Uknes S, Mauch-Mani B, Moyer M, Potter S, Williams S, Dincher S, Chandler D, Slusarenko A, Ward E, Ryals J: Acquired resistance in *Arabidopsis*. Plant Cell 4: 645–655 (1992).

132. Uknes S, Winter AM, Delaney T, Vernooij B, Morse A, Friedrich L, Nye G, Potter S, Ward E, Ryals J: Biological induction of systemic acquired resistance in *Arabidopsis*. Mol Plant-Microbe Interact 6: 692–698 (1993).
133. Umetani Y, Kodakari E, Yamamura T, Tanaka S, Tabata M: Glucosylation of salicylic acid by cell suspension cultures of *Mallotus japonicus*. Plant Cell Rep 9: 325–327 (1990).
134. van Damme EJM, Willems P, Torrekens S, van Leuven F, Peumans WJ: Garlic (*Allium sativum*) chitinases: characterization and molecular cloning. Physiol Plant 87: 177–186 (1993).
135. van Loon LC: The induction of pathogenesis-related proteins by pathogens and specific chemicals. Neth J Plant Path 89: 265–273 (1983).
136. van Loon LC, Antoniw JF: Comparison of the effects of salicylic acid and ethephon with virus-induced hypersensitivity and acquired resistance in tobacco. Neth J Plant Path 88: 237–256 (1982).
137. van de Rhee MD, Bol JF: Induction of the tobacco PR-1a gene by virus infection and salicylate treatment involves an interaction between multiple regulatory elements. Plant J 3: 71–82 (1993).
138. van de Rhee MD, Lemmers R, Bol JF: Analysis of regulatory elements involved in stress-induced and organ-specific expression of tobacco acidic and basic β-1,3-glucanase genes. Plant Mol Biol 21: 451–461 (1993).
139. van de Rhee MD, van Kan JAL, Gonzalez-Jaen MT, Bol JF: Analysis of regulatory elements involved in the induction of two tobacco genes by salicylate treatment and virus infection. Plant Cell 2: 357–366 (1990).
140. Vane JR: Inhibition of prostaglandin synthesis as a mechanism of action for aspirin-like drugs. Nature-New Biol 231: 232–235 (1971).
141. Vernooij B, Friedrich L, Morse A, Reist R, Kolditz-Jawhar R, Ward E, Uknes S, Kessmann H, Ryals J: Salicylic acid is not the translocated signal responsible for inducing systemic acquired resistance but is required in signal transduction. Plant Cell 6: 959–968 (1994).
142. Vigers AJ, Roberts WK, Selitrennikoff CP: A new family of plant antifungal proteins. Mol Plant-Microbe Interact 4: 315–323 (1991).
143. Ward ER, Uknes SJ, Williams SC, Dincher SS, Wiederhol DL, Alexander DC, Ahl-Goy P, Métraux JP, Ryals J: Coordinate gene activity in response to agents that induce systemic acquired resistance. Plant Cell 3: 1085–1094 (1991).
144. Weete JD: Induced systemic resistance to *Alternaria cassiae* in sicklepod. Physiol Mol Plant Path 40: 437–445 (1992).
145. Weissman G: Aspirin. Sci Am 264: 84–90 (1991).
146. Whalen MC, Innes RW, Bent AF, Staskawicz BJ: Identification of *Pseudomonas syringae* pathogens of *Arabidopsis* and a bacterial locus determining avirulence on both *Arabidopsis* and soybean. Plant Cell 3: 49–59 (1991).
147. White RF: Acetylsalicylic acid (aspirin) induces resistance to tobacco mosaic virus in tobacco. Virology 99: 410–412 (1979).
148. White RF: Serological detection of pathogenesis-related proteins. Neth J Plant Path 89: 311–317 (1983).
149. White RF, Antoniw JF: Virus-induced resistance responses in plants. Crit Rev Plant Sci 9: 443–455 (1991).
150. White RF, Rybicki EP, von Wechmar MB, Dekker JL, Antoniw JF: Detection of PR-1 type proteins in Amaranthaceae, Chemopodiaceae, Graminae and Solanaceae by immunoelectroblotting. J Gen Virol 68: 2043–2048 (1987).
151. Wilson DC, Thain JF, Minchin PEH, Gubb IR, Reilly AJ, Skipper YD, Doherty HM, O'Donnell PJ, Bowles DJ: Electrical signalling and systemic proteinase inhibitor induction in the wounded plant. Nature 360: 62–65 (1992).
152. Woloshuk CP, Meulenhoff JS, Sela-Buurlage M, van den Elzen PJM, Cornelissen BJC: Pathogen-induced proteins with inhibitory activity toward *Phytophthora infestans*. Plant Cell 3: 619–628 (1991).
153. Yalpani N, León J, Lawton MA, Raskin I: Pathway of salicylic acid biosynthesis in healthy and virus-inoculated tobacco. Plant Physiol 103: 315–321 (1993).
154. Yalpani N, Schulz M, Davis MP, Balke NE: Partial purification and properties of an inducible uridine 5'-diphosphate-glucose: salicylic acid glucosyltransferase from oat roots. Plant Physiol 100: 457–463 (1992).
155. Yalpani N, Shulaev V, Raskin I: Endogenous salicylic acid levels correlate with accumulation of pathogenesis-related proteins and virus resistance in tobacco. Phytopathology 83: 702 (1993).
156. Yalpani N, Silverman P, Wilson TMA, Kleier DA, Raskin I: Salicylic acid is a systemic signal and an inducer of pathogenesis-related proteins in virus-infected tobacco. Plant Cell 3: 809–818 (1991).
157. Yang SF, Pratt HK: The physiology of ethylene in wounded plant tissue. In: Wahl G (ed) Biochemistry of Wounded Plant Tissues, pp. 595–622. Walter de Gruyter, Berlin (1978).
158. Ye XS, Pan SQ, Kuć J: Pathogenesis-related proteins and systemic resistance to blue mould and tobacco mosaic virus induced by tobacco mosaic virus, *Peronspora tabacina* and aspirin. Physiol Mol Plant Path 35: 161–175 (1989).
159. Ye XS, Pan SQ, Kuć J: Specificity of induced systemic resistance as elicited by ethephon and tobacco mosaic virus in tobacco. Plant Sci 84: 1–9 (1992).
160. Yoshikawa M, Tsuda M, Takeuchi Y: Resistance to fungal diseases in transgenic tobacco plants expressing the phytoalexin elicitor-releasing factor, β-1,3-endoglucanase from soybean. Naturwissenschaften 80: 417–420 (1993).

Plant Molecular Biology **26**: 1459–1481, 1994.

Plant hormone conjugation

Günther Sembdner*, Rainer Atzorn and Gernot Schneider
Institut für Pflanzenbiochemie, Weinberg 3, D-06018 Halle, Germany (author for correspondence)*

Received and accepted 11 October 1994

Key words: plant hormone, conjugation, auxin, cytokinin, gibberellin, abscisic acid, jasmonate, brassinolide

Introduction

Plant hormones are an unusual group of secondary plant constituents playing a regulatory role in plant growth and development. The regulating properties appear in course of the biosynthetic pathways and are followed by deactivation via catabolic processes. All these metabolic steps are in principle irreversible, except for some processes such as the formation of ester, glucoside and amide conjugates, where the free parent compound can be liberated by enzymatic hydrolysis. For each class of the plant hormones so-called 'bound' hormones have been found. In the early literature this term was applied to hormones bound to other low-molecular-weight substances or associated with macromolecules or cell structures irrespective of whether structural elucidation had been achieved. After the characterization of the first gibberellin (GA) glucoside – GA_8-2-O-β-D-glucoside from maturing fruit of *Phaseolus coccineus* [175, 176] – the term GA conjugate was used for a GA covalently bound to another low-molecular-weight compound [184]. Subsequently, the term was extended to all other groups of plant hormones [178], including their precursors and metabolites as well as to secondary plant constituents in general.

Plant hormone conjugates have been studied intensively during the past decades and good progress was made concerning their chemistry (including structural elucidation, synthesis etc.) and, more recently, their biochemistry (including enzymes for conjugate formation or hydrolysis), and their genetical background. However, the most important biological question concerning the physiological relevance of plant hormone conjugation can so far be answered in only a few cases (see Conjugation of auxins). There is evidence that conjugates might act as reversible deactivated storage forms, important in hormone 'homeostasis' (i.e. regulation of physiologically active hormone levels). In other cases, conjugation might accompany or introduce irreversible deactivation. The difficulty in investigating these topics is, in part, a consequence of inadequate analytical methodology. However, the advent of analytical techniques such as HPLC-MS or capillary electrophoresis-MS may help to resolve matters.

Conjugation of auxins

It is the main intention in this section to review the conjugation of naturally occurring auxins; the numerous data on conjugates of synthetic auxins will not be discussed. In addition to biosynthesis, catabolism is another way to control the levels of free indole-3-acetic acid (IAA), and conjugation represents one important aspect of IAA catabolism. However, at least some IAA conjugates are not merely irreversibly deactivated end products of metabolism but instead act as temporary storage forms, from which IAA can be released via hydrolysis. Convincing data about IAA metabolism in *Zea mays* suggest that in seedlings conjugate hydrolysis in the endosperm represents the dominating source of free IAA in the coleoptile [5, 6, 7]. It is not known whether this mechanism is valid for higher plants in general.

After the first comprehensive review about 'bound auxins' in 1982 [27], the number of identified IAA catabolites increased, as documented by several subsequent reviews [4, 8, 88, 143, 152]. Major catabolic routes (Fig. 1) are (1) oxidative decarboxylation of IAA, (2) non-decarboxylative oxidative catabolism and (3) ester and amino acid conjugation. The latter can be divided into formation of conjugates from which IAA hydrolysis is still possible and into compounds where IAA was inactivated via oxidation after conjugate formation.

The formation and physiological significance of IAA conjugation is of primary importance in this article, but from a regulatory point of view it is necessary to discuss briefly alternative routes of IAA degradation. The oxidative decarboxylation pathway is catalysed by peroxidases, leading in several plant species to products such as indole-3-methanol [16, 149, 194] and indole-3-carboxylic acid [3, 151]. It has been proposed that a special 'IAA-oxidase' is responsible for these conversions [9, 50, 70]. However, there is a discrepancy between results of *in vitro* oxidation under different conditions and the relatively low occurrence of these catabolites in plant tissues [143].

The main products of the non-decarboxylation pathway, which appears to operate in many plant species, are oxindole-3-acetic acid and dioxindole-3-acetic acid [66, 141, 142]. In *Zea mays*, 7-hydroxylation and subsequent glucosylation of oxindole-3-acetic acid have also been observed [126]. The concentrations of both substances exceeded the levels of free IAA about ten-fold, implying that this is a major route for the inactivation of IAA. In contrast, there are many data indicating that the formation of both ester and amino acid IAA conjugates is associated with a transport function rather than modes of auxin inactivation [5, 12, 90].

Ester conjugates

Most of the available information on synthesis and hydrolysis of IAA esters (see Fig. 2) comes from experiments with *Zea mays* [4, 5, 7, 8, 27],

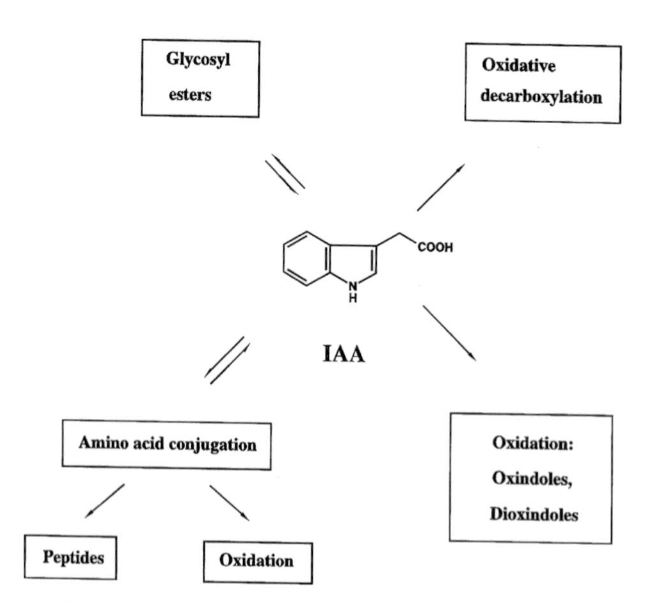

Fig. 1. Main routes of IAA metabolism in higher plants.

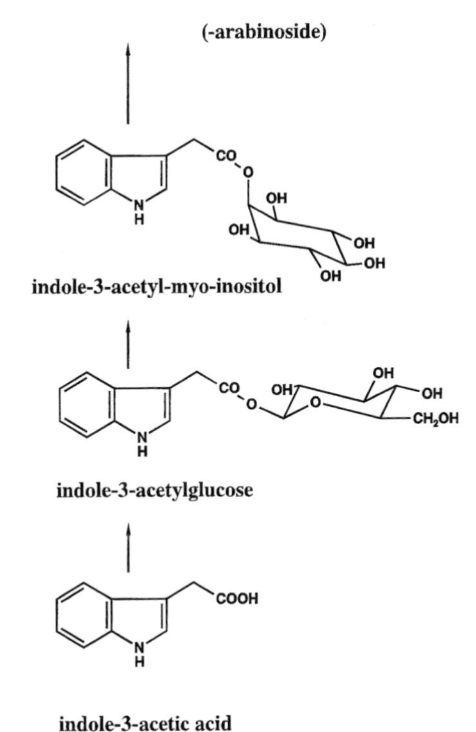

Fig. 2. Formation of IAA ester conjugates.

although IAA esters have been found in many other plant species [8, 21, 143]. The first evidence for an IAA-glucoside in plants was presented by Zenk in 1961 [233]. In *Zea mays* kernels, 1-O-(indole-3-acetyl)-β-D-glucose [43] and 2-O-(indole-3-acetyl)-myo-inositol have been detected as well as 5-O-β-1-arabinopyranosyl-2-O-(indole-3-acetyl)-myo-inositol and 5-galactopyranosyl-2-O-(indole-3-acetyl)-myo-inositol [25, 203, 204]. In addition, a high-molecular-weight IAA ester of a cellulosic glucan has been detected in extracts from maize [139].

The enzymology of IAA ester formation was studied by Michalczuk and Bandurski [104, 105], using crude cell-free preparations from immature kernels of sweet maize which converted (1) 2-^{14}C-IAA and UDPG to IAA-β-D-glucopyranoside and IAA-myo-inositol, and (2) UDP-galactose and IAA-myo-inositol to IAA-myo-inositol-galactose and IAA-myo-inositol-arabinose [4, 6, 30]. Typically not more than 20% of the IAA-myo-inositol was converted, which contrasts with the first glucosylation steps where there was almost always complete conversion of the substrate [4, 29, 30]. The conjugating enzymes are soluble and can be separated from each other by Sephadex G 150 chromatography [4]. However, further characterization of IAA ester-forming enzymes is still lacking.

Interesting information on IAA metabolism as an important regulative element has come from experiments with genetically manipulated plants where the auxin biosynthesis genes from the Ti plasmid of *Agrobacterium* were expressed to obtain auxin overproducing plants [e.g. 68, 186, 187, 189]. Quantitative determinations of bound and free IAA showed an increase of both forms, but often conjugates accumulated to a higher extent. In most cases the identity of the IAA conjugates was not determined, but there is some evidence [188] that they consist at least partly of ester compounds, although IAA amino acid conjugates were the main products. Experiments of this kind are a powerful way to show how plant cells can regulate the levels of active auxins and how they deal with excess production of IAA. It also opens possibilities for a better access to the metabolizing enzymes.

The release of IAA from ester conjugates has been studied extensively in the maize coleoptile [7, 8]. A combination of quantification and turnover studies revealed that most of the free IAA in the copeoptile tips of growing shoots did not originate from *de novo* synthesis but from ester hydrolysis in the endosperm. Similar studies with other species are not known, so whether this source of IAA is widespread in young seedlings remains to be determined.

Numerous feeding experiments in combination with biological activity determinations and subsequent analysis of metabolites [8, 28, 46, 125] indicate that the high physiological activity of IAA esters is indirect, resulting from release of free IAA. The state of knowledge about enzymes which can hydrolyse IAA from esters is once again confined to maize, and apart from studies using more or less crude enzyme preparations, not much information is available.

Amide conjugates

There are two types of amide conjugates formed with IAA in which either the indole ring of the IAA remains unchanged or oxindole or dioxindole derivatives are synthesized after formation of the peptide bond (Fig. 3). IAA-aspartate from seeds of soybean was the first amino acid conjugate to be identified conclusively [26]. This form of IAA conjugation occurs in legume seeds [4], but has been observed in other species too, for example in shoots of *Pinus silvestris* [1] and fruits of tomato [21]. IAA-glutamate conjugates are less common [45], and to date no other IAA amino acid conjugates have been detected in higher plants. However, there are several reports about larger amide conjugates. For instance, Bialek and Cohen [11] detected an IAA peptide with a molecular weight of about 5 kDa in *Phaseolus*, and the presence of an IAA glycoprotein conjugate has also been reported [137].

The function of amide conjugates is not fully understood. The peptide bond-forming enzymes are not, as yet, well-characterized, and it has not been possible to produce these conjugates *in vitro*. There is evidence from many experiments mostly confined to plant tissue cultures [e.g. 90] or seeds [e.g. 12] that hydrolysis of IAA-aspartate takes place to a high extent. This might explain its high biological activity, which also applies to other IAA amino acid conjugates [89]. Not much is known about the hydrolysing enzymes. There are only few data about isolation and characterization of a crude extract from *Phaseolus* [10, 143].

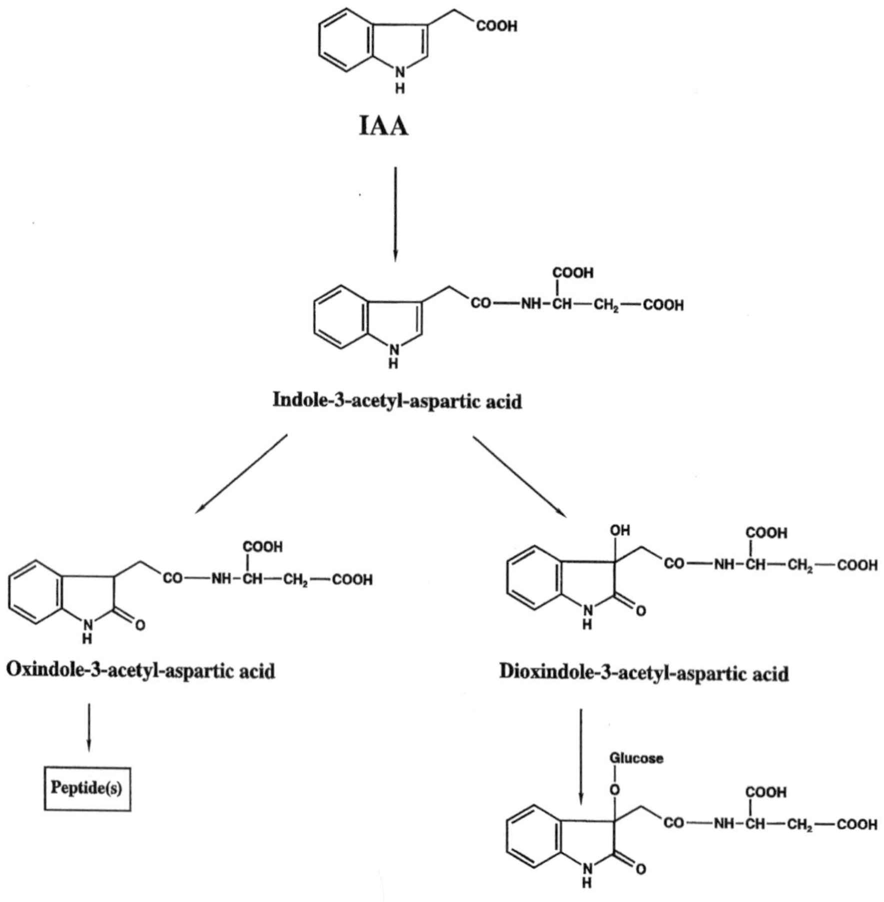

Fig. 3. Formation of IAA amino acids conjugates.

Interestingly, no common peptidases or proteinases are able to cleave the amide bond of such IAA conjugates, indicating that the enzyme involved is very specific. On the other hand, kinetic studies in soybean seedlings show that the IAA-aspartate pool increases during germination [4, 12], and its level found in the shoots of 7-day old seedlings is about twice the level found in the dry seed. The site of compartmentation of IAA amino acid conjugates is not known in general.

In recent years evidence arose about subsequent metabolization of IAA amino acid conjugates. Studies by Tsurumi and Wada [198, 199, 200, 201] have shown that oxidation of IAA-aspartate represents an important pathway of irreversible IAA inactivation in *Vicia*. The first step is conjugation with aspartate, followed by oxidation of the indole ring at two positions (Fig. 3), and subsequent glucosylation, but this last step does not seem to be obligatory. Metabolites of similar structures have been found in *Dalbergia* [116, 128] and tomato [21]. After feeding tritium-labelled IAA and IAA-aspartate to protonemata of the moss *Funaria hygrometrica* more than 80% of the radioactivity was found in compounds co-chromatographing with dioxindole aspartate [15]. Apart from dioxindole derivatives, a similar oxindole conjugate has been found in tomato [144]. The steps of synthesis are similar, starting with the peptide bond formation. The exact structure of the final product is not clear yet, but it seems to be a small peptide of still unknown amino acid sequence. In contrast to simpler amino acid conjugates, it is possible to synthesize the compound *in vitro* with a crude enzyme extract.

Conjugation of cytokinins

During the past decade, more progress was made in the field of cytokinin metabolism than in the field of cytokinin biosynthesis which is the subject of controversal discussion, as reflected in several reviews [62, 81, 96, 98]. Unlike to the situation for other plant hormone classes, the number of metabolites obtained after feeding of synthetic cytokinins such as benzyl adenine (BA) or kinetin to numerous plant tissues almost exceeds the number of known endogenous cytokinin metabolites. Therefore it is much more difficult to classify the many products of cytokinin metabolism. The scheme illustrated in Fig. 4 is an extension to proposals by Horgan [62].

One type of metabolism involves the cleavage of the N^6 side chain which results in a complete loss of biological activity. The enzyme involved in this reaction is called 'cytokinin oxidase' and has been characterized in various plant species [e.g. 19, 20, 22, 119, 133]. Since this type of metabolism is not a form of conjugation, it will not be discussed further. This applies also to other kinds of side-chain modification that do not involve conjugation.

The second type of metabolism comprises the interconversions of cytokinin bases, nucleosides and nucleotides. The 9-ribosides and their 5′-mono, di- and triphosphates are amongst the most abundant naturally occurring cytokinins and metabolites, and they exist in the plant cell in apparent equilibrium [62, 98]. Obviously several enzymes involved in adenylate metabolism will utilize the cytokinins as substrates and, so far, all of the detected enzymes exhibit lower affinities for the cytokinins than for adenine or adenosine. For instance, Chen and co-workers [24] in their studies on preparations from wheat germ cells discovered an adenosine phosphorylase which converts 2-isopentenyladenine (2iP) to 2-isopentenlyadenosine (2iPA), an adenosine kinase [23] which converts directly 2iPA to 2iPMP, and an adenosine phosphoribosyltransferase which converts directly 2iP to 2iPMP. The significance of such conversions is not completely understood, but Laloue and Pehte [75] have shown that tobacco cells are impermeable to cytokinin nucleotides but not to bases and ribosides.

Apart from the different types of cytokinin glucosylation, which are the most prominent cytokinin conjugates and will be discussed below in more detail, N-alanyl conjugation and O-acetylation [98] are also reported (see Fig. 4). After feeding of zeatin (Z) and BA, their alanyl conjugates have been found in *Lupinus* [133] and in immature apple seeds [44]. The alanyl conjugate

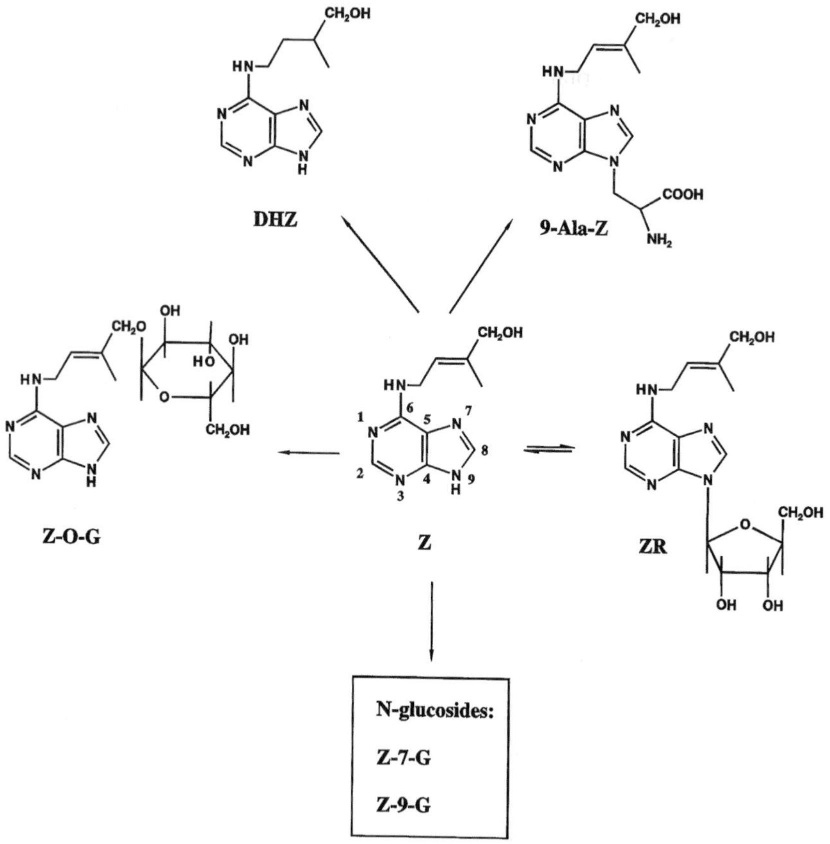

Fig. 4. Survey on conjugation of cytokinins, shown for zeatin.

of Z is also an endogenous compound in *Lupinus* [192]. An enzyme, β-(6-allylaminopurine-9-yl)-adenine synthase, has been partly purified from developing *Lupinus* seeds [44]. These conjugates are extremely stable [49, 130, 133], suggesting that their production represents a form of irreversible conjugation.

The cytokinin ribosides should not be regarded as real conjugates since it is still an open question whether they are active per se or via release of the free bases. A novel form of glycoside is Z-O-xyloside from *Phaseolus* [41]. The corresponding enzyme, O-xylosyl-transferase, is well-characterized and has been purified to homogenity; also a monoclonal antibody was raised against the enzyme [113, 114], and prospects of obtaining a cDNA clone would appear to be good.

Cytokinin glucosides

Cytokinin glucosides are of widespread distribution in many plant species [98], and glucosylation is possible at four positions (Fig. 4). Not all structures identified are reflected in this article so that the below discussion represents only a small facet of all known substances.

N-glucosides are known in 3-, 7-, and 9-position of the purine ring. Z-7-G was the only detectable cytokinin in radish seedlings [193], Z-9-G was the major compound in *Vinca rosea* crown gall tissue [177] and a minor compound in maize kernels [193]. After feeding of iP and iPA to cytokinin-dependent tobacco cells, iP-7-G was the major metabolite [77], as was BA-7-G after BA feedings [49, 76]. In de-rooted radish seedlings, three N-glucosides (BA-3-G, BA-7-G, BA-

9-G) were found [83]. BA-7-G was also found in corn tissue cultures [49, 132]. In common, N-glucosides are extremely stable in plant tissues [62, 98], and their biological activity is considerably lower than the activity of their free bases [82], perhaps of possible side chain cleavage by cytokinin oxidases. N-glucosylation might represent an irreversible form of cytokinin conjugation.

Side-chain glucosylation leads to the other form of cytokinin glucosides, the O-glucosides. The Z-O-glucosides are abundant in *Lupinus* [191, 192], *Phaseolus* [113, 114, 130], and *Vinca rosea* [177]. The O-glucosides are less stable than N-glucosides; for example they can be hydrolyzed by almond β-glucosidase (Emulsin) [98]. Z-O-G hydrolysis was observed after feeding in detached leaves of *Lupinus luteus* [133], primary leaves of *Phaseolus vulgaris* [130], and in *Vinca rosea* crown gall [63]. Interestingly, they cannot be inactivated by cytokinin oxidases [97, 177].

There are somewhat contrasting results about the biological activity of cytokinin-O-glucosides: Letham [82] found them as equally active as the free bases, whereas Kleczkowski *et al.* [67] reported higher activity for the bases. On the other hand, Mok and co-workers [112, 114] detected much higher biological activity for the glucosides, suggesting that they might be active per se and not via release. Most of the authors regard them as genuine cytokinin storage forms from which the bases are liberated and so regulate the levels of active cytokinins [e.g. 62]. In this context, the finding by Brzobohaty *et al.* [18] about a β-glucosidase from maize root meristem which is able to release active cytokinins from conjugates is of interest.

Much progress has been made in isolating and characterizing O-glucosyltransferase in *Phaseolus* [114]. The enzyme was purified to homogenity [41, 93], and it has high substrate specificity, utilizing *trans*-zeatin but neither dihydro-zeatin, *cis*-zeatin nor zeatin riboside [92, 115]. The molecular mass was about 50 kDa. As already mentioned for O-xylosyltransferase, a monoclonal antibody was raised against this enzyme.

In summary, the metabolic picture of cytokinins remains complex because of the large number of different endogenous cytokinins which can vary greatly from one plant species to another. But because of the substantial progress in recent years in isolating conjugating enzymes, as well as cytokinin oxidase, there are good prospects to elucidate the quantitative relationships between different metabolic routes of cytokinin conjugation. Also promising, are the increasing reports about cytokinin levels in transgenic plants where the isopentenyltransferase gene from *Agrobacterium tumefaciens* is overexpressed [e.g. 99, 123]. In general, such transformed plants show elevated total endogenous cytokinin levels and abnormal phenotypes [94]. However, not much is known yet about the rates of metabolism in such systems.

Conjugation of gibberellins

Since the structural identification of the first gibberellin (GA) glucoside, GA_8-2-O-β-D-glucoside (GA_8-2-O-G), from maturing fruits of *Phaseolus coccineus* [175, 176, 184], a series of GA glucosyl conjugates have been isolated and structurally elucidated; in addition, acyl and alkyl GA derivatives have been found. Today, the conjugation process is considered to be an important aspect of GA metabolism in plants. The field of GA conjugation has been reviewed previously in a general way [163, 168] as well as in the context of special biochemical and physiological processes [39, 78, 148, 155, 163, 182, 183].

The most common GA conjugates isolated from plants are those in which the GAs are connected to glucose. These conjugates can be divided into two groups: glucosyl ethers (or O-glucosides), where a hydroxy group of the GA skeleton is linked to the glucose, and glucosyl esters, in which the glucose is attached via the GA-C-7-carboxyl group. So far, the conjugating sugar moiety has had a β-D-glucopyranose structure. A summary of the isolated and structurally elucidated GA glucosyl conjugates and some additional conjugates is given in Table 1. In the case of the GA-O-glucosides the glucose moiety is linked either to 2-O-, 3-O-, 11-O-, 13-O- or 17-O-position of the parent GA (see Fig. 5). From

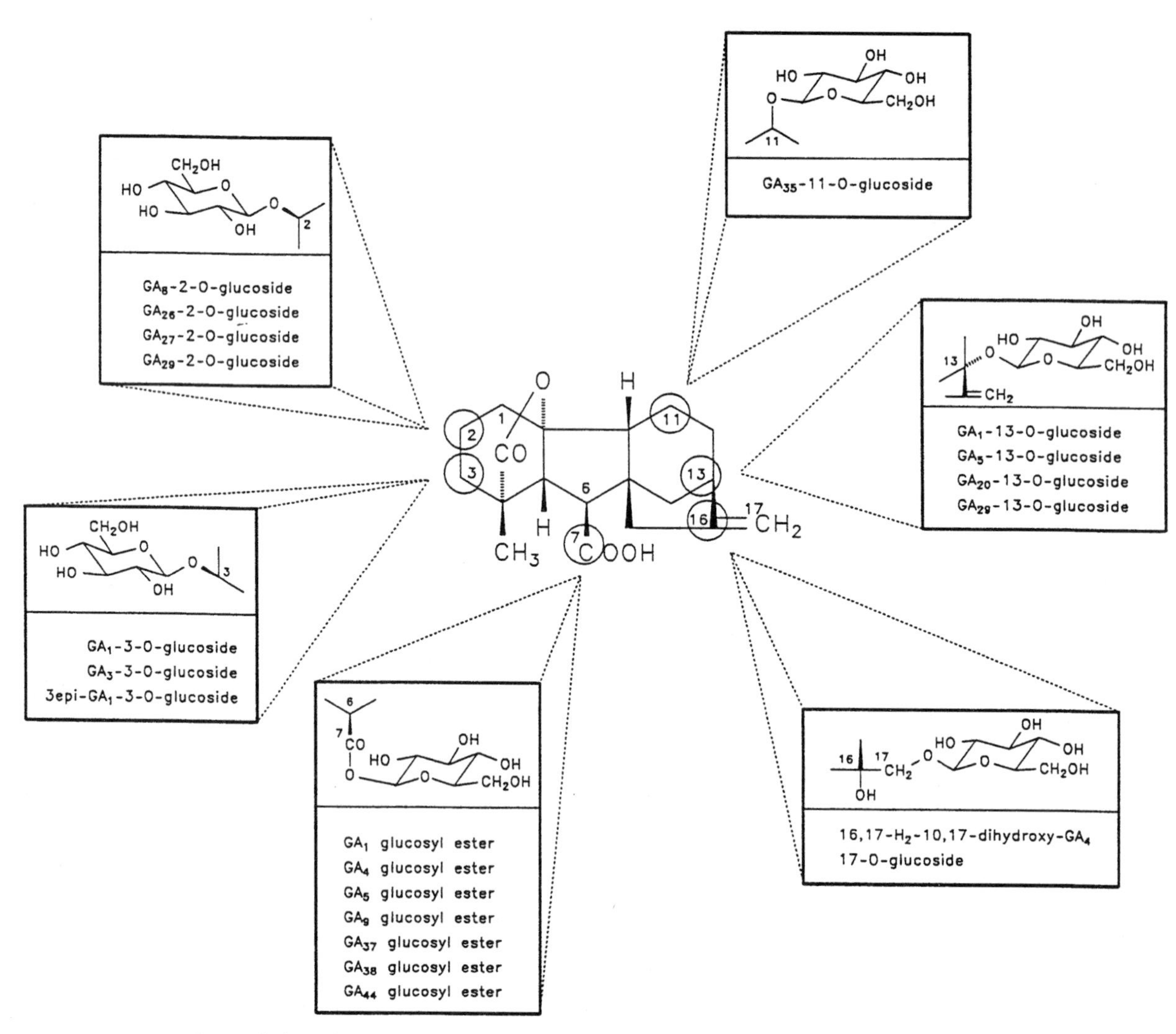

Fig. 5. Schematic structures of endogenously occurring gibberellin glucosyl conjugates.

the occurrence of GA conjugates in various species of higher plants it can be assumed that GA conjugates are distributed ubiquitously [121]. In addition to the GA conjugates shown in Table 1, there are many other reports of the occurrence of conjugates in which identification has been based solely on chromatographic parameters or the identification of the parent GA after hydrolysis. The same also applies to the numerous putative GA conjugates detected in metabolic studies [168]. Further characterization of these compounds is largely dependent upon access to appropriate standards and progress in analytical methodology. For the analysis of GA-O-glucosides (GA-O-G), GC-MS of permethylated derivatives has provided reliable data [145, 156, 160, 164, 166, 171], while with GA glucosyl esters (GA-GE), LC-MS is now the favoured approach [117, 118, 122]. Partial synthesis of numerous GA-O-glycosyl derivatives has provided both standards and labelled compounds for use both as internal standards and as substrates in metabolic studies [58, 162, 163, 170, 172, 173].

Knowledge of the enzymology of GA conjugation is still limited except for some data on the biosynthesis and metabolism of GA glucosyl conjugates. In maturing fruits of *Phaseolus coccineus* a GA glucosylating activity was found [69, 120,

Table 1. Naturally occurring GA conjugates.

Conjugate	Plant source/Reference
GA-O-glucosides	
Ga_1-3-O-glucoside	*Dolichos lablab* [220], *Hordeum vulgare* [169], *Phaseolus coccineus* [156, 158, 182], *Zea mays* [169]
GA_1-13-O-glucoside	*Phaseolus coccineus* [156]
3-epiGA_1-3-O-glucoside	*Phaseolus coccineus* [156]
GA_3-3-O-glucoside	*Pharbitis nil* [221, 223, 226], *Quamoclit pennata* [212]
16,17-H_2, 16,17-dihydroxy-GA_4-17-O-glucoside	*Oryza sativa* [211]
GA_5-13-O-glucoside	*Phaeolus coccineus* [156]
GA_8-2-O-glucoside	*Althea rosea* [52], *Hordeum vulgare* [169], *Pharbitis nil* [225, 226], *Phaseolus coccineus* [156], *Phaeolus vulgaris* [55, 57], *Zea mays* [169]
GA_{20}-13-O-glucoside	*Hordeum vulgare* [169], *Pisum sativum* [171], *Triticum aestivum* [80], *Zea mays* [167, 169]
GA_{26}-2-O-glucoside	*Pharbitis nil* [221, 225, 226]
GA_{27}-2-O-glucoside	*Pharbitis nil* [221, 225, 226]
GA_{29}-2-O-glucoside	*Hordeum vulgare* [169], *Pharbitis nil* [221], *Phaseolus coccineus* [156], *Pisum sativum* [171], *Zea mays* [169]
GA_{29}-13-O-glucoside	*Hordeum vulgare [169], Pisum sativum* [171], *Zea mays* [169]
GA_{35}-11-O-glucoside	*Cytisus scoparius* [216, 217]
GA B-D-glucopyranosyl esters	
GA_1 glucosyl ester	*Phaseolus vulgaris* [55, 56, 57]
GA_4 glucosyl ester	*Phaseolus vulgaris* [55, 56, 57]
GA_5 glucosyl ester	*Pharbitis purpurea* [210]
GA_9 glucosyl ester	*Picea sitchensis* [85, 117], *Pseudotsuga menziesii* [42, 103], *Pinus concorta* [42]
GA_{37} glucosyl ester	*Phaseolus vulgaris* [55, 56, 57]
GA_{38} glucosyl ester	*Phaseolus bulgaris* [55, 56, 57]
GA_{44} glucosyl ester	*Pharbitis purpurea* [210]
GA alkyl ester	
GA_1 n-propyl ester	*Cucumis sativus* [53]
GA_3 n-propyl ester	*Cucumis sativus* [53]
GA_1 methyl ester	*Lygodium japonicum* [215]
GA_{73} methyl ester	*Lygodium japonicum* [214, 215]
GA_{88} methyl ester	*Lygodium japonicum* [213]
GA acyl derivatives	
GA_3-3–O-acetate	*Gibberella fujikuroi* [174]
GA_{39}-3–O-isopentanoate	*Cucurbita maxima* [13]
GA-related conjugate	
Gibberethione	*Pharbitis nil* [227]

180], which has been shown to be a glucosyltransferase located exclusively in the pericarp. This enzyme preferentially utilizes UDP-glucose as a glucose donor and accepts GA_3 as well as, albeit less efficiently, GA_7 and GA_{30}, forming exclusively the 3-O-β-D-glucopyranosides [69]. This substrate specificity, however, contradicts the fact that GA_3 glucose conjugates have to date not been identified as endogenous constituents in this plant material. As a consequence, the physiological significance of the GA_3 glucosylation remains unclear. Enzyme preparations from cotyledons of 24 h imbibed seeds of *Phaseolus coccineus* transform labelled GA_4 to GA_1, GA_{34}, GA_4 glucosyl ester and a GA_{34} glucoside [35, 202]. Cell-free systems from germinating peas (*Pisum sativum*) metabolize labelled GA_{12}-aldehyde into a GA_{12}-aldehyde glucosyl ester conjugate [64]. Cytosolic enzyme fractions from cells of *Lycopersicon peruvianum* grown in suspension cultures have been shown to specifically transform GA_7 and GA_9 to the corresponding glucosyl esters in the presence of UDP-glucose [180].

Using radioactively labelled GA-O-G and GA-GE, the hydrolytic cleavage within several bioassay systems was found to parallel the biological activities of the conjugates [58, 84]. These findings have led to the suggestion that GA glucose conjugates per se are biologically inactive. Any response obtained in the assay then reflects the degree of hydrolysis and the activity of the released parent GA [179]. In such circumstances, the occurrence of a series of specific β-glucosidases might be anticipated. In keeping with this possibility GA_8-2-O-G is as active as GA_8 after leaf application to dwarf rice seedlings [34], whereas GA_3-3-O-G is much less active than its aglycon [222]. It is therefore of relevance that a β-glucosidase fraction from dwarf rice leaves hydrolyses GA_8-2-O-G 200 times faster than GA_3-3-O-G [153]. Furthermore, in extracts from maturing pods of *Phaseolus coccineus*, a soluble β-glucosidase has been detected which exhibits a high hydrolysing activity toward the endogenous GA_8-2-O-G. This GA_8-2-O-G hydrolysing activity decreases during pod maturation [154], which appears to be functionally related to the increase in GA glucosylating activity in the same tissue [69]. Fungal β-glucosidases, such as cellulase efficiently hydrolyse GA-13-O-glucosides [157]. In contrast, enzyme preparations from plants exhibit only low activity [153, 182] with this naturally occurring group of GA conjugates [156, 169, 171]. The ubiquitous occurrence of GA glucosyl conjugates and their facile metabolic formation provoke the assignment of some distinct physiological functions, which, due to a lack of convincing evidence, are still contradictory and speculative.The loss of biological activity in the course of the conjugation process and the increased polarity of GA glucosyl conjugates are considered to favour GA conjugates for being deposited into the vacuole. From the occurrence of GA conjugates in bleeding sap of trees, a possible function in the long-distance transport has been suggested [37, 38]. It also has been suggested that the glucosyl moiety of GA conjugates may cause a distorted orientation of the GA molecule within the membrane, which prohibits the appropriate binding to an assumed receptor [190]. Because of their preferential formation and accumulation during seed maturation it has been proposed that GA glucose conjugates may function as storage products [79, 80, 169]. This, however, applies only to conjugates of biologically active GAs, where hydrolysis, for example during early stages of seed germination, gives raise to free GAs prior to the onset of *de-novo* GA biosynthesis.

In the case of 2β-hydroxylated GAs, which are themselves biologically inactivated metabolites, conjugation may be a step within the process of further catabolism. The easy formation and hydrolysis of GA glucosyl conjugates, which means reversible deactivation/activation, is also discussed in connection with the regulation of free GA pools. The rapid exchange of pools of GA glucosyl ester, GA glucoside and free GAs has been shown in maize seedlings [164]. There are also indications that, in the case of *Brassica* mutants, different light conditions may influence GA metabolism including the formation of GA conjugates [147]. The tentative physiological roles of GA glucose conjugates will only be clarified if appropriate methods for identification and quantification of pool sizes become available and are used to investigate physiologically relevant processes. Special attention should be paid to the problem of compartmentalization, which invariably makes it always difficult to measure specific pools.

Conjugation of abscisic acid

Whereas the main route of abscisic acid (ABA) biosynthesis in higher plants was until recently a matter of discussion [32, 229, 231], a lot of information was available on ABA metabolites and conjugates [see 108, 109, 206] shortly after ABA was first discovered [31, 127]. Two methods of ABA degradation are known (Fig. 6). One possibility is the conversion to phaseic acid (PA) and dihydrophaseic acid (DPA), with subsequent conjugation, the other route is the direct formation of ABA conjugates. Interestingly, most of the presently known ABA metabolites had been characterized before 1984 (see [87]), and recent developments have been confined to their possible functions rather than the discovery of new structures.

Metabolism of ABA to phaseic acid and related compounds seems to be the main inactivation pathway [138, 232]. It leads over 6-hydroxymethyl-ABA to PA, DPA and some polar conjugates, with a side-branch from 6-hydroxymethyl-ABA to β-hydroxy-β-methyl-glutaryl-hydroxy-ABA [59]. PA is usually present in plant tissues in small amounts [87, 196], whereas accumulation of DPA and also its conjugates has been observed in many plants, especially in association with stress [144, 229] and at some stages of germination [36, 61]. To a lesser extent, PA and DPA can be conjugated to esters of the β-D-glucopyranoside type [60, 111], whereas glucose esters of these metabolites have not been found.

The earliest feeding experiments showed that considerable amounts of ABA were subjected to conjugation [108]. The glucose ester (ABAGE) was the first identified conjugate of ABA [72], and later investigations showed that it was synthesized in ripening fruits of several plant species [109, 123, 146]. ABA-β-glucopyranoside (ABAG) is another quantitatively important conjugate. After being originally isolated from apple seeds [86], it now seems to be ubiquitously distributed in germinating seeds. In germinating barley grain both esters account for up to 20% of the total metabolites [36, 61].

The knowledge of the enzymes involved in ABA conjugation is still very poor [87]. A glucosyltransferase has been described, but in no case a substantial release of ABA from conjugates has been detected. This indicates that ABA conjugation is probably an irreversible process which contrasts with the properties of similar conjugates of other plant hormones. This is in keeping with data on the biological activity of ABA conjugates. Whereas PA seems to have a similar activity to ABA in stomata closure [207] and inhibition of α-amylase synthesis [61], ABA conjugates are inactive.

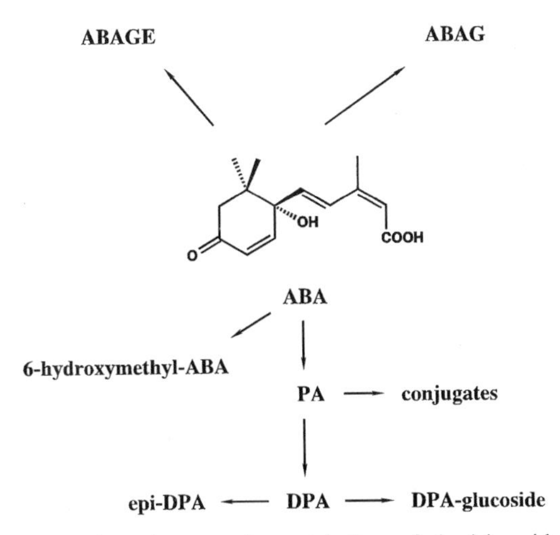

Fig. 6. Overview over the metabolism of abscisic acid.

Conjugation of jasmonates

Jasmonic acid ((−)-JA) and its stereo isomer (+)-7-iso-JA (synonymous with (+)-2-epi-JA) are the major representatives of a group of native plant bioregulators called jasmonates. They are widespread in the plant kingdom and exert various physiological activities when applied exogenously to plants [181]. Their functional role as native regulators is being studied intensively and evidence is given for their involvement in processes such as plant senescence [134, 140, 205] and the formation of vegetative storage organs [71, 94]. Even more striking is the potential role of jasmonates in the signalling of external stress impulses, such as herbivore [48] and pathogen [51] attack, mechanical forces (touch [208]) and osmotic stress [136], to give internal stress responses usually measured as activation or expression of specific genes and formation of characteristic proteins [181].

Metabolical formation of jasmonate conjugates

Investigations on the metabolic transformation of exogenously applied jasmonates using excised shoots of barley seedlings, tomato and potato plants [100, 101, 102], as well as cell suspension cultures of tomato, potato [65], and *Eschscholtzia* [209], showed that conjugation is common in plant tissues either without or after other metabolic transformations. According to these results, summarized in Fig. 7, major metabolic steps (others than conjugation) are hydroxylation at C11 (usually) or C12 yielding the 11-OH or 12-OH derivatives (tuberonic acid-related) and reduction of the C6 keto group resulting in cucurbic acid-related metabolites. Conjugation by O-glucosylation of the hydroxylated metabolites gives either 11-O- or 12-O-glucosides, or 6-O-glucosides of cucurbic acid-related structures.

Metabolic conjugation with amino acids of either non-metabolised jasmonates or of their side-chain hydroxylated derivatives is widespread in plants. In the case of barley shoots, valine, isoleucine, and leucine conjugates have been identified [100, 101, 102]. In cell suspension cultures of tomato and potato [65], instead of amino acid conjugate formation, conjugation at C1 with sugars took place, yielding JA glucosyl and gentiobiosyl ester as the major metabolites. In a cell culture suspension of *Eschscholtzia*, JA was metabolized to the 11-O-β-D-glucoside of 11-(R)-OH-JA [209].

Natural occurrence of jasmonate conjugates and their possible physiological role

With the exception of the sugar conjugates formed in cell suspensions, conjugates detected as

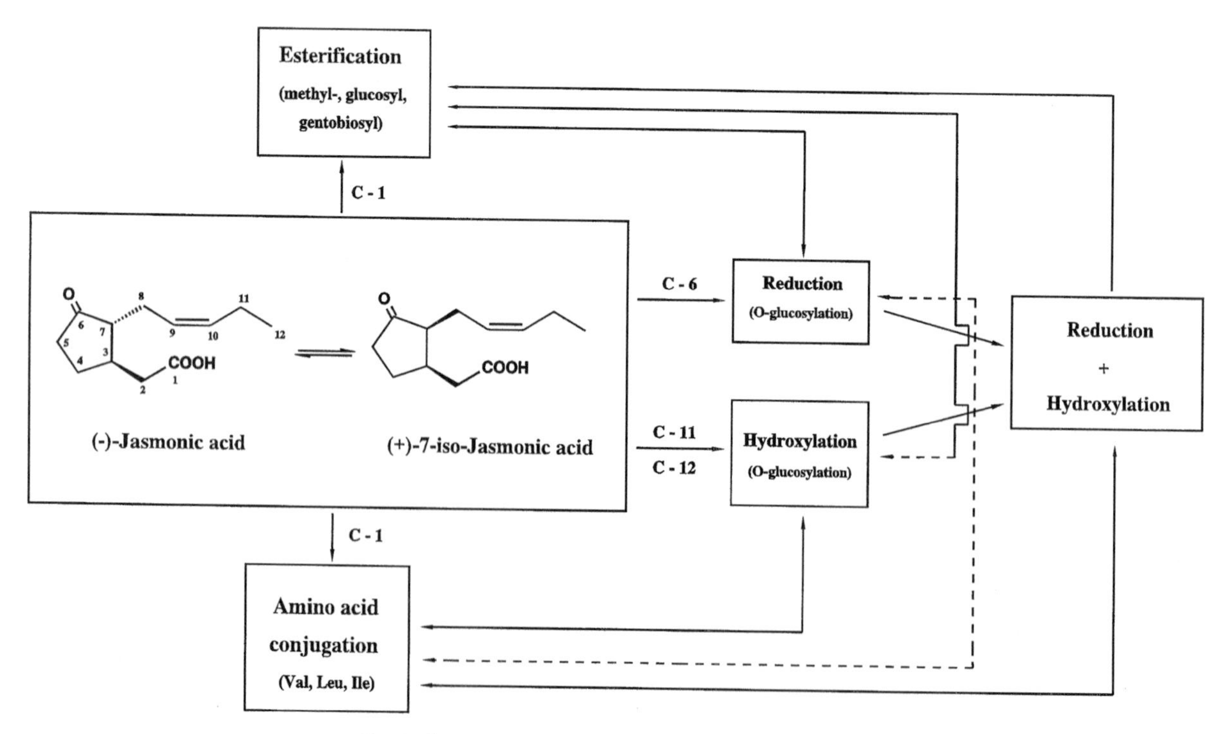

Fig. 7. Survey on metabolic routes of jasmonic acid.

metabolites after exogenous application of jasmonates are known to occur as native constituents in plants (Fig. 8). Thus, the 6-O-glucoside of cucurbic acid was found in pumpkin seeds [73] and the 12-O-β-D-glucopyranoside of 12-OH-JA, the aglucon of which is designated tuberonic acid, occurs as a native compound in potato [228]) and Jerusalem artichoke [95] that induces tuber formation [71]. Several (S)-amino acid conjugates of JA and other jasmonates have been found: tyrosine (Tyr), tryptophane (Trp), and phenylalanine (Phe) conjugates in flowers of the broad bean [17, 159], isoleucine (Ile) conjugates in fruits as well as Ile, leucine (Leu) and valine (Val) conjugates in young leaves of this plant [159]. Whether this distribution pattern of amino acid conjugates in different broad bean organs is related to any physiological role, is a matter for study. A phenylalanine conjugate of 12-acetoxy-JA is known to occur in *Praxelis* [14], and the isoleucine conjugates of (–)JA, (+)-7-iso-JA, and 9,10-dihydro-JA were isolated from the fungus *Gibberella fujikuroi* [33, 106]. The isoleucine conjugate of (–)-JA was found also in pollen of *Pinus mugo*. JA-Ile inhibits the pollen germination, whereas free JA is neither inhibiting nor stimulating (Knöfel, unpublished results). The results indicate speculation about the possible role of JA-Ile in regulation of this process; however, besides pollen germination also flower senescence [140] has to be considered as a process affected. Good evidence for a physiological role of jasmonate conjugates in stress signalling [135, 136, 181]) comes from the following results.

1. Like free jasmonates, in barley leaf segments jasmonate conjugates, such as the naturally occurring (–)-JA-(S)-Ile, were found to be active in inducing so-called jasmonate-inducible proteins (JIPs), whereas the (–)-JA-Trp is of very low activity [54].

2. In barley leaves osmotic stress by sorbitol, mannitol, sucrose, fructose etc. leads to JIPs

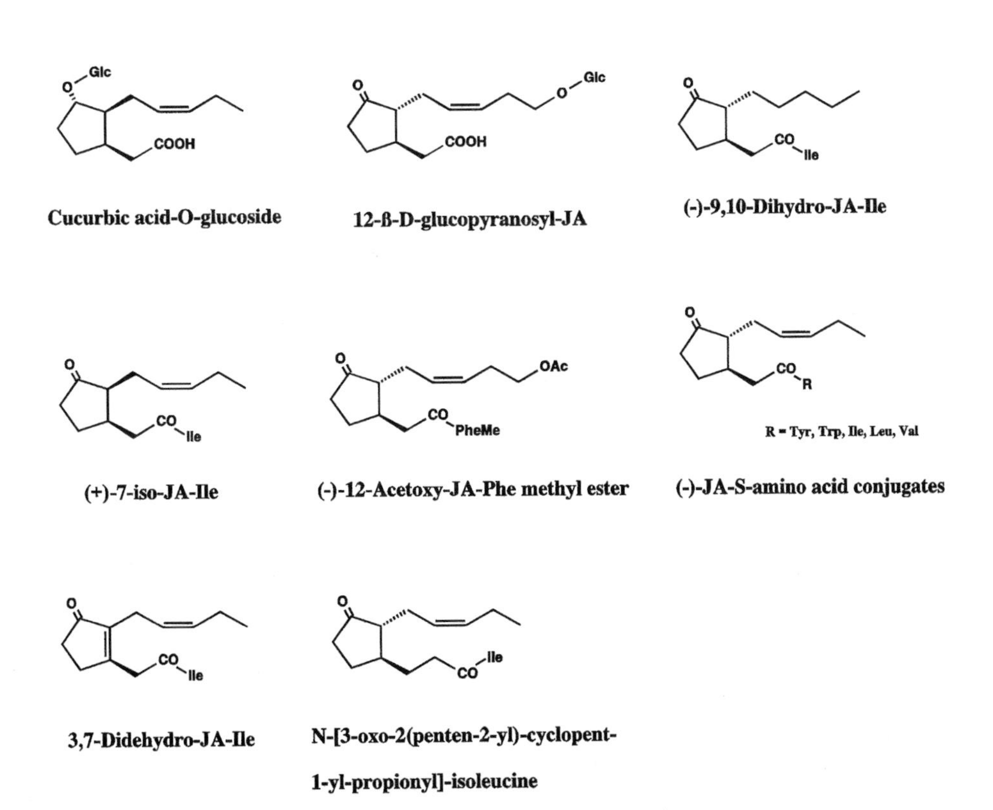

Fig. 8. Structures of endogenously occurring jasmonates.

together with the accumulation of both free acid and conjugated jasmonates like JA-Ile, JA-Leu, JA-Val [74].

According to the basic definition for conjugates (see Introduction), the methyl esters of jasmonates have also to be considered as conjugates, although they differ markedly from all other conjugate types. The JA methyl ester has been shown to possess physiological potencies of the same order, or even higher, than the free acid, probably depending on differences in the uptake and on the plant species used. The occurrence of JA methyl ester is well established in essential oils [40], but its physiologically relevance in plant tissues [197] must still be confirmed (use of methanol extraction could cause artefact formation from other endogenous ester conjugates). Nevertheless, volatile methyl jasmonate either applied through the atmosphere or released from *Artemisia* leaves was able to induce proteinase inhibitor (PI) [47, 48, 150]. Thus, methyl jasmonate might be a volatile signal in interplant communication, released in response to wounding or other stress situations.

In summary, jasmonates are, like classic phytohormones, transformed metabolically to conjugates, the types of which resemble those of auxins and gibberellins. Concerning jasmonate conjugation, apparently amino acid conjugates dominate. Some of them are of high activity when exogenously applied. They are widely distributed, and their endogenous levels increase rapidly in response to external stress. How they are involved in the transduction chain between external stress impulse and internal stress response has to be studied further. Even more speculative is the physiological role of JA amino acid conjugates in pollen germination and senescence processes. O-glucosides represent another important group of jasmonate conjugates they might be of physiological relevance in regulation of tuber formation. Whether they are special transport molecules remains an open question. Of further interest is the existence of jasmonate methyl ester conjugates. These volatile compounds exhibit high physiological potency and, thus, qualify as potential air-borne signals.

[236]

Conjugation of brassinosteroids

Only a few papers have so far dealt with conjugation of the brassinosteroids (Fig. 9). This field is in its very infancy [91] and the seemingly low concentration of brassinosteroid conjugates may well represent a major difficulty in their detection and analysis.

1 R_1 = OH; R_2 = H
2 R_1 = H; R_2 = OH

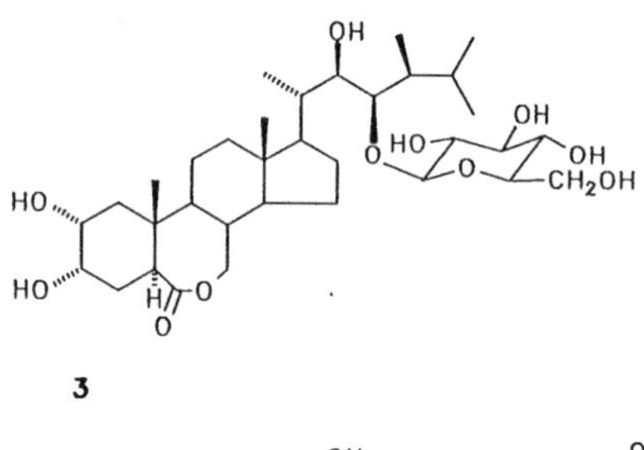

3

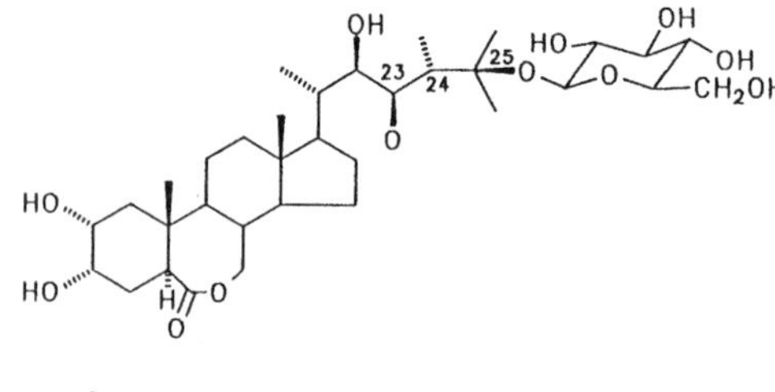

4

5 R = Louryl
6 R = Myristyl

Fig. 9. Structures of endogenously occurring brassinosteroid conjugates.

Although brassinosteroid molecules contain a series of functional groups, only hydroxyl groups at C-23 and C-25 have been found to be linked to glucosyl moieties. Extracts from seeds of *Phaseolus vulgaris* have been shown to contain 23-0-β-D-glucosyl-25-methyl dolichosterol (Fig. 9-1) and its 2β isomer (Fig. 9-2) [218, 219, 224].

Conjugates of brassinosteroids have also been detected in metabolic studies. After feeding brassinolide to mung beans, the corresponding 23-O-β-D-glucosyl conjugate (Fig. 9-3) was identified [195], while cell cultures of *Lycopersicon esculentum* transform 24-epi-brassinolide to the 25-O-β-D-glucosyloxy derivative (Fig. 9-4) [165]. Although 23-O-glucosyl brassinolide is as active as the free brassinolide in the rice lamina inclination test, there has been discussion suggesting that 23-O-glucosylation of brassinosteroids represents a regulatory deactivation step [195]. Recently, two acyl conjugates of brassinosteroids have been identified carrying the conjugation moieties at the 3-hydroxy group. These 3β-O-lauryl and 3β-O-myristyl derivatives of teasterone (Fig. 9-5 and 9-6) were isolated from pollen of *Lilium longifolium* [2].

References

1. Andersson B, Sandberg G: Identification of endogenous *N*-(3-indoleacetyl) aspartic acid in Scots pine (*Pinus silvestris* L.) by combined gas chromatography-mass spectrometry, using high-performance liquid chromatography for quantification. J Chromatogr 238: 151–156 (1982).
2. Asakawa S, Abe H, Natsume M: New acyl conjugated brassinosteroids from Lily pollen. XVth International Botany Congress (Yokohama 1993), Abstract 4163 (1993).
3. Badenoch-Jones J, Summons RE, Rolfe BG, Letham DS: Phytohormones, *Rhizobium* mutants, and nodulation in legumes. IV. Auxin metabolites in pea root nodules. J Plant Growth Regul 3: 23–29 (1984).
4. Bandurski RS: Metabolism of indole-3-acetic acid. In: Crozier A, Hillman JR (eds) The Biosynthesis and Metabolism of Plant Hormones, SEB-Series 23, pp. 183–200, Cambridge University Press, Cambridge (1984).
5. Bandurski RS, Desrosiers MF, Jensen P, Pawlak M, Schulze A: Genetics, chemistry, and biochemical physiology in the study of hormonal homeostasis. In: Karssen CM, van Loon LC, Vreugdenhil D (eds) Progress in Plant Growth Regulation, pp. 1–2, Kluwer Academic Publishers, Dordrecht (1992).
6. Bandurski RS, Schulze A: Concentrations of indole-3-acetic acid and its esters in *Avena* and *Zea*. Plant Physiol 54: 257–262 (1974).
7. Bandurski RS, Schulze A, Desrosiers M, Jensen P, Epel B: Relationship between stimuli, IAA and growth. In: Pharis RP, Rood SB (eds) Plant Growth Substances 1988, pp. 341–352, Springer-Verlag, Berlin/Heidelberg/New York (1990).
8. Bandurski RS, Schulze A, Domagalski W, Komoszynski M, Lewer P, Nonhebel H: Synthesis and metabolism of conjugates of indole-3-acetic acid. In: Schreiber K, Schütte HR, Sembdner G (eds) Conjugated Plant Hormones: Structure, Metabolism and Function, pp. 11–20. VEB Deutscher Verlag der Wissenschaften, Berlin (1987).
9. Beffa R, Martin HV, Pilet P-E: In vitro oxidation of indoleacetic acid by soluble auxin-oxidases and peroxidases from maize roots. Plant Physiol 94: 485–491 (1990).
10. Bialek K, Cohen JD: Hydrolysis of an indole-3-acetic acid amino acid conjugate by an enzyme preparation from *Phaseolus vulgaris*. Plant Physiol 75 (suppl): 108 (1984).
11. Bialek K, Cohen JD: Isolation and partial characterization of the major amide-linked conjugate of indole-3-acetic acid from *Phaseolus vulgaris*. Plant Physiol 88: 99–104 (1986).
12. Bialek K, Cohen JD: Free and conjugated indole-3-acetic acid in developing bean seeds. Plant Physiol 91: 775–779 (1989).
13. Blechschmidt S, Castel U, Gaskin P, Hedden P, Graebe JE, MacMillan J: GC/MS analysis of the plant hormones in seeds of *Cucurbita maxima*. Phytochemistry 23: 553–558 (1984).
14. Bohlmann F, Wegner P, Jakupovic J, King RM: Struktur und Synthese von N-(Acetoxy)-jasmonoylphenylalaninmethylester aus *Praxelis clematidea*. Tetrahedron 40: 2537–2540 (1984).
15. Bopp M, Atzorn R: Hormonelle Regulation der Moosentwicklung. Naturwissenschaften 79: 337–346 (1992).
16. Brown BH, Crozier A, Sandberg G: Catabolism of indole-3-acetic acid in chloroplast fractions from light-grown *Pisum sativum* L. seedlings. Plant Cell Environm 9: 527–534 (1986).
17. Brückner C, Kramell R, Schneider G, Schmidt J, Preiss A, Sembdner G, Schreiber K: N-[(−)jasmonoyl]-S-tryptophan and a related tryptophane conjugate from *Vicia faba*. Phytochemistry 27: 275–276 (1988).
18. Brzobohaty B, Moore I, Kristoffersen P, Bako L, Campos N, Schell J, Palme K: Release of active cytokinin by

a β-glucosidase localized to the maize root meristem. Science 262: 1051–1054 (1993).
19. Burch LR, Horgan R: The purification of cytokinin oxidase from *Zea mays* kernels. Phytochemistry 28: 1313–1319 (1989).
20. Burch LR, Horgan R: Cytokinin oxidase and the degradative metabolism of cytokinins. In: Kaminek M, Mok D, Zazimalova E (eds) Physiology and Biochemistry of Cytokinins in Plants, pp. 29–32. SPB Academic Publishing, The Hague (1992).
21. Catala C, Östin A, Chamarro J, Sandberg G, Crozier A: Metabolism of indole-3-acetic acid by pericarp discs from immature and mature tomato (*Lycopersicon esculentum* Mill). Plant Physiol 100: 1457–1463 (1992).
22. Chaffield JM, Armstrong DJ: Cytokinin oxidase from *Phaseolus vulgaris* callus cultures. Affinity for concanavalin A. Plant Physiol 88: 245–247 (1988).
23. Chen C-M, Eckert RL: Phosphorylation of cytokinin by adenosine kinase from wheat germ. Plant Physiol 59: 443–447 (1977).
24. Chen C-M, Petschow B: Metabolism of cytokinin: ribosylation of cytokinin bases by adenine phosphorylase from wheat germ. Plant Physiol 62: 871–874 (1978).
25. Chisnell JR: Myo-inositol esters of indole-3-acetic acid are endogenous components of *Zea mays* L. shoot tissue. Plant Physiol 74: 278–283 (1984).
26. Cohen JD: Identification and quantification analysis of indole-3-acetyl-aspartate from seeds of *Glycine max* L. Plant Physiol 70: 749–753 (1982).
27. Cohen JD, Bandurski RS: Chemistry and physiology of the bound auxins. Annu Rev Plant Physiol 33: 403–430 (1982).
28. Cohen JD, Bialek K: The biosynthesis of indole-3-acetic acid in higher plants. In: Crozier A, Hillman JR (eds) The Biosynthesis and Metabolism of Plant Hormones, SEB-Series 23, pp. 165–181, Cambridge University Press, Cambridge (1984).
29. Corcuera LJ, Bandurski RS: Biosynthesis of indol-3-yl-acetyl-myo-inositol arabinoside in kernels of *Zea mays* L. Plant Physiol 70: 1664–1666 (1982).
30. Corcuera LJ, Michalczuk L, Bandurski RS: Enzymic synthesis of indol-3-yl-acetyl-myo-inositol galactoside. Biochem J 207: 283–290 (1982).
31. Cornforth JW, Milborrow BV, Ryback G: Synthesis of (+) abscisin II. Nature 205: 1269–1270 (1965).
32. Creelman RA: Abscisic acid physiology and biosynthesis in higher plants. Physiol Plant 75: 31–36 (1989).
33. Cross BE, Webster GRB: New metabolites of *Gibberella fujikuroi*. Part XV. N-jasmonoyl- and N-dihydrojasmonoyl-isoleucine. J Chem Soc Commun 1970: 1839–1842 (1970).
34. Crozier A, Kuo CC, Durley RC, Pharis RP: The biological activities of 26 gibberellins in nine plant bioassays. Can J Bot 48: 867–877 (1970).
35. Crozier A, Turnbull CGN, Malcolm JM, Graebe JE: Gibberellin metabolism in cell-free preparations from *Phaseolus coccineus*. In: Takahashi N, Phinney BO, MacMillan J (eds) Gibberellins, pp. 83–93, Springer-Verlag, New York (1991).
36. Dashek WV, Singh BN, Walton DC: Abscisic acid localisation and metabolism in barley aleurone layers. Plant Physiol 64: 43–48 (1979).
37. Dathe W, Sembdner G, Kefeli VI, Vlasov PV: Gibberellins, abscisic acid, and related inhibitors in branches and bleeding sap of birch (*Betula pubescens* Ehrh.). Biochem Physiol Pflanzen 173: 238–248 (1978).
38. Dathe W, Sembdner G, Yamaguchi I, Takahashi N: Gibberellins and growth inhibitors in spring bleeding sap, roots and branches of *Juglans regia* L. Plant Cell Physiol 23: 115–123 (1982).
39. Davies PJ: Plant Hormones and Their Role in Plant Growth and Development. Martinus Nijhoff, Dordrecht/Boston/Lancaster (1987).
40. Demole E, Lederer E, Mercier D: Isolement et determination da la structure du jasmonate de methyle, constituant odorant characteristique de l'essence de jasmin. Helv Chim Acta 45: 675–685 (1962).
41. Dixon SC, Martin RC, Mok MC, Shaw G, Mok DWS: Zeatin glycosylation enzymes in *Phaseolus*. Isolation of O-glucoslytransferase from *P. lunatus* and comparison to O-xylosyltransferase from *P. vulgaris*. Plant Physiol 90: 1316–1321 (1989).
42. Doumas P, Imbault N, Moritz T, Oden PC: Detection and identification of gibberellins in douglas fir (*Pseudotsuga menziesii*) shoots. Physiol Plant 85: 489–494 (1992).
43. Ehmann A: Identification of 2-O-(indole-3-acetyl)-D-glucopyra-4-O-(indole-3-acetyl)-D-glucopyranose and 6-O-(indole-3-acetyl)-D-glucopyranose from kernels of *Zea mays* by gas-liquid chromatography-mass spectrometry. Carbohydr Res 34: 99–114 (1974).
44. Entsch B, Letham DS, Parker CW, Summons RE, Gollnow BE: Metabolites of cytokinins. In: Skoog F (ed) Plant Growth Regulation 1979, pp. 109–118, Springer-Verlag, Berlin (1979).
45. Epstein E, Baldi BG, Cohen JD: Identification of indole-3-acetylglutamate from seeds of *Glycine max* L. Plant Physiol 80: 256–258 (1986).
46. Epstein E, Cohen JD, Bandurski RS: Concentration and metabolic turnover of indoles in germinating kernels of *Zea mays*. Plant Physiol 65: 415–421 (1980).
47. Farmer EE, Johnson RR, Ryan CA: Regulation of expression of proteinase inhibitor genes by methyl jasmonate and jasmonic acid. Plant Physiol 98: 995–1002 (1992).
48. Farmer EE, Ryan CA: Interplant communication: airborne methyl jasmonate induces synthesis of proteinase inhibitors in plant leaves. Proc Natl Acad Sci USA 87: 7713–7716 (1990).
49. Gawer M, Laloue M, Terrine C, Guern J: Metabolism and biological significance of benzyladenine-7-glucoside. Plant Sci Lett 8: 262–274 (1977).

50. Grambow HJ, Langenbeck-Schwich B: The relationship between oxidase activity, peroxidase activity, hydrogen peroxide, and phenolic compounds in the degradation of indole-3-acetic acid in vitro. Planta 157: 131–137 (1983).
51. Gundlach H, Müller MJ, Kutchan TM, Zenk MH: Jasmonic acid is a signal transducer in elicitor-induced plant cell cultures. Proc Natl Acad Sci USA 89: 2389–2393 (1992).
52. Harada H, Yokota T: Isolation of gibberellin A_8-glucoside from shoot apices of *Althaea rosea*. Planta 92: 100–104 (1970).
53. Hemphill DD, Baker LR, Sell HM: Isolation of novel conjugated gibberellins from *Cucumis sativus* seed. Can J Biochem 51: 1647–1653 (1973).
54. Herrmann G, Kramell H-M, Kramell R, Weidhase RA, Sembdner G: Biological activity of jasmonic acid conjugates. In: Schreiber K, Schütte H-R, Sembdner G (eds) Conjugated Plant Hormones: Structure, Metabolism and Function, pp. 315–322. VEB Deutscher Verlag der Wissenschaften, Berlin (1987).
55. Hiraga K, Kawabe S, Yokota T, Murofushi N, Takahashi N: Isolation and characterization of plant growth substances in immature seeds and etiolated seedlings of *Phaseolus vulgaris*. Agric Biol Chem 38: 2521–2527 (1974).
56. Hiraga K, Yokota T, Murofushi N, Takahashi N: Isolation and characterization of a free gibberellin and glucosyl esters of gibberellins in mature seeds of *Phaseolus vulgaris*. Agric Biol Chem 36: 345–347 (1972).
57. Hiraga K, Yokota T, Murofushi N, Takahashi N: Isolation and characterization of gibberellins in mature seeds of *Phaseolus vulgaris*. Agric Biol Chem 38: 2511–2520 (1974).
58. Hiraga K, Yokota T, Takahashi N: Biological activity of some synthetic gibberellin glucosyl esters. Phytochemistry 13: 2371–2376 (1974).
59. Hirai N, Fukui H, Koshimizu K: A novel abscisic acid metabolite from seeds of *Robinia pseudacacia*. Phytochemistry 17: 1625–1627 (1978).
60. Hirai N, Koshimizu K: A new conjugate of dihydrophaseic acid from avocado fruit. Agric Biol Chem 47: 365–371 (1983).
61. Ho THD, Uknes SJ: Regulation of abscisic acid metabolism in the aleurone layers of barley seeds. Plant Cell Rep 1: 270–273 (1982).
62. Horgan R: Present and future prospects for cytokinin research. In: Kaminek M, Mok D, Zazimalova E (eds) Physiology and Biochemistry of Cytokinins in Plants, pp. 3–14. SPB Academic Publishing, The Hague (1992).
63. Horgan R, Palni LMS, Scott IM, McGaw BA: Cytokinin biosynthesis and metabolism in *Vinca rosea* crown gall tissue. In: Guern J, Peaud-Lenoel C (eds) Metabolism and Molecular Activities of Cytokinins, pp. 56–65. Springer-Verlag, Berlin (1981).
64. Kamiya JE, Graebe JE: The biosynthesis of all major pea gibberellins in a cell free system from *Pisum sativum*. Phytochemistry 22: 681–689 (1983).
65. Kehlen A: Untersuchungen zum Metabolismus von Jasmonsäure. Ph.D. thesis, Universität Halle (1991).
66. Kinashi H, Suzuki Y, Takeuchi S, Kawarada A: Possible metabolic intermediates from IAA to β-acid in rice bran. Agric Biol Chem 40: 2465–2470 (1976).
67. Kleczkowski K, Spanier K, Schell J: Cytokinins, their O-glucosides and riboside phosphates in the shoot-inducing mutants of tobacco crown gall tissue culture. In: Schreiber K, Schütte HR, Sembdner G (eds) Conjugated Plant Hormones: Structure, Metabolism and Function, pp. 138–152. VEB Deutscher Verlag der Wissenschaften, Berlin (1987).
68. Klee H, Horsch RB, Hinchee MA, Hoffmann NL: The effects of overproduction of two *Agrobacterium tumefaciens* T-DNA auxin biosynthesic gene products in transgenic *Petunia* plants. Genes Devel 1: 86–96 (1987).
69. Knöfel H-D, Schwarzkopf E, Müller P, Sembdner G: Enzymic glucosylation of gibberellins. J Plant Growth Regul 3: 127–140 (1984).
70. Kobayashi S, Sugioka K, Nakamo M, Tero-Kubota S: Analysis of stable end products and intermediates of oxidative decarboxylation of indole-3-acetic acid by horseradish peroxidase. Biochemistry 23: 4589–4597 (1984).
71. Koda Y: The role of jasmonic acid and related compounds in the regulation of flower development. Int Rev Cytol. 135: 155–199 (1992).
72. Koshimizu K, Fukui H, Mitsui T: Isolation of (+) abscisyl-β-glucopyranoside from immature fruit of *Lupinus luteus*. Agric Biol Chem 30: 941–943 (1968).
73. Koshimizu K, Fukui H, Usuda S, Mitsui T: Plant growth inhibitors in seeds of pumpkin. In: Plant Growth Substances 1973, pp. 86–92. Hirokawa, Tokyo (1973).
74. Kramell R, Atzorn R, Brückner C, Lehmann J, Schneider G, Sembdner G, Parthier B: Effects of osmotic stress on endogenous jasmonates in barley. I. Isolation and identification of jasmonic acid and its conjugates with amino acids as induced metabolites. J Plant Growth Regul (submitted).
75. Laloue M, Pehte C: Dynamics of cytokinin metabolism in tobacco cells. In: Wareing P (ed) Plant Growth Substances 1982, pp. 215–223, Academic Press, London (1982).
76. Laloue M, Pehte-Terrine C, Guern J: Uptake and metabolism of cytokinins in tobacco cells: studies in relation to the expression of their biological activities. In: Guern J, Peaud-Lenoel C (eds) Metabolism and Molecular Activities of Cytokinins, pp. 80–96. Springer-Verlag, Berlin (1981).
77. Laloue M, Terrine C, Guern J: Cytokinins: Metabolism and biological activity of N6-(d2-isopentenyl) adenine in tobacco cells and callus. Plant Physiol 59: 478–483 (1977).
78. Lehmann H, Sembdner G: Plant hormone conjugates.

In: Purohit SS (ed) Hormonal Regulation of Plant Growth and Development, vol. 3, pp. 245–310. Agro Botanical Publications, Bikaner (1986).
79. Lenton JR, Appleford NEJ: Gibberellin production and action during germination of wheat. In: Takahashi N, Phinney BO, MacMillan J (eds) Gibberellins, pp. 125–135. Springer-Verlag, New York (1991).
80. Lenton JR, Appleford NEJ, CrokersJ: Gibberellin-dependent α-amylase production in germinating wheat (*Triticum aestivum*) grain. In: Frontiers of Gibberellin Research 1993, Abstr. 19. Tokyo Riken (1993).
81. Letham DS, Palni LMS: The biosynthesis and metabolism of cytokinins. Annu Rev Plant Physiol 34: 163–197 (1983).
82. Letham DS, Palni LMS, Tao GQ, Gollnow Bl, Bates CM: Regulators of cell division in plant tissues. XXIX. The activities of cytokinin glucosides and alanine conjugates in cytokinin bioassay. J Plant Growth Regul 2: 3–17 (1983).
83. Letham DS, Tuo GQ, Parker CW: An overview of cytokinin biosynthesis. In: Wareing PF (ed) Plant Growth Substances 1982, pp. 143–153. Academic Press, London (1982).
84. Liebisch HW: Uptake, translocation and metabolism of GA_3 glucosyl ester. In: Schreiber K, Schütte HR, Sembdner G (eds) Biochemistry and Chemistry of Plant Growth Regulators, pp. 109–113. Institute for Plant Biochemistry, Academy of Sciences of the GDR, Halle (1974).
85. Lorenzi R, Horgan R, Heald JK: Gibberellin A9 glucosyl ester in needles of *Picea sitchensis*. Phytochemistry 15: 789–790 (1976).
86. Loveys BR, Milborrow BV: Isolation and characterisation of 1′-O-abscisic acid-β-glucopyranoside from vegetative tomato tissue. Aust J Plant Physiol 8: 571–589 (1981).
87. Loveys BR, Milborrow BV: Metabolism of abscisic acid. In: Crozier A, Hillman JR (eds) The Biosynthesis and Metabolism of Plant Hormones, SEB-Series 23, pp. 71–103. Cambridge University Press, Cambridge (1984).
88. Magnus V: Auxin conjugation. In: Schreiber K, Schütte HR, Sembdner G (eds) Conjugated Plant Hormones: Structure, Metabolism and Function, pp. 31–40. VEB Deutscher Verlag der Wissenschaften, Berlin (1987).
89. Magnus V, Hangarter RP, Good NE: Interaction of free indole-3-acetic acid and its amino acid conjugates in tomato hypocotyl cultures. J Plant Growth Regul 11: 67–75 (1992).
90. Magnus V, Nigovic B, Hangarter RP, Good NE: N-(Indol-3-ylacetyl)amino acids as sources of auxin in plant tissue culture. J Plant Growth Regul 11: 19–28 (1992).
91. Marquardt V, Adam G: Recent advances in brassinosteroid research. In: Ebing W (ed) Chemistry of Plant Protection, pp. 103–139. Springer-Verlag, Berlin/Heidelberg/New York (1991).
92. Martin RC, Martin RR, Mok MC, Mok DWS: A monoclonal antibody specific to zeatin O-glycosyltransferase of *Phaseolus*. Plant Physiol 94: 1290–1294 (1990).
93. Martin RC, Mok MC, Shaw G, Mok DWS: An enzyme mediating the conversion of zeatin to dihydrozeatin in *Phaseolus* embryos. Plant Physiol 90: 1630–1635 (1989).
94. Martineau B, Houck CM, Sheehy RE, Hiatt WR: Fruit-specific expression of the *A. tumefaciens* isopentenyl transferase gene in tomato: effects of fruit ripening and defence-related gene expression in leaves. Plant J 5: 11–19 (1994).
95. Matsuura H, Yoshihara T, Ichihara A, Kikuta Y, Koda Y: Tuber-forming substances in Jerusalem artichoke (*Helianthus tuberosus* L.). Biosci Biotech Biochem 57: 1253–1256 (1993).
96. McGaw BA: Cytokinin biosynthesis and metabolism. In: Davies PJ (ed) Plant Hormones and Their Role in Plant Growth and Development, pp. 76–93. Martinus Nijhoff, Dordrecht (1987).
97. McGaw BA, Horgan R: Cytokinin catabolism and cytokinin oxidase. Phytochemistry 22: 1103–1105 (1983).
98. McGaw BA, Horgan R: Cytokinin biosynthesis and metabolism. In: Crozier A, Hillman JR (eds) The Biosynthesis and Metabolism of Plant Hormones, SEB-Series 23, pp. 105–133, Cambridge University Press, Cambridge (1984).
99. McGaw BA, Horgan R, Heald JK, Wullems GJ, Schilperoort RA: Mass-spectrometric quantitation of cytokinins in tobacco crown-gall tumours induced by mutated octopine Ti plasmids of *Agrobacterium tumefaciens*. Planta 176: 230–234 (1988).
100. Meyer A, Gross D, Schmidt J, Jensen E, Vorkefeld S, Semdner G: Cucurbic acid-related metabolites of the plant growth regulator dihydrojasmonic acid in barley (*Hordeum vulgare*). Biochem Physiol Pflanzen 187: 401–408 (1991).
101. Meyer A, Gross D, Vorkefeld S, Kummer M, Schmidt J, Sembdner G, Schreiber K: Metabolism of the plant growth regulator dihydrojasmonic acid in barley shoots. Phytochemistry 28: 1007–1011 (1989).
102. Meyer A, Schmidt J, Gross D, Jensen E, Rudolph A, Vorkefeld S, Sembdner G: Amino acid conjugates as metabolites of the plant growth regulator dihydrojasmonic acid in barley (*Hordeum vulgare*). J Plant Growth Regul 10: 17–25 (1991).
103. Meyer A, Schneider G, Sembdner G: Endogenous gibberellins and inhibitors of the douglas fir. Abstract 51, International Symposium on Plant Growth Regulators, Liblice (1984).
104. Michalczuk L, Bandurski RS: In vitro biosynthesis of esters of indole-3-acetic acid Plant Physiol 65: 157 (suppl.) (1980).
105. Michalczuk L, Bandurski RS: Enzymic synthesis of 1-O-indole-3-acetyl-β-D-glucose and indole-3-acetyl.-myo-inositol. Biochem. J. 207: 273–283 (1982).
106. Miersch O, Brückner C, Schmidt J, Sembdner G: Cy-

clopentane fatty acids from *Gibberella fujikuroi*. Phytochemistry 31: 3835–3937 (1992).
107. Miersch O, Herrmann G, Kramell H-M, Sembdner G: Biological acitivity of jasmonic acid glucosyl ester. Biochem Physiol Pflanzen 182: 425–428 (1987).
108. Milborrow BV: The identification of (+)-abscisin II [(+)-dormin] in plants and measurement of its concentration. Planta 76: 93–113 (1967).
109. Milborrow BV: The metabolism of abscisic acid. J Exp Bot 21: 17–29 (1970).
110. Milborrow BV: The chemistry and physiology of abscisic acid. Annu Rev Plant Physiol 25: 259–307 (1974).
111. Milborrow BV, Vaughan G: Characterisation of dihydrophaseic acid 4′-O-β-D-glucopyranoside as a major metabolite of abscisic acid. Aust J Plant Physiol 9: 361–372 (1982).
112. Mok MC, Mok DWS, Marsden KE, Shaw G: The biological activity and metabolism of a novel cytokinin metabolite, O-xylosylzeatin, in callus tissue of *Phaseolus vulgaris* and *P. lunatus*. J Plant Physiol 130: 423–431 (1987).
113. Mok DWS, Mok MC, Martin RC, Bassil NV, Lightfoot DA: Zeatin metabolism in *Phaseolus*: enzymes and genes. In: Karssen CM, van Loon LC, Vreugdenhil D (eds) Progress in Plant Growth Regulation, pp. 597–606. Kluwer, Dordrecht (1992).
114. Mok DWS, Mok MC, Shaw G: Cytokinin activity, metabolism and function in *Phaseolus*. In: Kaminek M, Mok D, Zazimalova E (eds) Physiology and Biochemistry of Cytokinins in Plants, pp. 41–46. SPB Academic Publishing, The Hague (1992).
115. Mok DWS, Mok MC, Shaw G, Dixon SC, Martin RC: Genetic differences in the enzymatic regulation of zeatin metabolism in *Phaseolus* embryos. In: Pharis RP, Rood SB (eds) Plant Growth Substances 1988, pp. 267–274, Springer-Verlag, Berlin/Heidelberg/New York (1990).
116. Monteiro AM, Crozier A, Sandberg G: The biosynthesis and conjugation of indole-3-acetic acid in germinating seed and seedlings of *Dalbergia dolichopetala*. Planta 174: 561–568 (1988).
117. Moritz T: The use of combined capillary liquid chromatography/mass spectrometry for the identification of a gibberellin glucosyl conjugate. Phytochem Anal 3: 32–37 (1992).
118. Moritz T, Schneider G, Jensen E: Capillary liquid chromatography/fast atom bombardment mass spectrometry of gibberellin glucosyl conjugates. Biol Mass Spectrom 21: 554–559 (1992).
119. Motyka V, Kaminek M: Characterization of cytokinin oxidase from tobacco and poplar callus cultures. In: Kaminek M, Mok D, Zazimalova E (eds) Physiology and Biochemistry of Cytokinins in Plants, pp. 33–39. SPB Academic Publishing, The Hague (1992).
120. Müller P, Knöfel H-D, Sembdner G: Studies on the enzymatical synthesis of gibberellin-O-glucosides. In: Schreiber K, Schütte HR, Sembdner G (eds) Conjugated Plant Hormones: Structure, metabolism and function, pp. 115–119, VEB Deutscher Verlag der Wissenschaften, Berlin (1987).
121. Murakami Y: Distribution of bound gibberellin in higher plants and its hydrolysis by enzymes from different sources. Bull Nat Inst Agric Sci Ser D 36: 69–123 (1985).
122. Murofushi N, Yang Y-Y, Yamaguchi I, Schneider G, Kato Y: Liquid chromatography/atmospheric pressure chemical ionization mass spectrometry of gibberellin conjugates. In: Karssen CM, van Loon LC, Vreugdenhil D (eds) Progress in Plant Growth Regulation, pp. 900–904. Kluwer Academic Publishers, Dordrecht (1992).
123. Naumann R, Dörffling K: Variation of free and conjugated abscisic acid, phaseic acid and dihydrophaseic acid levels in ripening barley grains. Plant Sci Lett 27: 111–117 (1982).
124. Nilsson O, Moritz T, Imbault N, Sandberg G, Olsson O: Hormonal characterization of transgenic tobacco plants expressing the *rolC* gene of *Agrobacterium rhizogenes* TI-DNA. Plant Physiol 102: 363–371 (1982).
125. Nonhebel HM, Cooney TP: Measurement of the in vitro rate of indole-3-acetic acid turnover. In: Pharis RP, Rood SB (eds) Plant Growth Substances 1988, pp. 333–340. Springer-Verlag, Berlin/Heidelberg/New York (1990).
126. Nonhebel HM, Kruse LI, Bandurski RS: Indole-3-acetic acid catabolism in *Zea mays* seedlings. Metabolic conversion of oxindole-3-acetic acid to 7-hydroxy-2-oxindole-3-acetic acid-7′-O-β-D-glucopyranoside. J Biol Chem 260: 12685–12689 (1985).
127. Ohkuma K, Addicott FT, Smith OE, Thiessen WE: The structure of abscisin II. Tetrahedron Lett 29: 2529–2535 (1965).
128. Östin A, Monteiro AM, Crozier A, Jensen E, Sandberg G: Analysis of indole-3-acetic acid metabolites from *Dalbergia dolichopetala* by high-performance liquid chromatography-mass spectrometry. Plant Physiol 100: 63–68 (1992).
129. Palmer MV, Horgan R, Wareing PF: Cytokinin metabolism in *Phaseolus vulgaris*. I. Variation in cytokinin levels in leaves of decapitated plants in relation to bud outgrowth. J Exp Bot 32: 1231–1241 (1981).
130. Palmer MV, Horgan R, Wareing PF: Cytokinin metabolism in *Phaseolus vulgaris*. II. Comparative metabolism of exogenous cytokinins by detached leaves. Plant Sci Lett 22: 187–195 (1981).
131. Parker CW, Letham DS: Regulators of cell division in plant tissues, XVI. Metabolism of zeatin by radish cotyledons and hypocotyls. Planta 114: 199–218 (1973).
132. Parker CW, Letham DS: Regulators of cell division in plant tissues. XVII. Metabolism of zeatin in *Zea mays* seedlings. Planta 115: 337–344 (1974).
133. Parker CW, Letham DS, Gollnow BI, Summons RE,

Duke CC, McLeod JK: Regulators of cell division in plant tissues. XXV. Metabolism of zeatin in lupin seedlings. Planta 142: 239–251(1978).
134. Parthier B: Jasmonates: hormonal regulators or stress factors in leaf senescence? J Plant Growth Regul 9: 1–7 (1990).
135. Parthier B: Jasmonates, new regulators of plant growth and development: many facts and few hypotheses on their actions. Bot Acta 104: 446–454 (1991).
136. Parthier B, Brückner C, Dathe W, Hause B, Hermann G, Knöfel HD, Kramell H-M, Kramell R, Lehmann J, Miersch O, Reinbothe S, Sembdner G, Wasternack C, zur Nieden U: Jasmonates: metabolism, biological activities, and modes of action in senescence and stress responses. In: Karssen CM, van Loon LC, Vreugdenhil D (eds) Progress in Plant Growth Regulation, pp. 276–288, Kluwer Academic Publishers, Dordrecht (1992).
137. Percival FW, Bandurski RS: Esters of indole-3-acetic acid from *Avena* seeds. Plant Physiol 58: 60–67 (1986).
138. Pierce M, Raschke K: Synthesis and metabolism of abscisic acid in detached leaves of *Phaseolus vulgaris* L. after loss and recovery of turgor. Planta 153: 156–165 (1981).
139. Piskornik Z, Bandurski RS: Purification and partial characterization of a glucan containing indole-3-acetic acid. Plant Physiol 50: 176–182 (1972).
140. Porat R, Borochov A, Halevy AH: Enhancement of *Petunia* and *Dendrobium* flower senescence by jasmonic acid methyl ester is via the promotion of ethylene production. Plant Growth Regul 13: 297–301 (1993).
141. Reinecke DM, Bandurski RS: Oxindole-3-acetic acid, an catabolite in *Zea mays*. Plant Physiol 71: 211–213 (1983).
142. Reinecke DM, Bandurski RS: Oxidation of indole-3-acetic acid to oxindole-3-acetic acid by an enzyme preparation from *Zea mays* seedlings. Plant Physiol 75 (suppl): 108 (1984).
143. Reinecke DM, Bandurski RS: Auxin biosynthesis and metabolism. In: Davies PJ (ed) Plant Hormones and their Role in Plant Growth and Development, pp. 24–42. Martinus Nijhoff, Dordrecht/Boston/Lancaster (1987).
144. Riov J, Bangerth F: Metabolism of auxin in tomato fruit tissue: formation of high molecular weight conjugates of oxindole-3-acetic acid via the oxidation of indole-3-acetylaspartic acid. Plant Physiol 100: 1396–1402 (1992).
145. Rivier L, Gaskin P, Albone KS, MacMillan J: GC-MS Identification of endogenous gibberellins and gibberellin conjugates as their permethylated derivatives. Phytochemistry 20: 687–692 (1981).
146. Rock CD, Zeevaart JAD: Abscisic (ABA)-aldehyde is a precursor to, and 1′,4′-trans-ABA-diol a catabolite of, ABA in apple. Plant Physiol 93: 915–923 (1990).
147. Rood SB: Genetic and environmental control of gibberellin physiology in *Brassica*. In: Frontiers of Gibberellin Research 1993, Abstract 38. Tokyo Riken (1993).
148. Rood SB, Pharis RP: Evidence for reversible conjugation of gibberellins in higher plants. In: Schreiber K, Schütte HR, Sembdner G (eds) Conjugated Plant Hormones: Structure, Metabolism and Function, pp. 183–190. VEB Deutscher Verlag der Wissenschaften, Berlin (1987).
149. Ros Barcelo A, Pedreno MA, Ferrer MA, Sabater F, Munoz R: Indole-3-methanol is the main product of the oxidation of indole-3-acetic acid catalyzed by two cytosolic basic isoperoxidases from *Lupinus*. Planta 181: 448–450 (1990).
150. Ryan CA: The search for the proteinase inhibitor-inducing factor. PIF. Plant Mol Biol 19: 123–133 (1992).
151. Sandberg G, Jensen E, Crozier A: Analysis of 3-indole carboxylic acid in *Pinus silvestris* needles. Phytochemistry 23: 99–102 (1984).
152. Sandberg G, Crozier A, Ernstsen A: Indole-3-acetic acid and related compounds. In: Rivier L, Crozier A (eds) The Principles and Practice of Plant Hormone Analysis, vol. 2, pp. 169–301. Academic Press, London (1987).
153. Schliemann W: Hydrolysis of conjugated gibberellins by β-glucosidases of dwarf rice (*Oryza sativa* L. cv. 'Tanginbozu'). J Plant Physiol 116: 123–132 (1984).
154. Schliemann W: β-Glucosidase with gibberellin A8-2-O-glucoside hydrolysing activity from pods of runner beans. Phytochemistry 27: 689–692 (1988).
155. Schliemann W: Zum Konzept der reversiblen Konjugation bei Phytohormonen. Naturwissenschaften 78: 392–401 (1991).
156. Schliemann W, Schaller B, Jensen E, Schneider G: Native gibberellin-O-glucosides from mature seeds of *Phaseolus coccineus*. Phytochemistry 35: 35–38 (1994).
157. Schliemann W, Schneider G: Untersuchungen zur enzymatischen Hydrolyse von Gibberellin-O-glucosiden. I. Hydrolysegeschwindigkeiten von Gibberellin-13-O-glucosiden. Biochem Physiol Pflanzen 174: 738–745 (1979).
158. Schliemann W, Schneider G: Metabolic formation and occurrence of gibberellin A_1-3-O-β-D-glucopyranoside in immature fruits of *Phaseolus coccineus* L. Plant Growth Regul 8: 85–90 (1989).
159. Schmidt J, Kramell R, Brückner C, Sembdner G, Schreiber K, Stach J, Jensen E: Gas chromatographic/mass spectrometric and tandem mass spectrometric investigations of synthetic amino acid conjugates of jasmonic acid and enogenously occurring related compounds from *Vicia faba* L. Biomed Environm Mass Spectrom 19: 327–338 (1990).
160. Schmidt J, Schneider G, Jensen E: Capillary gas chromatography/mass spectrometry of permethylated gibberellin glucosides. Biomed Environm Mass Spectrom 17: 7–13 (1988).
161. Schneider G: Über strukturelle Einflüsse bei der Gluco-

sylierung von Gibberellinen. Tetrahedron 37: 545–549 (1981).

162. Schneider G: Gibberellin conjugates. In: Crozier A (ed) The Biochemistry and Physiology of Gibberellins, vol 1, pp. 389–456. Praeger Publishers, New York (1983).
163. Schneider G: Gibberellin conjugation. In: Schreiber K, Schütte HR, Sembdner G (eds) Conjugated Plant Hormones: Structure, Metabolism and Function, pp. 158–166. VEB Deutscher Verlag der Wissenschaften, Berlin (1987).
164. Schneider G, Jensen E, Spray CR, Phinney BO: Hydrolysis and reconjugation of gibberellin A_{20} glucosyl ester by seedlings of *Zea mays* L. Proc Natl Acad Sci USA 89: 8045–8048 (1992).
165. Schneider B, Kolbe A, Porzel A, Adam G: A novel metabolite of 24-epi-brassinolide in cell suspension culture of *Lycopersicon esculentum*. Phytochemistry 36: 319–321 (1994).
166. Schneider G, Schaller B, Jensen E: RP-HPLC Separation of permethylated free and glucosylated gibberellins: a method for the analysis of gibberellin metabolites. Phytochem Anal, submitted (1994).
167. Schneider G, Schliemann W: The occurrence of gibberellin-O-glucosides in mature seeds of Gramineae and Leguminosae. XVth International Botany Congress,Yokohama, 1993, Abstract 4160 (1993).
168. Schneider G, Schliemann W: Conjugation of gibberellins: an overview. Plant Growth Regul, in press (1994).
169. Schneider G, Schliemann W, Schaller B, Jensen E: Identification of native gibberellin-O-glucosides in *Zea mays* L. and *Hordeum vulgare* L. In: Karssen CM, van Loon LC, Vreugdenhil D (eds) Progress in Plant Growth Regulation, pp. 566–570. Kluwer Academic Publishers, Dordrecht (1992).
170. Schneider G, Schreiber K, Jensen E, Phinney BO: Synthesis of gibberellin A_{29}-β-D-glucosides and β-D-glucosyl derivatives of [17-^{13}C, T2]gibberellin A_5, A_{20}, and A_{29}. Liebigs Ann Chem 1990: 491–494 (1990).
171. Schneider G, Sembdner G, Jensen E, Bernhard U, Wagenbreth D: GC-MS identification of native gibberellin-O-glucosides in pea seeds. J Plant Growth Regul 11: 15–18 (1992).
172. Schneider G, Sembdner G, Schreiber K: Synthese von O(3)- und O(13)-glucosylierten Gibberellinen. Tetrahedron 33: 1391–1397 (1977).
173. Schneider G, Sembdner G, Schreiber K, Phinney BO: Partial synthesis of some physiologically relevant gibberellin glucosyl conjugates. Tetrahedron 45: 1355–1364 (1989).
174. Schreiber K, Schneider G, Sembdner G, Focke I: Isolierung von O(2)-Acetyl-Gibberellinsäure als Stoffwechselprodukt von *Fusarium moniliforme* Sheld. Phytochemistry 5: 1221–1225 (1966).
175. Schreiber K, Weiland J, Sembdner G: Isolierung und Struktur eines Gibberellinglucosides. Tetrahedron Lett 1967: 4285–4288 (1967).
176. Schreiber K, Weiland J, Sembdner G: Isolierung von Gibberellin-A_8-O(3)-β-D-glucopyranosid aus Früchten von *Phaseolus coccineus*. Phytochemistry 9: 189–198 (1970).
177. Scott IM, Martin GC, Horgan R, Heald JK: Mass spectrometric measurement of zeatin glucoside levels in *Vinca rosea* L. crown gall tissue. Planta 154: 273–276 (1982).
178. Sembdner G: Conjugates of plant hormones. In: Schreiber K, Schütte HR, Sembdner G (eds) Biochemistry and Chemistry of Plant Growth Regulators, pp. 283–302. Institute for Plant Biochemistry Academy of Sciences of the GDR, Halle (1974).
179. Sembdner G, Groß D, Liebisch H-W, Schneider G: Biosynthesis and metabolism of plant hormones. In: MacMillan J (ed) Encyclopedia of Plant Physiology, New Series, vol 9, pp. 281–444. Springer-Verlag, Berlin/Heidelberg/New York (1980).
180. Sembdner G, Knöfel H-D, Schwarzkopf E, Liebisch HW: In vitro glucosylation of gibberellins. Biol Plant 27: 231–236 (1985).
181. Sembdner G, Parthier B: The biochemistry and the physiology and molecular actions of jasmonates. Annu Rev Plant Physiol Mol Biol 44: 569–589 (1993).
182. Sembdner G, Schliemann W, Schneider G: Biochemical and physiological aspects of gibberellin conjugation. In: Takahashi N, Phinney BO, MacMillan J (eds) Gibberellins, pp. 249–263, Springer-Verlag, New York (1991).
183. Sembdner G, Schneider G: Gibberellin conjugation: a physiologically relevant process in hormone metabolism of plants. In: Kutacek M, Elliott MC, Machackova I (eds) Molecular Aspects of Hormonal Regulation of Plant Development, Proceedings 14th Biochemical Congress Prague 1988, pp. 151–173, SPB Academic Publishers, The Hague (1990).
184. Sembdner G, Weiland J, Aurich O, Schreiber K: Isolation, structure and metabolism of a gibberellin glucoside. In: Plant Growth Regulators, pp. 70–86. SCI Monograph 31, London (1968).
185. Sharkey TD, Raschke K: Effects of phaseic acid and dihydrophaseic acid on stomata and the photosynthetic apparatus. Plant Physiol 65: 291–297 (1980).
186. Sitbon F, Edlund A, Gardestrom P, Olsson O, Sandberg G: Compartmentation of indole-3-acetic acid metabolism in protoplasts isolated from leaves of wild-type and IAA-overproducing transgenic tobacco plants. Planta 191: 274–279 (1993).
187. Sitbon F, Hennion S, Sundberg B, Little CHA, Olsson O, Sandberg G: Transgenic tobacco plants coexpressing the *Agrobacterium tumefaciens iaaM* and *iaaH* genes display altered growth and indoleacetic acid metabolism. Plant Physiol 99: 1062–1069 (1992).
188. Sitbon F, Östin A, Olsson O, Sandberg G: Conjugation of indole-3-acetic acid (IAA) in wild-type and IAA-overproducing transgenic tobacco plants, and identifi-

cation of the main conjugates by frit-fast atom bombardment liquid chromatography-mass spectrometry. Plant Physiol 101: 313–320 (1993).
189. Sitbon F, Sundberg B, Olsson O, Sandberg G: Free and conjugated indoleacetic acid (IAA) contents in transgenic tobacco plants expressing the *iaaM* and *iaaH* IAA biosynthesis genes from *Agrobacterium tumefaciens*. Plant Physiol 95: 480–485 (1991).
190. Stoddart JL, Venis MA: Molecular and subcellular aspects of hormone action. In: MacMillan J (ed) Encyclopedia of Plant Physiology, New Series, vol 9, pp. 445–510. Springer-Verlag, Berlin/Heidelberg/New York (1980).
191. Summons RE, Entsch B, Parker CW, Letham DS: Mass spectrometric analysis of cytokinins in plant tissues. III. Quantitation of the cytokinin glucoside complex of lupin pods by saturable isotope dilution. FEBS Letta 107: 21–25 (1979).
192. Summons RE, Letham DS, Gollnow Bl, Parker CW, Entsch B, Johnson LP, McLeod JK, Rolfe BG: Cytokinin translocation and metabolism in species of the Leguminosae: studies in relation to shoot and nodule development. In: Guern J, Peaud-Lenoel C (eds) Metabolism and Activity of Cytokinins, pp. 69–80, Springer-Verlag, Berlin (1981).
193. Summons RE, McLeod JK, Parker CW, Letham DS: The occurrence of raphanatin as an endogenous cytokinin in radish seed: identification and quantitation by GC-MS using deuterium internal standards. FEBS Lett 82: 211–214 (1977).
194. Sundberg B, Sandberg G, Jensen E: Identification and quantification of indole-3-methanol in etiolated seedlings of Scots pine (*Pinus silvestris* L.). Plant Physiol 77: 952–955 (1985).
195. Suzuki H, Kim SK, Takahashi N, Yokota T: Metabolism of castasterone and brassinolide in mung bean explant. Phytochemistry 33: 1361–1367 (1993).
196. Tinelli ET, Sondheimer E, Walton DC: Metabolites of 2-^{14}C-abscisic acid. Tetrahedron Lett 2: 139–140 (1973).
197. Tsurumi S, Asahi Y: Identification of jasmonic acid in *Mimosa pudica* and its inhibitory efect on auxin- and light-induced opening of the pulvinules. Physiol Plant 64: 207–211 (1985).
198. Tsurumi S, Wada S: Metabolism of indole-3-acetic acid and natural occurrence of dioxindole-3-acetic acid derivatives in *Vicia* roots. Plant Cell Physiol 21: 1515–1525 (1980).
199. Tsurumi S, Wada S: Identification of 3-(O-β-glucosyl)-2-indolone- 3-acetylaspartic acid as a new indole-3-acetic acid metabolite in *Vicia* seedlings. Plant Physiol 79: 667–671 (1985).
200. Tsurumi S, Wada S: Identification of 3-hydroxy-2-indolone-3-acetylaspartic acid as a new indole-3-acetic acid metabolite in *Vicia* roots. Plant Cell Physiol 27: 559–562 (1986).
201. Tsurumi S, Wada S: Oxidation of indole-3-acetylaspartic acid in *Vicia*. In: Pharis RP, Rood SB (eds) Plant Growth Substances 1988, pp. 353–359. Springer-Verlag, Berlin/ Heidelberg/New York (1990).
202. Turnbull CGN, Crozier A: Metabolism of [1,2-^{3}H]gibberellin A_4 by epicotyls and cell-free preparations from *Phaseolus coccineus* L. seedlings. Planta 178: 267–274 (1989).
203. Ueda M, Bandurski RS: Structure of indole-3-acetic acid myoinositol esters and pentamethyl-myoinositols. Phytochemistry 13: 243–253 (1974).
204. Ueda M, Ehmann A, Bandurski RS: Gas-liquid chromatographic analysis of indole-3-acetic acid myoinositol esters in maize kernels. Plant Physiol 46: 715–719 (1970).
205. Ueda J, Kato J: Promotive effect of methyl jasmonate on oat leaf senescence in the light. Z Pflanzenphysiol 103: 357–359 (1981).
206. Walton DC: Biochemistry and physiology of abscisic acid. Annu Rev Plant Physiol 31: 453–489 (1980).
207. Walton DC: Structure-activity relationships of abscisic acid analogs and metabolites. In: Addicott FT (ed) Abscisic Acid, pp. 113–146, Praeger Scientific, New York (1983).
208. Weiler EW: Octadecanoid-derived signalling molecules involved in touch perception in a higher plant. Bot Acta 106: 2–4 (1993).
209. Xia Z-Q, Zenk MH: A new metabolite of the plant growth regulator jasmonic acid. Poster, 18th IUPAC Symposium Chemical and Natural Products, Strassbourg (1992).
210. Yamaguchi I, Kobayashi M, Takahashi N: Isolation and characterization of glucosyl esters of gibberellin A_5 and A_{44} from immature seeds of *Pharbitis purpurea*. Agric Biol Chem 44: 1975–1977 (1980).
211. Yamaguchi I, Yokei M, Nishizawa M, Yang YY, Chinio M, Murofushi N: Immunological technique in the research of gibberellins. XVth International Botany Congress, Yokohama, 1993, Abstract 4.3.1.3 (1993).
212. Yamaguchi I, Yokota T, Yoshida S, Takahashi N: High pressure liquid chromatography of conjugated gibberellins. Phytochemistry 18: 1699–1702 (1979).
213. Yamane H: Antheridiogens and gibberellins in Schizaeaceous ferns. XVth International Botany Congress, Yokohama, 1993, Abstract 4.3.2.3 (1993).
214. Yamane H, Sato Y, Nohara K, Nakayama M, Murofushi N, Takahashi N, Takeno K, Furuya M, Furber M, Mander LN: The methyl ester of a new gibberellin, GA_{73}: the principal antheridiogen in *Lygodium japonicum*. Tetrahedron Lett 29: 3959–3962 (1988).
215. Yamane H, Takahashi N, Takeno K, Furuya M: Identification of gibberellin A_9 methyl ester as a natural substance regulating formation of reproductive organs in *Lygodium japonicum*. Planta 147: 251–256 (1979).
216. Yamane H, Yamaguchi I, Murofushi N, Takahashi N: Isolation and structure of gibberellin A_{35} and its gluco-

side from immature seed of *Cytisus scoparius*. Agric Biol Chem 35: 1144–1146 (1971).

217. Yamane H, Yamaguchi I, Murofushi N, Takahashi N: Isolation and structures of gibberellin A_{35} and its glucoside from immature seed of *Cytisus scoparius*. Agric Biol Chem 38: 649–655 (1974).
218. Yokota T, Kim SK, Kosaka Y, Ogino Y, Takahashi N: Conjugation of brassinosteroids. In: Schreiber K, Schütte HR, Sembdner G (eds) Conjugated Plant Hormones: Structure, Metabolism and Function, pp. 288–296. VEB Deutscher Verlag der Wissenschaften, Berlin (1987).
219. Yokota T, Kim SK, Ogino Y, Takahashi N: Various brassinisteroids from *Phaseolus vulgaris* seeds: structure and biological activity. Proc 14th Annu Plant Growth Regulator Soc America Meeting, Honolulu, pp. 28–29 (1987).
220. Yokota T, Kobayashi S, Yamane H, Takahashi N: Isolation of a novel gibberellin glucoside, 3-O-β-D-glucopyranosyl gibberellin A_1 from *Dolichos lablab* seed. Agric Biol Chem 42: 1811–1812 (1978).
221. Yokota T, Murofushi N, Takahashi N: Structure of new gibberellin glucoside in immature seeds of *Pharbitis nil*. Tetahedron Lett 1970: 1489–1491 (1970).
222. Yokota T, Murofushi N, Takahashi N, Katsumi M: Biological activities of gibberellins and their glucosides in *Pharbitis nil*. Phytochemistry 10: 2943–2949 (1971).
223. Yokota T, Murofushi N, Takahashi N, Tamura S: Gibberellins in immature seeds of *Pharbitis nil*. III. Isolation and structures of gibberellin glucosides. Agric Biol Chem 35: 583–595 (1971).
224. Yokota T, Ogino Y, Suzuki H, Takahashi N, Saimoto H, Fujioka S, Sakurai A: Metabolism and biosynthesis of brassinosteroids. In: Cutler HC, Yokota T, Adam G (eds) Brassinosteroids: Chemistry, Bioactivity and Applications, pp. 86–96. ACS Symposium Ser 474, American Chemical Society, Washington (1991).
225. Yokota T, Takahashi N, Murofushi N, Tamura S: Isolation of gibberellins A_{26} and A_{27} and their glucosides from immature seeds of *Pharbitis nil*. Planta 87: 180–184 (1969).
226. Yokota T, Takahashi N, Murofushi N, Tamura S: Structures of new gibberellin glucosides in immature seeds of *Pharbitis nil*. Tetrahedron Lett 1969: 2081–2084 (1969).
227. Yokota T, Yamazaki S, Takahashi N, Iitaka Y: Structure of pharbitic acid, a gibberellin-related diterpenoid. Tetrahedron Lett 1974: 2957–2960 (1974).
228. Yoshihara T, Omer EA, Koshino H, Sakamura S, Kikuta Y, Koda Y: Structure of a tuber-inducing stimulus from potato leaves (*Solanum tuberosum* L.). Agric Biol Chem 53: 2835–2837 (1989).
229. Zeevaart JAD, Creelman RA: Metabolism and physiology of abscisic acid. Annu Rev Plant Physiol Plant Mol Biol 39: 439–473 (1988).
230. Zeevaart JAD, Gage DA, Creelman RA: Recent studies of the metabolism of abscisic acid. In: Pharis RP, Rood SB (eds) Plant Growth Substances 1988, pp. 233–240, Springer-Verlag, Berlin/Heidelberg/New York (1990).
231. Zeevaart JAD, Heath TG, Gage DA: Evidence for a universal pathway of abscisic acid biosynthesis in higher plants from ^{18}O incorporation patterns. Plant Physiol 91: 1594–1601 (1989).
232. Zeevaart JAD, Rock CD, Fantauzzo F, Heath TG, Gage DA: Metabolism of ABA and its physiological implications. In: Davies WJ, Jones HG (eds) Abscisic Acid: Physiology and Biochemistry, pp. 39–52. Bios Scientific Publications, Oxford (1991).
233. Zenk MH: 1-(Indole-3-acetyl)-β-D-glucose, a new compound in the metabolism of indole-3-acetic acid in plants. Nature 191: 493–494 (1961).

Plant Molecular Biology **26**: 1483–1497, 1994.

Cytokinin metabolism: implications for regulation of plant growth and development

Břetislav Brzobohatý [1,2,*], Ian Moore [2,3] and Klaus Palme [2]
[1] *Institute for Biophysics, AS CR, Královopolská 135, CZ-61265 Brno, Czech Republic (* author for correspondence);* [2] *Max-Planck-Institut für Züchtungsforschung, Carl-von-Linné Weg 10, D-50829 Köln, Germany;* [3] *Department of Plant Sciences, University of Oxford, South Parks Rd., Oxford, OX1 3RB, UK*

Received and accepted 11 October 1994

Key words: plant hormone, cytokinin, metabolism, conjugation, β-glucosidase, transgenic plants

Introduction

This review describes recent advances in the study of cytokinin metabolism. It highlights how plant development is influenced by cytokinin synthesis, conjugation and conjugate hydrolysis, and what has been learned of the enzymes that regulate these processes. Although cytokinin metabolism and physiology are complex issues, some of the key enzymatic players are now being identified. This holds out the prospect of rapid progress in the near future. Just as much of what we know about the control of animal cell proliferation was learned by studying the cellular counterparts of viral oncogenes, so important information about the control of plant development by phytohormones has come from studying the genes of bacterial pathogens that subvert host phytohormone metabolism to their own advantage. We will focus on what has been learned from the use of such genes, and describe progress in identifying their functional counterparts in plants.

Cytokinins

Plant growth and differentiation relies on two fundamental cell activities – cell elongation and division. As early as in the end of the past century the concept of two separate and specific factors controlling cell enlargement and division was formulated. The first experimental evidence of chemical control of cell division was provided by Haberlandt in 1913 [32]. He demonstrated that phloem diffusates could stimulate parenchymatous potato tuber cells to convert to mitotic cells. More than 40 years later, Skoog and co-workers were able to show that while auxin induced cell enlargement in tobacco pith tissue cells, a simultaneous addition of the cytokinin kinetin (6-furfurylamino purine) was necessary to induce cell division [56, 57]. The first naturally occurring compound able to induce plant cell divisions was purified by Letham in 1963 and identified as 6-(4-hydroxy-3-methylbut-*trans*-2-enylamino) purine, commonly known as zeatin [46]. Since that time a number of compounds which stimulate plant cell division in combination with auxin have been isolated and named cytokinins (Fig. 1).

Skoog's classical experiments led not only to the discovery of the first cytokinin, but simultaneously demonstrated a general feature of phytohormone action, namely involvement of more than one phytohormone in promoting a biological response. Later, it was established that phytohormones can interact both synergistically and antagonistically (e.g. cytokinins act together with auxin to stimulate callus cell division but oppose auxin with regard to stimulation of lateral bud formation).

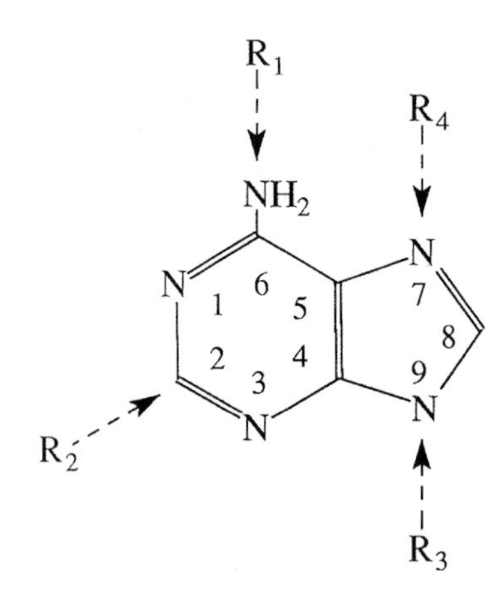

R_1	R_2	R_3	R_4	Trivial name
	H	H	-	$N^6(\Delta^2$-isopentenyl) adenine
	H	ribosyl	-	$N^6(\Delta^2$-isopentenyl) adenosin
	H	riboside	-	$N^6(\Delta^2$-isopentenyl) adenosine-5´-monophosphate
	H	-	glucosyl	$N^6(\Delta^2$-isopentenyl) adenine-7-glucoside
	H	H	-	*trans*-zeatin
	H	ribosyl	-	*t*-zeatin riboside
OH	H	glucosyl	-	*t*-zeatin-9-glucoside
	H	-	glucosyl	*t*-zeatin-7-glucoside
	H	alanyl	-	lupinic acid
	H	ribotide	-	*t*-zeatin riboside-5´-monophosphate
OG	H	H	-	zeatin-O-glucoside
	H	ribosyl	-	zeatin riboside-O-glucoside
	H	H	-	dihydrozeatin
	H	ribosyl	-	dihydrozeatin riboside
OH	H	glucosyl	-	dihydrozeatin-9-glucoside
	H	-	glucosyl	dihydrozeatin-7-glucoside
	H	alanyl	-	dihydrolupinic acid
	H	ribotide	-	dihydrozeatin riboside-5´-monophosphate
OG	H	H	-	dihydrozeatin-O-glucoside
	H	ribosyl	-	dihydrozeatin riboside-O-glucoside
	H	H	-	N^6(benzyl) adenine
	H	ribosyl	-	N^6(benzyl) adenosine
	H	glucosyl	-	N^6(benzyl) adenine-9-glucoside
	H	-	glucosyl	N^6(benzyl) adenine-7-glucoside

Fig. 1. Cytokinin structure and nomenclature. Glucosyl (G) and ribosyl groups refer to the *β-D*-glucopyranosyl and *β-D*-ribofuranosyl group.

Although cytokinins were first discovered by their ability to induce cell division, they are now known to act in combination with other phytohormones to regulate diverse responses in plants, including seed germination, *de novo* bud formation, release of buds from apical dominance, leaf expansion, reproductive development, and senescence.

Significant effort has been dedicated to identification of molecular mechanisms of cytokinin signal transduction. More than twenty years ago a search for cytokinin receptors was initiated [29]. However, despite the isolation of a number of cytokinin binding proteins from different plant sources [3, 23, 34, 43, 62, 65, 75, 77, 79, 80, 99, 100] no functional cytokinin receptor has been reported yet. The ability of cytokinins to stimulate transcription has been demonstrated in several experimental systems and cytokinin regulated promoters have been identified [1, 12, 15, 82, 98]. Although it is possible that upon cytokinin binding some soluble cytokinin binding proteins may regulate the activity of these promoters in a manner similar to animal steroid receptors, the involvement of receptor linked signalling cascades cannot be excluded. Membrane-bound receptor-like proteins have recently been cloned from plants and were shown to belong to the group of receptor kinases or the two-component signal receptors (see Theologis and Walker, this volume). However, the involvement of G-protein-linked receptors or other so far unknown receptor systems in phytohormone signalling cannot be excluded. An indication suggesting that phosphorylation is involved in cytokinin signalling pathway came from the demonstration that transcription of a cytokinin-inducible gene could be suppressed by a protein kinase inhibitor [17]. Further support for a role of protein phosphorylation in cytokinin response was provided recently by the demonstration that a wheat protein kinase homologue was transcriptionally regulated by cytokinins [81]. This finding may further support the notion of a link between protein phosphorylation and other players in cytokinin signal transduction, i.e. changes in calmodulin and calcium levels [37].

Understanding molecular mechanisms of cytokinin action will require both precise knowledge of cytokinin signal transduction pathways and an understanding of how active cytokinin levels are modulated. An intriguing feature of cytokinins, as well as other phytohormones, is the complexity of metabolic transformations to which they can be subjected in the plant cell. It is conceivable that metabolic modifications may be important in modulating cytokinin activity, compartmentation and transportability. Thus the level of active cytokinin at a particular site of action may be influenced by a large number of factors: *de novo* synthesis; oxidative degradation; formation and hydrolysis of inactive conjugates, transport into and out of particular cells; and subcellular compartmentalisation to or away from sites of action. On top of this, there is also the possibility that physiological responses may be modulated by variations in the ability of cells to respond to a particular concentration of free cytokinin. In this paper we will concentrate principally on *de novo* biosynthesis, degradation, conjugate formation, and conjugate hydrolysis, which will be dealt with in that order.

De novo *synthesis of cytokinins*

Cytokinin biosynthesis in developed plants takes place mainly in roots [19, 36, 94, 107], although smaller amounts can be synthesized by the shoot apex and some other plant tissues. Two biosynthetic pathways have been reported for cytokinins: the *de novo* biosynthetic pathway [11, 101] and the tRNA pathway [33, 88]. It is very likely that the *de novo* biosynthetic pathway accounts for the majority of newly synthesized cytokinin. The key step in *de novo* cytokinin biosynthesis is the formation of N^6-(Δ^2-isopentenyl) adenosine-5′-monophosphate from Δ^2-isopentenyl pyrophosphate and adenosine-5′-monophosphate catalysed by isopentenyltransferase (IPT). Although IPT activity has been detected in plants, attempts to purify the enzyme to homogeneity have not yet been successful [10, 76]. It may be that a gene-tagging approach such as that of Walden *et al.* (this volume) will provide a faster route to the identification of cytokinin biosynthetic enzymes. Recently, an *Arabidopsis* mutant, *amp1*, demonstrating various abnormalities reminiscent of cytokinin action, was shown to have a six-fold elevated cytokinin content. The elevated cytokinin content was proposed to be the primary effect of *amp1* mutation, and one of the possible roles for wild-type AMP1 is as a negative regulator of cytokinin biosynthesis [9].

Bacterial isopentenyltransferase

The soil phytophatogen *Agrobacterium tumefaciens* causes crown gall disease on several plant species by transforming the host with a bacterial DNA sequence (T-DNA) that leads to increased phytohormone biosynthesis by the transformed cells, which in turn promotes cell proliferation. Molecular analysis of the genes on the T-DNA allowed the identification of one (*tmr* or *ipt*) coding for an isopentenyltransferase. Several groups have used the gene to study the effects of enhanced cytokinin biosynthesis on plant growth and development. Tobacco calli overexpressing the *ipt* gene from a strong promoter (CaMV 35S) initiated shoots more frequently, rapidly and profusely than did calli containing the *ipt* gene expressed from its own promoter [91]. Similarly, the chimaeric CaMV 35S-*ipt* construct was able to induce shoot proliferation on undifferentiated cucumber tissues, whereas the *ipt* gene expressed from its own promoter was not. These changes correlated well with the measured increases in endogenous cytokinin level in these tissues [91], and mimiked the classic effects of cytokinin on plant tissue cultures described by Skoog and Miller [87].

Constitutive overproduction of cytokinin leads to almost complete suppression of root formation and, consequently, prevents plant regeneration. Controlled *ipt* expression in transgenic plants was first achieved by placing *ipt* under the control of heat-shock-inducible promoters (*hsp*) [54, 84, 89, 90]. Experiments with these transgenic plants led to a number of important conclusions about cytokinin action. Several morphological consequences of introducing an *hsp-ipt* gene into tobacco were observed. The transgenic plants displayed reduced height (to ca. 50% of the control). This reduction was due to a decrease in internode length rather than node number. Further, the transgenic plants also have smaller leaves, which show delayed senescence [89], a greatly reduced root system, a less developed vascular system, and reduced apical dominance.

Apical dominance has also been modified through the use of bacterial genes that alter free auxin levels. When the free IAA content was increased 10-fold by expression of the *A. tumefaciens* T-DNA auxin biosynthetic genes *iaaM* and *iaaH*, the transformed plants are almost completely apically dominant [42]. In addition, a gene encoding an IAA-lysine synthetase of *Pseudomonas savastanoi* has been used to reduce the free IAA content of transgenic plant material. When free IAA content was reduced by 10- to 20-fold, lateral buds are also released from apical dominance [78, 96]. Thus, it appeared that the apical dominance is not controlled by auxin or cytokinin alone, but by the ratio of the two growth factors. A high auxin/cytokinin ratio will result in the suppression of growth of lateral shoots, while a low ratio results in growth.

Interestingly, the morphological changes induced in plants containing the *hsp-ipt* construct were observed even in non-heat-treated plants. Although no *ipt* mRNA could be detected in such plants, the cytokinin levels were increased between twofold for zeatin riboside and up to sevenfold for zeatin riboside 5′-monophosphate [54]. Whilst a single heat treatment dramatically increased the abundance of *ipt* mRNA and the cytokinin content (63-fold for zeatin, 24-fold for zeatin riboside and 3-fold for zeatin riboside 5′-monophosphate), no further phenotypic alterations were observed [54]. Smart and co-workers [89] described similar differences in appearance between transformed and untransformed plants at the control temperature, the transformed plants being shorter and having larger axillary shoots than untransformed plants. In addition, after multiple heat shocks the *hsp-ipt* transformed plants exhibited an abundance of tiny shoots at the apex and a release of lateral buds, which was not found in the untransformed heat-shocked plant.

These studies parallel those of Kares and co-workers [41] who found that low to moderate increases in endogenous levels of IAA were sufficient to induce physiological responses *in planta*. This study used a heat-shock inducible system for increasing the endogenous IAA levels consisting of the auxin biosynthetic genes *iaaM* under the control of a heat shock promoter and *iaaH* transcribed from its own promoter. Interestingly, endogenous IAA in uninduced transformed plant material reached levels up to ten times higher than untransformed plants. This increase was able to promote root induction in three different plant systems. Heat-shock treatment caused a further increase in IAA levels in the range from 11–20 times compared to untransformed controls and increased dramatically the extent of root formation.

Hence, relatively minor changes in cytokinin and auxin levels are sufficient to cause the observed developmental alterations. Indeed, these are just two of several examples with transgenic plants where relatively small changes of 3- to 10-fold in measured hormone concentration (at least in whole tissue homogenates) are sufficient to significantly alter quantitative aspects of plant development [e.g. 26, 42, 96]. Trewavas [103] argued that a true understanding of the control exerted by plant hormones over a given process *in planta* can come only from studying the effects of small fluctuations in hormone concentration around endogenous levels – dramatic increases or reductions in hormone levels can show only that a hormone is in some way able to promote a particular process, or that it is necessary for that process, without revealing anything about the control exerted by the hormone over that process in non-experimental material. In view of disagreements in the past about the relevance of some responses that are elicited with broad dose-response curves by application of exogenous phytohormone, it is gratifying that much of our classical understanding of phytohormone action is supported by small fluctuations in measured phytohormone concentration generated from within.

One other point that emerges from these studies is the lack of homeotic, or qualitative developmental changes in transgenic plants in response to changes in phytohormone content of up to 80-fold. The changes observed in *hsp-ipt* plants for example are essentially quantitative in nature. Tissue culture experiments clearly indicate the potential of phytohormones provided in the medium to greatly alter the developmental fate of cells [87]. Results obtained with exogenous cytokinin application have been extended by increasing endogenous cytokinin levels with help of *ipt* gene in transgenic calli. As low as a six-fold increase in endogenous cytokinin level (compared to a leaf cytokinin content) in *N. plumbaginifolia* led to induction of shoot formation. An approximately threefold increase in endogenous cytokinin level has been associated with callus formation upon transformation in cucumber hypocotyles, whereas a 172-fold increase in endogenous cytokinins (compared to normal hypocotyl levels) was necessary to induce shoots on cucumber calli [91]. These levels of endogenous cytokinins are in the range achieved in *hsp-ipt* plants or only slightly exceed them. Thus, these results indicate the importance of the tissue in determining the nature of the response to a particular hormone content.

This type of morphogenic response has only been observed once in regenerated transgenic plants. In this case plants were regenerated which contained the *ipt* gene separated from the 35S CaMV promoter by the transposable element Ac ('*35S-Ac-ipt*' [27]). As a consequence of transposition in certain cells of the transgenic plants, the *ipt* gene was activated, and expressed to high levels producing sectors of high cytokinin concentration. This led to vivipary, that is, the formation of adventitious shoots derived from vascular parenchymal cells in the leaf. Thus, endogenous cytokinin can alter the developmental fate of certain cells in *ex vitro* cultured plants. However, adventitious shoots were observed to emerge only from the vascular parenchyma, whilst the activation of *ipt* gene by transposition occurs in numerous cell types, so the response of cells to cytokinin appears to be determined by cell type as much as by cytokinin concentration.

The local cytokinin concentrations induced by transposition in these *35S-Ac-ipt* plants were much higher than in the non-heat-treated *hsp-ipt* which may account for the different morphogenic responses in these two systems. However, the highest cytokinin levels were detected in the aberrant shoots of *35S-Ac-ipt* plants, and, surprisingly, they were not much higher than in heat-shocked leaves containing the *hsp-ipt* gene fusion. As several of the adventitious shoots emerged from sites on a leaf quite distant from the site of transposition (and consequently of cytokinin biosynthesis) it may be the gradient of cytokinin which is important, or as suggested by Estruch *et al.* [27], transport to the vascular parenchyma cells may cause a local accumulation of cytokinin to high levels. Later in development, *35S-Ac-ipt* plants formed epiphylous flower buds on leaf tips [28]. Some of these buds were morphologically normal while a fraction of them showed abnormalities such as organ transitions and fusion between organs located in different whorls. Interestingly, free cytokinin levels in the normal epiphyllous buds were only moderately (not more than ten times) higher compared to normal leaves or flower buds. In contrast, 100–1000-fold higher cytokinin levels were found in the abnormal epiphyllous flower buds. The increase in cytokinin levels was accompanied by a strongly reduced expression of tobacco homologues of floral homeotic genes *DEFA*, *GLO* and *PLENA* of *Antirrhinum majus* [28].

Thus, despite the lack of a well characterized plant isopentenyltransferase, the availability of a bacterial *ipt* has allowed some analysis of the consequences of deregulated cytokinin synthesis in plants. The results obtained in experiments involving small (5–10-fold) alterations in cytokinin content have confirmed much of the classical view of cytokinin action in plants, while the significance of developmental alterations associated with considerably larger changes in cytokinin content remains to be properly clarified.

Cytokinin deactivation

Attenuation of hormone-induced cellular responses requires an efficient way of hormone inactivation. Active cytokinins can be inactivated by degradation or conjugation to different low-molecular-weight metabolites, such as sugars and amino acids.

Degradation

At present the only plant enzymes known to catalyse the degradation of cytokinins to inactive products that lack the N^6-side-chain are cytokinin oxidases. An enzyme activity converting N^6-(Δ^2-isopentenyl) adenosine to adenosine was demonstrated in a cell-free system for the first time by Paces *et al.* [66]. Whitty and Hall [111] were the first to use the term 'cytokinin oxidase' for a similar enzyme activity detected in maize kernels. Cytokinin oxidase activities have now been isolated from a number of higher-plant sources [5, 8, 40, 44, 51, 63, 66, 67, 111]. N^6-(Δ^2-isopentenyl) adenosine is the preferred substrate for all cytokinin oxidase activities studied. Zeatin and zeatin riboside were also substrates in all cases studied. However, enzymes from certain sources exhibit clear preference for N^6-(Δ^2-isopentenyl) adenosine compared to zeatin [8, 40, 63, 67]. The presence and position of the double bond in the isoprenoid side-chain appears to be important to substrate activity. Thus, dihydrozeatin is resistant to cytokinin oxidase action *in vitro* [8, 40, 51, 63, 111] and appears to be a major cytokinin in tissues with a high cytokinin oxidase activity [64]. Similarly, *O*-glucosylation of the side chain protects cytokinins from the *in vitro* attack of cytokinin oxidase [51]. On the other hand, *N*-glucosylation or other ring substitutions may decrease the affinity of cytokinin oxidase for a given cytokinin, but do not necessarily eliminate substrate activity [51]. Cytokinin oxidase cannot accept cytokinin nucleotides as substrates [44]. Plant cytokinin oxidases exert very low or no activity against N^6-benzyladenin and kinetin *in vitro* [8, 40, 44, 51, 63, 111] and these cytokinins are degraded only very slowly *in vivo* [18, 108].

Cytokinin oxidase activity and cytokinin degradation can be regulated on several levels. Tissue culture experiments have demonstrated that cytokinin oxidase activity can be induced by transient increases in the cytokinin supply available in the culture medium [8, 40, 63, 69, 102]. A possible role of cytokinin oxidase compartmentalization in modulation of cytokinin degradation by *P. vulgaris* and *P. lunatus* calli has been suggested by Kaminek and Armstrong [40]. *Phaseolus vulgaris* calli can degrade exogenously supplied cytokinins much faster than *P. lunatus* calli. The cytokinin oxidase isolated from *P. lunatus* was apparently not glycosylated and exhibited a pH optimum of 8.4, features typical of intracellular enzymes. In contrast, the cytokinin oxidase from *P. vulgaris* was glycosylated and exhibited a pH optimum of 6.5, features often detected in secreted proteins. Thus, the faster degradation of cytokinins in tissue culture media by *P. vulgaris* calli could be explained if the *P. vulgaris* cytokinin oxidase is secreted into the cell exterior and consequently has a better access to exogenously supplied cytokinins than the *P. lunatus* cytokinin oxidase sequestered in an internal cell compartment. Cytokinin degradation *in vivo* can be influenced by other phytohormones. Thus, auxins stimulated cytokinin degradation in several experimental systems [35, 53, 70, 74, 112, 113], although not necessarily by increasing the level of cytokinin oxidase protein [63]. On the other hand, abscisic acid caused suppression of kinetin degradation in lettuce seeds [55] and the conversion of zeatin to dihydrozeatin in *P. vulgaris* [93]. Limited attention has been paid to the role of cytokinin degradation in the regulation of plant growth and development. However, from the data available it appears that cytokinin degradation varies in different tissues and in the same tissue during its development [39, 86, 104, 111] suggesting that cytokinin oxidase may have a significant role in the control of cytokinin levels during plant development.

Significant progress in understanding the role of cytokinin oxidase in the regulation of plant development can be expected from the application of molecular techniques. Therefore, the recent success in obtaining antibodies against maize cytokinin oxidase and their use in the isolation of a cDNA clone [6] are of particular importance.

Conjugation

A common feature of plant hormones is that they frequently occur in the cell as conjugates to various sugars and amino acids. These conjugates usually possess low intrinsic physiological activity, and what activity they do show correlates with their rate of hydrolysis in plant tissues. Conjugation is often seen as a way of removing free, active hormones from a tissue, however the conjugation process is often reversible, and, as conjugates can frequently accumulate in great excess over free forms of phytohormone, the conjugate pools must also be considered as sources of free hormone and may represent storage or inactive transportable forms of the hormone. This section considers the biochemistry and physiology of conjugate formation after which we discuss the physiological role of conjugate formation and hydrolysis.

Cytokinins occur frequently as *N*-glucosides and *O*-glucosides. They can be *N*-glucosylated on the purine ring and *O*-glucosylated on the N^6-substituted side-chain. *N*-glucosylation at the 3, 7, or 9 position of zeatin and dihydrozeatin have been identified in tobacco, radish, and lupin [14, 31, 52]. The radish cytokinin-7-glucosyltransferase was studied in more detail [20, 21]. The enzyme catalysed the formation of both 7- and 9-glucosides, but favoured the 7 position. Both UDP-glucose and TDP-glucose served as glucosyl donor. The enzyme can use *trans*-zeatin (K_m 0.15 mM) as well as other natural and synthetic cytokinins as substrates.

N-glucosides of cytokinins are biologically inactive and very stable *in vivo* where there do not appear to be hydrolases that can release the free phytohormone from the conjugate. Thus it is suspected that *N*-glucosylation may be important in regulating levels of active cytokinin through inactivation [47]. Indeed, increased cytokinin production upon heat shock treatment in *hsp-ipt* plants (see above) was accompanied by accumulation of zeatin-N^7-β-glucoside at high levels [54]. Similarly, after external cytokinin application a significant portion of the internalized cytokinin was converted to *N*-glucosides [52, 71, 72, 73]. Nevertheless, the precise function of these compounds in normal plant development is not known, but it has been proposed that they play a role in *Agrobacterium rhizogenes* pathogenicity (see below).

Cytokinins such as zeatin and dihydrozeatin with hydroxylated N^6-side-chain are often glucosylated to form the *O*-glucosides. Cytokinin-*O*-glucosides have been identified in many plant species [see 47]. The enzyme mediating *O*-glucosylation, *O*-glucosyltransferase, was isolated from *P. lunatus* [16]. The *O*-glucosyltransferase has a distinct substrate specificity. The enzyme recognizes *trans*-zeatin, but not dihydrozeatin, *cis*-zeatin, or ribosylzeatin. Both UDP-glucose and UDP-xylose serve as glycosyl donors for the enzyme *in vitro*, although UDP-glucose was the favoured substrate. Immunologically related proteins can be detected in both dicots and monocots [59, 60]. It is possible that the *amp1* mutant of *Arabidopsis thaliana* described above is defective in some aspect of cytokinin glycosylation, leading to increased accumulation of the free forms [9].

The function of *O*-glucosyl derivatives of zeatin is not clear. However, their accumulation when cytokinin accumulates in tissues and their decrease during phases of active growth have been taken as an evidence in favour of a storage role [47].

Other types of conjugates

While cytokinin glucosides appear to be ubiquitous cytokinin conjugates, cytokinin-*O*-xylosides seem to be specific for *Phaseolus vulgaris* [45, 105]. Interestingly, although *O*-xylosylzeatin has been found to be even more active than zeatin in some biossays, it was converted to zeatin only slowly in the same tissue. Therefore, it is possible that *O*-xylosylzeatin represents a conjugate with intrinsic cytokinin activity [58]. The *O*-xylosyltransferase has been purified and characterized [105]. The enzyme utilizes *trans*-zeatin as the preferred substrate and UDP-xylose as the donor of the xylosyl moiety. The enzyme also accepts dihydrozeatin as a substrate, but not *cis*-zeatin or ribosylzeatin. The enzyme is specifically expressed in endosperm [49, 50, 60]. Interestingly, the zeatin-*O*-xylosyltransferase was found to be associated with both the nucleus and cytoplasm [50]. The biological significance of this finding is unclear, but the authors suggested possible role(s) for the enzyme in transport or targeting of cytokinins or cytokinin-related molecules between the nucleus and cytoplasm [50]. Interestingly, cytokinins have recently been localized to both nucleus and cytoplasm in developing somatic embryos of *Dactylis glomerata* L. [38]. These findings will certainly increase interest in studying the role of subcellular compartmentalization in phytohormone action.

In addition to glycosylated forms of cytokinin, conjugates of cytokinins with amino acids have been identified. Alanine can be conjugated to the nitrogen 9 of the cytokinin purine ring. An enzyme catalysing this reaction, β-(9-cytokinin) alanine synthase, was purified from immature *Lupinus luteus* seeds. The product derived from zeatin was identified as lupinic acid [22]. As with *N*-glucosylation of cytokinins, alanyl conjugation probably represents a process of cytokinin inactivation since the alanine conjugates are also biologically inactive and stable.

Conjugate hydrolysis – bacterial enzymes

RolC

The study of phytohormone conjugates was given a boost by the discovery a few years ago that the plant pathogenic soil bacterium *Agrobacterium rhizogenes* may exploit conjugate pools during pathogenesis. *A. rhizogenes* induces the hairy root disease in infected plants, resulting in neoplastic root proliferation. These neoplasias result from activities of the *rol* genes (mainly *rolA*, *B* and *C*) which are transferred to the plant genome by the T-DNA of *A. rhizogenes* Ri plasmids. In an attempt to decipher the role of individual *rol* genes and to get an insight into the molecular basis of the hairy root disease, individual *rol* genes have been transformed into various plant species. Distinct pleiotropic morphological changes in transgenic plants expressing each of the genes demonstrated that the products of each *rol* gene can interfere with normal plant developmental processes independently and in distinct ways.

Recently, the gene product of *rolC* has been shown to release cytokinins from their *N*-glucosides [26]. Transgenic tobacco plants expressing the *rolC* gene from its own promoter resulted in reduced apical dominance and internodal distance, altered leaf morphology, small flowers and reduced fertility [83]. More severe phenotypes were observed in transgenic plants expressing the *rolC* gene under the control of the strong 35S CaMV promoter. These plants were very small, the apical dominance and internode length were drastically reduced, leaves were pale green and lanceolate, and flowers were male-sterile [30, 83]. An interesting aspect of *rolC* gene action is its cell-autonomy both in root induction on leaf discs [83] and in the reduction of leaf pigment content [95]. This observation has been interpreted as evidence that the *rolC* biological effect is not due to the *rolC*-mediated synthesis of a growth factor that is transported in leaf tissue [95].

The glycoconjugate hydrolysing activity of ROLC protein may cause increased levels of cytokinins in *rolC* transgenics [26, 85]. However, only some of the morphological and physiological changes associated with the *rolC* phenotype, namely reduction of apical dominance and height, and enhanced tuber formation in potatoes, can be explained by simply postulating increased cytokinin levels. Stimulation of root initiating and growth, and reduction in chlorophyll content which are typical of *rolC* transgenic plants are not observed either upon application of exogenous cytokinin or in plants expressing the *ipt* gene. In addition, the cell-autonomous behaviour of *rolC* contrasts with the ability of cytokinins generated by *IPT* to act at a distance [27]. It may be that some process such as intra- and intercellular transport of cytokinin is altered when their glycosylation is disturbed, leading to accumulation of free cytokinin within the cells. However, an unambiguous interpretation of data from such transgenics can be complicated by the fact that free hormone levels are influenced by numerous input and output pathways and these can be altered during tissue development. Only when these are better understood will we be able to interpret effects such as those of *rolC* with confidence.

Conjugate hydrolysis – plant enzymes

The stability of cytokinin-*O*-glucosides in different tissues varies markedly. While they accumulate to relatively high levels in tissues such as endosperm many plant tissues contain glycosidase activities that can release active cytokinin. The ability of plant tissues to hydrolyse the cytokinin-*O*-glucosidic bond contrasts with their apparent inability to hydrolyse cytokinin-*N*-glucosides to free cytokinin. Consequently, cytokinin-*O*-glucosides are frequently active in bioassays, and they have been suggested to be a storage or transport form of cytokinins. The biological activity exhibited by cytokinin-*O*-glucosides is associated with the release of free cytokinins, so cytokinin-*O*-glucosides are considered to be intrinsically inactive forms of cytokinins which can be easily converted to free active cytokinins by action of β-glucosidases. Cytokinin-*O*-glucoside levels have been observed to decrease rapidly during some phases of plant development, for example during germination in maize seeds [92], lateral bud development in bean plants [68], and the breaking of dormancy and apical bud growth in potato tubers [106]. Regulated release of free cytokinins from their glucosides may require the action of specific β-glucosidases. Although glycosidase activities were initially thought to have broad substrate specificities, it has appeared that when single glycosidase species are investigated they often show considerable selectivity for the sugar and aglycone portion of their substrates [13].

Recently, a novel 60 kDa β-glucosidase (p60) has been identified by photoaffinity labelling with azido-IAA in maize coleoptiles [7]. The purified enzyme exhibited activity towards general β-glucosidase substrates (e.g. *p*-nitrophenyl-β-*D*-glucopyranoside or 6-bromo-2-naphthyl-β-*D*-glucopyranoside), and the activity of p60 on various auxin and cytokinin conjugates was investigated. p60 was able to hydrolyse cytokinin-*O*- and -*N*3-glucosides but not cytokinin-*N*7- or -*N*9-glucosides. It also hydrolysed indoxyl-glucoside but not the auxin conjugates IAA-aspartate, IAA-glycine or IAA-myo-inositol, or IAA glucose ester, consistent with the specificity of the enzyme for the glucosidic bond. It had no or very low activity on several other naturally occurring glucosides (e.g. salicin, cellobiose and laminiaribose) which were cleaved efficiently by a broad-specificity β-glucosidase from *Caldocellum saccharolyticum* [7]. Interestingly, expression of this enzyme is tightly regulated during the maize life cycle with the highest levels being found in the developing seedling. One possible role for a β-glucosidase specific for this developmental stage might be to release free phytohormones from their conjugates which are transported from the endosperm to the growing regions.

The activity of p60 *in vitro* on cytokinin-glucosides indicated that it has the potential to modulate the intracellular concentration of active cytokinin. However, as mentioned previously, the kinetics of cytokinin accumulation are likely to be highly complex, and assays of activity *in vitro* do not clarify what impact a particular enzyme will have on the equilibria that control the metabolism and transport of cytokinin in the highly compartmentalized environment of the cell. The physiological relevance of p60 in cytokinin metabolism has been investigated in a biological assay. Firstly, amino acid sequence analysis of p60 allowed isolation of a cDNA, *Zm-p60.1*, coding for a protein closely related to p60 and with the same enzymatic properties. It was found that tobacco protoplasts that expressed Zm-p60.1 either transiently or constitutively could use inactive cytokinin glucosides such as zeatin-*O*-glucoside and kinetin-*N*3-glucoside to initiate cell division in the absence of any other exogenous cytokinin source [4].

In the roots of maize seedlings *Zm-p60.1* was found to be located in the meristematic cells. It was suggested that in young maize root meristems *Zm-p60.1* liberates free cytokinins from the exogenous supply of cytokinin-*O*-glucosides arriving from the endosperm and consequently helps to maintain meristem activity [4]. Thus, *Zm-p60.1* may be one of the key enzymes involved in regulation of equilibria between free phytohormones and their conjugates, and a valuable tool to study and manipulate these equilibria.

It is noteworthy that the endogenous cytokinin glucosidase, Zm-p60, and the bacterial enzyme, ROLC, have different substrate specificities, each of which appears tailored to its biological role. Thus Zm-p60 hydrolyses *O*-glucosides which are thought to be transient storage forms of cytokinin, but it does not hydrolyse N7- or N9-glucosides which appear to be terminally inactivated forms of cytokinin. The ROLC protein, however, is able to hydrolyse these N7- and N9-glucosides in addition to *O*-glucosides. Consequently, during pathogenesis, the bacterium is able to exploit the normally inactive N7- and N9-glucoside pools (which can accumulate to considerable levels in tissues of certain species [47]) to increase cytokinin levels, and by reversing *de novo* glycosylation on N7 and N9, it simultaneously antagonizes one of the pathways available to plant cells to compensate for the increased accumulation of free cytokinin.

Interestingly, IAA glucose ester was found to be a particularly efficient inhibitor of cytokinin glucoside hydrolysis by Zm-p60.1. This may explain why this protein was labelled by IAA azido derivative. The inhibitory effect of IAA glucose ester on cytokinin glucoside hydrolysis raises the possibility of interaction between auxin and cytokinin conjugate metabolism. Evidence in favour of mutual interactions between phytohormones at the metabolic level has been reported; IAA and its derivatives were found to be competitive inhibitors of cytokinin-alanine synthase *in vitro* [74], cytokinins inhibited 2,4-D conjugation to amino acids in tissue culture [61], ABA stimulated formation of cytokinin-*O*-glucosides from free cytokinins in tobacco plants [110] and *ipt* gene expression in *N. glutinosa* tumours resulted in both cytokinin and auxin autonomy accompanied by IAA accumulation [2]. Interestingly, transgenic plants expressing the *ipt* gene under the control of an auxin-inducible *SAUR* gene promoter exhibited increased tolerance to the toxic effects of exogenously applied auxins and auxin transport inhibitors [48].

Indirect evidence supporting an important role for Zm-p60.1 in maize development comes from genetic studies. In maize, a single locus (*Glu1*) coding for β-glucosidase activity has been identified up to now, and a large number of alleles have been described at this locus [97]. As the *Glu1* locus was characterized on the basis of β-glucosidase activity found in coleoptiles and since only one β-glucosidase has been identified in this material [25] it is likely that *Glu1* contains the Zm-p60 gene family. Lines originally reported to carry null alleles have now been shown to contain the enzyme in amounts similar to that of normal genotypes, but in a poorly soluble complex that did not lend itself to the original zymogram assay method. Thus, of the numerous alleles that are known at the *Glu1* locus, no null allele has yet been identified [24], supporting the proposition that the GLU1 glucosidase activity is critical for maize development.

Conclusions

The study of bacterial enzymes that modify cytokinin metabolism has allowed progress to be made in elucidating the physiological roles of cytokinins in numerous plant systems, and has highlighted the potential importance of the conjugate pools. There has also been recent progress in identifying some of the enzymatic activities that are likely to prove central to the control of cytokinin synthesis, compartmentalization, and inactivation. These initial studies must now be followed through with purification of the enzymes themselves, and isolation of the corresponding cDNA clones, through genetics, T-DNA tagging, or classical biochemistry. With these tools in hand, perhaps the most urgent questions to address will be the tissue and subcellular localization of each activity, and the consequences of overexpression, ectopic expression or antisense expression of the cDNAs. In this way a picture should begin to emerge of the spatial and temporal control of cytokinin levels in plants. This will provide a valuable basis for understanding the other important aspects of cytokinin action, namely its perception and signal transduction, two processes that are currently poorly understood.

Acknowledgements

B.B. is supported by a grant from the Grant Agency of the Czech Republic (204/93/0350) and K.P. by a grant from the DFG. We wish to thank Dr Csaba Koncz and Dr Christopher Redhead for critically reading the manuscript.

References

1. Banowetz GM: The effects of endogenous cytokinin content on benzyladenine-enhanced nitrate reductase induction. Physiol Plant 86: 341–348 (1992).
2. Binns AN, Labriola J, Black RC: Initiation of auxin autonomy in *Nicotiana glutinosa* cells by the cytokinin-biosynthesis gene from *Agrobacterium tumefaciens*. Planta 171: 539–548 (1987).
3. Brinegar AC, Blumenthal S, Cooper G: Photoaffinity labeling of mung bean mitochondrial proteins using (^{3}H)-2-azido-6-benzylaminopurine. In: Kaminek M, Mok DWS, Zazimalova E (eds), Physiology and Biochemistry of Cytokinins in Plants, pp. 301–307. SPB Academic Publishers, The Hague (1992).
4. Brzobohaty B, Moore I, Kristoffersen P, Bako L, Campos N, Schell J, Palme K: Release of active cytokinin by a β-glucosidase localized to the maize root meristem. Science 262: 1051–1054 (1993).
5. Burch LR, Horgan R: The purification of cytokinin oxidase from *Zea mays* kernels. Phytochemistry 28: 1313–1319 (1989).
6. Burch LR, Horgan R: Cytokinin oxidase and the degradative metabolism of cytokinins. In: Kaminek M, Mok

DWS, Zazimalova E (eds), Physiology and Biochemistry of Cytokinins in Plants, pp. 29–32. SPB Academic Publishers, The Hague (1992).
7. Campos N, Bako L, Feldwisch J, Schell J, Palme K: A protein from maize labeled with azido-IAA has novel β-glucosidase activity. Plant J 2: 675–684 (1992).
8. Chatfield JM, Armstrong DJ: Regulation of cytokinin oxidase activity in callus tissues of *Phaseolus vulgaris* L. cv. Great Northern. Plant Physiol 80: 493–499 (1986).
9. Chaudhury AM, Letham S, Craig S, Dennis ES: *ampl* – a mutant with high cytokinin levels and altered embryonic pattern, faster vegetative growth, constitutive photomorphogenesis and precocious flowering. Plant J 4: 907–916 (1993).
10. Chen CM: Cytokinin biosynthesis in cell-free systems. In: Wareing PF (ed.), Plant Growth Substances, pp. 155–164. Academic Press, London (1982).
11. Chen CM, Melitz DK: Cytokinin biosynthesis in a cell-free system from cytokinin-autotrophic tobacco tissue cultures. FEBS Lett 107: 15–20 (1979).
12. Claes B, Smalle J, Dekeyser R, Van Montagu M, Caplan A: Organ-dependent regulation of a plant promoter isolated from rice by 'promoter-trapping' in tobacco. Plant J 1: 15–26 (1991).
13. Conn EE: β-Glycosidases in plants: substrate specificity. In: Essen A. (ed.), β-Glucosidases: Biochemistry and Molecular Biology. ACS Symposium Series 533, pp. 15–26. Maple Press, York, PA (1993).
14. Cowly DE, Duke CC, Liepa AJ, Mac Leod JK, Letham DS: The structure and synthesis of cytokinin metabolites. I. The 7- and 9-β-glucofuranosides and pyranosides of zeatin and 6-benzylaminopurine. Aust J Chem 31: 1095–1111 (1978).
15. Dehio C, de Bruijn FJ: The early nodulation gene *SrEnod2* from *Sesbania rostrata* is inducible by cytokinin. Plant J 2: 117–128 (1992).
16. Dixon SC, Martin RC, Mok MC, Shaw G, Mok DWS: Zeatin glycosylation enzymes in *Phaseolus*: isolation of *O*-glucosyltransferase from *P. lunatus* and comparison to *O*-xylosyltransferase from *P vulgaris*. Plant Physiol 90: 1316–1321 (1989).
17. Dominov JA, Stenzler L, Lee S, Schwarz JJ, Leisner S, Howell SH: Cytokinins and auxins control the expression of a gene in *Nicotiana plumbaginifolia* cells by feedback regulation. Plant Cell 4: 451–461 (1992).
18. Doree M, Guern J: Short-term metabolism of some exogenous cytokinins in *Acer pseudoplatanus* cells. Biochim Biophys Acta 304: 611–622 (1993).
19. Engelbrecht L: Cytokinins in leaf-cuttings of *Phaseolus vulgaris* L. during their development. Biochem Physiol Pflanzen 163: 335–343 (1972).
20. Entsch B, Parker CW, Letham DS, Summons RE: Preparation and characterization, using high performance liquid chromatography, of an enzyme forming glucosides of cytokinins. Biochim Biophys Acta 570: 124–139 (1979).
21. Entsch B, Letham DS, Parker CW, Summons RE, Gollnow BI: Metabolites of cytokinins. In: Skoog F (ed), Plant Growth Substances 1979, pp. 109–118. Springer-Verlag, Berlin (1980).
22. Entsch B, Parker CW, Letham DS: An enzyme from lupine seeds forming alanine derivatives of cytokinins. Phytochemistry 22: 375–381 (1983).
23. Erion JL, Fox JE: Purification and properties of a protein which binds cytokinin-active 6-substituted purines. Plant Physiol 67: 156–162 (1981).
24. Esen A, Cokmus C: Maize genotypes classified as null at the Glu locus have β-glucosidase activity and immunoreactive protein. Biochem Genet 28: 319–336 (1990).
25. Esen A: Purification and partial characterization of maize (*Zea mays* L.) β-glucosidase. Plant Physiol 98: 174–182 (1992).
26. Estruch JJ, Chriqui D, Grossmann K, Schell J, Spena A: The plant oncogene *rolC* is responsible for the release of cytokinins from glucoside conjugates. EMBO J 10: 2889–2895 (1991).
27. Estruch JJ, Prinsen E, Van Onckelen H, Schell J, Spena A: Viviparous leaves produced by somatic activation of an inactive cytokinin-synthesizing gene. Science 254: 1364–1367 (1991).
28. Estruch JJ, Granell A, Hansen G, Prinsen E, Redig P, Van Onckelen H, Schwarz-Sommer Z, Sommer H, Spena A: Floral development and expression of floral homeotic genes are influenced by cytokinins. Plant J 4: 379–384 (1993).
29. Firn RD: Too many binding proteins, not enough receptors? In: Klämbt D (ed), Plant Hormone Receptors. pp 1–11. NATO ASI Series H: Cell Biology Vol. 10. Springer-Verlag, Berlin/Heidelberg (1987).
30. Fladung M: Transformation of diploid and tetraploid potato clones with the *rolC* gene of *Agrobacterium rhizogenes* and characterization of transgenic plants. Plant Breed 104: 295–304 (1990).
31. Fox JE, Cornette J, Deleuze G, Dyson W, Giersak G, Niu P, Zapata J, McChesney J: The formation, isolation and biological activity of a cytokinin 7-glucoside. Plant Physiol 52: 627–632 (1973).
32. Haberlandt G: Zur Physiologie der Zellteilungen. Sitzungsber K Preuss Akad Wiss: 318–345 (1913).
33. Hall RH: N^6-(Δ^2-isopentenyl) adenosine: chemical reactions, biosynthesis, metabolism and significance to the structure and function of tRNA. In: Davidson JN, Cohn WE (eds) Progress in Nucleic Acid Research and Molecular Biology, Vol 10, pp. 57–86. Academic Press, New York (1970).
34. Hamaguchi N, Iwamura H, Fujita T: Fluorescent anticytokinins as a probe for binding. Isolation of cytokinin-binding proteins from the soluble fraction and identification of a cytokinin-binding site on ribosomes of tobacco (*Nicotiana tabacum* cultivar Wisconsin No. 38) callus cells. Eur J Biochem 153: 565–572 (1985).
35. Hansen CE, Meins FJr, Milani A: Clonal and physi-

ological variation in the cytokinin content of tobacco cell lines differing in cytokinin requirements and capacity for neoplastic growth. Differentiation 29: 1–6 (1985).

36. Henson I.E., Wareing PF: Cytokinins in *Xanthium strumarium* L.: Distribution in the plant and production in the root system. J Exp Bot 27: 1268–1278 (1976).
37. Hepler PK, Wayne RO: Calcium and plant development. Ann Rev Plant Physiol 36: 397–439 (1985).
38. Ivanova M, Todorov IT, Atanassova L, Dewitte W, Van Onckelen HA: Co-localization of cytokinins with proteins related to cell proliferation in developing somatic embryos of *Dactylis glomerata* L. J Exp Bot 45: 1009–1017 (1994).
39. Jones RJ, Schreiber BM, McNeil K, Brenner ML, Foxon G: Cytokinin levels and oxidase activity during maize kernel development. In: Kaminek M, Mok DWS, Zazimalova E (eds) Physiology and Biochemistry of Cytokinins in Plants, pp. 235–239. SPB Academic Publishers, The Hague (1992).
40. Kaminek M, Armstrong DJ: Genotypic variation in cytokinin oxidase from *Phaseolus* callus cultures. Plant Physiol 93: 1530–1538 (1990).
41. Kares C, Prinsen E, Van Onckelen H, Otten L: IAA synthesis and root induction with *iaa* genes under heat shock promoter control. Plant Mol Biol 15: 225–236 (1990).
42. Klee HJ, Horsch RB, Hinchee MA, Hein MB, Hoffmann NL: The effects of overproduction of two *Agrobacterium tumefaciens* T-DNA auxin biosynthetic gene products in transgenic petunia plants. Genes Devel 1: 86–96 (1987).
43. Kobayashi K, Zbell B, Reinert J: A high affinity binding site for cytokinin to a particulate fraction in carrot suspension cells. Protoplasma 106: 145–155 (1981).
44. Laloue M, Fox JE: Cytokinin oxidase from wheat: partial purification and general properties. Plant Physiol 90: 899–906 (1989).
45. Lee YH, Mok MO, Mok DWS, Griffin DA, Shaw G: Cytokinin metabolism in *Phaseolus* embryos. Plant Physiol 77: 635–641 (1985).
46. Letham DS: Zeatin, a factor inducing cell division from *Zea mays*. Life Sci 8: 569–573 (1963).
47. Letham DS, Palni LMS: The biosynthesis and metabolism of cytokinins. Ann Rev Plant Physiol 34: 163–197 (1983).
48. Li Y, Shi X, Strabala TJ, Hagen G, Guilfoyle TJ: Transgenic tobacco plants that overproduce cytokinins show increased tolerance to exogenous auxin and auxin transport inhibitors. Plant Sci 100: 9–14 (1994).
49. Martin RC, Martin RR, Mok MC, Mok DWS: A monoclonal antibody specific to zeatin O-glycosyltransferases of *Phaseolus*. Plant Physiol 94: 1290–1294 (1990).
50. Martin RC, Mok MC, Mok DWS: Cytolocalization of zeatin O-xylosyltransferase in *Phaseolus*. Proc Natl Acad Sci USA 90: 953–957 (1993).
51. McGaw BA, Horgan R: Cytokinin oxidase from *Zea mays* kernels and *Vicia rosea* crown-gall tissue. Planta 159: 30–37 (1983).
52. McGaw BA, Heald JK, Horgan R: Dihydrozeatin metabolism in radish seedlings. Phytochemistry 23: 1373–1377 (1984).
53. McGaw BA, Horgan R, Heald JK, Wullems GJ, Schilperoort RA: Mass-spectrometric quantitation of cytokinins in tobacco crown-gall tumours by mutated octopine Ti plasmids of *Agrobacterium tumefaciens*. Planta 176: 230–234 (1988).
54. Medford JI, Horgan R, El-Sawi Z, Klee HJ: Alterations of endogenous cytokinins in transgenic plants using a chimeric isopentenyl transferase gene. Plant Cell 1: 403–413 (1989).
55. Miernyk JA: Abscisic acid inhibition of kinetin nucleotide formation in germinating lettuce seeds. Physiol Plant 45: 63–66 (1979).
56. Miller CO, Skoog F, Saltza MH von, Strong FM: Kinetin, a cell division factor from deoxyribonucleic acid. J Am Chem Soc 77: 1329–1334 (1955).
57. Miller CO, Skoog F, Okomura FS, von Saltza MH, Strong FM: Isolation, structure and synthesis of kinetin, a substance promoting cell division. J Am Chem Soc 78: 1345–1350 (1956).
58. Mok MC, Mok DWS, Marsden KE, Shaw G: The biological activity and metabolism of a novel cytokinin metabolite, O-xylosylzeatin, in callus tissue of *Phaseolus vulgaris* and *P. lunatus*. J Plant Physiol 130: 423–431 (1987).
59. Mok DWS, Mok MC, Martin RC, Bassil NV, Lightfoot DA: Zeatin metabolism in *Phaseolus*: enzymes and genes. In: Karsen CM, van Loon LC, Vreugdenhil D (eds) Progress in Plant Growth Regulation. pp. 597–606, Kluwer Academic Publishers, Dordrecht (1992).
60. Mok DWS, Mok MC, Martin RC, Bassil N, Shaw G: Immuno-analysis of zeatin metabolic enzymes of *Phaseolus*. In: Kaminek M, Mok DWS, Zazimalova E (eds) Physiology and Biochemistry of Cytokinins in Plants, pp. 17–23. SPB Academic Publishers, The Hague (1992).
61. Montague MI, Enns RK, Siegel NZ, Jaworski EG: Inhibition of 2,4-dichlorophenoxyacetic acid conjugation to amino acids by treatment of cultured soybean cells with cytokinins. Plant Physiol 67: 701–704 (1981).
62. Moore FH: A cytokinin-binding protein from wheat germ. Isolation by affinity chromatography and properties. Plant Physiol 64: 594–599 (1979).
63. Motyka V, Kaminek M: Characterization of cytokinin oxidase from tobacco and poplar callus cultures. In: Kaminek M, Mok DWS, Zazimalova E (eds) Physiology and Biochemistry of Cytokinins in Plants, pp. 33–39. SPB Academic Publishers, The Hague (1992).
64. Nandi SK, De Klerk GJM, Parker CW, Palni LMS: Endogenous cytokinin levels and metabolism of zeatin riboside in genetic tumour tissues and non-tumorous tissues of tobacco. Physiol Plant 78: 197–204 (1990).

65. Olsen KW, Zaluzec EJ, Zaluzec MM, Fernandez EJ, Pavkovic SF: Crystallographic and binding studies on a cytokinin peanut agglutinin complex. Biophys J 59: 296A (1991).
66. Paces V, Werstiuk E, Hall RH: Conversion of N^6-(Δ^2-isopentenyl) adenosine to adenosine by enzyme activity in tobacco tissue. Plant Physiol 48: 775–778 (1971).
67. Paces V, Kaminek M: Effect of ribosylzeatin on the enzymatic degradation of N^6-(Δ^2-isopentenyl) adenosine. Nucl Acids Res 3: 2309–2314 (1976).
68. Palmer MV, Horgan R, Wareing PF: Cytokinin metabolism in *Phaseolus vulgaris* L. I. Variations in cytokinin levels in leaves of decapitated plants in relation to lateral bud outgrowth. J Exp Bot 32: 1231–1241 (1981).
69. Palmer MV, Palni LMS: Substrate effects on cytokinin metabolism in soybean callus tissue. J Plant Physiol 126: 365–371 (1987).
70. Palni LMS, Burch L, Horgan R: The effect of auxin concentration on cytokinin stability and metabolism. Planta 174: 231–234 (1988).
71. Parker CW, Letham DS: Regulators of cell division in plant tissues. XVI. Metabolism of zeatin by radish cotyledons and hypocotyls. Planta 114: 199–218 (1973).
72. Parker CW, Wilson MM, Letham DS, Cowley DE, MacLeod JK: The glucosylation of cytokinins. Biochem Biophys Res Commun 55: 1370–1376 (1973).
73. Parker CW, Letham DS: Regulators of cell division in plant tissues. XVIII. Metabolism of zeatin in *Zea mays* seedlings. Planta 115: 337–344 (1974).
74. Parker CW, Entsch B, Letham DS: Inhibitors of two enzymes which metabolize cytokinins. Phytochemistry 25: 303–310 (1986).
75. Polya GM, Davis AW: Properties of a high affinity cytokinin-binding protein from wheat germ. Planta 139: 139–147 (1978).
76. Reinecke DM, Brenner ML, Rubenstein I: Cytokinin biosynthesis in developing *Zea mays* kernels. Plant Physiol 99 (Suppl): 66 (1992).
77. Roberts DD, Goldstein IJ: Adenine binding sites of the lectin from lima beans (*Phaseolus lunatus*). J Biol Chem 258: 13820–13824 (1983).
78. Romano C, Hein M, Klee H: Inactivation of auxin in tobacco transformed with the indole acetic acid-lysine synthetase gene of *Pseudomonas savastanoi*. Genes Devel 5: 438–446 (1991).
79. Romanov GA, Taran VY, Chvojka L, Kulaeva ON: Receptor-like cytokinin-binding protein(s) from barley leaves. J Plant Growth Regul 7: 1–7 (1988).
80. Romanov GA, Taran VY, Venis MA: Cytokinin-binding protein from maize shoots. J Plant Physiol 136: 208–212 (1990).
81. Sano H, Youssefian S: Light and nutritional regulation of transcripts encoding a wheat protein kinase homolog is mediated by cytokinins. Proc Natl Acad Sci USA 91: 2582–2586 (1994).
82. Schmitt JM, Piepenbrock M: Regulation of phosphoenolpyruvate carboxylase and crassulacean acid metabolism induction in *Mesembryanthenum crystallinum* L. by cytokinin. Modulation of leaf gene expression by roots? Plant Physiol 99: 1664–1669 (1992).
83. Schmülling T, Schell J, Spena A: Single genes from *Agrobacterium rhizogenes* influence plant development. EMBO J 2621–2629 (1988).
84. Schmülling T, Beinsberger S, De Greef J, Schell J, Van Onckelen H, Spena A: Construction of heat-inducible chimaeric gene to increase the cytokinin content in transgenic plant tissue. FEBS Lett 2: 401–406 (1989).
85. Schmülling T, Fladung M, Grossman K, Schell J: Hormonal content and sensitivity of transgenic tobacco and potato plants expressing single rol genes of *Agrobacterium rhizogenes* T-DNA. Plant J 3: 371–382 (1993).
86. Singh S, Palni LMS, Letham DS: Cytokinin biochemistry in relation to leaf senescence. V. Endogenous cytokinin levels and metabolism of zeatin riboside in leaf discs from green and senescent tobacco (*Nicotiana rustica*) leaves. J Plant Physiol 139: 279–283 (1992).
87. Skoog F, Miller CO: Chemical regulation of growth and organ formation in plant tissues cultured in vitro. Symp Soc Exp Biol 11: 118–130 (1957).
88. Skoog F, Armstrong DJ: Cytokinins. Ann Rev Plant Physiol 21: 359–384 (1970).
89. Smart C, Scofield S, Bevan M, Dyer T: Delayed leaf senescence in tobacco plants transformed with tmr, a gene for cytokinin production in *Agrobacterium*. Plant Cell 3: 647–656 (1991).
90. Smigocki AC: Cytokinin content and tissue distribution in plants transformed by a reconstructed isopentenyl transferase gene. Plant Mol Biol 16: 105–115 (1991).
91. Smigocki AC, Owens LD: Cytokinin gene fused with a strong promoter enhances shoot organogenesis and zeatin levels in transformed plant cells. Proc Natl Acad Sci USA 85: 5131–5135 (1988).
92. Smith AR, Van Staden J: Changes in endogenous cytokinin levels in kernels of *Zea mays* L. during imbibition and germination. J Exp Bot 29: 1067–1075 (1978).
93. Sondheimer E, Tzou D: The metabolism of 8-^{14}C-zeatin in bean axes. Plant Physiol 47: 516–520 (1971).
94. Sossountzov L, Maldiney R, Sotta B, Sabbagh I, Habricot Y, Bonnet M, Miginiac E: Immunocytochemical localization of cytokinins in Craigella tomato and a sideshootless mutant. Planta 175: 291–304 (1988).
95. Spena A, Aalen RB, Schulze SC: Cell autonomous behavior of the *rolC* gene of *Agrobacterium rhizogenes* during leaf development: a visual assay for transposon excision in transgenic plants. Plant Cell 1: 1157–1164 (1989).
96. Spena A, Prinsen E, Fladung M, Schulze SC, Van Onckelen H: The indoleacetic acid-lysine synthetase gene of *Pseudomonas syringae* subsp. *savastanoi* induces developmental alterations in transgenic tobacco and potato plants. Mol Gen Genet 227: 205–212 (1991).
97. Stuber CW, Goodmann MM, Johnson FM: Genetic

control and racial variation of β-glucosidase isozymes in maize (*Zea mays* L.). Biochem Genet 15: 383–394 (1977).
98. Sugiharto B, Burnell JN, Sugiyama T: Cytokinin is required to induce the nitrogen-dependent accumulation of mRNAs for phosphoenolpyruvate carboxylase and carbonic anhydrase in detached maize leaves. Plant Physiol 100: 153–156 (1992).
99. Takegami T, Yoshida K: Isolation and purification of cytokinin binding protein from tobacco leaves by affinity column chromatography. Biochem Biophys Res Commun 67: 782–789 (1975).
100. Taran VY, Romanov GA, Venis MA: Purification of soluble zeatin-binding proteins from maize shoots. In: Kaminek M, Mok DWS, Zazimalova E (eds) Physiology and Biochemistry of Cytokinins in Plants, pp. 165–167. SPB Academic Publishers, The Hague (1992).
101. Taya T, Tanaka Y, Nishimura S: 5′ AMP is a direct precursor of cytokinin in *Dictyostelium discoideum*. Nature 271: 545–547 (1978).
102. Terrine C, Laloue M: Kinetics of N^6-(Δ^2-isopentenyl) adenosine degradation in tobacco cells. Plant Physiol 65: 1090–1095 (1980).
103. Trewavas AJ: Growth substances in context: a decade of sensitivity. Biochem Soc Transact 20: 102–108 (1992).
104. Turner JE, Mok MC, Mok DWS: Zeatin metabolism in fruits of *Phaseolus*: comparison between embryos, seedcoat, and pod tissue. Plant Physiol 79: 321–322 (1985).
105. Turner JE, Mok DWS, Mok MC, Shaw G: Isolation and partial purification of an enzyme catalyzing the formation of O-xylosylzeatin in *Phaseolus vulgaris* embryos. Proc Natl Acad Sci USA 84: 3714–3717 (1987).
106. Van Staden J, Dimalla GG: Endogenous cytokinins and the breaking of dormancy and apical dominance in potato tubers. J Exp Bot 29: 1077–1084 (1978).
107. Van Staden J, Smith AR: The synthesis of cytokinins in excised roots of maize and tomato under aseptic conditions. Ann Bot 42: 751–753 (1978).
108. Van Staden J, Mooney PA: The effect of cytokinin preconditioning on the metabolism of adenine derivatives in soybean callus. J Plant Physiol 133: 466–469 (1988).
109. Wang TL, Thompson AG, Horgan R: A cytokinin glucoside from leaves of *Phaseolus vulgaris* L. Planta 135: 285–288 (1977).
110. Whenam RJ: Effect of systemic tobacco mosaic virus infection on endogenous cytokinin concentration in tobacco (*Nicotiana tabacum* L.) leaves: consequences for the control of resistance and symptom development. Physiol Mol Plant Pathol 35: 85–95 (1989).
111. Whitty CD, Hall RH: A cytokinin oxidase in *Zea mays*. Can J Biochem 52: 781–799 (1974).
112. Wyndaele R, Christiansen J, Horseele R, Rudelsheim P, Van Onckelen H: Functional correlation between endogenous phytohormone levels and hormone autotrophy of transformed and habituated soybean cell lines. Plant Cell Physiol 29: 1095–1101 (1988).
113. Zhang R, Letham DS, Wong OC, Nooden LD, Parker CW: Cytokinin biochemistry in relation to leaf senescence. II. The metabolism of benzylaminopurine in soybean leaves and the inhibition of conjugation. Plant Physiol 83: 334–340 (1987).

Plant Molecular Biology **26**: 1499–1519, 1994.

Molecular genetics of auxin and cytokinin

Lawrence Hobbie[+], Candace Timpte[+] and Mark Estelle*
*Department of Biology, Indiana University, Bloomington, IN 47405, USA (*author for correspondence)*

Key words: auxin, cytokinin, plant hormones, plant development

Introduction

The indeterminate nature of plant development requires that plants continuously regulate cell division and elongation in response to genetic, as well as a wide variety of different environmental signals [53]. Since their discovery, many physiological studies have shown that the plant hormones auxin and cytokinin have fundamental roles in the regulation of cell growth [30, 43]. Both compounds appear to be essential for growth of plant cells in culture. Organogenesis in tissue culture is regulated by altering the concentration of auxin and cytokinin. In the intact plant, auxin and cytokinin are apparently involved in a bewildering array of growth processes. One explanation for this complexity is that most and perhaps all changes in cell division or elongation are mediated by cytokinin or auxin. According to this view, developmental responses to environmental stimuli such as light or temperature are mediated by these hormones. Thus, auxin and cytokinin may occupy a central position in the regulation of cell growth. Despite the importance of auxin and cytokinin we still know very little about their molecular mode of action. In this review, we summarize recent progress in the field with an emphasis on molecular and genetic studies.

Auxin

Auxin is involved in diverse aspects of plant growth and development, including phototropism, gravitropism, elongation growth, vascular differentiation, and lateral branching of roots and shoots [33]. Although the molecular details of auxin action remain largely unknown, recent molecular and genetic studies have provided a number of interesting new results. Within the past five years, several auxin-binding proteins have been cloned and characterized, auxin has been shown to have a rapid effect on several ion channels, and new information on the function of auxin-induced genes has been obtained. Genetics is also beginning to contribute to our understanding of auxin action, with the recent cloning of the affected gene in an auxin-resistant mutant. We can anticipate that in the next five years some of this new information will be incorporated into a more complete picture of auxin action. A number of recent reviews also discuss aspect of auxin and cytokinin action [54, 56, 109, 153].

Cell elongation and the plasma membrane H^+-ATPase

One of the best studied auxin responses is the rapid elongation of excised hypocotyl, stem, or coleoptile sections. According to the acid growth hypothesis, cell elongation in stem sections is due to auxin-induced acidification of the cell wall, which in an unknown manner causes 'loosening' of the wall itself, allowing turgor pressure to produce cell expansion [116]. Experimental challenges to the acid growth theory [79], have led to

[+] The two first authors are co-first authors of the paper.

a vigorous debate [117, 132] which will not be summarized here. Recently it was shown that auxin application to intact pea stems can also induce elongation [162]. This *in vivo* result supports the physiological relevance of auxin during elongation growth. However, it remains unclear how much 'acid growth' contributes to auxin-induced elongation *in vivo*.

Acidification of the cell wall is thought to be caused by the activation of the plasma membrane H^+-ATPase. Physiological concentrations of auxin (micromolar and below) increase the activity of this enzyme in maize coleoptiles [58]. This increase in H^+-ATPase activity results in an increase in proton efflux and therefore a more negative cellular membrane potential (hyperpolarization). However, inactive auxins also produced hyperpolarization in this system [47] presumably by activating the H^+-ATPase. As the inactive auxins did not induce elongation growth, this result suggests that activation of the H^+-ATPase alone is not sufficient for such growth.

In early experiments, auxin-induced hyperpolarization was measured by impalement of tissue sections [13]; in such samples, the resting potential of the cell is between −100 and −130 mV, and the cells hyperpolarize by about −25 mV. More recently, tobacco mesophyll protoplasts have been used to investigate the electrophysiology of auxin action [36]. Perhaps because of damage during protoplast isolation or electrode impalement [151], these cells have an unusually low resting potential of about −5 mV. Auxin still caused a rapid hyperpolarization of these cells by −7 to −8 mV, as measured by impalement [11]. This effect was confirmed by whole-cell patch clamping [122]. However, the objection may be raised that these cells do not represent a physiologically relevant system because of their unusual resting potential. An alternative, and more widely accepted system for plant electrophysiology is guard cells from either *Vicia faba* or *Commelina commensis*. These cells are useful because of their experimental accessibility and high density of ion channels. They have been studied as isolated protoplasts [88, 91] or in epidermal peels (16). A 5 μM concentration of 2,4-dichlorophenoxyacetic acid (2,4D) caused hyperpolarization of guard cell protoplasts due to activation of the H^+-ATPase [88]. Maximum hyperpolarization occurred after 15 to 20 min. A similar effect was observed in *V. faba* mesophyll protoplasts, but the magnitude of the current induced was only half that seen in the guard cells. Thus, auxin appears to have qualitatively similar effects in guard cells and mesophyll cells.

The biochemical basis for auxin induction of H^+-ATPase activity is unclear. Auxin has been shown to decrease the K_m of the ATPase for ATP [50, 124]. However, this change would only affect ATPase activity if ATP is limiting, a condition which has not been demonstrated. Auxin application also results in increased synthesis and exocytosis of the ATPase. In maize coleoptiles, the amount of plasma membrane-localized ATPase increased significantly within 10 min of indole-3-acetic acid (IAA) addition [58]. Within 30 min the level of ATPase reached the maximum, 180% of the control value. Furthermore, this ATPase was rapidly turned over: inhibition of protein or RNA synthesis produced a rapid drop in ATPase levels. Other modes of regulation of the ATPase are also possible, such as phosphorylation [127].

Auxin-induced changes in ion channel activity

Recently, several intriguing examples of auxin-mediated changes in ion channel activity have been reported. In *Vicia faba* guard cells, analyzed using the whole cell patch-clamp configuration, auxin did not alter the properties of inward potassium channels, but it did affect anion channels [90]. Application of 1-naphthalene acetic acid (NAA) at concentrations of 5 to 100 μM shifted the membrane potential at which a guard cell anion channel was activated. In the absence of auxin, the activation potential was about −30 mV. When auxin was applied to the outside but not the inside of the cell, the activation potential shifted toward the physiological resting potential (usually −100 to −160 mV) by 10 to 35 mV (depending on the NAA concentration). Experiments on membrane patches showed that auxin applied to the original outside of the patch gave similar effects, indicating that all the signalling machinery necessary to produce the change in channel properties is present in the membrane itself. These effects were produced within 15 s of auxin application. The physiological relevance of this rapid effect is unclear, as the observed effect of auxin appears to be insufficient to lead to anion channel activation. Perhaps in combination with other effects on channel activity, or on overall cellular electrochemical balance, this effect of auxin on anion channels will indeed be important.

Various compounds known to affect anion channel function in other systems, 'channel blockers', were tested on guard cell anion channels. Compounds unrelated to auxin structurally as well as those with some similarities to auxin were found to have effects on the anion channel activation potential similar to that seen with active auxins [91]. Other carboxylic acids and anions, including malate and chloride, also affected anion channel function [60]. Blatt and Thiel [16] suggest that this non-specificity in auxin's effect on the anion channel calls into question the physiological relevance of these effects. The issue of specificity could be addressed in the future by testing whether antibodies against the auxin-binding protein ABP1 (see below) could block auxin's effect on anion channels.

Marten *et al.* [91] used affinity chromatography to enrich for guard cell proteins that bound to one of the auxin-related channel blockers. They identified in the enriched fraction a 60 kDa protein that cross-reacted with antibodies against mammalian anion channel proteins. However, in mesophyll cells anion channels could not be identified electrophysiologically or immunologically. Thus, channel protein expression appears to be tissue specific. It would be interesting to look at the distribution of auxin-responsive ion channels in various tissues, especially those involved in rapid elongation growth such as etiolated hypocotyls or the elongation zone of roots. As many plant tissues are difficult to analyze electrophysiologically, these studies will require molecular and immunological probes for various plant ion channels.

Inward and outward potassium channels are thought to be very important for the electrical properties of plant cells. In *Vicia faba* guard cells, studied using an impalement technique, low concentrations of auxin (1–10 μM) increased the inward potassium current, whereas higher concentrations of auxin (30–100 μM) decreased it [16]. This response by the inward potassium channels was almost instantaneous ($t_{1/2}$ of 100–350 ms). By contrast, the outward potassium current rose monotonically in response to increasing concentrations of auxin; this response took 1–3 min to develop. It is unclear why these authors observed this strong effect of auxin on the potassium channels when Marten *et al.* [90] did not. The two studies used different techniques: impalement of guard cells in epidermal peels [16] vs. whole-cell patch clamping of guard cell protoplasts [90] and buffers with a variety of small differences in ionic concentrations. In addition, Marten *et al.* apparently tested only 100 μM IAA (inhibitory in [16]) and did not test the lower IAA concentrations that activated the inward potassium channel in Blatt and Thiel's study [16]. It is hoped that additional studies using the patch clamping technique will resolve these contradictions.

Auxin activation of the inward K-current appears to account for the initial depolarization of the membrane potential observed by many investigators [9, 13, 16, 46]. This small initial depolarization was induced by 1–10 μM auxin, whereas 100 μM auxin induced hyperpolarization.

Further advances in understanding the role of ion channels in auxin response can be expected as more of the channel proteins are subjected to molecular analysis. Two putative *Arabidopsis* potassium channels were cloned by complementation of yeast mutants defective in potassium uptake [7, 134]. One of these, KAT1, was characterized in *Xenopus* oocytes and shown to encode a potassium influx channel [125].

Possible second messengers

Products of phospholipase A_2

In animals, phospholipase A_2 action produces arachidonic acid, a precursor for the important signalling molecules prostaglandins and leukotrienes [105]. In plants, the products of phospholipase A_2 action on phosphatidylcholine are lysophosphatidylcholine and free fatty acids. Lysophophatidylcholine stimulated the H^+-ATPase in *Avena* plasma membrane vesicles [110]. Active auxins increased the activity of phospholipase A_2 in zucchini microsomal vesicles [8] and in cultured soybean cell microsomes [128]. Thus, stimulation of the phospholipase activity could be an important intermediate in auxin activation of the plasma membrane ATPase. Scherer and Andre [128] hypothesized that lysophosphatidylcholine could stimulate the ATPase by directly binding to it, by activating a membrane-associated protein kinase that could phosphorylate it, or both.

Phosphatidylinositol

Phospholipase C-catalyzed hydrolysis of PIP_2 produces the important second messengers IP_3 and diacylglycerol (DAG). Diacylglycerol activates protein kinase C and IP_3 triggers calcium release. Their importance in plants is uncertain [35], but at least two reports indicate an effect of auxin on phosphatidylinositol metabolism. When auxin was added to growth-arrested cultured *Catharanthus roseus* cells that depended on auxin for division, there was a rapid turnover of phosphatidylinositols, including a transient 2- to-3-fold rise of IP_2 and IP_3, within one minute [42]. Addition of 1 μM IAA to microsomal membranes from carrot cell suspension cultures appeared to increase levels of IP_2 and IP_3 within 30 s after addition [164]. Thus, changes in inositol phospholipids may be important for auxin signal transduction. It will be interesting to analyze this pathway in systems where other aspects of the auxin pathway are also studied.

Calcium

The importance of calcium as a second messenger in plants is becoming better established [22, 114]. Early experiments in a variety of systems using calcium-depleted conditions, calcium chelators, and calmodulin inhibitors implicated calcium in the auxin signal transduction chain [21]. Gonzalez *et al.* [55] recently found that some but not all calmodulin inhibitors tested could inhibit auxin-dependent medium acidification by *Avena sativa* coleoptile segments. Similarly, Brock *et al.* [20] showed that calcium channel blockers inhibited IAA-induced growth of *Avena pulvini*, although EGTA (a calcium chelator) did not.

Recently, more direct data on the effect of auxin on calcium levels have been obtained using calcium-sensitive fluorescent dyes or Ca^{+2}-sensitive microelectrodes. In two instances, auxin was observed to cause a decrease in calcium levels [46, 150], and in three instances, it caused an increase in calcium levels [9, 51, 69]. These different results may be due to differences in species, experimental tissue, or the concentration of auxin applied. Calcium-measuring techniques using the calcium-sensitive luminescent protein aequorin promise many advantages for studies of calcium in plants [78]. Unfortunately, no reports of auxin effects on calcium levels using this approach have been yet published.

pH

Changes in cytoplasmic pH also can act as a second messenger in animal systems [101]. Several reports document auxin effects on pH. Auxin has been found to cause either acidification [46, 150] or alkalinization [51, 69], in all cases within minutes of application. Again, differences in experimental systems and protocols presumably explain the different effects seen. It is possible that a low concentration of auxin may cause acidification and a higher level causes alkalinization.

Phosphorylation state

Protein phosphorylation is a very important regulatory mechanism in microbial and animal systems, but the evidence for a role for protein phosphorylation in auxin signal transduction has until recently been virtually non-existent. Many of the necessary components, such as kinases and phosphatases, have been shown to exist in plants

[153], but auxin's effect on their activity is largely unknown. Recently two MAP kinase genes were cloned from *Arabidopsis* using a PCR approach [99]. One of the MAP kinase proteins was phosphorylated by an extract of auxin-activated tobacco BY-2 cells that had been treated with 2,4-D for 5 or 10 min. This phosphorylation is a potential activation signal for the MAP kinase. In addition, they observed a rapid and transient activation of a kinase activity in BY-2 cells following 2,4-D addition. This kinase had a molecular weight and substrate specificity characteristic of MAP kinases, although it was not shown directly to be identical to either of the two cloned kinases. These experiments support the model that auxin rapidly activates a MAP kinase kinase, which phosphorylates and activates a MAP kinase, which in turn regulates many downstream substrates including those necessary for cell division [71]. This pathway is already known to be largely conserved between animals and fungi, and it would not be surprising if it were also conserved in plants.

Auxin-induced genes

Many auxin-induced genes have been isolated and characterized. As auxin is known to regulate cell division and differentiation, many of the genes induced by auxin may be part of general cellular pathways of growth control and metabolism, and not specific to the early steps of auxin action. Such genes are generally induced over a period of hours after auxin addition. Examples include ARG1 from mung bean, identified as a fatty acid desaturase [161], α-amylase from *Pisum* cotyledons [65], ascorbate oxidase from pumpkin [37], enzymes of polyamine biosynthesis [111], and the *Arabidopsis dbp* gene [6]. The *dbp* gene encodes a protein with similarity to histone H1 and binds double-stranded DNA.

Several genes down-regulated by auxin have also been identified. Expression of the proteinase inhibitor 2 gene of potato was inhibited by auxin [75]. Reddy and Poovaiah [118] cloned an auxin down-regulated gene from strawberry that encoded a small proline- and threonine-rich protein of unknown function. Sucrose induction of the soybean vegetative storage protein was inhibited by auxin [31]. Datta *et al.* [29] identified three families of genes from soybean that were down-regulated by auxin, probably due to destabilization of the mRNAs. No function for the down-regulated genes is yet known.

Some genes are induced by other stimuli in addition to auxin, indicating that they do not have a specific function in auxin action. One example is ARG2, from mung bean, which is related to the LEA5-A gene of cotton and responds to water potential as well as to auxin [161]. A second example are the GH2/4 genes from soybean, which are induced by heavy metals and other stresses as well as by auxin and are homologous to glutathione *S*-transferases (GST) [56]. Droog *et al.* [33] showed that a previously identified multigene family of auxin-induced genes from tobacco, with homologues in potato and soybean, was also related to GST. One member of this family showed GST activity *in vitro*. Auxins and GSTs will be discussed in more detail in the section on auxin-binding proteins.

The p34 (cdc2) protein kinase is a key regulatory component in cell cycle control. Expression of the plant homologue of cdc2 has been localized to meristems and regions of actual or potential cell proliferation [61, 92]. In *Arabidopsis* and soybean, treatment of tissues with auxin induced CDC2 expression [61, 92, 96]. Martinez *et al.* noted that increased CDC2 expression in radish roots upon auxin treatment was due to an increase in the number of lateral root initials (which express CDC2) and not to increased expression in existing initials [92]. John *et al.* [70] found that induction of CDC2 occurred within 10 min of auxin treatment of pea seedling roots. However, much of the regulation of CDC2 activity occurs post-translationally and not at the level of gene expression. Increased CDC2 activity in excised tobacco pith culture required both auxin and cytokinin, although auxin alone induced an increase in CDC2 protein level [70]. The strong correlation seen between CDC2 expression and competence for cell division suggests that at least part of auxin's effect on cell division may result from changes in expression of the CDC2 genes. The rapidity of auxin induction of CDC2 mRNA led John *et al.* [70] to speculate that this induction is a direct auxin response rather than a secondary effect.

Several gene families are induced very rapidly and specifically by auxin and may therefore be directly involved in auxin action. This class includes the SAUR genes [93], the GH1 and GH3 genes [57], pJCW1 and 2 [156], and the pIAA genes [146]. The functions of these genes are unknown, but some clues are beginning to emerge.

SAUR genes comprise a family of small genes, first identified in soybean [93] and since cloned from mung bean [160], *Arabidopsis* [52, 66], and pea [66]. Expression of SAUR mRNAs is induced 25–60-fold within 2.5 min of applying auxin to auxin-depleted tissue. This induction is at least partially due to increased transcription [49, 93]. SAUR genes are also induced by treatment with protein synthesis inhibitors such as cycloheximide, but this induction seemed not to be due to an increase in the rate of transcription [49]. It could not be determined if cycloheximide treatment caused an increase in SAUR transcript stability. Treatment with auxin and cycloheximide caused superinduction of the SAUR genes [49]. These findings suggest that the SAUR genes may be controlled by a rapidly turned-over repressor or that the SAUR transcript may be degraded by a short-lived ribonuclease. In soybean the SAUR genes are found as a cluster. SAUR transcripts have been localized to regions of rapid elongation, for example to the elongating side of gravitropically stimulated shoots [83], but the function of the encoded proteins remains unknown.

The GH1 and 3 genes are induced within 10–20 min of auxin application. GH1 expression is induced 3–10-fold by auxin, while GH3 is induced 25–60-fold. Franco *et al.* [49] found that GH3 expression was not induced by cycloheximide.

Several recent results provide important clues to the function of the pIAA 4/5 and 6 genes, originally isolated from pea [146]. In pea and in *Arabidopsis*, these genes are members of a large family (at least 14 homologues in *Arabidopsis*) of genes that encode short-lived nuclear proteins [1]. One domain is predicted to form an amphipathic β-α-α structure, which is similar to the DNA-binding domain from certain prokaryotic repressors [107]. These data suggest that these proteins function as repressors or activators of gene expression. The gene family defined by the presence of this putative DNA-binding domain includes the auxin-induced genes ARG3 and ARG4 [160], GH3 [57], Aux22 and Aux28 [4] and their *Arabidopsis* homologues AtAux2-11 and AtAux2-27 [27]. As with the SAUR genes, cycloheximide induced expression of the IAA4/5 and 6 genes [146], but this induction was reported to be due to both increased transcription and mRNA stabilization [10]. Auxin induction of these same genes was due to increased transcription only. These findings again suggest the existence of a short-lived repressor protein that controls the expression of auxin-regulated genes. While the initial pathway of induction of the IAA4/5 and 6 genes remains unknown, these data do suggest a plausible pathway for signal transduction after these genes are turned on: they autoregulate synthesis to higher levels and regulate the expression of the numerous downstream genes through direct interactions with their respective promoters.

Ballas *et al.* [10] identified an 'auxin-responsive element' in the promoter of the IAA4/5 gene from pea. Deletion of this element abolished responsiveness of the IAA4/5 to auxin, and addition of this element to a CaMV 35S promoter conferred auxin responsiveness. This element is conserved in several other auxin-inducible genes. 'Footprinting' experiments demonstrated that nuclear proteins bind to parts of the auxin-responsive element. However, there was no difference in protein binding when nuclear extracts from auxin-treated or untreated cells were used. This argues against models in which an auxin-receptor complex binds directly to target genes or in which an auxin-responsive repressor binds directly to part of the promoter. Rather, there must be some less direct protein-protein interaction occurring to cause changes in transcriptional activity. According to one model, in the absence of auxin, a repressor would bind a promoter-bound activator and prevent its interaction with other necessary transcription factors. In the presence of auxin, the repressor is modified and can no longer inhibit the activator's action [10]. Identification of the auxin-responsive element should facilitate identification of upstream steps in the signal transduction chain.

Potential auxin receptors

ABP1 and its isoforms
Much effort has been devoted to isolating and characterizing auxin-binding proteins, with the hope that such proteins might be functional auxin receptors [72, 103, 109, 121]. The best characterized of these auxin-binding proteins is the maize 'ABP1' and its homologues, proteins of 20–23 kDa. This protein was originally purified [86, 87, 135] by classical biochemical techniques, using auxin binding as an assay and cloned [63, 68, 48]. Much correlative evidence suggests that ABP1 functions as a receptor. Initially, such evidence came largely from studies using antibodies against ABP1 to block auxin responses in various assays. Löbler and Klämbt [87] showed that anti-ABP1 IgG blocked auxin-induced corn coleoptile elongation. Subsequently, Barbier-Brygoo *et al.* [11] showed that auxin-induced hyperpolarization of tobacco mesophyll protoplasts was inhibited by the anti-ABP1 antibody. Surprisingly, addition of exogenous maize ABP1 to tobacco cells enhanced the auxin response [12]. Additional immunological evidence for the function of ABP1 was obtained using whole-cell patch clamp analysis of maize protoplasts [122]. Auxin was shown to induce an outward-directed positive current, presumed to be due to activation of the proton ATPase, that led to hyperpolarization of the membrane potential. Again, an anti-ABP1 polyclonal antibody blocked this response. In addition, an antipeptide antibody, D16, raised against the putative auxin-binding region of ABP1, was shown to act as an auxin agonist. Thus, there is strong evidence that a protein with at least a great deal of antigenic similarity to ABP1 is important in mediating these auxin responses.

Additional evidence in support of ABP1's function was obtained using a peptide corresponding to a putative external segment near the carboxyl-terminus of ABP1. This peptide, but not those corresponding to other parts of the protein, was found to inactivate the inward potassium channel when applied to *Vicia faba* guard cells [147]. Inactivation of the inward potassium channel corresponds to the effect seen with higher ($>30\ \mu$M) concentrations of auxin, as described above [16]. The inactivation by the peptide could be prevented by blocking cytoplasmic acidification, as found also for the inactivation of this channel by auxin. Direct measurement using the pH-sensitive dye BCECF confirmed that the carboxyl-terminal peptide caused alkalinization of the guard cell cytoplasm. This result suggests that this electrical response may normally be produced by that region of ABP1 interacting with components of a signal transduction chain that cause cellular pH changes and thereby affect channel activity.

ABP1 homologues have now been cloned from several species, including strawberry [80] and *Arabidopsis* [108, 136]. In *Arabidopsis*, ABP1 appears to be encoded by a single gene [108], whereas in maize and strawberry it consists of a small multigene family, with five genes identified to date in maize [133, 163]. No striking differences in sequence or expression pattern have been identified among the different isoforms that would suggest functional specificity. Maize ABP1 is a 43 kDa dimer of 22–23 kDa subunits. The proteins possess a signal sequence, suggesting that they are cotranslationally inserted through the membrane of the endoplasmic reticulum (ER), and a carboxyl-terminal Lys-Asp-Glu-Leu (KDEL) sequence which causes ER localization. A single site for N-linked glycosylation is located at amino acid 98 in the mature 163 amino acid protein. A conserved hexapeptide sequence sixty residues from the N-terminus has been identified as the probable auxin binding site [152]. Apart from these features, the sequence does not provide immediate insight into protein function.

One major puzzle with respect to ABP1 function is its cellular location. Since auxin is a lipophilic molecule and known to enter the cell by diffusion, a plasma membrane localized receptor is not required. Experiments using 1-naphthylphthalamic acid (NPA, an inhibitor of the auxin efflux carrier) to raise the intracellular auxin concentration indicated that auxin may act at an internal cellular site [154]. In contrast, the finding that auxin conjugated to large impermeant molecules can cause protoplast hyperpolarization by acting at the plasma membrane suggests a plasma membrane-localized receptor [151]. Consistent with the function of the KDEL sequence, the majority of auxin-binding activity in earlier experiments ('site 1', [115]) was localized to ER-derived membranes. However, the studies in which anti-ABP1 antibodies were shown to inhibit auxin responses indicate that at least one species of auxin receptor is localized to the plasma membrane. Jones and Herman [73] used an anti-ABP1 antibody to show that ABP1 in maize coleoptile cells and in cultured cells is found throughout the endomembrane system (especially in the ER and Golgi) and also in the plasma membrane and cell wall. The extent of labelling in the latter two structures was surprising considering the earlier biochemical studies, but consistent with the physiological results. Napier and Venis [103] showed that the interaction of a monoclonal antibody that recognized an epitope near the carboxyl-terminus of ABP was reduced in the presence of auxin, indicating that auxin caused a conformational change in that region. Two attractive models can be proposed: auxin could cause secretion of ABP1 by inducing sequestration of the KDEL sequence, or by causing the KDEL to be clipped off. However, neither model seems tenable, as depleting cultured maize cells of auxin was shown to enhance secretion of ABP1 [73]. In addition, the secreted ABP1 was shown to contain KDEL.

As is mentioned above, the ABP1 sequence is not informative with respect to function. There is a report that purified ABP1 can bind to and stimulate the activity of RNA polymerase II [123], but the *in vivo* significance of this observation is unclear. One model for ABP1 function states that a secreted form of ABP1 binds to a plasma membrane 'docking protein' in an auxin-dependent manner. Binding initiates signal transduction by an unknown mechansim [28]. ABP1's native form is thought to be a homodimer, with one auxin binding site per dimer. Perhaps the docking protein is induced to dimerize by binding the dimerized ABP1; dimerization of transmembrane receptors is frequently associated with their activation. Until such a docking protein is identified and its function analyzed this model must remain an intriguing speculation.

Other auxin-binding proteins

Photoaffinity labelling, using azido-derivatives of IAA, is a relatively recent advance in auxin receptor studies. A number of groups have identified at least 4 different auxin-binding proteins using this technique. Hicks *et al.* [64] applied this approach to zucchini plasma membrane preparations and detected labelling of 40 and 42 kDa proteins which appeared to be subunits of native complexes of 87 and 300 kDa. Peptide mapping indicated that the 40 and 42 kDa proteins were related to each other. Intriguingly, these proteins partitioned into Triton X-114 after extraction, normally a characteristic of integral membrane proteins. Further characterization of these proteins awaits the purification of amounts adequate for peptide sequencing and cloning.

Macdonald *et al.* [89] identified 3 azido-IAA labeled proteins in the soluble fraction of cultured *Hyoscyamus muticus* cells, with molecular weights of 31, 25, and 24 kDa. The 31 and 24 kDa proteins showed many similarities to each other and were found to be antigenically related to the basic form of β-1,3-glucanase. Authentic β-1,3-glucanase behaved identically to these proteins in the azido-IAA labelling assay and immunological tests, confirming the identity of β-1,3-glucanase as an auxin-binding protein. The function of the auxin binding is unknown. The 25 kDa protein is discussed below.

Feldwisch *et al.* [45] identified 3 azido-IAA-labelled proteins of 60, 58, and 23 kDa in maize coleoptile plasma membrane preparations, and an additional protein of 24 kDa (pm24) in microsomal preparations. The 60 kDa protein was subsequently identified as a β-glucosidase [23].

The 23 kDa protein (pm23) was characterized in some detail. Binding of azido-IAA to pm23 was competed by active auxins but not by inactive structurally-related compounds. pm23 was extracted from membranes by TX-114. It was recognized by an anti-ABP1 antibody raised against recombinant ABP1 from *E. coli*, but not by other anti-ABP1 antibodies. This antigenic similarity suggests that pm23 could be an isoform of ABP1, although it must be more diverged than the isoforms so far described to explain the lack of recognition by the other antibodies. Palme [109] indicated that pm23 showed no homology to other known proteins. The microsomal protein pm24, however, is not recognized by any of the anti-ABP1 antibodies. It is unclear if pm24 is homologous to the 24 kDa *Arabidopsis* protein described by Zettl *et al.* [165].

At least two other auxin-binding proteins have been identified as glutathione-S-transferases (GSTs). Bilang *et al.* [14] purified a soluble 25 kDa protein from *Hyoscyamus muticus* cells (the same protein previously identified in [89]) using azido IAA labelling, and from a partial peptide sequence showed it to be a GST. Zettl *et al.* [165] also used IAA photoaffinity labelling on *Arabidopsis* plasma membrane and microsomal vesicles to purify and clone a 24 kDa protein that is a GST. GSTs are widely distributed enzymes that transfer glutathione (γ-Glu-Cys-Gly) to a variety of electrophilic substrates. In plants the best known function of GSTs is detoxification of herbicides [149]. Bilang *et al.* [14] and Zettl *et al.* [165] proposed several possibilities for the role of GST in auxin binding. One possibility is that GST could conjugate glutathione to IAA, thereby modulating IAA activity or levels. This model was not supported by preliminary *in vitro* experiments [165]. A second possibility is that IAA could bind to GST at a second non-enzymatic site and thereby modulate GST activity. Some species of GST from animals are known to have a second site for steroid binding. Third, GST could affect the levels of growth regulators, such as jasmonic acid, by modifying intermediates in the biosynthetic pathway. Indeed, the purified ABP/GST was found to have activity on compounds similar to intermediates in the pathway of jasmonic acid synthesis [165]. A final possibility is that GSTs, by controlling the cellular level of glutathione, could affect the cellular redox potential and thus influence plant development. Earnshaw and Johnson [34] found that GSH addition to carrot suspension cells could switch the cells between somatic embryogenesis and cell proliferation. It is interesting to note that auxin induces the expression of several GST genes [33, 145]. This may be because the plant cells are trying to cope with a large amount of a foreign compound ('defense response'), or, if GST is somehow involved in auxin signal transduction, then induction of the GST gene upon auxin treatment could be a way to amplify the hormone signal.

Insights from genetic approaches

Experiments using mutants affected in response to auxin, and transgenic plants with altered auxin levels have begun to contribute to our understanding of auxin biology and mechanism of action. This topic has been recently reviewed by Klee and Estelle [77] and Hobbie and Estelle [67], and so only a few selected important findings will be discussed here.

Transgenic plants with bacterial genes that reduce or elevate internal auxin levels are very useful tools for understanding the physiological role of auxin. The IAA-Lys gene encodes an activity which conjugates auxin to lysine, forming a molecule without auxin activity and thus reducing the internal levels of auxin in transgenic plants by up to 10-fold [119, 143]. The IAA-M gene converts tryptophan to an auxin precursor which is converted to IAA by an endogenous activity, thus raising auxin levels up to 10-fold [77, 137]. In both cases, the phenotypes of the transgenic plants are consistent with previous physiological studies. It is also interesting to note that the transgenic plants were able to partially compensate for altered auxin levels by either synthesizing more auxin (in the IAA-Lys plants) or by conjugating more auxin to inactive forms (in the IAA-M plants).

A large number of mutants with alterations in auxin sensitivity or auxin metabolism has been isolated from a variety of species and used to investigate the physiological role of auxin. For example, an auxin-resistant mutant isolated from *N. tabacum* by Muller *et al.* [102] was very useful in validating the assay for auxin action developed by Ephritikhine *et al.* [36]. Protoplasts from the auxin-resistant mutant were shown to require higher concentrations of auxin to reach the same degree of hyperpolarization as wild-type protoplasts. In another series of studies, a number of auxin-auxotrophic mutants were isolated from *N. plumbaginifolia* tissue culture [17, 48]. Auxin levels in the mutants were approximately normal, indicating that the true defect in these mutants might lie in compartmentalization or some more subtle aspect of auxin metabolism. Attempts to regenerate mature plants from the culture lines were unsuccessful either in the presence or absence of auxin, suggesting that auxin auxotrophy is lethal.

One of the major advantages of a genetic approach to auxin action is the ability to isolate genes of unknown biochemical function. It is likely that the molecular characterization of these genes and their products will help us fill in many of the gaps in our knowledge, and perhaps reveal unanticipated molecular mechanisms. The most progress in molecular characterization of auxin-related mutants is being made in *Arabidopsis*. Using a straightforward screen for mutant seedlings able to elongate roots on concentrations of auxins inhibitory to wild-type root growth, our group and others have isolated auxin-resistant mutants that define at least five loci [67]. Common phenotypes among these mutants include defects in root gravitropism, organ elongation, and cross-resistance to additional plant hormones. In this review we will discuss the single locus that has been studied at the molecular level.

Mutants at the *AXR1* locus are characterized by a reduction in stature, increased lateral branching of the shoot, decreased lateral root branching, and wrinkled leaves [85]. The phenotype is consistent with a reduction in auxin sensitivity in all tissues of the plant. *axr1* mutant plants are also resistant to cytokinin and ethylene.

The *AXR1* gene was cloned using a map-based strategy [82]. It encodes a protein of 540 amino acids with a predicted molecular weight of 60 kDa. The AXR1 protein is similar to the N-terminal portion of the ubiquitin-activating enzyme, E1. This enzyme is responsible for the first step in the ubiquitin-conjugation pathway [62]. The most frequent outcome of protein ubiquitination is rapid degradation of the protein, but some proteins are known to be stably or reversibly ubiquitinated, indicating that ubiquitination may also serve as a regulatory modification. AXR1's role in the ubiquitin pathway is unknown. and it is possible that it does not act in this pathway at all. It is unlikely that it acts as an E1 enzyme, as AXR1 is about half the size of known E1s and lacks an active-site cysteine residue. Experiments are in progress to test possible AXR1 functions related to the ubiquitin pathway. It may act to modify or target the action of enzymes in the ubiquitin pathway in some manner. At least one AXR1-like gene has been identified in *Arabidopsis* (C. Timpte, unpublished results), indicating that there is a small *AXR1* gene family. Low-stringency Southern blots suggest that *AXR1* genes are widely distributed among plants (D. Lammer, unpublished). The relationship between AXR1 and E1, and the fact that some of the auxin-induced genes both encode and may be regulated by short-lived proteins [1, 82], suggests a role for rapid protein turnover in control of the auxin response pathway.

A second promising approach was employed by Hayashi *et al.* [59; Walden *et al.*, this volume]. They mutagenized tobacco protoplasts with a T-DNA-derived plasmid that contained multiple enhancer sequences near the right border, and identified a clone in which insertion of the plasmid conferred auxin-independent growth. Plants regenerated from these protoplasts were morphologically normal. Using plasmid rescue, they isolated the gene whose expression had been activated in the mutant protoplasts. It encoded a novel protein, highly basic and 570 amino acids in length. Further application of this technique should allow identification of additional genes involved in auxin-regulated growth.

The cytokinins

The cytokinins were first identified by Skoog and Miller [138] as an essential factor for plant cell division in culture. Subsequent studies have suggested a role for cytokinins in a wide variety of developmental processes including photomorphogenesis, chloroplast biogenesis and maintenance, and senescence [74]. In addition, many experiments, beginning with Skoog and Miller [138], have demonstrated that auxin and cytokinin interact to regulate several aspects of growth and development, including organogenesis in tissue culture, and apical dominance in mature plants. Unfortunately, there has been little progress in our understanding of either cytokinin physiology or mechanism of action during the past several years. In this article, we describe several recent studies which illustrate new genetic and molecular approaches to cytokinin biology. The reader should consult McGaw [94], Kaminek [74] and Chen *et al.* [25] for recent comprehensive reviews.

Manipulating cytokinin levels in transgenic plants

Cytokinin deficient mutants have not been identified in any plant. Consequently, much of our knowledge of the physiological role of cytokinins has come from experiments in which cytokinin has been applied to isolated organs or tissues. In order to circumvent some of the problems associated with this approach, several groups have generated transgenic plants with elevated cytokinin levels by introducing the isopentenyl transferase (*ipt*) gene from *Agrobacterium tumefaciens*. The enzyme encoded by the *ipt* gene (also known as *tmr* or gene 4) catalyses the condensation of isopentenyl pyrophosphate with adenosine monophosphate to produce isopentenyl AMP (iPMP), a precursor of several cytokinins [5, 81]. The *ipt* gene under control of its own promoter or fused to several different promoters has been introduced into petunia, potato, tobacco, and *Arabidopsis*. In most respects, the results of these experiments are consistent with earlier physiological studies. Elevated cytokinin levels resulted in a reduction in stature, the release of apical dominance, changes in vascular development, and in some cases, an inhibition of root growth.

In several studies, high cytokinin levels were achieved by fusing the *ipt* gene to the CaMV 35S promoter and introducing this construct into potato [106], cucumber [142], and several *Nicotiana* species [15, 142]. All of these transformants displayed an extreme shooty phenotype with little or no root development. In some instances, cell lines transformed with a 35S-*ipt* gene were auxin independent despite showing no change in IAA levels. This interesting result suggests that high cytokinin levels may alter auxin sensitivity. Surprisingly, tobacco plants expressing *ipt* from its bacterial promoter also accumulated high levels of zeatin compared to control plants [129] and were unable to form roots. When these shoots were grafted onto normal stems, they developed into stunted and highly branched mature plants.

In order to permit regeneration of mature plants and exert more control over cytokinin levels, several researchers fused the *ipt* gene to an inducible promoter and introduced this construct into either tobacco or *Arabidopsis*. Heat shock promoters from *Drosophila* [129, 143], maize [95], and soybean [4, 140] have been used to regulate *ipt* gene expression. The *ipt* gene under control of a maize hsp70 promoter was introduced into tobacco and *Arabidopsis* by Medford *et al.* [95]. This promoter was constituitively active at a low level, since non-heat induced tobacco plants had a three-fold increase in zeatin riboside and seven-fold increase in zeatin monophosphate over control plants. This increase in basal cytokinin levels affected plant development dramatically. Transgenic plants were reduced in stature, had released axillary buds, smaller stem and leaf area, reduced xylem and a reduced root system with wider and shorter roots than control plants. Surprisingly, heat induction did not lead to further developmental alterations beyond those described, despite dramatic increases in cytokinin levels. With heat induction, zeatin levels increased 30-fold, while zeatin riboside and zeatin monoriboside were stimulated 50-fold and 20-fold respectively. These results suggest that cytokinin response may be saturated even with relatively modest increases in cytokinin levels. Similar results were obtained by Smigocki [140] using a *Drosophila* hsp70 promoter fusion and by Smart *et al.* [139] using a heat-inducible soybean promoter. In the experiments with the soybean promoter, transgenic plants displayed delayed leaf senescence in addition to the other changes. Ainley *et al.* [4] used a more tightly regulated soybean heat shock promoter to regulate the *ipt* gene in tobacco. Transcription from this promoter was undetectable at control temperature. One of the classic cytokinin responses is the generation of shoots from leaf explants or from stem or leaf callus tissue. Leaves excised from control and transgenic plants were cultured on hormone free media. Essentially all of the heat shocked transgenic leaf explants produced shoots while only 1–2% of the control explants produced shoots. None of the transgenic leaf explants produced shoots in the absence of heat treatment. Intact transgenic plants subjected to daily heat treatments demonstrated morphological changes [4]. Although zeatin riboside levels increased only 5-fold, the plants had reduced node elongation, crinkled and downfolded leaves with increased chlorosis, and enlarged stems with expanded xylem, cortex, and pith cells. Additionally, flower bud development was delayed, resulting in late or aborted flowering.

The maize hsp70 fusion was also introduced into *Arabidopsis* by Medford *et al.* [95]. Plants grown in long days (16 h) showed no morphological changes even with heat treatments. However, under short day and heat treated conditions, root growth was affected and xylem proliferation decreased. Cytokinin levels were not reported.

In an interesting variation on these approaches, Li *et al.* [84] fused *ipt* to the auxin-inducible SAUR promoter. This promoter is primarily active in elongating tissue, presumably in response to auxin. Thus, transgenic plants which carry the SAUR-*ipt* fusion should have elevated cytokinin levels in tissues which are responding to auxin. The overall phenotype of these transgenic tobacco plants resembles those previously described: reduced stature, release of axillary buds, reduction in root initiation and growth, aberrant tissue development in stems. Additionally, these plants had adventitious shoots on leaves and petioles and exhibited complex and variable changes in senescence. The authors also demonstrate increased nutrient translocation to tissues which are expressing *ipt* (presumably tissues with high cytokinin levels), and suggest that differences in nutrient distribution may contribute to the transgenic phenotype. This suggestion is supported by experiments which indicate that applied cytokinin application increases the sink strength of treated tissue [155].

Estruch *et al.* [40] used the transposable element *Ac* to generate somatic sectors with high cytokinin levels. *Ac* was cloned into the untranslated leader of the *ipt* gene under control of the CaMV 35S promoter, thus inactivating the gene. Somatic transposition of *Ac* is expected to restore gene activity. Early excision events following transformation would result in undifferentiated callus tissue, but excisions occurring later should result in mosiac plants composed of cells expressing the 35S-*ipt* gene and cells which do not express the gene. Local effects of increased cytokinin can thus be studied throughout plant development. Mosiac plants exhibited loss of apical dominance and bulges on the leaf blade. These bulged regions contained higher levels of cytokinin than the adjacent normal leaf tissue. Furthermore, adventitious buds formed on the tips of the leaves, possibly because cytokinin is transported to, and accumulates at, the leaf tip. Thus, increased local cytokinin appears to override the normal developmental pattern, resulting in adventitious buds [41].

Recent experiments suggest that the *rolC* gene from *Agrobacterium rhizogenes* may function by altering cytokinin metabolism. Transgenic tobacco plants expressing *rolC* either from the endogenous promotor or the constituitive CaMV 35S promoter have phenotypic changes which include dwarfing, reduced apical dominance, increased root branching, and thin lanceolate leaves with reduced chlorophyll content [131]. Some of these changes might be explained by changes in cytokinin levels or sensitivity. Indeed, the *rolC* gene encodes a protein which is capable of hydrolyzing cytokinin glucosides *in vitro* [40, 41]. Thus the *rolC* gene product may hydrolyze cytokinin-glucosides, and thereby release active forms of cytokinin. However, recent analysis of hormone levels in *rolC* transgenic plants suggests that their phenotype is not solely due to increased cytokinin levels [130]. Levels of auxin and abscisic acid were also significantly altered in *rolC* transgenic tobacco and potato plant tissues. *RolC* transgenic plants also displayed altered sensitivity to auxin, abscisic acid, cytokinin, ACC (an ethylene precursor), and TIBA (an auxin transport inhibitor). Moreover, when *ipt* and *rolC* transgenic plants are crossed, the progeny display a *rolC* phenotype and not the *ipt* phenotype [130]. These experiments demonstrate that the regulation of free hormone levels in plants is complex. Furthermore, regulation of plant hormone activity can also be accomplished by altering the sensitivity of different tissues to hormones.

Cytokinin-related mutants

When grown on medium containing a low concentration of cytokinin (2.5 μM BA), *Arabidopsis* seedlings exhibit the 'cytokinin root syndrome' which consists of an inhibition of primary root growth and stimulation of root hair elongation. Su and Howell [144] used this response to screen for cytokinin-resistant mutants in *Arabidopsis*. Five independent recessive alleles of a new locus called *CKR1* were identified. The phenotype of *ckr1* mutant plants is relatively subtle. In the absence of cytokinin, seedling roots are longer and root hairs are shorter than those of wild-type plants. Rosette leaves are slightly cupped and more yellow than wild type when grown in constant illumination. This yellowing is absent when plants are grown in shorter days. Mature *ckr1* plants have normal, fertile flowers and are similar in stature to wild-type plants. Additionally, the *ckr1* mutation is specific to cytokinin since resistance to dihydrozeatin, kinetin and zeatin was observed, but cross-resistance to auxin was not found. Recent experiments have shown that the *ckr1* mutants are affected in other cytokinin responses. For example the mutants have an attenuated response to infection by *Agrobacterium* strains (Su and Howell, personal communication). Although clearly altered in cytokinin response, the morphology of *ckr1* plants is not severely affected. The authors suggest that either the *ckr1* mutants are leaky or that more than one cytokinin-response pathway exists.

Several *Arabidopsis* mutants that were isolated in screens for auxin resistance also have altered sensitivity to cytokinin. Both the *axr1* and *aux1* mutations confer decreased sensitivity to cytokinin (C. Lincoln, unpublised; [38]). This cross-resistance is particularly interesting because auxin and cytokinin cooperate to regulate many aspects of plant development. Growth processes such as apical dominance appear to be regulated by the ratio of auxin to cytokinin rather than the absolute concentration of either hormone. It is possible that auxin-cytokinin cross resistance in *aux1* and *axr1* plants is related to this interaction.

A cytokinin-resistant mutant has also been identified in *Nicotiana plumbaginifolia* [18]. The phenotype of this recessive mutation, originally called *ckr1*, includes resistance to cytokinin during seedling development, a reduction in root branching, and wiltiness of the shoot. Later studies demonstrated that the *ckr1* mutant is deficient in abscisic acid biosynthesis and is allelic to two other mutations called I217 and Esg152 [120]. These mutants define a single, recessive locus and all share reduced seed dormancy and a wilty phenotype. Additionally, the mutants have increased auxin tolerance and significantly lower endogenous levels of abscisic acid. Parry *et al.* [112] showed that the defect in *ckr1* affects the final step in the ABA biosynthetic pathway, the conversion of ABA-aldehyde to ABA. Consequently the mutant locus was renamed *Aba1* [120]. Cytokinin resistance may be a secondary effect of changes in abscisic acid levels. These results demonstrate the difficulty of selecting for mutations which affect a specific hormone response and emphasize the complexity of hormone action and interaction.

Another way to identify mutants affected in hormone physiology is to screen for plants with characteristic changes in morphology. The *amp1* (*a*ltered *m*eristem *p*rogram) mutants of *Arabidopsis* were isolated by screening for plants with an altered number of cotyledons and/or changes in leaf growth [24]. Mutant seedlings have a variable number of cotyledons and mature plants display reduced apical dominance, altered vegetative phyllotaxy as well as altered floral morphology and delayed senescence. Root explants from mutant plants also regenerate shoots at a higher rate than wild-type explants. Cytokinin analysis indicated that *amp1* plants had six-fold higher levels of zeatin riboside and four-fold higher levels of dihydrozeatin than wild-type plants. Elevated cytokinin levels could be caused by an increase in cytokinin biosynthesis or decreased cytokinin degradation. Since the mutation is recessive, the *AMP1* gene may function in cytokinin breakdown or as a negative regulator of cytokinin biosynthesis. Most aspects of the *amp1* phenotype are consistent with studies of transgenic plants with higher cytokinin levels. Furthermore, the phenotype suggests that cytokinin plays an important role in determining both the embryonic pattern and the vegetative pattern of organ initiation. Like the *ckr1* mutants, the *amp1* mutation also affects photomorphogenesis. When grown in the dark, mutant plants display a de-etiolated phenotype, including a shortened hypocotyl, loss of the apical hook and the inititation of leaf development [24]. These results, together with results summarized below, suggest that cyokinin acts to facilitate part of the de-etiolation program. Interestingly, the *amp1* mutation partially suppresses the phenotype of the phytochrome-deficient mutant, *hy2*. Thus, phytochrome may act, in part, through cytokinin.

Chory *et al.* [26] have recently obtained additional evidence for a role for cytokinin in photomorphogenesis. These researchers find that micromolar concentrations of cytokinin cause de-etiolation of dark-grown *Arabidopsis* seedlings similar to that observed in the *det1* and *det2* mutants. These results imply that an increase in cytokinin is sufficient to override a light requirement for leaf and chloroplast development in *Arabidopsis*. Significant differences in cytokinin levels between dark-grown *det1* mutants and wild-type plants were not observed although slight increases in zeatin (2.9 ×) and isopentenyladenosine (3.4 ×) were noted in light-grown *det1* seedlings. Since both *det1* and *det2* are recessive, the wild-type genes may act as negative regulators in a cytokinin signal transduction pathway that is coupled to photoreception. Chory [26] proposes a model in which light and cytokinin act independently or sequentially through common signal transduction intermediates such as *det1* and *det2* to control downstream light-regulated responses.

Other *Arabidopsis* mutants with possible defects in cytokinin metabolism or cytokinin action include the *su*per *r*oot (*sur*) mutant of *Arabidopsis* and the adenine phosphoribosyltransfrase (APRT)-deficient mutant BM3 (100). The *sur* mutant produces excessive lateral roots and does not develop past the four leaf stage [19]. However, normal shoots and flowers will regenerate from root explants treated with cytokinin. The APRT-deficient mutant was used to show that APRT functions to convert benzyladenine to the nucleotide form and therefore may be important for cytokinin interconversion [100].

Hormone mutants have also been described in the moss *Physcomitrella patens*. Application of cytokinin to wild-type *P. patens* protonema results in the overproduction of gametophytes, suggesting that cytokinin plays a role in regulating gametophyte production. A large number of gametophore overproducing mutants (OVE) have been isolated and recent genetic analysis has assigned seven of these mutants to three complementation groups [44]. Earlier studies revealed that mutant plants export the cytokinins isopentenyladenine and zeatin into the liquid growth medium [157, 158]. The accumulation of cytokinin is 100-fold greater from the OVE mutants than the wild type [113]. Physiological studies of cytokinin metabolism suggest that high cytokinin levels are not due to decreased cytokinin degradation in the OVE mutants OVE A78, OVE A102, OVE A201, and OVE B300 [113]. All but one of the OVE mutants are recessive. Based on these data, the authors propose that the OVE mutants are cytokinin overproducers in which the OVE genes have a negative regulatory function. A variety of additional cytokinin-related mutants have been isolated and characterized in *P. patens* including a cytokinin sensitive chloroplast mutant [2, 159]. With the recent advances in transformation techniques, moss may prove to be an excellent model system for studying cytokinin action [126].

Molecular aspects of cytokinin action

The molecular mechanism of cytokinin action is completely unknown. As for other plant growth regulators, cytokinin is thought to interact with a specific protein receptor, thus initiating a signal transduction cascade, perhaps resulting in changes in gene expression. Several cytokinin-binding proteins have been identified, but there is little evidence that any of these proteins function as receptors [109]. Most recently, Mitsui *et al.* [97, 98] have shown that one peptide component of a cytokinin-binding activity isolated from tobacco is similar to S-adenosyl-L-homocysteine hydrolase. Since S-adenosyl-L-homocysteine is an inhibitor of methyltransferase, these authors speculate that some of the physiological effects of cytokinin are mediated through changes in methylation. However, at present there is no other evidence for this view.

Cytokinin regulation of gene expression has been reviewed recently by Chen *et al.* [25] and only a few studies will be mentioned here. One of the best characterized examples of cytokinin regulated genes is the pLS216 gene from *Nicotiana plumbaginifolia* [32]. This gene is a member of a family of environmentally responsive genes, and therefore has been renamed *msr1*, for mutliple stimulus response. These genes encode proteins related to the stringent starvation response protein from *E. coli*. Dominov *et al.* [32] showed that the *msr1* mRNA accumulated in response to both cytokinin and auxin although the kinetics of accumulation differed for the two hormone treatments. The response to auxin peaked between 3–6 h, while the cytokinin response peaked in 24–48 h. In addition, cytokinin appeared to sensitize cells to subsequent auxin treatment. Based on these results, these workers propose a feedback mechanism to control gene expression. In their model, cytokinin stimulates *msr1* expression and either acts in a positive manner to sensitize the auxin response or blocks the feedback inhibition of the auxin response.

In addition to the traditional approach of exogenous cytokinin application, several groups have used transgenic plants to study the role of cytokinin in regulation of gene expression. For example, Estruch *et al.* [41] used *Ac* to generate somatic sectors expressing the *ipt* gene (described above). These plants also developed adventitious buds on the leaves. After floral initiation on the main stem, the ectopic meristems which form on the upper leaves are floral, forming both normal and abnormal floral buds [39]. The abnormal flower buds are characterized by homeotic transformation and/or organ fusion. These abnormalities correlate with high levels of *ipt* transcript and increased cytokinin content of the buds. In addition, increased cytokinin levels were correlated with decreased accumulation of homologues of three floral homeotic genes, DEFA, GLO, and PLENA from *Antirrhinum*. These researchers postulate that cytokinin fulfills some of the requirements of a morphogen since it has the following effects: (1) triggers ectopic developmental programs, (2) modifies a genetically well-defined developmental program, (3) alters apical development and causes stem fasciation [39].

Concluding remarks

The plant hormones are one of the most interesting and challenging areas of plant biology. Ubiquitous in their occurence and bewildering in the diversity of their responses, it is clear that a complete understanding of their biological role and mechanism of action will not come easily. However, there is plenty of cause for optimism. The development of new biochemical and molecular approaches will certainly lead to exciting new insights in the coming years. In addition, genetic approaches, particularly those involving *Arabidopsis*, are beginning to have a major impact on the field. The next five years should be an exciting time for researchers interested in plant hormones.

Note added in proof

Recent studies indicate that the *ckr1* mutations are alleles of the previously identified *EIN2* locus (G. Roman and J. Ecker, personal communication).

References

1. Abel S, Oeller PW, Theologis A: Early auxin-induced genes encode short-lived nuclear proteins. Proc Natl Acad Sci USA 91: 326–330 (1994).
2. Abel WO, Knebel W, Koop HU, Marienfeld JR, Quader H, Reski R, Schnepf E, Spoerlein B: A cytokinin rensitive mutant of the moss *Physcomitrella patens*, defective in chloroplast division. Protoplasma 152: 1–13 (1989).
3. Ainley WM, Walker JC, Nagao RT, Key JL: Sequence and characterization of two auxin-regulated genes from soybean. J Biol Chem 263: 10658–10666 (1988).
4. Ainley WM, McNeil KJ, Hill JW, Lingle WL, Simpson RB, Brenner ML, Nagao RT, Key JL: Regulatable endogenous production of cytokinins up to 'toxic' levels in transgenic plants and plant tissues. Plant Mol Biol 22: 13–23 (1993).
5. Akiyoshi D, Klee H, Amasino R, Nester EW, Gordon MP: T-DNA of *Agrobacterium tumefaciens* encodes an enzyme of cytokinin biosynthesis. Proc Natl Acad Sci USA 81: 5994–5998 (1984).
6. Alliotte T, Tire C, Engler G, Peleman J, Caplan A, Van Montagu M, Inze D: An auxin-regulated gene of *Arabidopsis thaliana* encodes a DNA-binding protein. Plant Physiol 89: 743–752 (1989).
7. Anderson JA, Huprikar SS, Kochian LV, Lucas WJ, Gaber RF: Functional expression of a probable *Arabidopsis thaliana* potassium channel in *Saccharomyces cerevisiae*. Proc Natl Acad Sci USA 89: 3736–3740 (1992).
8. Andre B, Scherer GFE: Stimulation by auxin of phospholipase A in membrane vesicles from an auxin-sensitive tissue is mediated by an auxin receptor. Planta 185: 209–214 (1991).
9. Ayling SM, Brownlee C, Clarkson DT: The cytoplasmic streaming response of tomato root hairs to auxin; observations of cytosolic calcium levels. J Plant Physiol 143: 184–188 (1994).
10. Ballas N, Wong L-M, Theologis A: Identification of the auxin-responsive element, *AuxRE*, in the primary indoleacetic acid-inducible gene, *PS-1AA4/5*, of pea (*Pisum sativum* L.). J Mol Biol 233: 580–596 (1993).
11. Barbier-Brygoo H, Ephritikhine G, Klambt D, Ghislain M, Guem J: Functional evidence for an auxin receptor at the plasmalemma of tobacco mesophyll protoplasts. Proc Natl Acad Sci USA 86: 891–895 (1989).
12. Barbier-Brygoo H, Ephritikhine G, Klambt D, Maurel C, Palme K, Schell J, Guern J: Perception of the auxin signal at the plasma membrane of tobacco mesophyll protoplasts. Plant J 1: 83–93 (1991).
13. Bates GW. Goldsmith MHM: Rapid response of the plasma-membrane potential in oat coleoptiles to auxin and other weak acids. Planta 159: 231–237 (1983).
14. Bilang J, Macdonald H, King PJ, Sturm A: A soluble auxin-binding protein from *Hyoscyamus muticus* is a glutathione *S*-transferase. Plant Physiol 102: 29–34 (1993).
15. Binns AN, Labliola J, Black RC: Initiation of auxin autonomy in *Nicotiana glutinosa* cells by the cytokinin-biosynthesis gene from *Agrobacterium tumefaciens*. Planta 171: 539–548 (1987).
16. Blatt MR, Thiel G: K^+ channels of stomatal guard cells: bimodal control of the K^+ inward-rectifier evoked by auxin. Plant J 5: 55–68 (1994).
17. Blonstein AD, Vahala T, Koornneef M, King PJ: Plants regenerated from auxin-auxotrophic variants are inviable. Mol Gen Genet 215: 58–64 (1988).
18. Blonstein AD, Parry AD, Horgan R, King PJ: A cytokinin-resistant mutant of *Nicotiana plumbaginifolia* is wilty. Planta 183: 244–250 (1991).
19. Boerjan W, Den Boer B, Van Montagu M: Molecular genetic approaches to plant development. Int J Devel Biol 36: 59–66 (1992).
20. Brock TG, Burg J, Ghosheh NS, Kaufman PB: The role of calcium in growth induced by indole-3-acetic acid and gravity in the leaf-sheath pulvinus of oat (*Avena sativa*). J Plant Growth Regul 11: 99–103 (1992).
21. Brummell DA, Hall JL: Rapid cellular responses to auxin and the regulation of growth. Plant Cell Environ 10: 523–543 (1987).
22. Bush DS: Regulation of cytosolic calcium in plants. Plant Physiol 103: 7–13 (1993).
23. Campos N, Bako L, Feldwisch J, Schell J, Palme K: A protein from maize labeled with azido-IAA has novel beta-glucosidase activity. Plant J 2: 675–684 (1992).
24. Chaudhury AM, Letham S, Craig S, Dennis ES: AMP1-a mutant with high cytokinin levels and altered embryonic pattern, faster vegetative growth, constituitive photomorphogenesis and precocious flowering. Plant J 4: 907–916 (1993).

25. Chen C-M, Jin G, Anderson BR, Ertl J: Modulation of plant gene expression by cytokinins. Aust J Plant Physiol 20: 609–619 (1993).
26. Chory J, Reinecke D, Sim S, Washburn T, Brenner M: A role for cytokinins in de-etiolation in *Arabidopsis*. Plant Physiol 104: 339–347 (1994).
27. Conner TW, Goekjian VH, LaFayette PR, Key JL: Structure and expression of two auxin-inducible genes from *Arabidopsis*. Plant Mol Biol 15: 623–632 (1990).
28. Cross JW: Cycling of auxin-binding protein through the plant cell: pathways in auxin signal transduction. New Biol 3: 813–819 (1991).
29. Datta N, LaFayette PR, Kroller PA, Nagao RT, Key JL: Isolation and characterization of three families of auxin down-regulated cDNA-clones. Plant Mol Biol 21: 859–869 (1993).
30. Davies PJ: The plant hormones: Their nature, occurrence, and functions. In: Davies PJ (ed) Plant Hormones and their Role in Plant Growth and Development, pp. 1–11. Kluwer Academic Publishers, Dordrecht (1988).
31. DeWald DB, Sadka A, Mullet JE: Sucrose modulation of soybean *Vsp* gene expression is inhibited by auxin. Plant Physiol 104: 439–444 (1994).
32. Dominov JA, Stenzler L, Lee S, Schwarz JJ, Leisner S, Howell SH: Cytokinins and auxins control the expression of a gene in *Nicotiana plumbaginifolia* cells by feedback regulation. Plant Cell 4: 451–461 (1992).
33. Droog FNJ, Hooykaas PJJ, Libbenga KR, van der Zaal EJ: Proteins encoded by an auxin-regulated gene family of tobacco share limited but significant homology with glutathione-*S*-transferases and one member indeed shows *in vitro* GST activity. Plant Mol Biol 21: 965–972 (1993).
34. Earnshaw BA, Johnson MA: The effect of glutathione on development in wild carrot suspension cultures. Biochem Biophys Res Comm 133: 988–993 (1985).
35. Einspahr KJ, Thompson GA Jr: Transmembrane signalling via phosphatidylinositol 4,5-bisphosphate hydrolysis in plants. Plant Physiol 93: 361–355 (1990).
36. Ephritikhine G, Barbier-Brygoo H, Muller J-F, Guern J: Auxin effects on the transmembrane potential difference of wild-type and mutant tobacco protoplasts exhibiting a differential sensitivity to auxin. Plant Physiol 83: 801–804 (1987).
37. Esaka M: Regulation of ascorbate oxidase gene expression in higher plants. Vitamins 67: 301–310 (1993).
38. Estelle M, Klee HJ: Auxin and cytokinin in *Arabidopsis*. In: Meyerowitz E, Somerville C, (eds), Cold Spring Harbor Press, Cold Spring Harbor, NY, (1994).
39. Estruch JJ, Granell A, Hansen G, Prinsen E, Redig P, Vanonckelen H, Schwarzsommer Z, Sommer H, Spena A: Floral development and expression of floral homeotic genes are influenced by cytokinins. Plant J 4: 379–384 (1993).
40. Estruch JJ, Parets-Soler A, Schmulling T, Spena A: Cytosolic localization in transgenic plants of the rolC peptide from *Agrobacterium rhizogenes*. Plant Mol Biol 17: 547–550 (1991).
41. Estruch JJ, Prinsen E, VanOnckelen H, Schell J, Spena A: Viviparous leaves produced by somatic activation of an inactive cytokinin-synthesizing gene. Science 254: 1364–1367 (1991).
42. Ettlinger C, Lehle L: Auxin induces rapid changes in phosphatidylinositol metabolites. Nature 331: 176–178 (1988).
43. Evans ML: Functions of hormones at the cellular level of organization. In: TK Scott (ed), Hormonal Regulation of Development II. Encyclopedia of Plant Physiology, vol, 10, pp. 23–79. Springer-Verlag, Berlin (1984).
44. Featherstone DR, Cove DJ, Ashton NW: Genetic analysis by somatic hybridization of cytokinin overproducing developmental mutants of the moss *Physcomitrella patens*. Mol Gen Genet 222: 217–224 (1990).
45. Feldwisch J, Zette R, Hesse F, Schell J, Palme K: An auxin-binding protein is localized to the plasma membrane of maize coleoptile cells: Identification by photoaffinity labeling and purification of a 23-kDa polypeptide. Proc Natl Acad Sci USA 89: 475–479 (1992).
46. Felle H: Auxin causes oscillations of cytosolic free calcium and pH in *Zea mays* coleoptiles. Planta 174: 495–499 (1988).
47. Felle H, Peters W, Palme K: The electrical response of maize to auxins. Biochim Biophys Acta 1064: 199–204 (1991).
48. Fracheboud Y, King PJ: An auxin-auxotrophic mutant of *Nicotiana plumbaginifolia*. Mol Gen Genet 227: 397–400 (1991).
49. Franco AR, Gee MA, Guilfoyle TJ: Induction and superinduction of auxin-responsive mRNAs with auxin and protein synthesis inhibitors. J Biol Chem 265: 15845–15849 (1990).
50. Gabathuler R, Cleland RE: Auxin regulation of a proton translocating ATPase in pea root plasma membrane vesicles. Plant Physiol 79: 1080–1085 (1985).
51. Gehling CA, Irving HR, Parish RW: Effects of auxin and abscisic acid on cytosolic calcium and pH in plant cells. Proc Natl Acad Sci USA 87: 9645–9649 (1990).
52. Gil P, Liu Y, Orbovic V, Verkamp E, Poff KL, Green PJ: Characterization of the auxin-inducible SAUR-AC1 gene for use as a molecular genetic tool in *Arabidopsis*. Plant Physiol 104: 777–784 (1994).
53. Goldberg, R: Plants: Novel developmental processes. Science 240: 1460–1467 (1988).
54. Goldsmith MHM: Cellular signalling: new insights into the action of the plant growth hormone auxin. Proc Natl Acad Sci USA 90: 11442–11445 (1993).
55. Gonzalez-Daros F, Carrasco-Luna J, Calatayud A, Salguero J, del Valle-Tascon S: Effects of calmodulin antagonists on auxin-stimulated proton extrusion in *Avena sativa* coleoptile segments. Physiol Plant 87: 68–76 (1993).

56. Guilfoyle TJ, Hagen G, Li Y, Ulmasov T, Liu Z, Strabala T, Gee M: Auxin-regulated transcription. Aust J Pant Physiol 20: 489–502 (1993).
57. Hagen G, Kleinschmidt A, Guilfoyle T: Auxin-regulated gene expression in intact soybean hypocotyl and excised hypocotyl sections. Planta 162: 147–153 (1984).
58. Hager A, Debus G, Edel HG, Stransky H, Serrano R: Auxin induces exocytosis and the rapid synthesis of a high-turnover pool of plasma-membrane H^+-ATPase. Planta 185: 527–537 (1991).
59. Hayashi H, Czaja I, Lubenow H, Schell J, Walden R: Activation of a plant gene by T-DNA tagging: auxin-independent growth in vitro. Science 258: 1350–1353 (1992).
60. Hedrich R, Marten I: Malate-induced feedback regulation of plasma membrane anion channels could provide a CO_2 sensor to guard cells. EMBO J 12: 897–901 (1993).
61. Hemerly AS, Ferreira P. Engler JdeA, Van Montagu M, Engler G, Inze D: *cdc2a* expression in *Arabidopsis* is linked with competence for cell division. Plant Cell 5: 1711–1723 (1993).
62. Hershko A, Ciechanover A: The ubiquitin system for protein degradation. Annu Rev Biochem 61: 69–102 (1992).
63. Hesse T, Feldwisch J, Balshusemann D, Bauw G, Puype M, Vandekerckhove J, Lobler M, Klambt D, Schell J, Palme K: Molecular cloning and structural analysis of a gene from *Zea mays* (L.) coding for a putative receptor for the plant hormone auxin. EMBO J 8: 2453–2461 (1989).
64. Hicks GR, Rice MS, Lomax TL: Characterization of auxin-binding proteins from zucchini plasma membrane. Planta 189: 83–90 (1993).
65. Hirasawa E, Yamamoto S: Properties and synthesis de novo of auxin-induced alpha-amylase in pea cotyledons. Planta 184: 438–442 (1991).
66. Ho T-hD, Hagen G: Hormonal regulation of gene expression. J Plant Growth Regul 12: 197–205 (1993).
67. Hobbie L, Estelle M: Genetic approaches to auxin action. Plant Cell Environ 17: 525–540 (1994).
68. Inhohara N, Shimomura S, Fukui T, Futai M: Auxin-binding protein located in the endoplasmic reticulum of maize shoots: Molecular cloning and complete primary structure. Proc Natl Acad Sci USA 86: 3564–3568 (1989).
69. Irving HR, Gehring CA, Parish RW: Changes in cytosolic pH and calcium of guard cells precede stomatal movements. Proc Natl Acad Sci USA 89: 1790–1794 (1992).
70. John PCL. Zhang K, Dong C, Diedelich L, Wightman F: $p34^{cdc2}$ related proteins in control of cell cycle progression, the switch between division and differentiation in tissue development, and stimulation of division auxin and cytokinin. Aust J Plant Physiol 20: 503–526 (1993).
71. Jonak C, Heberle-Bors E, Hirt H: MAP kinases: universal multi-purpose signalling tools. Plant Mol Biol 24: 407–416 (1994).
72. Jones AM: Do we have the auxin receptor yet? Physiol Plant 80: 154–158 (1990).
73. Jones AM, Herman EM: KDEL-containing auxin-binding protein is secreted to the plasma membrane and cell wall. Plant Physiol 101: 595–606 (1993).
74. Kaminek M: Progress in cytokinin research. Trends Biotechnol 10: 159–164 (1992).
75. Kernan A, Thornburg RW: Auxin levels regulate the expression of a wound-inducible proteinase inhibitor II-chloramphenicol acetyl transferase gene fusion *in vitro* and *in vivo*. Plant Physiol 91: 73–78 (1989).
76. Klee HJ, Horsch RB, Hinchee MA, Hein MB, Hoffmann NL: The overproduction of two *Agrobacterium tumefaciens* T-DNA auxin biosynthetic gene products in transgenic petunia plants. Genes Devel 1: 86–96 (1987).
77. Klee H, Estelle M: Molecular genetic approaches to plant hormone biology. Annu Rev Plant Physiol Plant Mol Biol 42: 529–551(1991).
78. Knight MR. Read NC, Campbell AK, Trewavas AJ: Imaging calcium dynamics in living plants using semi-synthetic recombinant aequorins. J Cell Biol 121: 83–90 (1993).
79. Kutschera U, Schopfer P: Evidence against the acid-growth theory of auxin action. Planta 163: 483–493 (1985).
80. Lazarus CM, Napier RM, Yu L-X, Lynas C, Venis MA: Auxin-binding proteins-antibodies and genes. In: Jenkins GI, Schuch W (eds) Molecular Biology of Plant Development, pp. 129–148, Company of Biologists Ltd, Cambridge, UK, (1991).
81. Letham DS. Palni LMS: The biosynthesis and metabolism of cytokinins. Ann Rev Plant Phys 34: 163–197 (1983).
82. Leyser HMO, Lincoln CA, Timpte C, Lammer D, Turner J, Estelle M: *Arabidopsis* auxin-resistance gene AXR1 encodes a protein related to ubiquitin-activating enzyme E1. Nature 364: 161–164 (1993).
83. Li Y, Hagen G, Guilfoyle TJ: An auxin-responsive promoter is differentially induced by auxin gradients during tropisms. Plant Cell 3: 1167–1175 (1991).
84. Li Y, Hagen G, Guilfoyle TJ: Altered morphology in transgenic tobacco plants that overproduce cytokinins in specific tissues and organs. Devel Biol 153: 386–395 (1992).
85. Lincoln C, Britton JH, Estelle M: Growth and development of the *axr1* mutants of *Arabidopsis*. Plant Cell 2: 1071–1080 (1990).
86. Löbler M, Klämbt D: Auxin-binding protein from coleoptile membranes of com (*Zea mays* L.), I. Purification by immunological methods and characterization. J Biol Chem 260: 9848–9853 (1985).
87. Löbler M, Klämbt D: Auxin-binding protein from co-

leoptile membranes of com (*Zea mays* L.). II. Localization of a putative auxin receptor. J Biol Chem 260: 9854–9859 (1985).
88. Lohse G, Hedrich R: Characterization of the plasma membrane H^+-ATPase from *Vicia faba* guard cells. Modulation by extracellular factors and seasonal changes. Planta 188: 206–214 (1992).
89. MacDonald H, Jones AM, King PJ: Photoaffinity labeling of soluble auxin-binding proteins. J Biol Chem 266: 7393–7399 (1991).
90. Marten I, Lohse G, Hedrich R: Plant growth hormones control voltage-dependent activity of anion channels in plasma membrane of guard cells. Nature 353: 758–762 (1991).
91. Marten I, Zeilinger C, Redhead C, Landry DW, Al-Awqati Q, Hedrich R: Identification and modulation of a voltage-dependent anion channel in the plasma membrane of guard cells by high-affinity ligands. EMBO J 11: 3569–3575 (1992).
92. Martinez MC, Jørgensen J-E, Lawton MA, Lamb CJ, Doerner PW: Spatial patterns of *cdc2* expression in relation to meristem activity and cell proliferation during plant development. Proc Natl Acad Sci USA 89: 7360–7364 (1992).
93. McClure BA, Guilfoyle T: Characterization of a class of small auxin-inducible soybean polyadenylated RNAs. Plant Mol Biol 9: 611–623 (1987).
94. McGaw BA: Cytokinin biosynthesis and metabolism. In: Davies PJ (ed) Plant Hormones and their Role in Plant Growth and Development, pp. 76–93, Kluwer Academic Publishers, Dordrecht (1988).
95. Medford J, Horgan R, El-Sawi Z, Klee H: Alterations of endogenous cytokinins in transgenic plants using a chimeric isopentenyl transferase gene. Plant Cell 1: 403–413 (1989).
96. Miao G-H, Hong Z, Velma DPS: Two functional soybean genes encoding $p34^{cdc2}$ protein kinases are regulated by different plant developmental pathways. Proc Natl Acad Sci USA 90: 943–947 (1993).
97. Mitsui S, Wakasugi T, Sugiura M: A cDNA encoding the 57 kDa subunit of a cytokinin-binding protein complex trom tobacco -- the subunit has high homology to *S*-adenosyl-homocysteine hydrolase. Plant Cell Physiol 34: 1089–1096 (1993).
98. Mitsui S, Sugiura M: Purification and properties of a cytokinin-binding protein from tobacco leaves. Plant Cell Physiol 34: 543–547 (1993).
99. Mizoguchi T, Gotoh Y, Nishida E, Yamaguchi-Shinozaki K, Hayashida N, Iwasaki T, Kamada H, Shinozaki K: Characterization of two cDNAs that encode MAP kinase homologues in *Arabidopsis thaliana* and analysis of the possible role of auxin in activating such kinase activities in cultured cells. Plant J 5: 111–122 (1994).
100. Moffatt B, Pethe C, Laloue M: Metabolism of benzyladenine is impaired in a mutant of *Arabidopsis thaliana* lacking adenosine phosphoribosyltransferase activity. Plant Physiol 95: 900–903 (1991).
101. Moolenaar WH, Mummery CL, van der Saag PT, de Laat SW: Rapid ionic events and the initiation of growth in serum-stimulated neuroblastoma cells. Cell 23: 789–798 (1981).
102. Muller J-F, Goujaud J, Caboche M: Isolation *in vitro* of naphthaleneacetic acid-tolerant mutants of *Nitotiana tabacum* which are impaired in root morphogenesis. Mol Gen Genet 199: 194–200 (1985).
103. Napier RM, Venis MA: Monoclonal antibodies detect an auxin-induced conformational change in the maize auxin-binding protein. Planta 182: 313–318 (1990).
104. Napier RM, Venis MA: Receptor for plant growth regulators: recent advances. J Plant Growth Regul 9: 113–126 (1990).
105. Needleman P, Turk J, Jakschik B, Morrison AR, Lefkowith JB: Arachidonic acid metabolism. Annu Rev Biochem 55: 69–102 (1986).
106. Ooms G. Kaup A, Roberts J: From tumor to tuber; tumor cell characteristics and chromosome numbers of crown gall derived tetraploid potato plants. Theor Appl Genet 66: 169–172 (1983).
107. Pabo CT, Sauer R: Transcription factors: structural families and principles of DNA recognition. Annu Rev Biochem 61: 1053–1095 (1992).
108. Palme K, Hesse T, Campos N, Garbers C, Yanofsky MF, Schell J: Molecular analysis of an auxin binding protein gene located on chromosome 4 of *Arabidopsis*. Plant Cell 4: 193–201 (1992).
109. Palme K: From binding proteins to hormone receptors? J Plant Growth Regul 12: 171–178 (1993).
110. Palmgren MG, Sommarin M: Lysophosphatidylcholine stimulates ATP dependent proton accumulation in isolated oat root plasma membrane vesicles. Plant Physiol 90: 1009–1014 (1989).
111. Park KY, Lee SH: Effects of ethylene and auxin on polyamine levels in suspension-cultured tobacco cells. Physiol Plant 90: 382–290 (1994).
112. Parry AD, Blonstein AD, Babiano MJ, King PJ, Horgan R: Abscisic acid metabolism in a wilty mutant of *Nicotiana plumbaginifolia* Planta 183: 237–243 (1991).
113. Perry CD, Cove DJ: Transfer RNA pool sizes and half lives in wild-type and cytokinin overproducing strains of the moss, *Physcomitrella patens*. Physiol Plant 67: 680–684 (1986).
114. Poovaiah BW, Reddy ASN: Calcium and signal transduction in plants. Crit Rev Plant Sci 12: 185–211 (1993).
115. Ray PM: Auxin-binding sites of maize coleoptiles are localized on membranes of the endoplasmic reticulum. Plant Physiol 59: 594–599 (1977).
116. Rayle DL, Cleland R: Control of plant cell enlargement by hydrogen ions. Curr Top Devel Biol 11: 187–214 (1977).
117. Rayle DL, Cleland R: The acid growth theory of auxin-

induced cell elongation is alive and well. Plant Physiol 99: 1271–1274 (1992).

118. Reddy ASN, Poovaiah BW: Molecular cloning and sequencing of a cDNA for an auxin-repressed mRNA: correlation between fruit growth and repression of the auxin-regulated gene. Plant Mol Biol 14: 127–136 (1990).
119. Romano CP, Hein MB, Klee HJ: Inactivation of auxin in tobacco transformed with the indoleacetic acid-lysine synthetase gene of *Pseudomonas savastanoi*. Genes Devel 5: 438–446 (1991).
120. Rousselin P, Kraepiel Y, Maldiney R, Miginiac E, Caboche M: Characterization of three hormone mutants of *Nicotiana plumbaginifolia*: evidence for a common ABA deficiency. Theor Appl Genet 85: 213–221 (1992).
121. Rubery PH: Auxin receptors. Annu Rev Plant Physiol 32: 569–596 (1981).
122. Ruck A, Palme K, Venis MA, Napier RM, Felle HH: Patch-clamp analysis establishes a role for an auxin binding protein in the auxin stimulation of plasma membrane current in *Zea mays* protoplasts. Plant J 4: 41–46 (1993).
123. Sakai S, Kikuchi M, Nakajima N: Interaction between auxin-binding protein-I and RNA polymerase II. Biosci Biotech Biochem 56: 1225–1229 (1992).
124. Santoni V, Vansuyt, G, Rossignol, M: The changing sensitivity to auxin of the plasma-membrane H^+-ATPase: relationship between plant development and ATPase content of membranes. Planta 185: 227–232 (1991).
125. Schachtman DP, Schroeder JI, Lucas WJ, Anderson JA, Gaber RF: Expression of an inward-rectifying potassium channel by the *Arabidopsis KAT1* cDNA. Science 258: 1654–1658 (1992).
126. Schaefer D. Zyrd JP, Knight CD, Cove DJ: Stable transformation of the moss *Physcomitrella patens*. Mol Gen Genet 226: 418–424 (1991).
127. Schaller GE, Sussman, MR: Phosphorylation of the plasma-membrane H^+-ATPase of oat roots by a calcium stimulated protein kinase. Planta 173: 509–518 (1988).
128. Scherer GFE, Andre B: Stimulation of phospholipase A_2 by auxin in microsomes from suspension-cultured soybean cells is receptor-mediated and influenced by nucleotides. Planta 191: 515–523 (1993).
129. Schmulling T, Beinsberger S, DeGreet J, Schell J, Van Onckelen H, Spena A: Construction of a heat inducible chimeric gene to increase cytokinin content in transgenic plant tissue. FEBS Lett 249: 401–406 (1989).
130. Schmulling T, Fladung M, Grossman K, Schell J: Hormonal content and sensitivity of transgenic tobacco and potato plants expressing single rol genes of *Agrobacterium rhizogenes* T-DNA. Plant J 3: 371–382 (1993).
131. Schmulling T, Schell J, Spena A: Single genes from *Agrobacterium rhizogenes* influence plant development. EMBO J 7: 2621–2629 (1988).
132. Schopfer P: Determination of auxin-dependent pH changes in coleoptile cell walls by a null-point method. Plant Physiol 103: 351–357 (1993).
133. Schwob E, Choi S-Y, Simmons, C, Migliaccio F, Ilag L, Hesse T, Palme K, Soll D: Molecular analysis of three maize 22 kDa auxin-binding protein genes-transient promoter expression and regulatory regions. Plant J 4: 423–432 (1993).
134. Sentenac H, Bonneaud N, Minet M, Lacroute F, Salmon J-M, Gaymard F, Glignon C: Cloning and expression in yeast of a plant potassium ion transport system. Science 256: 663–665 (1992).
135. Shimomura S, Sotobayash T, Futai M, Fukui T: Purification and properties of an auxin-binding protein from maize shoot membranes. J Biochem 99: 1513–1524 (1986).
136. Shimomura S. Liu W, Inohara N, Watanabe S, Futai M: Structure of the gene for an auxin-binding protein and a gene for 7SL RNA from *Arabidopsis thaliana*. Plant Cell Physiol 34: 633–637 (1993).
137. Sitbon F, Ostin A. Sundberg B, Olsson O, Sandberg G: Conjugation of indole-3-acetic acid (IAA) in wild-type and IAA-overproducing transgenic tobacco plants, and identification of the main conjugates by frit-fast atom bombardment liquid chromatography-mass spectrometry. Plant Physiol 101: 313–320 (1993).
138. Skoog F, Miller C: Chemical regulation of growth and organ formation in plant tissues cultured *in vitro*. Soc Expt Biol Symp 11: 188–231 (1957).
139. Smart CM, Scofield SR, Bevan MW, Deyer TA: Delayed leaf senescence in tobacco plants transformed with *tmr*, a gene for cytokinin production in *Agrobacterium*. Plant Cell 3: 647–656 (1991).
140. Smigocki AC: Cytokinin content and tissue distribution in plants transformed by a reconstructed isopentenyl transferase gene. Plant Mol Biol 16: 105–115 (1991).
141. Smigocki AC, Owens LD: Cytokinin to auxin ratios and morphology of shoots and tissues transformed by chimeric isopentenyltransferase gene. Plant Physiol 91: 808–811 (1989).
142. Spena A, Prinsen E, Fladung M, Schulze SC, Van Onckelen H: The indoleacetic acid-lysine synthetase gene of *Pseudomonas syringae* subsp. *savastanoi* induces developmental alterations in transgenic tobacco and potato plants. Mol Gen Genet 227: 205–212 (1991).
143. Su W, Howell S: A single genetic locus, CKR1, defines *Arabidopsis* mutants in which root growth is resistant to low concentrations of cytokinin. Plant Physiol 99: 1569–1574 (1992).
144. Takahashi Y, Nagata T: pal B: An auxin-regulated gene encoding glutathione-*S*-transferase. Proc Natl Acad Sci USA 89: 56–59 (1992).
145. Theologis A, Huynh TV, Davis RW: Rapid induction of specific mRNAs by auxin in pea epicotyl tissue. J Mol Biol 183: 53–68 (1985).
146. Thiel G, Blatt MR, Fricker MD, White IR, Millner P: Modulation of K^+-channels in *Vicia* stomatal guard cells

by peptide homologues to the auxin-binding protein C terminus. Proc Natl Acad Sci USA 90: 1493–1497 (1993).
147. Tillmann U, Viola G, Kayser B, Siemeister G, Hesse T, Palme K, Lobler M, Klambt D: cDNA clones of the auxin-binding protein from com coleoptiles (*Zea mays* L.): isolation and characterization by immunological methods. EMBO J 8: 2463–2467 (1989).
148. Timmerman KP: Molecular characterization of corn glutathione-*S*-transferase isozymes involved in herbicide detoxification. Physiol Plant 77: 465–471 (1989).
149. Tretyn, A, Wagner G, Felle HH: Signal transduction in *Sinapsis alba* root hairs: auxins as external messengers. J Plant Physiol 139: 187–193 (1991).
150. Venis MA, Thomas EW, Barbier-Brygoo H, Ephritikhine G, Guem J: Impermeant auxin analogues have auxin activity. Planta 182: 232–235 (1990).
151. Venis MA, Napier RM, Barbier-Brygoo H, Maurel C, Perrot-Rechenmann C, Guem J: Antibodies to a peptide from the maize auxin-binding protein have auxin agonist activity. Proc Natl Acad Sci USA 98: 7208–7212 (1992).
152. Verhey SD, Lomax TL: Signal transduction in vascular plants. J Plant Growth Regul 12: 179–195 (1993).
153. Vesper MJ, Kuss CL: Physiological evidence that the primary site of auxin action is an intracellular site. Planta 182: 486–491 (1991).
154. Waring PF, Phillips IDJ: Growth and differentiation in plants. Pergamon Press, New York (1981).
155. Walker JC, Key JL: Isolation of cloned cDNAs to auxin-responsive poly(A)$^+$ RNAs of elongating soybean hypocotyl. Proc Natl Acad Sci USA 9: 7185–7189 (1982).
156. Wang TL, Horgan R, Cove D: Cytokinins from the moss *Physcomitrella patens*. Plant Physiol 68: 735–738 (1981).
157. Wang TL, Beutelmann P, Cove D: Cytokinin biosynthesis in mutants of the moss *Physcomitlella patens*. Plant Physiol 68: 739–744 (1981).
158. Wang TL, Futers TS, McGeory F, Cove DJ: Moss mutants and the analysis of cytokinin metabolism. In: Crozier A, Hillman JR (eds) The Biosynthesis and Metabolism of Plant Hormones. Society of Experimental Biologists Seminar Series 23, pp. 135–164, Cambridge University Press, Cambridge (1984).
159. Yamamoto KT, Mori H, Imaseki H: cDNA cloning of indole-3-acetic acid-regulated genes: Aux22 and SAUR from mung bean (*Vigna radiata*) hypocotyl tissue. Plant Cell Physiol 33: 93–97 (1992).
160. Yamamoto KT: Further characterization of auxin-regulated mRNAs in hypocotyl sections of mung bean (*Vigna radiata* (L.) Wilczek): Sequence homology to genes for fatty-acid desaturases and atypical late-embryogenesis-abundant protein, and the mode of expression of the mRNAs. Planta 192: 359–364 (1994).
161. Yang T, Law DM, Davies PJ: Magnitude and kinetics of stem elongation induced by exogenous indole-3-acetic acid in intact light-growth pea seedlings. Plant Physiol 102: 717–724 (1993).
162. Yu L-X, Lazarus CM: Structure and sequence of an auxin-binding protein gene from maize (*Zea mays* L.). Plant Mol Biol 16: 925–930 (1991).
163. Zbell B, Walter-Back C: Signal transduction of auxin on isolated plant cell membranes: indications for a rapid polyphosphoinositide response stimulated by indoleacetic acid. J Plant Physiol 133: 353–360 (1988).
164. Zettl R. Schell J, Palme K: Photoaffinity labeling of *Arabidopsis thaliana* plasma membrane vesicles by 5-azido-[7-^{3}H]indole-3-acetic acid: Identification of a glutathione *S*-transferase. Proc Natl Acad Sci USA 91: 689–693 (1994).

Plant Molecular Biology **26**: 1521–1528, 1994.

Activation tagging: a means of isolating genes implicated as playing a role in plant growth and development

Richard Walden [1,*], Klaus Fritze [1], Hiroaki Hayashi [1,3], Edvins Miklashevichs [2], Hinrich Harling [1] and Jeff Schell [1]
[1] *Max Planck Institut für Zuchtungsforschung, Carl von Linné Weg 10, D-50829 Köln, Germany (* author for correspondence);* [2] *Institute of Microbiology, Kleisti, LV 1067 Riga, Latvia;* [3] *Current address: Faculty of Agriculture, University of Tokyo, Bunkyo-ku, Tokyo 113, Japan*

Received 21 April 1994; accepted 26 April 1994

Key words: activation tagging, *Arabidopsis*, auxin, cytokinin, polyamines, signal transduction, T-DNA tagging

Abstract

Activation T-DNA tagging has been used to generate a variety of tobacco cell lines selected by their ability to grow either in the absence of auxin or cytokinin in the culture media, or under selective levels of an inhibitor of polyamine biosynthesis. The majority of the cell lines studied in detail contain single T-DNA inserts genetically co-segregating with the selected phenotype. While most of the plants regenerated from the mutant cell lines appear phenotypically normal, several display phenotypes which could be inferred to result from disturbances in the content, or the metabolism, of auxins and cytokinins, or polyamines. The tagging vector is designed to allow the isolation of tagged plant genes by plasmid rescue. Confirmation that the genomic sequence responsible for the selected phenotype has indeed been isolated is provided by PEG-mediated protoplast DNA uptake of rescued plasmids followed by selection for protoplast growth under the original selective conditions. Several plasmids have been rescued from the mutant lines which confer on transfected protoplasts the ability to grow either in the absence of auxin or cytokinin in the culture media, or under selective levels of an inhibitor of polyamine biosynthesis. This review describes the background to activation tagging and our progress in characterizing the genes that have been tagged in the mutant lines we have generated.

Introduction

The power of mutant analysis in plants as a means of dissecting complex biochemical and developmental processes has only recently been fully realized by molecular biologists. While the biochemical analysis of mutant plant lines has provided valuable insight into the consequences of the mutation, it is now becoming feasible to isolate and characterize the mutated genes themselves, hence allowing an understanding of the molecular basis of the mutation itself [4, 13, 11]. Transposon and T-DNA tagging, as well as chromosome walking, have all proved to be valuable means of isolating mutated plant genes [7]. In particular, T-DNA tagging in *Arabidopsis* has proven to be an especially powerful means of gene isolation and an ever increasing number of genes tagged in this manner are being described [6, 15]. Central to each of these methods of mutant generation is the practical requirement that mutations, which generally involve disruption of a transcriptional unit, result routinely in a recessive mutation. This means that the mutant phenotype can only be visualised following selfing of the mutated individuals and this demands not an insubstantial amount of effort. Moreover, the generation of a specific mutation is fortuitous. We have developed an alternative method of T-DNA tagging where the mutation is a consequence not of gene disruption, rather the activation of expression of the tagged gene. This has come to be known as activation tagging [30, 31]. Because the mutation itself results from the activation of gene expression the resulting phenotype is dominant. This allows a direct selection for a desired phenotype from amongst the population of primary transformants.

To achieve the deregulation of expression of flanking plant genes upon insertion into the plant genome we have engineered a T-DNA tag to contain the transcriptional enhancer sequence of the cauliflower mosaic virus (CaMV) 35S RNA promoter cloned as a tandem tetramer near the border of the T-DNA. Mutant cell lines are generated by *Agrobacterium* protoplast co-cultivation. Thus a large number of mutant individuals can be generated with ease and a chemical selection can be applied to the primary transformants. Mutant cell lines generated in this manner can be cultured further and ultimately, plants regenerated.

We have used activation tagging to generate mutants affected in processes which have been previously proved to be relatively difficult to mount a genetic analysis, namely, the action of the plant growth substances auxin and cytokinin, as well as the polyamines. The notion underlying these experiments is that growth factors are likely to exert their effects through the induction of gene expression. By the activation tagging of a gene which normally becomes active only in the presence of a particular growth factor and whose action results in cell growth and division, we might isolate genes that play a role in a specific signal transduction pathway. In the case of isolating genes whose overexpression results in callus growth in the absence of growth substances we are, in effect, isolating plant cellular protooncogenes.

Activation tagging

The development of activation tagging exploits several experimental observations. First, protoplast *Agrobacterium* co-cultivation is an effective means of producing extremely large numbers of transgenic plant cells with relative ease [3, 23]. Second, during the transformation process the T-DNA of *Agrobacterium tumefaciens* inserts preferentially into potentially transcribed regions of the plant genome [16]. Third, there is a multiplying effect on activity when transcriptional enhancer sequences derived from a promoter sequence are cloned as multimers and they can activate gene expression distant from their site of insertion [12]. Finally, central to the use of activation tagging is the notion that the majority of mutations produced by overexpression of a gene will be dominant which in turn allows direct selection for a specific phenotype from amongst the population of primary transformants.

The tagging vector we have developed is based on the pPCV series of plant transformation plasmids [17]. The plasmid backbone contains origin of replication and mobilisation sequences allowing maintenance in *Agrobacterium* and left (LB) and right (RB) border sequences delimiting the T-DNA containing a hygromycin resistance gene at the LB sequence driven by the nopaline synthase promoter, the high-copy-number *Escherichia coli* plasmid pIC19H [22], and located 20 bp from the RB sequence the transcriptional enhancer sequence (-90 to -420) [27] of the 35S RNA promoter of CaMV cloned as a tetramer (Fig. 1). The idea of using such a T-DNA as a tag is that following insertion into the plant genome the multiple enhancers will activate the expression of flanking plant genes and that the plasmid allows rescue of flanking sequences with ease. The production of a dominant mutation has a further advantage in that functional testing of flanking plant sequences can be carried out by transfecting protoplasts with the rescued DNA

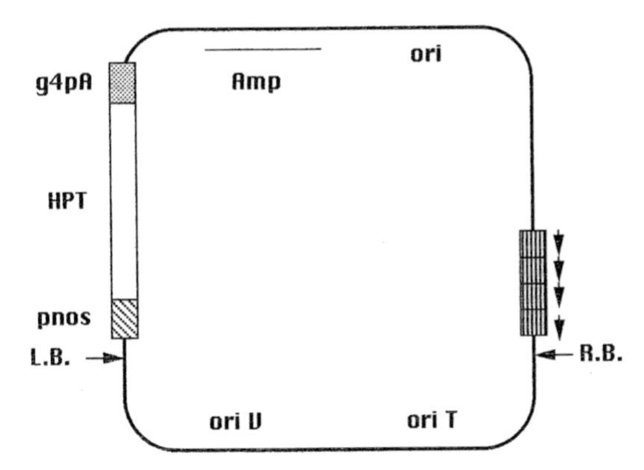

Fig. 1. Structure of the activation tagging vector pPCVICEn4HPT. Functional regions of the plasmid are shown. *oriV* and *oriT* are responsible for stable maintenance in *Agrobacterium* while between the left (L.B.) and right (R.B.) borders of the T-DNA is the hygromycin resistance gene (HPT) fused with the nopaline synthase promoter and the poly A sequences of gene 4 of the *Agrobacterium* T-DNA, an ampicillin resistance gene (Amp) and origin of replication (ori) functional in *E. coli* and the enhancer sequence of the 35S RNA promoter cloned in tandem (arrows).

and selecting for growth under the original selection pressure. This then provides the ultimate proof that the tagged gene responsible for the selected phenotype has indeed been isolated and opens the way for its localization by further testing of deletion derivatives of the rescued plasmid.

We have used tobacco protoplast *Agrobacterium* co-cultivation to generate mutant cell lines transformed with the activation tag which have the ability to grow in tissue culture in the absence of either auxin or cytokinin in the culture media, or in the presence of MGBG, an inhibitor of polyamine biosynthesis. By this means we aim to create novel mutants modified in the metabolism or signal transduction of these plant growth substances, or polyamines.

Auxin-independent mutants

Although auxin has been long recognized as playing a central role in plant growth and development, the molecular basis of its action is little understood [25, 26]. While several experimental systems are available to study auxin action [2, 10, 29], debate remains concerning how auxin modulates its differing effects, such as cell elongation, cell division, side shoot growth and lateral root initiation and growth, either upon external application, or through the expression of transgenes involved in auxin biosynthesis [1 ,14, 28]. There is, however, general agreement that auxin is required in the culture media for protoplasts to divide and form callus [18, 24, 25]. We decided to exploit this in generating tagged mutants that might be modified in auxin biosynthesis, perception, or signal transduction. With this in mind, following tobacco protoplast tranformation with agrobacteria containing the activation gene tag, we selected for growth of transgenic cells in the absence of auxin in the culture media. In a typical experiment, using 30×10^6 protoplasts and with a transformation frequency of about 20–30%, we obtained 13 calli growing under these conditions. The majority of these calli regenerated into plants. Some of the regenerated plants display phenotypic changes such as increased frequency of side shoot formation, reduction of root initiation and growth as well as premature senescence, all of which could be considered as indicative of changes in auxin metabolism and action. The majority of regenerated plants however displayed no obvious phenotypic changes. Nevertheless, in all examples studied to date, protoplasts reisolated from the plant lines divide in culture in the absence of auxin in the media, and this characteristic genetically co-segregates with a single T-DNA insert. By Southern analysis, we judge that at least eight differing genes have been tagged in separate individuals.

We have studied one plant line, *axi 159* in most detail ([9], Walden *et al.*, submitted). In *axi 159* the T-DNA has inserted ca. 6000 bp downstream from the transcriptional start site of the gene responsible for auxin-independent growth, *axi 1*. *axi 1* itself is a 4111 bp gene split by nine introns and apparently is a member of a small multigene family in tobacco. Northern analysis indicates that in untransformed tobacco plants *axi 1* transcripts accumulate to highest levels in root tissue. In the tagged plant line, *axi 1* transcripts are detectable in all tissue tested, confirming the notion of the experiment that upon insertion of the tag into the plant gene the expression of *axi 1* has become deregulated. In freshly isolated protoplasts from untransformed plants *axi 1* expression requires auxin in the culture media. In protoplasts isolated from *axi 1* plants, *axi 1* transcripts accumulate both in the presence and absence of auxin in the culture media. The coding sequence of the gene displays no obvious similarities to genes that have been previously characterized. Nevertheless, preliminary work suggests that *axi 1* expression has an effect on gene expression in both transient assays and in intact plants (Walden, Cjaza and Bongartz, unpublished). Naturally our interest currently is focused on determining how direct, or indirect, this effect might be. Similarly, a number of other genes have been isolated from tobacco lines that are able to promote auxin-independent growth (Harling, Miklashevichs and Walden, unpublished).

Cytokinin independent mutants

Similar to the experiments described above we have generated tagged cell lines that have the ability to grow in the culture media in the absence of cytokinin. One of these lines regenerates plants that are dwarfed only with difficulty. The others, although able to regenerate, are all male- and female-sterile. This obviously precludes genetic analysis but in two examples plant DNA flanking the T-DNA insert has been rescued and, intriguingly, these are able to promote growth in the absence of both auxin and cytokinin (Miklashevichs and Walden, unpublished).

Mutants modified in polyamine metabolism

In the previous examples selection was based on the generation of mutants able to grow in the absence of either auxin, or cytokinin, in the culture media. However, with such a tagging system dominant selection, i.e. growth under normally toxic levels of a particular compound, is also feasible. To test this, and to establish a means of generating mutations in polyamine metabolism or perception, we have used the system to generate cell lines selected to grow in culture in the presence of selective levels of MGBG, an inhibitor of polyamine synthesis. The aim here was to generate mutants in polyamine metabolism, so that the role of polyamines in plant growth and development, an area that remains a matter of debate, might be assessed [5].

Methylglyoxal-bis(guanylhydrazone) (MGBG), is an inhibitor of *S*-adenosylmethionine decarboxylase (SAMdc), a key enzyme in the polyamine biosynthetic pathway being the rate-limiting step in the conversion of putrescine to spermidine and spermine. Previously, Malmberg and coworkers have generated through UV mutagenesis a variety of tobacco lines resistant to MBGB [20]. Plants regenerated from these mutant cell lines displayed a variety of phenotypes including aberrant flower formation with altered patterns of morphogenesis including organ replacement and sexual inversion. In some cases the biochemical basis of the mutation were partially characterized [21]. However, the severity of the phenotypes displayed by the regenerated plants precluded detailed genetic analysis, though in two examples the phenotypes observed appeared to segregate as a single dominant genetic locus [21, 19].

Building on these observations, we decided to apply the strategy of T-DNA activation tagging in an attempt not only to generate mutants modified in polyamine metabolism, but also to isolate the genes involved. We carried out activation tagging and recovered eight independent mutant lines from a tagging experiment involving 30×10^6 protoplasts. All lines regenerated into plants without apparent difficulty. Two of the lines displayed dramatic phenotypic changes including reduced internode length and leaf twisting. However, the most characteristic changes were seen in flower morphology where malformed flowers were male-sterile and possessed elongated styles. In one line parthenocarpy was frequently observed in unpollinated flowers. Two of the plant lines studied in detail contained increased levels of SAMdc ac-

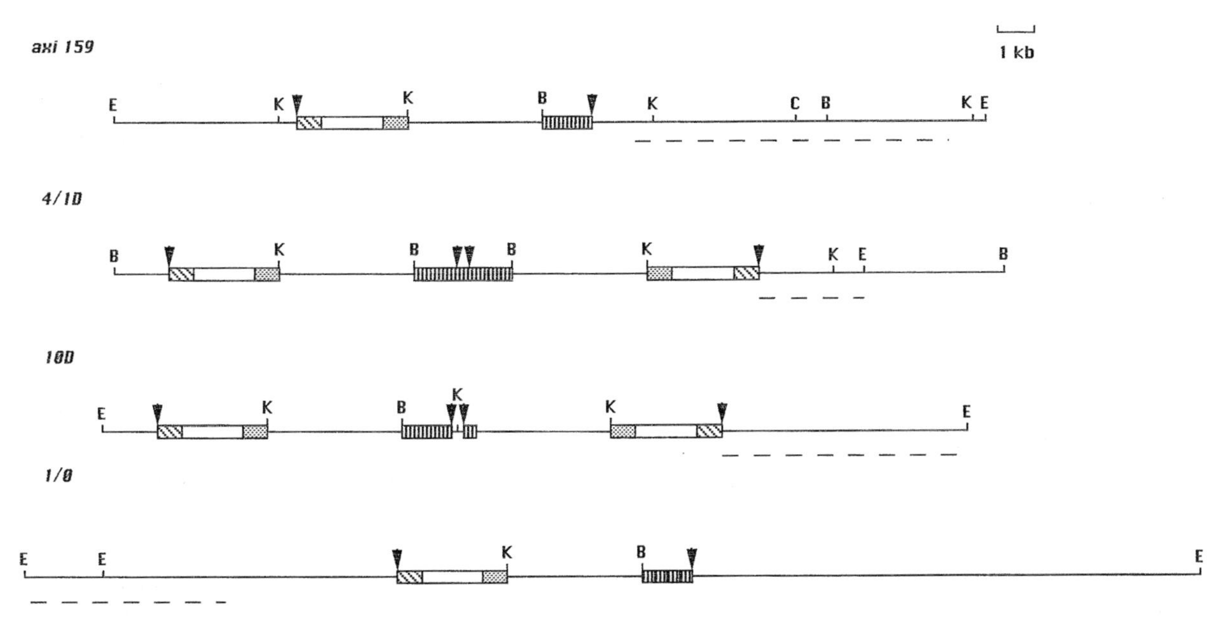

Fig. 2. Organization of the activation T-DNA tag in the genome of differing mutant lines. The organization of the T-DNA in three lines selected for growth in the absence of auxin (*axi159*, 4/1D and 10D), or in the absence of cytokinin (1/0) in the culture media are shown. The borders of the T-DNA are indicated by the inverted arrows and the internal organisation of the T-DNA is as in Fig. 1. In two cases the T-DNA has inserted as a complete (4/1D) or partial (10D) inverted dimer. The position of the region of the plant DNA demonstrated to direct either auxin, or auxin and cytokinin independent growth following plasmid rescue is indicated in each case by the broken line. Restriction enzyme sites are: *Bam* HI (B), *Cla* I (C), *Eco* RI (E) and *Kpn* I (K).

tivity compared with untransformed tobacco. In one plant line this increase in enzymatic activity is coupled with an approximate doubling in the level of putrescine and also significantly higher levels of spermidine compared with untransformed plants. These increases in free polyamines are accompanied by increases also in conjugated polyamines. Intriguingly, in the other plant line the levels of both free and conjugated putrescine, as well as free spermidine, are approximately the same as in untransformed callus but the levels of conjugated spermidine are reduced by about 50%. Plasmid rescue was used to recover the T-DNA and flanking plant sequences from one of the mutant lines and this, when transfected into protoplasts, confers MGBG-resistant growth (Fritze, Cjaza and Walden, submitted). Northern analysis indicates that the rescued plant DNA is overexpressed in the mutant plant line when compared with wild type tobacco and work is currently in progress to clone cDNAs corresponding to the rescued genomic sequences.

Discussion

One of the most direct ways of dissecting a complex biological process is mutant generation and analysis. Gene tagging provides a means not only of generating mutations but also of isolating the genes involved in a specific process. Routinely, mutants resulting from tagging, be it transposon or T-DNA tagging, are recessive making recovery of specific mutants fortuitous. Activation tagging, through the generation of dominant mutations, allows a means of selecting for a specific mutation. The selection schemes we have used, growth in the absence of a necessary growth substance, or in the presence of a selective chemical, relies on little prior knowledge of the process under study. Nevertheless, the mutations that are generated have one characteristic in common: the expression of the tagged gene allows growth under selective conditions. Thus, mutations that we recover could result from a variety of events including: the activation of a specific developmental pathway, overexpression of the target of selective pressure as well as detoxification, increased turnover or reduced uptake of the selective compound. While a selection is likely to generate mutations changed in a specific response or event, we have currently no means of predetermining the step in a process in which a mutation might be generated.

We have used T-DNA tagging as the initial step in a genetic analysis of the molecular basis of the action of the plant growth substances auxin and cytokinin, as well as polyamines, substances that have long been implicated as playing an important role in plant growth and development. Such an approach is likely to be fruitful in these areas of research because, despite intensive biochemical analysis, little is known of the molecular basis of auxin/cytokinin action and the exact role of polyamines in plant growth and development remains a matter of debate.

Here, we have summarized our initial results. Currently, we have studied in some detail about 10 mutants selected for growth in the absence of auxin, five selected for growth in the absence of cytokinin and eight selected for growth in the presence of MGBG. To ease analysis, we have only studied in detail those that contain single- or low-copy T-DNA inserts. Although still at a very preliminary stage, several points emerged.

1. Flexibility of the tagging event. We deliberately used transcriptional enhancers lacking a transcriptional start site so that in the activation of expression of plant genes we were released from the need for the T-DNA to insert upstream from an open reading frame which had no intervening start codons. This precluded us from being able to (1) generate antisense mutations following insertion of the tag downstream from a target gene, (2) recover the tagged gene from the plant genome by inverse PCR. In practice, however, it has allowed us to recover overexpression mutants where the tag has inserted upstream or downstream form the target gene. For example, in the case of *axi 159* we know that the tag has inserted downstream from *axi 1* ca. 6 kb from the transcriptional start site. In addition, preliminary analysis of several of the other tagged mutants that have been obtained indicates that the region of the plant genome responsible for the selected phenotype does not necessarily need to flank the right-border sequence of the T-DNA.

2. Cells able to grow in culture in the absence of auxin or cytokinin are able to regenerate into plants. The selection for growth in the absence of growth substances is based firmly on the observations, dating back at least 50 years, that neoplastic growth as a result of *Agrobacterium* transformation, or habituation, occurred under similar conditions. Such cultures are unable to regenerate into plants and at the outset we thought that the same would be true in our experiments. Although we have generated cell lines that have difficulty in regenerating, the majority have regenerated into plants able to progress through a normal growth cycle. While it may be premature to reach definitive conclusions before the functions of the tagged gene products are identified, it does suggest that the plant cell may have a variety of default systems so that overexpression of these genes does not disrupt normal developmental pathways. Thus, the types of gene tagged in these experiments contrast with the master genes of specific developmental pathways, most notably these of flower development, which apparently tolerate no such default pathways.

3. Dominant mutations ease functional testing. The ultimate proof that a specific gene has indeed been tagged and rescued is functional rescue. With recessive mutations this involves complementation of the mutant line with the wild-type allele of the mutated gene. In the case of dominant mutations, as described here, this involves the transformation of wild-type protoplasts with rescued plasmid sequences followed by screening for growth under selective conditions. This has proved to be a rapid and simple means of not only judging whether a gene has been rescued but also determining with deletion derivates where the gene is located. Routinely, a result is obtained 2 to 3 weeks at the most after protoplast transformation.

In this review we have described our preliminary work involving activation tagging. We think that this form of T-DNA tagging may provide a means of isolating genes involved in complex biochemical and morphological pathways. The selection schemes that we have adopted were deliberately aimed at yielding mutants affected in processes that are currently little understood at the molecular level. While in the short term our goal is to characterize the action of the genes that we have isolated, the numbers of apparently differing mutants that we have generated should allow us, in the long term, to dissect genetically the systems underlying the action of plant growth substances and polyamines.

Acknowledgements

We thank the undergraduate students who worked with the principle author in establishing the conditions for the experiments described here: Ursula Uwer, Michaela Dehio and Christel Schipmann. We have been technically assisted by Inge Czaja and Elke Bongartz. The work described had been funded by the Max Planck Gesellschaft, the Alexander von Humboldt Stiftung and the Graduiertenkolleg of the University of Cologne.

References

1. Cline MG: The role of hormones in apical dominance. New approaches to an old problem in plant development. Physiol Plant 90: 230–237 (1994).
2. Davies PJ (ed): Plant Hormones and their Role in Plant Growth and Development. Martinus Nijhoff, Dordrecht (1987).
3. Depicker AG, Herman L, Jacobs A, Schell J, Van Montagu M: Frequencies of simultaneous transformation with different T-DNA and their relevance to Agrobacterium/plant cell interaction. Mol Gen Genet 210: 477–484 (1987).
4. Estelle M: The plant hormone auxin: Insight in sight. Bioessays 14: 439–444 (1992).
5. Evans PT, Malmberg RL: Do polyamines have roles in plant development? Annu Rev Plant Physiol Plant Mol Biol 40: 235–269.
6. Feldman K: T-DNA insertional mutagenesis in *Arabidopsis*: mutational spectrum. Plant J 1: 71–82 (1992).
7. Gibson, S, Somerville CR: Isolating plant genes. Trends Biotechnol 11: 306–313 (1993).
8. Goldsmith MHM: Cellular signalling: new insights into the action of the plant growth hormone auxin. Proc Natl Acad Sci USA 90: 11442–11445 (1993).
9. Hayashi H, Czaja I, Schell J, Walden R: Activation of a plant gene implicated in auxin signal transduction by T-DNA tagging. Science 258: 1350–1353 (1992).
10. Jacobs WP: Plant Hormones and Development. Cambridge University Press, Cambridge (1979).
11. Jones AM: Surprising signals in plant cells. Science 263: 183–184 (1994).
12. Kay R, Chan A, Daly M, McPherson J: Duplication of CaMV 35S promoter sequence creates a strong enhancer for plant genes, Science 236: 1299–1302 (1987).
13. King PJ: Plant hormone mutants. Trends Genet 229: 181–188 (1988).
14. Klee H, Estelle M: Molecular genetic approaches to plant hormone biology. Annu Rev Plant Physiol 42: 529–551 (1991).
15. Koncz C, Németh K, Rédei G, Schell, J: T-DNA mutagenesis in *Arabidopsis*. Plant Mol Biol 20: 963–976 (1992).
16. Koncz C, Martini N, Mayerhofer R, Koncz-Kalman Z, Körber H, Rédei GP, Schell J: High-frequency T-DNA mediated gene tagging in plants. Proc Natl Acad Sci USA 86: 8467–8471 (1989).
17. Koncz C, Schell J: The promoter of the TL-DNA gene 5 controls the tissue specific expression of chimeric genes carried by a novel type of *Agrobacterium* binary vector. Mol Gen Genet 204: 383–396 (1985).
18. Linsmaier EL, Skoog F: Organic growth factor requirements of tobacco tissue cultures. Physiol Plant 18: 100–127 (1965).
19. Malmberg RL, McIndoo J: Ultraviolet mutagenesis and genetic analysis of resistance to methlyglyoxal-bis(guanylhydrazone). Mol Gen Genet 196: 28–34 (1984).
20. Malmberg RL, McIndoo J, Hiatt AC, Lowe Ba: Genetics of polyamine biosynthesis in tobacco: developmental switches in the flower. Cold Spring Harbor Symp 50: 475–482 (1985).
21. Malmberg RL, Rose DG: Biochemical genetics of resistance to MGBG in tobacco: mutants that alter SAM-decarboxylase or polyamine ratios, and floral morphology. Mol Gen Genet 207: 9–14 (1987).
22. Marsh JL, Erfle M, Wykes, EJ: The pIC plasmid and phage vectors with versatile cloning sites for recombinant selection by insertional inactivation. Gene 32: 481–485 (1984).
23. Marton L, Wullems GJ, Molendijk L: *In vitro* transformation of cultured cells from *Nicotiana tabacum* by *Agrobacterium tumefaciens*. Nature 277: 129–131 (1979).
24. Murashige T, Skoog F: A revised medium for rapid growth and bioassays with tobacco tissue cultures. Physiol Plant 15: 473–497 (1962).
25. Nagata T, Takebe I: Cell wall regeration and cell division in isolated tobacco mesophyll protoplasts. Planta 92: 301–308 (1990).
26. Napier RM, Venis MA: From auxin binding protein to plant hormone receptor? Trends Biol Sci 16: 72–75 (1991).

27. Odell JT, Nagy F, Chua N-H: Identification of DNA sequences required for the activity of the cauliflower mosaic virus 35S promoter. Nature 313: 810–812 (1985).
28. Spena A, Estruch JJ, Schell J: On microbes and plants: new insights in phytohormonal research. Curr Opin Biotechnol 3: 159–163 (1992).
29. Theologis A: Rapid gene regulation by auxin. Annu Rev Plant Physiol Plant Mol Biol 37: 407–438 (1986).
30. Walden R, Hayashi H, Schell J: T-DNA as a gene tag. Plant J 1: 281–288 (1991).
31. Walden R, Fritze K, Harling H: Induction of signal transduction pathways through promoter activation. Meth Cell Biol Plant Cell Biol, in press (1994).

Plant Molecular Biology **26**: 1529–1555, 1994.

Gibberellins: perception, transduction and responses

Richard Hooley
IACR Long Ashton Research Station, Department of Agricultural Sciences, University of Bristol, Long Ashton, Bristol BS18 9AF, UK

Received and accepted 20 August 1994

Key words: gibberellin, growth, development, perception, receptor, gene expression, signal transduction, response mutant, calcium

Introduction

Gibberellins (GAs) are a class of plant hormones that exert profound and diverse effects on plant growth and development. The chemistry and metabolism of GAs have been studied for several decades and this has led to a detailed understanding of the pathways involved in their biosynthesis and catabolism. Attempts to understand the perception and mechanism of action of GAs have been based heavily on studies with the cereal aleurone as a model system and have drawn extensively on parallels with the molecular mechanism of action of mammalian steroid hormones. During the past few years, the established view of GA-perception has been challenged and new techniques for identifying GA receptors have been developed. As our understanding of GA-regulated events in aleurone cells has advanced through molecular and cell biology approaches, other GA-responsive plant tissues have also proved to be tractible for studying GA-action. This has led to a greater awareness of the diversity of cellular events that can be modulated by GAs. These currently range from ion channel activity to gene expression and present multiple targets for GA-signalling. Complementary to these approaches, molecular genetic analyses of GA-response mutants seems poised to provide insight into the identity of genes involved in GA signal transduction. While understanding of the GA perception-transduction-response pathway is in its infancy, there are promising prospects for understanding the molecular mechanisms whereby GAs invoke a variety of cellular and developmental events in plants.

GAs and plant development

It is clear from studies of GA-deficient mutants, and from the effects of the application of exogenous GAs to plants or plant parts, that this class of plant hormones are essential and potent regulators of growth and development. They affect a broad range of events during the normal growth and development of higher plants and, for the purpose of this article, I have divided the types of response of plant cells and tissues to GAs into three categories: (1) cell growth in vegetative tissues, (2) flower and fruit development and (3) seed reserve mobilisation by aleurone cells (Fig. 1). While there may very well be common elements between these different classes of response, the division provides a useful framework within which to consider the GA perception-transduction-response pathways. It is important to emphasise, however, that GAs are not unique regulators of these events. Other plant hormones and environmental stimuli affect plant cell growth

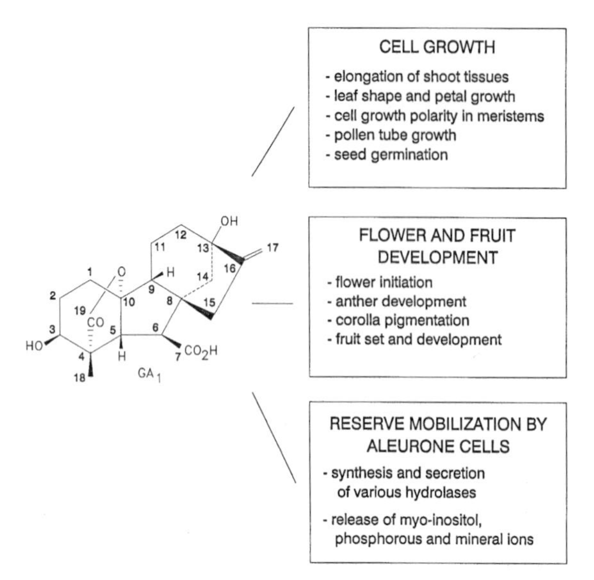

Fig. 1. Gibberellins and plant development. Structure of GA_1 with carbon atoms numbered and a summary of the three GA-response categories.

and the development of flowers and fruit. Seed reserve mobilisation, on the other hand, is under a more specific positive regulation by GAs.

Cell growth in vegetative tissues

Cell growth responses to GAs occur in a range of vegetative tissues. This can be confirmed by comparing the phenotype of a GA-deficient mutant with the wild-type or GA-treated mutant. Perhaps the most obvious of the GA-regulated growth responses in vegetative tissues is the elongation growth of shoots, particularly stem internodes, as well as hypocotyls, coleoptiles, mesocotyls and epicotyls. Other cell growth responses to GA are more subtle and occur only during specific stages of growth of the plant. For example, the transition from juvenile to adult leaf shape is affected by GA in a range of plant species, such as ivy [120], sweet pea [123] and tomato [71]. Cell growth polarity in root tips and shoot apical meristems is also markedly influenced by GAs [5]. During fruit set, GA may stimulate pollen germination and tube growth [106]. In petunia flowers, GA stimulates the growth of petals [154].

GAs promote elongation growth through effects on both cell division and expansion, although the latter process makes a greater contribution to increases in plant stature. The biophysics of GA promotion of elongation growth is complex and not well understood. It probably involves a combination of changes in the hydraulic properties of the cell and alterations in cell wall extensibility. However, the precise nature of these changes, their coordination and the primary action of GA on these processes are not understood [76].

The regulation of orientation of cell wall cellulose microfibril deposition is of primary importance in determining cell growth polarity and, therefore, a potential target for GA-action. The alignment of newly synthesised cellulose microfibrils appears to be mediated by the arrangement of underlying cortical microtubules. GA has been shown to induce transverse orientation of microtubules in a range of different cell types, including some in which the hormone does not affect the rate of elongation growth [5, 130]. GA regulation of the orientation of microtubules is an auxin-dependent process, although the nature of this cooperative response is not clear. Possible mechanisms underlying microtubule reorientation are being examined at the biochemical level and indicate that early changes in microtubule dynamics in response to GA are associated with modification of a specific tubulin isotype [25].

Changes in the expression of a number of genes are now known to accompany GA-induced elongation growth and present molecular targets for studies of GA-regulation of cell growth [95, 115, 129].

Exogenous GA is required for germination of GA-deficient mutants of tomato and Arabidopsis, suggesting that GA may be required for germination of wild-type seed of these species [86, 87]. In dicots, the radicle emerges through the endosperm and there is good evidence that GA promotes radicle emergence in tomato seed by enhancing the growth capacity of the embryo and by weakening the endosperm. The latter effect is caused by GA-induction and stimulation of the cell wall hydrolases endo-β-mannanase and mannohydrolase [104].

Flower and fruit development

Recent studies with developmental mutants of tomato and *Arabidopsis*, combined with the use of GAs and inhibitors of GA biosynthesis in horticulture and citriculture, implicated GAs as important regulators of flower and fruit development.

Flower induction
The transition from vegetative growth to flowering in long-day or cold-requiring plants and in conifers can be promoted by application of GAs. In fact, it is thought that vernalisation and long-day photoperiod promote stem elongation and flowering by stimulating GA biosynthesis [11]. Whether or not GAs influence stem elongation and floral initiation independently is still the subject of some debate, although the balance of evidence currently favours independent regulation of these events [11]. A definitive role for GA during flower induction has been established in *Arabidopsis* through studies with the extreme GA-deficient mutant *ga1-3* [161]. In short days, this mutant does not flower unless GA is applied. Because other less extreme GA-deficient mutants show only a delay in, rather than prevention of, flowering under short days, yet still have a dwarf phenotype, it seems possible that flower induction may have a lower GA response threshold than stem elongation. Whether or not GAs are required for flowering of *Arabidopsis* in long days is not clear. A delay in flowering in the extreme GA-deficient *ga1-3* grown under long days may be interpreted in favour of this, although the phenotype of other GA-deficient mutants and the GA-insensitive *gai* mutant do not. Leakiness of mutants can, however, cloud a definitive implication of cause and effect.

Photoperiod also influences floral development in *Arabidopsis* mutants that have lesions in a set of genes that collectively determine floral meristem development [108]. Recent evidence suggests that, in these mutants, GA can overcome the effect of short days on floral meristem identity and development [108]. Thus, in short days, GAs are important regulators of *Arabidopsis* flower initiation and may influence the expression of floral homeotic genes, such as *LFY* and *AG*.

Flower development
GAs affect the development of anthers which are a source of GAs that appear to influence corolla development and pigmentation [94, 114]. Inhibition of GA biosynthesis can delay anthesis, while exogenous GAs have the opposite effect. The GA-deficient *gib-1* and *gib-2* mutants of tomato show normal floral initiation but are arrested in anther development at a premeiotic G1 phase [67]. The extreme GA-deficient mutant of *Arabidopsis*, *ga1-3*, is male-sterile and has poorly developed stamens and petals [161]. GA-deficient mutants of maize exhibit abnormal flower development. In the terminal flowers, stamen development is not suppressed (as it is in the wild type), resulting in bisexual flowers with pistils and anthers. Flowers of lateral shoots develop anthers but not pistils. Male sterility in GA-deficient mutants of *Arabidopsis* and maize does not appear to be caused by the same arrest in anther development exhibited by the *gib-1* and *gib-2* mutants of tomato, since pollen is formed in these mutants but is not released [67].

In a number of species, the GA content of stamens increases prior to anthesis [114]. In petunia, removal of stamens or anthers at an early stage in corolla development causes a marked inhibition in the pigmentation of corolla tissues. GA applied to the site of stamen removal can, to a large extent, overcome this effect [154].

Fruit development
Together with auxin and cytokinins, GAs appear to be involved at a number of stages in fruit development [35]. Exogenous GAs can promote fruit set and influence development in a range of horticultural species [41, 114]. Developing seeds contain the highest cellular concentrations of GAs in plants. Whether or not these endogenous GAs are essential for fruit growth and development is not entirely clear. There is evidence for promotive effects on cell division and expansion, and on the development of ovaries [6, 35, 114]. The clearest evidence that endogenous GAs are important for

seed development has come from the lh^i mutant of pea which is GA-deficient in the shoot and developing seeds. Seeds of the lh^i mutant abort unless their endogenous GA level is restored to near that of the wild type [118]. At the present time, little insight into GA-perception and signal transduction has come from studies with developing seeds and fruits.

Seed reserve mobilisation by aleurone cells

Aleurone cells are the most widely used experimental system for studying the cellular and molecular biology of GA-action. Aleurone cells of the Gramineae are a highly specialised tissue that differentiates from peripheral endosperm cells during seed development and forms a layers one or three cells thick enclosing the endosperm. Shortly after the seed has germinated, aleurone cells begin to synthesise and secrete a variety of hydrolytic enzymes that participate in the systematic breakdown of starch and protein reserves of the endosperm to provide nutrients for the growing seedling. In addition, aleurone cells contain the majority of the seed's stored reserves of myo-inositol, phosphorous and mineral cations, such as K^+ and Mg^{2+}. These are released into the endosperm after germination and provide the growing seedling with carbohydrate for cell wall synthesis, phosphorous and essential cations. Reserve mobilisation by aleurone cells appears to be coordinated to a large extent by GA produced by the embryo [31, 73].

Cellular responses to GAs

It is clear that GAs, like other plant hormones, have pleiotropic effects during plant growth and development. This variety is also reflected at the cellular level. In attempting to understand GA signal transduction, it is relevant to consider the full spectrum of cellular responses that GAs are able to invoke. The diversity of known cellular responses to GAs illustrates the heterogeneity of signalling possibilities that can be propagated by GAs (Fig. 2).

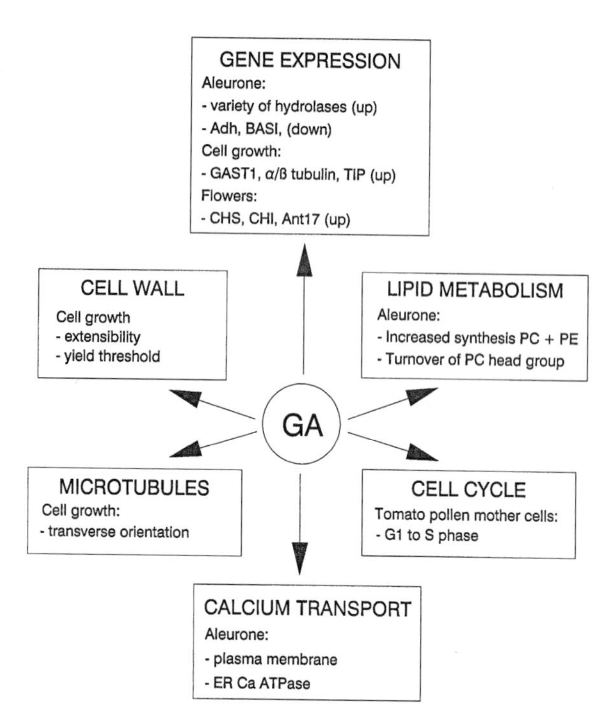

Fig. 2. Cellular responses to gibberellins. Different types of cellular responses to GAs together with brief examples that are elaborated further in the text.

GA structure-activity relations

Although 94 different GAs have been identified in higher plants and fungi, the majority of these are intermediates in a number of closely related biosynthetic pathways and only a relatively small number are biologically active *per se*. Structure-activity analysis of GAs has relied on bioassays in which exogenous GAs are applied to plants, or to parts thereof, and to characterisation of GA biosynthesis mutants. Bioassays suffer from the disadvantage that the ligand is usually applied at a position remote from its site of action. Thus, uptake, transport and metabolism can complicate assessment of the intrinsic activity. However, aleurone protoplasts are a particularly sensitive bioassay system since the GA is applied at, or very near to, its site of action [55, 136].

The biological activities of GAs measured in a number of assay systems are remarkably consistent, with only a handful displaying high biological activity [53, 127, 128, 136]. GAs that are highly active in promoting stem or hypocotyl elongation are almost always highly active in other cell growth responses and in inducing α-amylase gene expression in aleurone. Similarly, GAs that are inactive in cell growth responses are also inactive in aleurone.

The carboxyl group on C-6 of the B-ring, the 19,10-lactone and the hydroxylation status at C-2 and C-3 are particularly important in determining high biological activity of GAs. 3β hydroxylation is generally associated with high biological activity, while 2β-hydroxylated GAs are inactive in all assays. The hydroxylation status at other positions in the molecule are less critical in determining biological activity. It appears, therefore, that structural features in the A/B ring area of GAs are important for determining high biological activity.

An exception to this principle appears to be the induction of flowering in long-day plants. Bioassay data with *Lolium temulentum* suggest that GAs with high florigenic activity are less effective at stimulating elongation growth, and *vice versa*. A double bond in the A ring, either at C-1,2 or C-2,3, is essential for florigenic activity. While 3β hydroxylation is of major importance in promoting activity of GAs for stem elongation and α-amylase induction, 3β-hydroxylated GAs are less active than non-3β-hydroxylated GAs in flower induction in *Lolium*. Hydroxylation at C-12, -13 and -15 enhances flowering relative to stem growth [26, 27]. These observations raise the possibility that GA perception associated with flower induction may involve receptors with specificities for GAs different from those involved in cell growth and seed reserve mobilisation responses. Alternatively, the possibility that uptake, transport and metabolism may obscure the intrinsic activity of GAs is a problem that is difficult to resolve [26].

Except for this example, the balance of evidence suggests fairly conserved structure-activity relationships for all classes of responses to GAs. This can be interpreted in favour of a degree of conservation in the GA perception mechanism and, consequently, diversity in at least some of the downstream signalling components. This situation contrasts quite markedly with other classes of hormones. For example, mammalian steroid hormones show substantially greater diversity in their range of biologically active structures than do GAs. Nevertheless, in spite of this diversity, steroids mediate the majority of their responses through specific interactions with members of a receptor superfamily [111]. Based on our current knowledge of the structure-activity relationships of GAs, it seems likely that, with the possible exception of flower induction, structural features of the receptor(s) involved in the molecular recognition of ligand will be highly conserved between the different classes of response to the hormone. This does not, however, exclude the possibility that other features of the receptor(s) which may, for example, be involved in coupling to signal transduction pathways, might show structural diversity related to the response chain, or chains, in which they are involved.

Cellular site of GA perception

The structural similarity between GAs and mammalian steroid hormones has led to the suggestion that GAs might act on plant cells by a mechanism similar to steroids in mammals [137]. There is good evidence that at least some components of the steroid-signalling pathway are either present in plant cells, or that proteins which can substitute for these functions are [125]. Nevertheless, the question of whether GAs signal through such a pathway is not clear.

GAs are hydrophobic weak acids with pK_as in the range 4.0 to 4.2 [147]. They can enter plant cells by passive partitioning of undissociated molecules that become trapped as relatively impermeant anions in alkaline compartments within the cell. GA_1 uptake by spinach cell suspension cultures appears to have, in addition, a carrier-mediated component [107]. Whether or not GA transport proteins occur in other plant tissues is a question that has not been examined in detail. GA uptake by plant cells will be reduced substantially as the pH of the medium bathing them is increased above the pK_a [107]. Notwithstanding the complication of GA metabolism [98, 102], it can be argued that, if GAs are perceived by receptors inside plant cells, conditions that reduce GA uptake should decrease the response to the hormone. Some evidence supports this theory [44, 134], while other information opposes it [40, 136]. The interpretation of data produced under physiological extremes of pH however, may be complicated by nonspecific effects on the physiology of cells. For example, extracellular pH has a substantial effect on cytoplasmic Ca^{2+} in barley aleurone protoplasts [15].

GA perception at the plasma membrane

Membrane impermeant GAs are active in aleurone
We have attempted to identify the site of GA perception in aleurone protoplasts of *Avena fatua* with an impermeant derivative of GA_4 [55] using a principle originally pioneered with insulin [18]. Elements of the D ring of the GA molecule are not of major importance in determining biological activity of GAs [127, 128] and we have exploited this by synthesizing a GA_4 derivative with a thiol-containing addition on the D ring. This compound (GA_4-17-S-$(CH_2)_3$-SH) induces α-amylase in aleurone protoplasts of *A. fatua*, although its biological activity is approximately two orders of magnitude lower than that of GA_4 [55]. The derivative has been coupled, using a hydrophobic spacer arm, to 120 μm diameter beads of Sepharose 6B, such that elements of the GA_4 molecule that are most likely to confer biological activity are exposed, but can extend no further than ca. 1.95 nm from the surface of the Sepharose 6B. GA_4-17-Sepharose beads are not likely to cross the plasma membrane of aleurone cells. However, the exposed regions of the GA_4 molecule will be able to interact with both the surface of the plasma membrane and components within the membrane to a depth of ca. 1.95 nm. GA_4-17-Sepharose will not be able to penetrate the aleurone cell wall and this property has been used to monitor for any free GA_4 that might be released from the GA_4-17-Sepharose. When aleurone layers were co-incubated with aleurone protoplasts and the amounts of α-amylase mRNA induced in each tissue by the GA_4-Sepharose was determined, the impermeant GA_4 was found to induce high level α-amylase gene expression in aleurone protoplasts and only very low levels of α-amylase mRNA in aleurone cells [55]. These observations led us to propose that GA_4 can be perceived at the aleurone plasma membrane and is coupled to the regulation of α-amylase gene expression by an, as yet, undefined signal transduction pathway [55].

Another method of reducing the membrane permeability of GAs has been explored by the synthesis of GA_4-17-sulphonic acid [10]. The pK_a of the sulphonic acid group is 0.6 and, therefore, it will be ionised, and hence membrane-impermeant, at all physiological pH values. Nevertheless, GA_4-17-sulphonic acid is biologically active in aleurone layers and protoplasts of *A. fatua* and has the same activity as a similar derivative that contains a short hydrophobic addition at C-17 in place of the sulphonic acid group [10]. In the absence of evidence for carrier-mediated GA uptake in aleurone cells, the fact that two GA_4 derivatives that differ profoundly in their ability to cross the plasma membrane by passive diffusion have identical activities in aleurone layers and protoplasts again argues for GA perception at the external face of the plasma membrane.

GA microinjection
The theory that GA may be perceived at the aleurone plasma membrane [55] has recently been supported by the results of experiments [38] in which either GA_3 or GA_3 plus ABA were microinjected into single barley aleurone protoplasts embedded in thin films of agarose [51]. The responses were assessed by three criteria: stimulation of vacuolation, secretion of α-amylase into a starch-agarose film and transient GUS expression driven by 1.8 kb of the barley *α-Amy1/6-4* promoter. None of these responses was stimulated when GA_3 was microinjected to internal concentrations of between 1 and 250 μM, while subsequent addition of GA_3 to the external medium stimulated the responses. Similarly, external ABA antagonised the GA_3-stimulated re-

sponses while microinjection of up to 250 μM ABA did not. Microinjection most likely delivered the hormones directly into the cytoplasm and it is clear, therefore, that, when GA and ABA are presented to aleurone protoplasts in this way, they are not active [38].

A substantial body of evidence now supports a plasma membrane location for GA perception in aleurone cells and for ABA perception in both aleurone and stomatal guard cells [1]. The cellular location of GA receptors has not been examined in other tissues. Nevertheless, until receptors for these plant hormones are identified definitively it may be prudent to anticipate mechanisms of plant hormone perception that have evolved to meet the needs of higher plants and which may, therefore, be unique to them, rather than to adopt whole-heartedly preconceptions derived from models of mammalian hormone action. The possibility that there may be more than one class of GA receptor should also be considered. This principle has been established for steroid hormones, since there are now known to be specific steroid-binding sites in neuronal plasma membranes, which include GABA receptors [109].

GA-binding proteins

Since the mid 1970s, with the availability of high specific activity (30–50 Ci/mmol) [^{3}H]GA_1 and [^{3}H]GA_4, GA-binding activity has been sought in a range of plant tissues. Initial studies were based on *in vivo* assays analogous to those used for defining mammalian steroid hormone binding. Subsequently, *in vitro* binding assays were developed and used to define and to partially purify GA-binding activities, particularly in soluble protein fractions prepared from tissues that elongate in response to GAs. More recently, GA affinity chromatography and photoaffinity labelling with high-specific-activity radioiodinated GA derivatives have been developed and used to define specific GA-binding proteins. At the present time, however, there is no well characterised GA-binding activity that can be considered a firm candidate for a GA receptor.

Soluble GA-binding proteins and elongation growth

GA-binding activity has been detected in cytosol from pea epicotyls, cucumber hypocotyls, maize leaf sheaths and mung bean hypocotyls, tissues that show an elongation growth response to GAs. The existence of these soluble proteins that specifically bind biologically active GAs has been interpreted in favour of a steroid-like signalling mechanism for GAs [137]. The possibility that soluble GA receptors mediate the action of this class of plant hormones through a transduction pathway unrelated to a steroid mechanism of action seems to have been ignored.

The first clear demonstration of GA-binding activity came from studies with the highly GA-sensitive epicotyl hook of a dark-grown GA-deficient dwarf mutant of pea [138] and was subsequently characterised further by Keith and Srivastava [79]. In both of these investigations, [^{3}H]GA_1 was supplied to intact tissue. After homogenization, cytosol was recovered and binding activity was monitored in fractions eluting from a Sephadex gel filtration column. [^{3}H]GA_1 was associated with a high-molecular-weight fraction and a 30 kDa fraction. About 50% of the [^{3}H]GA_1 associated with the 30 kDa fraction had been metabolised, by 2β hydroxylation, to GA_8. [^{3}H]GA_1 associated with the high-molecular-weight fraction had not been metabolised. The binding activity in the high-molecular-weight fraction appears to have been due to both soluble and microsomal proteins, although the investigations concentrated on defining the former rather than the latter activity.

In vivo GA binding was reduced by biologically active, but not by inactive, GAs. However, it was not possible to exchange [^{3}H]GA_1 *in vitro* with excess GA_1 GA_3 or GA_8, nor was the binding pH-sensitive. These are not encouraging features as far as the biological relevance of the binding activity is concerned. Nevertheless, this work did spur *in vitro* GA-binding assays with cucumber hypocotyl cytosol [80] and a soluble GA_4-binding activity was demonstrated to be saturable, reversible, pH-sensitive with an optimum of 7.5 and displacable by active, but not inactive, GAs.

Because the $[^3H]GA_4$ dissociated from its binding protein more rapidly than the activity detected in pea epicotyls, a rapid DEAE-cellulose filter-binding assay was devised [81] and used to characterise the binding activity in greater detail. A single class of binding sites with a K_D of about 70 nM, a half life of dissociation of 6–7 min at 0–2 °C and an abundance of about 0.4 pmol per mg soluble protein was described. There was a good correlation between the binding affinity for different GAs and GA derivatives and their biological activity in this tissue [81, 167].

It was recognised that the soluble GA-binding activities were labile and easily lost during extraction. Improved methods of protein extraction combined with either the DEAE filter paper or gel-filtration assays were used to search for GA-binding activity *in vitro* in cytosol from pea epicotyls [90, 92]. The DEAE filter assay revealed a saturable, exchangeable and specific $[^3H]GA_4$ binding activity in both dwarf and tall pea. The K_D for $[^3H]GA_4$ was 130 nM in dwarf pea and 70 nM in tall pea, with the number of binding sites estimated as 0.66 and 0.43 pmol per mg soluble protein in the dwarf and tall, respectively. Specific binding of $[^3H]GA_1$ to pea proteins could not be demonstrated using the DEAE filter binding assay but Lashbrook *et al.* [90] were able to measure some, apparently specific, saturable and pH-sensitive $[^3H]GA_1$ binding to pea cytosol proteins. However, this was against a high background of non-specific binding.

GA-deficient mutants of *Zea mays* are highly responsive to GA_1, which is believed to be the endogenous GA regulating elongation growth in this, and probably other, plant species. Nevertheless, a detailed investigation of $[^3H]GA_1$ binding by leaf sheath cytosol from wild-type, GA-deficient dwarfs and the GA-insensitive *D8* mutant did not reveal any binding activity that was specific for biologically active GAs [82, 83].

Taken together, these observations indicate that specific binding of the endogenous growth-promoting GA_1 is very difficult to demonstrate in either *in vivo* or, particularly, *in vitro* binding assays with soluble proteins extracted from GA-responsive elongating tissues. Under certain conditions of extraction and assay, specific, reversible, saturable and high-affinity binding of $[^3H]GA_4$ can be demonstrated with soluble proteins prepared from some, but by no means all, tissues that show GA-regulated elongation growth. Attempts to purify these soluble GA-binding activities have met with limited success [137]. These proteins might be receptors or enzymes involved in GA metabolism. Evidence for and against a receptor role has been discussed [137, 168]. However, assignment of a definitive role must await purification and identification of these binding activities.

GA affinity chromatography has recently been used to purify GA-binding proteins [103]. A crude protein fraction containing cytosol and other proteins was prepared from mung bean hypocotyls. Non-specific binding proteins were removed by passage through 2,4-D-Sepharose before allowing proteins to bind to a GA_3-Sepharose matrix. Because of difficulties in preparing the affinity matrix, GA_3 was linked to the Sepharose through the carboxyl group on C-6. This is not an ideal choice, considering the importance of this functionality for biological activity. Nevertheless, the decision to use this type of matrix was also based on the observation, in small scale experiments, that a GA-binding activity was recovered from both GA_3-7-Sepharose and the more difficult-to-prepare GA_4-17-Sepharose. GA-binding proteins were recovered by washing the affinity matrix with a high-salt pH 7.6 buffer and purified further by anion-exchange and gel filtration chromatography. GA-binding activity was monitored by a simple ammonium sulphate precipitation assay using $[^3H]GA_4$ [103]. It is, perhaps, unfortunate that the authors chose to use a binding assay that has been so heavily discredited [149]. Nonetheless, a specific $[^3H]GA_4$ binding activity was observed in the initial crude protein fraction which became enriched during subsequent chromatography steps. The native protein was estimated, by gel filtration, to be between 150 and 200 kDa and, on the basis of SDS-PAGE analysis of protein fractions recovered from native PAGE that were shown to bind $[^3H]GA_4$, it was suggested that the protein might be heteroligomeric with subunits of 23 and 35 kDa. Notwithstanding the potential difficulties outlined above, the binding appeared to be specific for GA_4, GA_7 and GA_9. While this binding activity does not appear to have the characteristics expected of a receptor, the technique of GA affinity chromatography appears to have considerable potential for the purification of GA-binding proteins.

GA-binding activity in lettuce hypocotyl cell wall fractions

One other example of binding of the endogenous growth-promoting GA_1 with plant material has come from studies with lettuce hypocotyl. Tissue sections undergo elongation growth when exposed to exogenous biologically active GAs. Of the $[^3H]GA_1$ taken up by hypocotyl sections, some 2–5% was associated with a cell wall-containing fraction. Association of $[^3H]GA_1$ with cell wall material increased with time and was dependent on protein synthesis [139, 140, 141] but could only partly be displaced by prolonged chasing with unlabelled GA_1. The available evidence points strongly to a covalent association between the $[^3H]GA_1$ and components of the cell wall that might include carbohydrates. Studies of the temporal relationship between cell extension and $[^3H]GA_1$ association with cell wall material under conditions permissive and inhibitory for elongation growth indicated that the phenomenon was not simply a consequence of growth [142]. It has been suggested that interaction between GA_1 and cell wall components might interfere with the action of cross-linking enzymes that limit cell wall expansion [142]. This hypothesis has not been tested further.

GA-binding activity in aleurone

GA-regulation of α-amylase gene expression in aleurone cells of members of the Gramineae is, perhaps, the most extensively studied response to this plant hormone. Aleurone cells are a uniform cell type that are highly responsive to GA and, as such, should be a good source of receptors for this plant hormone. GA-binding activity in aleurones was first reported in an *in vitro* binding study with wheat subcellular fractions in which $[^3H]GA_1$ binding was assessed using a centrifugation assay [70]. Non-aqueous subcellular fractionation techniques led to the recovery of a fraction enriched in aleurone grains which bound $[^3H]GA_1$. However, the binding activity was competed for by ABA [70] and this raises some uncertainty about its possible biological relevance. Use of the *in vivo* GA-binding technique was complicated by rapid metabolism of GAs to inactive 2-β-hydroxylated and glycosylated forms. An attempt to define *in vivo* $[^3H]G_1$binding in barley aleurone overcame the problem of metabolism by incubation at 0–1.5 °C [78]. At equilibrium, $[^3H]GA_1$ was present at a higher concentration inside the cells than in the medium and, because no significant metabolism could be detected, this suggested that the $[^3H]GA_1$ had bound to subcellular components. A small proportion of the $[^3H]GA_1$ binding could be competed by a 50-fold molar excess of unlabelled GA_1, but not by the inactive GA_8 [78]. These observations provided evidence of specific GA binding in aleurone cells but the nature of the interaction has not been elaborated any further.

GA photoaffinity labelling

Photoaffinity labelling is a powerful technique for identifying hormone-binding proteins. Photoaffinity probes have been developed for, and used to identify and characterise, a number of proteins that appear to be involved in plant hormone sequestration, transport and action [13, 28, 30, 50, 57, 169].

In order to identify GA-binding proteins that may be involved in GA perception in aleurone, several aryl azido GA photoaffinity reagents have been developed [9, 54]. Initially, a tritiated reagent $[^3H]GA_4$-S-ABA [9] was used to photoaffinity label *A. fatua* aleurone protoplasts but it was not possible to detect any specific GA-binding proteins with this probe [54]. One shortcoming of this reagent was the long exposure times required for fluorographs. These difficulties were overcome by preparing GA_4 photoaffinity reagents radiolabelled to higher specific activity with ^{125}I. The GA_4 analogue (GA_4-17-yl-1′-(1′thia)-propan-3′-ol-4-azido-5-^{125}iodosalicylate), hereafter referred to as [^{125}I]GA_4-O-ASA, was synthesised and radioiodinated to >120 Ci/mmol [9]. The non-radiolabelled, iodinated GA_4-O-ASA is biologically active, inducing high-level α-amylase gene expression in *A. fatua* aleurone layers and protoplasts. It can be photolysed within seconds on exposure to intense visible light >300 nm [9].

The effectiveness and specificity of [^{125}I]GA_4-O-ASA as a photoaffinity probe for a model GA_4-binding protein has been confirmed by assessing covalent attachment of the reagent to a number of anti-GA monoclonal antibodies [150]. Further modification of the substrate has led to the preparation of a closely related GA_4 analogue GA_4-17-sulphoxyethyl-*p*-azido-salicylate (GA_4-17-SE-*p*-ASA) that can be radioiodinated to a specific activity of greater than 1000 Ci/mmol [56]. In the absence of an established *in vitro* GA-binding activity from aleurone cells, *in vivo* GA photoaffinity labelling has been performed as an extension of the *in vivo* GA-binding assay principle described above under conditions where [^{125}I]GA_4-O-ASA and [^{125}I]GA_4-17-SE-*p*-ASA are inducing α-amylase gene expression and, therefore, are likely to be interacting with a GA receptor. Numerous aleurone polypeptides are labelled by the photoaffinity probes, several approximately in proportion to their abundance [56, 152]. This non-specific labelling may be a feature of the photoaffinity probe that can be improved on by further refinements in its design. Nonetheless, judicious use of non-radiolabelled GAs as competitors allowed specific and non-specific interactions to be discriminated and led to the identification of two GA-binding proteins in aleurone cells of *A. fatua*, a 50 kDa polypeptide recovered in a soluble fraction and a 60 kDa polypeptide recovered in a microsomal membrane-containing fraction [56, 152]. Although the identities of these two GA-binding proteins are not yet known, the fact that they can be identified after SDS-PAGE means that it may be possible to purify and sequence them and thereby gain insight into their possible functions.

Attempts to detect specific GA binding in *in vitro* GA photoaffinity labelling experiments have so far met with only limited success [152], perhaps indicating the labile nature of specific binding observed *in vitro*. Photoactive derivatives of N-substituted phthalimides that have GA-agonist activity are also being developed as photoaffinity probes for GA-binding proteins [144].

Interaction of GAs with liposomes

The permeability of soybean lecithin liposomes to both neutral and charged molecules can be enhanced by GA_3, which lowers the phase transition temperature of the liposomes in a concentration-dependent manner over the range 250 μM to 2.5 mM [164, 165]. However, biologically inactive GA_8 had very similar effects. This is consistent with the suggestion, based on nuclear magnetic resonance studies [166], that the carboxyl group on C-6 associates with the quaternary nitrogen of phospholipid head groups. The effect of GAs on the phase-transition temperature of phospholipid liposomes has also been studied by electron spin resonance and differential scanning calorimetry [112]. A GA_4/GA_7 mixture interacted with the liposome surface and lowered the phase-transition temperature. In contrast to the observations of Wood and Paleg [164, 165], however, GA_8 was found to be inactive. The interaction between $GA_{4/7}$ and the liposome surface was enhanced at low pH presumably through minimising charge repulsion at the carboxyl group on C-6.

Whilst these observations with artificial membranes suggest that interactions between GAs and plant membranes may indeed occur and lead to changes in physicochemical properties, the biological significance of these *in vivo* is questionable. The specificity of the interactions for biologically active GAs is still controversial and the concentration range over which the responses are observed suggest affinities several orders of magnitude lower than the biological activity of the hormone *in vivo* [105].

Low-temperature preincubation has been reported to increase the sensitivity, measured as the amount of α-amylase synthesised, of aleurone layers from *Rht3* dwarf wheat to GA [131]. Changes in aleurone phospholipids, in particular the head group and acyl content of phosphatidylcholine and phosphatidylethanolamine, were correlated with changes in sensitivity in response to low-temperature treatment and this has been interpreted as indicating that membrane lipids are primary sites of GA_3 perception [131, 132, 133]. Whilst phospholipid composition may well influence the interaction of GAs with membranes [112], the available evidence is at best only correlative and provides no direct evidence for membrane lipids being the site of GA perception. In addition, a major difficulty with the interpretation of the studies of Singh and Paleg [131, 132, 133] is that phospholipid analyses were performed on total aleurone phospholipids. Clearly, as far as GA perception mechanisms are concerned, it would be more pertinent to analyse and quantify changes in the lipid composition of defined membrane fractions.

GA-response mutants

Single gene response mutants provide a means of identifying genes whose products may encode receptors or components of signalling pathways. They can be generated by either chemical mutagenesis, by ionising radiation or by insertional mutagenesis using transposons or *Agrobacterium tumefaciens* T-DNA. The mutated gene can be identified by positional cloning if sufficient markers are nearby, genomic subtraction if it comprises a large deletion, or, in the case of insertional mutants, using specific hybridisation probes. A number of higher-plant genes that encode, or are thought to encode, components of hormone signalling pathways have been identified by these means and others are certain to follow [22, 39, 77, 91, 93, 96].

Molecular genetic analysis of ethylene-signalling mutants has provided evidence that the regulation of growth by this plant hormone may involve a protein kinase cascade. A genetic model based on the epistatic relationships between ethylene-resistant and constitutive response mutants has been defined and molecular cloning has revealed the identity of the products of two genes in this pathway. The first, *ETR1*, encodes a histidine kinase with sequence similarity to two-component regulators, while the second, *CTR1*, encodes a Raf-like serine/threonine kinase [22, 77]. Components of ABA-signalling pathways have also been identified by molecular genetic approaches. A transcriptional activator which appears to interact with transcription factors involved in ABA-regulated gene expression and is confined to seed-specific ABA signalling [39, 93] has been cloned from maize and *Arabidopsis*. A component common to all ABA-signalling pathways has been identified as a novel type of protein serine/threonine phosphatase of the 2C type [91, 96].

A major factor in determining the success of the molecular genetic approach to studying signal transduction is the stringency of the screen used to select the desired mutant. This is based on an anticipated phenotype. For GA-response mutants, two strategies have been adopted, each based almost exclusively on plant stature, and these have led to the selection of two classes of GA-response mutants: (1) GA-insensitive dwarfs or semi-dwarfs and (2) constitutive GA responders (slender mutants). In view of these rather general selection criteria, perhaps it is not surprising that GA-response mutants are probably the most common plant hormone mutants [118].

Dwarf and semi-dwarf GA-insensitive mutants

A range of genetic lesions could be expected to give rise to plants with dwarf phenotypes and reduced sensitivity to GA. The majority of these will have nothing to do with GA perception or signal transduction. The strategy for identifying putative GA perception or transduction mutants has been to analyse the phenotypes in detail and identify those which very closely resemble GA-deficient mutants in all aspects except sensitivity to exogenous GA. Although extreme response capacity mutants could fall into this class, they may nevertheless help provide insight into later stages in transduction pathways and components of the responses *per se*. The recessive *lka* and *lkb* dwarf GA-insensitive mutants of pea, for example, have been found to have substantially higher cell wall yield thresholds than wild-type plants [118] and may, therefore, be useful tools with which to study the effects of GA on cell wall physicochemical properties during cell growth responses.

Dwarf and semi-dwarf GA-insensitive mutants have been characterised in a number of species [118]. Those that have received most attention are the *D8*, *D9* and *Mpl1* mutants of maize [45, 162], *gai Arabidopsis* [113] and *Rht3* wheat [34]. All are dominant or semi-dominant mutations. Based on the similarity in phenotype to GA-deficient mutants, it has been argued that the most likely candidates for receptor or signal transduction pathway mutants are maize *D8*, *Mpl1* and *Arabidopsis gai*.

Dominant mutations that confer insensitivity to GA might act through one of two mechanisms, depending on the function of the wild-type gene product. In the first mechanism, the wild-type gene product would be a negative regulator of the GA-response pathway and be inactivated by GA. A mutant in which the negative regulator could no longer be inactivated by GA would be a dominant gain-of-function mutant. In the second mechanism [48], the wild-type gene product would be a positive regulator of the response pathway but would function only as a multimer. A mutant in which this positive regulator loses its function would be dominant if the defective protein disrupted or inactivated the multimer necessary for GA action, a phenomenon also referred to as complex poisoning. Dominant mutations conferring insensitivity to other plant hormones have been identified. The *ETR1* and *ABI1* mutants of *Arabidopsis* are currently the best examples, since the gene products for both of these mutants have been identified and the amino acid substitutions conferring the mutant phenotype identified [22, 91, 96]. In neither of these cases is it yet possible to state whether the mutations are gain- or loss-of-function, but this should become clearer after detailed biochemical analyses.

Genetic analysis of the *D8* and *Mpl1* mutants of maize suggest that they are gain-of-function mutations that are largely independent of the wild-type gene products, although they could also result from overproduction or ectopic expression of the wild-type gene product. Clonal analysis of X-ray-induced wild-type somatic sectors in *D8* and *Mpl1* backgrounds indicates that, in certain tissues, the mutations are cell-autonomous [45].

The *gai* mutant of *Arabidopsis* has a phenotype similar in many respects to the *D8* and *Mpl1* mutants of maize and, because it is in a genetically tractible species, efforts have been focused on cloning this gene [113]. *GAI* is located on the top arm of chromosome 1 of *Arabidopsis* at map position 21.8 and has been placed some 0.6 centimorgans (cM) distal to the GA-deficient mutant *GA4* [113]. RFLP mapping of *GAI* has been hampered by difficulties in scoring *gai* segregation in the ecotypes used although *ga4* has been located between markers m219 and g2359 and it seems likely that *gai* will be in this vicinity too [113]. Derivative mutations of *gai* have been generated by γ-ray irradiation and an M1 screening procedure used to select for mutants with a wild-type phenotype. Although this strategy was aimed at generating second site suppressor mutations, segregation analysis of three derivative alleles recovered showed that they are most likely to be intragenic derivative alleles of *gai*. A fourth appears to be a large deletion or rearrangement [113].

It seems most likely that the derivative alleles of *gai* are loss-of-function mutations rather than restoration of wild-type since chance would favour the former. Peng and Harberd [113] have argued that the balance of evidence suggests that *gai* is most likely to be a gain-of-function mutation, although they are cautious to point out that this cannot be concluded until the *gai* and its derivative alleles are identified. Nevertheless, the observations suggest that the wild-type *GAI* may be dispensible. This might be because a similar protein, perhaps encoded by a related member of a multigene family, can substitute for the GAI product. Clear interpretation of the available data is not yet possible. However, the prospect that *gai* will soon be identified is good since it has been located to an approximately 50 kb region of genomic DNA (N. P. Harberd, personal communication).

Constitutive GA-response mutants

Mutants that have a phenotype similar to the wild-type plant that has been treated repeatedly with GA are referred to as slender mutants. Where it can be demonstrated that this phenotype has not been caused by overproduction of GA, it can be considered to be a GA constitutive response mutant. These are rare by comparison with dwarf and semi-dwarf mutants. The best characterised of this class of mutants are the *cry*^{s}*la* mutant of *Pisum sativum* [117], *sln1* in barley [32], the *Spy* mutants of *Arabidopsis* [68] and the *procera* mutant of tomato [71]. These mutations are all recessive. It is likely that slender mutants are loss-of-function mutations of a negative regulator, or regulators, of the GA perception-transduction pathway. Furthermore, because slender mutants are pleiotropic they may affect a step in the GA perception-transduction pathway that is common to all responses to GA. This class of mutant, therefore, provides further evidence of conservation in the mechanism of action of GA, in spite of heterogeneity in the classes of response mediated by the hormone. In addition, slender mutants also suggest that at least one element of the GA perception-response pathway, or an unrelated pathway that cross-talks strongly with it, has a repressor-like function.

The combined effect of mutations at two loci, *cry*^{s} and *la*, in pea produce a slender phenotype with parthenocarpic fruit development. Because the phenotype of plants with a mutation at only one, rather than both, of these loci is wild-type, it may be that the products of these two genes can substitute for one another in the GA perception-transduction pathway [117]. This slender phenotype appears to be independent of endogenous GA since it is unaffected by the presence of an additional mutation *na* that reduces endogenous GAs [117].

Another slender mutant of pea, *sln*, is of interest because it illustrates an unusual way in which a slender phenotype can arise [119]. At the seedling stage *sln* has a phenotype very similar to *cry*^{s}*la* and is insensitive to early-stage GA biosynthesis inhibitors. However, this mutant is not a GA-independent constitutive responder. The phenotype appears to be caused by abnormally high levels of the inactive GA_{20} in *sln* seed which are converted to active GA_1 in the young seedling. During development of wild-type seeds GA_{20} is catabolised to GA_{29} and GA_{29} catabolite, substrates that cannot be converted to active GAs. Seed of the *sln* mutant, therefore, have an unusually large reserve of potentially active GAs. Because these conversions take place, at least in part, in the maternal tissue of the testa, the *sln* phenotype has an unusual pattern of inheritence [119].

The *sln* mutant of barley is affected in cell growth, flower development and seed reserve mobilisation responses to GA [20, 32, 88]. The phenotype of *sln* barley is unaffected by GA biosynthesis inhibitors [24, 88] suggesting that the GA perception-transduction pathway is constitutively activated in this mutant. Nevertheless, ABA inhibits GA-induction of α-amylase, protease and nuclease in aleurone layers of the mutant, suggesting either that ABA negatively regulates the GA-perception-transduction pathway downstream of *sln* or that it operates through a parallel, but unrelated, pathway. ABA-induction of gene expression is unaffected in the mutant, an mRNA inducible by ABA, and dehydration in roots showed similar response to these treatments in the *sln* mutant and wild type [20].

Three independent recessive mutations with a slender GA-constitutive response phenotype have been generated in *Arabidopsis* by EMS mutagenesis combined with a screening strategy based on resistance to the GA biosynthesis inhibitor, paclobutrazol [68]. These mutations are at the same locus which has been named *SPINDLY* (*SPY*) and they, like the other slender mutants, affect all responses to GAs ranging from hypocotyl and stem elongation, anther development, flowering and fruit development. The phenotype of double mutants between *spy-1* and the GA-deficient dwarf *gal-2* indicates that *spy-1* suppresses all of the growth and development effects associated with GA deficiency.

Unlike *cry^s la* pea and *sln* barley mutants, the phenotype of *spy* mutants is influenced by GA levels. Hypocotyl length in the *spy 1 gal-2* double mutants increases in response to exogenous GA_3 and the dose-response curve parallels that of the single *gal-2* mutant. These observations have been interpreted as indicating that, in a GA-deficient background, *spy-1* and exogenous GA_3 can interact additively [68]. Two models explaining the function of spy have been suggested [68]. In the first, the GA perception-transduction pathway is branched and *spy* constitutively activates one of the branches allowing GA to signal through the other and augment the response. In the second model, the wild-type gene product of *SPY* negatively regulates cross-talk between the GA transduction pathway and an unrelated pathway.

Whether or not the slender mutants all affect the same component of the GA perception-transduction pathway is not entirely clear. The distinction between GA-responsive and GA-insensitive slender phenotypes might be related to the strength of the different alleles as suggested by Jupe *et al.* [74]. However, this may not be the case for *spy* mutants [68].

Relationships between GA-response mutants

In view of the pleiotropic nature of responses of plants to GAs, it is likely that response mutants that are either specific for different tissues or classes of response or that are affected at different points in the perception-transduction chain will be isolated. The dominant GA-insensitive dwarfs and recessive constitutive responders described above are probably affected at different points in the GA perception-transduction pathway. However, since both of these classes of mutant appear to be affected in every response to GA, these mutations must be in elements common to all GA responses throughout the plant.

Evidence of GA-response mutations that are either tissue- or response-specific is very limited. Barley mutants generated by sodium azide treatment were screened for those which had altered GA and ABA sensitivity in a half-seed assay that measured α-amylase production. Several response mutants were isolated, two of which were characterised and found to have reduced sensitivity to GA in the half-seed assay, producing less α-amylase and phosphatase in response to GA_3. They each had normal, or near-normal, growth phenotypes [52]. These mutants have not been characterised in any detail and it is not known where in the perception-transduction-response pathway they are affected.

Response mutants of other plant hormones are restricted to particular tissues and/or types of response. The ABA-insensitive mutants of *Arabidopsis ABI1* and *ABI3* provide a good example of this where the transcriptional activator encoded by *ABI3* is involved in seed-specific gene expression, while the protein phosphatase 2C-like gene product of *ABI1* appears to be involved in all ABA responses. If a larger number of GA–response mutants were generated, it might be possible to position them on both the common and distinct parts of GA perception-transduction pathways involved in the multiple responses of plant cells to GAs. However, there is very little evidence at present that it will be possible to separate cell growth, seed reserve mobilisation and fruit and flower development response categories with specific mutations.

Response regulation of GA biosynthesis

The GA-insensitive dominant or semi-dominant dwarf mutants of *Arabidopsis*, wheat and maize provide an interesting contrast to the positive correlation between GA concentration and growth observed in wild-type plants where elongation growth is associated with elevated levels of biologically active GAs. In these GA-insensitive mutants an inverse correlation exists and the levels of GA_1 and its precursor GA_{20} are elevated while GA_{19} is reduced [2, 33, 47, 145]. It has been argued that this is evidence that a negative feed back regulation of GA metabolism is defective in the mutants as a consequence of impairment in GA-perception or transduction. This phenomenon is thought to exist also with ethylene [43]. GA-insensitive mutants, such as the recessive *lk*, *lka* and *lkb* pea mutants that are thought to be affected in their response capacity, do not show the same correlation, suggesting that the feedback regulation is not directly influenced by growth rate *per se*. The endogenous GAs in slender barley are consistent with the response regulation theory since GA_1 and GA_{20} are lower than in the wild type, while GA_{19} is higher [24]. Certainly, it appears that one component of the GA perception-transduction pathway impacts on GA biosynthesis, probably through influencing the activity of GA 20 oxidase [2, 47].

Phospholipid metabolism: a lipid-based GA-signalling pathway?

Changes in phospholipid and choline metabolism are among the earliest effects of GA on wheat aleurone cells and provide support, for a lipid-based GA-signalling pathway. Within 2 h of exposure to GA_3, the activities of two enzymes involved in phospholipid biosynthesis, phosphorylcholine-cytidyl transferase and phosphorylcholine-glyceride transferase increase substantially [148]. During the lag phase between GA_3 addition and the induction of α-amylase, there is a marked change in the rate of turnover of *N*-methyl and methylene carbons of the phosphatidyl choline (PC) head groups [49, 148]. In non-GA_3-treated controls, *N*-methyl carbons turn over more rapidly than methylene carbons. GA_3 promotes phospholipid breakdown and changes the pattern of PC metabolism such that the whole choline head group, rather than *N*-methyl carbons, turn over. It has been suggested that this change in PC turnover may be a component of a lipid-based GA signal transduction pathway in aleurone [49, 148]. However, until further supporting evidence for this comes to light, the theory must remain speculative. In fact, GA_3-regulated PC turnover and phospholipid metabolism have also been shown to be associated with the aleurone spherosome fraction and, therefore, might be involved in endomembrane assembly [3, 160].

A further rapid effect of GA on aleurone lipids has been demonstrated by Fernandez and Staehelin [29], who found that GA induces the transfer of lipase activity from a subcellular fraction enriched in aleurone storage protein bodies to lipid bodies. This response is quite rapid, occurring between 45 and 120 min after GA treatment. It is partially overcome by ABA.

Calcium and calmodulin: components of the GA transduction-response pathway?

Many of the environmental and hormonal stimuli that act on plant cells induce changes in cytosolic Ca^{2+} (for recent reviews see [14, 116]). These changes are quite rapid and fall into three basic patterns [14]: (1) large, transient increases, (2) small, steady-state increases or decreases and (3) oscillatory changes with regular or irregular periodicities. GA has been shown to cause small steady-state increases in cytosolic Ca^{2+} in aleurone cells [15, 36]. It is important to consider how GA-induced changes in cytoplasmic Ca^{2+} come about, whether or not this category of cytosolic Ca^{2+} modulation has a signalling function and, if so, on what targets it has an impact.

A sizeable body of evidence indicates that Ca^{2+}, possibly in conjunction with calmodulin or a calmodulin-like protein, appears to be involved in coordinating the response of aleurone cells to GA and ABA [12, 17, 36, 37].

Cytosolic Ca^{2+} in aleurone protoplasts

The cytosolic Ca^{2+} status of barley aleurone protoplasts has been investigated by acid-loading of indo-1 and fluo-3 combined with fluorescence ratio analysis and confocal microscopy [15, 36]. Cytosolic Ca^{2+} increased from 50 to 150 nM between 4 and 6 h after treating barley aleurone protoplasts with GA. The increase was not uniform across the cytoplasm. Highest levels (700 nM) were observed at the periphery of the protoplasts and may have been associated with cortical ER [36, 37]. The GA-induced increase in cytosolic Ca^{2+} preceded, by several hours, GA-induced synthesis and secretion of α-amylase. When the concentration of Ca^{2+} in the medium bathing aleurone protoplasts was reduced below 1 mM, both the increase in cytosolic Ca^{2+} and α-amylase secretion were reduced, suggesting that Ca^{2+} influx at the aleurone plasma membrane sustains both these processes [36].

ABA overcomes the GA elevation of cytoplasmic Ca^{2+} within 3 h, restoring near basal levels some 2 h before it inhibits GA stimulation of α-amylase secretion [36]. This response of aleurone cells to ABA contrasts with the increase in cytoplasmic Ca^{2+} that ABA causes in stomatal guard cells [126].

There are three mechanisms by which GA (and ABA) might regulate Ca^{2+} influx at the aleurone plasma membrane. Firstly, the GA receptor may be a ligand-gated anion channel, secondly, GA may be perceived by an intracellular receptor which indirectly modulates the activity of a plasma membrane Ca^{2+} channel and, thirdly, a GA receptor at the aleurone plasma membrane [55] may modulate the activity of a plasma membrane Ca^{2+} channel. Plant plasma membrane Ca^{2+}-permeable channels have been characterised [126, 146] and, in stomatal guard cells, can be activated by ABA, leading to the elevation of cytoplasmic Ca^{2+} [126].

Research aimed at identifying, characterising and studying the regulation of Ca^{2+} channels in aleurone plasma membrane may give insight into GA and ABA signalling in this tissue. In addition, the observation that ABA elevates cytosolic Ca^{2+} in stomatal guard cells [126] while reducing it in aleurone cells [36] suggests that some tissue-specific regulation may be uncovered.

While it is clear that GA and ABA coordinately regulate cytoplasmic Ca^{2+} in aleurone cells, leading to small sustained changes in concentration, there have been few attempts to define biochemical targets for such changes. One candidate may be a homologue of the putative Ca^{2+}-modulated protein phosphatase 2C that has been identified as a component in ABA signalling in *Arabidopsis* [91, 96].

Ca^{2+}-calmodulin regulation of ion channel activity

Another target that has been suggested for Ca^{2+} and calmodulin regulation in aleurone cells is the tonoplast slow vacuolar (SV) ion channel. This channel is probably involved in the mobilisation of K^{+} from phytate stored in aleurone protein bodies [12]. SV channel activity has been detected by patch-clamp analysis of the membrane (tonoplast) of barley aleurone storage protein bodies [12]. Higher current densities were detected in protein bodies from GA-treated protoplasts compared with ABA or minus GA treatments, although single channel recordings were similar between these treatments. Channel opening was stimulated by Ca^{2+} and Ca^{2+} sensitivity was enhanced by calmodulin. Calmodulin antagonists, W7 and trifluoperazine, reduced channel activity. These observations provide evidence for Ca^{2+}-calmodulin regulation of the activity of this SV channel and this may be related to GA-induced increases in both of these signalling molecules in aleurone cells [36, 37]. However, the sensitivity of the SV channel to Ca^{2+} *in vitro* is outside the range of concentrations of cytoplasmic Ca^{2+} occurring in aleurone protoplasts [36]. Clearly, further investigation is required before Ca^{2+} and calmodulin can be placed with any confidence on a signal transduction pathway linking GA with SV channel activity *in vivo*.

Ca^{2+} transport and protein secretion in aleurone

Certainly one of the fates of cytoplasmic Ca^{2+} in aleurone cells is to join a substantial flux out of the cell via the endomembrane system. α-Amylase is a Ca^{2+}-containing metalloenzyme that requires at least equimolar bound Ca^{2+} for activity [16]. It can comprise as much as 70% of newly synthesised secreted protein in GA-treated aleurone cells and its flux through the endomembrane system demands maintenance of ER lumenal Ca^{2+} levels in excess of 5 μM [16]. Because cytoplasmic Ca^{2+} in GA-treated aleurone cells is ca. 150 nM, Ca^{2+} has to accumulate into the ER against a concentration gradient. This is achieved by a Ca^{2+}-ATPase located in the ER membrane [16]. Studies of ^{45}Ca transport into barley aleurone microsomal membrane fractions enriched in ER indicate that the transport of Ca^{2+} into ER is some 5-fold higher in GA-treated aleurone, compared with non-GA-treated controls. This is due to GA stimulating the activity of the ER Ca^{2+}-ATPase, an effect which is overcome by ABA [16, 17].

The effects of both GA and ABA on the activity of the ER Ca^{2+}-ATPase are slow, developing over 4 to 8 h, and, therefore, are unlikely to involve a direct action of the hormones on the activity of pre-existing transporters [17]. One mechanism of regulation appears to be stimulation by a calmodulin-like protein which increases in barley aleurone layers treated with GA [37]. Calmodulin stimulates Ca^{2+} uptake into ER membrane vesicles isolated from barley aleurone that has not been treated with GA to levels comparable with ER membranes from GA-treated tissue. In addition, the calmodulin antagonists, W5 and W7, inhibit Ca^{2+} uptake into ER membrane vesicles prepared from GA-treated aleurone [37]. GA also increases the level of another calcium binding protein, BiP, in the ER of aleurone cells and this is likely to be involved in protein folding [72].

Another role suggested for the GA- and ABA-induced changes in cytoplasmic Ca^{2+} in aleurone cells is the regulation of exocytosis, possibly by a mechanism similar to that in mammalian cells [170].

Calcium and cell growth responses to GA

There is no clear evidence at present to support a role for cytoplasmic Ca^{2+} in GA-induced stem elongation. Although a variety of Ca^{2+} channel blockers and antagonists has been shown to inhibit the growth of GA-treated stem segments of *A. sativa*, they did not affect the early growth in response to the hormone, even when the tissue was preincubated with millimolar concentrations of verapamil for 12 h prior to GA addition. Rather, the Ca^{2+} antagonists inhibited sustained growth of the tissue segments. These findings have been interpreted in support of a role for Ca^{2+} movement in maintaining, but not initiating, GA-induced growth [97]. Clearly there are limitations to the deductions that one may make on the basis of treating whole tissues with Ca^{2+} agonists and approaches that monitor cytoplasmic Ca^{2+} during GA-induced stem elongation may cast further light on this issue. It is not known whether cytosolic Ca^{2+} or calmodulin are involved in other plant cell responses to GA.

GA regulation of gene expression

Gene expression and cell growth responses to GA

Auxin can stimulate cell elongation in excised tissue segments in less than 20 to 30 min. In advance of this, there are rapid and specific auxin-induced changes in the transcription of primary response genes [4]. Elongation growth in response to GA is a much slower response, taking place over hours rather than minutes. In GA-deficient mutants of pea (*le*) and maize (*d5*), increases and decreases in the steady-state levels of *in vitro* translatable mRNAs have been reported as early as 30 minutes after applying GA, even though elongation responses were not visible until several hours later [23]. At the present time these are among the fastest responses yet reported to GA. Although the genes encoding these transcripts have not been identified, it is possible that they may be primary response genes, analogous to those identified in auxin-induced elongation growth.

In the GA-deficient *gib-1* mutant of tomato, decreases in the abundance of several translatable mRNAs have been reported, but only after

6 h of GA treatment. Other mRNAs decreased by 24 h [69]. cDNA clones corresponding to three of these mRNAs and showing both sustained and transient decreases in abundance have been isolated [69]. The first gene shown to be upregulated by GA during GA-stimulated stem elongation (*GAST1*) was identified in shoots of the GA-deficient *gib-1* mutant of tomato and encodes a 12.8 kDa polypeptide of unknown function [129]. An increase in *GAST1* mRNA is detectable 2 h after GA treatment and, by 12 h, levels are >20-fold higher than those in untreated controls. Thereafter, transcript levels decline to untreated levels by 48 h. GA regulation of expression appears to be at both transcriptional and post-transcriptional levels and ABA partially inhibits GA regulation of the gene [129]. In 244 nucleotides upstream of the start of transcription of *GAST1*, there do not appear to be sequences similar to the elements identified in α-*Amy* promoters which are known to be important for high-level GA-regulated expression. The significance of an inverted pyrimidine box sequence in the 3′-untranslated region is not known.

The first genes with a known function to be shown to be modulated by GA during elongation growth were tubulin genes [95]. Tubulin gene expression is enhanced in a concentration-dependent manner during GA_3-induced internode elongation in *A. sativa*. Levels of both α- and β-tubulin transcripts increase 5–6-fold 6 h after GA_3 treatment, a time when GA stimulation of growth is just detectable. Transcript levels continue to increase for up to 24 h. Elongation growth and GA stimulation of tubulin mRNA could be inhibited by ABA and cycloheximide. These observations suggest a close correlation between tubulin gene expression and GA-regulated elongation growth but also indicate that GA regulation of tubulin gene expression requires sustained protein synthesis. Tubulin genes therefore do not have the characteristics of primary response genes directly influenced by GA, suggesting that GA regulation of their expression might be through the hormone inducing the production of a regulatory protein that may be a component of the perception-transduction pathway.

Using a PCR-based subtractive hybridisation technique, Phillips and Huttly [115] have identified two cDNAs that are up-regulated after GA treatment of the GA-deficient *ga1* mutant of *Arabidopsis*. One of these cDNAs encodes tonoplast intrinsic protein (γ-TIP) present in vegetative tissues that is thought to function as a passive water channel. Gamma-TIP mRNA increases substantially between 8 and 24 h after GA treatment, although whether or not this increase is confined to extending tissue is not clear. It has been suggested [115] that an increase in γ-TIP may promote water transfer into the vacuole that could either generate turgor as a driving force for cell expansion or restore turgor lost through GA-induced stress relaxation of the cell wall. The other GA-stimulated cDNA encodes a proline- and glycine-rich protein with a potential secretory signal peptide. GA stimulates the abundance of this transcript between 8 and 24 h after treatment and, based on its expression in wild-type Landsberg *erecta*, it may be restricted to flowers [115].

Gene expression during flower and fruit development

Flower induction

A preliminary report suggests that GA may directly or indirectly regulate the expression of floral homeotic genes [108].

Anther development

GA-application to *gib-1* tomato plants rescues stamen development and gives rise to both increases and decreases in the abundance of a number of *in vitro* translation products [69] Within 8 h of GA application, increases and decreases in specific translation products could be seen. Changes in other translation products were not detected until 24 or 48 h. By comparing the *in-vitro* translation products from stamen and shoot poly(A)$^+$ RNA of GA-treated and untreated plants, it was possible to identify translation products that were modulated by GA and appeared to be specific to stamens. Because some of these altered in abundance before morphological changes associated with the rescue of stamen development became apparent, it is possible that they may be involved in GA-regulated stamen development [69]. The cDNA cloning and identification of the genes encoding these translation products has potential to give insight into GA regulation of gene expression during stamen development.

Anthocyanin biosynthesis

The anthers of petunia flowers are a source of GAs that promote the pigmentation and growth of tissues of the corolla [154]. GA induces the anthocyanin production necessary for corolla pigmentation in petunia when applied to emasculated flowers or to isolated petals incubated in the presence of sucrose [154, 156, 157]. Exogenous GA induces the expression of a number of anthocyanin biosynthetic genes. Chalcone synthase (CHS) is normally expressed in corollas. When they are detached from the flower and cultured without GA, CHS mRNA decreases but can be restimulated by GA. Nuclear run-on transcription experiments confirm that this effect is at the level of transcription and has a lag time of some 10 h. The fact that GA regulation of CHS transcription is sensitive to cycloheximide suggests that CHS is not a primary response gene regulated by GA [156, 157].

Other genes of the anthocyanin biosynthesis pathway are regulated by GA in petunia corollas. On the basis of the expression of chalcone isomerase (CHI) promoter-GUS fusions in transgenic petunia flowers, it seems likely that CHI may also be regulated at the level of transcription [156]. Two genes involved in later stages of the anthocyanin biosynthesis pathway, dihydroflavonol 4-reductase [155] and *ant17* [158], are also regulated by GA.

The regulation of expression of anthocyanin biosynthesis genes in petals is complex and appears to be controlled, at least in part, by regulatory genes that encode transcription factors of the basic helix-loop-helix and c-*myb* classes [94]. The possibility that GA may coordinately regulate genes of the anthocyanin biosynthesis pathway in petunia by acting on regulatory genes encoding these types of transcription factors [94, 157] is an attractive concept that should be tested. A perception-transduction-response cascade that has one or more regulatory genes as central components would be an effective mechanism for a single ligand to regulate coordinately the expression of multiple genes in a tissue-specific manner.

GA regulation of gene expression in aleurone

Aleurone cells are an excellent system for studying GA and ABA regulation of gene expression. Genes encoding α-amylase [73], a number of proteases [19, 85, 153, 159], (1-3,1-4)-β-glucanase [163] and *BEG1* (globulin-1) [46] are upregulated by GA and, in all cases, this effect is repressed by ABA. The expression of genes encoding alcohol dehydrogenase and an α-amylase/protease inhibitor is down regulated by GA [73].

The regulation of α-amylase gene expression in aleurone cells of wheat, barley and oat is the most intensively studied effect of this plant hormone on gene expression. Aleurone cells and protoplasts synthesise α-amylase in response to GA and this synthesis is accompanied by an increase in the steady-state levels of α-amylase mRNA [73]. Run-on transcriptions with nuclei isolated from barley and wild-oat aleurone protoplasts have demonstrated that the GA-induced increase in α-amylase mRNA is due primarily to increased transcription of α-amylase genes which can be overcome by ABA [65, 171].

In common with all other GA-responsive genes characterised to date, α-amylase expression increases only after several hours of exposure of aleurone cells to GA. GA induction of α-amylase mRNA is sensitive to cycloheximide and the amino acid analogue aminoethyl-*L*-cysteine [99, 100, 101], suggesting that protein synthesis is required before α-amylase gene transcription can be stimulated. Inhibition of protein synthesis after 2 or 4 h treatment with GA did not substantially inhibit α-amylase mRNA, suggesting that the protein(s) important for GA-regulated α-amylase gene expression had been synthesised during this period. It is likely that one or more of these proteins may be components of the GA perception-transduction pathway, or may be involved in an unrelated signal transduction pathway that cross-talks strongly with it.

In Gramineae, α-amylase is encoded by a multigene family that gives rise to several classes of isozymes [58, 60]. These genes are not all subject to the same regulation by GA. Some are insensitive to GA and are subject to developmental or tissue-specific regulation. Those which are GA-responsive can differ temporally and quantitatively in their expression, at both mRNA and protein levels. To a large extent this differential regulation of expression can be correlated with groups of α-amylase genes located on particular chromosomes and encoding proteins that can be grouped according to their isoelectric points (pIs) [7, 21, 59, 61, 62, 75].

Transient expression analysis of *α-Amy* promoter-reporter gene constructs in aleurone protoplasts and layers has been used to identify *cis* elements important for high-level GA- and ABA-regulated expression of wheat and barley α-amylase genes. Analysis of the promoter of the wheat low pI α-amylase gene (*α-Amy2/54*) in oat aleurone protoplasts revealed that elements involved in directing GA- and ABA-regulated expression lie within 289 bp upstream of the start of transcription [63]. Similar analysis of a barley high pI *Amy pHV19* gene promoter showed that the main elements important for GA and ABA regulation lie between 174 and 41 bp upstream from the start of transcription [66].

Transient expression analysis has also been used to assess the activity of multimerised individual elements that are known to be highly conserved in α-amylase promoters [58, 135]. Multiple copies of a 21 bp sequence from the barley *α-Amy1/6-4* promoter confer GA-inducible and ABA-repressible expression on a minimal promoter-reporter gene fusion. It has been suggested that a GA-response element (GARE) resides within these 21 bp [135] and the alignment of GA-regulated α-amylase gene promoters [58, 121] reveals a conserved motif UTAACAUANTCYGG (where U = A/G, Y = C/T and N = A/C/T/G) within this region. Sequences similar to this conserved motif are also present in the promoters of a thiol protease gene [159], a carboxypeptidase gene [8] and a gene encoding (1-3,1-4)-β-glucanase isoenzyme II [163], all of which are GA-regulated in aleurone cells. However, the promoter of a cathepsin B-like gene that is GA-regulated in wheat aleurone does not contain the GARE motif and a different conserved element has been shown to be important for GA-regulated expression [19]. The *GAST1* gene which is GA-regulated in tomato shoots does not contain the GARE motif [129].

Evidence of greater complexity in the GA regulation of α-amylase genes has emerged from subsequent studies. A deletion and mutation analysis within the 289 bp region of the wheat *α-Amy2/54* promoter identified at least three regions, one of which probably corresponds to the GARE area, necessary for high-level GA- and ABA-regulated expression. These elements appeared to operate in concert [64]. A similar linker-scanning mutation and deletion analysis of the barley low-pI *α-Amy32b* gene promoter [89] identified the GARE and a putative Opaque-2-binding sequence (O2S) as being essential for expression. A pyrimidine box [58] and a separate motif, TATCCATGCAGTG, were found to be important in determining absolute levels of expression [89]. It has been suggested that GARE cooperates with other *cis* elements, in particular O2S, thereby functioning as a GA-response complex (GARC) in which the pyrimidine box may facilitate interactions between proteins binding to O2S and GARE [89, 121].

Both the GARE and a TATCCAC motif are important for GA- and ABA-regulated expression of the high-pI barley *Amy pHV19* gene promoter. The pyrimidine box was found not to be important for expression and whether or not GARE cooperates with neighbouring *cis* elements is debatable [42, 122]. In view of the expression patterns of high- and low-pI α-amylase genes, differences between their promoters may not be unexpected and this may be evidence of such variation.

Cis elements are assumed to be sites with which

transcription factors interact in a sequence-specific manner. There is now convincing evidence, from gel retardation and DNase 1 footprinting analyses, that aleurone nuclear proteins interact specifically with DNA sequence elements in GA- and ABA-regulated α-amylase promoters [84, 110, 124, 143]. There are good correlations between *cis* elements defined by transient expression analysis and sequence elements shown to bind nuclear proteins. Clear interactions have been demonstrated between nuclear proteins and GARE, O2S, the TATCCAT motif, cAMP- and phorbol ester-like response elements and the pyrimidine box. Some evidence suggests that protein binding to GARE and the TATCCAT motif may be induced by GA [143], although other reports show no evidence for GA-inducible interactions [84, 124].

Future prospects

Evidence that GA can be perceived at the plasma membrane of aleurone cells has led to a reassessment of previous models of GA signal transduction. However, until a definitive identification can be made of GA receptors, their subcellular location is likely to remain controvertial. Recently developed techniques of GA photoaffinity labelling and GA affinity chromatography appear to offer powerful and specific methods for identifying GA-binding proteins, although whether or not they will help identify the elusive GA receptor remains to be seen. The recent successful isolation of high-purity aleurone plasma membrane [151] may help greatly in identifying GA-binding activity at this subcellular site.

It has become clear that Ca^{2+} and calmodulin play important roles in coordinating the responses of aleurone cells to GA and ABA. A better understanding of the temporal and spatial dynamics of cytoplasmic and ER Ca^{2+} may help elucidate the multifunctional role of Ca^{2+} in these cells. It will also be of value to determine whether or not GA modulates cytoplasmic Ca^{2+} in other GA-responsive tissues where Ca^{2+} flux associated with α-amylase secretion is not present. Biochemical approaches may identify specific targets for Ca^{2+} and calmodulin regulation.

It is reasonable to anticipate that genes encoding proteins which bind specifically to GARE and other sequence elements in α-amylase promoters will be identified soon. These may encode transcription factors that are components of the GA perception-transduction-response pathway. They might even be primary response genes regulated by GA. Similar detailed analysis of the promoters of other genes upregulated by GA during each of the three classes of response to this plant hormone may lead to the identification of other *cis* elements and *trans*-acting factors involved in GA-regulated gene expression.

Molecular genetic analysis of GA-response mutants seems well placed to identify genes responsible for insensitivity to GA and it is likely that the first of these to be identified will be the *gai* locus in *Arabidopsis*. A greater range of GA-response mutants is needed, in particular for mutants affected in specific, rather than all, responses to GAs. These may well come from refined screening procedures. We can also anticipate insight into GA signal transduction coming from molecular genetic studies of other plant hormones. Cross-talk between signalling pathways for different plant hormones seems very likely and may involve proteins that play central roles in plant cell signalling.

Acknowledgement

I am grateful to Anne Berry for her assistance in preparing this manuscript.

References

1. Anderson BE, Ward JM, Schroeder JI: Evidence for an extracellular reception site for abscisic acid in *Commelina* guard cells. Plant Physiol 104: 1177–1183 (1994).
2. Appleford NEJ, Lenton JR: Gibberellins and leaf expansion in near-isogenic wheat lines containing *Rht1* and *Rht3* dwarfing alleles. Planta 183: 229–236 (1991).
3. Arnalte M-E, Cornejo M-J, Bush DS, Jones RL: Gibberellic acid stimulates lipid metabolism in barley aleurone protoplasts. Plant Sci 77: 223–232 (1991).

4. Ballas N, Wong L-M, Theologis A: Identification of the auxin-responsive element, *AuxRE*, in the primary indoleacetic acid-inducible gene, *PS-IAA4/5*, of pea (*Pisum sativum*). J Mol Biol 233: 580–596 (1994).
5. Baluska F, Parker JS, Barlow PW: A role for gibberellic acid in orienting microtubules and regulation cell growth polarity in the maize root cortex. Planta 191: 149–157 (1993).
6. Bardense GWM, Karssen CM, Koornneef M: Role of endogenous gibberellins during fruit and seed development. In: Takahashi N, Phinney BO, MacMillan J (eds) Gibberellins, pp. 179–187. Springer-Verlag, Berlin (1991).
7. Baulcombe DC, Huttly AK, Martienssen RA, Barker RF, Jarvis MG: A novel wheat α-amylase gene (α-Amy3). Mol Gen Genet 209: 33–40 (1987).
8. Baulcombe DC, Barker RF, Jarvis MG: A gibberellin responsive wheat gene has homology to yeast carboxypeptidase. J Biol Chem 262: 13726–13735 (1987).
9. Beale MH, Hooley R, Smith SJ, Walker RP: Photoaffinity probes for gibberellin-binding proteins. Phytochemistry 31: 1459–64 (1992).
10. Beale MH, Ward JL, Smith SJ, Hooley R: A new approach to gibberellin perception in aleurone: novel, hydrophylic, membrane-impermeant, GA-sulphonic acid derivatives induce α-amylase formation. Physiol Plant 85: A136 (1992).
11. Bernier G, Havelange A, Houssa C, Petitjean A, Lejeune P: Physiological signals that induce flowering. Plant Cell 5: 1147–1155 (1993).
12. Bethke PC, Jones RL: Ca^{2+}-calmodulin modulates ion channel activity in storage protein vacuoles of barley aleurone cells. Plant Cell 6: 277–285 (1994).
13. Brinegar AC, Cooper G, Stevens A, Hauer CR, Shabanowitz J, Hunt DF, Fox EJ: Characterization of a benzyladenine binding-site peptide isolated from a wheat cytokinin-binding protein: sequence analysis and identification of a single affinity-labelled histidine residue by mass spectrometry. Proc Natl Acad Sci USA 85 5927–5931 (1988).
14. Bush DS: Regulation of cytosolic calcium in plants. Plant Physiol 103: 7–13 (1993).
15. Bush DS, Jones RL: Cytoplasmic calcium and α-amylase secretion from barley aleurone protoplasts. Eur J Cell Biol 46: 466–469 (1988).
16. Bush DS, Biswas AK, Jones RL: Gibberellic-acid-stimulated Ca^{2+} accumulation in endoplasmic reticulum of barley aleurone: Ca^{2+} transport and steady-state levels. Planta 178: 411–420 (1989).
17. Bush DS, Biswas AK, Jones RL: Hormonal regulation of Ca^{2+} transport in the endomembrane system of the barley aleurone. Plant Physiol 189: 507–515 (1993).
18. Cautrecasas P: Interaction of insulin with the cell membrane: the primary action of insulin. Proc Natl Acad Sci USA 63: 450–457 (1969).
19. Cejudo FJ, Ghose TK, Stabel P, Baulcombe DC: Analysis of the gibberellin-responsive promoter of a cathepsin B-like gene from wheat. Plant Mol Biol 20: 849–856 (1992).
20. Chandler PM: Hormonal regulation of gene expression in the 'slender' mutant of barley (*Hordeum vulgare* L.). Planta 175: 115–120 (1988).
21. Chandler PM, Jacobsen JV: Primer extension studies on α-amylase mRNAs in barley aleurone. II. Hormonal regulation of expression. Plant Mol Biol 16: 637–645 (1991).
22. Chang C, Kwok SF, Bleecker AB, Meyerowitz EM: *Arabidopsis* ethylene-response gene *ETR1*: similarity of product to two-component regulators. Science 262: 539–544 (1993).
23. Chory J, Voytas DF, Olszewski NE, Ausubel FM: Gibberellin-induced changes in the populations of translatable mRNAs and accumulated polypeptides in dwarfs of maize and pea. Plant Physiol 83: 15–23 (1987).
24. Croker SJ, Hedden P, Lenton JR, Stoddart JL: Comparison of gibberellins in normal and slender barley seedlings. Plant Physiol 94: 194–200 (1990).
25. Duckett CM, Lloyd CW: Gibberellic acid-induced microtubule reorientation in dwarf peas is accompanied by rapid modification of an α-tubulin isotype. Plant J 5: 363–372 (1994).
26. Evans LT, King RW, Chu A, Mander LN, Pharis RP: Gibberellin structure and florigenic activity in *Lolium temulentum*, a long day plant. Planta 182: 97–106 (1990).
27. Evans LT, King RW, Mander LN, Pharis RP: The relative significance for stem elongation and flowering in *Lolium temulentum* of 3β-hydroxylation of gibberellins. Planta 192: 130–136 (1994).
28. Feldwisch J, Zettl R, Hesse F, Schell J, Palme K: An auxin-binding protein is located to the plasma membrane of maize coleoptile cells: identification by photoaffinity labeling and purification of a 23-kDa polypeptide. Proc Natl Acad Sci USA 89: 475–479 (1992).
29. Fernandez DE, Staehelin LE: Does gibberellic acid induce the transfer of lipase from protein bodies to lipid bodies in barley aleurone cells? Plant Physiol 85: 487–496 (1987).
30. Feyerabend M, Weiler EW: Photoaffinity labeling and partial purification of the putative plant receptor for the fungal wilt-inducing toxin, fusicoccin. Planta 178: 282–290 (1989).
31. Fincher GB: Molecular and cellular biology associated with endosperm mobilization in germinating cereal grains. Annu Rev Plant Physiol Plant Mol Biol 40: 305–346 (1989).
32. Foster CA: Slender: an accelerated extension growth mutant of barley. Barley Genet Newslett 7: 24–27 (1977).
33. Fujioka S, Yamane H, Spray CR, Katsumi M, Phinney BO, Gaskin P, MacMillan J, Takahashi N: The dominant non-gibberellin-responding dwarf mutant (*D8*) of maize accumulates native gibberellins. Proc Natl Acad Sci USA 85 9031–9035 (1988).

34. Gale MD, Marshall GA: The nature and genetic control of gibberellin insensitivity in dwarf wheat grain. Heredity 35: 55–65 (1975).
35. Gillaspy G, Ben-David H, Gruissem W: Fruits: a developmental perspective. Plant Cell 5: 1439–1451 (1993).
36. Gilroy S, Jones RL: Gibberellic acid and abscisic acid coordinately regulate cytoplasmic calcium and secretory activity in barley aleurone protoplasts. Proc Natl Acad Sci USA 89: 3591–3595 (1992).
37. Gilroy S, Jones RL: Calmodulin stimulation of unidirectional calcium uptake by the endoplasmic reticulum of barley aleurone. Planta 190: 289–298 (1993).
38. Gilroy S, Jones RL: Perception of gibberellin and abscisic acid at the external face of the plasma membrane of barley (*Hordeum vulgare* L.) aleurone protoplasts. Plant Physiol 104: 1185–1192 (1994).
39. Giraudat J, Hauge BM, Valon C, Smalle J, Parcy F, Goodman HM: Isolation of the *Arabidopsis ABI3* gene by positional cloning. Plant Cell 4: 1251–1261 (1992).
40. Goodwin PB, Carr DJ: The induction of amylase synthesis in barley aleurone layers by gibberellic acid I. Response to temperature. J Exp Bot 23: 1–7 (1972).
41. Greenberg J, Goldschmidt EE: Acidifying agents, uptake, and physiological activity of gibberellin A_3 in *Citrus*. HortScience 24: 791–793 (1989).
42. Gubler F, Jacobsen JV: Gibberellin-responsive elements in the promoter of a barley high-pI α-amylase gene. Plant Cell 4: 1435–1441 (1992).
43. Guzman P, Ecker JR: Exploiting the triple response of *Arabidopsis* to identify ethylene-related mutants. Plant Physiol 93: 907–914 (1990).
44. Hamabata A, Rodriguez E, Garcia-Maya M, Bernal-Lugo I: Effect of pH on the GA_3 induced α-amylase synthesis. J Plant Physiol 143: 349–352 (1994).
45. Harberd NP, Freeling M: Genetics of dominant gibberellin-insensitive dwarfism in maize. Genetics 121: 827–838 (1989).
46. Heck GR, Chamberlain AK, Ho DT-H: Barley embryo globulin-1 gene, *BEG1*-characterization of cDNA, chromosome mapping and regulation of expression Mol Gen Genet 239: 209–218 (1993).
47. Hedden P, Croker SJ: Regulation of gibberellin biosynthesis in maize seedlings. In: Karssen CM, Van Loon LC, Vreugdenhil D (eds) Progress in Plant Growth Regulation, pp. 534–544. Kluwer Academic Publishers, Dordrecht, Netherlands (1992).
48. Herskowitz I: Functional inactivation of genes by dominant negative mutations. Nature 329: 219–222 (1987).
49. Hetherington PR, Laidman DL: Influence of gibberellic acid and the *Rht3* gene on choline and phospholipid metabolism in wheat aleurone tissue. J Exp Bot 42: 1357–1362 (1991).
50. Hicks GR, Rayle DL, Jones AM, Lomax TL: Specific photoaffinity labeling of two plasma membrane polypeptides with an azido auxin. Proc Natl Acad Sci USA 86: 4948–4952 (1989).
51. Hillmer S, Gilroy S, Jones RL: Visualizing enzyme secretion from individual barley (*Hordeum vulgare*) aleurone protoplasts. Plant Physiol 102: 279–286 (1992).
52. Ho DT-H, Shih S-C, Kleinhofs A: Screening for barley mutants with altered hormone sensitivity in their aleurone layers. Plant Physiol 66: 153–157 (1980).
53. Hoad GV, Phinney BO, Sponsel VM, MacMillan J: The biological activity of sixteen gibberellin A_4 and gibberellin A_9 derivatives using seven bioassays. Phytochemistry 20: 703–713 (1981).
54. Hooley R, Beale MH, Smith SJ, MacMillan J: Novel affinity probes for gibberellin receptors in aleurone protoplasts of *Avena fatua*. In: Pharis RP, Rood SB (eds) Plant Growth Substances 1988, pp. 145–153. Springer-Verlag, Berlin/Heidelberg/New York (1990).
55. Hooley R, Beale MH, Smith SJ: Gibberellin perception at the plasma membrane of *Avena fatua* aleurone protoplasts. Planta 183: 274–280 (1991).
56. Hooley R, Smith SJ, Beale MH, Walker RP: *In vivo* photoaffinity labelling of gibberellin-binding proteins in *Avena fatua* aleurone. Aust J Plant Physiol 20: 573–584 (1993).
57. Hornberg C, Weiler EW: High-affinity binding sites for abscisic acid on the plasmalemma of *Vicia faba* guard cells. Nature 310: 321–324 (1984).
58. Huang N, Sutliff TD, Litts JC, Rodriguez RL: Classification and characterization of the rice α-amylase multigene family. Plant Mol Biol 14: 655–668 (1990).
59. Huang N, Koizumi N, Reinl S, Rodriguez RL: Structural organization and differential expression of rice α-amylase genes. Nucl Acids Res 18: 7007–7014 (1990).
60. Huang N, Stebbins GL, Rodriguez RL: Classification and evolution of α-amylase genes in plants. Proc Natl Acad Sci USA 89: 7526–7530 (1992).
61. Huang N, Reinl SJ, Rodriguez RL: *RAmy2A*: a novel α-amylase-encoding gene in rice. Gene 111: 223–228 (1992).
62. Huttly AK, Martienssen RA, Baulcombe DC: Sequence heterogeneity and differential expression of the α-Amy2 gene family in wheat. Mol Gen Genet 214: 232–240 (1988).
63. Huttly AK, Baulcombe DC: A wheat α-Amy2 promoter is regulated by gibberellin in transformed oat aleurone protoplasts. EMBO J 8: 1907–1913 (1989).
64. Huttly AK, Phillips AL, Tregear JW: Localisation of *cis* elements in the promoter of a wheat α-Amy2 gene. Plant Mol Biol 19: 903–911 (1992).
65. Jacobsen JV, Beach RL: Control of transcription of α-amylase and rRNA genes in barley aleurone protoplasts by gibberellic acid and abscisic acid. Nature 316: 275–277 (1985).
66. Jacobsen JV, Close TJ: Control of transient expression of chimaeric genes by gibberellic acid and abscisic acid in protoplasts prepared from mature barley aleurone layers. Plant Mol Biol 16: 713–724 (1991).
67. Jacobsen SE, Olszewski NE: Characterization of the

arrest in anther development associated with gibberellin deficiency of the *gib-1* mutant of tomato. Plant Physiol 97: 409–414 (1991).
68. Jacobsen SE, Olszewski NE: Mutations at the *SPINDLY* locus of *Arabidopsis* alter gibberellin signal transduction. Plant Cell 5: 887–896 (1993).
69. Jacobsen SE, Shi L, Xin X, Olszewski NE: Gibberellin-induced changes in the translatable mRNA populations of stamens and shoots of gibberellin-deficient tomato. Planta 192: 372–378 (1994).
70. Jelsema CL, Ruddat M, Morre DJ, Williamson FA: Specific binding of gibberellin A_1 to aleurone grain fractions from wheat endosperm. Plant Cell Physiol 18: 1009–1019 (1977).
71. Jones MG: Gibberellins and the *procera* mutant of tomato. Planta 172: 280–284 (1987).
72. Jones RL, Bush DS: Gibberellic acid and abscisic acid regulate the level of a BiP cognate in the endoplasmic reticulum of barley aleurone cells. Plant Physiol 97: 456–459 (1991).
73. Jones RL, Jacobsen JV: Regulation of synthesis and transport of secreted proteins in cereal aleurone. Int Rev Cytol 126: 49–88 (1991).
74. Jupe SC, Causton DR, Scott IM: Cellular basis of the effects of gibberellin and the *pro* gene on stem growth in tomato. Planta 174: 106–111 (1988).
75. Karrer EE, Litts JC, Rodriguez RL: Differential expression of α-amylase genes in germinating rice and barley seeds. Plant Mol Biol 16: 797–805 (1991).
76. Katsumi M, Ishida K: Gibberellin control of cell elongation. In: Takahashi N, Phinney BO, MacMillan J (eds) Gibberellins, pp. 211–219. Springer-Verlag, New York (1991).
77. Kieber JJ, Rothenberg M, Roman G, Feldmann K, Ecker JR: CTR1, a negative regulator of the ethylene response pathway in *Arabidopsis*, encodes a member of the Raf family of protein kinases. Cell 72: 427–441 (1993).
78. Keith B, Boal R, Srivastava LM: On the uptake, metabolism and retention of $[^3H]GA_1$ by barley aleurone layers at low temperatures. Plant Physiol 66: 956–961 (1980).
79. Keith B, Srivastava LM: *In vivo* binding of gibberellin A_1 in dwarf pea epicotyls. Plant Physiol 66: 962–967 (1980).
80. Keith B, Foster NA, Bonettemaker M, Srivastava LM: *In vitro* gibberellin A_4 binding to extracts of cucumber hypocotyls. Plant Physiol 68: 344–348 (1981).
81. Keith B, Brown S, Srivastava LM: *In vitro* binding of gibberellin A_4 to extracts of cucumber measured by using DEAE-cellulose filters. Proc Natl Acad Sci USA 79: 1515–1519 (1982).
82. Keith B, Rappaport L: *In vitro* $[^3H]$ gibberellin A_1 binding to soluble proteins from GA-sensitive and GA-insensitive dwarf maize mutants. In: Fox EJ, Jacobs M (eds) Molecular Biology of Plant Growth Control, pp. 289–298. Alan R. Liss, New York (1987).
83. Keith B, Rappaport L: *In vitro* gibberellin A_1 binding in *Zea mays* L. Plant Physiol 85: 934–941 (1987).
84. Kim J-K, Cao J, Wu R: Regulation and interaction of multiple protein factors with the proximal promoter regions of a rice high pI α-amylase gene. Mol Gen Genet 232: 383–393 (1992).
85. Koehler SM, Ho DT-H: Hormonal regulation, processing, and secretion of cysteine proteases in barley aleurone layers. Plant Cell 2: 769–783 (1990).
86. Koorrnneef M, Van der Veen JH: Induction and analysis of gibberellin-sensitive mutants in *Arabidopsis thaliana* (L.) Heynh. Theor Appl Genet 58: 257–263 (1980).
87. Koornneef M, Bosma TDG, Hanhart CJ, Van der Veen JH, Zeevaart JAD: The isolation and characterization of gibberellin-deficient mutants in tomato. Theor Appl Genet 80: 852–857 (1990).
88. Lanahan MB, Ho DT-H: Slender barley: a constitutive gibberellin-response mutant. Planta 175: 107–114 (1988).
89. Lanahan MB, Ho DT-H, Rogers SW, Rogers JC: A gibberellin response complex in cereal α-amylase gene promoters. Plant Cell 4: 203–211 (1992).
90. Lashbrook CC, Keith B, Rappaport L: *In vitro* gibberellin A_1 binding to a soluble fraction from dwarf pea epicotyls. In: Fox EJ, Jacobs M (eds) Molecular Biology of Plant Growth Control, pp. 299–308. Alan R. Liss, New York (1987).
91. Leung J, Bouvier-Durand M, Morris P-C, Guerrier D, Chefdor F, Giraudat J: *Arabidopsis* ABA response gene *ABI1*: features of a calcium-modulated protein phosphatase. Science 264: 1448–1452 (1994).
92. Liu Z-H, Srivastava LM: *In vitro* binding of gibberellin A_4 in epicotyls of dwarf pea and tall pea. In: Fox EJ, Jacobs M (eds) Molecular Biology of Plant Growth Control, pp. 315–322. Alan R. Liss, New York (1987).
93. McCarty DR, Hattori T, Carson CB, Vasil V, Lazar M, Vasil LK: The *Viviparous-1* developmental gene of maize encodes a novel transcriptional activator. Cell 66: 895–905 (1991).
94. Martin C, Gerats T: Control of pigment biosynthesis genes during petal development. Plant Cell 5: 1253–1264 (1993).
95. Mendu N, Silflow CD: Elevated levels of tubulin transcripts accompany the GA_3-induced elongation of oat internode segments. Plant Cell Physiol 34: 973–983 (1993).
96. Meyer K, Leube MP, Grill E: A protein phosphatase 2C involved in ABA signal transduction in *Arabidopsis thaliana*. Science 264: 1452–1455 (1994).
97. Montague MJ: Calcium antagonists inhibit sustained gibberellic acid-induced growth of *Avena* (Oat) stem segments. Plant Physiol 101: 399–405 (1993).
98. Musgrave A, Kays SE, Kende H: Uptake and metabolism of radioactive gibberellins by barley aleurone layers. Planta 102: 1–10 (1972).
99. Muthukrishnan S, Chandra GR, Maxwell EA: Hor-

mone-induced increases in levels of functional mRNA and α-amylase mRNA in barley aleurones. Proc Natl Acad Sci USA 76: 6181–6185 (1979).
100. Muthukrishnan S, Chandra GR, Maxwell EA: Hormonal control of α-amylase gene expression in barley studies using a cloned cDNA probe. J Biol Chem 258: 2370–2375 (1983).
101. Muthukrishnan S, Chandra GR, Albaugh GP: Modulation by abscisic acid and *S*-2-aminoethyl-*L*-cysteine of α-amylase mRNA in barley aleurone cells. Plant Mol Biol 2: 249–258 (1983).
102. Nadeau R, Rappaport L, Stolp CF: Uptake and metabolism of ^{3}H-gibberellin A_1 by barley aleurone layers: response to abscisic acid. Planta 107: 315–324 (1972).
103. Nakajima M, Sakai S, Kanazawa K, Kizawa S, Yamaguchi I, Takahashi N, Murofushi N: Partial purification of a soluble gibberellin-binding protein from mung bean hypocotyls. Plant Cell Physiol 34: 289–296 (1993).
104. Ni B-R, Bradford KJ: Germination and dormancy of abscisic acid- and gibberellin-deficient mutant tomato (*Lycopersicon esculentum*) seeds. Sensitivity of germination to abscisic acid, gibberellin, and water potential. Plant Physiol 101: 607–617 (1993).
105. Nissen P: Dose responses of gibberellins. Physiol Plant 72: 197–203 (1988).
106. Nitsch J: Hormonal factors in growth and development. In: Hume AC (ed) The Biochemistry of Fruits and their Products, vol. II, pp. 427–472. Academic Press, London (1970).
107. Nour JM, Rubery PH: The uptake of gibberellin A_1 by suspension-cultured *Spinacia oleracea* cells has a carrier-mediated component. Planta 160: 436–443 (1984).
108. Okamuro JK, den Boer BGW, Jofuku KD: Regulation of *Arabidopsis* flower development. Plant Cell 5: 1183–1193 (1993).
109. Orchinik M, Murray TF, Moore FL: Steroid modulation of GABA(A) receptors in an amphibian brain. Brain Res 646: 258–266 (1994).
110. Ou-Lee T-M, Turgeon R, Wu R: Interaction of a gibberellin-induced factor with the upstream region of an α-amylase gene in rice aleurone tissue. Proc Natl Acad Sci USA 85: 6366–6369 (1988).
111. Parker MG: Steroid and related receptors. Curr Opin Cell Biol 5: 499–504 (1993).
112. Pauls KP, Chambers JA, Dumbroff EB, Thompson JE: Perturbation of phospholipid membranes by gibberellins. New Phytol 91: 1–17 (1982).
113. Peng J, Harberd NP: Derivative alleles of the *Arabidopsis* gibberellin-insensitive (*gai*) mutation confer a wild-type phenotype. Plant Cell 5: 351–360 (1993).
114. Pharis RP, King RW: Gibberellins and reproductive development in seed plants. Annu Rev Plant Physiol Plant Mol Biol 36: 517–568 (1985).
115. Phillips AL, Huttly AK: Cloning of two gibberellin-regulated cDNAs from *Arabidopsis thaliana* by subtractive hybridization: expression of the tonoplast water channel, γ-TIP, is increased by GA_3. Plant Mol Biol 24: 603–615 (1994).
116. Poovaiah BW, Reddy ASN: Calcium and signal transduction in plants. Crit Rev Plant Sci 12: 185–211 (1993).
117. Potts WC, Reid JB, Murfet IC: Internode length in *Pisum*. Gibberellins and the slender phenotype. Physiol Plant 63: 357–364 (1985).
118. Reid JB: Plant hormone mutants. J Plant Growth Regul 12: 207–226 (1994).
119. Reid JB, Ross JJ, Swain SM: Internode length in *Pisum* a new slender mutant with elevated levels of C_{19} gibberellins. Planta 188: 462–467 (1992).
120. Robbins WJ: Gibberellic acid and the reversal of adult *Hedera* to a juvenile state. Am J Bot 44: 743–746 (1957).
121. Rogers JC, Rogers SW: Definition and functional implications of gibberellin and abscisic acid *cis*-acting hormone response complexes. Plant Cell 4: 1443–1451 (1992).
122. Rogers JC, Lanahan MB, Rogers SW: The *cis*-acting gibberellin response complex in high-pI α-amylase gene promoters. Plant Physiol 105: 151–158 (1994).
123. Ross JJ, Murfet IC, Reid JB: Distribution of gibberellins an *Lathyrus odoratus* L. and their role in leaf growth. Plant Physiol 102: 603–608 (1993).
124. Rushton PJ, Hooley R, Lazarus CM: Aleurone nuclear proteins bind to similar elements in the promoter regions of two gibberellin-regulated α-amylase genes. Plant Mol Biol 19: 891–901 (1992).
125. Schena M, Lloyd AM and Davis RW: A steroid-inducible gene expression system for plant cells. Proc Natl Acad Sci USA 88: 10421–10425 (1991).
126. Schroeder JI, Hagiwara S: Repetitive increases in cytosolic Ca^{2+} of guard cells by abscisic acid activation of nonselective Ca^{2+} permeable channels. Proc Natl Acad Sci USA 87: 9305–9309 (1990).
127. Serebryakov EP, Agnistikova VN, Suslova LM: Growth-promoting activity of some selectively modified gibberellins. Phytochemistry 23: 1847–1854 (1984).
128. Serebryakov EP, Epstein NA, Yasinskaya NP, Kaplun AB: A mathematical additive model of the structure-activity relationships of gibberellins. Phytochemistry 23: 1855–1863 (1984).
129. Shi L, Gast RT, Gopalraj M, Olszewski NE: Characterization of a shoot-specific, GA_3- and ABA-regulated gene from tomato. Plant J 2: 153–159 (1992).
130. Shibaoka H: Plant hormone-induced changes in the orientation of cortical microtubules: alterations in the cross-linking between microtubules and the plasma membrane. Annu Rev Plant Physiol Plant Mol Biol 45: 527–544 (1994).
131. Singh SP, Paleg LG: Low temperature-induced GA_3 sensitivity of wheat II. Changes in lipids associated with the low temperature-induced GA_3 sensitivity of isolated aleurone of wheat. Plant Physiol 76: 143–147 (1984).
132. Singh SP, Paleg LG: Low temperature-induced GA_3

sensitivity of wheat. IV. Comparison of low temperature effects on the phospholipids of aleurone tissue of dwarf and tall wheat. Aust J Plant Physiol 12: 277–289 (1985).
133. Singh SP, Paleg LG: Low temperature-induced GA_3 sensitivity of wheat. VI. Effect of inhibitors of lipid biosynthesis on α-amylase production by dwarf (*Rht3*) and tall (*rht*) wheat, and on lipid metabolism of tall wheat aleurone tissue. Aust J Plant Physiol 13: 409–416 (1986).
134. Sinjorgo KMC, de Vries MA, Heistek JC, van Zeijl MJ, van der Veen SW, Douma AC: The effect of external pH on the gibberellic acid response of barley aleurone. J Plant Physiol 142: 506–509 (1993).
135. Skriver K, Olsen FL, Rogers JC, Mundy J: *Cis*-acting elements responsive to gibberellin and its antagonist abscisic acid. Proc Natl Acad Sci USA 88: 7266–7270 (1991).
136. Smith SJ, Walker RP, Beale MH, Hooley R: Biological activity of some gibberellins and gibberellin derivatives in aleurone cells and protoplasts of *Avena fatua*. Phytochemistry 33: 17–20 (1993).
137. Srivastava LM: The gibberellin receptor. In: Klambdt D (ed), Plant Hormone Receptors, pp. 192-227. Springer-Verlag, Berlin/Heidelberg (1987).
138. Stoddart JL, Breidenbach W, Nadeau R, Rappaport L: Selective binding of $[^3H]$ gibberellin A_1 by protein fractions from dwarf pea epicotyls. Proc Natl Acad Sci USA 71: 3255f–3259 (1974).
139. Stoddart JL: Interaction of $[^3H]$ gibberellin A_1 with a sub-cellular fraction from lettuce (*Lactuca sativa* L.) hypocotyls. I. Kinetics of labelling. Planta 146: 353–361 (1979).
140. Stoddart JL: Interaction of $[^3H]$ gibberellin A_1 with a sub-cellular fraction from lettuce (*Lactuca sativa* L.) hypocotyls. II. Stability and properties of the association. Planta 146: 363–368 (1979).
141. Stoddart JL, Williams PD: Interaction of $[^3H]$ gibberellin A_1 with a sub-cellular fraction from lettuce (*Lactuca sativa* L.) hypocotyls. Requirement for protein synthesis. Planta 147: 264–268 (1979).
142. Stoddart JL, Williams PD: Interaction of $[^3H]$ gibberellin A_1 with a sub-cellular fraction from lettuce (*Lactuca sativa* L.) hypocotyls. The relationship between growth and incorporation. Planta 148: 485–490 (1980).
143. Sutliff TD, Lanahan MB, Ho DT-H: Gibberellin treatment stimulates nuclear factor binding to the gibberellin response complex in a barley α-amylase promoter. Plant Cell 5: 1681–1692 (1993).
144. Suttle JC, Hultstrand JF, Tanaka FS: The biological activities of five azido N-substituted phthalimides: potential photoaffinity reagents for gibberellin receptors. Plant Growth Regul 11: 311–318 (1992).
145. Talon M, Koornneef M, Zeevaart JAD: Accumulation of C_{19}-gibberellins in the gibberellin-insensitive dwarf mutant *gai* of *Arabidopsis thaliana* (L.) Heynh. Planta 182: 501–505 (1990).
146. Thuleau P, Ward JM, Ranjeva R, Schroeder JI: Voltage-dependent calcium-permeable channels in the plasma membrane of a higher plant cell. EMBO J 13: 2970–2975 (1994).
147. Tidd BK: Dissociation constants of the gibberellins. J Chem Soc 1521–1523 (1964).
148. Vakharia DN, Brearley CA, Wilkinson MC, Gaillard T, Laidman DL: Gibberellin modulation of phosphatidylcholine turnover in wheat aleurone tissue. Planta 172: 502–507 (1987).
149. Venis M: Hormone-binding studies and the misuse of precipitation assays. Planta 162: 502–505 (1984).
150. Walker RP, Beale MH, Hooley R: Photoaffinity labelling of MAC182, a gibberellin-specific monoclonal antibody. Phytochemistry 31: 3331–3335 (1992).
151. Walker RP, Waterworth WM, Hooley R: Preparation and polypeptide composition of plasma membrane and other subcellular fractions from wild oat (*Avena fatua*) aleurone. Physiol Plant 89: 388–398 (1993).
152. Walker RP, Waterworth WM, Beale MH, Hooley R: Gibberellin-photoaffinity labeling of wild oat (*Avena fatua* L.) aleurone protoplasts. Plant Growth Regul 12: 1–9 (1994).
153. Watanabe H, Abe K, Emori Y, Hosoyama H, Arai S: Molecular cloning and gibberellin-induced expression of multiple cysteine proteinases of rice seeds (oryzains). J Biol Chem 266: 16897–16902 (1991).
154. Weiss D, Halevy AH: Stamens and gibberellins in the regulation of corolla pigmentation and growth in *Petunia hybrida*. Planta 179: 89–96 (1989).
155. Weiss D: Regulation of corolla pigmentation and development in petunia flower. Ph. D. thesis, The Hebrew University of Jerusalem, Israel (1990).
156. Weiss D, van Tunen AJ, Halevy AH, Mol JNM, Gerats AGM: Stamens and gibberellic acid in the regulation of flavonoid gene expression in the corolla of *Petunia hybrida*. Plant Physiol 94: 511–515 (1990).
157. Weiss D, van Blokland R, Kooter JM, Mol JNM, van Tunen AJ: Gibberellic acid regulates chalcone synthase gene transcription in the corolla of *Petunia hybrida*. Plant Physiol 98: 191–197 (1992).
158. Weiss D, van der Luit AH, Kroon JTM, Mol JNM, Kooter JM: The petunia homologue of the *Antirrhinum majus candi* and *Zea mays A2* flavonoid genes; homology to flavanone 3-hydroxylase and ethylene-forming enzyme. Plant Mol Biol 22: 893–897 (1993).
159. Whittier RF, Dean DA, Rogers JC: Sequence analysis of α-amylase and thiol protease genes that are hormonally regulated in barley aleurone cells. Nucl Acids Res 15: 2515–2535 (1987).
160. Wilkinson MC, Laidman DL, Galliard T: Two sites of phosphatidyl choline synthesis in the wheat aleurone cell. Plant Sci Lett 35: 195–199 (1984).
161. Wilson RN, Heckman JW, Somerville CR: Gibberellin is required for flowering in *Arabidopsis thaliana* under short days. Plant Physiol 100: 403–408 (1992).

162. Winkler RG, Freeling M: Physiological genetics of the dominant gibberellin-nonresponsive maize dwarfs, *Dwarf8* and *Dwarf9*. Planta 193: 341–348 (1994).
163. Wolf N: Structure of the genes encoding *Hordeum vulgare* (1-3,1-4)-β-glucanase isoenzymes I and II and functional analysis of their promoters in barley aleurone protoplasts. Mol Gen Genet 234: 33–43 (1992).
164. Wood A, Paleg LG: The influence of gibberellic acid on the permeability of model membrane systems. Plant Physiol 50: 103–108 (1972).
165. Wood A, Paleg LG: Alteration of liposomal membrane fluidity by gibberellic acid. Aust J Plant Physiol 1: 31–40 (1974).
166. Wood A, Paleg LG, Spotswood TM: Hormone-phospholipid interaction: a possible hormonal mechanism of action in the control of membrane permeability. Aust J Plant Physiol 1: 167–169 (1974).
167. Yalpani N, Srivastava LM: Competition for *in vitro* [^{3}H] gibberellin A_4 binding in cucumber by gibberellins and their derivatives. Plant Physiol 79: 963–967 (1985).
168. Yalpani N, Suttle JC, Hultstrand JF, Rodaway SJ: Competition for *in vitro* [^{3}H] gibberellin A_4 binding in cucumber by substituted phthalamides. Plant Physiol 91: 823–828 (1989).
169. Zettl R, Feldwisch J, Boland W, Schell J, Palme K: 5′-Azido-[3,6-$^3H_2$]-1-naphthylphthalamic acid, a photoactivatable probe for naphthylphthalamic acid receptor proteins from higher plants: identification of a 23-kDa protein from maize coleoptile plasma membranes. Proc Natl Acad Sci USA 89: 480–484 (1992).
170. Zorec R, Tester M: Cytoplasmic calcium stimulates exocytosis in a plant secretory cell. Biophys J 63: 864–867 (1992).
171. Zwar JA, Hooley R: Hormonal regulation of α-amylase gene transcription in wild oat (*Avena fatua* L.) aleurone protoplasts. Plant Physiol 80: 459–463 (1986).

Plant Molecular Biology **26**: 1557–1577, 1994.

Current advances in abscisic acid action and signalling

Jérôme Giraudat*, François Parcy, Nathalie Bertauche, Françoise Gosti, Jeffrey Leung, Peter-Christian Morris, Michelle Bouvier-Durand and Nicole Vartanian
Institut des Sciences Végétales, Centre National de la Recherche Scientifique UPR 40, 91198 Gif-sur-Yvette Cedex, France (author for correspondence)*

Received and accepted 21 June 1994

Key words: abscisic acid, gene regulation, mutants, seeds, stomata, stress

Abstract

Abscisic acid (ABA) participates in the control of diverse physiological processes. The characterization of deficient mutants has clarified the ABA biosynthetic pathway in higher plants. Deficient mutants also lead to a revaluation of the extent of ABA action during seed development and in the response of vegetative tissues to environmental stress. Although ABA receptor(s) have not yet been identified, considerable progress has been recently made in the characterization of more downstream elements of the ABA regulatory network. ABA controls stomatal aperture by rapidly regulating identified ion transporters in guard cells, and the details of the underlying signalling pathways start to emerge. ABA actions in other cell types involve modifications of gene expression. The promoter analysis of ABA-responsive genes has revealed a diversity of *cis*-acting elements and a few associated *trans*-acting factors have been isolated. Finally, characterization of mutants defective in ABA responsiveness, and molecular cloning of the corresponding loci, has proven to be a powerful approach to dissect the molecular nature of ABA signalling cascades.

Introduction

Abscisic acid (ABA) is a naturally occurring plant hormone (or growth regulator) that was identified in the 1960s (see [2] for a review of the early studies on ABA). ABA is probably present in all higher plants, and has been implicated in the control of a wide range of essential physiological processes including seed development and plant adaptation to environmental stress.

The present review emphasizes the recent advances made in elucidating the ABA signal transduction pathways. Identification of plant responses that are regulated by ABA *in vivo* is a prerequisite to studies aiming at characterizing the underlying regulatory pathways. The first part of this article is thus devoted to an update of certain aspects of ABA physiology and in particular of the information derived from the characterization of ABA-biosynthetic mutants. This brief overview is only meant to provide the necessary background to studies on ABA signalling; further details on ABA functions can be found in several books and reviews [1, 18, 64, 166, 175]. In the second part, we then analyse more extensively the respective contribution of various experimental approaches to our present knowledge of the molecular cascades mediating ABA effects at the cellular level.

Recent advances from the analysis of ABA-deficient mutants

As for many other biologically active substances, the possible roles of ABA in plants were initially investigated by monitoring endogenous ABA contents and by analysing the effects of exogenously applied ABA or biosynthetic inhibitors (reviewed in [166]). Mutants deficient in ABA synthesis provide a means to more directly assess the roles of ABA under physiological conditions. As will be described below, such mutants lead us to a reevaluation of the extent of ABA action during seed development and in the response of vegetative tissues to various stress conditions. Deficient mutants have also contributed extensively to the clarification of the biosynthetic pathways of ABA.

The ABA biosynthetic pathway in plants

Mutants that are deficient in ABA have been isolated in a variety of species (see Table 1). Among the best characterized are the maize *viviparous* (*vp*) [114, 134], tomato *flacca, sitiens* and *notabilis* [153, 156], *Arabidopsis aba* [79], potato *droopy* [131], pea *wilty* [168] and *Nicotiana plumbaginifolia aba1* [123, 139] mutants. These mutants display various abnormalities that, as expected for mutants deficient in the hormone, can be restored to wild type by exogenous supply of ABA.

ABA is a sesquiterpenoid with mevalonic acid as its precursor. While phytopathogenic fungi synthesize ABA by a direct pathway (also known as the C_{15} pathway) from farnesyl pyrophosphate, numerous distinct lines of evidence now support that higher plants rather synthesize ABA by an indirect (or C_{40}) pathway from xanthophylls (see [155, 167, 175, 176] for detailed reviews). In particular, the biosynthetic defects identified in ABA-deficient mutants are congruent with the indirect pathway schematically depicted in Fig. 1. The maize *vp2*, *vp5*, *vp7* and *vp9* mutants are blocked in the early stages of carotenoid biosynthesis [108, 114]. The *Arabidopsis aba* mutants are impaired in the epoxidation reaction convert-

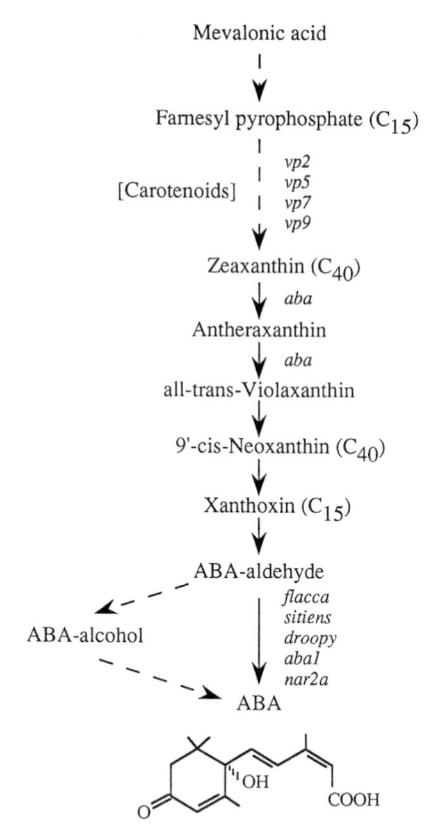

Fig. 1. Simplified pathway of ABA biosynthesis in higher plants. The metabolic blocks in various ABA-deficient mutants are indicated. Adapted from [155, 167, 175, 176].

ing zeaxanthin to antheraxanthin and most likely in the subsequent one leading to violaxanthin [26, 137]. The tomato *flacca* and *sitiens* [156], potato *droopy* [27] and *N. plumbaginifolia aba1* [123, 139] mutants are all blocked in the final step(s) of ABA biosynthesis, oxidation of ABA aldehyde to ABA. This step might be catalyzed by an enzyme that requires a molybdenum cofactor, as suggested in particular by the barley *nar2a* mutant [165]. Defects in such a cofactor might conceivably affect additional reactions unrelated to ABA biosynthesis and thus possibly explain the pleiotropic effects of the barley *nar2a* [165], tomato *flacca* [152] and *N. plumbaginifolia aba1* [123, 139] mutations. In addition, the existence of a minor shunt pathway that involves reduction of ABA aldehyde to ABA alcohol and oxidation of ABA alcohol to ABA (most likely via a cytochrome P-450 mono-oxygenase) was uncovered by the tomato *flacca* and *sitiens* [136] and other mutations (reviewed in [136, 155, 176]). Other ABA-biosynthetic mutations have not yet been unambiguously related to a particular step of the pathway (discussed in [155]).

As illustrated below for some ABA-response mutants, techniques are now available to clone a gene simply on the basis of its associated mutant phenotypes. The available ABA-biosynthetic mutants might thus lead in the near future to the identification of genes encoding various enzymes involved in ABA biosynthesis. These would be valuable tools to improve our understanding of the regulation of ABA biosynthesis but also to better identify which cells synthesize ABA.

Roles of ABA during seed development

ABA has been proposed to be an essential regulator of various processes occurring during roughly the last two thirds of seed development, i.e. during the 'maturation' and 'post-abscission' phases in the nomenclature proposed by Galau *et al.* [38]. In a number of mono- and dicotyledonous species, endogenous ABA content has indeed been shown to peak during this period before returning to low levels in the dry seed (for reviews see [6, 132]).

When removed from the ovule after the end of pattern formation, most embryos display the ability to germinate precociously on culture medium. Such precocious germination could be prevented

Table 1. Characteristics of the various ABA mutants. Additional details on the mutant phenotypes can be found in the text. The loci that have been cloned are in bold, and the potential function of the encoded protein is indicated.

Species	Name	Defect	References
ABA-deficient mutants			
Z. mays	*vp2*	carotenoid biosynthesis	[108, 114, 134]
	vp5	carotenoid biosynthesis	[108, 114, 134]
	vp7	carotenoid biosynthesis	[108, 114, 134]
	vp9	carotenoid biosynthesis	[108, 114, 134]
	vp8	?	[108, 114, 134]
A. thaliana	*aba*	epoxidation of xanthophylls	[26, 79, 137]
L. esculentum	*notabilis*	?	[153]
	flacca	oxidation of ABA aldehyde to ABA	[153, 156]
	sitiens	oxidation of ABA aldehyde to ABA	[153, 156]
S. phureja	*droopy*	oxidation of ABA aldehyde to ABA	[27, 131]
P. sativum	*wilty*	?	[27, 168]
N. plumbaginifolia	*aba1/ckr1*	oxidation of ABA aldehyde to ABA	[123, 139]
H. vulgare	*nar2a*	oxidation of ABA aldehyde to ABA (deficient in molybdenum cofactor)	[165]
ABA-insensitive mutants			
A. thaliana	***abi1***	ABA responsiveness	[80]
		(Ca^{2+}-modulated protein phosphatase)	[86, 106]
	abi2	ABA responsiveness	[80]
	abi3	ABA responsiveness, seed-specific	[80, 112, 119]
		(transcription activator)	[45]
	abi4	ABA responsiveness (seed-specific?)	[34]
	abi5	ABA responsiveness (seed-specific?)	[34]
Z. mays	***vp1***	ABA responsiveness, seed-specific	[100, 134, 135]
		(transcription activator)	[102]
H. vulgare	*cool*	ABA sensitivity in guard cells	[133]

in many cases by adding ABA to the culture medium, thereby suggesting that the increased seed ABA levels play a similar role *in vivo* (for reviews see [6, 132]). The various nonallelic ABA-biosynthetic *viviparous* mutants identified in maize indeed display precocious germination on the mother plant, i.e. vivipary [114, 134]. During wild-type *Arabidopsis* seed development, there is only a transient period during which freshly harvested developing seeds can readily germinate upon water imbibition. This germination capacity is then lost in later stages: seeds develop primary dormancy. In contrast, seeds of the ABA-deficient *aba* mutants remain non dormant until ripeness [75, 79]. Reciprocal crosses between wild type and the *aba* mutant further demonstrated that embryonic rather than maternal ABA participates in dormancy induction [75]. Reduced seed dormancy has also been reported for *N. plumbaginifolia* ABA-deficient mutants [139].

At the molecular level, seed developmental stages are characterized by the accumulation of distinct sets of mRNAs and corresponding proteins in the embryo and endosperm (see [22, 49, 68, 132, 159] for reviews). At least in dicots, storage proteins are characteristic markers of the 'maturation' stage [38]. Their accumulation is followed by that of various classes of late embryogenesis-abundant (LEA) proteins thought to participate in desiccation tolerance (for reviews on the structural characteristics of LEA proteins, see [28, 29]). Since several of these molecular markers have been reported to be precociously inducible by exogenous ABA in cultured embryos, developmental variations in seed ABA levels have been considered as candidate endogenous signals controlling the corresponding gene expression programs (see [69, 132] for references). However, several genes from rapeseed [35], sunflower [48] or wheat [109, 110] are inducible in excised immature embryos by both exogenous ABA and osmotic stress, but in the latter case without any significant increase in endogenous ABA content. The role of endogenous ABA has been further challenged by a systematic analysis of gene expression during cotton embryogenesis. By monitoring the expression pattern of a large set of marker mRNAs both *in vivo* [68] and under various culture conditions [69], Hughes and Galau have identified several classes of coordinately expressed mRNAs. Their expression patterns can be explained as unique combinations of a few temporal programs of gene expression. However, these programs appear to be mainly controlled by as yet unidentified 'maturation' and 'post-abscission' developmental factors distinct from variations in ABA levels [38, 69].

While no ABA-mutant is available in cotton, the effect of ABA-biosynthetic mutations on seed gene expression has been analysed in *Arabidopsis* and maize. Most strikingly, such mutations inhibit only slightly, if at all, the *in vivo* accumulation of storage protein mRNAs both in *Arabidopsis* ([121]; F. Parcy, C. Valon and J. Giraudat, unpublished results) and in maize [120]. These mutant seeds nevertheless contain severely reduced ABA levels and in particular do not display the peaks of ABA content observed during wild-type seed development [75, 114, 120]. In agreement with the conclusions of Hughes and Galau, the wild-type variations in bulk ABA content thus do not appear to be the major developmental signal controlling expression of the storage protein genes ('maturation' program). Although the accumulation of various maize [120, 128, 129, 164, 169] and *Arabidopsis* ([33]; F. Parcy, C. Valon and J. Giraudat, unpublished results) LEA mRNAs is significantly reduced in ABA-biosynthetic mutant seeds, in most cases the extent of this inhibition does not seem to be linearly correlated to the reduction in ABA content. It has also been outlined that in wild type, LEA mRNAs reach their maximal abundance at the very end of seed development whereas ABA content simultaneously decreases [68]. Expression of LEA genes is thus probably controlled by some additional developmental factor(s) (such as the 'post-abscission factor' of Hughes and Galau), even though endogenous ABA levels seem to play a significant role in modulating the intensity of this expression.

Similarly, the *aba* mutation alone does not prevent the acquisition of desiccation tolerance during *Arabidopsis* seed development. Endogenous

ABA nevertheless appears to participate in this developmental process since seeds of the *aba,abi3* digenic mutant (see below) remain desiccation-intolerant, a phenotype that can be reversed to wild type by exogenous supply of ABA [78, 105].

The above data on ABA-biosynthetic mutants indicate that wild-type developmental variations in bulk ABA content appear to control the induction of seed dormancy but are not the primary regulators of the other responses analysed. Since ABA-dependent regulatory pathway(s) appear nevertheless involved, these latter responses might possibly be controlled by developmental variations in their sensitivity to ABA [160]. Such possible variations in ABA sensitivity are however apparently not attributable to developmental variations in the expression of the maize *VP1* [102] and *Arabidopsis ABI3* (F. Parcy, C. Valon and J. Giraudat, unpublished results) genes which encode putative elements of seed ABA-signalling pathways (see below).

Roles of ABA in response to environmental stress conditions

During vegetative growth, ABA has been proposed for a long time to be an essential mediator in triggering plant responses to various adverse environmental conditions such as drought, high salinity or cold. This conclusion was initially supported by the observed stress-induced increases in endogenous ABA levels [15, 16, 151] and by the ability of exogenously applied ABA to mimic many of the plant morphological and physiological responses to these environmental stimuli (for reviews see [19, 94, 161, 175]).

In particular, ABA is thought to be responsible for triggering stomatal closure under conditions of water deficiency. The effects of exogenous ABA on stomatal guard cells are now documented by a wealth of electrophysiological studies (see below). Most of the available ABA-deficient mutants display an increased tendency to wilt and/or enhanced water loss in excised aerial parts, suggestive of a defect in stomatal regulation [79, 113, 123, 131, 153, 168]. These defects have however not yet been traced at the cellular level, for instance by electrophysiological recordings in guard cells. ABA-biosynthetic mutants have so far only rarely been analysed for other global stress responses. The *Arabidopsis aba* mutation, however, affects the development of freezing tolerance [41, 62] and impairs the production of characteristic root structures in response to progressive drought [163].

In the past few years, the possible role of ABA in mediating stress responses has been further strengthened by molecular studies. A large number of genes which are similarly regulated by stress and exogenous ABA have indeed been identified in a variety of plant species (for reviews see [13, 22, 70, 149]). Although the documented or putative molecular functions of the proteins encoded by these stress-regulated genes is obviously of great biological interest, this point is beyond the scope of the present review. The use of ABA-biosynthetic mutants and/or ABA biosynthesis inhibitors has demonstrated that endogenous ABA indeed contributes to the regulation of several ABA-responsive genes by desiccation or drought [16, 50, 83, 117, 128, 129], cold [83, 116, 117] and salt [11]. Interestingly, ABA also participates in the accumulation of ferritin mRNA in response to iron stress [88].

The characterization of ABA-biosynthetic mutants however also revealed that the ability of a given gene to respond to exogenous ABA under non-stress conditions does not necessarily imply that this gene is actually regulated by endogenous ABA upon stress. Several *Arabidopsis* genes which are inducible by exogenous ABA are regulated by cold in an ABA-independent manner [41, 116].

Additional evidence demonstrate the existence of ABA-independent pathways mediating gene regulation in response to stress. For instance, genes unresponsive to ABA represent a significant proportion of those recovered from screens on the basis of differentially expressed mRNA upon water stress [50, 55, 171]. Although ABA-independent regulation might take place prior to the stress-induced *de novo* synthesis of ABA [54–56, 173], such ABA-independent regulation does not occur exclusively in the early stages following the onset of stress conditions [50]. A *cis*-acting element involved in ABA-independent stress regulation of gene expression has been recently characterized [174]. As shown in Table 2, this DRE element does not resemble any of the known ABA-responsive elements (see below).

In conclusion, available data do support a role for ABA in mediating various environmental stress. However, instead of acting as the central co-ordinator of all aspects of the plant response, ABA seems required for only some of the regulatory pathways involved. These pathways are most likely integrated into a more complex regulatory network. Diverse types of experiments indeed support that ABA-dependent and ABA-independent pathways interact in regulating the expression of certain genes in response to drought [116, 173] or osmotic [11] stress.

Roles of ABA in wound response

Recently, ABA has also been implicated in mediating local and systemic wound response. After mechanical wounding, a specific set of proteins (thought to be defense-related) accumulate both at the local wound site and systemically throughout the plant. By far the best characterized of these induced proteins are protease inhibitors I and II of tomato and potato (reviewed in [140]). However, other proteins such as those with homology to cathepsin D inhibitor, threonine deaminase, or leucine aminopeptidase are also induced [65] and many plant families other than the Solanaceae are now known to possess a systemic wound response (reviewed in [140]).

Several lines of evidence support that ABA contributes to transducing the wound response in potato and tomato [126, 127]. In the absence of wounding, applied ABA results in high levels of protease inhibitor II (Pin II) gene induction in both species. Mechanical wounding of potato leaves increases ABA levels both locally and systemically. More importantly, ABA-deficient mutants of tomato (*sitiens*) and potato (*droopy*) show much lower local and systemic levels of Pin II gene induction in response to wounding [126, 127]. Similar observations were later made for additional mRNA markers of the potato wound response [65]. Conditions of water stress (shown to enhance endogenous ABA levels) did not induce Pin II nor other wound-responsive genes, which suggests that in potato plants two independent transduction mechanisms regulate the ABA-dependent wound and water stress responses respectively [65]. More recently it has been demonstrated that the potato Pin II promoter is inducible by wounding and exogenously applied ABA in transgenic rice, demonstrating that the basic induction machinery is conserved between monocots and dicots [170].

Although the results summarized above support a role for ABA in mediating the systemic wound response, definitive proof of ABA as the systemic messenger is lacking. Furthermore, a number of other molecules or signals seem to participate in promulgating the systemic wound response (reviewed in [140]). In particular, jasmonic acid (which has effects similar to those of

Table 2. Sequence motifs of the various *cis*-acting elements mentioned in the text.

Element	Gene	Reference		Sequence
DRE	*RD29A*	[173]	–167	TACCGACAT
Em1A	*Em*	[95]	–153	GGACACGTGGC
Motif I	*rab16A*	[111]	–186	CCGTACGTGGCGC
hex-3	(synthetic)	[82]		GGACGCGTGGC
Sph	*C1*	[60]	–145	TCCATGCATGCAC
?	*CDeT27-45*	[115]	–361	AAGCCCAAATTTCACAGCCCGATAACCG
GARE	*Amy1/6-4*	[150]	–148	GGCCGATAACAAACTCCGGCC
GARE	*Amy32b*	[138]	–120	GTAACAGAGTCTGG

ABA on various other physiological processes reviewed in [124]) appears to be of central importance. Exogenous applications of jasmonic acid (JA) [31, 65] or of its biosynthetic intermediates [32] induce wound-responsive genes. JA seems to act downstream of ABA in regulating the wound response [65, 125]. Several lines of evidence suggest that wounding induces lipase activity which releases linolenic acid, the precursor to JA, from the plasma membrane and then several oxygenation steps subsequently lead to JA biosynthesis [32, 125]. One might then envisage that in the first stage of the wound response, JA synthesis is ABA-dependent and that ABA, for instance, controls the early induction of one of the JA biosynthetic enzymes [104].

Signal transduction

The ultimate objective of studies on signal transduction is a molecular description of the regulatory network that coordinates perception of the signal to cellular responses. Chemical compounds such as ABA are generally thought to elicit a cascade of events by interacting with a specific receptor site(s). Saturable ABA binding sites have been described [66, 67], but the corresponding proteins have not been further identified. In contrast, significant progress has been made in the identification of more downstream elements that contribute to the ABA regulation of ionic currents in stomatal guard cells or of gene expression in various tissues.

ABA signalling pathways in stomatal guard cells

The aperture of stomatal pores is controlled by changes in the turgor of the two surrounding guard cells. In order to optimize CO_2 and water vapour exchanges with the atmosphere, guard cell volume responds within minutes to a variety of signals (reviewed in [77]). In particular, during conditions of water stress, the increased ABA levels in guard cells [58] are thought to reduce water loss through transpiration by promoting stomatal closure. As already mentioned, the 'wilty' phenotype displayed by most ABA-deficient mutants provide suggestive evidence for an action of endogenous ABA on stomatal regulation.

Guard cell volume is controlled osmotically, mainly by large influx (stomatal opening) or efflux (stomatal closure) of K^+, balanced by flux of anions. Tracer flux studies on isolated epidermal strips revealed that externally applied ABA evokes the efflux of K^+ and anions from the guard cells; the released ions originating both from the cytoplasm and the vacuole [90–92]. These studies also provided evidence for the involvement of second messengers in the ABA effects since a 2 min ABA pulse was sufficient to elicit the full K^+ ($^{86}Rb^+$) efflux response which lasted ca. 20 min, and the rate of K^+ efflux continued to rise after ABA washout [91].

Electrophysiological techniques further dissected the above fluxes into their various ionic current components. The considerable progress made towards the deciphering of the electrical responses triggered by ABA in the plasma membrane of guard cells are detailed in several recent reviews [9, 92, 143]. Although several aspects such as the exact time sequence of events still remain to be clarified, a general scheme starts to emerge as briefly summarized here. The first electrical change detected after exposure to ABA is an initial depolarization which reflects a net influx of positive charges [158]. The respective contributions of anion efflux and of Ca^{2+} influx to this early depolarization response are still a matter of controversy [9, 147, 158]. Nevertheless, in either scenario this first event would then lead to the activation of two types of Ca^{2+}-sensitive and voltage-dependent anion conductances which is the basis of the long-term depolarization and large anion efflux observed in response to ABA. These two anionic currents are carried by two distinct channels (or two functional modes of a single channel type): R- and S-type [61, 144, 145]. The characteristics of the S-type anion channel identified in the plasma membrane of *Vicia faba* guard cells make it a good candidate for mediating most of the long-term anion efflux [146]. The above depolarization generates the driving force for K^+

efflux through outward rectifying K^+ channels, which provide the predominant pathway for long-term K^+ efflux. These K^+ currents are activated by ABA in a largely voltage-independent manner [7]. Conversely, ABA inhibits K^+ influx through inward rectifying K^+ channels that represent the major pathway for K^+ uptake into guard cells during stomatal opening [7, 8, 158].

Both Ca^{2+} and H^+ most likely participate as intracellular secondary messengers in mediating the above ABA effects on stomatal aperture and/or plasma membrane channels. An elevation of the cytoplasmic Ca^{2+} concentration (Ca_i) above ca. 600 nM by photolysis of caged Ca^{2+} suffices to produce stomatal closing [44]. Along with this observation, increases in Ca_i inhibit the inward rectifying K^+ channels and activate the voltage-dependent anion channels [61, 144]. The Ca^{2+}-induced inactivation of the inward rectifying K^+ channels appears to be mediated by a Ca^{2+}-dependent protein phosphatase related to the animal calcineurin [89]. ABA has been shown to induce an increase in guard cell Ca_i, as revealed by fluorescent indicators, which precedes stomatal closure [96]. There is however a considerable variability in the reported ABA-induced elevations of Ca_i, which might simply reflects technical limitations in monitoring Ca_i [98], or alternatively indicates that ABA treatment is not systematically accompanied by an increase in Ca_i [42]. The existence of an additional, Ca^{2+}-independent, pathway in ABA action is supported by several observations including the Ca^{2+}-independent activation of outward rectifying K^+ channels (reviewed in [9, 42, 143]). These K^+ channels are in contrast affected by the cytoplasmic pH (pH_i). ABA was shown to evoke an alkalinisation of the cytoplasm of guard cells [72] which is a necessary intermediate in the ABA activation of the outward rectifying K^+ channels [8]. Both of these pathways (Ca^{2+} and H^+) thus appear to be essential in mediating ABA-evoked stomatal closure; the details of their mutual interactions however remain to be determined.

The cellular origin of the ABA-induced cytosolic alkalinisation is unknown (discussed in [8]). The ABA-evoked rise in Ca_i is most likely contributed both by an influx of external Ca^{2+} (through a non-specific cation channel) as well as by Ca^{2+} release from intracellular stores [42, 97, 147]. Inositol (1,4,5)-trisphosphate (IP_3) is an attractive intermediate for triggering this intracellular Ca^{2+} mobilization. When released in the cytoplasm, IP_3 induces a rise in Ca_i (apparently from internal stores) followed by stomatal closure [44], and IP_3 inactivates the inward rectifying K^+ channel [10]. Activators of G-proteins have also been shown to inactivate these same channels, in a Ca^{2+}-dependent manner in the case of GTP-γ-S [30].

These various observations, together with evidence suggesting that ABA acts from the outside of guard cells [4, 59], are of course reminiscent of animal hormone-receptor/G-protein linked transduction cascades (see [23, 76] for reviews). One could thus speculate that by binding to a transmembrane receptor at the plasma membrane [67], ABA activates a G-protein, which then triggers the release of IP_3. IP_3 would then evoke efflux of Ca^{2+} from the vacuole [3] and/or endoplasmic reticulum into the cytosol. Numerous aspects of this minimal framework for the Ca^{2+}-mediated ABA effects however clearly await experimental confirmation. This analysis might be difficult since several lines of evidence suggest that guard cells contain another G-protein linked cascade which plays an antagonistic regulatory role, i.e. promotes stomatal opening [30, 84, 85].

Another key aspect of the above model remains uncertain, namely the location of the ABA reception site. Recent data indeed support that an extracellular reception site is critical in the ABA inhibition of stomatal opening [4] but that ABA acts from within guard cells to promote stomatal closure and the associated inhibition of inward K^+ currents [148]. The molecular identification of these sites of ABA action (receptors) clearly represents an exciting challenge for the future.

At least some parts of the ABA signalling pathways identified in guard cells might also be present in other cell types. For instance, ABA-induced membrane depolarizations have been observed in radish seedlings and in epidermal and mesophyll cells from tobacco (reviewed in [92]). Also, ABA-induced increases in Ca_i associated with an increase in pH_i have been observed in cells of corn roots, corn coleoptiles and parsley hypocotyls [40]. These various cellular events have however not yet been unambiguously related to ABA-regulated physiological processes.

In contrast, as described above, ABA actions both in seeds and in vegetative tissues involve modifications of gene expression. Analysing the promoter of such ABA-responsive genes thus represents a bottom-up approach to the characterization of the corresponding ABA regulatory pathways.

Promoter analysis of ABA-responsive genes

Over 150 genes from various species are known to be inducible by exogenous ABA (reviewed in [13, 22, 70, 149]. Most of these genes were originally identified as being expressed during late seed development and/or in the vegetative tissues of plants exposed to environmental stress. As discussed above, their responsiveness to applied ABA does not necessarily imply that all these genes are primarily regulated by endogenous ABA content *in vivo*. Nevertheless, these target genes are useful tools to investigate the cellular components involved in their ABA induction. In most cases ABA responsiveness has been monitored only by northern blot analysis of steady-state mRNA levels. Several genes have however been further demonstrated to be regulated by ABA at least in part at the transcriptional level. Analysing the promoters of these genes provide a powerful means to identify the terminal components of the ABA regulatory cascade, namely the *cis*-acting element(s) and *trans*-acting factor(s) involved in ABA responsiveness.

The best characterized class of *cis*-acting ABA-responsive elements (ABRE) is exemplified by the Em1a element from the wheat *Em* gene [95] and the Motif I element from the rice *rab16A* gene [111]. These two elements were identified and characterized by roughly similar experimental strategies. Upstream sequences from the *Em* or *rab16A* genes were fused to a reporter gene and the expression of these chimeric constructs were analysed by transient assays in protoplasts. Promoter regions involved in the induction by ABA were initially delimited by 5′ deletion analysis, and found to contain various sequence motifs (Em1a/b and Em2, Motif I and Motif IIa/b) conserved in the promoters of other ABA-responsive genes [95, 111]. Additional evidence was then obtained that at least the Em1a and Motif I elements directly contribute to ABA responsiveness. A 75 bp fragment of the *Em* promoter (containing the Em1a/b and Em2 motifs) confers ABA responsiveness to a truncated (−90) derivative of the cauliflower mosaic virus (CaMV) 35S promoter whereas two single-basepair changes in the Em1a motif decrease ABA induction from 12- to 2-fold [57]. Six tandemly repeated copies of the *rab16A* Motif I fused to a minimal (−46) CaMV 35S promoter provide induction by ABA [150].

Gel retardation DNA binding assays and footprinting experiments demonstrated that the Em1a and Motif I sequences interact with nuclear proteins [57, 111]. Complementary DNA clones that encode proteins with binding affinity for the Em1a (wheat EmBP-1 protein [57]) and Motif I (tobacco TAF-1 protein [118]) elements respectively have been isolated. Although their *in vivo* roles in ABA responses await further analysis, the EmBP-1 and TAF-1 proteins display the two adjacent domains characteristic of the bZIP (basic region-leucine zipper) transcription factors family.

As shown in Table 2, the Em1a and Motif I elements are similar to each other and to the palindromic CACGTG motif known as the G-box [46, 73]. This observation was initially intriguing since several motifs related to the G-box were known to participate in mediating the regulation of unrelated plant genes by a diversity of stimuli distinct from ABA such as light [24] or anaerobiosis [21]. It is now clear that the nucleotides flanking the ACGT core play a critical role in controlling the bZIP DNA binding specificity. Sequence elements with a ACGT core have been accordingly subdivided into three categories (G-, C- and A-box) and bZIP proteins classified into three groups depending on their respective affinity for G-box and C-box elements [73]. These differential binding specificities provide a possible explanation of how the diverse bZIP proteins and ACGT-containing elements are integrated into different *in vivo* regulatory networks. These results also outline that, as discussed for TAF-1 [141], a bZIP protein identified on the basis of its *in vitro* binding to a particular ACGT-containing element is not necessarily the endogenous factor that interacts *in vivo* with this element.

The experimental evidence summarized above support that the Em1a and Motif I elements at least contribute to the ABA responsiveness of the wheat *Em* and rice *rab16A* genes, respectively. Although related sequence motifs with a ACGT core have been found in a number of other ABA-responsive genes [70], their biological significance remains uncertain until functionally assessed. The CCACGTGG element seems to indeed participate in the ABA induction of the maize *rab28* gene [130]. In contrast, the Em1a-like motifs present in the maize *C1* and *Craterostigma plantagineum CDeT27-45* genes are not major determinants of the ABA responsiveness of these genes [60, 115].

Promoter deletion analysis showed that although the Em1a and Motif I elements are clearly necessary components, they do not account alone for the full transcriptional ABA induction of the *Em* and *rab16A* genes [95, 111]. Also, in gain of function experiments several tandem copies of Motif I were needed to confer ABA responsiveness to the minimal (–46) 35S promoter (that contains only a functional TATA box) whereas Motif I is not tandemly repeated in the native *rab16A* promoter [150]. Individual ABREs may thus require other element(s) to couple hormone effects to the transcriptional apparatus. Interestingly, such an abscisic acid responsive complex (ABRC) was artificially built by substituting a single copy of the Motif I ABRE to the gibberellin responsive element in a promoter fragment of the *Amy32b* barley α-amylase gene [138]. The single copies of the ABRE and of the otherwise non ABA-responsive O2S element were both required to provide ABA responsiveness, which indicates that the O2S-ABRE unit functioned as an ABRC. Endogenous coupling elements from native ABA-responsive promoters remain to be identified.

Additional types of *cis*-acting ABREs distinct from the above G-box-related ones have been characterized. A combination of deletion and point mutation analyses identified an element named Sph (Table 2) which is critical for the ABA-activation of the maize *C1* promoter [60]. Related motifs occur in the promoter regions of several other ABA-regulated genes but their functional significance is unknown [60]. Tetramers of a synthetic element named hex-3 (Table 2) confer transcriptional activity upon a truncated (–90) CaMV 35S promoter, and this activity can be enhanced by ABA [82]. An element which appears required, although alone insufficient, to confer responsiveness to ABA has been recently identified in the *Craterostigma plantagineum CDeT27-45* gene [115]. Binding of this element (Table 2) to nuclear factor(s) is ABA-inducible [115], which is not the case for the binding activities to the G-box related Em1a and Motif I elements [57, 111]. Finally, preliminary evidence suggests that the ABA-inducible ATMYB2 homologue of the MYB transcription factor [162] might contribute to the induction of the *Arabidopsis RD22* gene [172] by ABA.

All elements described so far participate in transcriptional activation by ABA. In contrast, a few other elements are known to mediate transcriptional repression by ABA. In particular, the gibberellin-responsive elements (GARE) shown in Table 2 are essential for the gibberellin-inducible, ABA-repressible expression of the barley α-amylase *Amy1/6-4* [150] and *Amy32b* [138] genes. Recent data support that in barley aleurone protoplasts, ABA can regulate gene expression by acting from the external face of the plasma membrane [43]. This indicates the existence of intracellular cascade(s) linking this external site to nuclear transcription.

In conclusion, molecular dissection of ABA-regulated promoters has already revealed a diversity of *cis*-acting sequences that represent likely end-points of ABA-regulatory pathways. This inventory is possibly still far from complete. Identifying the endogenous *trans*-acting factors that bind to these *cis* elements, and understanding how they are connected to more upstream elements of the ABA-signalling cascade(s) will be the next logical (and probably most difficult) steps in this bottom-up approach. As in many other cases (reviewed in [71]), protein kinases and/or phosphatases (see *ABI1* gene below) are likely to be involved in the ABA-pathways regulating transcription. Additional potential candidates start to emerge. Putative transcription activators such as the VP1 and ABI3 (see below), and GF14 [20] proteins can potentially participate in ABA-related transcriptional complexes. Also, the maize Rab17 protein might play a role in the ABA regulation of nuclear protein transport [47].

During the past few years, molecular techniques have emerged to clone genes identified only on the basis of their associated mutant phenotypes. Characterization of mutants defective in ABA responsiveness thus potentially represents a powerful approach to dissect the molecular nature of ABA-signalling cascades, as illustrated in the next section.

Characterization of ABA-response mutants

Mutants that are impaired in their responsiveness to ABA have been described in several plant species, including maize [134], *Arabidopsis* [34, 51, 80, 112, 119] and barley [133] (see Table 1). These mutants are distinct from ABA-biosynthetic mutants in that they do not have reduced endogenous ABA levels and their phenotypes cannot be reversed to wild type by exogenous supply of ABA. These ABA-response mutants are generally pleiotropic in phenotypes and are thus believed to unravel components of signal transduction chains. Since only the *Arabidopsis abi1*, *abi2* and *abi3* [80] and the maize *vp1* [134] mutants have already been analyzed to a substantial extent, and are targets for molecular studies, they will be the main subjects of the discussion below.

The seed-specific Arabidopsis ABI3 *and maize* VP1 *genes*

The *Arabidopsis ABA-INSENSITIVE-3* (*ABI3*) and maize *VIVIPAROUS-1* (*VP1*) loci are both active only in seeds. Phenotypically, no alteration was detected in the vegetative tissues of *abi3* mutant plants [37, 41, 50, 80, 116]. Also, *vp1* mutations were shown to inhibit the anthocyanin biosynthetic pathway only in embryo and aleurone tissues [25, 134]. Molecular cloning of the *VP1* [102] and *ABI3* [45] genes provided a simple explanation for this specificity in that neither gene is found to be expressed in vegetative tissues [45, 101, 102, 122]. The *ABI3* gene is transiently expressed beyond seed germination in young seedlings, but this expression is strictly confined to the organs of embryonic origin (cotyledons and hypocotyl) [122]. This residual expression might possibly explain the few abnormal phenotypes described for young *abi3* mutant seedlings [37].

The *Arabidopsis abi3* and maize *vp1* mutants display some common ABA-related phenotypes. *Abi3* mutants were initially recovered by selecting for seeds capable of germinating in the presence of inhibitory ABA concentrations [80]. Mature seeds of the severe *abi3-3* [112], *abi3-4* and *abi3-5* [119] mutant alleles are several orders of magnitude less sensitive to the inhibition of germination by exogenous ABA. Unlike in the *Arabidopsis aba* biosynthetic mutants, the endogenous ABA levels are not reduced in the *abi3* (nor in the non-allelic *abi1* and *abi2*) mutants [80]. Developing *abi3* embryos fail to become dormant [80, 112, 119], similar to the phenotype described for the *aba* mutant [75, 79]. This deficiency however does not lead to vivipary unless *Arabidopsis* plants are grown under high-humidity conditions [112]. In contrast, like other maize *viviparous* (*vp*) mutants, the *vp1* mutations were originally identified as leading to precocious germination of the embryo while still attached to the mother plant (vivipary) [134]. Whereas most of the other *VP* loci affect early steps in the biosynthesis of carotenoids and ABA, *vp1* embryos do not have reduced ABA content [114]. Excised immature *vp1* embryos rather exhibit somewhat reduced sensitivity to growth inhibition by exogenous ABA in culture [135].

Further evidence supporting that the ABI3 and VP1 proteins can participate in ABA signalling was obtained with the use of the cloned genes. A transcriptional fusion between the CaMV 35S promoter and the *VP1* cDNA was electroporated into maize suspension culture protoplasts together with a construct carrying the GUS reporter placed under the control of the wheat *Em* promoter [95]. In this transient assay system, VP1 overexpression synergistically enhances the transcriptional activation of the heterologous *Em* promoter by exogenous ABA [102]. More recently, *Arabidopsis* plants were stably transformed with transcriptional fusions between the double enhanced CaMV 35S promoter and the ABI3 cDNA (2 × 35S::ABI3). In these 2 × 35S::ABI3 transgenic lines, ectopic expression of ABI3 conferred to plantlets both an increased sensitivity to the inhibition of root growth by exogenous ABA and the ability to accumulate the normally seed-specific *At2S3* [53], *CRC* [121] and *AtEm1* [39] endogenous mRNAs in response to applied ABA (F. Parcy and J. Giraudat, unpublished results). The ectopically expressed ABI3 protein can thus functionally interact with ABA-regulatory cascade(s) present in differentiated vegetative tissues.

Both *abi3* and *vp1* mutations inhibit the *in vivo* accumulation of various endogenous mRNA species characteristic of developmental stages occurring during the last two thirds of seed development, as discussed above. In *Arabidopsis* seeds, these ABI3-dependent mRNAs were initially shown to include cruciferin and napin storage protein mRNAs as well as the late embryogenesis-abundant (LEA) *AtEm6* mRNA [33, 37, 112, 121]. In a recent and more systematic analysis, 19 cDNA probes have been used to compare the kinetics of expression of the corresponding mRNAs throughout silique development in the wild-type and in the severe *abi3-4* mutant [122]. This study demonstrated that the *abi3-4* mutation markedly inhibits the accumulation of multiple transcripts (including the above *At2S3*, *CRC* and *AtEm1* mRNAs) throughout the last two thirds of *Arabidopsis* seed development. This mutation however does not globally disrupt the various temporal programs of gene expression characteristic of late seed development since several mRNA markers retained near wild-type expression patterns in *abi3-4* [122]. Maize *vp1* mutant embryos have been examined at various selected time points and found to similarly contain severely reduced endogenous levels of several globulin storage protein and LEA-type mRNAs [101, 120, 129, 169].

In developing seeds of both species however, the abundance of at least some of these marker mRNAs remains markedly higher in ABA-deficient mutants than in *abi3* or *vp1* [33, 129, 169]. When the abundance of four marker mRNAs that are totally repressed in *vp1* mutant embryos was systematically compared in the various *viviparous* biosynthetic mutants, substantial correspondence between transcript level and ABA content was observed only for the *Em* mRNA [120]. Similarly, the *Arabidopsis* napin *At2S3* and cruciferin *CRC* mRNA levels are slightly if at all reduced in the ABA-deficient *aba-1* mutant whereas the *abi3-4* mutation markedly inhibits the accumulation of both mRNAs (F. Parcy, C. Valon and J. Giraudat, unpublished results). As discussed above, the expression of such genes is most likely controlled primarily by developmental factors distinct from variations in ABA levels. Mutant phenotypes nevertheless indicate that the ABI3 and VP1 proteins are essential for the regulation of gene expression by these unidentified factors.

Additional observations suggest that ABI3 and VP1 roles are not confined to ABA signalling. Accumulation of seed storage lipids is inhibited in *abi3-1* but not in *aba-1 Arabidopsis* mutants [37]. Unlike *aba* mutants, embryos of the strong *abi3* mutant alleles fail to lose chlorophyll and to acquire desiccation tolerance during seed development [112, 119]. Maize *vp1* seeds are defective in anthocyanin accumulation, a phenotype displayed by none of the ABA-deficient *vp* mutants [134]. This colorless phenotype results from the failure to express the *C1* regulatory gene in *vp1* seed tissues and interestingly, partially distinct *cis*-acting sequences mediate activation of the *C1* promoter by VP1 and exogenous ABA, respectively [60].

Altogether the above data indicate that *in vivo* the ABI3 and VP1 proteins interact with ABA-signalling cascades (for instance those controlling dormancy) but also with distinct regulatory pathways. Available evidence support that these proteins are transcriptional activators. As shown in Fig. 2, the primary structures of the VP1 [102] and ABI3 [45] proteins display a similar arrangement of domains with distinct biochemical characteristics. Some of these domains further correspond to discrete regions of remarkable amino acid conservation [45]. No significant sequence similarities to other known proteins were found, and no typical motifs associated with DNA-binding were detected. However, several regions of the polypeptide chains present features previously described in transcriptional activation domains [45, 102]. In particular, experimental evidence support that the N-terminal acidic domain of VP1 can indeed participate in transcriptional activation [102]. As transcriptional activators, the ABI3 and VP1 proteins could control the intensity of gene expression during seed development by interacting with various transcription factors related to distinct regulatory pathways. Such molecular interactions however remain to be experimentally demonstrated.

From the wealth of data accumulated in the past few years on these two systems, it emerges that the ABI3 and VP1 proteins play a much more complex role than initially anticipated. In the future, combined genetic and molecular approaches should unravel further details about the exact function of these proteins in the regulatory networks controlling seed development in mono- and dicotelydonous species respectively. In this respect, several recent *Arabidopsis* mutants represent promising tools since they share several phenotypes with *abi3* mutants but apparently do not display reduced ABA responsiveness [14, 81, 103].

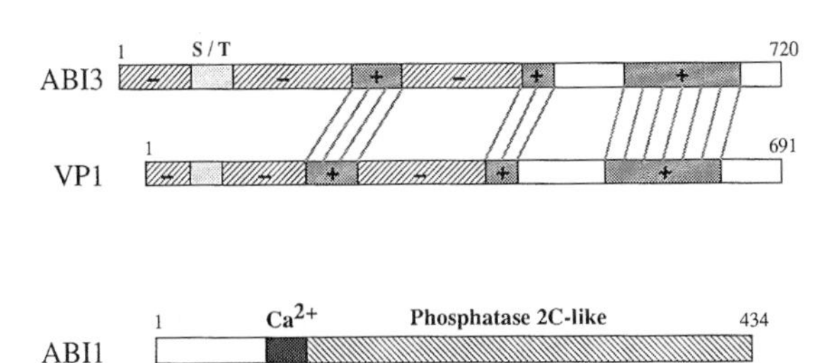

Fig. 2. Schematic diagrams of the architecture of the ABI3, VP1 and ABI1 proteins. Top: the *Arabidopsis* ABI3 and maize VP1 proteins display a similar arrangement of acidic (–), basic (+) and serine/threonine rich (S/T) domains. The three basic domains further correspond to regions of high amino acid sequence identity. See [45, 102] for additional details. Bottom: the *Arabidopsis* ABI1 protein displays a novel amino-terminal domain containing an EF-hand Ca^{2+}-binding motif (Ca^{2+}), and a carboxy-terminal domain homologous to the 2C class of serine/threonine protein phosphatases. See [86, 106] for additional details.

Additional *Arabidopsis* ABA-insensitive mutants have been recently isolated by similar means [36]. These mutants correspond to at least four new *Arabidopsis* loci, two of which (named *ABI4* and *ABI5*) have been characterized to some extent [34]. Available *abi4* and *abi5* mutant alleles display rather weak phenotypes, some of which (for instance reduced *AtEm6* mRNA levels in dry seeds) are also found in *abi3*. From these data and from the characterization of various *abi* digenic mutants, it has been proposed that ABI4 and ABI5 might act in the same pathway than ABI3 [34]. Further experiments and/or stronger alleles might help to firmly assess the direct contribution of ABI4 and ABI5 to ABA-signalling *per se* as well as their exact relationship with ABI3.

The Arabidopsis abi1 *and* abi2 *mutants*

A. thaliana is the only plant species where ABA-response mutants displaying phenotypes both in seeds and in vegetative tissues have been identified. Like *abi3*, the *abi1* and *abi2* mutants were initially selected for their reduced sensitivity to the inhibition of seed germination by exogenous ABA [80]. These mutants also share the reduced seed dormancy phenotype [80], similar to that described for the *aba* mutants [75, 79]. However, available *abi1* and *abi2* mutant alleles do not display the additional *abi3* seed phenotypes described above [33, 37, 78].

Do the ABI1 and/or ABI2 proteins interact with the same seed ABA-regulatory cascade(s) as ABI3? In a germination assay, the *abi3-1,abi1-1* and *abi3-1,abi2-1*/+ digenic mutants are markedly more resistant to ABA than any of these monogenic mutants [37]. Also, whereas *in vivo* accumulation of the *AtEm6* mRNA level is inhibited in *abi3-1*, but not *abi1-1* nor *abi2-1* mature seeds, this mRNA level is further reduced in *abi3-1,abi1-1* and *abi3-1,abi2-1/+* digenic mutants [33]. These additive effects are ambiguous in terms of epistatic interactions since a single phenotype was scored in each set of data and none of the mutant alleles used has been proven to be null (*abi3-1* in particular is definitely a much weaker allele than e.g. *abi3-4*) [5]. Homozygous *abi2-1* and *abi3-1* monogenic mutants are viable but the homozygous *abi2-1,abi3-1* double mutant appears lethal [37]. This synthetic lethal phenotype [52] might indicate that the ABI2 and ABI3 proteins belong to the same seed response pathway(s).

In addition to the above seed phenotypes, the *abi1* and *abi2* mutants are defective in numerous ABA-responses during vegetative growth. Like the *aba* biosynthetic mutants [79], *abi1* and *abi2* plants display disturbed water relations as shown by their increased tendency to wilt [80]. In the case of *abi1-1* (hereafter simplified as *abi1*) which has been analysed in more detail, this wilty phenotype has been traced to improper regulation of stomatal aperture on the abaxial (lower) surface of the leaf, which is on average twice as wide in the mutant as that in the wild type [86].

Whereas *aba* mutants are impaired in cold acclimation, *abi1* nor *abi2-1* (hereafter simplified as *abi2*) mutations do not alter this process [41, 116]. In contrast to these common phenotypes, *abi1* and *abi2* mutations differentially affect the ABA-dependent morphological [163] and molecular [50] responses of *Arabidopsis* plants to progressive drought stress. The *abi1* mutation also affects the ABA-dependent accumulation of other mRNAs in response to rapid desiccation and/or cold [83, 117, 173], the *abi2* mutant was not analysed in these studies.

In addition to the above *in vivo* ABA-dependent processes, the *abi1* and *abi2* mutants have also been characterized for various responses to exogenously applied ABA. *Abi1* and *abi2* seedling growth [37, 80], including root development [86, 106], is more resistant to ABA inhibition than wild type. In particular, root meristematic cells of *abi1* plantlets retain their ability to passage through the S phase of the mitotic cycle in the presence of inhibitory ABA concentrations [86]. *Abi1* and *abi2* mutations impair the root hair deformation induced by applied ABA [142]. These mutations also inhibit ABA-induced proline accumulation and changes in protein synthesis [37]. The *abi1* mutation has been further shown to interfere with the ABA-induced accumulation of all identified mRNAs tested [41, 50, 83, 116, 117, 173]. Unfortunately, the *abi1* and *abi2* mutants have been only rarely [41, 50] compared in these studies. Interestingly, the ABA-induced accumulation of various cold-responsive genes is impaired in *abi1* but not *abi2* [41].

Although essentially all the above phenotypic analyses have been performed on single *abi1* and *abi2* mutant alleles, available data already indicate that the ABI1 and ABI2 proteins contribute to many ABA-regulated responses in vegetative tissues. The relationship between the ABI1 and ABI2 proteins in the ABA-signalling network nevertheless remains unclear. The common mutant phenotypes suggest that these proteins may both regulate certain processes. However, the differential effects of the *abi1* and *abi2* mutations on several responses suggest that the ABI1 and ABI2 proteins might belong to distinct branches of the ABA-signalling network. Additional *Arabidopsis* loci have been recently identified by selecting for mutants with reduced sensitivity to the ABA inhibition of seedling growth [51]. Some of these mutants also display reduced seed dormancy and/or disturbed regulation of leaf water status, and should thus help to further decipher the branching of the ABA-signalling network.

The *ABI1* locus has been cloned recently independently by us [86] and by Meyer *et al.* [106]. The sequence of this gene predicts that it encodes a protein of 434 amino acids that shares sequence similarity (35% identity, 55% similarity) in its carboxyl-terminus with the 2C class of serine/threonine protein phosphatases (PP2Cs) identified in rat [154] and yeast [93]. However, in contrast to these classical PP2Cs, which are Mg^{2+}-

or Mn^{2+}-requiring enzymes [17], the ABI1 protein is appended with a novel amino-terminal domain containing an EF-hand Ca^{2+}-binding site [107]. The combination of these two motifs (Fig. 2) suggests that the ABI1 protein is a modified phosphatase 2C which may have acquired an ability to interact with Ca^{2+}.

The structural features of the ABI1 protein evoke several intriguing possibilities regarding its role in ABA signalling, particularly with regards to stomatal aperture and cellular division in the root meristem. However, the direct involvement of the ABI1 protein in these processes would still need to be verified by further biochemical, physiological and genetic analysis. ABA is known to induce an increase in cytoplasmic Ca^{2+} in a variety of cell types [99]. Moreover, exogenous ABA inhibits cell division by arresting nuclei preferentially in the G1 phase [12, 87]. The $p34^{cdc2}$ gene, which is required for the G1/S transition and the entry into mitosis has been cloned from *Arabidopsis* [63]. Its expression was found to be completely inhibited in the lateral root tips and decreased over the vascular cylinder of the entire root by exogenous ABA. Although these results are not directly comparable to ours because of different experimental criteria employed, it is conceivable that ABI1 in response to ABA or associated Ca^{2+} changes could antagonize the phosphorylation events necessary for the synthesis and activity of similar cell cycle components controlling entry into S-phase [63, 74].

As mentioned above, Ca^{2+} is also strongly implicated as one of the second messengers involved in stomatal response [9, 99, 143]. Recent physiological studies with kinase and phosphatase inhibitors further suggest that stomatal movements as well as some of the electrogenic units involved (for example, plasma membrane H^+ pump, voltage-independent inward- and outward-rectifying K^+ channels) are sensitive to protein phosphorylation [85, 89, 157]. ABI1, as a potential calcium-modulated phosphatase, could couple ABA-stimulus response by modifying the phosphorylation states of these target proteins. Further, although the analysis of *abi1* mutant has so far been focused on ABA sensitivity, the features of the protein suggest that it may have a more versatile role. The protein might serve to cross-talk and integrate ABA and other Ca^{2+}-dependent stimuli that converge on phosphorylation-regulated signalling pathways. The nature of these integrated pathways should now become accessible for systematic investigation with the cloned gene available.

Conclusion

We are still ignorant of the identity and functions of many of the elements involved in ABA signalling, but have arrived at an exciting edge where pieces of the puzzle are emerging at an increasing rate. The impressive progress made in the last few years already provide conceptual frameworks for further studies. The combined use of physiological, genetic and molecular approaches will undoubtedly continue to unravel exciting and possibly unexpected aspects of ABA-signal transduction pathways in plant cells.

Acknowledgements

We thank Dorothea Bartels, Hélène Barbier-Brygoo and Michel Delseny for critical comments on this manuscript, and the numerous colleagues who provided us with reprints and preprints of their publications. Work in our laboratory is funded by the Centre National de la Recherche Scientifique, the European Economic Community and the Ministère de la Recherche et de la Technologie.

References

1. Addicott FT (ed): Abscisic Acid. Praeger Scientific, New York (1983).
2. Addicott FT, Carns HR: History and introduction. In: Addicott FT (ed) Abscisic Acid, pp. 1–21. Praeger Scientific, New York (1983).
3. Alexandre J, Lassalles JP, Kado RT: Opening of Ca^{2+} channels in isolated red beet root vacuole membrane by inositol 1,4,5-trisphosphate. Nature 343: 567–569 (1990).
4. Anderson BE, Ward JM, Schroeder JI: Evidence for an extracellular reception site for abscisic acid in *Com-*

melina guard cells. Plant Physiol 104: 1177–1183 (1994).
5. Avery L, Wasserman S: Ordering gene function: the interpretation of epistasis in regulatory hierarchies. Trends Genet 8: 312–316 (1992).
6. Black M: Involvement of ABA in the physiology of developing and mature seeds. In: Davies WJ, Jones HG (eds) Abscisic Acid Physiology and Biochemistry, pp. 99–124. BIOS Scientific Publishers, Oxford (1991).
7. Blatt M: Potassium channel currents in intact stomatal guard cells: rapid enhancement by abscisic acid. Planta 180: 445–455 (1990).
8. Blatt MR, Armstrong F: K^+ channels of stomatal guard cells: abscisic acid-evoked control of the outward rectifier mediated by cytoplasmic pH. Planta 191: 330–341 (1993).
9. Blatt MR, Thiel G: Hormonal control of ion channel gating. Annu Rev Plant Physiol Plant Mol Biol 44: 543–567 (1993).
10. Blatt MR, Thiel G, Trentham DR: Reversible inactivation of K^+ channels of *Vicia* stomatal guard cells following the photolysis of caged inositol 1,4,5-trisphosphate. Nature 346: 766–769 (1990).
11. Bostock RM, Quatrano RS: Regulation of *Em* gene expression in rice. Interaction between osmotic stress and abscisic acid. Plant Physiol 98: 1356–1363 (1992).
12. Bouvier-Durand M, Real M, Côme D: Changes in nuclear activity upon secondary dormancy induction by abscisic acid in apple embryo. Plant Physiol Biochem 27: 511–518 (1989).
13. Bray EA: Molecular responses to water deficit. Plant Physiol 103: 1035–1040 (1993).
14. Castle LA, Meinke DW: A *FUSCA* gene of *Arabidopsis* encodes a novel protein essential for plant development. Plant Cell 6: 25–41 (1994).
15. Chen H-H, Li PH, Brenner ML: Involvement of abscisic acid in potato cold acclimation. Plant Physiol 71: 362–365 (1983).
16. Cohen A, Bray EA: Characterization of three mRNAs that accumulate in wilted tomato leaves in response to elevated levels of endogenous abscisic acid. Planta 182: 27–33 (1990).
17. Cohen P: The structure and regulation of protein phosphatases. Annu Rev Biochem 58: 453–508 (1989).
18. Davies WJ, Jones HG (eds): Abscisic acid physiology and biochemistry. BIOS Scientific Publishers, Oxford (1991).
19. Davies WJ, Tardieu F, Trejo CL: How do chemical signals work in plants that grow in drying soil? Plant Physiol 104: 309–314 (1994).
20. de Vetten NC, Lu G, Ferl RJ: A maize protein associated with the G-box binding complex has homology to brain regulatory proteins. Plant Cell 4: 1295–1307 (1992).
21. DeLisle AJ, Ferl RJ: Characterization of the *Arabidopsis Adh* G-box binding factor. Plant Cell 2: 547–557 (1990).
22. Delseny M, Gaubier P, Hull G, Saez-Vasquez J, Gallois P, Raynal M, Cooke R, Grellet F: Nuclear genes expressed during seed desiccation: relationship with response to stress. In: Basra AS (ed) Stress-Induced Gene Expression in Plants, pp. 25–59. Harwood Academic Publishers, Reading, UK (1994).
23. Dohlman HG, Thorner J, Caron MG, Lefkowitz RJ: Model systems for the study of seven-transmembrane-segment receptors. Annu Rev Biochem 60: 653–688 (1991).
24. Donald RGK, Cashmore AR: Mutation of either G box or I box sequences profoundly affects expression of the *Arabidopsis rbcS-1A* promoter. EMBO J 9: 1717–1726 (1990).
25. Dooner HK: *Viviparous-1* mutation in maize conditions pleiotropic enzyme deficiencies in the aleurone. Plant Physiol 77: 486–488 (1985).
26. Duckham SC, Linforth RST, Taylor IB: Abscisic acid deficient mutants at the *aba* gene locus of *Arabidopsis thaliana* are impaired in the epoxidation of zeaxanthin. Plant Cell Environ 14: 601–606 (1991).
27. Duckham SC, Taylor IB, Linforth RST, Al-Naieb RJ, Marples BA, Bowman WR: The metabolism of *cis* ABA-aldehyde by the wilty mutants of potato, pea and *Arabidopsis thaliana*. J Exp Bot 40: 901–905 (1989).
28. Dure III L: The Lea proteins of higher plants. In: Verma DPS (ed) Control of Plant Gene Expression, pp. 325–335. CRC Press, Boca Raton, FL (1993).
29. Dure III L, Crouch M, Harada J, Ho T-HD, Mundy J, Quatrano R, Thomas T, Sung ZR: Common amino acid sequence domains among the LEA proteins of higher plants. Plant Mol Biol 12: 475–486 (1989).
30. Fairley-Grenot K, Assmann S: Evidence for G-protein regulation of inward K^+ channel current in guard cells of fava bean. Plant Cell 3: 1037–1044 (1991).
31. Farmer EE, Ryan CA: Interplant communication: airborne methyl jasmonate induces synthesis of proteinase inhibitors in plant leaves. Proc Natl Acad Sci USA 87: 7713–7716 (1990).
32. Farmer EE, Ryan CA: Octadecanoid precursors of jasmonic acid activate the synthesis of wound-inducible proteinase inhibitors. Plant Cell 4: 129–134 (1992).
33. Finkelstein RR: Abscisic acid-insensitive mutations provide evidence for stage-specific signal pathways regulating expression of an *Arabidopsis* late embryogenesis-abundant (lea) gene. Mol Gen Genet 238: 401–408 (1993).
34. Finkelstein RR: Mutations at two new *Arabidopsis* ABA response loci are similar to the *abi3* mutations. Plant J 5: 765–771 (1994).
35. Finkelstein RR, Crouch ML: Rapeseed embryo development in culture on high osmoticum is similar to that in seeds. Plant Physiol 81: 907–912 (1986).
36. Finkelstein RR, Doyle MP: Molecular genetic analysis of abscisic acid signal transduction in *Arabidopsis*. Fifth International Conference on *Arabidopsis* Research, Columbus, OH (1993).

37. Finkelstein RR, Somerville CR: Three classes of abscisic acid (ABA)-insensitive mutations of *Arabidopsis* define genes that control overlapping subsets of ABA responses. Plant Physiol 94: 1172–1179 (1990).
38. Galau GA, Jakobsen KS, Hughes DW: The controls of late dicot embryogenesis and early germination. Physiol Plant 81: 280–288 (1991).
39. Gaubier P, Raynal M, Hull G, Huestis GM, Grellet F, Arenas C, Pagès M, Delseny M: Two different *Em*-like genes are expressed in *Arabidopsis thaliana* seeds during maturation. Mol Gen Genet 238: 409–418 (1993).
40. Gehring CA, Irving HR, Parish RW: Effects of auxin and abscisic acid on cytosolic calcium and pH in plant cells. Proc Natl Acad Sci USA 87: 9645–9649 (1990).
41. Gilmour SJ, Thomashow MF: Cold acclimation and cold-regulated gene expression in ABA mutants of *Arabidopsis thaliana*. Plant Mol Biol 17: 1233–1240 (1991).
42. Gilroy S, Fricker MD, Read ND, Trewavas AJ: Role of calcium in signal transduction of *Commelina* guard cells. Plant Cell 3: 333–344 (1991).
43. Gilroy S, Jones RL: Perception of gibberellin and abscisic acid at the external face of the plasma membrane of barley (*Hordeum vulgare* L.) aleurone protoplasts. Plant Physiol 104: 1185–1192 (1994).
44. Gilroy S, Read ND, Trewavas AJ: Elevation of cytoplasmic calcium by caged calcium or caged inositol trisphosphate initiates stomatal closure. Nature 343: 769–771 (1990).
45. Giraudat J, Hauge BM, Valon C, Smalle J, Parcy F, Goodman HM: Isolation of the *Arabidopsis ABI3* gene by positional cloning. Plant Cell 4: 1251–1261 (1992).
46. Giulano G, Pichersky E, Malik VS, Timko MP, Scolnik PA, Cashmore AR: An evolutionary conserved protein binding sequence upstream of a plant light-regulated gene. Proc Natl Acad Sci USA 85: 7089–7093 (1988).
47. Goday A, Jensen AB, Culianez-Macia FA, Alba MM, Figueras M, Serratosa J, Torrent M, Pages M: The maize abscisic acid-responsive protein Rab17 is located in the nucleus and interacts with nuclear localization signals. Plant Cell 6: 351–360 (1994).
48. Goffner D, This P, Delseny M: Effects of abscisic acid and osmotica on helianthinin gene expression in sunflower cotyledons *in vitro*. Plant Sci 66: 211–219 (1990).
49. Goldberg RB, Barker SJ, Perez-Grau L: Regulation of gene expression during plant embryogenesis. Cell 56: 149–160 (1989).
50. Gosti F, Bertauche N, Vartanian N, Giraudat J: Abscisic acid-dependent and -independent regulation of gene expression by progressive drought in *Arabidopsis thaliana*. Mol Gen Genet, in press (1994).
51. Grill E, Ehrler T, Meyer K, Leube M: Steps of abscisic acid action. Fifth International Conference on *Arabidopsis* Research, Columbus, OH (1993).
52. Guarente L: Synthetic enhancement in gene interaction: a genetic tool come of age. Trends Genet 9: 362–366 (1993).
53. Guerche P, Tire C, Grossi de Sa F, De Clercq A, Van Montagu M, Krebbers E: Differential expression of the *Arabidopsis* 2S albumin genes and the effect of increasing gene family size. Plant Cell 2: 469–478 (1990).
54. Guerrero F, Mullet JE: Increased abscisic acid biosynthesis during plant dehydration requires transcription. Plant Physiol 80: 588–591 (1986).
55. Guerrero FD, Jones JT, Mullet JE: Turgor-responsive gene transcription and RNA levels increase rapidly when pea shoots are wilted. Sequence and expression of three inducible genes. Plant Mol Biol 15: 11–26 (1990).
56. Guerrero FD, Mullet JE: Reduction of turgor induces rapid changes in leaf translatable RNA. Plant Physiol 88: 401–408 (1988).
57. Guiltinan MJ, Marcotte WR, Quatrano RS: A plant leucine zipper protein that recognizes an abscisic acid response element. Science 250: 267–271 (1990).
58. Harris MJ, Outlaw Jr. WH: Rapid adjustement of guard-cell abscisic acid levels to current leaf-water status. Plant Physiol 95: 171–173 (1991).
59. Hartung W: The site of action of abscisic acid at the guard cell plasmalemma of *Valerianella locusta*. Plant Cell Environ 6: 427–428 (1983).
60. Hattori T, Vasil V, Rosenkrans L, Hannah LC, McCarty DR, Vasil IK: The *Viviparous-1* gene and abscisic acid activate the *C1* regulatory gene for anthocyanin biosynthesis during seed maturation in maize. Genes Devel 6: 609–618 (1992).
61. Hedrich R, Busch H, Raschke K: Ca^{2+} and nucleotide dependent regulation of voltage dependent anion channels in the plasma membrane of guard cells. EMBO J 9: 3889–3892 (1990).
62. Heino P, Sandman G, Lang V, Nordin K, Palva ET: Abscisic acid deficiency prevents development of freezing tolerance in *Arabidopsis thaliana* (L.) Heynh. Theor Appl Genet 79: 801–806 (1990).
63. Hemerly AS, Ferreira P, de Almeida Engler J, Van Montagu M, Engler G, Inzé D: *cdc2a* expression in *Arabidopsis* is linked with competence for cell division. Plant Cell 5: 1711–1723 (1993).
64. Hetherington AM, Quatrano RS: Mechanisms of action of abscisic acid at the cellular level. New Phytol 119: 9–32 (1991).
65. Hildmann T, Ebneth M, Pena-Cortes H, Sanchez-Serrano JJ, Willmitzer L, Prat S: General roles of abscisic and jasmonic acids in gene activation as a result of mechanical wounding. Plant Cell 4: 1157–1170 (1992).
66. Hocking TJ, Clapham KJ, Cattell KJ: Abscisic acid binding to subcellular fractions from leaves of *Vicia faba*. Planta 138: 303–304 (1978).
67. Hornberg C, Weiler EW: High-affinity binding sites for abscisic acid on plasmalemma of *Vicia faba* guard cells. Nature 310: 321–324 (1984).
68. Hughes DW, Galau GA: Temporally modular gene ex-

pression during cotyledon development. Genes Devel 3: 358–369 (1989).
69. Hughes DW, Galau GA: Developmental and environmental induction of *Lea* and *LeaA* mRNAs and the postabscission program during embryo culture. Plant Cell 3: 605–618 (1991).
70. Hull G, Gaubier P, Delseny M, Casse-Delbart F: Abscisic acid inducible genes and their regulation in higher plants. Current Top Mol Genet (Life Sci Adv) 1: 289–305 (1993).
71. Hunter T, Karin M: The regulation of transcription by phosphorylation. Cell 70: 375–387 (1992).
72. Irving HR, Gehring CA, Parish RW: Changes in cytosolic pH and calcium of guard cells precede stomatal movements. Proc Natl Acad Sci USA 89: 1790–1794 (1992).
73. Izawa T, Foster R, Chua N-H: Plant bZIP protein DNA binding specificity. J Mol Biol 230: 1131–1144 (1993).
74. Jacobs T: Control of the cell cycle. Devel Biol 153: 1–15 (1992).
75. Karssen CM, Brinkhorst-van der Swan DLC, Breekland AE, Koornneef M: Induction of dormancy during seed development by endogenous abscsic acid: studies on abscisic acid deficient genotypes of *Arabidopsis thaliana* (L.) Heynh. Planta 157: 158–165 (1983).
76. Kaziro Y, Itoh H, Kozasa T, Nakafuku M, Satoh T: Structure and function of signal-transducing GTP-binding proteins. Annu Rev Biochem 60: 349–400 (1991).
77. Kearns EV, Assmann SM: The guard cell environment connection. Plant Physiol 102: 711–715 (1993).
78. Koornneef M, Hanhart CJ, Hilhorst HWM, Karssen CM: *In vivo* inhibition of seed development and reserve protein accumulation in recombinants of abscisic acid biosynthesis and responsiveness mutants in *Arabidopsis thaliana*. Plant Physiol 90: 463–469 (1989).
79. Koornneef M, Jorna ML, Brinkhorst-van der Swan DLC, Karssen CM: The isolation of abscisic acid (ABA) deficient mutants by selection of induced revertants in non-germinating gibberellin sensitive lines of *Arabidopsis thaliana* (L.) Heynh. Theor Appl Genet 61: 385–393 (1982).
80. Koornneef M, Reuling G, Karssen CM: The isolation and characterization of abscisic acid-insensitive mutants of *Arabidopsis thaliana*. Physiol Plant 61: 377–383 (1984).
81. Kraml M, Keith K, McCourt P: A non-dormant *Arabidopsis* mutant which is sensitive to ABA. Fifth International Conference on *Arabidopsis* Research, Columbus, OH (1993).
82. Lam E, Chua N-H: Tetramer of a 21-base pair synthetic element confers seed expression and transcriptional enhancement in response to water stress and abscisic acid. J Biol Chem 266: 17131–17135 (1991).
83. Lang V, Palva ET: The expression of a *rab*-related gene, *rab18*, is induced by abscisic acid during the cold acclimation process of *Arabidopsis thaliana* (L.) Heynh. Plant Mol Biol 20: 951–962 (1992).
84. Lee HJ, Tucker EB, Crain RC, Lee Y: Stomatal opening is induced in epidermal peels of *Commelina communis* L. by GTP analogs or pertussis toxin. Plant Physiol 102: 95–100 (1993).
85. Lee Y, Assmann SM: Diacylglycerols induce both ion pumping in patch-clamped guard-cell protoplasts and opening of intact stomata. Proc Natl Acad Sci USA 88: 2127–2131 (1991).
86. Leung J, Bouvier-Durand M, Morris P-C, Guerrier D, Chefdor F, Giraudat J: *Arabidopsis* ABA-response gene *ABI1*: features of a calcium-modulated protein phosphatase. Science 264: 1448–1452 (1994).
87. Levi M, Brusa P, Chiatante D, Sparvoli E: Cell cycle reactivation in cultured pea embryo axes. Effect of abscisic acid. In Vitro Cell Devel Biol 29: 47–50 (1993).
88. Lobréaux S, Hardy T, Briat J-F: Abscisic acid is involved in the iron-induced synthesis of maize ferritin. EMBO J 12: 651–657 (1993).
89. Luan S, Li W, Rusnak F, Assmann SM, Schreiber SL: Immunosuppressants implicate protein phosphatase regulation of K^+ channels in guard cells. Proc Natl Acad Sci USA 90: 2202–2206 (1993).
90. MacRobbie EAC: Effects of ABA in isolated guard cells of *Commelina communis* L. J Exp Bot 32: 563–572 (1981).
91. MacRobbie EAC: Calcium-dependent and calcium-independent events in the initiation of stomatal closure by abscisic acid. Proc R Soc Series B 241: 214–219 (1990).
92. MacRobbie EAC: Effect of ABA on ion transport and stomatal regulation. In: Davies WJ, Jones HG (eds) Abscisic Acid Physiology and Biochemistry, pp. 153–168. BIOS Scientific Publishers, Oxford (1991).
93. Maeda T, Tsai AYM, Saito H: Mutations in a protein tyrosine phosphatase gene (*PTP2*) and a protein serine/threonine phosphatase gene (*PTC1*) cause a synthetic growth defect in *Saccharomyces cerevisiae*. Mol Cell Biol 13: 5408–5417 (1993).
94. Mansfield TA: Hormones as regulators of water balance. In: Davies RD (ed) Plant Hormones and their Role in Plant Growth and Development, pp. 411–430. Martinus Nijhoff, Dordrecht (1988).
95. Marcotte WR, Russell SH, Quatrano RS: Abscisic acid-responsive sequences from the *Em* gene of wheat. Plant Cell 1: 969–976 (1989).
96. McAinsh MR, Brownlee AM, Hetherington AM: Abscisic acid-induced elevation of guard cell calcium precedes stomatal closure. Nature 343: 186–188 (1990).
97. McAinsh MR, Brownlee C, Hetherington AM: Partial inhibition of ABA-induced stomatal closure by calcium-channel blockers. Proc R Soc Series B 243: 195–201 (1991).
98. McAinsh MR, Brownlee C, Hetherington AM: Visualizing changes in cytosolic-free Ca^{2+} during the response

of stomatal guard cells to abscisic acid. Plant Cell 4: 1113–1122 (1992).
99. McAinsh MR, Brownlee C, Sarsag M, Webb AAR, Hetherington AM: Involvement of second messengers in the action of ABA. In: Davies WJ, Jones HG (eds) Abscisic Acid Physiology and Biochemistry, pp. 137–152. BIOS Scientific Publishers, Oxford (1991).
100. McCarty DR, Carson CB, Lazar M, Simonds SC: Transposable element-induced mutations of the *viviparous-1* gene in maize. Devel Genet 10: 473–481 (1989).
101. McCarty DR, Carson CB, Stinard PS, Robertson DS: Molecular analysis of *viviparous-1*: an abscisic acid-insensitive mutant of maize. Plant Cell 1: 523–532 (1989).
102. McCarty DR, Hattori T, Carson CB, Vasil V, Lazar M, Vasil IK: The *viviparous-1* developmental gene of maize encodes a novel transcriptional activator. Cell 66: 895–905 (1991).
103. Meinke DW: A homeotic mutant of *Arabidopsis thaliana* with leafy cotyledons. Science 258: 1647–1650 (1992).
104. Melan MA, Dong X, Endara ME, Davis KR, Ausubel FM, Peterman TK: An *Arabidopsis thaliana* lipoxygenase gene can be induced by pathogens, abscisic acid, and methyl jasmonate. Plant Physiol 101: 441–450 (1993).
105. Meurs C, Basra AS, Karssen CM, van Loon LC: Role of abscisic acid in the induction of desiccation tolerance in developing seeds of *Arabidopsis thaliana*. Plant Physiol 98: 1484–1493 (1992).
106. Meyer K, Leube MP, Grill E: A protein phosphatase 2C involved in ABA signal transduction in *Arabidopsis thaliana*. Science 264: 1452–1455 (1994).
107. Moncrief ND, Kretsinger RH, Goodman M: Evolution of EF-hand calcium-modulated proteins. I. Relationships based on amino acid sequences. J Mol Evol 30: 522–562 (1990).
108. Moore R, Smith JD: Graviresponsiveness and abscisic acid content of roots of carotenoid-deficient mutants of *Zea mays* L. Planta 164: 126–128 (1985).
109. Morris PC, Kumar A, Bowles DJ, Cuming AC: Osmotic stress and abscisic acid induce expression of the wheat *Em* genes. Eur J Biochem 190: 625–630 (1990).
110. Morris PC, Weiler EW, Maddock SE, Jones MGK, Lenton JR, Bowles DJ: Determination of endogenous abscisic acid levels in immature cereal embryos during *in vitro* culture. Planta 173: 110–116 (1988).
111. Mundy J, Yamaguchi-Shinozaki K, Chua N-H: Nuclear proteins bind conserved elements in the abscisic acid-responsive promoter of a rice *rab* gene. Proc Natl Acad Sci USA 87: 1406–1410 (1990).
112. Nambara E, Naito S, McCourt P: A mutant of *Arabidopsis* which is defective in seed development and storage protein accumulation is a new *abi3* allele. Plant J 2: 435–441 (1992).
113. Neill SJ, Horgan R: Abscisic acid production and water relations in wilty tomato mutants subjected to water deficiency. J Exp Bot 36: 1222–1231 (1985).
114. Neill SJ, Horgan R, Parry AD: The carotenoid and abscisic acid content of viviparous kernels and seedlings of *Zea mays* L. Planta 169: 87–96 (1986).
115. Nelson D, Salamini F, Bartels D: Abscisic acid promotes novel DNA-binding activity to a desiccation-related promoter of *Craterostigma plantagineum*. Plant J 5: 451–458 (1994).
116. Nordin K, Heino P, Palva ET: Separate signal pathways regulate the expression of a low-temperature-induced gene in *Arabidopsis thaliana* (L.) Heynh. Plant Mol Biol 16: 1061–1071 (1991).
117. Nordin K, Vahala T, Palva ET: Differential expression of two related, low-temperature-induced genes in *Arabidopsis thaliana* (L.) Heynh. Plant Mol Biol 21: 641–653 (1993).
118. Oeda K, Salinas J, Chua N-H: A tobacco bZip transcription activator (TAF-1) binds to a G-box-like motif conserved in plant genes. EMBO J 10: 1793–1802 (1991).
119. Ooms JJJ, Léon-Kloosterziel KM, Bartels D, Koornneef M, Karssen CM: Acquisition of desiccation tolerance and longevity in seeds of *Arabidopsis thaliana*. A comparative study using abscisic acid-insensitive *abi3* mutants. Plant Physiol 102: 1185–1191 (1993).
120. Paiva R, Kriz AL: Effect of abscisic acid on embryo-specific gene expression during normal and precocious germination in normal and *viviparous* maize (*Zea mays*) embryos. Planta 192: 332–339 (1994).
121. Pang PP, Pruitt RE, Meyerowitz EM: Molecular cloning, genomic organization, expression and evolution of 12S seed storage protein genes of *Arabidopsis thaliana*. Plant Mol Biol 11: 805–820 (1988).
122. Parcy F, Valon F, Raynal M, Gaubier P, Delseny M, Giraudat J: Regulation of gene expression by the *Arabidopsis ABI3* (abscisic acid-insensitive) gene: analysis of mutant and transgenic plants. Fifth International Conference on *Arabidopsis* Research, Columbus, OH (1993).
123. Parry AD, Blonstein AD, Babiano MJ, King PJ, Horgan R: Abscisic acid metabolism in a wilty mutant of *Nicotiana plumbaginifolia*. Planta 183: 237–243 (1991).
124. Parthier B: Jasmonates, new regulators of plant growth and development: many facts and few hypotheses on their actions. Bot Acta 104: 446–454 (1991).
125. Pena-Cortes H, Albrecht T, Prat S, Weiler EW, Willmitzer L: Aspirin prevents wound-induced gene expression in tomato leaves by blocking jasmonic acid biosynthesis. Planta 191: 123–128 (1993).
126. Pena-Cortés H, Sanchez-Serrano JJ, Mertens R, Willmitzer L, Prat S: Abscisic acid is involved in the wound-induced expression of the proteinase inhibitor II gene in potato and tomato. Proc Natl Acad Sci USA 86: 9851–9855 (1989).
127. Pena-Cortes H, Willmitzer L, Sanchez-Serrano JJ: Abscisic acid mediates the wound induction but not

developmental-specific expression of the proteinase inhibitor II gene family. Plant Cell 3: 963–972 (1991).
128. Pla M, Goday A, Vilardell J, Gomez J, Pagès M: Differential regulation of ABA-induced 23-25kDa proteins in embryo and vegetative tissues of the *viviparous* mutants of maize. Plant Mol Biol 13: 385–394 (1989).
129. Pla M, Gomez J, Goday A, Pagès M: Regulation of the abscisic acid-responsive gene *rab28* in maize *viviparous* mutants. Mol Gen Genet 230: 394–400 (1991).
130. Pla M, Vilardell J, Guiltinan MJ, Marcotte WR, Niogret M-F, Quatrano RS, Pagès M: The *cis*-regulatory element CCACGTGG is involved in ABA and water-stress responses of the maize gene *rab28*. Plant Mol Biol 21: 259–266 (1993).
131. Quarrie S: *Droopy*: a wilty mutant of potato deficient in abscisic acid. Plant Cell Environ 5: 23–26 (1982).
132. Quatrano RS: The role of hormones during seed development. In: Davies PJ (ed) Plant Hormones and their Role in Plant Growth and Development, pp. 494–514. Kluwer Academic Publishers, Dordrecht (1988).
133. Raskin I, Ladyman JAR: Isolation and characterization of a barley mutant with abscisic acid-insensitive stomata. Planta 173: 73–78 (1988).
134. Robertson DS: The genetics of vivipary in maize. Genetics 40: 745–760 (1955).
135. Robichaud C, Sussex IM: The response of *viviparous-1* and wild-type embryos of *Zea mays* to culture in the presence of abscisic acid. J Plant Physiol 126: 235–242 (1986).
136. Rock CD, Heath TG, Gage DA, Zeevaart JAD: Abscisic alcohol is an intermediate in abscisic acid biosynthesis in a shunt pathway from abscisic aldehyde. Plant Physiol 97: 670–676 (1991).
137. Rock CD, Zeevaart JA: The *aba* mutant of *Arabidopsis thaliana* is impaired in epoxy-carotenoid biosynthesis. Proc Natl Acad Sci USA 88: 7496–7499 (1991).
138. Rogers JC, Rogers SW: Definition and functional implications of gibberellin and abscisic acid *cis*-acting hormone response complexes. Plant Cell 4: 1443–1451 (1992).
139. Rousselin P, Kraepiel Y, Maldiney R, Miginiac E, Caboche M: Characterization of three hormone mutants of *Nicotiana plumbaginifolia*: evidence for a common ABA deficiency. Theor Appl Genet 85: 213–221 (1992).
140. Ryan CA: The search for the proteinase inhibitor-inducing factor, PIIF. Plant Mol Biol 19: 123–133 (1992).
141. Salinas J, Oeda K, Chua N-H: Two G-box related sequences confer different expression patterns in transgenic tobacco. Plant Cell 4: 1485–1493 (1992).
142. Schnall JA, Quatrano RS: Abscisic acid elicits the water-stress response in root hairs of *Arabidopsis thaliana*. Plant Physiol 100: 216–218 (1992).
143. Schroeder J: Plasma membrane ion channel regulation during abscisic acid-induced closing of stomata. Phil Trans R Soc Lond B 338: 83–89 (1992).
144. Schroeder J, Hagiwara S: Cytosolic calcium regulates ion channels in the plasma membrane of *Vicia faba* guard cells. Nature 338: 427–430 (1989).
145. Schroeder J, Keller BU: Two types of anion channel currents in guard cells with distinct voltage regulation. Proc Natl Acad Sci USA 89: 5025–5029 (1992).
146. Schroeder J, Schmidt C, Sheaffer J: Identification of high-affinity slow anion channel blockers and evidence for stomatal regulation by slow anion channels in guard cells. Plant Cell 5: 1831–1841 (1993).
147. Schroeder JI, Hagiwara S: Repetitive increases in cytosolic Ca^{2+} of guard cells by abscisic acid activation of nonselective Ca^{2+}-permeable channels. Proc Natl Acad Sci USA 87: 9305–9309 (1990).
148. Schwartz A, Wu W-H, Tucker EB, Assmann SM: Inhibition of inward K^+ channels and stomatal response by abscisic acid: an intracellular locus of phytohormone action. Proc Natl Acad Sci USA 91: 4019–4023 (1994).
149. Skriver K, Mundy J: Gene expression in response to abscisic acid and osmotic stress. Plant Cell 2: 503–512 (1990).
150. Skriver K, Olsen FL, Rogers JC, Mundy J: *Cis*-acting DNA elements responsive to gibberellin and its antagonist abscisic acid. Proc Natl Acad Sci USA 88: 7266–7270 (1991).
151. Stewart CR, Voetberg G: Relationship between stress-induced ABA and proline accumulations and ABA-induced proline accumulation in excised barley leaves. Plant Physiol 79: 24–27 (1985).
152. Tal M, Imber D, Erez A, Epstein E: Abnormal stomatal behavior and hormonal imbalance in *flacca*, a wilty mutant of tomato. V. Effect of abscisic acid on indoleacetic acid metabolism and ethylene evolution. Plant Physiol 63: 1044–1048 (1979).
153. Tal M, Nevo Y: Abnormal stomatal behaviour and root resistance, and hormonal imbalance in three wilty mutants of tomato. Biochem Genet 8: 291–300 (1973).
154. Tamura S, Lynch KR, Larner J, Fox J, Yasui A, Kikuchi K, Suzuki Y, Tsuiki S: Molecular cloning of rat type 2C (1A) protein phosphatase mRNA. Proc Natl Acad Sci USA 86: 1796–1800 (1989).
155. Taylor IB: Genetics of ABA synthesis. In: Davies WJ, Jones HG (eds) Abscisic Acid Physiology and Biochemistry, pp. 23–37. BIOS Scientific Publishers, Oxford (1991).
156. Taylor IB, Linforth RST, Al-Naieb RJ, Bowman WR, Marples BA: The wilty tomato mutants *flacca* and *sitiens* are impaired in the oxidation of ABA-aldehyde to ABA. Plant Cell Environ 11: 739–745 (1988).
157. Thiel G, Blatt MR: Phosphatase antagonist okadaic acid inhibits steady state K^+ currents in guard cells of *Vicia faba*. Plant J 5: 727–733 (1994).
158. Thiel G, MacRobbie EAC, Blatt MR: Membrane transport in stomatal guard cells: the importance of voltage control. J Membrane Biol 126: 1–18 (1992).
159. Thomas TL: Gene expression during plant embryogen-

esis and germination: an overview. Plant Cell 5: 1401–1410 (1993).
160. Trewavas A: How do plant growth substances work? II. Plant Cell Environ 14: 1–12 (1991).
161. Trewavas AJ, Jones HG: An assessment of the role of ABA in plant development. In: Davies WJ, Jones HG (eds) Abscisic Acid: Physiology and Biochemistry, pp. 169–188. BIOS Scientific Publishers, Oxford (1991).
162. Urao T, Yamaguchi-Shinozaki K, Urao S, Shinozaki K: An Arabidopsis *myb* homolog is induced by dehydration stress and its gene product binds to the conserved MYB recognition sequence. Plant Cell 5: 1529–1539 (1993).
163. Vartanian N, Marcotte L, Giraudat J: Drought rhizogenesis in *Arabidopsis thaliana*. Differential responses of hormonal mutants. Plant Physiol 104: 761–767 (1994).
164. Vilardell J, Martinez-Zapater JM, Goday A, Arenas C, Pagès M: Regulation of the *rab17* gene promoter in transgenic *Arabidopsis* wild-type, ABA-deficient and ABA-insensitive mutants. Plant Mol Biol 24: 561–569 (1994).
165. Walker-Simmons M, Kudrna DA, Warner RL: Reduced accumulation of ABA during water stress in a molybdenum cofactor mutant of barley. Plant Physiol 90: 728–733 (1989).
166. Walton DC: Biochemistry and physiology of abscisic acid. Annu Rev Plant Physiol Plant Mol Biol 31: 453–489 (1980).
167. Walton DC: Abscisic acid biosynthesis and metabolism. In: Davies PJ (ed) Plant Hormones and their Role in Plant Growth and Development, pp. 113–131. Kluwer Academic Publishers, Dordrecht (1988).
168. Wang T, Donkin M, Martin E: The physiology of a wilty pea: abscisic acid production under water stress. J Exp Bot 35: 1222–1232 (1984).
169. Williams B, Tsang A: A maize gene expressed during embryogenesis is abscisic acid-inducible and highly conserved. Plant Mol Biol 16: 919–923 (1991).
170. Xu D, McElroy D, Thornburg RW, Wu R: Systemic induction of a potato *pin2* promoter by wounding, methyl jasmonate, and abscisic acid in transgenic rice plants. Plant Mol Biol 22: 573–588 (1993).
171. Yamaguchi-Shinozaki K, Koizumi M, Urao S, Shinozaki K: Molecular cloning and characterization of 9 cDNAs for genes that are responsive to desiccation in *Arabidopsis thaliana*: sequence analysis of one cDNA clone that encodes a putative transmembrane channel protein. Plant Cell Physiol 33: 217–224 (1992).
172. Yamaguchi-Shinozaki K, Shinozaki K: The plant hormone abscisic acid mediates the drought-induced expression but not the seed-specific expression of *rd22*, a gene responsive to dehydration stress in *Arabidopsis thaliana*. Mol Gen Genet 238: 17–25 (1993).
173. Yamaguchi-Shinozaki K, Shinozaki K: Characterization of the expression of a desiccation-responsive *rd29* gene of *Arabidopsis thaliana* and analysis of its promoter in transgenic plants. Mol Gen Genet 236: 331–340 (1993).
174. Yamaguchi-Shinozaki K, Shinozaki K: A novel *cis*-acting element in an *Arabidopsis* gene is involved in responsiveness to drought, low-temperature, or high-salt stress. Plant Cell 6: 251–264 (1994).
175. Zeevaart JAD, Creelman RA: Metabolism and physiology of abscisic acid. Annu Rev Plant Physiol Plant Mol Biol 39: 439–473 (1988).
176. Zeevaart JAD, Rock CD, Fantauzzo F, Heath TG, Gage DA: Metabolism of ABA and its physiological implications. In: Davies WJ, Jones HG (eds) Abscisic Acid Physiology and Biochemistry, pp. 39–52. BIOS Scientific Publishers, Oxford (1991).

Plant Molecular Biology **26**: 1579–1597, 1994.

Ethylene biosynthesis and action: a case of conservation

Thomas I. Zarembinski and Athanasios Theologis*
Plant Gene Expression Center, Albany, CA 94710, USA (author for correspondence)*

Received and accepted 1 July 1994

Key words: aminotransferase, dioxygenase, ethylene, kinase cascade, raf kinase, two-component system

Introduction

Ethylene is one of the simplest organic molecules with biological activity. At concentrations as low as 0.1 ppm in air, it has been shown to have dramatic effects on plant growth and development [1]. Neljubov [78] was the first to show that ethylene has three major effects in etiolated pea seedlings called the triple response: (1) diageotropic growth, (2) thickening of stem and inhibition of stem elongation, and (3) exaggeration of apical hook curvature. Since then, numerous ethylene effects have been described in light-grown plants such as sex determination in curcurbits, fruit ripening in climacteric fruits, epinastic curvature, flower senescence, and root initiation [1]. Interestingly, ethylene has also been shown to have opposite effects in some plants; for instance, it inhibits stem elongation in most dicots, whereas in some aquatic dicots and rice, it stimulates growth [1, 45, 72]. Such growth is essential for the survival of such plants so as to keep its foliage above water [45, 72].

Until the early 1970s very little was known about how ethylene is biosynthesized, how its production is regulated, and how a plant perceives its presence in nanoliter quantities. Since then, a large amount of biochemical and genetic data has been gathered, indicating that every component of the ethylene production and action pathways in plants so far studied has a homologue found in other prokaryotic or eukaryotic systems (see Table 1). Many excellent reviews on ethylene production and perception have been published [23, 27, 46, 47, 48, 108, 109, 122, 123] and the reader is encouraged to read them for additional information. This review will focus on data obtained over the past five years during which some of the genes responsible for ethylene biosynthesis and perception were cloned.

Ethylene biosynthesis

Introduction

The pathway for ethylene biosynthesis was elucidated by Shang Fa Yang and his collaborators in the late 1970s [123] and has provided the basis for all subsequent biochemical and molecular genetic analysis of the pathway (see Fig. 1). Methionine is the biological precursor of ethylene; it is converted to *S*-adenosylmethionine (AdoMet) by the enzyme methionine adenosyl transferase

Table 1. Ethylene biosynthetic and signal transduction proteins and their homologues.

Gene	Homologue	References
ACS	Aspartate aminotransferase	95, 104
ACO	2-oxoglutarate-dependent dioxygenases	38, 121
ETR1	prokaryotic 2-component modules	16, 82
CTR1	Raf protein kinases	32, 49

ACS, ACC synthase; ACO, ACC oxidase.

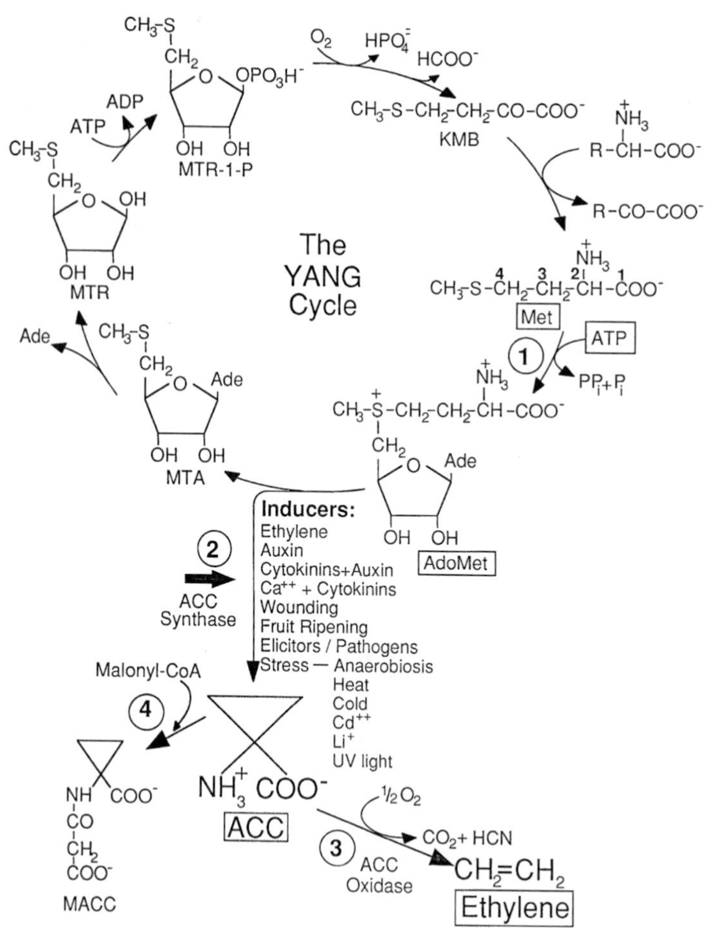

Fig. 1. The ethylene biosynthetic pathway of higher plants. AdoMet, *S*-adenosyl-*L*-methionine; ACC, 1-aminocyclopropane-1-carboxylic acid; KMB, 2-keto-4-methylthiobutyrate; MACC, malonyl-ACC; MTA, 5′-methylthioadenosine; MTR, 5′-methylthioribose; MTR-1-P, MTR-1-phosphate (after [123]).

(step 1 in Fig. 1). The rate-limiting step is the conversion of AdoMet to 1-aminocyclopropane-1-carboxylate (ACC) and methylthioadenosine (MTA) which is catalyzed by ACC synthase (step 2, Fig. 1). ACC is then converted to either ethylene, CO_2, and HCN by ACC oxidase (step 3 in Fig. 1, an O_2-dependent process) or N-malonyl-ACC (MACC) by malonyl transferase (step 4, Fig. 1). The latter reaction constitutes a possible regulatory step by inactivating ACC [123]. Interestingly, methionine is recycled through the pathway by converting methylthioadenosine to methionine. The net result is that the ribose moiety of ATP is converted to methionine from which ethylene is derived; the CH_3-S group of MTA is conserved for continued regeneration of methionine. Thus, given a constant pool of CH3-S group and available ATP, a high rate of ethylene production can be achieved without high intracellular concentrations of methionine (a less abundant amino acid).

The enzymes that catalyze steps 3 and 4 in Fig. 1 have been purified to homogeneity [20, 26, 31, 67, 87]. ACC synthase however has only partially been purified because of its low abundance and lability [46]. It was a combination of molecular biological approaches and heterologous expression that allowed the isolation and identification of the genes encoding AdoMet [85, 117], ACC synthase [98, 116], and ACC oxidase [34, 102].

ACC synthase

ACC synthase is a cytosolic enzyme which catalyzes the first committed step in the ethylene biosynthetic pathway. Its half-life is short; the $t1/2$ of tomato ACC synthase is 58 min [51]. There are a multitude of both internal cues and external inducers which elicit *de novo* synthesis of the enzyme [46, 123]. Taken together, ACC synthase represents the key regulatory enzyme in the pathway.

The first ACC synthase cDNA was cloned from zucchini using a novel experimental approach [98]. Antibodies to partially purified enzyme from zucchini fruit tissue treated with IAA + LiCl were purified on an affinity matrix containing total proteins from uninduced zucchini fruit. The purified antibodies were highly enriched for those recognizing ACC synthase and were used to screen an expression library. The authenticity of the isolated clones was verified by expression in *Escherichia coli* and yeast [98]. Subsequently, Van Montagu and his colleagues cloned two ACC synthase cDNAs from ripe tomato fruit. ACC synthase was purified from 200 kg of tomatoes and degenerate oligonucleotides from peptide sequences of the purified enzyme were used to screen a tomato fruit cDNA library [115, 116]. Two partial cDNA clones were obtained corresponding to two genes now known as *LE-ACS2* and *LE-ACS4* [116]. Since these initial reports, ACC synthase cDNAs and genomic sequences have been cloned from numerous plant species such as apple [21], zucchini [41], tomato [59, 80, 96, 125], *Arabidopsis* [58, 114], winter squash [76, 77], rice [126], orchid (S.D. O'Neill, GenBank accession number L07882, unpublished), carnation [82], mungbean [8, 9, 10, 50], soybean [62] and tobacco [5]. The emerging picture is that ACC synthase is encoded by a divergent multigene family where each gene is differentially regulated by a different subset of inducers. For example, in tomato the enzyme is encoded by at least nine genes and six of them are induced by auxin (Kawakita and Theologis, unpublished). An interesting aspect of ACC synthase gene expression is its inducibility by protein synthesis inhibitors such as cycloheximide. All the rice [126], *Arabidopsis* [58] and tomato (Kawakita and Theologis, unpublished) genes cloned so far are induced by cycloheximide. Cycloheximide inducibility is the hallmark of primary responsive genes [110]. These results suggest that the expression of the ACC synthase gene may be under the control of a labile repressor(s) molecule or that their transcripts are labile and cycloheximide simply stabilizes them by removing a labile nuclease. Some of the ACC synthase genes have been mapped in tomato [96], *Arabidopsis* [58, 114], and rice [126]; their map positions do not correspond to any known ethylene biosynthesis or action mutants.

Comparison of the primary sequences of ACC synthase genes cloned so far reveals a great deal about the enzyme's evolution. Phylogenetic analysis has shown that there are three major branches in the phylogenetic tree, indicating three major classes of ACC synthase polypeptides [59]. Furthermore, this trifurcation had to occur before the divergence of monocots and dicots, since monocot sequences exist in two separate branches [58, 59, 126].

Biochemical and sequence comparisons show that all ACC synthases have striking similarities to two aminotranferases in particular, *Bacillus* sp. strain YM-2 aspartate [105] and rat tyrosine [37, 69] aminotransferases (see Fig. 2). Like the aminotransferases, ACC synthase is a pyridoxal phosphate-dependent enzyme [123] that functions as a dimer in zucchini [98], winter squash [99], and tomato [57]. Sequence analysis of the ACC synthase genes as well as sequencing of the dodecapeptide containing the lysine residue which forms a Schiff base with pyridoxal phosphate [124] has indicated high sequence similarity to the same region in aminotransferases [77, 107]. There is also extensive conservation between the predicted secondary structure of ACC synthase and that of the aspartate aminotransferase obtained by X-ray crystallography [127, 128]. It has been previously found that eleven out of the twelve residues conserved among all aminotransferases are also present in all ACC synthases [41, 96]. However, on the basis of a recent comprehensive alignment of 51 aminotransferases, it was found

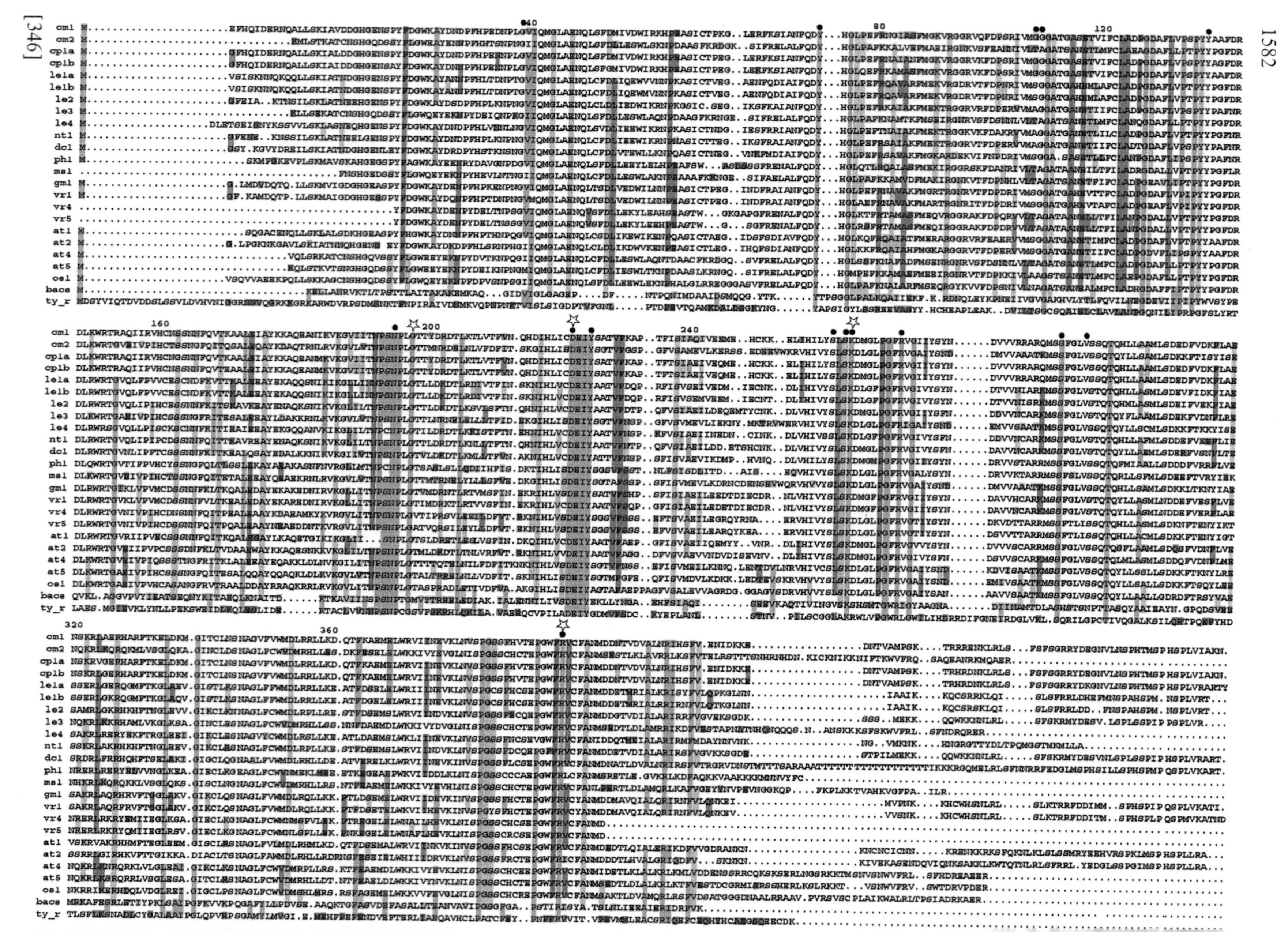

that only four residues are invariant (Gly-197, Asp-222, Lys-258, Arg-386) [70]. All four are also present in all ACC synthases corresponding to residues Gly-212, Asp-237, Lys-278, and Arg-412 (see Fig. 2). Interestingly, the bacillus aspartate aminotransferase is more similar to ACC synthase (19–28% identity) than to the *E. coli*, pig cytosolic and mitochondrial enzymes (13–14% identity). Despite these similarities, ACC synthase and aspartate aminotransferases have distinct substrate specificities determined by complementation of *E. coli* aminotransferase mutants (Zarembinski and Theologis, unpublished).

Closer examination of all ACC synthase amino acid sequences shows that while they share significant sequence similarity, their carboxyl termini are quite divergent (Fig. 2). This hypervariable positively charged carboxyl terminus of ACC synthase is critical for dimerization and for determining some kinetic parameters of the enzyme [57]. There are at least two domains in the carboxyl terminus of ACC synthase that influence its activity. One of them is responsible for the substrate-based (AdoMet) inhibition [57]. Furthermore, ACC synthase appears to be active as a dimer and as a monomer; the wild-type tomato LE-ACS2 isoenzyme is a dimer whereas a carboxyl terminal deletion mutant (last 52 residues deleted) has a nine-fold higher V_{max} and functions as a monomer [57]. Contrastingly, it has been previously reported that wild-type tomato ACC synthase is a dimer when expressed in *E. coli*, but is a monomer when purified from tomato pericarp tissue [99]. A possible explanation of these results is that during purification there is proteolysis of the carboxyl terminus [96, 97], thus producing monomers like the above deletion mutant.

ACC oxidase

ACC oxidase was far more difficult to study than ACC synthase because an *in vitro* enzyme assay was missing [47]. It was only after cloning its gene and discovering sequence similarity to the iron and ascorbate-dependent dioxygenases [35] it was deduced that the enzyme probably requires cofactors iron and ascorbate for its activity [118]. Since then, ACC oxidase has been purified and biochemically characterized from apple [20, 26, 87] and avocado [67]. Biochemical experiments have confirmed that ACC oxidase requires iron, ascorbate, and CO_2 for activity [20, 26, 67, 87]. ACC oxidase activity is not as highly regulated as that of ACC synthase. It is constitutive in most vegetative tissues [123] but it is induced during fruit ripening [68], senescence [120] and wounding [14], and by fungal elicitors [102]. Its subcellular location is still a point of controversy. The primary sequence suggests ACC oxidase to be a cytosolic enzyme since it does not contain putative membrane-spanning domains or a signal peptide [34]. However, there is a large body of data indicating that the enzyme is either associated with the plasma membrane [47] or that it is apoplastic [4, 55, 94]. Like ACC synthase, it is an unstable enzyme; the $t1/2$ of apple ACC oxidase is 2 h [87].

The isolation of the first ACC oxidase gene was somewhat fortuitous; it was isolated in tomato by differential screening [100]. Its authenticity was verified by a combination of antisense experiments [35] and *in vivo* expression in two heterologous systems: *Xenopus* oocytes [102] and yeast [34, 121]. Subsequently, ACC oxidase was cloned from numerous plants (see legend to Fig. 3) and,

Fig. 2. Amino acid sequence alignments of ACC synthases and aminotransferases. ACC synthases: cm1 [77], cm2 [76], cp1a [97], cp1b [41], le1a, le1b, le2, le3, le4 [96], nt1 [5], dc1 [82], ph1 (S. O'Neill, unpubl), ms1 [21], gm1 [62], vr1, vr4, vr5 [8, 9, 10], at1 (X. Liang and A. Theologis, unpublished), at2 [58], at4 (S. Abel and A. Theologis, unpublished), at5 (X. Liang and A. Theologis, unpublished), os1 [126]. Aminotransferases: bacs [105], ty–r [37]. Residues conserved between both aminotransferases and ACC synthases are shaded in blue. Residues conserved between only *Bacillus* sp. strain YM-2 aspartate aminotransferase (bacs) or rat tyrosine aminotranferase (ty–r) and the ACC synthases are shaded in orange and red, respectively. The filled stars designate the residues which represent the active site residues that play functional and structural roles as described on the basis of the X-ray structure of vertebrate aspartate aminotransferases [105]. Unfilled stars represent the four invariant residues present in all aminotransferases and ACC synthases. The numbering is with respect to the pig cytosolic aspartate aminotransferase.

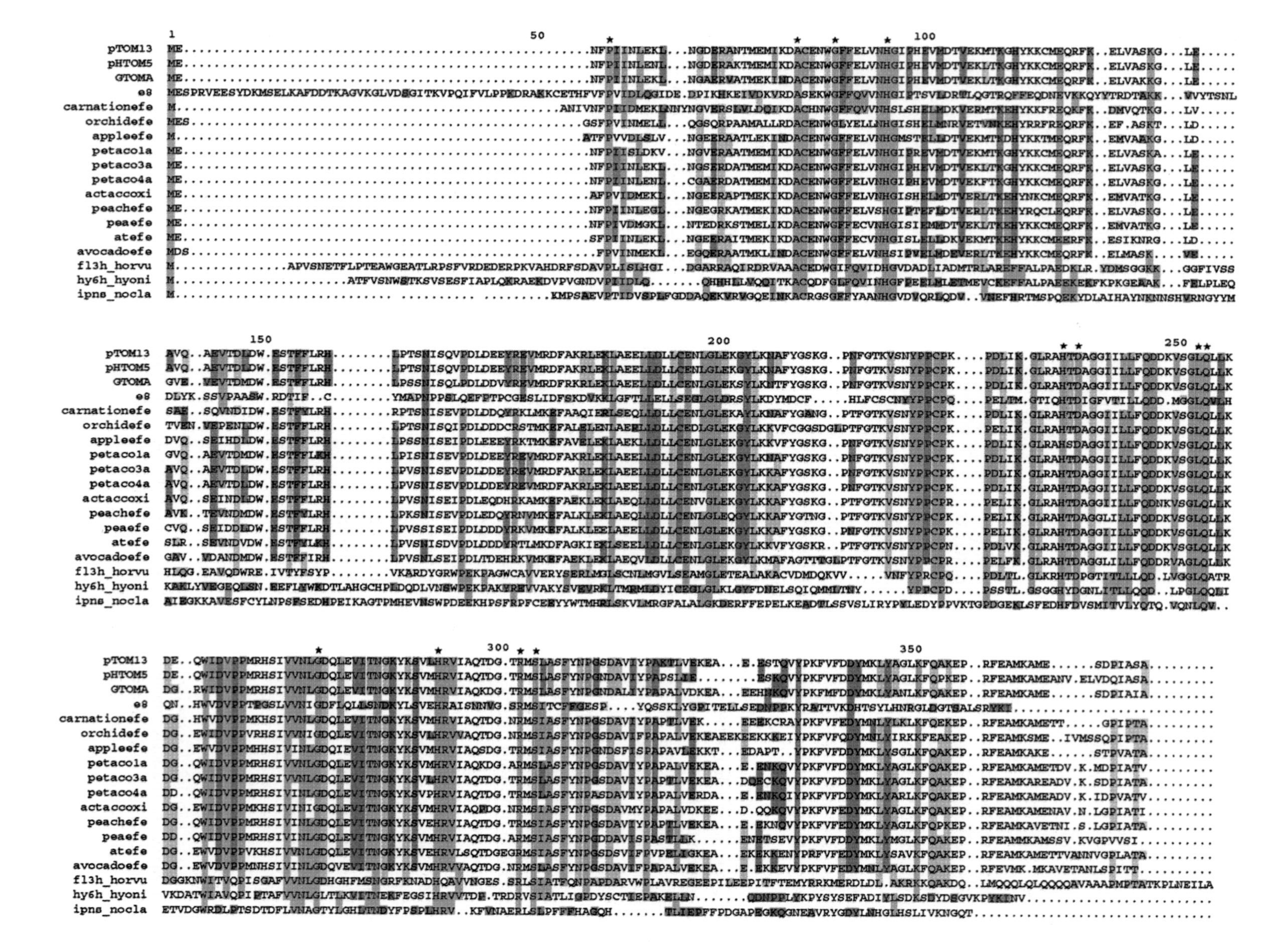
1
50
100
pTOM13
pHTOM5
GTOMA
e8
carnationefe
orchidefe
appleefe
petaco1a
petaco3a
petaco4a
actaccoxi
peachefe
peaefe
atefe
avocadoefe
fl3h_horvu
hy6h_hyoni
ipns_nocla
150
200
250
300
350

like ACC synthase, is encoded by a multigene family but with limited divergence (90% sequence similarity in the petunia four-member family [106] and 88% similarity between the two functional tomato ACC oxidases pHTOM5 and pTOM13 [102]). Primary sequence comparison (Fig. 3) has shown that ACC oxidase is a member of the family of iron- and ascorbate-dependent dioxygenases ([38, 106], Fig. 3).

Ethylene and fruit ripening

Fruit ripening in the climacteric tomato fruit has been one of the most intensely studied ethylene-mediated developmental processes [27, 111]. The reason is two-fold: first, a large number of dramatic changes occur within a short period of time in the tomato fruit during ripening, many of which are under ethylene control or are initiated by ethylene exposure. Autocatalysis of ethylene production is a characteristic feature of ripening fruits (including tomato) and other senescing tissues in which a massive increase in ethylene production is triggered by exposure to ethylene. Color changes, softening, and conversion of starch to sugar are also associated with the ripening process. Thus, ripening fruit represents an interesting model system to study ethylene biosynthesis and perception. Second, ripening fruit has economic importance; billions of dollars worth of fruits and vegetables rot (or overripen) before they can reach the consumer. Therefore, an understanding of the ripening process is important for learning how to control it during shipping and storage [107].

Until recently, there existed a great deal of discussion as to whether ethylene is the trigger for ripening in climacteric fruits, or is simply a by-product of the ripening process [35, 52, 79]. Recent data using antisense technology in tomato show that ethylene is the controlling factor for fruit ripening. The best example of such an experiment was done by driving the expression of the tomato ACC synthase *LE-ACS2* cDNA in its antisense orientation with the CaMV 35S promoter in transgenic tomato [79]. The LE-ACS2 antisense fruits produce large amounts of *LE-ACS2* antisense mRNA which completely inhibit the expression of the ACC synthase genes, *LE-ACS2* and *LE-ACS4*, expressed during fruit ripening [79]. More importantly, the antisense fruits produce less than 0.1 nl $g^{-1} h^{-1}$ ethylene, do not show a climacteric rise in respiration, and never ripen. The fruits ripen only when treated for at least six days with exogenous ethylene or propylene (an ethylene analogue) [79]. This result indicates first that the lesion is specific to ethylene production. Second, ethylene is not a trigger, but a rheostat for fruit ripening such that ethylene must be present continuously to induce a rapidly turning-over set of mRNAs and proteins which initiate the ripening process [79, 107]. Other successful attempts to reduce ethylene production in tomato include ACC oxidase antisense experiments [35] and overexpression of the *Pseudomonas syringae* gene encoding ACC deaminase. This enzyme metabolizes ACC before it can be converted to ethylene [52].

The LE-ACS2 antisense plants have also provided a valuable tool to study ethylene-dependent and -independent gene expression during fruit ripening. Northern analysis of antisense and wild-type fruits show that there are at least two signal transduction pathways important for fruit ripening: an ethylene-independent (developmentally controlled) and an ethylene-dependent path-

Fig. 3. Amino acid sequence alignments of ACC oxidases and dioxygenases. ACC oxidases: pTOM13 [53], pHTOM5 [102], GTOMA [39], e8 [18], carnationefe [119], orchidefe (S.D. O'Neill, GenBank accession number L07912, unpublished), appleefe [22], petacola, petaco3a, petaco4a [106], acataccoxi [63], peachefe [14], peaefe (S.C. Peck, D.C. Olson and H. Kende, unpublished), atefe (M. A. Gomez-Lim, unpublished), avocadoefe [68]. Dioxygenases: f13h–horvu (M. Meldgaard, unpublished), hy6h–hyoni [66], ipns–nocla [17]. Residues conserved between ACC oxidases and at least two dioxygenases are shaded in blue. Residues conserved between only flavanone-3-hydroxylase (f13h–horvu), hyoscyamine-6-dioxygenase (hy6h–hyoni), or isopenicillin N synthase (ipns–nocla) and the ACC oxidases are shaded in orange, red, and green respectively. The stars designate amino acids that are conserved across all members of the Fe(II) and ascorbate requiring superfamily of enzymes [106]. The numbering is with respect to the E8 amino acid sequence.

way [111]. Furthermore, ethylene not only affects transcription but translation as well. The gene encoding polygalacturonase (PG) is known to be developmentally regulated during tomato fruit ripening [111]. While antisense fruits express PG mRNA, they do not accumulate PG polypeptide unless continuous ethylene or propylene is added, indicating that either the translatability of the PG mRNA or the turnover rate of the PG polypeptide is under ethylene control [111]. Similar conclusions have also been reached using transgenic *rin* plants that express PG from the E8 promoter [30].

Genetic analysis of ethylene biosynthesis

Several mutants have been isolated or constructed using antisense technology which overproduce ethylene in both tomato and *Arabidopsis*. They can be divided into two major classes: the first class are ethylene overproducers with lesions that affect the activity of the ethylene biosynthetic pathway, whereas the second class are ethylene perception mutants that overproduce ethylene and will be discussed in the next section. The major representatives of the first class are the *eto1*, *eto2*, and *eto3* (*Arabidopsis*) mutations. The *eto1* mutation is recessive and is responsible for a ten-fold ethylene overproduction in dark-grown *Arabidopsis* seedlings [33]. The *eto2* and *eto3* mutations are dominant which cause 20- and 100-fold higher ethylene production in etiolated *Arabidopsis* seedlings, respectively [48]. Interestingly, the Eto$^-$ phenotype is specific to etiolated seedlings; light-grown plants do not overproduce ethylene [48]. Therefore, it is of great interest that *ACS2* promoter-GUS fusions in wild-type *Arabidopsis* show higher *ACS2* gene expression in light than in dark wild-type *Arabidopsis* seedlings [93]. Also, the Eto$^-$ phenotype does not affect ACC oxidase activity [33]. ETO1 is probably a regulatory protein since it does not map to any of the five known *Arabidopsis* ACC synthase genes [58]. The possibility exists that ETO1 is a negative regulator of ethylene biosynthesis, probably at the level of ACC synthase gene expression. Interestingly, cycloheximide enhances the level of mRNA from all five *Arabidopsis* ACC synthase genes which suggests that ETO1 may be a labile repressor of ACC synthase whose synthesis is blocked by protein synthesis inhibition [58].

Recently, it was shown that transgenic tomato fruit expressing antisense E8 mRNA produces ten-fold higher levels of ethylene [86]. These results indicate that E8 is a negative regulator of ethylene biosynthesis. Since E8 has homology to ACC oxidase (Fig. 3) the results also indicate that one of the components of the regulatory machinery responsible for monitoring ethylene production requires a redox reaction for its activity. It has been postulated that one or more of the *eto* mutations may be E8 mutations [107].

Ethylene perception

Introduction

Until very recently, the understanding of the ethylene sensing apparatus has lagged behind that of the ethylene biosynthetic pathway. This was due to the inability of classical biochemical techniques to shine light on the ethylene signal transduction pathway and to the absence of good plant genetic models to study ethylene action. Classic physiological studies have pointed out that the ethylene receptor is very particular in the type of ligand it will accept. The preferred ligand is a small aliphatic (two carbons with no large side groups is the best), unsaturated (double bond is preferred) molecule free of resonance forms and with its terminal carbon free of positive charge [13]. Furthermore, it has also been postulated that the ethylene receptor probably contains a metal ion since unsaturated aliphatic molecules bind metals readily [13]. The metal is probably Zn^{2+} since zinc-deficient tomato plants are ethylene-insensitive. Copper and iron-deficient plants still show strong epinasty in the presence of exogenous ethylene [13]. Carbon dioxide appears to be a competitive inhibitor of the ethylene reception site [13]. Finally, oxygen is required for ethylene action, indicating that the receptor's metal ion must be oxidized by molecular oxygen either directly or indirectly before reception of and activation by ethylene can occur [13]. Ethylene has been shown to act via calcium and protein phosphorylation; both have been shown to be essential for ethylene-dependent expression of pathogenesis-related (PR) proteins [89, 90]. As it will be discussed later, molecular genetic evidence strongly supports the phosphorylation aspect of the ethylene perception pathway.

It was however genetic analysis of the ethylene perception using the small crucifer *Arabidopsis thaliana* as a model system that established the nature of the ethylene-sensing machinery. Etiolated *Arabidopsis* seedlings show each aspect of the triple response clearly and reproducibly. The small size and short generation time of *Arabidopsis* allows the screening of thousands of seedlings very rapidly. Mutants that either (a) fail to respond to exogenous ethylene (ethylene-insensitive (*ein*), ethylene-resistant (*etr*), and ACC-insensitive (*ain*) mutants) or (b) constitutively display the triple response in the absence of hormone (constitutive triple response (*ctr1*) and ethylene-overproducing (eto)) were isolated [7, 23, 33, 49, 113].

Ethylene-insensitive mutants

Etiolated seedlings of the ethylene-insensitive mutants do not show the triple response in the presence of ethylene; they are tall with open hooks as compared to their wild-type counterparts which are thick and short. Eight distinct ethylene-insensitive mutants have been isolated so far [7, 23, 48, 113]. Three of them, *etr1*, *ein2*, and *ein3*, have been recently cloned [23]. The ETR1 locus is interesting since all of its alleles isolated so far are dominant, suggesting that the mutant proteins encoded by these alleles may either inhibit a complex that ETR1 is part of (dominant negative) or constitutively suppress the ethylene response by locking ETR1 into a particular conformational state [16]. Genetic data indicate that ETR1 acts very early in the ethylene signal transduction pathway, either as the ethylene receptor itself, or as a protein which interacts with the receptor [16]. A second ethylene-insensitive mutation known as *ein2* is recessive and not allelic to *etr1*. Strong alleles to *ein2* are pleiotropic and do not show all known ethylene responses [23, 33]. Additional Ein$^-$ mutants have been isolated that show weak triple response. A representative member of this class is the recessive mutation *ein3* [48]. The *ain* mutants are another class that was isolated during a screen for ACC insensitivity (i.e. long, etiolated plants among a short wild-type background of seedlings). Six of the alleles are due to a single recessive mutation and confer ethylene insensitivity in seedlings at concentrations of ethylene as high as 100 ppm. Interestingly, Ain$^-$ mutants retain their apical hook unlike the Ein$^-$ mutants. In light-grown adult plants, *ain* mutants show decreased ethylene sensitivity with respect to leaf senescence but show no differences compared to wild-type plants when exposed to biotic and abiotic stresses known to elicit ethylene production. Unlike the *ein* loci, the *ain1-1* allele produces three-fold less ethylene. This result suggests the AIN1 protein positively regulates ethylene production.

In tomato there is a ripening mutant *nr* (never ripen) which was isolated about 40 years ago [91] and has recently been found to be ethylene-insensitive. The *nr* mutation is semidominant and pleiotropic, blocking senescence and abscission of flowers, epinasty, fruit ripening, and the triple response in etiolated seedlings [54]. It has been recently shown that NR is the tomato homologue of ETR1 (M. Lanahan and H. Klee, personal communication).

Constitutive triple-response mutants

The second class of ethylene perception mutants has the opposite phenotype of their ethylene-insensitive counterparts; they show the triple response in the absence of exogenous ethylene. All these mutants are recessive and fall into a single complementation group called *ctr1*. The Ctr$^-$ phenotype is pleiotropic like other ethylene perception mutants and mimics wild-type plants grown in 10 ppm ethylene. The apical hook remains closed longer when plants are transferred from dark to light. The leaves, inflorescence, and root system are much more compact than in wild-type plants, due to the smaller cell size in the Ctr$^-$ plants [49]. The *ctr1* mutation also lengthens the time needed for bolting [49]. The Ctr$^-$ phenotype is insensitive to inhibitors of ethylene biosynthesis and action, indicating that CTR1 is involved in ethylene perception [49]. The data suggest that CTR1 represses the ethylene signal transduction pathway which is constitutively active, and ethylene relieves the inhibition.

In tomato there is a semidominant mutation, *epi*, which confers a constitutive ethylene response (epinastic leaves, swelling of stem and petiolar cortex, and abundant lateral roots) [28, 29]. Treatment of Epi$^-$ seedlings with inhibitors of ethylene biosynthesis or action fail to normalize the Epi$^-$ phenotype [28]. The possibility exists that the EPI protein is a homologue of CTR1.

Genetic and biochemical model for the ethylene-sensing apparatus

Double-mutant analysis (epistasis) has established the following genetic model for the ethylene signal transduction pathway:

ETO1, ETO2, ETO3→ETR1→CTR1→EIN2→EIN3→triple response.

The ETR1 and EIN2 are placed in the same pathway rather than in separate pathways because the effects of the *etr1* and *ein2* mutations are not additive [23]. The precise position of *ein2* in this pathway has been recently determined to be downstream of CTR1 (G. Roman and J. Ecker, personal communication). Recently, the cloning of ETR1 [16] and CTR1 [49] has indicated that some of the components of the pathway are protein kinases suggesting that plants sense ethylene via a kinase cascade.

The nature of the ETR1 protein

The amino acid sequence of ETR shows striking similarity to a superfamily of prokaryotic proteins which are components of a basic communication module known as the two-component system. There are as many as fifty different types of two-component systems within a prokaryotic cell [103]. Each two-component system consists of two separate proteins: a sensor and an associated response regulator [83, 103]. The sensor has two domains: an extracellular input and a cytoplasmic histidine kinase domain. The response regulator is composed of a receiver module and typically an output domain (transcriptional activation). This arrangement offers a highly efficient mechanism by which prokaryotes respond to changes in their environment, such as nitrogen availability, chemical signals, osmotic stress, and oxygen tension.

ETR1 shares highest similarity to a subset of the bacterial two-component proteins that contain both histidine kinase (sensor) and receiver (response regulator) domains on the same polypeptide (Fig. 4; [16]). The N-terminus of ETR1 contains three putative transmembrane domains (Fig. 4). ETR1 lacks the variable carboxyl-terminal domain which is present in the response regulator of most two-component members and

Fig. 4. Amino acid sequence alignments of ETR1 with all known eukaryotic two-component system homologues and various bacterial two-component sensors containing both a histidine kinase and receiver domain. ETR1 [16], ARCB [44], BARA [75], RCSC [104], BVGS [3], LEMA [40], SLN1 [81], SSK1 [65], 282, 80 [101]. 282 and 80 represent two consensus amino acid domains for prokaryotic sensor and response regulators obtained from the ProDom protein domain database [101]. Residues conserved between ETR1 and other two component proteins are shaded in blue. The dots correspond to the highly conserved amino acids involved in phosphotransfer in all two-component systems. Motifs and residues conserved between all histidine kinase and receiver domains are shaded in orange [84, 103]. The three hydrophobic regions that compose the putative transmembrane domain are shaded in red. The thick lines designate residues which form the hydrophobic core of the response regulator, CheY [103]. The open boxes denote the motif characteristic of ATP-binding proteins [81]. The numbering is with respect to the ETR1 amino acid sequence.

```
      1                                                50                                                100                                                                                    150
ETR1  ..MEVCNCIEPQWPADELLMKYQYISDFFIAIAYFSIPLELIYFVKKSAVFPYRWVLVQFGAFIVLCGATHLINLWTFTTHSRTVALVMTTAKVLTAVVSCATALMLVH.........IIPDLLSVKTRELFLKNKAAELDREMGLIRTQEETGRHVRML/THEIRSTLDRHTILKT......
ARCB  ..........................MKQIRLLAQYYVDLMMKLGLVRFSMLLALALVVLAIVVQMAVTMVLHGQVESIDVIRSIFFGLLITP.........WAVYFLSVVVEQ..LEESRQRLSRLVQKLEEMRE.......RDLSLNVQLKDNIAQLN......
BARA  ......................................MTNYSLRARMMILILAPTVLIGLLLSIFFVVHRYNDLQRQLEDAGASIIEPLAVSTEYGMSL.........QNRESIGQLISV..LHRRHSDIVRAISVYDENNRLFVTSNFHLDPSSMQLGSNVPFPR......
RCSC  FSAAYDLNRVFLIGSDNLCMANFGLRDMPVERDTALKALHERINKYRNAPQDDSGSNLYWISEGPRPGVGYFYALTPVYLANRLQALLGVEQTIRMENFFLPGTLPMGV.........TILDENGHTLIS..LTGPESKIKGDPRWMQERSWFGYTEGFRELVLKKNLPPSS..LS......
BVGS  IGVDTVEELIAKLRSGEADMAGALFVNSARESFLSFSRPYVRNGMVIVTRQDPDAPVDADHLDGRTVALVRNSAAIPLLQRRYPQAKVVTADNPSEAMLMVANGQADAVVQTQISASYYVNRYFAGKLRIASALDLPPAEIALATTRGQTELMSILNKALYSISNDELASIISRWRGSDGDPRT
LEMA  .......................................MLLL/TILPASLMAAMLGGYFTWMQLSELQSQLLQRGEMIAQDLAPLAANALGR.........KDKVLLSRIAT....QTLEQTDVRAVSFLDTDRTVLAHAGPTMISPS.PIGSGSQL.........
SLN1  ALEHSTIAIISAVYNSQGKASGYHFVFPPYGSRSDLPQKVFSIKNDTFISSAFRNGKGG.SLKQTNILSTRNTALGYSPCSFNLVNWVAIVSQPESVFLSPATKLAKIITGTVIAIGVFVILLTLPLAHWAVQPIVRLQKATELITEGRGLRPSTPRTISRASSFKRGFSSGFAVPSS......
SSK1  ........................................................................................................................................................................................
282   IGVDTVEELIAKLRSGEADMAGALFVNAARESFLSFSRPYVRNGMVIVTRQDPDAPVDADHLDGRTIAMVRNSAAIPLLQRRYPQAKVVTADNPTEAMLMVADGQADAVVQTQISASYYVNRYFAGKLRIASALDLPPAEIALATTRGQTELMSILNKALYSISNDELASIISRWRGSDGDPRT
80    ........................................................................................................................................................................................

                                   200                                               250                                              300
ETR1  .........TLVELGRTLALEECALWMPTRTGLELQLSYTLRHQHPVEYTVPIQLPVINQVFGTSRAV.KISPNSPVARLRPVSGKYMLGEVVAVRVPLLHLSNFQINDWPELSTKRYALMVLMLPSDSARQWHVHELELVEVVADQVAVALSHAAILEESMRARDLLMEQNVALDLARREAET
ARCB  .........QEIAVREKAEAELQETFGQLKIEIKEREETQIQLEQQSSFLRSFLDASPDLVFYRNEDK.EFSGCNRAMELLTGKSEKQLVHLKPADVYSPEAAAKVI....ETDEKVFRHNVSLTYEQWLDYPDGRKACF...EIRKVPYYDRVGK.....RHGLMGFGRDITERKRYQDALER
BARA  .........QLTVTRDGDIMILRTPIISESYSPDESPSSDA.KNSQNMLGYIALELDLKSVRLQQYKE.IFISSVMMLFCIGIALIFGWRLMRDVTGPIRNMVNTVDRIRRGQLDSRVEGFMLGELDMLKNGINSMAMSL...AAYHEEMQHNIDQATSDLRETLEQMEIQNVELDLAKKRAQE
RCSC  .........IVYSVPVDKVLERIRMVILNAILLNVLAGAAL.FTLARMYERRIFIPAESDALRLEEHE.QFNRKIVASAPVGICI...LRTADGVNILSNELAHTYLNML/THEDRQRL/TQIICGQQVNFVDVL/TSNNTNL...QISFVHSRYRNENVAICVLVDVSSRVKMEESLQEMAQAAEQ
BVGS  WYAYRNEIYLLIGLGLLSALLFLSWIVYLRRQIRQRKRAERALNDQLEFMRVLIDGTPNPIYVRDKEGRMLLCNDAYLDTFGVTADAVLGKTIPEANVVGDPA..LAREMHEFLL/TRVAAEREPRFEDRDVTLHGRTRHVYQWTIPYGDS.LGELKGIIGGWIDITERAELLRELHDAKESADA
LEMA  ..........LSSTGTDATRYLLPVFGSQRHLTSPIIPAEA....DTLLGWVELEISHNGTLLRGYRS.LFASLLLIL/TGLAFTATLAVRMSRTINGPMSQIKQAVSQLKDGNLETRLPPLGSRELDELASGINRMAATL...QNAQEELQLSIDQATEDVRQNLETIEIQNIELDLARKEALE
SLN1  ........LLQFNTAEAGSTTSVSGHGGSGHGSGAAFSANSSMKSAINLGNEKMSPPEEENKIPNNHTDAKISMDGSLN.HDLLGPHSLRHNDTDRSSNRSHIL/TTSANL/TEARLPDYRRLFSDELSDL/TETFNTMTDALDQHYALLEERVRARTK..............QLEAAKIEAEA
SSK1  ........................................................................................................................................................................................
282   WYAYRNEIYLLIGLGLLSALLFLSWIVYLRRQIRQRKRAERALNDQLEFMRVLIDGTPNPIYVRDKEGRMLLCNDAYLDTFGVTADAVLGKTIPEANVVGDPA..LAREMHEFLL/TRMAAEREPRFEDRDVTLHGRTRHVYQWTIPYGDS.LGELKGIIGGWIDITERAEMLRELHDAKESADE
80    ........................................................................................................................................................................................

          350 ●                                          400                                               450
ETR1  AIRARNDFLAVMNHEMRTPMHAIIALSSLLQ..ETELTPEQRLMVETILKSSNLLATLMNDVLDLSRLEDGSLQLELGTFNLHTLFREVLNLIKPIAVVKKLPITLNLAPDLPEFVV..GDEKRLMQIILNIVGNAVKFSK.QGSISVTALVTKSDTR..........................
ARCB  ASRDKTTFISTISHELRTPLNGIVGLSRILL..DTELTAEQEKYLKTIHVSAVTLGNIFNDIIDMDKMERRKVQLDNQPVDFTSFLADLENLSALQAQQKGLRFNLEPTLPLPHQVI..TDGTRLRQILWNLISNAVKFTQ.QGQVTVRVRYDEG....D........................
BARA  AARIKSEFLANMSHELRTPLNGVIGFTRL/TL..KTELTPTQRDHLNTIERSANNLLAIINDVLDFSKLEAGKLILESIPFPLRSTLDEVVTLLAHSSHDKGLEL/TLNIKSDVPDNVI..GDPLRLQQIITNLVGNAIKFTE.NGNIDILVEKRALSNTKV........................
RCSC  ASQSKSMFLATVSHELRTPLYGIIGNLDLLQ..TKELPKGVDRLVTAMNNSSSLLLKIISDILDFSKIESEQLKIEPREFSPREVMNHITANYLPLVVRKQLGLYCFIEPDVPVALN..GDPMRLQQVISNLLSNAIKFTD.TGCIVLHVRADGDY..........................
BVGS  ANRAKTTFLATMSHEIRTPMNAIIGMLELALLRPTDQEPDRQS.IQVAYDSARSLLELIGDILDIAKIEAGKFDLAPVRTALRVLPEGAIRVFDGLARQKGIELVLKTDIVGVDDVL..IDPLRMKQVLSNLVGNAIKFTT.EGQVVLAVTARPDGDA.A........................
LEMA  ASRIKSEFLANMSHEIRTPLNGILGFTHLLQ..KSELTPRQFDYLGTIEKSADNLLSIINEILDFSKIEAGKLVLDNIPFNLRDLLQDTL/TILAPAAHAKQLELVSLVYRDTPLALS..GDPLRLRQIL/TNLVSNAIKFTR.EGTIVARAMLEDETEEHA........................
SLN1  ANEAKTVFIANISHELRTPLNGILGMTAISM..EETDVNKIRNSLKLIFRSGELLLHILTELLTFSKNVLQRTKLEKRDFCITDVALQIKSIFGKVAKDQRVRLSISLFPNLIRTMVLWGDSNRIIQIVMNLVSNALKFTPVDGTVDVRMKLLGEYDKELSEKKQYKEVYIKKGTEVTENLETT
SSK1  ........................................................................................................................................................................................
282   ANRAKTTFLATMSHELRTPMNGIIGMLQLLL..NTELTPGQDKYLNTIHRSARSLLEIIGDILDFAKIEAGKFDLEPIPFALRALMEDVIRVFDGLARQKGLELVLNIDIVVPDDVI..IDPMRLKQVLSNLVNNAIKHTP.EGG...SITISVQADE.D........................
80    ........................................................................................................................................................................................

                                                                                     500                                                   550
ETR1  ...............................................................................AVKDSGAGINPQDIPKIFTKFAQTQ.SLATRSSGGSGLGLAISKRFVNLMEGNIWIESDGLGKGCTAIFDVKL
ARCB  ............................................................................MLHFEVEDSGIGIPQDELDKIFAMYYQVKDSHGGKPATGTGIGLAVSRRLAKNMGGDITVTSEQ.GKGSTFTL/TIHA
BARA  ............................................................................QIEVQIRDTGIGIPERDQSRLFQAFRQADAS.ISRRHGGTGLGLVITQKLVNEMGGDISFHSQP.NRGSTFWFHINL
RCSC  .............................................................................LSIRVRDTGVGIPAKEVVRLFDPFFQVGTG.VQRNFQGTGLGLAICEKLISMMDGDISVDSEP.GMGSQF..TVRI
BVGS  ............................................................................HVQFSVSDTGCGISEADQRQLFKPFSQVGGSAEAGPAPGTGLGLSISRRLVELMGGTLVMRSAP.GVGTTVSVDLRL
LEMA  ............................................................................QLRISVQDTGIGLSSQDVRALFQAFSQADNS.LSRQPGGTGLGLVISKRLIEQMGGEIGVDSTP.GEGSEFWISLKL
SLN1  DKYDLPTLSNHRKSVDLESSATSLGSNRDTSTIQEEITKRNTVANESIYKKVNDREKASNDDVSSIVSTTTSSYDNAIFNSQFNKAPGSDDEEGGNLGRPIENPKTWVISIEVEDTGPGIDPSLQESVFHPFVQGDQT.LSRQYGGTGLGLSICRQLANMMHGTMKLESK.VGVGSKFTFTLPL
SSK1  ........................................................................................................................................................................................
282   ............................................................................NIQVEVSDSGIGIPEEQQDRLFEPFV......TSKESGMGLGLSISRNIIDQHGGMITIGNRE.QGG.........
80    ........................................................................................................................................................................................
                                                                                     ▭                             ▭

                                                                                                                                                                     600
ETR1  GISER....SNESKQS..........................................................................................................GIPKVPAIPRHSNFTGLKVLVMDENGVSRMVTKGLLVHL
ARCB  PSVAE....EVDDAFDEDDMP..........................................................................................................................LPALNVLLVEDIELNVIVARSVLEKL
BARA  DLNPN....IIIEGPSTQCLAGKRLAYVEPNSAAAQCTLDILSETPLEVV....YSPTFSALPPAHYDMMLLGIAVTFREPL/TMQHERLAKAVSMTDFL....MLALPCHAQVNAEKLKQDGIGACLLKPL/TPTRLLPAL/TEFCHHKQNTLLPVTDESKLAMTVMAVDDNPANLKLIGALLEDM
RCSC  PLYGA....QYPQKKGVEGLSGKRCWLAVRNASLCQFLETSLQRSGIVVTTYEGQEPTPEDV...............LITDEVVSKKWQGRAVVTFCRRHIGIPLEEAPGEWVHSVAAPHELPALLARIYLIEMESDDPANALPSTDKAVSDNDDMMILVVDDHPINRRLLADQLGSL
BVGS  TMVEK....SVQAA...................................................................................................................PPAAATAATPSKPQVSLRVLVVDDHKPNLMLLRQQLDYL
LEMA  PKARE....DKEESLNIP.LGGLRAAVLEHHDLARQALEHQLEDCGLQTIVFNNLENLLNGVTAAHETPAAIDLAVLGVTALEISPERLRQHIWDLENLNCKVMVLCPTTEHALFQLAVHDVYTQLQAKPACTRKLQKALSELIAPRAVRADIGPPLSSRAPRVLCVDDNPANLLLVQTLLEDM
SLN1  NQTKEISFADMEFPFEDEFNPESRKNRRVKFSVAKSIKSRQSTSSVATPATNRSSL/TNDVLPEVRSKGKHETKDVGNPNMGREEKNDNGGLEQLQEKNIKPSICL/TGAEVNEQNSLSSKHRSRHEGLGSVNLDRPFLQSTGTATSSRNIPTVKDDDKNETSVKILVVEDNHVNQEVIKRMLNLE
SSK1  .......................................................................................................................................FPKINVLIVEDNVINQAILGSFLRKH
282   ........................................................................................................................................................................................
80    .....................................................................................................................................................DDDAAIRRVLQQMLENA
                                                                                                                                                     ▬▬      ▪  ▬   ▬

                        ●   650                                                                          700
ETR1  GCE.VTTVSSNEECLRVVSH.....EH.KVVFMDVCMPGVENYQIALRIH.................EKFTKQRHQRPL..LVALSGNTDKSTKEKCMSFGLDGVLLKPVSLDNIRDVLSDLLEPRVLYEGM..........................
ARCB  GNS.VDVAMTGKAALEMFKP.....GEYDLVLLDIQLPDMTGLDISREL/T.................KRYPREDLP..P..LVAL/TANVL.KDKQEYLNAGMDDVLSKPLSVPALTAMIKKFWDTQDDEESTVTTEENSKSEALLDIPMLEQY.....LELVGPKLITDGLAVFEK
BARA  VQH.VELCDSGHQAVERAKQ.....MPFDLILMDIQMPDMDGIRACELIH.................QL...PHQQQTP..VIAVTAHAMAGQKEKLLGAGMSDYLAKPIEEERLHNLLLRYKPGSGISSRVVTPEVNEIVVNPNAT...LDWQLALRQAAGKTDLARDMLQMLLD
RCSC  GYQ.CKTANDGVDALNVLSK.....NHIDIVLSDVNMPNMDGYRLTQRTR.................QLG.....LTLP..VIGVTANALAEEKQRCLESGMDSCLSKPVTLDVIKQTLTVYAERVRKSRES................................
BVGS  GQR.VIAADSGEAALALWRE.....HAFDVVITDCNMPGISGYELARRIR.................AAEAAPGYGRTRCILFGFTASAQMDEAQACRAAGMDDCLFKPIGVDALRQRLNEAVARAAL..PTPPSPQAAAPATDDATPTAFSAESILALTQNDEALIRQLLEEVIR
LEMA  GAE.VVAVEGGYAAVNAVQQ.....EAFDLVLMDVQMPGMDGRQATEAIR.................AWEAERNQSSLP..IVALTAHAMANEKRSLLQSGMDDYLTKPISERQLAQVVLKW...TGLALRNPAPERQNEALEVHVGPLVLDHEEGLRLAAGKADLAADMLAMLLA
SLN1  GIENIELACDGQEAFDKVKELTSKGENYNMIFMDVQMPKVDGLLSTKMIR.........................RDLGYTSPIVALTAFADDSNIKECLESGMNGFLSKPIKRPKLKTILTEFCAAYQGKKNNK............................
SSK1  KIS.YKLAKNGQEAVNIWKE.....GGLHLIFMDLQLPVLSGIEAAKQIRDFEKQNGIGIQKSLNNSHSNLEKGTSKRFSQAPVIIVALTASNSQMDKRKALLSGCNDYLTKPVNLHWLSKKITEWGCMQALIDFDSWKQGESRMTDSVLVKSPQKPIAPSNPHSFKQATSMTPTHSPVR
282   ........................................................................................................................................................................................
80    GYQ.VTAADDGQEALNLLQN.....NNPDVVIMDINMPGMDGLELLRQLR.................QQNP...MQQLP..IIIMTAHGDVADAVQALEAGADDYIAKPFDPDEL......IARIR.........................................
                              ▬▬           ▪  ▬   ▪                                                ▬▬                       ▪    ▪   ▬
```

serves as a transcriptional activation domain [103]. All four mutations in ETR1 are clustered in the putative transmembrane domains [16], suggesting that insertion of ETR1 into the plasma membrane is impaired in these mutants, or that the ethylene signal somehow cannot be relayed from the N-terminal input domain to the histidine kinase domain due to conformational constraints. However, it has been recently shown that over-expression of the ETR1 polypeptide mutated at the critical His-353 and Asp-642 residues (see Fig. 4) confers ethylene insensitivity (C. Chang and E. Meyerowitz, personal communication). This result suggests that the *etr1* acts as dominant negative mutation.

By analogy to the bacterial two-component system, one can visualize ETR1 as the ethylene sensor. It may sense ethylene by an extracellular metal-containing input domain and transduces the signal through autophosphorylation of its histidine kinase domain. Subsequent phosphotransfer occurs first to the aspartate residue of the *cis* response regulator and then to the *trans* cognate response regulator domain to indirectly alter gene expression. Such phosphotransfer routes are seen in bacterial two-component systems in which the sensor has both a histidine kinase and a receiver module on the same protein [43, 112].

The nature of the CTR1 protein

The amino acid sequence of CTR1 shows that its carboxyl terminus shares significant similarity to the Raf family of serine/threonine protein kinases [49]. Raf was originally isolated as a key retroviral protein (v-raf) which tranforms embryo fibroblasts and epithelial cells in culture [32]. Since then, cellular homologues of v-raf have been isolated in mammals, *Drosophila*, and chicken [15, 32]. The Raf proteins mediate dramatic changes in cell growth and differentiation by transducing signals from cell-surface receptors to transcription factors [32]. Raf has been shown to affect dorsoventral patterning and R7 photoreceptor differentiation in *Drosophila* [11, 19], meiotic maturation and mesoderm development in *Xenopus* oocytes [25, 64], and vulval development in *C. elegans* [36]. CTR1 shares the same tridomain structure of all known Raf proteins [32]: Its carboxyl terminus shares high sequence similarity to the conserved kinase domain of Raf-1 [49]; The N-terminal half of CTR1 contains both the conserved cysteine motif and serine/threonine rich tract found in Raf proteins and probably acts to regulate the C-terminal kinase domain [49]. Interestingly, two point mutations found in the *CTR1* gene change invariant residues of the kinase domain, indicating that phosphorylation of downstream target proteins by the CTR1 suppresses ethylene's effects (Fig. 5).

A model

One model that incorporates what we have learned from the genetics of ethylene signalling

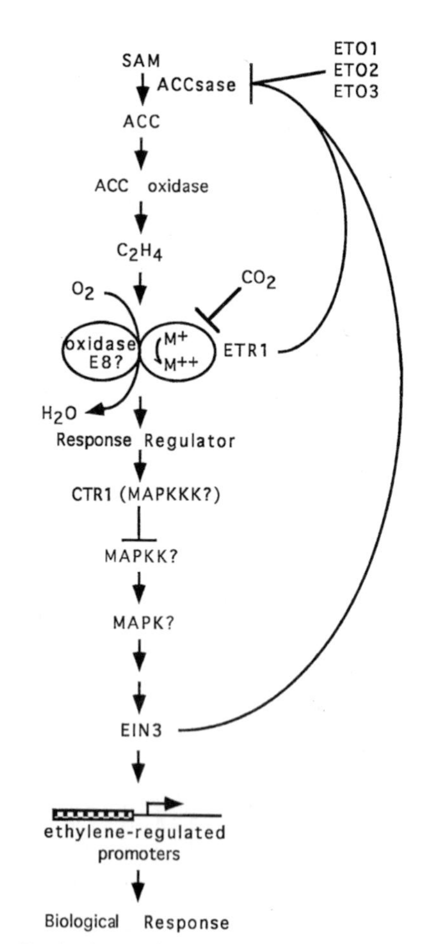

Fig. 1. A putative ethylene-sensing pathway.

and the putative nature of ETR1 and CTR1 proteins is presented in Fig. 5. Recently it has been shown in yeast that the homologue of ETR1, SLN1, has both a sensor domain and a response regulator domain on the same polypeptide [65, 81]. SLN1 is thought to phosphorylate a eukaryotic response regulator, SSK1, which in turn regulates a MAP kinase kinase kinase (MAPKKK) [42, 65]. This kinase then phosphorylates a MAPKK (PBS2 kinase) which regulates a MAPK (HOG1) which controls genes important for the osmolarity response [42, 65]. From work in mammalian cells, the CTR1 homologue, Raf-1, is a MAPKKK which directly phosphorylates and activates the MAPKK *in vitro* [32]. Since EIN3 is not a MAPKK or MAPK (J. Ecker, personal communication), it may act downstream of these kinases (Fig. 5). If CTR1 is in fact a Raf kinase, it is expected that two more classes of *ctr* mutations should be isolated: Class 1 would be lesions in the kinases downstream of CTR1 (such as MAPKK and MAPK), whereas Class 2 would be specific serine/threonine and tyrosine phosphatases which would neutralize the above kinases. Such classes of mutants have been isolated in an analogous yeast signal transduction pathway that senses osmolarity [42, 65].

According to the model presented in Fig. 5 which is based on the genetic evidence and our knowledge of signalling systems in other species, the ETR1 protein is a kinase which is active in the absence of ethylene, and inactive in its presence. The expectation would therefore be that if CTR1 is an indirect target of ETR1, it would be phosphorylated and active in the absence of ethylene, and dephosphorylated and inactive in its presence. Activated CTR1 may eventually phophorylate EIN3, which may be a transcription factor that is inactivated by phosphorylation and only active when the ETR1 and CTR1 kinases are switched off in the presence of ethylene.

In Fig. 5, the E8 is viewed as the putative oxidase of the ethylene receptor postulated by Stanley Burg [13]. This view is based on the observation that E8 shows sequence similarity to dioxygenases (Fig. 3) and its inactivation leads to ethylene overproduction [86]. We suggest that interference with the ethylene sensor that results in lower levels of reception is interpreted by the cell as an absence of the hormone leading to ethylene overproduction.

Ethylene-mediated changes in gene expression

Ethylene effects are believed to be mediated by transcriptional activation of a large set of genes [12, 24, 61]. Putative ethylene-regulated genes have been cloned and studied in order to understand ethylene-regulated processes such as fruit ripening [60, 68, 100], defense response to pathogens [12, 92], and senescence [88]. Unfortunately, there is a scarcity of information as to which *cis*-acting promoter elements in these genes confer ethylene inducibility. Deletion analysis has defined small (<100 bp) promoter fragments conferring ethylene responsiveness in the bean chitinase 5B gene [12, 92], the tobacco PR-1B gene [24] and the tomato E4 gene [74]. While there is a 11 bp sequence conserved between the PR-1B gene and other ethylene-regulated pathogenesis-related (PR) genes [24], the E4 sequence is different [74], suggesting that there are at least two signal transduction pathways which transcriptionally activate ethylene-responsive genes, or that there is a promiscuous transcription factor which recognizes different *cis*-acting elements [73]. Several DNA-binding proteins have been detected that interact with these elements but none of them have been purified [24, 71, 74].

Ethylene and disease resistance

Pathogen attack typically has one of two outcomes on plants. If the bacteria contain an avirulence gene (i.e. avirulent bacteria) which corresponds to a particular plant resistance gene, a localized cell death patch occurs (hypersensitive response, HR) and bacteria fail to spread to other parts of the plant. This reaction is known as the resistant response. If such a match does not occur, the bacteria are considered virulent, cause necrotic lesions and spread systemically. This is known as the susceptible response [6]. It has been proposed that ethylene plays a role in both responses [6] but genetic evidence is lacking. The question arises as to whether ethylene mediates the responses to the inducer (pathogen attack) or its production is a by-product of the defense response. Recent experimental evidence using the ethylene-insensitive mutants *etr1* and *ein2* and the ethylene overproducer *eto1* has indicated that ethylene is not essential. None of the mutants affect the resistant response in *Arabidopsis* and both the HR response and inhibition of bacterial growth occur when the mutants are infected with avirulent bacteria [6]. Interestingly, all mutants behave as wild-type plants when infected with virulent bacteria except for the *ein2* mutant. Ein2$^-$ shows far fewer necrotic lesions than its wild-type counterpart and no restriction in bacterial proliferation. The ethylene overproducer *eto1* neither confer greater disease resistance with avirulent bacteria nor greater disease susceptibility with virulent bacteria. These results indicate that EIN2 plays an essential role in causing necrotic lesions during pathogen attack [6].

Many plants respond to pathogen infection by inducing long-lasting, broad-spectrum resistance also called systemic acquired resistance (SAR) [95]. During this resistance response, numerous PR proteins are induced, one of which is known to confer increased tolerance to pathogen infection [2]. Using the ethylene-insensitive mutants *etr1* and *ein2*, it was shown that the development of SAR response does not require ethylene [56].

Conclusions and future directions

Studies on ethylene have led the way in advancing our understanding of the biosynthesis of a plant hormone at the biochemical and molecular level and they now lead our attempts to understand the biochemical machinery responsible for the perception of a plant hormone. Understanding the tissue and cell-specific expression of the ACC synthase and ACC oxidase multigene families during plant development will offer new knowledge of the role of ethylene as a signalling molecule. We must also understand the regulation of the ethylene signalling pathway at the biochemical level. It will require the isolation of all the components of the pathway and the development of the appropriate biochemical experimental system. The cloning of ethylene signalling homologues in yeast raises the possibility that this microorganism may become the system of choice for both biochemical and genetic analysis of the ethylene signal transduction pathway. We envision that not far in the future, it will be possible to construct an ethylene-sensing yeast strain with all or part of the *Arabidopsis* ethylene sensing components. The current knowledge also has the potential to elucidate the molecular details of the autocatalytic ethylene production. Finally, the spectacular advances in ethylene research and its applications to world agriculture offer the best example that fundamental research is the only tool for solving 'mission-oriented' and 'strategic importance' applied agronomical problems. This view is supported by the recent cloning of the *Arabidopsis* ethylene perception genes in tomato, indicating that the fundamental knowledge obtained from a weed can be effectively used to control senescense of agronomical important plants.

Acknowledgements

This work was supported by grants to A. T. from the NSF (DCB-8645952, 8819129, -8916286, and MCB-8316475), and the USDA (5835-21430-002-00D, 5335-21430-003-00D). We thank Alice Tarun and Claudia Kohler for reading the manuscript.

References

1. Abeles FB, Morgan PW, Saltveit ME, Jr: Ethylene in Plant Biology. Academic Press, New York (1992).
2. Alexander D, Goodman RM, Gut-Rella M, Glascock C, Weymann K, Friedrich L, Maddox D, Ahl Goy P, Luntz T, Ward E, Ryals J: Increased tolerance to two oomycete pathogens in transgenic tobacco expressing pathogenesis-related protein 1a. Proc Natl Acad Sci USA 90: 7327–7331 (1993).

3. Arico B, Miller JF, Roy C, Stibitz S, Monack D, Falkow S: Sequences required for expression of *Bordetella pertussis* virulence factors share homology with prokaryotic signal transduction proteins. Proc Natl Acad Sci USA 86: 6671–6675 (1989).
4. Ayub RA, Rombaldi C, Petitprez M, Latche A, Pech JC, Lelievre JM: Biochemical and immunocytological characterization of ACC oxidase in transgenic grape cells. In: Pech JC, Latche A, Balague C (eds) Cellular and Molecular Aspects of the Plant Hormone Ethylene, pp. 98–99. Kluwer Academic Publishers, Dordrecht, Netherlands (1992).
5. Bailey BA, Avni A, Li N, Mattoo AK: Nucleotide sequence of the *Nicotiana tabacum* cv Xanthi gene encoding 1-aminocyclopropane-1-carboxylate synthase. Plant Physiol 100: 1615–1616 (1992).
6. Bent AF, Innes RW, Ecker JR, Staskawicz BJ: Disease development in ethylene-insensitive *Arabidopsis thaliana* infected with virulent and avirulent *Psedomonas* and *Xanthomonas* pathogens. Mol Plant-Microbe Interact 5: 372–378 (1992).
7. Bleecker AB, Estelle MA, Somerville G, Kende H: Insensitivity to ethylene conferred by a dominant mutation in *Arabidopsis thaliana*. Science 241: 1086–1089 (1988).
8. Botella JR, Arteca JM, Schlagnhaufer CD, Arteca RN: Identification and characterization of a full-length cDNA encoding for an auxin-induced 1-aminocyclopropane-1-carboxylate synthase from etiolated mung bean hypocotyl segments and expression of its mRNA in response to indole-3-acetic acid. Plant Mol Biol 20: 425–436 (1992).
9. Botella JR, Schlagnhaufer CD, Arteca JM, Arteca RN: Identification of two new members of the 1-aminocyclopropane-1-carboxylate synthase-encoding multigene family in mung bean. Gene 123: 249–253 (1993).
10. Botella JR, Schlagnhaufer CD, Arteca RN, Phillips AT: Identification and characterization of three putative genes for 1-aminocyclopropane-1-carboxylate synthase from etiolated mung bean hypocotyl segments. Plant Mol Biol 18: 793–797 (1992).
11. Brand AH, Perrimon N: Raf acts downstream of the EGF receptor to determine dorsoventral polarity during *Drosophila* oogenesis. Genes Devel 8: 629–639 (1994).
12. Broglie KE, Biddle P, Cressman R, Broglie R: Functional analysis of DNA sequences responsible for ethylene regulation of a bean chitinase gene in transgenic tobacco. Plant Cell 1: 599–607 (1989).
13. Burg SP, Burg EA: Molecular requirements for the biological activity of ethylene. Plant Physiol 42: 144–152 (1967).
14. Callahan AM, Morgens PH, Wright P, Nichols Jr KE: Comparison of Pch313 (pTOM13 homolog) RNA accumulation during fruit softening and wounding of two phenotypically different peach cultivars. Plant Physiol 100: 482–488 (1992).
15. Calogeraki I, Barnier JV, Eychene A, Felder MP, Calothy G, Marx M: Genomic organization and nucleotide sequence of the coding region of the chicken c-R*mil* (B-*raf*-1) proto-oncogene. Biochem Biophys Res Comm 193: 1324–1331 (1993).
16. Chang C, Kwok SF, Bleecker AB, Meyerowitz EM: *Arabidopsis* ethylene-response gene *ETR1*: similarity of product to two-component regulators. Science 262: 539–544 (1993).
17. Coque JJ, Martin JF, Calzada JG, Liras P: The cephamycin biosynthetic genes *pcbAB*, encoding a large multidomain peptide synthetase, and *pcbC* of *Nocardia lactamdurans* are clustered together in an organization different from the same genes in *Acremonium chrysogenum* and *Penicillium chrysogenum*. Mol Microbiol 5: 1125–1133 (1991).
18. Deikman J, Fischer RL: Interaction of a DNA binding factor with the 5′-flanking region of an ethylene-responsive fruit ripening gene from tomato. EMBO J 7: 3315–3320 (1988).
19. Dickson B, Sprenger F, Morrison D, Hafen E: Raf functions downstream of Ras1 in the Sevenless signal transduction pathway. Nature 360: 600–603 (1992).
20. Dong JG, Fernandez-Maculet JC, Yang SF: Purification and characterization of 1-aminocyclopropane-1-carboxylate oxidase from apple fruit. Proc Natl Acad Sci USA 89: 9789–9793 (1992a).
21. Dong JG, Kim WT, Yip WK, Thompson GA, Li L, Bennett AB, Yang SF: Cloning of a cDNA encoding 1-aminocyclopropane-1-carboxylate synthase and expression of its mRNA in ripening apple fruit. Planta 185: 38–45 (1991).
22. Dong JG, Olson D, Silverstone A, Yang S-F: Sequence of a cDNA Coding for a 1-aminocyclopropane-1-carboxylate oxidase homolog from apple fruit. Plant Physiol 98: 1530–1531 (1992).
23. Ecker JR, Theologis A: Ethylene: a unique signalling molecule. In: Somerville C, Meyerowitz E (eds), *Arabidopsis* 485–521. Cold Spring Harbor Laboratory Press, Cold Spring Harbor, N.Y. (1994).
24. Eyal Y, Meller Y, Lev-Yadun S, Fluhr R: A basic-type PR-1 promoter directs ethylene responsiveness, vascular and abscission zone-specific expression. Plant J 4: 225–234 (1993).
25. Fabian JR, Morrison DK, Daar IO: Requirement for raf and map kinase function during the meiotic maturation of *Xenopus* oocytes. J Cell Biol 122: 645–652 (1993).
26. Fernandez-Maculet JC, Dong JG, Yang SF: Activation of 1-aminocyclopropane-1-carboxylate oxidase by carbon dioxide. Biochem Biophys Res Comm 193: 1168–1173 (1993).
27. Fray RG, Grierson D: Molecular genetics of tomato fruit ripening. Trends Genet 9: 438–443 (1993).
28. Fujino DW, Burger DW, Bradford KJ: Ineffectiveness of ethylene biosynthetic and action inhibitors in phenotypically reverting the epinastic mutant of tomato

(*Lycopersicon esculentum* Mill.). J Plant Growth Regul 8: 53–61 (1989).

29. Fujino DW, Burger DW, Yang S-F, Bradford KJ: Characterization of an ethylene overproducing mutant of tomato (*Lycopersicon esculentum* Mill. Cultivar VFN8). Plant Physiol 88: 774–779 (1988).
30. Giovannoni JJ, DellaPenna D, Lashbrook CC, Bennett AB, Fischer RL: Expression of a chimeric polygalacturonase gene in transgenic *rin* (ripening inhibitor) tomato fruit. In: Bennett AB, O'Neill SD (eds) Horticultural Biotechnology, pp. 217–227. Wiley-Liss, New York (1990).
31. Guo L, Arteca RN, Phillips AT, Liu Y: Purification and characterization of 1-aminocyclopropane-1-carboxylate N-malonyltranferase from etiolated mung bean hypocotyls. Plant Physiol 100: 2041–2045 (1992).
32. Gupta Williams N, Roberts TM: Signal transduction pathways involving the Raf proto-oncogene. Cancer Metastasis Rev 13: 105–116 (1994).
33. Guzman P, Ecker JR: Exploiting the triple response of *Arabidopsis* to identify ethylene-related mutants. Plant Cell 2: 513–523 (1990).
34. Hamilton AJ, Bouzawen M, Grierson D: Identification of a tomato gene for the ethylene-forming enzyme by expression in yeast. Proc Natl Acad Sci USA 88: 7434–7437 (1991).
35. Hamilton AJ, Lycett GW, Grierson D: Antisense gene that inhibits synthesis of the hormone ethylene in transgenic plants. Nature 346: 284–287 (1990).
36. Han M, Golden A, Han Y, Sternberg PW: *C. elegans* lin-45 raf gene participates in let-60 ras-stimulated vulval differentiation. Nature 363: 133–140 (1993).
37. Hargrove JL, Scoble HA, Mathews WR, Baumstark BR, Biemann K: The structure of tyrosine aminotransferase: evidence for domains involved in catalysis and enzyme turnover. J Biol Chem 264: 45–53 (1989).
38. Hedden P: 2-Oxoglutarate-dependent dioxygenases in plants: mechanism and function. Biochem Soc Trans 20: 373–376 (1992).
39. Holdsworth MJ, Schuch W, Grierson D: Nucleotide sequence of an ethylene-related gene from tomato. Nucl Acids Res 15: 10600 (1987).
40. Hrabak EM, Willis DK: The lemA gene required for pathogenicity of *Pseudomonas syringae* pv. *syringae* on bean is a member of a family of two-component regulators. J Bact 174: 3011–3020 (1992).
41. Huang P-L, Parks JE, Rottmann WH, Theologis A: Two genes encoding 1-aminocyclopropane-1-carboxylate synthase in zucchini (*Cucurbita pepo*) are clustered and similar, but differentially expressed. Proc Natl Acad Sci USA 88: 7021–7025 (1991).
42. Hughes DA: Histidine kinases hog the limelight. Nature 369: 187–188 (1994).
43. Iuchi S: Phosphorylation/dephosphorylation of the receiver module at the conserved aspartate residue controls transphosphorylation activity of histidine kinase in sensor protein ArcB of *Escherichia coli*. J Biol Chem 268: 23972–23980 (1993).
44. Iuchi S, Matsuda Z, Fujiwara T, Lin EC: The *arcB* gene of *Escherichia coli* encodes a sensor-regulator protein for anaerobic repression of the arc modulon. Mol Microbiol 4: 715–727 (1990).
45. Jackson MB: Ethylene and responses of plants to soil waterlogging and submergence. Annu Rev Plant Physiol 36: 145–174 (1985).
46. Kende H: Enzymes of ethylene biosynthesis. Plant Physiol 91: 1–4 (1989).
47. Kende H: Ethylene biosynthesis. Annu Rev Plant Physiol Plant Mol Biol 44: 283–307 (1993).
48. Kieber JJ, Ecker JR: Ethylene gas: it's not just for ripening any more! Trends Genet 9: 356–362 (1993).
49. Kieber JJ, Rothenberg M, Roman G, Feldmann KA, Ecker JR: CTR1, a negative regulator of the ethylene response pathway in *Arabidopsis*, encodes a member of the Raf family of protein kinases. Cell 72: 427–441 (1993).
50. Kim WT, Silverstone A, Yip WK, Dong JG, Yang SF: Induction of 1-aminocyclopropane-1-carboxylate synthase mRNA by auxin in mung bean hypocotyls and cultured apple shoots. Plant Physiol 98: 465–471 (1992).
51. Kim WT, Yang SF: Turnover of 1-aminocyclopropane-1-carboxylic acid synthase protein in wounded tomato fruit tissue. Plant Physiol 100: 1126–1131 (1992).
52. Klee HJ, Hayford MB, Kretzmer KA, Barry GF, Kishore GM: Control of ethylene synthesis by expression of a bacterial enzyme in transgenic tomato plants. Plant Cell 3: 1187–1193 (1991).
53. Kock M, Hamilton A, Grierson D: *ETH1*, a gene involved in ethylene synthesis in tomato. Plant Mol Biol 17: 141–142 (1991).
54. Lanahan MB, Yen H-C, Giovannoni JJ, Klee HJ: The *Never Ripe* mutation blocks ethylene perception in tomato. Plant Cell 6: 521–530 (1994).
55. Latche A, Dupille E, Rombaldi C, Cleyet-Marel JC, Lelievre JM, Pech JC: Purification, characterization and subcellular localization of ACC oxidase from fruits. In: Pech JC, Latche A, Balague C (eds) Cellular and Molecular Aspects of the Plant Hormone Ethylene, pp. 39–45. Kluwer Academic Publishers, Dordrecht, Netherlands (1992).
56. Lawton KA, Potter SL, Uknes S, Ryals J: Acquired resistance signal transduction in *Arabidopsis* is ethylene independent. Plant Cell 6: 581–588 (1994).
57. Li N, Mattoo AK: Deletion of the carboxyl-terminal region of 1-aminocyclopropane-1-carboxylic acid synthase, a key protein in the biosynthesis of ethylene, results in catalyticaly hyperactive, monomeric enzyme. J Biol Chem 269: 6908–6917 (1994).
58. Liang X, Abel S, Keller JA, Shen NF, Theologis A: The 1-aminocyclopropane-1-carboxylate synthase gene family of *Arabidopsis thaliana*. Proc Natl Acad Sci USA 89: 11046–11050 (1992).

59. Lincoln JE, Campbell AD, Oetiker J, Rottmann WH, Oeller PW, Shen NF, Theologis A: LE-ACS4, a fruit ripening and wound-induced 1-aminocyclopropane-1-carboxylate synthase gene of tomato (*Lycopersicon esculentum*). J Biol Chem 268: 19422–19430 (1993).
60. Lincoln JE, Cordes S, Read E, Fischer RL: Regulation of gene expression by ethylene during *Lycopersicon esculentum* (tomato) fruit development. Proc Natl Acad Sci USA 84: 2793–2797 (1987).
61. Lincoln JE, Fischer RL: Diverse mechanisms for the regulation of ethylene-inducible gene expression. Mol Gen Genet 212: 71–75 (1988).
62. Liu D, Li N, Dube S, Kalinski A, Herman E, Mattoo AK: Molecular characterization of a rapidly and transiently wound-induced soybean (*Glycine max* L.) gene encoding 1-aminocyclopropane-1-carboxylate synthase. Plant Cell Physiol 34: 1151–1157 (1993).
63. MacDiarmid CWB, Gardner RC: A cDNA sequence from kiwifruit homologous to 1-aminocyclopropane-1-carboxylic acid oxidase. Plant Physiol 101: 691–692 (1993).
64. Macnicol AM, Muslin AJ, Williams LT: Raf-1 kinase is essential for early *Xenopus* development and mediates the induction of mesoderm by FGF. Cell 73: 571–583 (1993).
65. Maeda T, Wurgler-Murphy SM, Saito H: A two-component system that regulates an osmosensing MAP kinase cascade in yeast. Nature 369: 242–245 (1994).
66. Matsuda J, Okabe S, Hashimoto T, Yamada Y: Molecular cloning of hyoscyamine-6-β-hydroxylase, a 2-oxoglutarate-dependent dioxygenase, from cultured roots of *Hyoscyamus niger*. J Biol Chem 266: 9460–9464 (1991).
67. McGarvey DJ, Christoffersen RE: Characterization and kinetic parameters of ethylene-forming enzyme from avocado fruit. J Biol Chem 267: 5964–5967 (1992).
68. McGarvey DJ, Yu H, Christoffersen RE: Nucleotide sequence of a ripening-related cDNA from avocado fruit. Plant Mol Biol 15: 165–167 (1990).
69. Mehta PK, Christen P: Homology of 1-aminocyclopropane-1-carboxylate synthase, 8-amino-7-oxononanoate synthase, 2-amino-6-caprolactam racemase, 2,2-dialkylglycine decarboxylase, glutamate-1-semialdehyde 2,1-aminomutase and isopenicillin-N-epimerase with aminotransferases. Biochem Biophys Res Comm 198: 138–143 (1994).
70. Mehta PK, Hale TI, Christen P: Aminotransferases: demonstration of homology and division into evolutionary subgroups. Eur J Biochem 214: 549–561 (1993).
71. Meller Y, Sessa G, Eyal Y, Fluhr R: DNA-protein interactions on a cis-DNA element essential for ethylene regulation. Plant Mol Biol 23: 453–463 (1993).
72. Metraux J-P, Kende H: The role of ethylene in the growth response of submerged deep water rice. Plant Physiol 72: 441–446 (1983).
73. Montgomery JR: Regulation of gene expression during tomato fruit ripening. Ph. D. thesis, University of California, Berkeley (1993).
74. Montgomery J, Goldman S, Deikman J, Margossian L, Fischer RL: Identification of an ethylene-responsive region in the promoter of a fruit ripening gene. Proc Natl Acad Sci USA 90: 5939–5943 (1993).
75. Nagasawa S, Tokishita S, Aiba H, Mizuno T: A novel sensor-regulator protein that belongs to the homologous family of signal-transduction proteins involved in adaptive responses in *Escherichia coli*. Mol Microbiol 6: 799–807 (1992).
76. Nakagawa N, Mori H, Yamazaki K, Imaseki H: Cloning of a complementary DNA for auxin-induced 1-aminocyclopropane-1 carboxylate synthase and differential expression of the gene by auxin and wounding. Plant Cell Physiol 32: 1153–1163 (1991).
77. Nakajima N, Mori H, Yamazaki K, Imaseki H: Molecular cloning and sequence of a complementary DNA encoding 1-aminocyclopropane-1-carboxylate synthase induced by tissue wounding. Plant Cell Physiol 31: 1021–1029 (1990).
78. Neljubov D: Uber die horizontale Mutation der Stengel von *Pisum sativum* und einiger anderer. Pflanzen Beih Bot Zentralbl 10: 128–239 (1901).
79. Oeller PW, Wong LM, Taylor LP, Pike DA, Theologis A: Reversible inhibition of tomato fruit senescence by antisense RNA. Science 254: 437–439 (1991).
80. Olson DC, White JA, Edelman L, Harkins RN, Kende H: Differential expression of two genes for 1-aminocyclopropane-1-carboxylate synthase in tomato fruits. Proc Natl Acad Sci USA 88: 5340–5344 (1991).
81. Ota IM, Varshavsky A: A yeast protein similar to bacterial two-component regulators. Science 262: 566–569 (1993).
82. Park KY, Drory A, Woodson WR: Molecular cloning of an 1-aminocyclopropane-1-carboxylate synthase from senescing carnation flower petals. Plant Mol Biol 18: 377–386 (1992).
83. Parkinson JS: Signal transduction schemes of bacteria. Cell 73: 857–871 (1993).
84. Parkinson JS, Kofoid EC: Communication modules in bacterial signalling proteins. Annu Rev Genet 26: 71–112 (1992).
85. Peleman J, Boerjan W, Engler G, Seurinck J, Botterman J, Alliotte T, Van Montagu M, Inze D: Strong cellular preference in the expression of a housekeeping gene of *Arabidopsis thaliana* encoding S-adenosylmethionine synthetase. Plant Cell 1: 81–93 (1989).
86. Penarrubia L, Aguilar M, Margossian L, Fischer RL: An antisense gene stimulates ethylene hormone production during tomato fruit ripening. Plant Cell 4: 681–687 (1992).
87. Pirrung MC, Kaiser LM, Chen J: Purification and properties of the apple fruit ethylene-forming enzyme. Biochemistry 32: 7445–7450 (1993).
88. Raghothama KG, Lawton KA, Goldsbrough PB,

Woodson WR: Characterization of an ethylene-regulated flower senescence-related gene from carnation. Plant Mol Biol 17: 61–71 (1991).
89. Raz V, Fluhr R: Calcium requirement for ethylene-dependent responses. Plant Cell 4: 1123–1130 (1992).
90. Raz V, Fluhr R: Ethylene signal is transduced via protein phosphorylation events in plants. Plant Cell 5: 523–530 (1993).
91. Rick CM, Butler L: Phytogenetics of the tomato. Adv Genet 8: 267–382 (1956).
92. Roby D, Broglie K, Gaynor J, Broglie R: Regulation of a chitinase gene promoter by ethylene and elicitors in bean protoplasts. Plant Physiol 97: 433–439 (1991).
93. Rodrigues-Pousada RA, Rycke RD, Dedonder A, Caeneghem WV, Engler G, Van Montagu M, Van Der Straeten D: The *Arabidopsis* 1-aminocyclopropane-1-carboxylate synthase gene 1 is expressed during early development. Plant Cell 5: 897–911 (1993).
94. Rombaldi C, Petitprez M, Cleyet-Marel JC, Rouge P, Latche A, Pech JC, Lelievre JM: Immunocytolocalisation of ACC oxidase in tomato fruits. In: Pech JC, Latche A, Balague C (eds) Cellular and Molecular Aspects of the Plant Hormone Ethylene, pp. 96–97. Kluwer Academic Publishers, Dordrecht, Netherlands (1992).
95. Ross AF: Localized acquired resistance to plant virus infection in hypersensitive hosts. Virology 14: 329–339 (1961).
96. Rottmann WH, Peter GF, Oeller PW, Keller JA, Shen NF, Nagy BP, Taylor LP, Campbell AD, Theologis A: 1-Aminocyclopropane-1-carboxylate synthase in tomato is encoded by a multigene family whose transcription is induced during fruit and floral senescence. J Mol Biol 222: 937–961 (1991).
97. Sato T, Oeller PW, Theologis A: The 1-aminocyclopropane-1-carboxylate synthase of *Cucurbita*. J Biol Chem 266: 3752–3759 (1990).
98. Sato T, Theologis A: Cloning the mRNA encoding 1-aminocyclopropane-1-carboxylate synthase, the key enzyme for ethylene biosynthesis in plants. Proc Natl Acad Sci USA 86: 6621–6625 (1989).
99. Satoh S, Mori H, Imaseki H: Monomeric and dimeric forms and the mechanism-based inactivation of 1-aminocyclopropane-1-carboxylate synthase. Plant Cell Physiol 34: 753–760 (1993).
100. Slater A, Maunders MJ, Edwards K, Schuch W, Grierson D: Isolation and characterization of cDNA clones for tomato polygalacturonase and other ripening-related proteins. Plant Mol Biol 5: 137–147 (1985).
101. Sonnhammer ELL, Kahn D: Modular arrangement of proteins as inferred from analysis of homology. Prot Sci 3: 482–492 (1994).
102. Spanu P, Reinhardt D, Boller T: Analysis and cloning of the ethylene-forming enzyme from tomato by functional expression of its mRNA in *Xenopus laevis* oocytes. EMBO J 10: 2007–2013 (1991).
103. Stock JB, Ninfa AJ, Stock AM: Protein phosphorylation and regulation of adaptive responses in bacteria. Microbiol Rev 53: 450–490 (1989).
104. Stout V, Gottesman S: RcsB and RcsC: a two-component regulator of capsule synthesis in *Escherichia coli*. J Bact 172: 659–669 (1990).
105. Sung M, Tanizawa K, Tanaka H, Kuramitsu S, Kagamiyama H, Hirotsu K, Okamoto A, Higuchi T, Soda K: Thermostable aspartate aminotransferase from a thermophilic bacillus species: gene cloning, sequence determination, and preliminary X-ray characterization. J Biol Chem 266: 2567–2572 (1991).
106. Tang X, Wang H, Brandt AS, Woodson WR: Organization and structure of the 1-aminocyclopropane-1-carboxylate oxidase gene family from *Petunia hybrida*. Plant Mol Biol 23: 1151–1164 (1993).
107. Theologis A: One rotten apple spoils the whole bushel: the role of ethylene in fruit ripening. Cell 70: 181–184 (1992).
108. Theologis A: What a gas! Curr Biol 3: 369–371 (1993).
109. Theologis A: Control of ripening. Curr Opin Biotechnol 5: 152–157 (1994).
110. Theologis A, Huynh TV, Davis RW: Rapid induction of specific mRNAs by auxin in pea epicotyl tissue. J Mol Biol 183: 53–68 (1985).
111. Theologis A, Oeller PW, Wong LM, Rottmann WH, Gantz DM: Use of a tomato mutant constructed with reverse genetics to study fruit ripening, a complex development process. Devel Genet 14: 282–295 (1993).
112. Uhl MA, Miller JF: Autophosphorylation and phosphotransfer in the *Bordetella pertussis* BvgAS signal transduction cascade. Proc Natl Acad Sci USA 91: 1163–1167 (1994).
113. Van der Straeten D, Djudzman A, Vancaeneghem W, Smalle J, Van Montagu M: Genetic and physiological analysis of a new locus in *Arabidopsis* that confers resistance to 1-aminocyclopropane-1-carboxylic acid and ethylene and specifically affects the ethylene signal-transduction pathway. Plant Physiol 102: 401–408 (1993).
114. Van Der Straeten D, Rodrigues-Pousada RA, Villarroel R, Hanley S, Van Montagu M: Cloning, genetic mapping, and expression analysis of an *Arabidopsis thaliana* gene that encodes 1-aminocyclopropane-1-carboxylate synthase. Proc Natl Acad Sci USA 89: 9969–9973 (1992).
115. Van Der Straeten D, Van Wiemeersch L, Goodman HM, Van Montagu M: Purification and partial characterization of 1-aminocyclopropane-1-carboxylate synthase from tomato pericarp. Eur J Biochem 182: 639–647 (1989).
116. Van Der Straeten D, Wiemeersch LV, Goodman HM, Van Montagu M: Cloning and sequence of two different cDNAs encoding 1-aminocyclopropane-1-carboxylate synthase in tomato. Proc Natl Acad Sci USA 87: 4859–4863 (1990).

117. Van Doorsselaere J, Gielen J, Van Montagu M, Inze D: A cDNA encoding S-adenosyl-L-methionine synthetase from poplar. Plant Physiol 102: 1365–1366 (1993).
118. Ververidis P, John P: Complete recovery *in vitro* of ethylene-forming enzyme activity. Phytochemistry 30: 725–727 (1991).
119. Wang H, Woodson WR: A flower senescence-related mRNA from carnation shares sequence similarity with fruit ripening-related mRNAs involved in ethylene biosynthesis. Plant Physiol 96: 1000–1001 (1991).
120. Wang H, Woodson WR: Nucleotide sequence of a cDNA encoding the ethylene-forming enzyme from petunia corollas. Plant Physiol 100: 535–536 (1992).
121. Wilson ID, Zhu YL, Burmeister DM, Dilley DR: Apple ripening-related cDNA clone PAP4 confers ethylene-forming ability in transformed *Saccharomyces cerevisiae*. Plant Physiol 102: 783–788 (1993).
122. Yang SF, Dong JG: Recent progress in research of ethylene biosynthesis. Bot Bull Acad Sin 34: 89–101 (1993).
123. Yang SF, Hoffman NE: Ethylene biosynthesis and its regulation in higher plants. Annu Rev Plant Physiol 35: 155–189 (1984).
124. Yip W-K, Dong J-G, Kenny JW, Thompson GA, Yang SF: Characterization and sequencing of the active site of 1-aminocyclopropane-1-carboxylate synthase. Proc Natl Acad Sci USA 87: 7930–7934 (1990).
125. Yip W-K, Moore T, Yang SF: Differential accumulation of transcripts for four tomato 1-aminocyclopropane-1-carboxylate synthase homologs under various conditions. Proc Natl Acad Sci USA 89: 2475–2479 (1992).
126. Zarembinski TI, Theologis A: Anaerobiosis and plant growth hormones induce two genes encoding 1-aminocyclopropane-1-carboxylate synthase in rice (*Oryza sativa L.*). Mol Biol Cell 4: 363–373 (1993).
127. Jansonius JN, Eichele G, Ford GC, Picot D, Thaller C, Vincent GC: Spatial structure of mitochondrial aspartate aminotransferase. In: Christen P, Metzler DE (eds) Transaminases, pp. 109–138. John Wiley and Sons, New York (1985).
128. Rost B, Sander C: Improved prediction of protein secondary structure by use of sequence profiles and neural networks. Proc Natl Acad Sci USA 90: 7558–7562 (1993).

Plant Molecular Biology **26**: 1599–1609, 1994.

Structure and function of the receptor-like protein kinases of higher plants

John C. Walker
Division of Biological Sciences, University of Missouri, Columbia, MO 65211, USA

Received 9 March 1994; accepted 26 April 1994

Key words: Protein kinase, protein phosphorylation, receptor, self-incompatibility, signal transduction

Abstract

Cell surface receptors located in the plasma membrane have a prominent role in the initiation of cellular signalling. Recent evidence strongly suggests that plant cells carry cell surface receptors with intrinsic protein kinase activity. The plant receptor-like protein kinases (RLKs) are structurally related to the polypeptide growth factor receptors of animals which consist of a large extracytoplasmic domain, a single membrane spanning segment and a cytoplasmic domain of the protein kinase gene family. Most of the animal growth factor receptor protein kinases are tyrosine kinases; however, the plant RLKs all appear to be serine/threonine protein kinases. Based on structural similarities in their extracellular domains the RLKs fall into three categories: the S-domain class, related to the self-incompatibility locus glycoproteins of *Brassica*; the leucine-rich repeat class, containing a tandemly repeated motif that has been found in numerous proteins from a variety of eukaryotes; and a third class that has epidermal growth factor-like repeats. Distinct members of these putative receptors have been found in both monocotyledonous plants such as maize and in members of the dicotyledonous Brassicaceae. The diversity among plant RLKs, reflected in their structural and functional properties, has opened up a broad new area of investigation into cellular signalling in plants with far-reaching implications for the mechanisms by which plant cells perceive and respond to extracellular signals.

Introduction

The past decade has yielded profound new insights into how a eukaryotic cell transmits signals from the plasma membrane to the cell's interior. Both biochemical and molecular genetic analyses, in a variety of organisms, have shown that many signal transduction pathways are conserved throughout evolution and require the action of protein kinases. In plants, the past five years have witnessed an explosion of interest in the molecular analyses of protein kinases that may be involved in signal transduction pathways. The cloning of the first plant protein kinase gene was reported in 1989 [23] and a search of the DNA sequence databases [3] reveals that nearly one hundred protein kinase genes from plants have since been identified. One exciting observation from these results is the identification of new families of protein kinases (e.g. [14] and [47]). The availability of cDNA clones for these plant protein kinases has made it possible to perform a detailed comparative analysis of their structural characteristics, and to begin to dissect their biochemical properties and physiological functions. This review will focus on one family of plant protein kinases: the receptor-like protein kinases (RLKs).

Signal transduction by transmembrane protein kinases

In animal cells, common types of receptors for extracellular ligands are receptor tyrosine kinases. All of the proteins in this family have a large extracellular domain, span the plasma membrane once, and contain a tyrosine kinase catalytic domain [8]. Ligand binding to the extracellular domain induces receptor homodimerization and stimulates receptor autophosphorylation, an essential step in tyrosine kinase-mediated signal transduction. Autophosphorylation occurs at multiple sites and each potentially functions as a high-affinity binding site for a diverse array of proteins that contain a *src* homology-2 (SH2) domain [15, 31]. Proteins containing an SH2 domain are divided into two types: those with a catalytic activity and those that function as molecular adaptors. These molecular adaptors typically have an additional motif, the *src* homology-3 (SH3) domain that mediates interactions with other signalling molecules. Hence, ligand binding to a transmembrane receptor protein kinase can result in the activation of several intracellular events, from production of second messengers to initiation of a protein kinase cascade [15, 31].

Transmembrane signalling by protein kinases is not limited to tyrosine kinase receptors. Metazoan cells also utilize another family of protein kinase receptors. Although these receptors have the same architecture as the tyrosine kinase receptors – a large extracellular domain, a single transmembrane region and a cytoplasmic protein kinase domain – they autophosphorylate on serine and/or threonine, not on tyrosine. These receptors are members of the transforming growth factor β (TGF-β) receptor family [25]. TGF-β induced signalling requires the interaction of two distinct transmembrane serine/threonine proteins kinases known as the type I and type II receptors. The nature of the intracellular targets for this heterodimeric signalling complex is unknown.

Identification, classification and primary structure of the RLKs of plants

In plants cells the analysis of transmembrane signalling by protein kinases is still in its infancy. Much of the evidence gathered to date in support of receptor protein kinases is based on the predicted structures of proteins derived from cDNA clone sequences. The designation of this novel class of plant signalling molecules, the receptor-like protein kinases, is based on their structural similarity with the mammalian growth factor receptor protein kinases. Plant RLKs can be placed into three major classes on the basis of similarities in their extracytoplasmic domains (Fig. 1). Within each category they can be further discrimi-

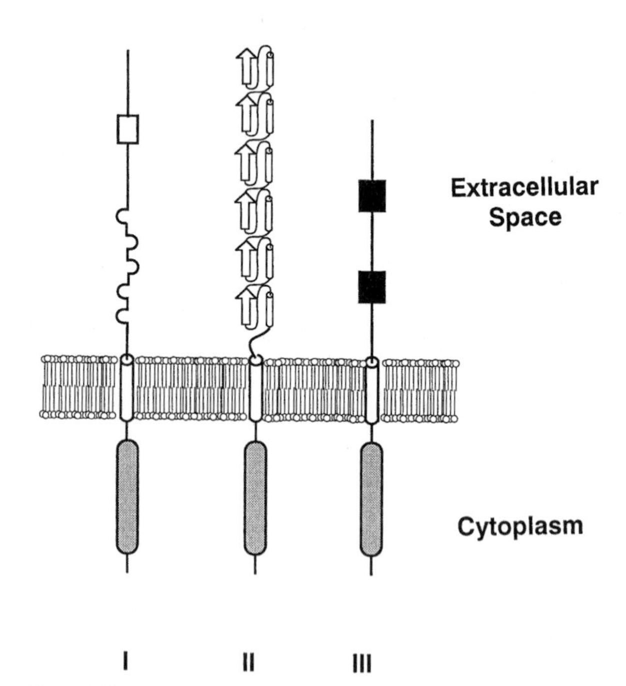

Fig. 1. The receptor-like protein kinases of plants. The structures of the three classes of plant RLKs is illustrated. Domains represented: protein kinase domain (gray oval), transmembrane domain (open cylinder), conserved cysteine rich region in the S-domain (semi-circles), conserved PTDT block of the S-domain (open box), leucine rich repeats (cylinder and arrow) and the epidermal growth factor repeats (black boxes). These illustrations are notdrawn to scale.

nated by differences in extracellular domains, protein kinase domains, and patterns of gene expression. This classification is useful because it illustrates that there is a large family of *RLK* genes in plants and that each gene encodes a distinct protein kinase that probably plays a unique role in cellular signalling.

The first class of RLKs are characterized by having an extracellular S-domain. The S-domain was originally described in the self-incompatibility-locus glycoproteins (SLGs) from the Brassicaceae. The SLGs are part of the sporophytic self-incompatibility response that normally inhibits self-pollination [28]. The distinguishing feature of the S-domain is an array of ten cysteine residues found in combination with other conserved motifs (Fig. 2). S-domain RLKs have been found in *Brassica* species, *Arabidopsis thaliana*, and maize. The first RLK found, ZmPK1, is a member of the S-domain class and was isolated by PCR using oligonucleotide primers designed to amplify protein kinase sequences from maize mRNA [47]. The S-domain of ZmPK1 is 27% identical to the *Brassica oleracea* SLG_{13} allele and contains the conserved amino acids characteristic of the S-domain family. Subsequently, an S-locus-linked RLK, designated *SRK* for S receptor kinase, was reported [40]. The S-domain of the SRK_6 is 89% identical to the SLG_6 allele but only 67% identical to the SLG_2 allele. By comparison, the S domain of SLG_2 is ca. 90% identical to the corresponding *SRK*2 gene. This

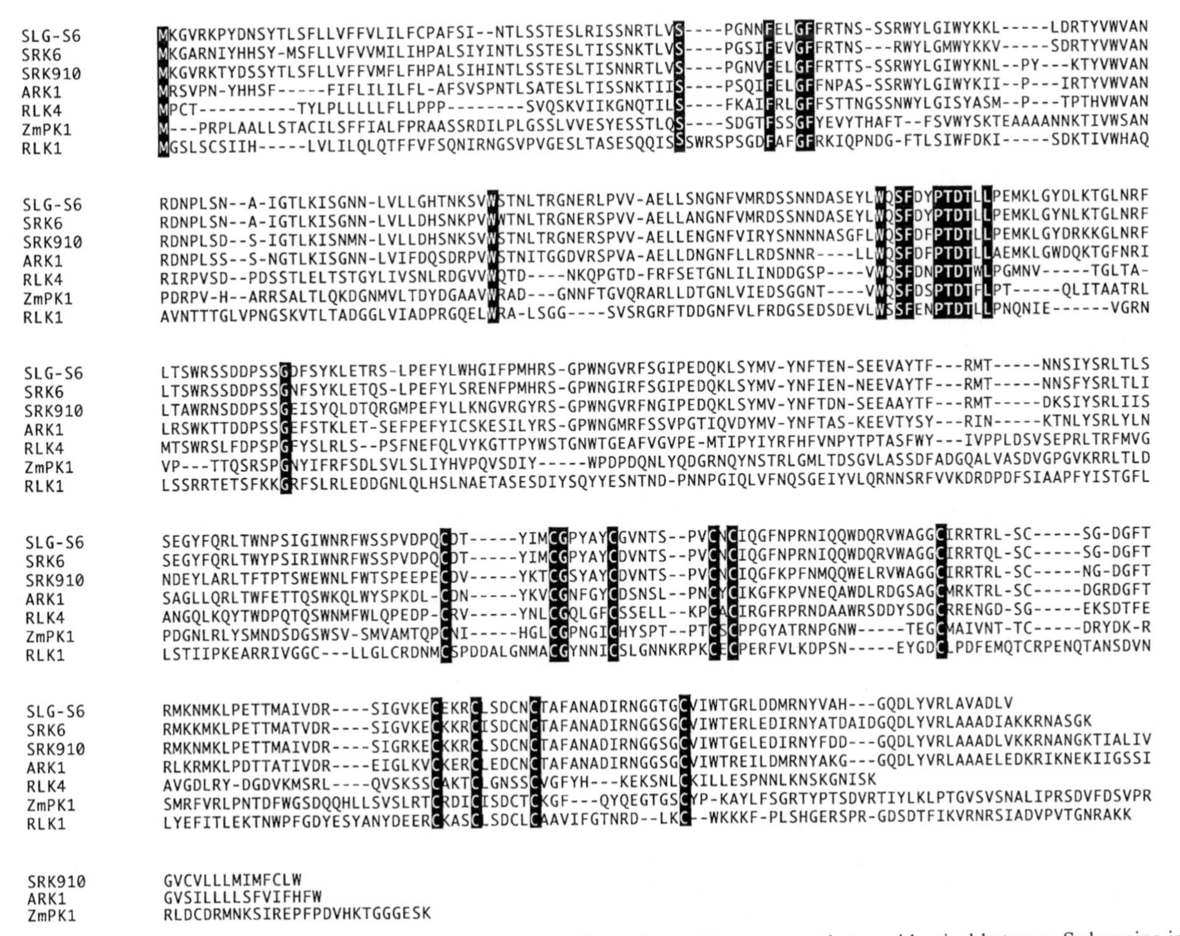

Fig. 2. Amino acid sequence alignments of RLK S-domains. The amino acid sequences that are identical between S-domains in the S-domain are shown outlined in black boxes. GenBank accession numbers: SLG-S6 (Y00268), SRK6 (M76647), SRK910 (M97667), ARK1 (M80238), RLK4 (M84659), ZmPK1 (X52384), RLK1(M84658).

high degree homology between genetically linked pairs of the *SLG* and *SRK* genes suggests that they are evolving in concert within a given S genotype [40]. Other *SRK* alleles have also been identified and characterized in *Brassica* species [11, 22].

Several members of the S-domain class of *RLK*s are clearly not analogous of the *SRK* genes. *ARK*1 from *Arabidopsis thaliana* shares sequence similarity with both *Zm*PK1 and the *SRK* genes [43]. *ARK*1 was isolated by screening a genomic library from *Arabidopsis* with *Brassica SLG* and *SRK* cDNA probes. The predicted ARK1 polypeptide is approximately 60% identical to the S domains of the *Brassica* SLGs and SRKs but only 23% identical to ZmPK1. Using a similar approach, the protein kinase domain of *Zm*PK1 was used to screen an *Arabidopsis* cDNA library in a search for additional *RLK* genes [46]. Three different *RLK* genes were characterized in this study; two of them, *RLK*1 and *RLK*4, belong to the S-domain family of receptor protein kinases. RLK1 and RLK4 share about 25% amino acid sequence identity with the SLGs and 22% identity with ZmPK1 in the S-domain. Thus, the S-domain *RLK* genes represent the largest family of transmembrane receptor protein kinases so far identified in higher plants, and the individual members within this class have substantial primary sequence divergence. This divergence and the differences in the patterns of expression of different S-domain genes (discussed below) suggest that each of these receptors plays a unique role in cellular signalling.

The second RLK category includes members of the leucine rich repeat (LRR) family of proteins (Fig. 3). The LRR motif is found in many proteins that participate in protein-protein interactions and have diverse cellular functions and locations. LRRs are usually found tandemly repeated and can occur in very divergent forms with gaps and insertions within or between repeats. Recently the basic tertiary structure of the LRR repeats has been determined to constitute a new class of α/β fold [19]. The most conserved element (Fig. 3) of the LRR repeat forms a β sheet thought to be an exposed face involved in protein-

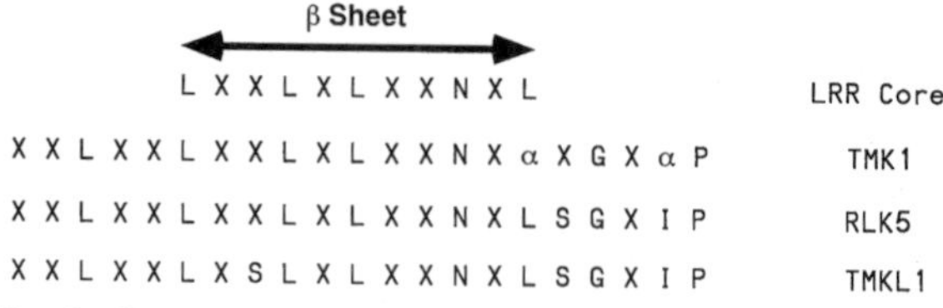

Fig. 3. Comparison of consensus sequences found in the LRR RLKs. The LRR core sequence that forms a beta sheet is shown at the top, with the consensus LRR of TMK1, TMKL1 and RLK5 aligned below. X indicates any amino acid and α indicates any amino acid with analiphatic side chain.

protein interactions [19]. Differences in this region among the repeats may encode the specificity of interaction. Outside of the consensus core of the LRR there is little sequence similarity in the extracellular domains of the LRR RLKs. To date, three LRR containing protein kinases have been described in *Arabidopsis thaliana*: TMK1, RLK5 and TMKL1 (Fig. 3). *TMK*1 and *TMKL*1 were isolated in the course of chromosomal walks in search of other genes [4, 45], and *RLK*5 was isolated by low-stringency hybridization with a ZmPK1 catalytic domain probe [46]. The organization of the LRRs in these three genes is very different: TMK1 has 11 LRRs arranged in two blocks, *TMKL*1 has seven LRR repeats in a single continuous block and *RLK*5 has 21 LRRs in a single continuous block. However, unlike other *RLK* genes, *TMKL*1 does not appear to encode an active protein kinase, since several of the conserved protein kinase subdomains are missing or altered (discussed below).

To date, *pro*25 from *Arabidopsis* is the sole example of a plant gene that contains an epidermal growth factor-like repeat. *pro*25 encodes a protein that interacts with the amino terminus of a light-harvesting chlorophyll *a*/*b* binding protein (LHCP) [21] and was selected in yeast on that basis. Deduced amino acid sequence of *pro*25 indicates that it encodes a protein kinase domain linked via a membrane-spanning region to a domain containing two epidermal growth factor-like repeats. Unlike other plant RLKs, the deduced amino acid sequence of *pro*25 does not appear to contain an amino terminal signal peptide.

The plant RLKs represent a unique subfamily of protein kinases. Plant *RLK* genes may have evolved from a common ancestral transmembrane protein kinase, or, alternatively, conserved features of these enzymes may have some functional significance. Despite their structural diversity and varied substrate specificity, the catalytic domains of most eukaryotic protein kinases contain eleven blocks (subdomains) of conserved amino acid sequences [12, 13]. This primary sequence conservation reflects secondary and tertiary structural conservation as well [42]. Predicted amino acid sequences of the plant RLKs suggest that these enzymes belong to the serine/threonine class of protein kinases (Fig. 4). Eukaryotic protein kinases are commonly classified as either serine/threonine-specific or tyrosine-specific, although dual-specificity protein kinases that can phosphorylate both serine/threonine and tyrosine residues have recently been reported. Two subdomains, VIb and VIII, are used to differentiate serine/threonine protein kinases from tyrosine kinases. In plant RLK sequences, subdomain VIb sequences (Fig. 4) more closely match the consensus DLKPEN found in serine/threonine kinases, as opposed to DLAARN common to tyrosine kinases. Subdomain VIII of plant RLKs matches the GTPXYIAPE motif of the serine/threonine kinases more closely than it does the FPIKWMAPE motif of the tyrosine kinases. TMKL1 is missing several essential residues required for protein kinase activity and therefore probably lacks phosphotransfer ability.

The protein kinase catalytic domain of RLKs is flanked on its amino terminal side by a juxtamembrane domain, which separates the transmembrane domain from the protein kinase catalytic core. Its sequence is highly variable among plant RLK family members and ranges in size from 35 amino acids in RLK5 to 85 amino acids

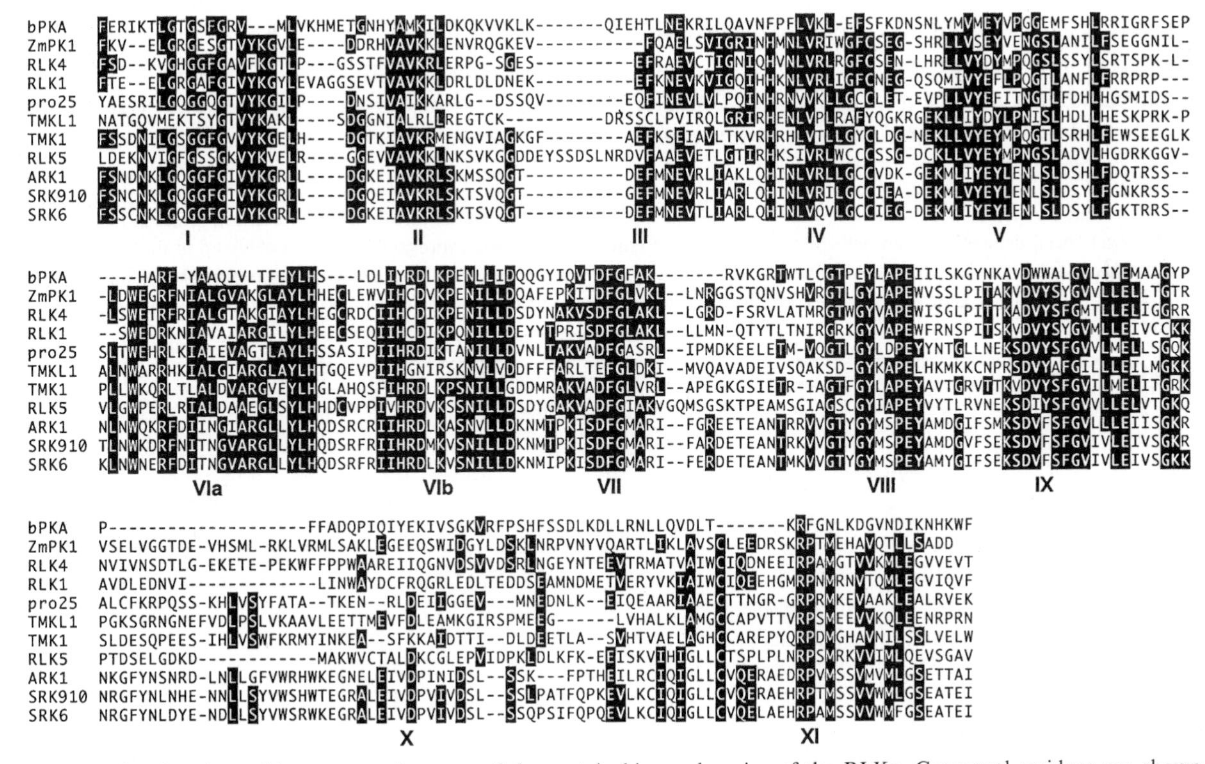

Fig. 4. Predicted amino acid sequence alignment of the protein kinase domains of the RLKs. Conserved residues are shown outlined in black boxes. The eleven conserved domains found in the protein kinase superfamily are indicated by Roman numerals below each region. The sequences of the plant RLKs are aligned relative to the bovine protein kinase A catalytic subunit (bPKA). GenBank accession numbers: bPKA (X67154), ZmPK1 (X52384), RLK4 (M84659), RLK1 (M84658), pro25 (L04999), TMKL1 (X72863), TMK1 (L00670), RLK5 (M84660), ARK1 (M80238), SRK910 (M97667),SRK6 (M76647).

in TMK1. On the other side of the protein kinase catalytic core resides the carboxy-terminal tail. The tail of plant RLKs is short relative to animal cell receptor tyrosine kinases and ranges from 0 to 70 amino acids. The function(s) of the juxtamembrane domain and carboxy-terminal tail in the plant RLKs are not known but these regions are important sites for regulation in the tyrosine kinase family of growth factor receptors [44] and may have similar roles in some plant RLKs.

With the exception of *pro*25 all plant RLK genes encode polypeptides with a hydrophobic amino terminus. This region is thought to act as a signal peptide and targets the proteins to the endoplasmic reticulum during synthesis. The transmembrane domains of the RLKs, which vary in length from 22 to 28 amino acids, have little in common beyond their hydrophobicity. All RLK transmembrane domains are bounded on their carboxyl terminal flanks by several basic amino acids. This organization – hydrophobic domain followed by basic residues – is typical of the Type I integral membrane proteins [37]. The basic amino acids act as a 'stop-transfer' signal which situates the amino terminal domain in the extracellular space and the carboxyl terminal domain on the cytoplasmic face of the lipid bilayer.

Expression of the RLK genes

Variations in structures among different plant *RLK* gene products may reflect functional diversity in this gene family. Diverse patterns of expression of the *RLK* genes (Table 1) further support the concept that these proteins may be involved in mediating a variety of cellular signalling processes.

The *SRK* genes, which may play a role in the pollen self-incompatibility response in the Brassicaceae, are expressed predominantly in pistils, although expression can also be detected in anthers [11, 40]. Expression of the *SRK* genes has not been found in leaf, root or petal tissues. This expression pattern closely resembles that observed for the *SLG* genes and is consistent with the function of SRK in controlling pollen self-incompatibility.

In *Arabidopsis* the S-domain protein kinases have quite distinct patterns of gene expression

Table 1. Classification and expression of the RLK genes. Pluses indicate the tissues in which the expression of the indicated RLK mRNAs has been detected. –, no detectable mRNA was found; (+), mRNA expression has been detected in these tissues, but al lower levels than in other expressing tissues; ND, not determined or data not available. Pluses in the center of the 'floral' column indicate that these RLK genes are expressed in floral tissue but that it has not been determined if the expression is in the anther, pistils or both.

<table>
<tr><th>RLK Class</th><th>Gene</th><th>Species</th><th colspan="4">Expression</th><th>Reference</th></tr>
<tr><th></th><th></th><th></th><th colspan="2">vegetative</th><th colspan="2">floral</th><th></th></tr>
<tr><th></th><th></th><th></th><th>leaf</th><th>root</th><th>anthers</th><th>pistils</th><th></th></tr>
<tr><td>S-domain</td><td>ZmPK1</td><td>Z. mays</td><td>+</td><td>+</td><td>–</td><td>(+)</td><td>47</td></tr>
<tr><td></td><td>RLK1</td><td>A. thaliana</td><td>+</td><td>–</td><td colspan="2">nd</td><td>46</td></tr>
<tr><td></td><td>RLK4</td><td>A. thaliana</td><td>(+)</td><td>+</td><td colspan="2">nd</td><td>46</td></tr>
<tr><td></td><td>ARK1</td><td>A. thaliana</td><td>+</td><td>–</td><td colspan="2">(+)</td><td>43</td></tr>
<tr><td></td><td>SRK6</td><td>B. oleracea</td><td>–</td><td>–</td><td>(+)</td><td>+</td><td>40</td></tr>
<tr><td></td><td>SRK910</td><td>B. napus</td><td>–</td><td>–</td><td>(+)</td><td>+</td><td>11</td></tr>
<tr><td>LRR</td><td>RLK5</td><td>A. thaliana</td><td>+</td><td>+</td><td colspan="2">nd</td><td>46</td></tr>
<tr><td></td><td>TMK1</td><td>A. thaliana</td><td>+</td><td>+</td><td colspan="2">+</td><td>4</td></tr>
<tr><td></td><td>TMKL1</td><td>A. thaliana</td><td>+</td><td>+</td><td colspan="2">+</td><td>45</td></tr>
<tr><td>EGF</td><td>pro25</td><td>A. thaliana</td><td>+</td><td>–</td><td>–</td><td>–</td><td>21</td></tr>
</table>

compared to the expression of the *Brassica* SRK genes. *ARK*1 expression is highest in leaf tissues and lower in floral buds and stems, with no mRNA detectable in roots [43]. In contrast, *RLK*1 mRNA appears to accumulate in leaf and stem tissue, while *RLK*4 expression is restricted primarily to roots [46]. The expression of *RLK*1 and *RLK*4 in flowers has not been examined.

*ZmPK*1 expression is primarily restricted to the early developmental stages of a maize seedling with a low level of expression also observed in the silks [47]. Immunoblot analyses and immunocytochemistry at the electron microscopic level suggest that ZmPK1 is localized in the plasma membrane, consistent with its predicted role as a cell surface receptor (V.P. Counihan and J.C. Walker, unpublished results).

The expression of the LRR RLK genes is widely distributed. *TMKL*1 mRNA is expressed in *Arabidopsis* seedlings, rosette leaves, stems and roots [45]. Expression is also observed in flower buds with a reduced level of mRNA found in succeedingly older siliques. *RLK*5 is expressed at low levels in roots and rosette leaves and stems [46]. However, the temporal pattern of *RLK*5 expression was not examined and it is not known if *RLK*5 is expressed in any other tissues in the *Arabidopsis* plant. Immunoblots using antibodies directed against the LRR domain of TMK1 shows that this protein is in roots, leaves, siliques and flowers, but not stems of the *Arabidopsis* plant [4]. The TMK1 polypeptide appears to be membrane-associated and can be solublized in 1% Triton X-100, consistent with the proposed role of TMK1 as a transmembrane protein kinase.

Transcripts of the EGF containing RLK, *pro*25, are detectable in light-grown leaf tissue and at a very low level in dark-adapted leaves, but not in roots, stems, flowers, siliques, or etiolated seedlings [21]. Antiserum raised to *Escherichia coli* expressed *pro*25 detects both a cell surface protein and one in the chloroplast [20]. *pro*25 was isolated in a selection designed to detect proteins that specifically interact with the 50 amino terminal residues of the LCHP protein and is unique among RLKs in that it lacks an amino terminal signal peptide. It has been suggested that *pro*25 encodes a plasma membrane protein that may share an epitope with another protein found in chloroplast membranes [20].

Genetic and biochemical analysis of RLK function

Important clues about the roles RLKs may play in plant cell regulation have come from the analysis of pollen self-incompatibility in *Brassica*. Self-incompatibility is controlled by the *S*-locus encoding at least two genes, *SLG* and *SRK*. These genes co-segregate with the self-incompatibility phenotype [28]. There are two reports that indicate that self-compatible lines of *Brassica* species have functional *SLG* genes, but non-functional *SRK* genes. In pollen self-compatible lines of *Brassica napus* the structures and expression of the *SLG* and *SRK* genes have been examined [10]. The sequence and expression of the *SLG* gene indicates that it encodes a functional gene. Therefore, a defect in the *SLG* gene is not thought to contribute to a failure in the recognition of self pollen in this *B. napus* line. Expression of the corresponding *SRK* also appears to the same as in self-incompatible lines. However, DNA sequence analysis of the *SRK* gene reveals a one base pair deletion within the S domain that would cause a frameshift and premature translation stop. Except for this deletion the transmembrane and protein kinase domains appear to be unaltered. This observation implies that this *SRK* allele yields a truncated protein that lacks protein kinase activity. The SRK protein remains to be analyzed, and it is not known if self-incompatibility is restored by transformation with a functional *SRK* allele. Nonetheless, these observations support the proposition that *SRK* genes play a role in mediating self-incompatibility.

The requirement for a functional *SRK* allele in the control of self-incompatibility is also supported by genetic evidence from a self-compatible *B. oleracea* line that expresses a functional SLG but not the corresponding *SRK* gene [29]. DNA sequence analysis shows that the promoter, S domain, and transmembrane region of the *SRK* gene are deleted in this line. Thus, the structure and lack of mRNA expression of this *SRK* allele demonstrates that this self-compatible line, which does not reject its own pollen, has a null *SRK* allele. Although other regions of the *S*-locus may be deleted in this line, these results offer further support for the role of the *SRK* genes in mediating recognition of self pollen in the Brassicaceae.

Biochemical analysis of RLK function has shown that these enzymes have intrinsic protein kinase activity. When the carboxy terminal half (the protein kinase domain) of *TMK*1 was fused to the maltose-binding protein and expressed in *E. coli* [4], the recombinant product demonstrated autophosphorylation activity. One dimensional phosphoamino acid analysis showed that phosphothreonine is predominant with some phosphoserine also detected. No phosphotyrosine was detected by this analysis. Using a similar approach a glutathione *S*-transferase fusion protein with the catalytic domain of the *SRK*-910 gene was shown to autophosphorylate [11]. An *in vitro* mutagenized form of the recombinant fusion protein, in which a critical, invariant lysine residue was substituted with alanine, was not active. Phosphoamino acids analysis of the autophosphorylated protein indicated that phosphorylation was predominantly on threonine with a low level of phosphoserine also detectable. As with TMK1 no phosphotyrosine was detected. The *SRK*6 [41] and *pro*25 (B.D. Kohorn, personal communication) genes have also been shown to encode functional serine/threonine protein kinases that autophosphorylate when expressed as recombinant proteins in *E. coli.*

The intrinsic biochemical properties of the RLK5 protein kinase have also been examined. The *RLK*5 catalytic domain was expressed as two different recombinant fusion proteins in *E. coli.* Both hybrid proteins have similar kinetic properties, autophosphorylate on serine and threonine residues at a ratio of 10:1 and have significantly greater activity in the presence of Mn^{2+} than Mg^{2+}, a property that is a characteristic of mammalian transmembrane protein kinases. A lysine to glutamic acid substitution, K711E, abolishes RLK5 protein kinase activity. The autophosphorylation reaction is first order with respect to enzyme concentration, and the catalytically active fusion protein will cross-phosphorylate the K711E protein. This suggests that the autophosphorylation can result from either an intramolecular reaction or by transphosphorylation. Tryptic cleavage of the autophosphorylated protein and two-dimensional thin-layer electrophoresis indicate that several sites in the catalytic domain are phosphorylated (M.A. Horn and J.C. Walker, unpublished observations).

Evidence for activity of native, yet unidentified, RLKs comes from analyses of protein kinases extracted from *Arabidopsis thaliana*. When a detergent-solublized, lectin agarose affinity-purified preparation from microsomal membranes was assayed for protein kinase activity [33] several polypeptides, ranging in size from 90 to 135 kDa, were capable of autophosphorylation. The protein kinase activity in these assays was higher in the presence of Mn^{2+} than Mg^{2+}. Phosphoamino acid analysis revealed phosphorylation primarily on serine and threonine with no phosphotyrosine detected. These experiments are consistent with there being multiple protein kinases in plant membranes with the biochemical attributes characteristic of receptor protein kinases.

Summary and speculations

The architecture of the *RLK* gene products strongly suggests that they function as cell surface receptors. The diversity of structure and array of gene expression patterns of different members of the *RLK* family further suggest that they respond to diverse extracellular signals and display different physiological functions. Biochemical experiments have shown that RLKs have an intrinsic protein kinase activity and autophosphorylate on serine and/or threonine. But what are the ligands that activate these receptors and what are the cellular components that carry the signal to the interior of the cell?

Recent progress in animal cell systems has revealed a variety of signalling molecules responsible for the transmission of information downstream from animal receptor tyrosine kinases. Studies of cell proliferation control have elucidated a diversity of downstream transducing proteins, including molecules that facilitate the reversible formation of protein complexes by recognizing only the activated forms of the receptor tyrosine kinases [15, 31], small GTP binding proteins such as *ras* [34], serine/threonine protein kinases such as *raf* [27], and the evolutionarily conserved MAP kinase phosphorylation cascade [1, 5, 30]. In addition to these protein kinase-mediated pathways, the complementary action of protein phosphatases is known to play a critical role in signalling from the membrane to the nucleus [9].

In higher plants, evidence has been obtained for the existence of a number of soluble, cytoplasmic serine protein kinases including *raf*-like protein kinases [17, 18], elements of the MAP kinase pathway [2, 7, 16, 26], protein phosphatases [38, 39] and G-proteins [24]. It is likely, therefore, that plant signal transduction pathways will eventually be found to incorporate some of these conserved elements. However, it remains to be determined whether the plant RLKs are components of signalling pathways similar to those described in animals, or whether they will be coupled to as yet undescribed signal transduction pathways unique to the plant kingdom.

It is also interesting to speculate about possible ligands for these candidate receptor protein kinases. Considering the close primary structural similarities between plant RLKs and mammalian growth factor receptors it is not unreasonable to presume that RLKs may operate by mechanisms similar to those of their animal counterparts. Without exception, all of the receptor protein kinases with known ligands are activated by polypeptides, independent of whether they are receptor tyrosine kinases or receptor serine/threonine kinases. The ligands for plant RLKs may also be polypeptides. One of the strongest arguments in support of this hypothesis is the relationship of the extracellular domains of the LRR and EGF RLKs to other proteins of known function. In LRR RLKs, the extracellular domain consists of several leucine rich repeats. Strong evidence from studies of several LRR proteins suggests that this motif is involved in protein-protein interactions [19, 35]. Thus, these repeats represent sites for potential interactions with polypeptide ligands. A similar argument can be made for the EGF RLK, pro25. Evidence suggests that the EGF-like motifs mediate interactions with other cellular proteins [6, 36]. However, the LRR and EGF-like motifs might just as well mediate receptor dimerization, a critical event in the activation of all known receptor protein kinases, and play no role in ligand recognition and binding. Given the paucity of known polypeptide signals in plants [32] plant RLKs may be unlike their mammalian counterparts and interact with oligosaccharides or other non-protein ligands. Regardless of the nature of the ligands that activates these plant cell receptors, their diversity in structure and diverse patterns of expression suggest that there are many signalling molecules and pathways yet to be discovered in plants.

Acknowledgements

I wish to thank Thomas Jacobs, Karen Cone, Bruce Kohorn and members of my laboratory for their valuable comments about the manuscript. Research in the author's laboratory has been supported by grants from the National Institutes of Health, the National Science Foundation and the University of Missouri Food for the Twenty First Century program.

References

1. Ahn NG, Seger R, Krebs EG: The mitogen-activated protein kinase activator. Curr Biol 4: 992–999 (1992).
2. Banno H, Hirano K, Nakamura T, Irie K, Nomoto S, Matsumoto K, Machida Y: NPK1, a tobacco gene that encodes a protein with a domain homologous to yeast BCK1, STE11, and Byr2 protein kinases. Mol Cell Biol 13: 4745–4752 (1993).
3. Benson D, Lipman DJ, Ostell J: GenBank. Nucl Acid Res 21: 2963–2965 (1993).
4. Chang C, Schaller GE, Patterson SE, Kwok SF, Meyerowitz EM, Bleecker AB: The *TMK*1 gene from *Arabi-*

dopsis codes for a protein with structural and biochemical characteristics of a receptor protein kinase. Plant Cell 4: 1263–1271 (1992).
5. Crews CM, Erikson RL: Extracellular signals and reversible protein phosphorylation: What to Mek of it all. Cell 74: 215–217 (1993).
6. Davis CG: The many faces of epidermal growth factor repeats. New Biol 2: 410–419 (1990).
7. Duerr B, Gawienowski M, Ropp T, Jacobs T: MsERK1: A mitogen-activated protein kinase from a flowering plant. Plant Cell 5: 87–96 (1993).
8. Fantl WJ, Johnson DE, Williams LT: Signalling by receptor tyrosine kinases. Annu Rev Biochem 62: 453–481 (1993).
9. Feng GS, Hui CC, Pawson T: SH2-containing phosphotyrosine phosphatase as a target of protein-tyrosine kinases. Science 259: 1607–1611 (1993).
10. Goring DR, Glavin TL, Schafer U, Rothstein SJ: An S receptor kinase gene in self-compatible *Brassica napus* has a 1-bp deletion. Plant Cell 5: 531–539 (1993).
11. Goring DR, Rothstein SJ: The S-locus receptor kinase gene in a self-incompatible *Brassica napus* line encodes a functional serine/threonine kinase. Plant Cell 4: 1273–1281 (1992).
12. Hanks SK, Quinn AM: Protein kinase catalytic domain sequence database: identification of conserved features of primary structure and classification of family members. Meth Enzymol 200: 38–61 (1991).
13. Hanks SK, Quinn AM, Hunter T: The protein kinase family: conserved features and deduced phylogeny of the catalytic domains. Science 241: 42–52 (1988).
14. Harper JF, Sussman MR, Schaller GE, Putnam-Evans C, Charbonneau H, Harmon AC: A calcium-dependent protein-kinase with a regulatory domain similar to calmodulin. Science 252: 951–954 (1991).
15. Heldin C-H: SH2 domains: elements that control protein interactions during signal transduction. Trends Biochem Sci 16: 450–452 (1991).
16. Jonak C, Pay A, Bogre L, Hirt H, Heberle-Bors E: The plant homologue of MAP kinase is expressed in a cell-cycle dependent and organ-specific manner. Plant J 3: 611–617 (1993).
17. Kieber JJ, Ecker JR: Ethylene gas: it's not just for ripening any more. Trends Genet 9: 356–362 (1993).
18. Kieber JJ, Rothenberg M, Roman G, Feldmann KA, Ecker JR: *CTR*1, a negative regulator of the ethylene response pathway in *Arabidopsis*, encodes a member of the *raf* family of protein kinases. Cell 72: 427–441 (1993).
19. Kobe B, Deisenhofer J: Crystal structure of porcine ribonuclease inhibitor, a protein with leucine-rich repeats. Nature 366: 751–756 (1993).
20. Kohorn BD, Fujiki M: A family of proteins containing a serine/threonine kinase and/or an EGF repeat has members in the plasma membrane and chloroplast. Fifth International Conference on Arabidopsis Research (1993).
21. Kohorn BD, Lane S, Smith TA: An *Arabidopsis* serine/threonine kinase homologue with an epidermal growth factor repeat selected in yeast for its specificity for a thylakoid membrane protein. Proc Natl Acad Sci USA 89: 10989–10992 (1992).
22. Kumar V, Trick M: Sequence complexity of the S-receptor kinase gene family in *Brassica*. Mol Gen Genet 241: 440–446 (1993).
23. Lawton MA, Yamamoto RT, Hanks SK, Lamb CJ: Molecular cloning of plant transcripts encoding protein kinase homologs. Proc Natl Acad Sci USA 86: 3140–3144 (1989).
24. Ma H: Protein phosphorylation in plants: enzymes, substrates and regulators. Trends Genet 9: 228–230 (1993).
25. Massague J: Receptors for the TGF-β family. Cell 69: 1067–1070 (1992).
26. Mizoguchi T, Gotoh Y, Nishida E, Yamaguchi-Shinozaki K, Hayashida N, Iwasaki T, Kamada H, Shinozaki K: Characterization of two cDNAs that encode MAP kinase homologues in *Arabidopsis thaliana* and analysis of the possible role of auxin in activating such kinase activities in cultured cells. Plant J 5: 111–122 (1994).
27. Moodie SA, Wolfman A: the 3Rs of life: Ras, Raf and growth regulation. Trends Genet 10: 44–48 (1994).
28. Nasrallah JB, Nasrallah ME: Pollen-stigma signalling in the sporophytic self-incompatibility response. Plant Cell 5: 1325–1335 (1993).
29. Nasrallah JB, Rundle SJ, Nasrallah ME: Genetic evidence for the requirement of the *Brassica* S-locus receptor kinase gene in the self-incompatibility response. Plant J 5: 373–384 (1994).
30. Neiman AM: Conservation and reiteration of a kinase cascade. Trends Genet 9: 390–394 (1993).
31. Pawson T, Gish GD: SH2 and SH3 domains: from structure to function. Cell 71: 359–362 (1992).
32. Pearce G, Strydom D, Johnson S, Ryan CA: A polypeptide from tomato leaves induces wound-inducible proteinase inhibitor proteins. Science 253: 895–898 (1991).
33. Schaller GE, Bleecker AB: Receptor-like kinase activity in membranes of *Arabidopsis thaliana*. FEBS Lett 333: 306–310 (1993).
34. Schlessinger J: How receptor tyrosine kinases activate ras. Trends Biochem Sci 18: 273–275 (1993).
35. Schneider R, Schweiger M: A novel modular mosaic of cell adhesion motifs in the extracellular domains of the neurogenic *trk* and *trkB* tyrosine kinase receptors. Oncogene 6: 1807–1811 (1991).
36. Siegelman MH, Cheng IC, Weissman IL, Wakeland EK: The mouse lymph node homing receptor is identical with the lymphocyte cell surface marker Ly22: role of the EGF domain in endothelial binding. Cell 61: 611–622 (1990).
37. Singer SJ: The structure and insertion of integral protein in membranes. Annu Rev Cell Biol 6: 247–296 (1990).
38. Smith RD, Walker JC: Isolation and expression of a maize type 1 protein phosphatase. Plant Physiol 97: 677–683 (1991).
39. Smith RS, Walker JC: Expression of multiple type 1

phosphoprotein phosphatases in *Arabidopsis thaliana*. Plant Mol Biol 21: 307–316 (1993).

40. Stein JC, Howlett B, Boyes DC, Nasrallah ME, Nasrallah JB: Molecular cloning of a putative receptor protein kinase gene encoded at the self-incompatibility locus of *Brassica oleracea*. Proc Natl Acad Sci USA 88: 8816–8820 (1991).
41. Stein JC, Nasrallah JB: A plant receptor-like gene, the S-locus receptor kinase of *Brassica oleracea* L., encodes a functional serine/threonine kinase. Plant Physiol 101: 1103–1106 (1993).
42. Taylor SS, Knighton DR, Zheng J, TenEyck LF, Sowadski JM: Structural framework for the protein kinase family. Annu Rev Cell Biol 8: 429–462 (1992).
43. Tobias CM, Howlett B, Nasrallah JB: An *Arabidopsis thaliana* gene with sequence similarity to the S-Locus receptor kinase of *Brassica oleracea*. Plant Physiol 99: 284–290 (1992).
44. Ullrich A, Schlessinger J: Signal transduction by receptors with tyrosine kinase activity. Cell 61: 203–212 (1990).
45. Valon C, Smalle J, Goodman HM, Giraudat J: Characterization of an *Arabidopsis thaliana* gene (TMKL1) encoding a putative transmembrane protein with an unusual kinase-like domain. Plant Mol Biol 23: 415–421 (1993).
46. Walker JC: Receptor-like protein kinase genes of *Arabidopsis thaliana*. Plant J 3: 451–456 (1993).
47. Walker JC, Zhang R: Relationship of a putative receptor protein kinase from maize to the S-locus glycoproteins of *Brassica*. Nature 345: 743–746 (1990).

Plant Molecular Biology **26**: 1611–1636, 1994.

GTP-binding proteins in plants: new members of an old family

Hong Ma
Cold Spring Harbor Laboratory, Cold Spring Harbor, NY 11724, USA

Received and accepted 13 June 1994

Key words: guanine nucleotide-binding proteins, heterotrimeric G proteins, small G proteins, biochemical detection, cDNAs, expression, yeast complementation, transgenic plants

Abstract

Regulatory guanine nucleotide-binding proteins (G proteins) have been studied extensively in animal and microbial organisms, and they are divided into the heterotrimeric and the small (monomeric) classes. Heterotrimeric G proteins are known to mediate signal responses in a variety of pathways in animals and simple eukaryotes, whiole small G proteins perform diverse functions including signal transduction, secretion, and regulation of cytoskeleton. In recent years, biochemical analyses have produced a large amount of information on the presence and possible functions of G proteins in plants. Further, molecular cloning has clearly demonstrated that plants have both heterotrimeric and small G proteins. Although the functions of the plant heterotrimeric G proteins are yet to be determined, expression analysis of an *Arabidopsis* Gα protein suggests that it may be involved in the regulation of cell division and differentiation. In contrast to the very few genes cloned thus far that encode heterotrimeric G proteins in plants, a large number of small G proteins have been identified by molecular cloning from various plants. In addition, several plant small G proteins have been shown to be functional homologues of their counterparts in animals and yeasts. Future studies using a number of approaches are likely to yield insights into the role plant G proteins play.

Introduction

GTP-binding regulatory proteins (for simplicity, referred to as G proteins hereafter) are members of a large family of guanine nucleotide binding proteins found in all eukaryotes. On the basis of subunit composition and size, G proteins have been classified as heterotrimeric or small (monomeric) G proteins. Heterotrimeric G proteins, consisting of α, β, and γ subunits, generally relay information from membrane receptors to intracellular effectors [16, 17, 59, 144, 155]. One of the best studied heterotrimeric G proteins is the G protein (G_s) that mediates the hormonal stimulation of adenylate cyclase, and is widespread in animal cells [59]. In this case, the binding of the signal epinephrine to its receptor triggers the activation of G_s, which then activates adenylate cyclase. The synthesis of cAMP in turn leads to a cascade of protein phosphorylations and dephosphorylations, which regulate the activity of a variety of proteins. Another well characterized example of heterotrimeric G protein is the transducins (G_t), which transmit visual signals in vertebrates [59]. The G_t-coupled receptor is the light-activated rhodopsin, and the G_t-activated effector is cGMP phosphodiesterase. Therefore, G_t mediates the light-activated hydrolysis of cGMP, which in turn regulates Na^+/Ca^{2+} channels in the photoreceptors, altering membrane potentials [160]. In animals, extensive biochemical and pharmacological studies have accumulated a large amount of information on many additional heterotrimeric G proteins and their functions [59, 144]. In recent years, molecular cloning has identified genes for previously known G proteins, as well as genes for new types of heterotrimeric G proteins in both mammals and other animals [87, 155]. Furthermore, a combination of genetic, biochemical and molecular analyses have uncovered G protein functions in simple eukaryotes [16, 17, 87, 155]. The sizes of the α subunits range from 35 to 45 kDa, those of the β and γ subunits are generally of 35–36 and 8–10 kDa, respectively. In addition, recent biochemical studies have uncovered novel proteins that are much larger (66–74 kDa) than known $G\alpha$'s, yet have some characteristics of $G\alpha$ subunits, such as GTP-binding, cross-reactivity with antisera against a conserved $G\alpha$ peptide, and interaction with G-protein-coupled membrane receptors [73, 79, 80, 125, 156]. It is not known, however, how structurally similar these new GTP-binding proteins are to known $G\alpha$ subunits, and whether they interact with either known or novel $G\beta\gamma$ subunits.

The small G proteins include a large number of molecules, from 20 to 30 kDa in size, and have very diverse functions [63, 64]. The discovery that the proto-oncogene *ras* encodes a small GTP-binding protein opened a new chapter in the history of G proteins. Although the precise biochemical function of mammalian *ras* is still not clear, increasing evidence suggests that it mediates signals received by membrane-receptor tyrosine kinase(s) and transmits them through other proteins to protein kinases, in particular the MAP kinase cascade [64, 109, 149, 164]. Genetic and molecular studies in invertebrate animals have also provided strong support for this signalling pathway, which is important for the regulation of specific cellular differentiations [145, 157]. Molecular, genetic and biochemical analyses of the Ras proteins in the yeast *Saccharomyces cerevisiae* have demonstrated an important role for Ras in the regulation of cellular growth; in particular, Ras regulate the activity of adenylate cyclase in yeast, and much (but not all) of the Ras function is mediated by cAMP in yeast [177, 178]. A variety of studies in recent years have identified a large number of different, but related, small G proteins functioning in a variety of processes, from signal transduction to controlling protein secretion, from regulating cytoskeletal functions to organizing membranes [63, 64].

Since the late 1980s, biochemical, physiological and molecular approaches have been successfully used to demonstrate the presence of G proteins in plants. For comparison, this article will summarize briefly G proteins in animals and simple eukaryotes, and focus the remainder of the discussion on the recent results on G proteins in plants. Other GTP-binding proteins, including tubulins and translation factors, will not be discussed here, although they perform important cellular functions. Other recent reviews on plant G proteins offer different emphases and perspectives [86, 94, 163].

G protein structures and functions in animals and simple eukaryotes

Heterotrimeric G proteins

Heterotrimeric G proteins were first identified in mammalian signal transduction pathways [87, 155]. In addition to the aforementioned G_s, there are three inhibitory G proteins ($G_{i(1-3)}$) which mediate the hormonal inhibition of adenylate cyclase activity. Furthermore, there are two transducins, G_{t1} and G_{t2}, which transmit visual signals to membrane potentials in rod and cone photoreceptor cells, respectively. The genes encoding the α subunits of these G proteins have been isolated [87]; the G_i's and G_t's are more similar to each other at the amino acid sequence level than they are to G_s (Fig. 1). Another $G\alpha$, G_o, was discovered which is most similar to G_i in sequence. Other mammalian $G\alpha$ genes have been isolated that are structurally and functionally re-

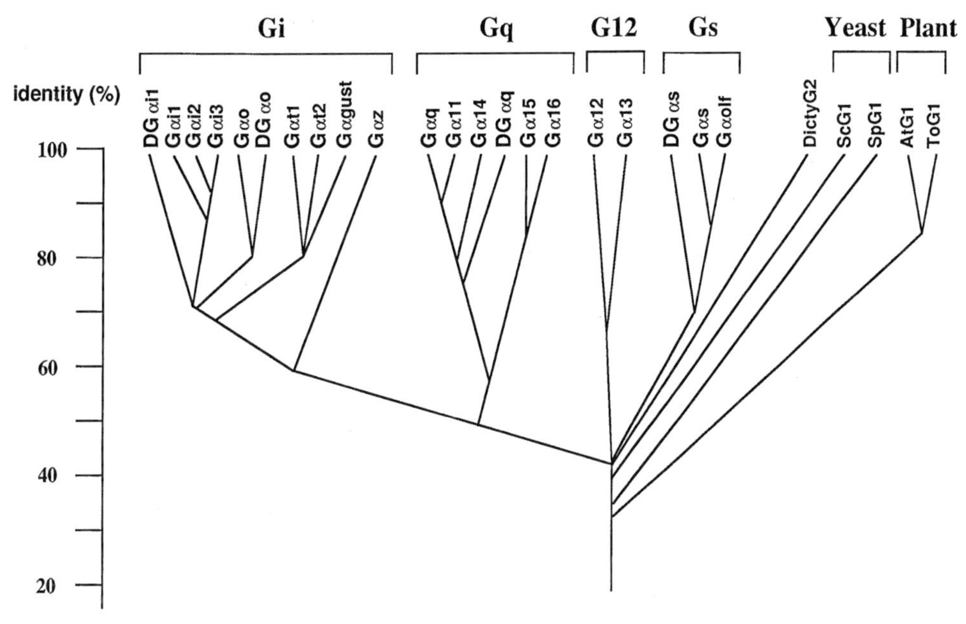

Fig. 1. Comparison of heterotrimeric G protein α subunits.The similarity tree is based on the percentage of amino acid sequence identity, and is modified from Fig. 2 of Simon *et al.* [155]; the additional amino acids in G_s, G_{olf}, and the yeast ScG1 were not included in the comparison, as described in Lochrie and Simon [103]. The DG's are from *Drosophila melanogaster*, DictyG2 is from *Dictyostelium discoideum*, ScG1 from *Saccharomyces cerevisiae*, SpG1 from *Schizosaccharomyces pombe*, AtG1 from *Arabidopsis thaliana*, ToG1 from tomato, and the other from mammals.

lated to G_s (G_{olf}, for olfactory response) or G_t (gustducin, for taste sensation) [111, 155]. Moreover, two new classes of Gα have been isolated recently: the G_q and the G_{12}/G_{13} classes [2, 158, 159, 179]. At least some of these new G proteins are involved in the regulation of phospholipase C [155]. Figure 1 shows the amino acid similarity between different Gα proteins. Genes encoding homologues of all major types of mammalian Gα proteins (Fig. 1) have also been isolated from *Drosophila* [36, 129, 132, 135–137, 158], and genes for the G_s and G_o types of Gα have been isolated from *Caenorhabditis elegans* [51, 102]. Among simple eukaryotes, the yeasts *S. cerevisiae* and *Schizosaccharomyces pombe* each have two Gα genes [40, 76, 115, 118, 119, 124]. *Dictyostelium* has as many as eight different α subunits, Gα1-Gα8 [52, 62, 92, 133, 182].

Heterotrimeric G protein α subunits have conserved consensus regions for GTP binding and GTPase activity [17, 87, 155]. A conserved arginine residue is found in all known Gα's, and it can serve as a site for ADP ribosylation by cholera toxin; however, only some Gα's are known to be modified by cholera toxin while others are not [155]. ADP ribosylation by cholera toxin blocks GTPase activity, resulting in an activated form of the α subunit (Fig. 2A). Some α subunits, including G_i's and G_t's, have a site (a cysteine residue near the C-terminus) for a similar modification by pertussis toxin. The ADP ribosylation by pertussis toxin interferes with the interaction between Gα and the receptor, and blocks the exchange of GDP for GTP, rendering the protein unable to be activated by the receptor (Fig. 2A, see below for the mechanism of G protein action).

Molecular cloning has identified several mammalian genes encoding at least 4 different β subunits and 5 γ subunits [74, 155]. Genes for β subunits have also been isolated from invertebrate animals squid [146], *Drosophila* [184, 185] and *C. elegans* [165]. In addition, the budding yeast *S. cerevisiae STE4* and *STE18* genes were found to encode β and γ subunits, respectively

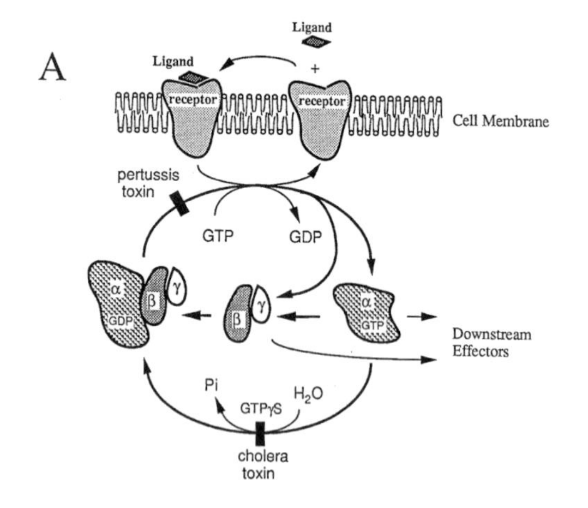

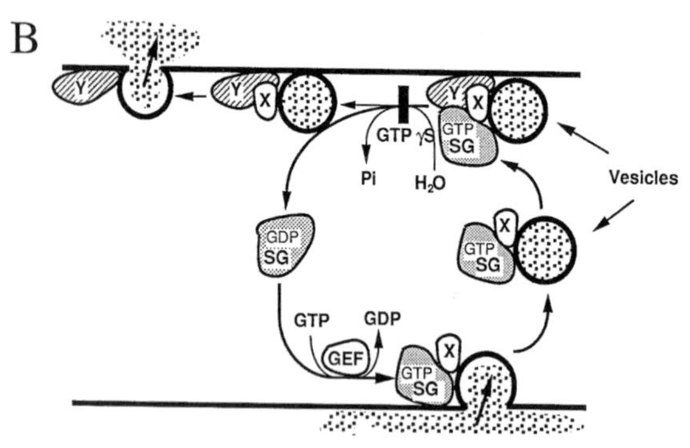

Fig. 2. Mechanisms of G protein actions. A. Heterotrimeric G proteins composed of the α, β and γ subunits. The α and/or $\beta\gamma$ subunits may interact with a variety of effectors. B. Small G proteins (SG) involved in secretion. Proteins X and Y are membrane-associated factors, and GEF is the guanine-nucleotide exchange factor, which interacts with the G protein. Solid bars indicate blockage by GTPγS or toxins.

[174], and *Dictyostelium* has one Gβ subunit involved in development [101]. The $\beta\gamma$ subunits have long been recognized to interact with the α subunit, thereby regulating its activity. In recent years, increasing evidence indicates that the $\beta\gamma$ subunits may also directly regulate effectors, and play a role in receptor interaction [74]. For example, studies indicate that G protein action in the pheromone response pathway in yeast involves the interaction of $\beta\gamma$ subunits with the effector [40, 174, 175]. In mammalian cells, the $\beta\gamma$ subunits have been shown to activate phospholipase A_2 in photoreceptors [78], K^+ channels in heart muscle [104], and the phospholipase C isoform β_2 from human granulocytes [22, 74, 84]. In addition, the $\beta\gamma$ subunits can activate type II adenylate cyclase in the presence of activated G_s, while they prevent the activated G_s from activating type I adenylate cyclase [98]. Not all of the β and γ subunits are functionally similar; in fact, different β and γ subunits act to distinguish different receptors under certain conditions [89, 90].

Receptors and effectors of the heterotrimeric G proteins

Heterotrimeric G proteins are coupled to receptors of the seven-transmembrane-segment class. A large number of receptors in this class have been identified using pharmacological, biochemical and molecular techniques [30, 42, 155]; they include the mammalian hormone and neurotransmitter receptors (adrenergic, serotoninergic, muscarinic, dopamine receptors, and so on), rhodopsin and the color vision opsins, a large family of odorant receptors, yeast pheromone receptors, and the *Dictyostelium* cAMP receptor. Since there are many more known receptors than known G proteins, different receptors likely interact with the same G protein. For example, there are three different color rhodopsins and only one color transducin, the signals received by all three rhodopsins are thought to be mediated by the same transducins. Furthermore, in the yeast mating response, both the a-factor and the α-factor receptors activate the same heterotrimeric G protein [70].

There are also numerous downstream effectors regulated by heterotrimeric G proteins [30, 155]. The first known G protein effectors are adenylate cyclases, which are regulated by G_s, and G_i. In addition, in olfactory epithelial cells, a specific isoform of adenylate cyclase [6, 130] acts downstream of G_{olf}, which is very similar to G_s [81]. As mentioned earlier, in the visual response, the effector of transducins is cGMP phosphodiesterase. Another class of effectors includes the isoforms of phospholipase C, which may be regulated by G_i, G_o and the G_q class of G proteins [155]. There has been evidence for regulation of phospholipase A_2 by transducins [78]. Finally, several types of K^+ channels are regulated by G_i's and G_o, some Ca^{2+} channels are regulated by G_s and G_o, and Na^+ channels are regulated by G_{i3} and G_s [18, 30]. In *Dictyostelium*, a guanylate cyclase has been shown to be regulated by $G\alpha 2$ [52]. Note that the same G protein may regulate different effectors: G_s can both activate adenylate cyclase and open Ca^{2+} channels, G_i can inhibit adenylate cyclase while opening K^+ channels, and *Dictyostelium* $G\alpha 2$ can activate both a guanylate cyclase and a phospholipase C [30, 52]. As many G protein-mediated responses are yet to be characterized at the molecular level, other effectors and G protein-effector interactions are certain to be uncovered.

Small G proteins

On the basis of amino acid sequence similarity, small G proteins have been grouped into several subfamilies [43, 63, 64]. All of the small G proteins have consensus regions for GTP-binding, which are related in varying degrees to those found in the α subunits of heterotrimeric G proteins [17]. Members of each subfamily share characteristic additional residues in the GTP binding consensus regions in addition to the residues that are conserved in all G proteins [17]. The most extensively studied small G proteins are in the ras subfamily which includes the products of several *ras* proto-oncogenes, and homologues in invertebrate animals and yeasts. The activity of the ras protein is directly regulated by guanine nucleotide exchange factors (GEFs), which stimulate the exchange of bound GDP for GTP, thus activating ras [11]. In addition, GTPase activating proteins (GAPs) stimulate the intrinsic GTPase activity of ras, leading to faster hydrolysis of the bound GTP and reducing ras activity [11]. Increasing evidence indicates that membrane receptor tyrosine kinases mediate the regulation of ras activity by extracellular signals, through the 'adaptor proteins' (e.g., the mammalian Grb2 protein) which bind to the activated, autophosphorylated receptor and to GEFs [149]. In addition, genetic, molecular and biochemical studies have uncovered a conserved kinase cascade (MAP kinase cascade) which functions downstream of ras [122]. Ras homologues in *Drosophila* and *C. elegans* are involved in signal transduction during development [145, 157]. In yeast, Ras proteins regulate cell growth by controlling the activity of adenylate cyclase and thus the level of cAMP [63, 178].

A second group of small G proteins consists of several mammalian rab proteins, the yeast YPT1 and SEC4 proteins, and related proteins; these are involved in vesicular transport in the secretory pathway [63, 64, 152]. Extensive biochemical and genetic studies indicate that the rab/ypt subfamily of small G proteins associate with membrane vesicles (probably interacting with membrane proteins) and shuttle between donor and acceptor membrane structures. It is believed that the binding of GDP or GTP stabilizes alternate conformations of these small G proteins, allowing them to associate with protein(s) on one structure or another. Furthermore, the bound GTP is hydrolyzed only when stimulated by the appropriate GAP protein. Therefore, GTP binding and hydrolysis drive a unidirectional transport of vesicle content (see below and Fig. 2B). The members of the Ypt/Rab subfamily are regulated by their cognate GEFs, GAPs, and a third type of regulator: the guanine-nucleotide dissociation inhibitors (GDIs), which inhibit GDP dissociation from rab [11].

In addition to the ras and rab/ypt subfamilies, another group includes the mammalian rho and rac, and the yeast CDC42 proteins, which function in cell polarity and cytoskeletal function [63, 64]. There are three other types of small G proteins [64]. One is represented by the mammalian and yeast ARFs, which are known to facilitate ADP ribosylation of α subunits by cholera toxin *in vitro*. More recently, ARF has been implicated in secretion, where its proposed function is to control the assembly of secretory vesicle coat proteins in a GTP-dependent manner [64]. In the yeast *S. cerevisiae*, another small G protein, CIN4, was found to be involved in chromosome segregation [14]. Although the sizes of the ARF and

CIN4 proteins lead to their classification as small G proteins, they resemble the larger heterotrimeric α subunits in two ways [14]. Firstly, the GTP-binding consensus regions of ARF and CIN4 are more similar to those of Gα's than to those of ras and rho. The second feature concerns the position and nature of lipid modification; ARF and CIN4 both contain a glycine as the second residue from the N terminus, where all known α subunits have a glycine. The glycine residues in ARF and some Gα's are known to be myristoylated [20, 82]. In contrast, most small G proteins such as ras do not have this glycine residue, but are isoprenylated (farnesylated for ras) at a cysteine residue very close to the C terminus [16]. Some small G proteins are also palmitoylated at a cysteine residue in the C-terminal half of the protein [21]. Other small G proteins include the mammalian Ran protein and its homologues and the *S. cerevisiae* Sar1 protein [17, 140]. The Ran protein is involved protein targeting to the nucleus and affects mitosis [116, 140]. The Ran protein lacks both the N-terminal glycine and the C-terminal cysteine residues; therefore, its post-translational modification, if any, is different from both Gα/ARF and ras. The Sar1 protein functions in vesicular transport [17].

Mechanisms of G protein activation

Much has been learned about the mechanisms of some mammalian heterotrimeric G proteins, particularly G_s and transducins [9, 16, 59, 155], which have served as models for other G proteins. Briefly, heterotrimeric G proteins function through a cycle of reactions and protein-protein interactions (Fig. 2A). At the resting (inactive) state, the α subunit binds GDP and associates with the βγ subunits to form a complex. Upon the binding of a ligand, a cell surface receptor is activated, and it catalyzes the exchange of the bound GDP on Gα for a GTP, which causes a conformational change of the α subunit, resulting in its dissociation from the βγ complex. The active GTP-bound α subunit then regulates its effector(s), such as adenylate cyclases or K^+ channels, leading to a cascade of downstream events. Often the GTP-bound α subunit is the known species which activates downstream events, while the βγ complex acts as an inhibitor of the α subunit. However, as mentioned before, several studies clearly demonstrate that the βγ subunits can also directly interact with downstream effectors [74, 98]. The hydrolysis of GTP to GDP and phosphate by the intrinsic GTPase of the α subunit returns the α subunit to its inactive conformation and the GDP-bound α subunit re-associates with the βγ subunits. Non-hydrolyzable GTP analogues, such as GTPγS, and mutations reducing the GTPase activity of the α subunit prolong the active state of heterotrimeric G proteins. In addition, pertussis toxin uncouples the receptor from its G protein and thus blocks signal transduction, and cholera toxin blocks the GTPase activity of the α subunit and fixes it in an activated form [87, 155].

Genetic and biochemical evidence suggests that the mechanisms of the ras proteins and close homologues are similar to those of the heterotrimeric G proteins [16, 17, 63, 178]. Mutations which lead to oncogenic and constitutively active forms of ras are found to alter two kinds of residues: (1) those important for GTP binding, and (2) those required for GTPase activity. The first class of mutant ras proteins are enhanced for GTP binding, and the second kinds of mutant ras proteins are defective in GTP hydrolysis [43, 63]. In other words, increased GTP binding by ras leads to a more active ras; therefore, the GTP-bound form is active, while the GDP-bound form of ras is inactive. However, the GDP/GTP exchange and GTP hydrolysis, that is the interconversion between the active and inactive states, is controlled differently for ras proteins than for heterotrimeric G proteins. Instead of activated transmembrane receptors as in the case of heterotrimeric G proteins, the GDP/GTP exchange of ras requires a different type of protein factor, known as the guanine nucleotide exchange factors, which are regulated by receptor tyrosine kinases. In addition, ras proteins have a very low intrinsic GTPase activity, which can be substantially increased by GAPs (GTPase-activating proteins). The larger

$G\alpha$ proteins, on the other hand, contain an insertion of more than 100 amino acid residues between the first and second consensus regions for GTP binding. Since $G\alpha$'s have a higher intrinsic GTPase activity than small G proteins, it has long been proposed that this extra region has a GTPase activating activity [17]. This has recently been demonstrated experimentally by using two truncated proteins: one contains just the 'insertion' region and can activate the GTPase activity of the other which contains the remainder of the protein [108]. Furthermore, the crystal structure of the transducin α subunit indicates that the GAP-like region folds into a domain separate from the GTPase domain of the protein [123].

In contrast to heterotrimeric G proteins and ras, another mechanism is likely to operate in the action of small G proteins involved in secretion, such as the yeast Sec4 and Ypt1, and mammalian rab proteins [15, 63, 166]. For these proteins, the GTP-bound form is required for a portion of a cyclical traffic of cellular membranes, and one GTP is required for one cycle (Fig. 2B). GTP hydrolysis is necessary for the cycle to be completed. Models have been proposed which postulate that the GTP-bound form of the small G protein associates with some factor(s) during part of the cycle, while the GDP-bound form associates with others. Since the normal function of these small G proteins requires the continuous cycling of GDP/GTP exchange and GTP hydrolysis, the binding of non-hydrolyzable GTP analogues, and mutations reducing GTPase activity (either in the G proteins or in the GAPs), inhibit the function of these small G proteins.

Biochemical studies of plant G proteins and their involvement in plant signalling

Detection of GTP-binding proteins

Plant cells respond to a variety of signals both from the environment (light, humidity, temperature, gravity and pathogens) and from other cells (hormones, nutrients). However, little is known about the mechanisms of signal transduction in plants. Many have thought that due to the conserved nature of G proteins, they may play important roles in plant signal transduction pathways, as they do in animals and simple eukaryotes. There are several ways to detect G proteins *in vitro*. One of the widely used methods to detect G proteins is a GTP-binding assay, usually using one of the non-hydrolyzable GTP analogues, such as GTPγS. In general, GTP-binding assays detect both heterotrimeric and small G proteins, and other GTP-binding proteins; therefore, binding activities present in plant extracts can be quite complex [46]. Consequently, they are usually used for preliminary studies, and additional, more specific assays are often needed to demonstrate that G proteins are present. Nevertheless, in a relatively well defined system, for example a partially purified membrane fraction, GTP-binding assays can still be very informative when one performs proper controls such as binding using ATP and treatment with specific signals. Furthermore, a filter assay for GTP binding by renatured proteins has been very successfully used to identify small G proteins. Since both the α subunit of heterotrimeric G proteins and small G proteins have intrinsic GTPase activities, the GTPase assay is another way these proteins may be detected. Indeed, assays for GTPase activity have been used to provide evidence for the presence of G proteins in plant cells (see below and Table 1). This method is also limited by its lack of specificity for particular types of G proteins.

In addition to GTP-binding and GTPase activity assays, a method for preferentially detecting heterotrimeric G proteins is the use of bacterial toxins that covalently link an ADP-ribose moiety to a particular amino acid residue (ADP ribosylation) of some α subunits of heterotrimeric G proteins. The most frequently used toxins are pertussis and cholera toxins [58]. The susceptible $G\alpha$ subunits may be labeled using these toxins and radioactive NAD^+, which serves as the donor of the ADP-ribose group. G protein α subunits sensitive to pertussis toxin, including G_i's and transducins, have a cysteine near the carboxy terminus (usually at the fourth position from the end), thus ADP ribosylation by pertussis toxin is

Table 1. Biochemical detection of GTP-binding proteins.

Plant	Tissue	Sizes[a]	Assays[b]	References
Arabidopsis thaliana	leaf, root	36, 31	anti-Gα, GTPγS (s)	[10]
Arabidopsis thaliana	leaf	33	anti-Gi	[173]
Avena sativa (oat)	etiolated seedling	24	GTP (f), anti-Gα, CTX	[143]
Chlamydomonas reinhardtii	eyespot	24	GTP (s), GTPase, anti-Gα	[91]
Cucurbita pepo (zucchini)	etiolated hypocotyl	33, 50	GTPγS (s), anti-Gs	[77]
Cucurbita pepo (zucchini)				[45]
Commelina communis	leaf, root	38, 34	anti-Gα	[10]
Dunaliella salina		28	GTP (f), anti-YPT1	[141]
Dunaliella saline		29, 30	GTP (f)	[141]
Glycine max (soybean)	cultured cells	45	anti-Gα, CTX	[99]
Hordeum vulgare (barley)	aleurone	32, 36	anti-Gα	[168]
Hordeum vulgare (barley)	aleurone	22, 24	GTP (f), anti-ras	[168]
Lemna paucicostata		?	GTP (s)	[65]
Oryza sativa (rice)	coleoptile	28, 30	GTP (f), anti-Gα	[189]
Pisum sativum (pea)	plumules	21	GTP (f), anti-ARF	[112]
Pisum sativum (pea)	plumule nucleus	27, 28, 30	GTP (f)	[29]
Pisum sativum (pea)	leaf chloroplast	24	GTP (f)	[148]
Pisum sativum (pea)	etiolated seedlings	25, 37	anti-Gi	[173]
Pisum sativum (pea)	etiolated seedlings	43	anti-GPA1	[173]
Pisum sativum (pea)	etiolated seedlings	40	anti-Gi/Go, PTX	[169]
Pisum sativum (pea)	etiolated epicotyl	?	GTP (s)	[67]
Spinacea oleracea (spinach)	leaf	?	GTPγS (s)	[114]
Vicia faba (broad bean)	leaf, root	37, 31	anti-Gα	[10]
Zea mays (maize)	root	27, 34	GTPγS (s), purification	[8]

[a] The sizes in kDa were estimated from protein gels; therefore, the sizes of proteins were not known when the GTP-binding activities detected in solution assays.

[b] GTP or GTPγS indicate the binding of labeled nucleotide on filter (f) or in solution (s); western experiments were indicated by the antiserum used: Gα, a conserved Gα peptide; GPA1, a peptide specific for GPA1 from *Arabidopsis*; YPT1, from *S. cerevisiae*; Gi, Go, Gs, and ARF are from mammals. CTX and PTX indicate labeling with cholera and pertussis toxins, respectively.

a sensitive assay for a subset of G protein α subunits. ADP ribosylation by cholera toxin is considerably more complex [58]. Although the arginine residue that is ADP-ribosylated is conserved among all known G protein α subunits, only $G\alpha_s$ has been well documented as being efficiently modified by cholera toxin, while others may be modified under some circumstances [58]. Furthermore, ADP ribosylation by cholera toxin is greatly stimulated by a soluble factor, called the ADP ribosylation factor (ARF), which is itself a small G protein, and is activated by the binding of GTP or GTP analogues [58]. Proteins other than G proteins can also serve as substrates for ADP ribosylation by cholera toxin, but they are ribosylated considerably more slowly than G_s and dominate the labeling pattern only if they are very abundant [58]. The effect of ARF and the possibility that proteins other than G proteins can be modified by cholera toxin make its use less desirable and more prone to artifacts. It must be emphasized that the toxins are only useful to identify the presence of certain classes of G proteins in an extract. Therefore, there might be G proteins which are not substrates for the toxin-catalyzed ADP ribosylations.

Immunoblot procedures using antisera against peptides from known G proteins have also been very powerful. A number of groups have used antibodies raised against a conserved peptide of the α subunit of heterotrimeric G proteins (see below). These analyses in combination with GTP-binding studies provide strong evidence for the presence of heterotrimeric G proteins in plants. The limitation of the antibody studies is that it requires cross-reactivity between the antibodies and the plant proteins. Since there are known animal α subunits which contain amino acid divergence in the conserved region used to raise the antibodies, it would not be surprising if some plant $G\alpha$ proteins also contain amino acid changes in the same region. Another problem is that not all cross-reacting proteins are G proteins. Nevertheless, with other independent assays, analyses using antibodies have produced very valuable information on potential plant G proteins.

Using one or more of these *in vitro* biochemical techniques, GTP-binding activities and proteins have been detected in a variety of plant species (Table 1). By using ^{35}S-labeled GTPγS binding in solution, GTP-binding activities were detected in the thylakoid membranes of spinach (*Spinacea oleracea*) leaves [114], and in membranes of rice coleoptile [190]. GTP-binding and GTPase activities were found in the eyespot of the green alga *Chlamydomonas reinhardtii* [91], and in membrane extracts from maize roots [8]. GTP-binding activities and substrates for ADP ribosylation catalyzed by pertussis toxin were detected in gel filtration fractions of extracts from pea (*Pisum sativum*) and *Lemna paucicostata* [65–67]. In addition, a filter assay for GTP binding has been used to uncover small G proteins in the green alga *Dunaliella salina* [113, 141], in microsomes of zucchini (*Cucurbita pepo*) hypocotyl [45], and in the chloroplast outer envelope membrane [148] and nuclear envelope from pea [29]. Furthermore, an ARF-like protein was detected in the cytosol of pea plumule cells [112].

The combination of GTP-binding studies and immunological analysis with antibodies raised against known G protein α subunits have detected potential heterotrimeric G protein subunits from *Arabidopsis*, broad bean (*Vicia faba*), *Commelina communis* [10], zucchini (*Cucurbita pepo*) [77], and barley (*Hordeum vulgare*) [168]. By using a GTP-binding assay, western blot analysis using anti-$G\alpha$ antibodies, and ADP ribosylation with cholera toxin, Romero *et al.* identified a 24 kDa GTP-binding protein in oat (*Avena sativa*) etiolated seedlings [143], although this protein is smaller than any known heterotrimeric G protein α subunits, and is within the size range of known small G proteins. Similarly, using anti-$G\alpha$ antibodies and a filter GTP-binding assay, two small GTP-binding proteins were identified in rice coleoptile membranes (28 and 30 kDa) [189], and in barley aleurone protoplasts (22 and 24 kDa, also recognized by anti-ras antibodies) [168]. It is possible that the anti-$G\alpha$ antibodies cross-react with small G proteins, and the conditions used for ADP ribosylation by cholera toxin allow the modification of small G proteins. The definitive identification and characterization of these proteins must await further molecular studies.

G protein involvement in plant signalling pathways

Studies of known G proteins indicate that the interaction of G proteins with receptors, effectors, and GTP/GDP occur at particular points of a cycle (Fig. 2); therefore, alteration at one point in the cycle affects the subsequent point(s). This property of G proteins has been exploited to learn possible G protein functions in individual signalling pathways. GTP analogues and the cholera and pertussis bacterial toxins described previously are useful tools to probe the involvement of G proteins in various cellular processes. GTP analogues, particularly GTPγS, are used frequently due to the relative ease with which they can probe G protein functions. However, GTP analogues affect both heterotrimeric and small G proteins; therefore, conclusions from studies using GTP analogues are not definitive. The bacterial toxins, particularly pertussis toxin, are more specific, and they are used for analyses of heterotrimeric G proteins. G proteins that are involved in mediating extracellular signals, such as the heterotrimeric G proteins, are usually activated by a receptor-ligand complex. The activation usually involves the exchange of a bound GDP for a GTP. Therefore, the presence of a G protein in a signalling pathway can often be detected as a stimulation of GTP binding by the signal. If at least a portion of the signalling pathway can be reconstituted *in vitro*, then the effect of the signal on GTP binding can be characterized using the GTP-binding assays described above. In addition, heterotrimeric G proteins are usually activated by binding to GTP. This activation is attenuated by the hydrolysis of GTP to GDP and phosphate due to the intrinsic GTPase activity of the α subunit. For small G proteins, the intrinsic GTPase activity is greatly stimulated by the GTPase activating protein (GAP). Because GTP analogues such as GTPγS and GMP-PNP are not hydrolyzable, they are more potent activators of G proteins. If a signal is known or suspected for a particular cellular process, such as response to light, then guanine nucleotides may be used to mimic the signal in generating the response.

The use of bacterial toxins can also probe G protein function in cellular processes. Cholera toxin can ADP-ribosylate both GTP- and GDP-bound forms [58], and cholera toxin-catalyzed ADP ribosylation inhibits intrinsic GTPase activity, prolonging the activated state of the α subunit. In contrast, pertussis toxin-catalyzed ADP ribosylation only occurs on the GDP-bound heterotrimeric form, and the modification uncouples the G protein from the receptor [59]. Therefore, pertussis toxin keeps the G protein in the inactive state. If a signal activates a G protein, then the ADP ribosylation catalyzed by pertussis toxin should be reduced. Further, a positive effect of cholera toxin on some cellular response would suggest that an activated G protein is involved in promoting the response, while a positive effect of pertussis toxin suggests that an activated G protein can inhibit the response.

A number of studies have implicated GTP-binding proteins in light-stimulated signalling pathways, using both GTP analogues and bacterial toxins. In *Lemna*, a single 8 h period of darkness induces flowering. It was found that, when the extracts were prepared from *Lemna* plants that had been in darkness for 8 h, GTP binding was inhibited by about 20% by red or far-red light, as compared to the binding in the dark, but not affected by blue light [66]. This suggests that the red/far-red receptor phytochromes may be involved in the regulation of one or more G proteins. However, it is puzzling that in this case similar inhibition was seen with red light and far-red light, which have opposite effects on phytochromes [134]. In more recent studies, it was found that GTP binding by pea nuclear membranes was stimulated by a 2 min exposure of red light, and such stimulation was eliminated if the red-light exposure was followed by a 4 min exposure of far-red light [29]. Similarly, far-red light-reversible, red light-stimulated GTP binding was observed with membranes from etiolated oat seedlings [143]. These results strongly suggest that one or more GTP-binding proteins are activated by red light via phytochrome, since the far-red reversibility is a characteristic of phytochrome signalling. In other studies, it was found that the GTP analogues GTPγS (30–100 μM intracellular) and Gpp(NH)p (50–100 μM intracellular) mimicked the effects of the light receptor phytochrome A on light-dependent synthesis of anthocyanin and the expression of a reporter gene (*GUS*) under the control of a light regulated *cab* gene promoter, suggesting that a G protein may be involved in phytochrome signal transduction [121]. In this case, cholera toxin alone had only a small effect on the light responses; however, cholera toxin in combination with a low concentration (1 μM) of GTPγS, which has no effect by itself, produced an effect similar to that of phytochrome A or 30–100 μM GTPγS [121]. These results are consistent with a role for a cholera toxin-sensitive G protein that is put into a prolonged activated state by ADP ribosylation. In addition to red light, blue light has also been observed to stimulate GTP binding in etiolated oat seedlings [143]. Moreover, blue light stimulates a GTPase activity, as well as a GTP-binding activity, in plasma membranes of etiolated pea seedlings [169]. A 40 kDa protein in these membranes was ADP-ribosylated by pertussis toxin in the absence but not presence of GTP and blue light; in addition, a protein of the same size (presumably the same protein) cross-reacted with antisera which detect transducin or G_i/G_o α subunits [169]. These results suggest that a heterotrimeric G protein in etiolated pea seedlings may be involved in blue light signal transduction.

Although the results discussed above certainly suggest that light signals can be mediated by heterotrimeric G proteins in plants, it is not known how light activates G proteins. The red light receptor phytochromes appear to be soluble proteins from their sequences [154]; furthermore, recent isolation of an *Arabidopsis* blue-light response gene (*HY4*) suggests that a blue light receptor is also a soluble protein that is similar to photolyases [1]. It is possible that these non-transmembrane photoreceptors interact with other proteins, possibly membrane-associated ones, which in turn interact with G proteins. Alternatively, the plant photoreceptors may directly interact with G proteins; this would represent the direct contact of G proteins with entirely new types of receptors. In either case, light signalling in plants is likely to provide new insights into G protein functions.

Plant cells also respond to a number of plant hormones. In one study, the auxin, indole-3-acetic acid (IAA), was observed to enhance GTPγS binding in rice coleoptile; further, GTPγS caused a reduction in auxin binding [190]. These findings suggest that auxin stimulates the exchange of the GDP bound on a G protein for a GTP. The effect of GTPγS may be explained in two ways: first, it is possible that the activated G protein due to the binding of GTPγS desensitizes the auxin receptor; alternatively, the auxin receptor may require the association with a GDP-bound G protein to interact with the ligand, auxin.

Biochemical studies have also suggested the involvement of GTP-binding proteins in the regulation of downstream events. It has been found that GTPγS affects K^+ currents in the guard cells of broad bean (*Vicia faba*) leaves [49]. In these guard cells, GTPγS was found to reduce an inward K^+ current, while GDPβS enhanced the current [49]. Since GTPγS activates while GDPβS inhibits G proteins, these results suggest that one or more G proteins negatively regulate K^+ inward currents. In addition, it was found that cholera toxin inhibits the inward K^+ current in guard cells, further supporting the idea that a G protein negatively regulates the K^+ currents [49]. Pertussis toxin also inhibits K^+ current in guard cells [49]. In general, pertussis toxin blocks the activation of G proteins; therefore, the inhibition by pertussis toxin on K^+ currents in guard cells suggests that a second G protein acts in these cells to positively regulate the K^+ current. Since reduced K^+ uptake inhibits stomatal opening, these results suggest that GTPγS and the toxins would inhibit stomatal opening in broad bean guard cells under these conditions. In contrast, both GTPγS and pertussis toxin induce stomatal opening in the epidermis of *Commelina communis* [97]. Again, similar effects of GTPγS and pertussis toxin suggest more than one protein may be involved. It is not clear why opposite effects were seen in these two systems; it is possible that the different conditions and techniques favor one G protein over another, since more than one G protein seems to be involved in both cases. However, GDPβS had no effect on stomatal opening in *Commelina communis* [97], indicating that the situation in this case is rather complex, and that additional studies are need before the involvement of G protein(s) can be ascertained. G proteins have also been implicated in the regulation of an outward K^+ current in broad bean mesophyll cells; GTPγS and cholera toxin, but not pertussis toxin, inhibit the outward K^+ current in these cells [100]. In addition, GTP and GTP analogues have been shown to affect swelling of wheat protoplasts [13] and the formation of inositol phosphate derivatives in *Acer pseudoplatanus* [41], indicating the possible involvement of GTP-binding proteins. In cultured French bean cells, both cholera and pertussis toxins enhance the response to a fungal elicitor by the cells [12], suggesting possible G protein participation in this signalling pathway. Recently, a 45 kDa protein in cultured soybean cells that is recognized by an anti-Gα antiserum and labeled by ADP ribosylation with cholera toxin has been suggested to be involved in the elicitation of the defense responses [99]. Finally, in cultured soybean cells, both cholera and pertussis toxins were found to stimulate the expression of a *cab* gene, which normally depends on phytochrome for expression, and rendered the *cab* gene expression light-independent [142].

In summary, biochemical and physiological experiments have produced an impressive amount of evidence for the involvement of G proteins in plant signalling processes. Even though more studies are needed before one can learn the nature of these G proteins, it is very encouraging that a variety of pathways seem to employ G proteins.

Molecular analyses of G proteins in plants

Isolation of genes encoding heterotrimeric G protein subunits

Although biochemical studies have produced much evidence for the existence of G proteins in plants and their involvement in plant signalling pathways, none of these proteins have been identified or purified. Therefore, little is known about these proteins at the molecular level. However, using molecular approaches, a number of cDNAs have been isolated that encode putative heterotrimeric G protein subunits (Table 2). Using PCR with degenerate oligonucleotides based on conserved peptides among known G protein α subunits, a gene (*GPA1*) was isolated from *Arabidopsis thaliana* which encodes a protein with 36% identity to mammalian G_i and transducins, and contains GTP-binding consensus regions for heterotrimeric G protein α subunits [107]. The isolation of *GPA1* provided a clear demonstration that heterotrimeric G protein(s) are present in plants. Subsequently, a homologue (*TGA1*) of *GPA1* was isolated from tomato using low-stringency hybridization with *GPA1* as the probe [106]; the two predicted proteins are 84% identical. Both *GPA1* and *TGA1* were shown to be single-copy genes by Southern analyses. Furthermore, PCR and low-stringency procedures have failed to identify additional genes encoding G protein α subunits (H. Huang and H. Ma, unpublished). More recently, a single homologue of *GPA1* has been isolated from each of soybean (L. Romero and E. Lam, pers. comm.), lotus (C. Poulsen, pers. comm.) and maize (C.D. Han and R. Martienssen, pers. comm.). These genes all encode proteins that are very similar (greater than 75% amino acid sequence identity) to GPα1, the product of *GPA1*. In addition, a 43 kDa protein was detected in pea membranes using an antiserum raised against a C-terminal peptide of GPα1 [173]. These results indicate that *GPA1*

Table 2. Plant heterotrimeric G proteins and related proteins identified by molecular cloning.

Plant[a]	Predicted proteins	Protein size (kDa)	Most similar[b] animal protein (%)	Expression and/or function	References
Heterotrimeric G proteins					
α subunits					
Arabidopsis thaliana	GPα1	44.6	Gt1 (36)	Expressed in all major organs	[107]
Lycopersicon esculentum (tomato)	TGα1	44.9	Gt1 (34)		[106]
β subunits					
Arabidopsis thaliana	AGβ1	41.0	β2 (44)	Expressed in many organs	[171]
Zea mays	ZGβ1	41.7	β2 (42)	Expressed in many organs	[171]
WD-40 proteins					
Arabidopsis thaliana	COP1	111.8	β3 (29)[c]	Light signal transduction	[38]
Chlamydomonas reinhardtii	Cblp	35.1	MHC12.3 (66)	Constitutively expressed	[150]
Nicotiana tabacum (tobacco)	arcA	35.8	MHC12.3 (67)	Auxin-regulated	[75]

[a] Homologues of GPα1 have been isolated from *Glycine max* (soybean; L. Romero and E. Lam, pers. comm.); *Lotus japonicus* (C. Paulsen, pers. comm.); and *Zea mays* (C.D. Han and R. Martienssen, pers. comm.).
[b] The Gt1 (rod transducin), β2 and β3 proteins are from man, and the MHC12.3 protein is from chicken.
[c] The percent identity is only for the WD-40 domain of COP1.

(and its homologues) is a conserved gene found in all flowering plants. Although the α subunits identified from plants are probably homologues of each other, their low levels of sequence identity to those from animals and simple eukaryotes make it unlikely that the plant Gα's are functional homologues of any of the non-plant ones. The apparent uniqueness of *GPA1* (and homologues) suggests that it has a non-redundant function in plants, and the fact that it is highly conserved in many plants suggest that its function is important. Furthermore, the uniqueness of *GPA1* and its homologues suggests that if there are other Gα genes in plants, they must be quite different from *GPA1*, such that they can not be detected through hybridization or PCR.

Do plants also have G protein β and γ subunits? One is tempted to say yes, since all other organisms that have α subunits also have β and γ subunits. All known Gβ's from animals and simple eukaryotes contain 7 repeats of a motif called WD-40, which is characterized by the dipeptide tryptophan-aspartate and is about 40 amino acids long [54]. The WD-40 motif is also found in a variety of proteins with diverse functions, including regulation of cell cycle [54], RNA splicing [33], cytoskeletal function [35], and transcriptional repression [180]. A chicken protein, MHC12.3, of unknown function also contains several WD-40 motifs [60]. Although there is no evidence that any of these non-Gβ WD-40 proteins interacts with Gα or Gγ, it is known that the yeast WD-40 protein TUP1 interacts with another transcriptional repressor SSN6 (also called CYC8) [181], which contains a different type of repeats [151]. Three genes have been isolated from plants which encode WD-40-containing proteins. Two of the predicted proteins (Cblp and arcA) are very similar to each other (68% identity), and to the chicken MHC12.3 protein (>65%), but have only about 25% of sequence identity to known β subunits [75, 150]. Therefore, these plant WD-40-containing proteins may have any of a number of functions that are not related to G proteins. The third plant WD-40 protein (COP1) has a large non-WD-40 N-terminal domain, including two zinc fingers, in addition to the C-terminal WD-40 domain [37]. These features and the phenotypes of *cop1* mutants have led to the hypothesis that COP1 may be a transcriptional repressor [37], as is the yeast TUP1 protein. Recently, cDNAs encoding proteins with a much higher degree of similarity (42% or more) to animal β subunits have been isolated from maize (*ZGB1*) and *Arabidopsis* (*AGB1*) [171]. This indicates that plants have at least one pair of α and β subunits. It is most likely that G protein γ subunit(s) is(are) also present in plants. Interestingly, like *GPA1*, both *ZGB1* and *AGB1* appear to be single-copy, suggesting that, if other Gβ genes exist in these plants, they also must have very different sequences.

A detailed analysis of the Arabidopsis GPA1 *expression pattern*

In order to gain more information on the function of GPα1, detailed analyses of its spatial and temporal expression were carried out using a fusion between *GPA1* and the reporter gene *uidA* (encoding a β-glucuronidase), and using immunolocalization studies with specific antibodies directed against a peptide from the C terminal region of GPα1 [72, 172]. The results show that the *GPA1* gene product is expressed in nearly all tissues examined and during all stages of plant development. The level of *GPA1* expression, however, varies in different tissues and at different stages. In germinating seeds, GPα1 level is high in the cotyledons and at the root tip. In young seedlings, the highest level of GPα1 is detected in the shoot and root apical meristems as well as the lateral root meristems and leaf primordia. As the plants develop vegetatively, GPα1 level remains very high in the meristems and primordia, and in the root elongation zones, and decreases as the rosette leaves and cauline leaves mature. In mature leaves and roots, the GPα1 levels are high in the vascular tissues, particular phloem, but lower in the leaf mesophyll cells, and not detectable in the epidermis. During early flower development, GPα1 is present at high levels in the floral meristem and floral organ primordia, and the level

Table 3. Plant small G proteins identified by molecular cloning.

Plant	Protein	Size[a] (kDa)	Homologue[b] (% identity)	Expression and/or function	References[c]
Arabidopsis thaliana	Ara	24.2	Rab11 (55)		[110]
	Ara2	24.0	Rab11 (63)		[3]
	Ara3	23.8	Rab8 (58)		[3]
	Ara4	24.0	Rab11 (57)		[3]
	Ara5	21.6*	Rab1 (75)		[3]
	Arf1	20.6	Arf1 (88)		[138]
	Rab2a		Rab2	Expressed preferentially in pollen	[127]
	Rab2b		Rab2		[127]
	Rab6	23.1	Rab6 (72)	Complements a yeast *ytp6* mutation	[7]
	Rab11	24.0	Rab11 (66)		[186]
	Rha1	21.7	Rab5 (62)	Expression high in root and callus	[4]
				Expressed primarily in guard cells and root tips	[162]
	Sar1	22.0	ScSar1 (63)	Suppresses a yeast *set12* mutation	[31]
Brassica napus	Bra	24.4	Rab11 (55)		(a)
Chlamydomonas reinhardtii	yptC1	22.6	Rab1 (81)	Complements a yeast *ypt1* mutation	(b)
	yptC4	23.6	Rab2 (79)		(b)
	yptC5	23.1	Rab7 (67)		(b)
	yptC6	24.2	Rab11 (76)		(b)
Glycine max	sRab1	22.4	Rab1 (75)	Complements a yeast *ypt1* mutation	[28]
				Membrane biogenesis during root nodulation	
	sRab7	23.1	Rab7 (61)		[28]
Lycopersicon esculentum	Rab1A	20.1*	Rab1 (77)		(c)
	Rab1B	22.5	Rab1 (74)	Complements a yeast *ypt1* mutation	(c)
	Rab1C	22.6	Rab1 (78)	Complements a yeast *ypt1* mutation	(c)
	Sar1	22.0	ScSar1 (62)	Expressed in several organs	[34]
	Ypt2	23.9	Rab8 (58)	Expressed in apical meristem	[53]
Nicotiana plumbaginifolia	Np-ypt3	24.2	Rab11 (68)	In stem and root, high in flowers	[32]
	Rhn	21.8	Rab5 (60)	High in roots, flowers, lower in stems	[161]
Nicotiana tabacum	Nt-rab5	22.0	Rab5 (62)	Expressed in stem and root, high in flowers	[32]
	Rgb1	22.4	Rab1 (79)		(d)
	Rgb2	21.8	Rab5 (62)		(d)
Oryza savita	Rgp1	24.9	Rab11 (55)	Expression reduced in 5azaC-induced dwarf	[147]
	Rgp2	23.9	Rab11 (63)	Expressed in several organs	[187]
	ric1	22.4	Rab1 (76)		[88]
	ric2	24.0	Rab11 (70)		[88]
Pisum sativum	pra1	23.9	Rab11 (56)	Expressed highly in leaves and roots	[117]
	pra2	24.6*	Rab11 (52)	Expressed at a low level in leaves	[117]
	pra3	24.9	Rab11 (57)	Expressed at moderate levels in leaves and roots	[117]
	pra4	24.0	Rab11 (67)	Expressed highly in roots, less in leaves	[117]
	pra5	24.1	Rab11 (67)	Expressed at low levels in leaves and roots	[117]
	pra6	24.0	Rab11 (66)	Expressed at moderate levels in leaves and roots	[117]
	pra7	24.2	Rab11 (64)	Expressed highly in roots, less in leaves	[117]
	pra8	22.4	Rab1 (75)	Expressed at moderate levels in leaves and roots	[117]
	pra9A	22.5	Rab1 (79)	Expressed highly in roots, less in leaves	[117]
	pra9B	22.5	Rab1 (77)	Expressed highly in roots, less in leaves	[117]
	pra9C	22.6	Rab1 (75)	Expressed highly in roots, less in leaves	[117]
	Rho1	22.5	Rac2 (59)	Expressed in all organs of seedling	[183]
	Rab	23.0	Rab7 (69)	Expressed in pod	[44]
Vicia faba	Gnrp1	22.6	Rab1 (75)		(e)
	Gnrp2	22.9*	Rab11 (69)		(e)

Table 3. (Continued)

Plant	Protein	Size[a] (kDa)	Homologue[b] (% identity)	Expression and/or function	References[c]
	Gnrp3	24.0	Rab11 (67)		(e)
	Gnrp4	25.2	Rab11 (56)		(e)
Vigna aconitifolia	vRab7	23.1	Rab7 (61)	Membrane biogenesis during root nodulation	[28]
Volvox carteri	yptV1	22.5	Rab1 (81)	Complements a yeast *ypt1* mutation Expressed throughout development	[48]
	yptV2	24.2	Rab8 (53)		[47]
	yptV3	22.3	None –	Expressed throughout development	[47]
	yptV4	23.7	Rab2 (79)	Expressed throughout development	[47]
	yptV5	23.1	Rab7 (66)	Expressed throughout development	[47]
Zea mays	yptm1	23.3	Rab1 (65)	Complements a yeast *ypt1* mutation	[128]
	yptm2	22.5	Rab1 (79)	Complements a yeast *ypt1* mutation	[128]
	yptm3	23.0	Rab2 (79)		[127]
	yptm4		Rab1		[127]

[a] If the MW value was provided by the reference, then it is used here; if it was not, then it was calculated using Intelligenetics Software. An asterisk next to a value indicates that the sequence was incomplete, and a blank indicates that the sequence was not available.

[b] The same mammalian or yeast protein is listed for each member of a subgroup. The following are the homologues (see Fig. 3 for references): human Rab1, Rab3, and Rab6; dog (*Canis familiaris*) Rab2, Rab4, Rab5, Rab7, Rab8, Rab11, Rac2 and Rho1; Bovine Arf1; and yeast (*Saccharomyces cerevisiae*) Sar1. The percent identity was calculated based on an alignment using the Pileup program of the GCG Package (version 7) [57].

[c] In addition to published references, others are listed as follows: (a) Y. Park, H. Kang, J. Kwak, H. Lee, and H. Nam, submitted to GenBank; (b) S. Fabry, pers. comm.; (c) A. Loraine, pers. comm.; (d) J.A. Napier and P.R. Shewry, submitted to GenBank; (e) G. Saalbach and J. Thielmann, submitted to GenBank.

decreases as the organs mature. Later during flower development, *GPA1* expression is concentrated in the vascular tissues, in the carpel wall, in the microspore tetrads, and in the ovules. During pollination, a high level of GPα1 is found in the growing pollen tube, and after pollination, *GPA1* is expressed highly in embryos until the late curved stage, but not in mature embryos. Therefore, GPα1 is present at high levels in all actively proliferating cells, as well as in the cells that have begun to differentiate, but lower in most fully differentiated cells. Among the differentiated cells, those in the vascular tissues have very high levels of GPα1, as do the carpel and silique walls. These tissues are all involved in nutrient transport.

The fact that GPα1 is present at very high levels in the undifferentiated cells of meristems and organ primordia, and in cells during the early phase of organ differentiation, and that its level reduces as organs become fully differentiated, suggests that GPα1 is involved in promoting active cell division, and that its function is reduced in differentiated cells. This is parallel to the function of some mammalian G proteins in regulating cell division and differentiation [61, 153]. Constitutively activated mutant G_s and G_i have been associated with human tumors [96, 105], indicating that active Gα promotes cell division. On the other hand, the level of the $G\alpha_{i2}$ protein has been observed to decline during differentiation of a cell line [56]. Furthermore, active $G\alpha_{i2}$ blocks while $G\alpha_{i2}$ antisense RNA stimulates this differentiation [170]. Other studies showed that $G\alpha_s$ antisense oligonucleotides stimulate the differentiation of fibroblasts to adipocytes [167]. G proteins (G_q and G_{i2}) have also been shown to mediate the effect of growth factors on DNA synthesis and calcium flux [95]. Recently, the human $G\alpha_{12}$ gene, when overexpressed, was found to cause transformation of NIH 3T3 cells [24]. Although it is premature to suggest that GPα1 functions by a mechanism similar to those of the mammalian

Gα proteins, it is not unreasonable to postulate that GPα1 may regulate cell division and/or differentiation in plants.

A shared property of the undifferentiated cells, which have high levels of GPα1, is that they all require high intake of energy and nutrients. It is possible that GPα1 regulates the uptake of nutrients by these cells. It is intriguing that most of the mature tissues that express GPα1 at high levels are involved in nutrient transport. G_s and G_i were first identified as required for the hormonal regulation of sugar metabolism, and G_s has recently been shown to regulate sugar uptake [71]. Furthermore, a fission yeast Gα protein, *GPA2*, has recently been demonstrated to function in nutrient sensing [76]. Therefore, it would not be surprising if GPα1 indeed is shown to be involved in the regulation of nutrient transport or metabolism in plant cells in the near future.

Isolation of genes encoding small G proteins

Many small G protein genes/cDNAs have been isolated using low-stringency hybridizations or PCR (Table 3). The first cloned plant small G protein gene (*ara*) was from *Arabidopsis thaliana*. Its product is related to a number of known small G proteins: 55% to rab11, 44% identical to YPT1, 31% to H-ras and yeast RAS1, 29% to rho, and 26% to ral [110]. Other *Arabidopsis* genes include *ara2* through *ara5* [3], *rab6* [7], and *rha1* [4], all of which encode proteins related to the rab/ypt proteins. Two other *Arabidopsis* genes were isolated which encode an ARF-related protein [138] and a Sar1 homologue (see below) [31].

A total of 13 genes have been identified in pea which encode small G proteins. One of these, *Rho1Ps*, encodes the first plant member of the rho/rac subfamily of small G proteins; the predicted protein is 59% identical to human rac2 [183]. When the pea rho protein was expressed in *Escherichia coli*, it was shown to be able to bind GTP in a filter assay. Other cloned pea small G protein genes are most similar to members of the rab/Ypt subfamily; these include *Psa-rab*, and *Pra1* through *Pra9A*, *Pra9B*, and *Pra9C* [44, 117]. Rab-related genes have also been isolated from other legumes such as soybean [28] (see below), from tomato [53], from tobacco [32, 161], and from monocots maize [128] and rice [88, 147, 187]. Five small G protein genes of the rab/ypt type were also isolated from the green alga *Volvox carteri*; three of these can bind GTP when expressed in *E. coli* [39, 47, 48].

A comparison of the plant small G proteins with some of those from animals and yeasts is shown in Fig. 3, and a summary of the properties of these proteins is shown in Table 3. Most of the isolated plant small G proteins are more similar to rab/Ypt proteins involved in protein and membrane trafficking than to any other small G proteins. The high levels of sequence similarity between the plant small G proteins with their animal and yeast homologues suggest that this group of proteins also perform highly conserved basic cellular functions in protein and membrane trafficking in plants [139]. It is likely that the genes listed here represent an incomplete set. Even with these cloned genes, there are more than one case where two or more genes from a single plant show high levels of similarity with the same mammalian protein. This could either be that some plants have more functionally redundant genes, or that these plants genes have evolved to carry out slightly different functions. Furthermore, one of the plant Ypt gene, *YptV3* from *Volvox*, is not very similar to any known Ypts/Rabs. As it was pointed by Fabry *et al.* [47], *YptV3* may play a role unique to plants (or algae); alternatively, its homologues in animals may be identified in the future. There are now only a few known plant genes in the other classes, Rho/Rac, Arf, and Sar1; this may represent the more recent history of these classes.

It is intriguing that small G proteins more similar to the true ras proteins have not been identified. It is possible that proteins similar to ras in function are present in plants, but they are not highly similar at the sequence level, and are yet to be isolated. It is known that ras proteins are involved in signal transduction [63, 149], as are heterotrimeric G proteins. The failure thus far to identify plant ras proteins, and the scarcity of isolated heterotrimeric G protein α and β sub-

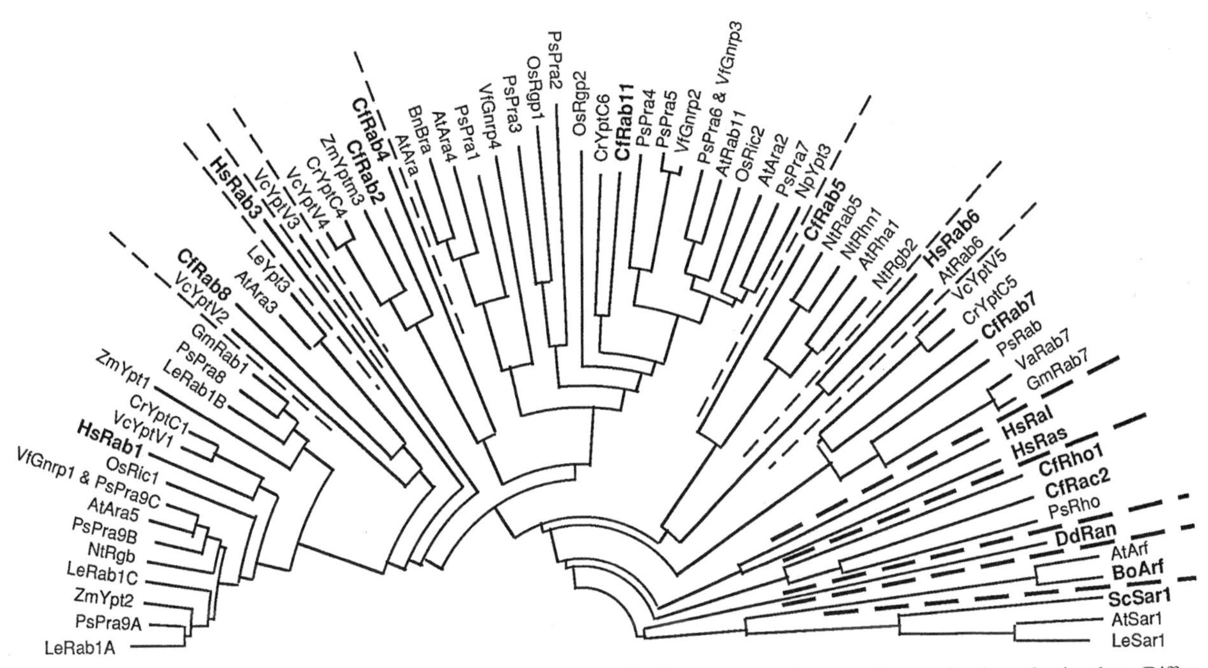

Fig. 3. Comparison of plant small G proteins with each other and with representative ones from animals and microbes. Different major subfamilies are separated by thick dashed lines, and the different type within the Rab/Ypt subfamily are separated by thin dashed lines. The names of each protein are preceded by two letters indicating the species. The species designations are: At, *Arabidopsis thaliana*; Bn, *Brassica napus*; Bo, bovine; Cf, *Canis familaris* (dog); Cr, *Chlamydomonas reinhardtii*; Dd, *Dictyostelium discoideum* (slime mold); Gm, *Glycine max* (soybean); Hs, *Homo sapiens*; Le, *Lycopersicon esculentum* (tomato); Np, *Nicotiana plumbaginifolia*; Nt, *Nicotiana tabacum*; Os, *Oryza sativa* (rice); Ps, *Pisum sativum* (pea); Sc, *Saccharomyces cerevisiae* (baker's yeast); Va, *Vigna aconitifolia*; Vc, *Volvox carteri*; Vf, *Vicia faba* (broad bean); Zm, *Zea mays* (maize). The sequences of VfGnrp1 and VfGnrp3 are identical to those of PsGBP11 and PsGBP6, respectively. References for the non-plant proteins are: bovine Arf [131]; CfRab2, CfRab5, and CfRab7 [26]; CfRab4, CfRab8, CfRab11, CfRho1 and CfRac2 [27]; HsRab1, HsRab3, and HsRab6 [188]; HsRas [23]; HsRalA [25]; DdRan [19]; and ScSar1 [120]; see Table 3 for references of plant proteins. Due to space constraints, most of the known yeast small G proteins are not shown here; the following are the known homologues between mammals and yeasts (Sp, *Schizosaccharomyces pombe*; references given for the yeast proteins): Rab1/ScYpt1 [55]; Rab6/SpRyh1 [69]; Rab7/ScYpt7 [176]; Rab5/SpYpt5 [5]; Rab8/SpYpt2 [68]; Rab11 /SpYpt3 [50]. The tree was drawn using MacDraft and MacDraw programs based on the output from the Molecular Evolutionary Genetics Analysis program (version 1.01) [93], following an alignment using the Pileup program of the GCG Package (version 7) [57].

units, suggest that proteins of these classes are not highly similar to those in animals and simple eukaryotes. This may reflect the fact that plant respond to environmental signals in very different ways than animals and microbes; such differences may account for the likely divergence of proteins involved in signal transduction pathways [85]. Because GTP-binding proteins have been implicated in a variety of plant signalling processes (see previous section), there probably exist additional G proteins in plants. These plant G proteins may be distant relatives of known heterotrimeric or small G proteins, or they may represent new families of GTP-binding proteins. Genetic and biochemical studies of plant signal transduction pathways should reveal the nature of these proteins in the near future.

Functional complementation of yeast mutants by plant small G proteins

One of the ways to test *in vivo* function of a protein encoded by a cloned cDNA is to introduce it into a yeast host which lacks a homologous function. This is feasible for many highly conserved proteins, including G proteins. This approach has been used for some of the isolated small G protein cDNAs. For example, a soybean homologue (*sRAB1*) of the budding yeast *YPT1* and mammalian *rab1* genes was isolated by PCR. The sRab1 protein is 75% identical to the rab1 protein. Furthermore, when the soybean *sRAB1* gene was fused to a yeast *GAL1* promoter and introduced into a yeast *ypt1* mutant, it complemented the cold-sensitive growth phenotype of the mutant [28]. Other plant Rab1/Ypt1 homologues, including the algal *YptV1* and *YptC1* (S. Fabry, pers. comm.), tomato *Rab1b* and *Rab1c* (A. Loraine and W. Gruissem, pers. comm.), and the maize *Yptm1* and *Yptm2* genes [127], have also been shown to complement a *S. cerevisiae ypt1* mutant. In another case, an *Arabidopsis* homologue (*AtRAB6*) of the fission yeast *Rhy1* and the human *rab6* genes was isolated using PCR, and shown to complement a budding yeast *ypt6* mutation (*YPT6* is a homologue of *Rhy1* and *rab6*) [7]. The predicted AtRab6 protein shares a high degree of similarity with the human and fission yeast homologues (>70% identity). When the *AtRAB6* cDNA was fused to the yeast *GAL10* promoter, and expressed in a temperature-sensitive *ypt6* null mutant of budding yeast, the *ypt6* mutant defective was corrected by the expression of *AtRAB6*. This complementation was specifically due to the production of a functional AtRab6 protein because a mutant *Atrab6* cDNA encoding a protein with a single amino acid change (Asn-122 → Ile) failed to complement.

Because highly conserved proteins can function in heterologous systems, functional complementation of yeast mutants can also be used as a means of isolating plant homologues of yeast genes. This approach has been successfully used to isolate an *Arabidopsis* homologue of the yeast *SAR1* gene, which encodes a distinct type of small G protein [31]. The yeast *SAR1* gene was identified as a clone which, when present at a high copy number, suppresses the temperature-sensitive phenotype of a *sec12* mutation. Since the *sec12* mutant is defective in formation of secretory vesicles from the endoplasmic reticulum, the *SAR1* gene is probably also involved in this process. The *Arabidopsis SAR1* cDNA was isolated by introducing a yeast expression cDNA library containing *Arabidopsis* cDNAs into a *sec12* mutant, and selecting for transformants which can grow at the restrictive temperature. The predicted budding yeast and *Arabidopsis* Sar1 protein sequences are 63% identical.

In these successful cases of functional complementation in yeast, the similarity between the homologues are about 60% or more. For many of the isolated small G proteins from plants, similarly high levels of sequence identity exist with particular animal/yeast homologues (Table 3); it is likely that these plant small G proteins are also able to complement the corresponding yeast mutations. For heterotrimeric G proteins, the similarity between the yeast proteins and the ones from multicellular organisms is much lower, ranging from 35% to 48% between the yeast GPA1 and various mammalian Gα's. It is not surprising, therefore, that only one of the mammalian Gα's (G_s) has been shown to complement the growth defect of the yeast *gpa1* mutant, but not its defect in pheromone response. These results once again suggest that G proteins for the fundamental cellular functions such as secretion are more conserved than G proteins involved in different signalling pathways.

Functional analysis using transgenic plants

In addition to biochemical studies and expression analysis, one important way of characterizing G protein functions is to examine the effects of altering gene expression in transgenic plants, including antisense, over-expression and ectopic expression. Recently, the effect of antisense RNA of two small G proteins from soybean (*sRAB1*) and another legume *Vigna aconitifolia* (*vRAB7*) was studied in transgenic soybean nodules [28]. The growth of nodules expressing the *sRAB1* antisense RNA were severely inhibited at an early stage, with only about 1/10 to 1/5 of the normal weight, and proportional reduction in nitrogenase activity. The yeast Ypt1 protein and its mammalian homologue rab1 are known to be involved in vesicular transport between ER and Golgi. Since the membranes (PBM) surrounding the symbiotic nitrogen-fixing bacteria are derived from the plasma membrane formed by fusion of vesicles, the phenotypes of *sRAB1* antisense nodules and the fact that *sRAB1* expression increases during nodulation suggest that *sRab1* is involved in the synthesis of the PBMs. A similar analysis was done with the *vRAB7* gene. In the nodules expressing antisense *vRAB7*, the infected cells had reduced number of bacteroids, and many small vesicles. In addition, there are some large vesicular structures which apparently contain degraded bacterial materials. In mammals, rab7 is localized to late endosomes, and the yeast homologue (Ypt7) is involved in the transport of vacuolar proteins. It seems that the legume rab7 homologue is required for the biogenesis of the PBM, which is similar to vacuoles in certain ways.

In another study, a rice gene, *rgp1*, encoding a rab-related protein having a 62% identity with the fission yeast Ypt3 protein, was introduced into tobacco plants by transformation in either the sense or the antisense orientation [83]. Both the transgenic plants carrying sense and the antisense constructs have a decrease in apical dominance and a dwarf phenotype. Since the *rgp1* antisense RNA apparently causes a reduction in the mRNA level of the tobacco homologue (*tgp1*), but *tgp1* is expressed normally in the sense transgenic plants, both a reduction and an increase in the gene function seem to disrupt the same process. Since reduction of apical dominance is also seen when the balance between cytokinin and auxin is shifted towards more cytokinin, the phenotype of the transgenic plants may be due to an alteration in hormonal balance. The transgenic plants also have abnormal floral phenotypes, with some homeotic organ conversion, suggesting either that the *rgp1* gene is involved in regulation of flower development, or that hormonal balance affects flower development. The latter possibility is supported by the recent observation that hormones affect the function of known *Arabidopsis* floral homeotic genes [126].

Future prospects

G proteins have been demonstrated to be important molecular switches in animals and simple eukaryotes. The spectrum of processes regulated by G proteins continues to widen, and the precise mechanisms by which G proteins function are becoming more and more clear. In plants, the presence of both heterotrimeric and small G proteins has be demonstrated. However, many questions remain to be answered. An immediate problem is the nature of the putative G proteins suggested by biochemical studies. Since pertussis toxin-sensitive G proteins seem to exist in plants, and the only known Gα protein, encoded by the *Arabidopsis GPA1* gene and homologues, does not have the conserved cysteine residue near the C terminus, plants are likely to have other heterotrimeric G proteins that are quite different from any known ones. Another puzzle is that known light receptors in plant are not like the classic G protein-coupled receptors that are characterized by seven transmembrane segments. In fact, no such membrane receptor has been identified in plants. An equally intriguing question is whether plants have true ras homologues. Certainly plants do not have genes that are highly similar to *ras*, as they do other types of small G proteins. It is still possible that plants have small G proteins which carry out signalling functions similar to ras proteins, but these plant proteins would be structurally divergent from the known ras proteins. As genetic, molecular, biochemical and physiological approaches continue to be employed for the analysis of plant signal transduction and G protein function, it is likely that new G proteins will be identified in plants, that the nature of G proteins implicated by biochemical studies will be determined, and that interaction between plant G proteins and receptors/effectors will be characterized. We are at the beginning of an exciting era of new discoveries and new insights about plant signal transduction and G protein functions.

Acknowledgements

The author thanks Drs S. Fabry, W. Gruissem, C. Han, E. Lam, A. Loraine, R. Martienssen, P. Millner, C. Poulsen, N. Raikhel, R. Reggiani, T. Reynolds, and L. Romero for communicating results before publication, M. Zhang for the computer analysis of the amino acid similarity tree for the small G proteins, and S. Assmann, C. Flanagan, A. Flint, D. Kostic, T. Stearns, T. Volpe, and C. Weiss for very helpful comments on this manuscript. The work in the author's laboratory was supported by the Cold Spring Harbor Laboratory Robertson Fund and by grants from the National Science Foundation (DCB-9004567 and MCB-9316048).

References

1. Ahmad M, Cashmore AR: *HY4* gene of *A. thaliana* encodes a protein with characteristics of a blue-light receptor. Nature 366: 162–166 (1993).
2. Amatruda TT III, Steele DA, Slepak VZ, Simon MI: Gα16, a G protein α subunit specifically expressed in hematopoietic cells. Proc Natl Acad Sci USA 88: 5587–5591 (1991).
3. Anai T, Hasegawa K, Watanabe Y, Uchimiya H, Ishizaki R, Matsui M: Isolation and analysis of cDNAs encoding small GTP-binding proteins of *Arabidopsis thaliana*. Gene 108: 259–264 (1991).
4. Anuntalabhochai S, Terryn N, Van Montagu M, Inzé D: Molecular characterization of an *Arabidopsis thaliana* cDNA encoding a small GTP-binding protein, Rha1. Plant J 1: 167–174 (1991).
5. Armstrong J, Craighead MW, Watson R, Ponnambalam S, Bowden S: *Schizosaccharomyces pombe* ypt5: a homologue of the rab5 endosome fusion regulator. Mol Biol Cell 4: 583–592 (1993).
6. Bakalyar HA, Reed RR: Identification of a specialized adenylyl cyclase that may mediate odorant detection. Science 250: 1403–1406 (1990).
7. Bennarek SY, Reynolds TL, Schroeder M, Grabowski R, Hengst L, Gallwitz D, Raikhel NV: A small GTP-binding protein from *Arabidopsis thaliana* functionally complements the yeast *ypt6* null mutant. Plant Physiol 104: 591–596 (1994).
8. Bilushi SV, Shebunin AG, Babakov AV: Purification and subunit composition of a GTP-binding protein from maize root plasma membranes. FEBS Lett 291: 219–221 (1991).
9. Birnbaumer L: Receptor-to-effector signalling through G proteins: roles for βγ dimers as well as α subunits. Cell 71: 1069–1072 (1992).
10. Blum W, Hinsch K-D, Schultz G, Weiler EW: Identification of GTP-binding proteins in the plasma membrane of higher plants. Biochem Biophys Res Commun 156: 954–959 (1988).
11. Boguski MS, McCormick F: Proteins regulating ras and its relatives. Nature 366: 643–654 (1993).
12. Bolwell GP, Coulson V, Rodgers MW, Murphy DL, Jones D: Modulation of the elicitation response in cultured French bean cells and its implication for the mechanism of signal transduction. Phytochemistry 30: 397–405 (1991).
13. Bossen ME, Kendrick RE, Vredenberg WJ: The involvement of a G-protein in phytochrome-regulated, Ca^{2+}-dependent swelling of etiolated wheat protoplasts. Physiol Plant 80: 55–62 (1990).
14. Botstein D, Segev N, Stearns T, Hoyt MA, Holden J, Kahn RA: Diverse biological functions of small GTP-binding proteins in yeast. Cold Spring Harbor Symp Quant Biol 53: 629–636 (1988).
15. Bourne HR: Do GTPases direct membrane traffic in secretion? Cell 53: 669–671 (1988).
16. Bourne HR, Sanders DA, McCormick F: The GTPase superfamily: a conserved switch for diverse cell functions. Nature 348: 125–132 (1990).
17. Bourne HR, Sanders DA, McCormick F: The GTPase superfamily: conserved structure and molecular mechanism. Nature 349: 117–127 (1991).
18. Brown AM, Birnbaumer L: Ionic channels and their regulation by G protein subunits. Annu Rev Physiol 52: 197–213 (1990).
19. Bush J, Cardelli JA: Molecular cloning and DNA sequence of a *Dictyostelium* cDNA encoding a ran/TC4 related GTP binding protein. GenBank (1993).
20. Buss JE, Munby SM, Casey PJ, Gilman AG, Sefton BM: Myristoylated α subunits of guanine nucleotide-binding regulatory proteins. Proc Natl Acad Sci USA 84: 7493–7497 (1987).
21. Buss JE, Sefton BM: Direct identification of palmitic acid as the lipid attached to p21 ras. Mol Cell Biol 6: 116–122 (1986).
22. Camps M, Carozzi A, Schnabel P, Scheer A, Parker PJ, Gierschik P: Isozyme-selective stimulation of phospholipase C-β2 by G protein βγ-subunits. Nature 360: 684–686 (1992).
23. Capon DJ, Chen EY, Levinson AD, Seeburg PH, Goeddel DV: The complete nucleotide sequences of the T24 human bladder carcinoma oncogene and its normal homologue. Nature 302: 33–37 (1983).
24. Chan AM-L, Fleming TP, McGovern ES, Chedid M, Miki T, Aaronson SA: Expression cDNA cloning of a transforming gene encoding the wild-type Gα12 gene product. Mol Cell Biol 13: 762–768 (1993).
25. Chardin P, Tavitian A: Coding sequences of human ralA and ralB cDNAs. Nucl Acids Res 17: 4380 (1989).
26. Chavrier P, Parton RG, Hauri HP, Simons K, Zerial M: Localization of low molecular weight GTP binding proteins to exocytic and endocytic compartments. Cell 62: 317–329 (1990).
27. Chavrier P, Vingron M, Sander C, Simons K, Zerial M: Molecular cloning of YPT1/sec4-related cDNAs from an epithelial cell line. Mol Cell Biol 10: 6578–6585 (1990).
28. Cheon C-I, Lee N.-G., Siddique A-BM, Bal AK, Verma

DPS: Roles of plant homologs of Rab1p and Rab7p in the biogenesis of the peribacteroid membrane, a subcellular compartment formed *de novo* during root nodule symbiosis. EMBO J 12: 4125–4135 (1993).
29. Clark GB, Memon AR, Tong C-G, Thompson GA Jr, Roux SJ: Phytochrome regulates GTP-binding protein activity in the envelope of pea nuclei. Plant J 4: 399–402 (1993).
30. Conklin BR, Bourne HR: Structural elements of Gα subunits that interact with G$\beta\gamma$, receptors, and effectors. Cell 73: 631–641 (1993).
31. d'Enfert C, Gensse M, Gaillardin C: Fission yeast and a plant have functional homologues of the Sar1 and Sec12 proteins involved in ER to Golgi traffic in budding yeast. EMBO J 11: 4205–4211 (1992).
32. Dallmann G, Sticher L, Marshallsay C, Nagy F: Molecular characterization of tobacco cDNAs encoding two small GTP-binding proteins. Plant Mol Biol 19: 847–857 (1992).
33. Dalrymple MA, Peterson-Bjorn S, Friessen JD, Beggs JD: The product of the *PRP4* gene of S. cerevisiae shows homology to β subunits of G proteins. Cell 58: 811–812 (1989).
34. Davies C: Cloning and characterization of a tomato GTPase-like gene related to yeast and *Arabidopsis* genes involved in vesicular transport. Plant Mol Biol 24: 523–531 (1994).
35. de Hostos EL, Bradtke B, Lottspeich F, Guggenheim R, Gerish G: Coronin, an actin binding protein of *Dictyostelium discoideum* localized to cell surface projections, has sequence similarities to g protein β subunits. EMBO J 10: 4097–4104 (1991).
36. de Sousa SM, Hoveland LL, Yarfitz S, Hurley JB: The *Drosophila* Go alpha-like G protein gene produces multiple transcripts and is expressed in the nervous system and in ovaries. J Biol Chem 264: 18544–18551 (1989).
37. Deng XW, Caspar T, Quail PH: *cop1*: a regulatory locus involved in light-controlled development and gene expression in *Arabidopsis*. Genes Devel 5: 1172–1182 (1991).
38. Deng XW, Matsui M, Wei N, Wagner D, Chu AM, Feldmann KA, Quail PH: *COP1*: an Arabidopsis regulatory gene, encodes a protein with both a zinc-binding motif and a Gβ homologous domain. Cell 71: 791–801 (1992).
39. Dietmaier W, Fabry S: Analysis of the introns in genes encoding small G proteins. Curr Genet 26: 497–505 (1994).
40. Dietzel C, Kurjan J: The yeast *SCG1* gene: a Gα-like protein implicated in the a- and α-factor response pathway. Cell 50: 1001–1010 (1987).
41. Dillenschneider M, Hetherington A, Graziana A, Alibert G, Berta P, Haiech J, Ranjeva R: The formation of inositol phosphate derivatives by isolated membranes from *Acer psudoplatanus* is stimulated by guanine nucleotides. FEBS Lett 208: 413–417 (1986).
42. Dohlman HG, Thorner J, Caron MG, Lefkowitz RJ: Model systems for the study of seven-transmembrane-segment receptors. Annu Rev Biochem 60: 653–688 (1991).
43. Downward J: The ras superfamily of small GTP-binding proteins. Trends Biochem Sci 15: 469–472 (1990).
44. Drew JE, Bown D, Gatehouse JA: Sequence of a novel plant *ras*-related cDNA from *Pisum sativum*. Plant Mol Biol 21: 1195–1199 (1993).
45. Drobak BK, Allan EF, Comerford JG, Roberts R, Dawson AP: Presence of guanine nucleotide-binding proteins in a plant hypoctyl fraction. Biochem Biophys Res Commun 150: 899–903 (1988).
46. Ephritikhine G, Pradier J-M, Guern J: Complexity of GTPγS binding to tobacco plasma membranes. Plant Physiol Biochem 31: 573–584 (1993).
47. Fabry S, Jacobsen A, Huber H, Palme K, Schmitt R: Structure, expression, and phylogenetic relationships of a family of *ypt* genes encoding small G-proteins in the green alga *Volvox carteri*. Curr Genet 24: 229–240 (1993).
48. Fabry S, Na N, Huber H, Palme K, Jaenicke L, Schmitt R: The *yptV1* gene encodes a small G-protein in the green alga *Volvox carteri*: gene structure and properties of the gene product. Gene 118: 153–162 (1992).
49. Fairley-Grenot K, Assmann SM: Evidence for G-protein regulation of inward K^+ channnel current in guard cells of fava bean. Plant Cell 3: 1037–1044 (1991).
50. Fawell E, Hook S, Sweet D, Armstrong J: Novel YPT1-related genes from *Schizosaccharomyces pombe*. Nucl Acids Res 18: 4264 (1990).
51. Fino SI, Plasterk RH: Characterization of a G-protein α-subunit gene from the nematode *Caenorhabditis elegans*. J Mol Biol 215: 483–487 (1990).
52. Firtel RA, van Haastert PJM, Kimmel AR, Devreotes PN: G protein linked signal transduction pathways in development: Dictyostelium as an experimental system. Cell 58: 235–239 (1989).
53. Fleming AJ, Mandel T, Roth I, Kuhlemeier C: The patterns of gene expression in tomato shoot apical meristem. Plant Cell 5: 297–309 (1993).
54. Fong HW, Hurley JB, Hopkins RS, Miake-Lye R, Johnson MS, Doolittle RF, Simon MI: Repetitive segmental structure of the transducin β subunit: homology with the *CDC4* gene and identification of related mRNAs. Proc Natl Acad Sci USA 83: 2162–2166 (1986).
55. Gallwitz D, Donath C, Sander C: A yeast gene encoding a protein homologous to the human *c-has/bas* proto-oncogene product. Nature 306: 704–707 (1983).
56. Galvin-Parton PA, Watkins DC, Malbon CC: Retinoic acid modulation of transmembrane signalling: analysis in F9 teratocarcinoma cells. J Biol Chem 265: 17771–17779 (1990).
57. Genetics Computer Group I: University Research Park, 575 Science Dr., Suite B, Madison, WI 53711, USA.
58. Gill DM, Woolkalis MJ: Cholera toxin-catalyzed

[^{32}P]ADP-ribosylation of proteins. Meth Enzymol 195: 267–280 (1991).

59. Gilman AG: G proteins: transducers of receptor-generated signals. Annu Rev Biochem 56: 615–649 (1987).
60. Guillemot F, Billault A, Auffray C: Physical linkage of a guanine nucleotide-binding protein-related gene to the chicken major histocompatibility complex. Proc Natl Acad Sci USA 86: 4594–4598 (1989).
61. Gupta SK, Gallego C, Johnson GL: Mitogenic pathways regulated by G protein oncogenes. Mol Biol Cell 3: 123–128 (1992).
62. Hadwiger JA, Wilkie TM, Strathmann M, Firtel RA: Identification of *Dictyostelium* G_α genes expressed during multicellular development. Proc Natl Acad Sci USA 88: 8213–8217 (1991).
63. Hall A: The cellular functions of small GTP-binding proteins. Science 249: 634–640 (1990).
64. Hall A: Ras-related proteins. Curr Opin Cell Biol 5: 265–268 (1993).
65. Hasunuma K, Funadera K: GTP-binding protein(s) in green plant, *Lemna paucicostata*. Biochem Biophys Res Commun 143: 908–912 (1987).
66. Hasunuma K, Furukawa K, Funadera K, Kubota M, Watanabe M: Partial characterization and light-induced regulation of GTP-binding proteins in *Lemna paucicostata*. Photochem Photobiol 46: 531–535 (1987).
67. Hasunuma K, Furukawa K, Tomita K, Mukai C, Nakamura T: GTP-binding proteins in etiolated epicotyls of *Pisum sativum* (Alaska) seedlings. Biochem Biophys Res Commun 148: 133–139 (1987).
68. Haubruck H, Engelke U, Mertins P, Gallwitz D: Structural and functional analysis of *ypt2*, an essential ras-related gene in the fission yeast *Schizosaccharomyces pombe* encoding a sec4 protein homologue. EMBO J 9: 1957–1962 (1990).
69. Hengst D, Lehmeier T, Gallwitz D: The *ryh1* gene in the fission yeast *Schizosaccharomyces pombe* encoding a GTP-binding protein related to ras, rho and ypt: structure, expression and identification of its human homologue. EMBO J 9: 1949–1955 (1990).
70. Herskowitz I: A regulatory hierarchy for cell specialization in yeast. Nature 342: 749–757 (1990).
71. Honnor RC, Naghshineh S, Cushman SW, Wolff J, Simpson IA, Londos C: Cholera and pertussis toxins modify regulation of glucose transport activity in rat adipose cell: evidence for mediation of a cAMP-independent process by G-proteins. Cell Signalling 4: 87–98 (1992).
72. Huang H, Weiss CA, Ma H: Regulated expression of the *Arabidopsis* Gα gene *GPA1*. Int J Plant Sci 155: 3–14 (1994).
73. Im M-J, Graham RM: A novel guanine nucleotide-binding protein coupled to the α_1-adrenergic receptor. J Biol Chem 265: 18944–18951 (1990).
74. Iñiguez-Lluhi J, Kleuss C, Gilman AG: The importance of G-protein $\beta\gamma$ subunits. Trends Cell Biol 3: 230–236 (1993).
75. Ishida S, Takahashi Y, Nagata T: Isolation of cDNA of an auxin-regulated gene encoding a G protein β subunit-like protein from tobacco BY-2 cells. Proc Natl Acad Sci USA 90: 11152–11156 (1993).
76. Isshiki T, Mochizuki N, Maeda T, Yamamoto M: Characterization of a fission yeast gene, *gpa2*, that encodes a Gα subunit involved in the monitoring of nutrition. Genes Devel 6: 2455–2462 (1992).
77. Jacobs M, Thelen MP, Farndale RW, Astle MC, Rubery PH: Specific guanine nucleotide binding by membranes from *Cucurbita pepo* seedlings. Biochem Biophys Res Commun 155: 1478–1484 (1988).
78. Jelsema CL, Axelrod J: Stimulation of phospholipase A2 activity in bovine rod outer segments by the $\beta\gamma$ subunits of transducin and its inhibition by the α subunit. Proc Natl Acad Sci USA 84: 3623–3627 (1987).
79. Jo H, Cha BY, Davis HW, McDonald JM: Identification, partial purification, and characterization of two guanosine triphosphate-binding proteins associated with insulin receptors. Endocrinology 131: 2855–2862 (1992).
80. Jo H, Radding W, Anantharamaiah GM, McDonald JM: An insulin receptor peptide (1135–1156) stimulates guanosine 5′-[γ-thio]triphosphate binding to the 67 kDa G protein associated with the insulin receptor. Biochem J 294: 19–24 (1993).
81. Jones DT, Reed RR: G_{olf}: an olfactory neuron specific-G protein involved in odorant signal transduction. Science 244: 790–795 (1989).
82. Kahn RA, Goddard C, Newkirk M: Chemical and immunological characterization of the 21 kDa ADP-ribosylation factor of adenylate cyclase. J Biol Chem 263: 8282–8287 (1988).
83. Kamada I, Yamauchi S, Toussefian S, Sano H: Transgenic tobacco plants expressing *rgp1*, a gene encoding a *ras*-related GTP-binding protein from rice, show distinct morphological characteristics. Plant J 2: 799–807 (1992).
84. Katz A, Wu D, Simon MI: Subunits $\beta\gamma$ of heterotrimeric G protein activate β2 isoform of phospholipase C. Nature 360: 686–689 (1992).
85. Kaufman LS: G proteins, paradigms, and plants. Int J Plant Sci 155: 1–2 (1994).
86. Kaufman LS: GTP-binding signalling proteins in higher plants. J Photochem Photobiol 22: 3–7 (1994).
87. Kaziro Y, Itoh H, Kozasa T, Nakafuku M, Satoh T: Structure and function of signal-transducing GTP-binding proteins. Annu Rev Biochem 60: 349–400 (1991).
88. Kidou S-I, Anai T, Umeda M, Aotsuka S, Tsuge T, Kato A, Uchimiya H: Molecular structure of ras-related small GTP-binding protein genes of rice plants and GTPase activities of gene products in *Escherichia coli*. FEBS Lett 332: 282–286 (1993).
89. Kleuss C, Scherübl H, Hescheler J, Schultz G, Wittig B:

Different β-subunits determine G-protein interaction with transmembrane receptors. Nature 358: 424–426 (1992).
90. Kleuss C, Scherübl H, Hescheler J, Schultz G, Wittig B: Selectivity in signal transduction determined by γ subunits of heterotrimeric G proteins. Science 259: 832–834 (1993).
91. Korolkov SN, Garnovskaya MN, Basov AS, Chunaev AS, Dumler IL: The detection and characterization of G-proteins in the eyespot of *Chlamydomonas reinhardtii*. FEBS Lett 270: 132–134 (1990).
92. Kumagai A, Pupillo M, Gundersen R, Miake-Lye R, Devreotes PN, Firtel RA: Regulation and function of $G\alpha$ protein subunits in Dictyostelium. Cell 57: 265–275 (1989).
93. Kumar S, Tamura K, Nei M: Molecular Evolutionary Genetics Analysis (version 1.01). Institute for Molecular and Evolutionary Genetics, Pennsylvania State University, University Park, PA 16802, USA.
94. Lam E: Heterotrimeric guanine nucleotide-binding proteins and light responses in higher plants. In: Raskin I, Schultz J (eds) Current Topics in Plant Physiology, vol. 11: Plant Signals in Interactions with Other Organisms, pp. 7–13. American Society for Plant Physiology, Rockville, TN (1993).
95. LaMorte VJ, Harootunian AT, Spiegel AM, Tsien RY, Feramisco JR: Mediation of growth factor induced DNA synthesis and calcium mobilization by G_q and G_{i2}. J Cell Biol 121: 91–99 (1993).
96. Landis CA, Masters SB, Spada A, Pace AM, Bourne HR, Vallar L: GTPase inhibiting mutations activate the α chain of G_s and stimulate adenylyl cyclase in human pituitary tumours. Nature 340: 692–696 (1989).
97. Lee HJ, Tucker EB, Crain RC, Lee Y: Stomatal opening is induced in epidermal peels of *Commelina communis* L. by GTP analogs or pertussis toxin. Plant Physiol 102: 95–100 (1993).
98. Lefkowitz RJ: The subunit story thickens. Nature 358: 372 (1992).
99. Legendre L, Heinstein PF, Low PS: Evidence for participation of GTP-binding proteins in elicitation of the rapid oxidative burst in cultured soybean cells. J Biol Chem 267: 20140–20147 (1992).
100. Li W, Assmann S: Characterization of a G-protein-regulated outward K^+ current in mesophyll cells of *Vicia faba* L. Proc Natl Acad Sci USA 90: 262–266 (1993).
101. Lilly P, Wu L, Welker DL, Devreotes PN: A G-protein β subunit is essential for *Dictyostelium* development. Genes Devel 7: 986–995 (1993).
102. Lochrie MA, Mendel JE, Sternberg PW, Simon MI: Homologous and unique G protein α subunits in the nematode *Caenorhabditis elegans*. Cell Regul 2: 135–54 (1991).
103. Lochrie MA, Simon MI: G protein multiplicity in eukaryotic signal transduction systems. Biochemistry 27: 4957–4965 (1988).
104. Logothetis DE, Kurachi Y, Galper J, Neer EJ, Clapham DE: The $\beta\gamma$ subunits of GTP-binding proteins activate the muscarinic K^+ channel in heart. Nature 325: 321–326 (1987).
105. Lyons J, Landis CA, Harsh G, Vallar L, Grunewald K, Feichtinger H, Duh Q-H, Clark OH, Kawasaki E, Bourne HR, McCormick F: Two G protein oncogenes in human endocrine tumors. Science 249: 655–659 (1990).
106. Ma H, Yanofsky MF, Huang H: Isolation and sequence analysis of *TGA1* cDNAs encoding a tomato G protein α subunit. Gene 107: 189–195 (1991).
107. Ma H, Yanofsky MF, Meyerowitz EM: Molecular cloning and characterization of *GPA1*, a G protein α subunit gene from *Arabidopsis thaliana*. Proc Natl Acad Sci USA 87: 3821–3825 (1990).
108. Markby DW, Onrust R, Bourne HR: Separate GTP binding and GTPase activating domains of a $G\alpha$ subunit. Science 262: 1895–1901 (1993).
109. Marshall MS: The effector interactions of $p21^{ras}$. Trends Biochem Sci 18: 250–254 (1993).
110. Matsui M, Sasamoto S, Kunieda T, Nomura N, Ryotaro I: Cloning of *ara*, a putative *Arabidopsis thaliana* gene homologous to the *ras*-related gene family. Gene 76: 313–319 (1989).
111. McLaughlin SK, McKinnon PJ, Margolskee RF: Gustducin is a taste-cell-specific G protein closely related to the transducins. Nature 357: 563–569 (1992).
112. Memon AR, Clark GB, Thompson GA Jr: Identification of an ARF type low molecular mass GTP-binding protein in pea (*Pisum sativum*). Biochem Biophys Res Comm 193: 809–813 (1993).
113. Memon AR, Herrin DL, Thompson GA Jr: Intracellular translocation of a 28 kDa GTP-binding protein during osmotic shock-induced cell volume regulation in *Dunaliella salina*. Biochim Biophys Acta 1179: 11–22 (1993).
114. Millner PA: Are guanine nucleotide-binding proteins involved in regulation of thylakoid protein kinase activity? FEBS Lett 226: 155–160 (1987).
115. Miyajima I, Nakafuku M, Nakayama N, Brenner C, Miyajima A, Kaibuchi K, Arai K, Kaziro Y, Matsumoto K: *GPA1*, a haploid-specific essential gene, encodes a yeast homolog of mammalian G protein which may be involved in mating factor signal transduction. Cell 50: 1011–1019 (1987).
116. Moore MS, Blobel G: The GTP-binding protein Ran/TC4 is required for protein import into the nucleus. Nature 365: 661–663 (1993).
117. Nagano Y, Murai N, Matsuno R, Sasaki Y: Isolation and characterization of cDNAs that encode eleven small GTP-binding protein from *Pisum sativum*. Plant Cell Physiol 34: 447–455 (1993).
118. Nakafuku M, Itoh H, Nakamura S, Kaziro Y: Occurrence in *Saccharomyces cerevisiae* of a gene homologous to the cDNA coding for the α subunit of mammalian G

proteins. Proc Natl Acad Sci USA 84: 2140–2144 (1987).

119. Nakafuku M, Obara T, Kaibuchi K, Miyajima I, Miyajima A, Itoh H, Nakamura S, Arai K-I, Matsumoto K, Kaziro Y: Isolation of a second yeast *Saccharomyces cerevisiae* gene (*GPA2*) coding for guanine nucleotide-binding regulatory protein: studies on its structure and possible function. Proc Natl Acad Sci USA 85: 1374–1378 (1988).
120. Nakano A, Muramatsu M: A novel GTP-binding protein, Sar1p, is involved in transport from the endoplasmic reticulum to the Golgi apparatus. J Cell Biol 109: 2677–2691 (1989).
121. Neuhaus G, Bowler C, Kern R, Chua N-H: Calcium/calmodulin-dependent and -indepenent phytochrome signal transduction pathways. Cell 73: 937–952 (1993).
122. Nishida E, Gotoh Y: The MAP kinase cascade is essential for diverse signal transduction pathways. Trends Biochem Sci 18: 128–131 (1993).
123. Noel JP, Hamm HE, Sigler PB: The 2.2 A crystal structure of transducin-α complexed with GTPγS. Nature 366: 654–663 (1993).
124. Obara T, Nakafuku M, Yamamoto M, Kaziro Y: Isolation and characterization of a gene encoding a G-protein α subunit from *Schizosaccharomyces pombi*: involvement in mating and sporulation pathways. Proc Natl Acad Sci USA 88: 5877–5881 (1991).
125. Ohmura T, Sakata A, Onoue K: A 68-kD GTP-binding protein associated with the T cell receptor complex. J Exp Med 176: 887–891 (1992).
126. Okamuro JK, den Boer BGW, Jofuku KD: Regulation of Arabidopsis flower development. Plant Cell 5: 1183–1193 (1993).
127. Palme K, Diefenthal T, Moore I: The *ypt* gene family from maize and Arabidopsis: structural and functional analysis. J Exp Bot 44: 183–195 (1993).
128. Palme K, Diefenthal T, Vingron M, Sander C, Schell J: Molecular cloning and structural analysis of genes from *Zea mays* (L.) coding for members of the *ras*-related *ypt* gene family. Proc Natl Acad Sci USA 89: 787–791 (1992).
129. Parks S, Wieschaus E: The *Drosophila* gastrulation gene concertina encodes a G_α-like protein. Cell 64: 447–458 (1991).
130. Pfeuffer E, Mollner S, Lancer D, Pfeuffer T: Olfactory adenylyl cyclase: identification and purification of a novel enzyme form. J Biol Chem 264: 18803–18807 (1989).
131. Price SR, Nightingale M, Tsai S-C, Williamson KC, Adamik R, Chen H-CC, Moss J, Vaughan M: Guanine nucleotide-binding proteins that enhance choleragen ADP-ribosyltransferase activity: nulceotide sequence and deduced amino acid sequence of an ADP-ribosylation factor cDNA. Proc Natl Acad Sci USA 85: 5488–5491 (1988).
132. Provost NM, Somers DE, Hurley JB: A *Drosophila melanogaster* G protein α subunit gene is expressed primarily in embryos and pupae. J Biol Chem 263: 12070–12076 (1988).
133. Pupillo M, Kumagai A, Pitt GS, Firtel RA, Devreotes PN: Multiple α subunits of guanine nucleotide-binding proteins in *Dictyostelium*. Proc Natl Acad Sci USA 86: 4892–4896 (1989).
134. Quail PH: Phytochrome: a light-activated moleculer switch that regulates plant gene expression. Annu Rev Genet 25: 389–409 (1991).
135. Quan F, Forte MA: Two forms of *Drosophila melanogaster* Gs α are produced by alternate splicing involving an unusual splice site. Mol Cell Biol 10: 910–917 (1990).
136. Quan F, Thomas L, Forte M: *Drosophila* stimulatory G protein α subunit activates mammalian adenylyl cyclase but interacts poorly with mammalian receptors: implications for receptor-G protein interaction. Proc Natl Acad Sci USA 88: 1898–1902 (1991).
137. Quan F, Wolfgang WJ, Forte MA: The *Drosophila* gene coding for the α subunit of a stimulatory G protein is preferentially expressed in the nervous system. Proc Natl Acad Sci USA 86: 4321–4325 (1989).
138. Regad F, Bardet C, Tremousaygue D, Moisan A, Lescure B, Axelos M: cDNA cloning and expression of an *Arabidopsis* GTP-binding protein of the ARF family. FEBS Lett 316: 133–136 (1993).
139. Reynolds TL, Raikhel NV: Targeting and trafficking of vacuolar proteins. In Tartakoff A (ed) Membranes: Specialized Functions in Plants. JAI Press, Greenwich, CT, in press (1994).
140. Roberge M: Checkpoint controls that couple mitosis to completion of DNA replication. Trends Cell Biol 2: 277–281 (1992).
141. Rodríguez-Rosales M, Herrin DL, Thompson GA Jr: Identification of low molecular mass GTP-binding proteins in membranes of the halotolerant alga *Dunaliella salina*. Plant Physiol 98: 446–451 (1992).
142. Romero LC, Lam E: Guanine nucleotide binding protein involvement in early steps of phytochrome-regulated gene expression. Proc Natl Acad Sci USA 90: 1465–1469 (1993).
143. Romero LC, Sommer D, Gotor C, Song P-S: G-protein in etiolated *Avena* seedlings: possible phytochrome regulation. FEBS Lett 282: 341–346 (1991).
144. Ross EM: Signal sorting and amplification through G protein-coupled receptors. Neuron 3: 141–152 (1989).
145. Rubin GM: Signal transduction and the fate of the R7 photoreceptor in *Drosophila*. Trends Genet 7: 372–377 (1991).
146. Ryba NJP, Pottinger JDD, Keen JN, Findlay JBC: Sequence of the β-subunit of the phosphatidylinositol-specific phospholipase C-directed GTP-binding protein from squid (*Loligo forbesi*) photoreceptors. Biochem J 273: 225–228 (1991).
147. Sano H, Youssefian S: A novel *ras*-related *rgp1* gene encoding a GTP-binding protein has reduced expression

in 5-azacytidine induced dwarf rice. Mol Gen Genet 228: 227–232 (1991).
148. Sasaki Y, Sekiguchi K, Nagano Y, Matsuno R: Detection of small GTP-binding proteins in the outer envelop membrane of pea chloroplasts. FEBS Lett 293: 124–126 (1991).
149. Schlessinger J: How receptor tyrosine kinases activate Ras. Trends Biochem Sci 18: 273–275 (1993).
150. Schloss JA: A chlamydomonas gene encodes a G protein β subunit-like polypeptide. Mol Gen Genet 221: 443–452 (1990).
151. Schultz DG, Carlson M: Molecular analysis of *SSN6*, a gene functionally related to the SNF1 protein kinase of *Saccharomyces cerevisiae*. Mol Cell Biol 7: 3637–3645 (1987).
152. Schwaninger R, Plutner H, Bokoch GM, Balch WE: Multiple GTP-binding proteins regulate vesicular transport from the ER to Golgi membranes. J Cell Biol 119: 1077–1096 (1992).
153. Seuwen K, Pouysségur J: G protein-controlled signal transduction pathways and the regulation of cell proliferation. Adv Cancer Res 58: 75–94 (1992).
154. Sharrock RA, Quail PH: Novel phytochrome sequences in *Arabidopsis thaliana*: structure, evolution, and differential expression of a plant regulatory photoreceptor family. Genes Devel 3: 1745–1757 (1989).
155. Simon MI, Strathmann MP, Gautam N: Diversity of G proteins in signal transduction. Science 252: 802–808 (1991).
156. Srivastava SK, Singh US: Insulin activates guanosine 5′-[γ-thio] triphosphate (GTPγS) binding to a novel GTP-binding protein, G_{IR}, from human placenta. Biochem Biophys Res Comm 173: 501–506 (1990).
157. Sternberg PW, Horvitz HR: Signal transduction during *C. elegans* vulval induction. Trends Genet 7: 366–371 (1991).
158. Strathmann M, Simon MI: G protein diversity: a distinct class of α subunits is present in vertebrates and invertebrates. Proc Natl Acad Sci USA 87: 9113–9117 (1990).
159. Strathmann MP, Simon MI: Gα12 and Gα13 subunits define a fourth class of G protein α subunits. Proc Natl Acad Sci USA 88: 5582–5586 (1991).
160. Stryer L: Cyclic GMP cascade of vision. Annu Rev Neurosci 9: 87–119 (1986).
161. Terryn N, Anuntalabhochai S, Van Montagu M, Inzé D: Analysis of a *Nicotiana plumbaginifolia* cDNA encoding a novel small GTP-binding protein. FEBS Lett 299: 287–290 (1992).
162. Terryn N, Arias MB, Engler G, Tiré C, Villarroel R, Van Montagu M, Inzé D: *rha1*, a gene encoding a small GTP binding protein from Arabidopsis, is expressed primarily in developing guard cells. Plant Cell 5: 1761–1769 (1993).
163. Terryn N, Van Montagu M, Inzé D: GTP-binding proteins in plants. Plant Mol Biol 22: 143–152 (1993).
164. Ullrich A, Schlessinger J: Signal transduction by receptors with tyrosine kinase activity. Cell 61: 203–212 (1990).
165. van der Voorn L, Gebbink M, Plasterk RH, Ploegh HL: Characterization of a G-protein β-subunit gene from the nematode *Caenorhabditis elegans*. J Mol Biol 213: 17–26 (1990).
166. Walworth NC, Goud B, Kabcenell AK, Novick PJ: Mutational analysis of *SEC4* suggests a cyclical mechanism for the regulation of vesicular traffic. EMBO J 8: 1685–1693 (1989).
167. Wang H, Watkins DC, Malbon CC: Antisense oligodeoxynucleotides to G_s protein α-subunit sequence accelerate differentiation of fibroblasts to adipocytes. Nature 358: 334–337 (1992).
168. Wang M, Sedee NJA, Heidekamp F, Snaar-Jagalska BE: Detection of GTP-binding proteins in barley aleurone protoplasts. FEBS Lett 329: 245–248 (1993).
169. Warpeha KMF, Hamm HE, Rasenick MM, Kaufman LS: A blue-light-activated GTP-binding protein in the plasma membranes of etiolated peas. Proc Natl Acad Sci USA 88: 8925–8929 (1991).
170. Watkins DC, Johnson GL, Malbon CC: Regulation of the differentiation of teratocarcinoma cell into primitive endoderm by $G\alpha_{i2}$. Science 258: 1373–1375 (1992).
171. Weiss CA, Garnaat CW, Mukai K, Hu Y, Ma H: Isolation of cDNAs encoding G protein β subunit homologues from maize *(ZGB1)* and *Arabidopsis* (*AGB1*). Proc Natl Acad Sci USA 91: 9554–9558 (1994).
172. Weiss CA, Huang H, Ma H: Immunolocalization of the G protein α subunit encoded by the *GPA1* gene in Arabidopsis. Plant Cell 5: 1513–1528 (1993).
173. White IR, Wise A, Finan PM, Clarkson J, Millner PA: GTP-binding proteins in higher plant cells. In: Cooke DT, Clarkson DT (eds) Transport and Receptor Proteins of Plant Membranes, pp. 185–192. Plenum Press, New York (1992).
174. Whiteway M, Hougan L, Dignard D, Thomas DY, Bell L, Saari GC, Grant FJ, O'Hara P, Mackay VL: The *STE4* and *STE18* genes of yeast encode potential β and γ subunits of the mating factor receptor-coupled G protein. Cell 56: 467–477 (1989).
175. Whiteway M, Hougan L, Thomas DY: Overexpression of the STE4 gene leads to mating response in haploid *Saccharomyces cerevisiae*. Mol Cell Biol 10: 217–222 (1990).
176. Wichmann H, Hengst L, Gallwitz D: Endocytosis in yeast: evidence for the involvement of a small GTP-binding protein (Ypt7p). Cell 71: 1131–1142 (1992).
177. Wigler M, Field J, Powers S, Broek D, Toda T, Cameron S, Nikawa J, Michaeli T, Colicelli J, Ferguson K: Studies of *RAS* function in the yeast *Saccharomyces cerevisiae*. Cold Spring Harbor Symp Quant Biol 53: 649–655 (1988).
178. Wigler MH: The RAS system in yeasts. In: Lacal JC,

McCormick F (eds) The *ras* Superfamily of GTPases, pp. 155–172. CRC Press, Boca Raton, FL (1993).
179. Wilkie TM, Scherle PA, Strathmann MP, Slepak VZ, Simon MI: Characterization of G-protein α subunits in the G_q class: expression in murine tissues and stromal and hematopoietic cell lines. Proc Natl Acad Sci USA 88: 10049–10053 (1991).
180. Williams FE, Trumbly RJ: Characterization of *TUP1*, a mediator of glucose repression in *Saccharomyces cerevisiae*. Mol Cell Biol 10: 6500–6511 (1990).
181. Williams FE, Varanasi U, Trumbly RJ: The CYC8 and TUP1 proteins involved in glucose repression in *Saccharomyces cerevisiae* are associated in a protein complex. Mol Cell Biol 11: 3307–3316 (1991).
182. Wu LJ, Devreotes PN: *Dictyostelium* transiently expresses eight distinct G-protein α-subunits during its developmental program. Biochem Biophys Res Commun 179: 1141–1147 (1991).
183. Yang Z, Watson JC: Molecular cloning and characterization of rho, a ras-related small GTP-binding protein from the garden pea. Proc Natl Acad Sci USA 90: 8732–8736 (1993).
184. Yarfitz S, Niemi GA, McConnell JL, Fitch CL, Hurley JB: A G β protein in the Drosophila compound eye is different from that in the brain. Neuron 7: 429–438 (1991).
185. Yarfitz S, Provost NM, Hurley JB: Cloning of a *Drosophila melanogaster* guanine nucleotide regulatory protein β-subunit gene and characterization of its expression during development. Proc Natl Acad Sci USA 85: 7134–7138 (1988).
186. Yi Y, Guerinot M: A new member of the small GTP-binding protein family in *Arabidopsis thaliana*. Plant Physiol 104: 295–296 (1994).
187. Youssefian S, Nakamura M, Sano H: Molecular characterization of *rgp2*, a gene encoding a small GTP-binding protein from rice. Mol Gen Genet 237: 187–192 (1993).
188. Zahraoui A, Touchot N, Chardin P, Tavitian A: The human rab genes encode a family of GTP-binding proteins related to yeast ypt1 and sec4 products involved in secretion. J Biol Chem 264: 12394–12401 (1989).
189. Zaina S, Breviario D, Mapelli S, Bertani A, Reggiani R: Two putative G-protein α subunits dissociate from rice coleoptile membranes after GTP stimulation. J Plant Physiol 143: 293–297 (1994).
190. Zaina S, Reggiani R, Bertani A: Preliminary evidence for involvement of GTP-binding protein(s) in auxin signal transduction in rice (*Oryza sativa* L.). J Plant Physiol 136: 653–658 (1990).

Plant Molecular Biology **26**: 1637–1650, 1994.

Green circuits – The potential of plant specific ion channels

Rainer Hedrich* and Dirk Becker
Institut für Biophysik, Herrenhäuser Strasse 2, 30419 Hannover, Germany (author for correspondence)*

Received 11 October 1994; accepted in revised form 11 October 1994

Introduction

In 1983 when visiting the German Botanical Congress in Vienna to present out initial studies on the identification of the first plant K^+ channel [73], the previous speaker finished his talk with the conclusion that unlike animals which use channels like drums, plants are more sophisticated since like playing a melody on a piano they are able to bring various carrier types and pumps into play.

In the light of the rapid progress in the field of plant membrane transport/biology during the past decade both statements turned out to be incorrect:

1) both animal and plant cells take advantage of ion channels, carriers, and pumps and
2) plant and animal K^+ channels are structurally closer related than was ever expected [93].

Because of this similarity on one hand and the presence of action potentials in both branches of the evolutionary tree on the other, one might suggest that with respect to function a common set of membrane elements has been evolved. The 'green circuits', however, differ anatomically from their animal counterparts and with regard to their composition in electrogenic elements. Although higher plants contain nerves, they are generally not concerned with the transmission of action potentials. These transport elements mediate the long distance water and nutrients supply of the various plant tissues such as roots, shoot, leaves, flowers, and developing buds.

The first models which have been constructed following electrophysiological recordings from 'sensitive' plants like *Mimosa pudica*, *Samanea saman*, or *Dionaea muscipula* to describe the peculiar fast transmission of electrical signals initiated at the site of stimulation spreading out all over the plant were applied to other plants, including vegetables. These models generated according to that of Hodgkin and Huxley [36] on the ionic basis of action potentials and its propagation within nervous systems involve conductive elements and shields. In nerve cells the cable is presented by the intracellular/-axonal electrolytes. The plasma membrane isolating the conductive cytoplasm is equipped with ion channels often located at specialized regions such as the nodes of Ranvier along the axon. Activation of voltage-dependent ion channels allows the permeation of charged molecules along their electrochemical gradients and thus propagation of an electrical signal.

In 1984 ion channels have been discovered in plants, too [54, 73]. Since then, however, these electroenzymes have neither been correlated to conductive macrostructures like the xylem or phloem nor have clusters of extremely high channel density been described in excitable plant cells.[1] Therefore four questions are still matter of debate:

1. Do electrical signals travel along the xylem and/or phloem or are other not yet defined structures involved?
2. Do isolators, functional equivalent to Schwann cells, such as polymeres (wax or

[1] Evidence for a high anion channel density has recently been reported for *Chara*, a giant green alga [96].

suberines) shield entire excitable plant cells or tissues to form a cable-like structure?

3. Is excitability in animals and plants founded by a similar set of ion channels or is there evidence for 'green circuits'?
4. Besides excitability, are 'green' ion channels required to fulfil plant specific tasks?

Summarizing the progress in the field of plant ion transport, this review will concentrate on 'green circuits' and plant specific ion channels or properties which trigger them.

A. The plant action potential

In contrast to animal cells Na^+ does not play a fundamental role in excitability; for non-halophytes Na^+ is even toxic with respect to growth and development. Instead of using Na^+ channels to depolarize the plasma membrane the 'green circuits' take advantage of voltage-dependent anion channels [7]. Furthermore, the uptake of sugars and amino acids or even ions in cotransport with Na^+ in their 'red' counterparts, is coupled in plants to the free enthalpie of the H^+ gradient [66, 23, 46, 99, 58, 69]. In the context of

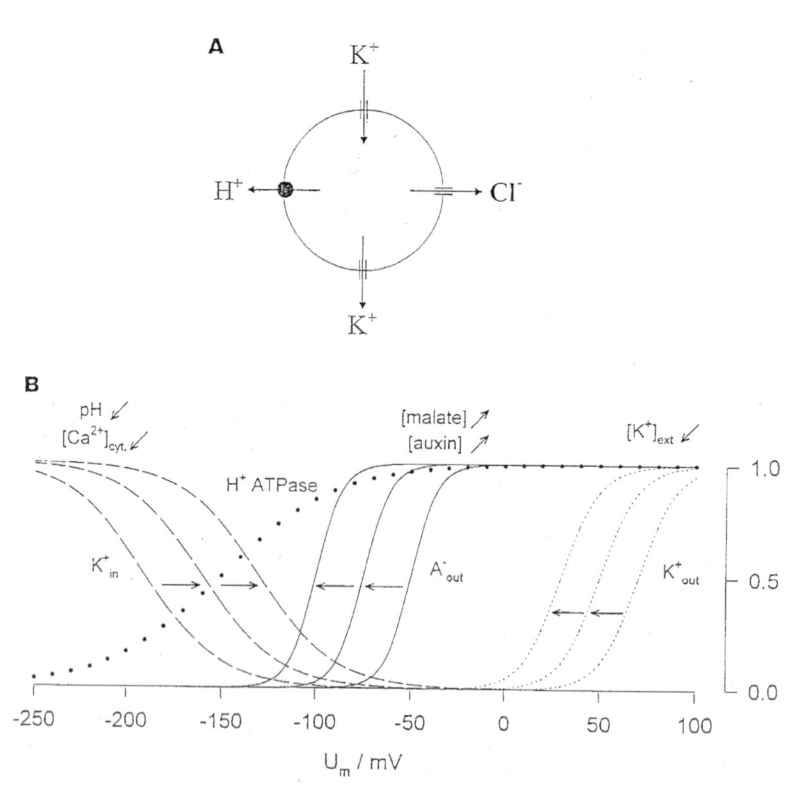

Fig. 1. Voltage dependent ion transporters in the plasma membrane of higher plant cells. A. Physiological direction of ion fluxes through K^+ uptake and K^+ release channels, anion channels and the H^+-ATPase. B. Activation curves, representing the relative conductance as a function of voltage for the ionic pathways shown in A (0 = closed channel; 1 = open channel). Voltage range fractionation, shift in the activation curve and its direction along the voltage axis is indicated by horizontal arrows. Upward and downward arrows on top of the activation curves for the individual ion channels behind the effectors indicate the direction of concentration change able to modify the membrane property in the given manner. Following resting levels for the various effectors were assumed: 30 mM K^+, pH 7.0, < 100 nM Ca^{2+}, 0 mM auxin and malate. Depending on the effector concentration the working range of guard cell anion channel 1 (GCAC1) is overlapping with K^+ uptake and K^+ release channels. Note, that the activation curve for the H^+-ATPase overlaps with each ion channel. Simplified activation curves were constructed from single Boltzman distributions which correlate quantitatively to data in the given literature.

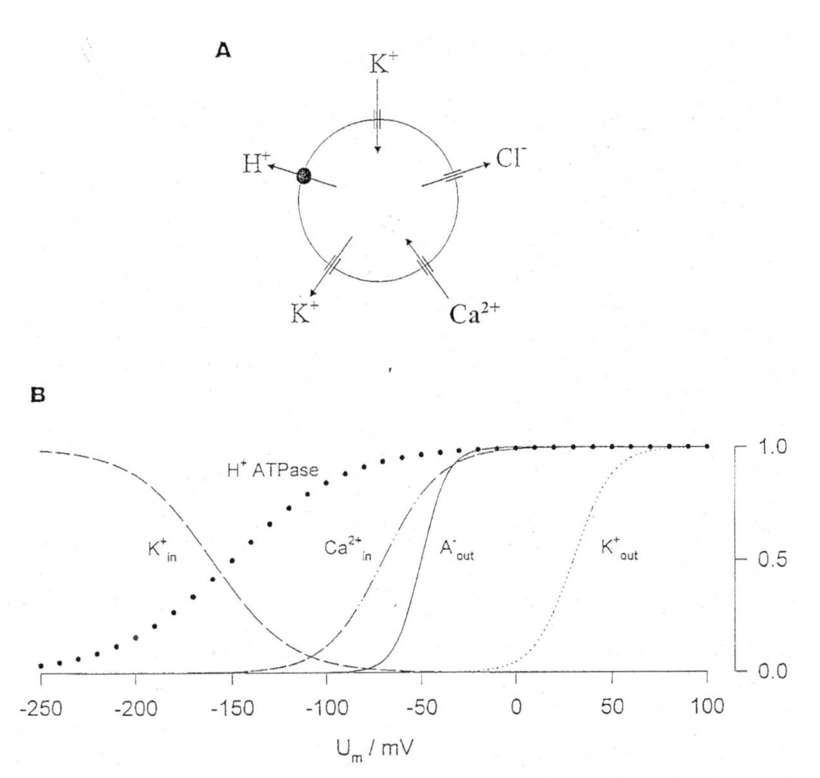

Fig. 2. Voltage dependent ion transporters in the plasma membrane of higher plant cells. A. Physiological direction of ion fluxes through K^+ uptake K^+ release channels, *Ca^{2+} channels*, anion channels and the H^+-ATPase. B. Activation curves, change in the relative conductance as a function of voltage for the ionic pathways given in A (0 = closed channel; 1 = open channel). In the absence of gating modifiers for the anion channel (cf. Fig. 1B), a Ca^{2+} channel with an activation threshold in the range of the resting potential upon stimulation could provide the initial depolarizing conductance of the plasma membrane. The resting potential is assumed to be located in the range of −250 to −150 mV.

cotransport it should be mentioned, however, that some archaebacteria can use the H^+ gradient as well as the Na^+ gradient to create energy [37].

The sequence of events during the plant action potential might include the activation of Ca^{2+}-, voltage-, and time-dependent anion channels [28, 33 and references therein]. Opening of anion channels will depolarize the plasma membrane towards the activation threshold of voltage-dependent K^+ channels which in turn will repolarize the membrane [78]. The shape of the action potential and its kinetics is determined by the relative contribution of the ionic conductances at rest – such as the inward-rectifying K^+ channels and the H^+ ATPase, responsible for the often very negative resting potentials. In the face of time-dependent (inactivating) anion channels [28] the termination of the action potential does, however, not consequently involve outward-rectifying K^+ channels. This is even more pronounced in the presence of hormones or signal metabolites [53, 29], which would separate the working range of anion channels from that of K^+ release channels.[2] This situation is given when ligands such as auxin or malate shift the activation curve of the *g*uard *c*ell *a*nion *c*hannel 1 (GCAC1) negative, towards the resting potential of the cell (Fig. 1B) and/or inactivation is completed before voltage- and time-dependent K^+ release channels open [75, 43].

[2] Note, that cell types such as guard cells or suspension cultured cells are equipped with anion channels which differ in voltage-dependence, unit conductance, and ligand sensitivity [108, 28, 53].

Table 1. K^+ uptake and release channels in the plant plasma membrane.

Plant	Tissue	Conductance	Selectivity	Modulation	Reference(s)
Vicia faba	guard cell	20 pS sym. K^+ [105 mM]	$K^+ > Na^+$ 17:1 $K^+ > Ca^{2+}$ 1:0.3 $K^+ > Rb^+ > Na^+ > Li^+ \gg Cs^+$	activation by hyperpolarization; block by Ba^{2+}, $Ca^{2+}_{ext.}$ and Al^{3+}; inhibition by GTPγS; Ca^{2+}_{cyt} and modulation by phosphorylation	[73, 74, 75, 76, 77, 79, 19, 20, 50]
Vicia faba	guard cell	20 pS sym. K^+ [105 mM]	$K^+ > Na^+$ 8:1 $K^+ > Rb^+ > Na^+ > Li^+ \gg Cs^+$	activation by depolarization ($V_m > -20$ mV); block by $Ba^{2+}_{ext.}$	[73, 74, 78]
Vicia faba	guard cell	40 pS $K^+_{ext.}$ [100 mM] $Na^+_{int.}$ [100 mM]	$K^+ > Na^+$ 20:1	activation by depolarization ($V_m > +70$ mV); activation by ABA; block by $TEA_{ext.}$ and $Cs^+_{int.}$	[70]
Vicia faba	guard cell	14 pS $K^+_{int.}$ [100 mM] $K^+_{ext.}$[10mM]		activation by depolarization, blocked by Protons	[39]
Zea mays	shoot suspension culture	40 and 125 pS $K^+_{ext.}$ [75 mM] cell attached	$K^+ > Cl^-$ 1:0.4	activation by depolarization, block by Cd^{2+}, verapamil and TEA, internal Ca^{2+} shifts voltage dependence	[18, 41, 42]
Arabidopsis thaliana	mesophyll	44, 66 and 109 pS $K^+_{int.}$ [220 mM] $K^+_{ext.}$ [50 mM]	$K^+ > Cl^-$	activation by depolarization and ATP	[89, 90]
Arabidopsis thaliana	tissue culture	63 pS sym. K^+ [105 mM]	$K^+ > Cl^-$	activation by depolarization ($V_m > 0$ mV), $K^+_{int.}$ modulates conductance	[47]
Asclepias tuberosa	suspension culture	40 pS $K^+_{ext.}$ [100 mM] $Na^+_{int.}$ [100 mM]	$K^+ > Cl^-$	activation by depolarization ($V_m > 0$ mV)	[71]
Samanea saman	extensor and flexor cells	20 pS $K^+_{int.}$ [125 mM] $K^+_{ext.}$ [25 mM]	$K^+ \gg Cl^-$ 100:3 $K^+ > Rb^+ > Na^+ = Li^+ = Cs^+$	activation by depolarization ($V_m > -30$ mV); block by TEA and quinine; voltage dependent block by Cs^+ and Ba^{2+}; block by Gd^{3+} and La^{3+}	[54, 55]

Table 1 (Continued).

Haemanthus albiflos, carriage return, H. katherinae, Clivia	endosperm	34 pS sym K^+ [100 mM]	$K^+ > Rb^+ = Na^+ = Li^+ = Cs^+$	activation by depolarization ($V_m > +80$ mV); activation by internal Ca^{2+} and Ba^{2+}	[92]
Dionaea muscipula	trap-lobe cells	3.3 pS sym. K^+ [30 mM]	$K^+ > Na^+$	activation by depolarization ($V_m > 0$ mV)	[38]
Pisum sativum	epidermal cells	35 pS sym. K^+ [100 mM]	$Na^+ > Li^+ > K^+$ 5:2:1	activation by internal Ca^{2+}	[17]
Plantago media	root	8 to 133 pS	$K^+ > Cl^-$	11 different cation-selective channel types 6 activated by depolarization, 5 activated by hyperpolarization	[101]
Hordeum vulgare	aleurone	35 pS $K^+_{int.}$ [100 mM] $K^+_{ext.}$ [10 mM]	$K > Na^+$ 35:1	activation by hyperpolarization	[14]
Hordeum vulgare	root xylem parenchyma	21 pS sym K^+ [100 mM]	$K^+ > Mg^{2+}$ $Ca^{2+} > K^+ \approx Na^+$	Two channel types, activated by depolarization; one channel type TEA insensitive	[103]
Hordeum vulgare	root xylem parenchyma	30 pS sym. K^+ [100 mM]	$K^+ > Rb^+ = Cs^+ > Li^+ = Na^+$	activation by hyperpolarization, permeable to Cs^+, voltage-dependent block by La^{3+}	[104]
Triticum spec.	root	32 pS sym. K^+ [100 mM]	$K^+ > Na^+$ 30:1	activation by depolarization	[67]
Triticum spec.	root	115 and 450 pS $K^+_{int.}$ [60 mM] $K^+_{ext.}$ [105 mM]	$K^+ > Cl^-$	activation by hyperpolarization	[22]
Avena sativa	mesophyll	15 pS sym. K^+ [100 mM]	$K^+ > Na^+$	activation by hyperpolarization, voltage-dependent block by Na^+ and Cs^+	[44]

Since voltage-dependent anion channels in plants require elevated cytoplasmic Ca^{2+} levels, the activation of Ca^{2+} permeable channels in the plasma membrane and/or the vacuolar membrane are proposed to represent an initial step within an action potential [33]. In line with this prediction several kinds of Ca^{2+} permeable channels have been found in both membranes, the activation of which is still under investigation [15, 102, 83, 98]. Equivalent to the heart muscle action potential voltage-dependent, slowly inactivating Ca^{2+} channels may dominate the depolarization phase. Indeed, evidence for the existence of L-type like Ca^{2+} channel in coexistence with a voltage-dependent anion channel has been reported recently for carrot cells [98, 5].

Voltage- and ligand-dependent anion channels in conjunction with K^+- and Ca^{2+} channels as well as an electrogenic H^+-ATPase may allow specialized plant cells or plant cells within their developmental program to respond to the variability in the environmental conditions by changes in the electrical activity of the plasma membrane (Fig. 2). The transduction of signals and information could hence be encoded by the shape of a single action potential or frequence of firing. Besides their macroscopic organization a characteristic feature of the 'green cables' is the absence of voltage-dependent Na^+ channels in favour of voltage- and ligand-sensitive Cl^- channels.

B. Plant specific properties of 'green ion channels'

Whereas excitability and consequently the presence of voltage-gated ion channels in animals is restricted to only a few, highly differentiated cell types, this class of channels was found throughout all plant species, cell types and developmental stages studied so far (Tables 1–3 and [27]). Compared to animals, plants are omnipotent and can adapt more easily to limitations in their neighbourhood since programs for pattern formation and e.g. differentiation are redundant. Therefore a plant is more robust, which is an important property in the light of their inability to flee from unfavourable environmental conditions. During their life cycle plants have to overcome periods where water supply is limiting (water/drought stress), the Na^+, Cl^- and pH in the soil is increasing (salinity stress) or is characterized by the presence of toxic cations released from heavy metall containing minerals by acid rain (e.g. Al^{3+}-toxicity). Omnipotence and adaption is therefore based on the ability of almost all plant cells and/or tissues to de-differentiate before individual clones start their developmental programs again. Germination, root- or shoot formation, and reproduction requires the differentiation into specialized cell types with distinct tasks such as

secretion of lytic enzymes, slime, sugars and salt	aleuron cells of growing seeds or cells in the root tip or gland cells
photosynthesis	mesophyll cells
movement (turgor-driven)	guard cells and cells in the pulvinus, modified leaf cells in carnivorous plants
microbe/pathogen interaction	root-hair cells
uptake, release and long distance transport of nutrients and water	cells in xylem and phloem

We have just started to gain new insights into the abundance, distribution, function and molecular structure of the various channel types in cells performing different tasks, we will focus on three voltage-dependent channel types for the following reasons:

K^+ channels	– a family of voltage-gated, inward-rectifying channels of known function and molecular structure
Anion channels	– a diverse class of ion channels where at least the functional properties of a voltage-gated one in guard cells has been investigated in detail.
Channels of the slow vacuolar (SV-) type	– a voltage-gated channel type found in all plants and cell types looked at.

K^+ uptake channels seem to represent a general feature of plant cells (see Table 1). Voltage-dependent K^+ uptake channels slowly activate when the plasma membrane is hyperpolarized towards potentials more negative than −80 to −100 mV [27, 4]. Upon prolonged stimulation by voltage this K^+ channel does not inactivate. Unlike its functional counterpart in animal cells and outward-rectifying depolarization activated potassium release channels in plants, the voltage-dependence of the inward rectifier is insensitive towards changes in extracellular K^+ concentration [35, 45, Bertl, pers. communication]. The substrate dependence of the current amplitude, however, is characterized by a Michaelis-Menten kinetics with an K_m of 3–4 mM [80].

Very negative membrane potentials and K^+ uptake are generally accompanied by a high H^+-ATPase activity [49], proton release and subsequent acidification of the extracellular/cell wall space [87]. In line with its supposed physiological function this channel is sensitive to pH changes. At neutral pH the K^+ current amplitude is small. Upon acidification the threshold potential of activation shifts more positive and consequently current amplitude and kinetics increase [12, 31, 56]. The plant-specific properties of this K^+ channel, provided by sustained activity upon voltage activation, K^+ selectivity within the range for the K^+ concentration found in the extracellular space of plants,[3] and linkage to the chemiosmotic motor via its voltage- and pH dependence, are in agreement with its physiological role: K^+ uptake and regulation of turgor and volume.

When the first molecular structures of K^+ uptake channels from *Arabidopsis thaliana* (KAT1, AKT1) appeared they where surprisingly homologous to those of the *Shaker* family of voltage-dependent outward-rectifying channels rather than to their physiological animal equivalents [86, 3, 35, 45]. Following functional expression in *Xenopus* oocytes, insect cell lines (e.g. *Sf9*) and yeast the gene product indeed carried the characteristic features of plant inward rectifiers [68, 31, 32]. Since its location within the plant is unknown its cellular function is still an open question. Because of the striking similarities in the electrophysiological fingerprint of KAT1 and the guard cell inward rectifier, KST1, its related gene in *Solanum tuberosum*, has been isolated from this cell type by heterologous screening [56]. Molecular localization and comparison of its *in vivo* (guard cell protoplasts) and *in vitro* (functional expressed in oocytes) properties indicated that inward K^+ currents in potato guard cells seem to result from the activity of the KST1 gene product only, even though a low expression of the potato guard cell AKT1-homologue (SKT1) could be determined (Müller-Röber, pers. communication). Future analysis of the phenotype of transgenic plants with regard to K^+ channel expression may allow a more detailed understanding of its physiological contribution during growth, development (specialization), movement, and reproduction. Analysis of the subunit composition (monomers, homo-, heterooligomers) and stochiometry or presence of regulators within an individual cell type or tissue together with its electrophysiological fingerprint should give new insights towards the understanding of the basis of functional diversity. The clarification of its plant cell-specific differences, such as selectivity, susceptibility to blockers, and threshold potential of voltage activation (Table 1) together with the analysis of mutants and structural chimera between 'red' and 'green' members of K^+ channel families, will provide the missing link between their structure and function. Within this context the question about the molecular structure of the 'green' outward rectifier and whether or not it is related to *Shaker* is of prime interest.

Anion channels are characterized by a great diversity in both branches of the phylogenetic tree of life (for plants see Table 2). This fact as well as the lack of any structural information prevents the identification of the individual counterparts in each phyla.

Nevertheless a plant specific anion channel represents the voltage- and Ca^{2+}-dependent *g*uard *c*ell *a*nion *c*hannel 1 (GCAC1). Its physiological role, a depolarizing activity (see above), and its electrophysiological properties resembles those of voltage-dependent Na^+ channels in animal nerve- or muscle cells [34, for review]. In guard cells and plant cells in general the opening of

[3] For K^+ uptake at nM K^+ concentrations from the soil see K^+/H^+ symporters; [69].

Table 2. Comparison of the basic characteristics of plant anion channels.

Plant	Membrane	Conductance	Selectivity	Activation	Reference(s)
Suspension cels *Asclepias tuberosa*	PM	100 pS	$Cl^- > K^+$	hyperpolarization	[71]
Suspension cells *Amaranthus tricolor*	PM	200 pS	$NO_3^- > Cl^- > K^+ > Asp^-$	hyperpolarization	[95]
Suspension cells *Nicotiana tabacum*	PM	15 pS		depolarization ATP	[108]
Roots *Triticum aestivum* *Triticum turgidum*	PM	4 pS	$NO_3^- \geq Cl^- > I^- \gg PO_4^{3-}, ClO_4^-$		[88]
Mesophyll cells *Peperomia metallica*	TM	65–150 pS	$NO_3^- > Cl^-$	depolarization	[72]
Cotyledons *Arabidopsis thaliana*	PM	5–40 pS		voltage-independent	[47]
Stem cells *Nicotiana tabacum*	PM	86; 146 pS	$Cl^- > K^+$	stretch	[21]
Epidermal cells *Pisum sativum*	PM	300	$NO_3^- > Cl^- = Br^- > I^- > F^- > Mal^{2-}$	hyperpolarization	[17]
Guard cells *Commelina communis*	PM	34 pS; 59 pS	$A^- > K^+$	stretch	[74]
Guard cells *Vicia faba* *Xanthium strumarium*	PM	24–39 pS	$NO_3^- \geq I^- > Br^- > Cl^- > Mal^{2-}$	depolarization Ca^{2+}, ATP	[40, 28, 48, 29, 16]
Guard cells *Vicia faba*	PM	1;33 pS		depolarization Ca^{2+}	[81, 82]
Guard cells *Vicia faba*	PM	27 pS; 13 pS	$Cl^- > K^+$	stretch depolarization	[15]

anion channels results in anion release from the cytoplasm into the extracellular space. Anion efflux is driven by the negative membrane potential and outward-directed anion gradient. The cytoplasmic concentration of e.g. Cl^- is in the order of 50 mM. In contrast to animal cells, where the Cl^- extracellular concentration exceeds that of the cytoplasm, plant cells are exposed to pond water-like media of low ionic strength (2–6 mM Cl^-; [91]). Therefore energy-coupled anion uptake systems, taking advantage of the plasma membrane proton gradient were postulated [94, for recent progress in the conformation of the cotransport hypothesis see 23, 64, 65]. Under conditions of extreme salt stress where the chloride concentration reaches sea water levels in conjunction with membrane potentials far more positive than −100 mV, anion influx mediated by anion channels is thermodynamically possible, only [100].

So far, a detailed analysis of cell-type specific functional properties of plant anion channels has only been provided for guard cells [for review see 33]. In this paragraph we will thus concentrate on GCAC1 located in the plasma membrane of guard cells. Pairs of this cell type, the stomata, form hydrodynamic, turgor-driven valves which are concerned in the control of water loss during photosynthetic CO_2 uptake [62]. Electrically and metabolically isolated from other cells, guard cells receive signals from the environment and within the plant (e.g. hormones and metabolites/ions, reflecting the growth rate, water status/salinity or metabolic status). Given the number and nature of stimuli affecting stomatal movement, guard cells have to perceive and integrate them, possibly through a change in electrical activity (movement of charges, excitability, or single transient/prolonged potential changes). Thereby coordinated changes in volume (mass flow of K^+ and anions) are used to adjust stomatal aperture to improve water use efficiency [for review see 62].

Within this circuit GCAC1 is supposed to present an essential element in membrane polarization as well as mediation of large and rapid anion fluxes.[4] This voltage-gated anion channel is modulated by extracellular hormones (auxin), the photosynthate malate [28] salinity changes (Cl^- concentration) [53, 29, 30] as well as cytoplasmic Ca^{2+} and nucleotides [28]. These ligands are capable to control the activation status (number of active channels and/or probability of opening), transport capacity (such as apoplastic Cl^- concentration affects its unit conductance), position of the voltage sensor, and consequently the voltage threshold of activation (auxin and malate; see Fig. 1).

The latter, modifiers of gating, enable guard cells to shift the working range of anion channel activity along the voltage axis. In this resting position the activity of GCAC1 overlaps with the voltage range of activity of K^+ release channels (Fig. 1B, dotted lines on the right hand side), allowing salt release and down regulation of turgor and volume, a pre-requisite for stomatal closure. Whereas the simultaneous voltage activation of both channels requires a pre-depolarization, such as opening of Ca^{2+} channels (Fig. 2), the presence of extracellular gating modifiers will shift the activation curve of GCAC1 towards the resting potential of the cell to activate this particular anion channel (Fig. 1). Separation of the activation curves for GCAC1 and the K^+ release channel will excite the plasma membrane, a property essential for rapid transduction of changes in the environmental conditions. Range fractionation, with respect to the membrane potential has also been found for K^+ uptake and K^+ release channels (Fig. 1). Triggers like changes in the cytoplasmic Ca^{2+} concentration, H^+-ATPase activity (ΔpH), and in the extracellular K^+ concentration shift the activation threshold of the individual channels along the voltage axis (Fig. 1B; [9, 76, 13, 33], Bertl pers. communication).

Activation and modulation of anion channels through modifiers of gating hence allow to repetitively or sustained interconvert the electrical properties and the resting potential of the plasma membrane from a hyperpolarized state (dominated by the K^+ uptake channel and the H^+-pump) into a depolarized state (dominated by GCAC1 and the K^+ release channel). The maxima and minima of the two extremes [97] might therefore depend on K^+ supply (nutrition), H^+-ATPase activity (energy charge [59]), and Ca^{2+} conductances [49].

Thus GCAC1 might be classified as a 'green' channel, since in contrast to animal ion channels it is gated by voltage as well as ligands the combination of which provides for its plant/guard cell specific properties.

SV-type channels are located in the membrane of vacuoles, the major intracellular store of plants for K^+, Na^+, Ca^{2+} salts, metabolites and lytic enzymes. This organelle with its transport systems embedded in the vacuolar membrane is involved in turgor-formation, the driving force for cell expansion, growth and development. In the vacuole as well as in the lysosomal compartments of 'red' cells V-type H^+-ATPases of highly conserved molecular structure have been detected [61, 57]. This finding, besides others, has led to the assumption that these endosomal organelles share functional properties. Because of the difference in size, up to 90% of the total cell volume

[4] For interconversion between rapid (R-type) and slow (S-type) gating modes of GCAC1 or different anion channels in favour of charge flow on one hand and mass flow on the other, see [48, 81, 16].

Table 3. Slow vacuolar SV-type channels* in the vacuolar membrane of various plant cells.

Plant	Tissue	Conductance	Solution/ mM	Selectivity	Modulation	Reference(s)
Hordeum vulgare	aleurone	26 pS	[100 KCl]	$K^+ \gg Cl^-$	Ca^{2+}- and CaM-activated; blocked by W-7 and TFP	[11]
Beta vulgaris	suspension culture	51–68 pS	[100 KCl]	$K^+ > Cl^-$	Ca^{2+}-activated	[60]
Nicotiana tabacum	mesophyll	60–80 pS	[100 KCl]			[26]
Beta vulgaris conditiva	hypocotyl root	65 pS	[100 KCl]	$K^+ > Cl^-$	Ca^{2+}-activated	[2]
Plantago media *Plantago maritima*	root	60–70 pS	[100 KCl]	$K^+ = Na^+ > Cl^-$	Ca^{2+}-activated	[51]
Chenopodium rubrum	suspension culture	70 pS	[100 KCl]	$K^+ > Cl^-$	Ca^{2+}-activated; blocked by CTX, (+)tubocurarine, W-7, W-5	[8, 105, 106, 107, 63]
Vigna unguiculata	stem	102 ± 4 pS	[100 KCl]	$K^+ \approx Na^+ > Cl^-$	blocked by a vacuolar factor	[52]
Riccia fluitans	thallus	120–140 pS	[200 KCl]			[26]
Beta vulgaris	taproot	120–160 pS	[200 KCl]	$K^+ = Na^+ > Ac^- > NO_3^- > Mal^{2-} > Cl^-$	Ca^{2+}-activated, blocked by cytosolic and vacuolar H^+; blocked by DIDS, Zn^{2+}	[24, 25, 83]
Allium cepa	guard cells	210 ± 17 pS	[200 KCl]	$Na^+ > K^+ > Rb^+ > Cs^+ \gg Cl^-$		[1]
Vicia faba	guard cells	281 ± 20 pS	[200 KCl]	$K^+ > TEA^+ > Cl^- \gg Ca^{2+} \gg Gluc^-$	Ca^{2+}-activated, blocked by cytosolic H^+; blocked by Zn^{2+}, W-7, TFP, calmidazolium	[83, 85]

* All SV-type channels are voltage-gated outward rectifiers.

in plants compared to 10–20% in some chromaffine cells but generally less than 10% in animal cells, patch clamp studies on the ion channel composition and the properties of individual channel types have been restricted to 'green' vacuoles. Consequently, we are unable to decide whether the features of the vacuolar ion channels correspond to the basic task of the lysosomal compartment or exhibit plant specific characters. In Table 3 we hence present cell specific differences of a channel common to plant vacuoles [26].

The slow vacuolar (SV-type) channel, named after its slow voltage-dependent kinetics [24], is activated by depolarized potentials (for use of the new convention see 10) in the presence of elevated cytoplasmic Ca^{2+} only (see [27] for summary). Depending on the cell type and experimental conditions this channel is permeable to cations such as K^+, Na^+, Ca^{2+} and even anions [24, 102, 83]. Even though patch-clamp studies on vacuoles released from their natural habitats became feasable [32], taking into account the activity and current direction of the H^+-pumping V-type ATPase and PP_iase, we still lack conclusive information about its short- and long-term electrical behaviour and the gradients for the various charge carriers *in vivo*. Therefore the alignment of channel properties to their physiological roles is still a problem. Depending on the plant, tissue or cell type, vacuolar ion fluxes may change direction and amplitude within minutes (guard cells, motor cells), hours to days and even month (storage cells in roots or fruits). We thus predicted that e.g. the guard cell plasma membrane and/or vacuolar membrane is equipped with ion transporters of high abundance or transport capacity [6, 83]. Indeed, when comparing the single channel amplitude (turn over) of the SV-type channels (Table 3) the guard cell representative is by far the most conductive. Permeable to ions stored in the vacuole, outward-rectifying SV-type channels might represent release channels for K^+ salts during stomatal closure. The serial arrangement of two membranes and the coordinated regulation of ion channels in the plasma membrane (K^+ release and anion channels) and in the vacuolar membrane (SV-channels) mediates transcellular (vacuolar to extracellular) ion efflux [84] and might display part of a green circuit as well.

Conclusion

To introduce the reader into the molecular biology and biophysics of plant ion channels we have selected three examples for primarily voltage-gated ion channels. Their 'green' features were discussed with respect to our current understanding of structure and function, physiology and plant/cell specificity. Unlike the voltage-gated ion channels in animal cells, plant channels are able to respond to changes within the cell, plant and the environment. Following the analysis of the electrical properties, channel structure and the identification of potential regulators, future studies will be directed towards the understanding of cell specific elements, the expression and assembly of different channel types or subunits to gain heterooligomerous channels with new properties.

Acknowledgements

We thank Heiner Busch, Petra Dietrich, Barbara Schulz-Lessdorf and Ingo Dreyer for assistance in preparing the manuscript. Experimental studies which provided the background for this review were supported by DFG grants to RH.

References

1. Adomeo G, Zeiger E: A cationic channel in the guard cell tonoplast of *Allium cepa*. Plant Physiol 105: 999–1006 (1994).
2. Alexandre J, Lassalles JP, Kado RT: Opening of Ca^{2+} channels in isolated red beet root vacuole membrane by inositol 1, 4, 5-trisphosphate. Nature 343: 567–570 (1990).
3. Anderson JA, Huprikar SS, Kochian LV, Lucas WJ, Gaber RF: Functional expression of a probable *Arabidopsis thaliana* potassium channel in *Saccharomyces cerevisiae*. Proc Natl Acad Sci USA 89: 3736–3740 (1992).
4. Assman SM: Signal transduction in guard cells. Annu Rev Plant Physiol 9: 345–375 (1993).
5. Barbara J-G, Stoeckel H, Takeda K: Hyperpolarisation-activated inward chloride current in protoplasts from suspension cultured carrot cells. Protoplasma 180: 136–144 (1994).
6. Becker D, Zeilinger C, Lohse G, Depta H, Hedrich R: Identification and biochemical characterization of the plasma-membrane H^+-ATPase in guard cells of *Vicia faba* L.. Planta 190: 44–50 (1993).
7. Beilby MJ: Electrophysiology of giant algal cells. Meth Enzymol, 174: 403–443 (1989).
8. Bentrup F-W: Cell physiology and membrane transport. Progress in Botany 51: 70–79 (1989).
9. Bertl A, Slayman CL: Complex modulation of cation channels in the tonoplast and plasma membrane of saccharomyces cerevisiae: Single-channel studies. J Exp Biol 172: 271–287 (1992).
10. Bertl A, Blumwald E, Coronado R, Eisenberg R, Findlay G, Gradmann D, Hille B, Köhler K, Kolb H-A, MacRobbie E, Meissner G, Miller C, Neher E, Palade P, Pantoja O, Sanders D, Schroeder J, Slayman C, Spanswick R, Walker A, and Williams A: Electrical measurements on endomembranes. Science Letters 258: 873–874 (1992).
11. Bethke PC, Jones RL: Ca^{2+}-Calmodulin modulates ion channel activity in storage protein vacuoles of barley aleurone cells. Plant Cell 6: 277–285 (1994).
12. Blatt MR: Ion channel gating in plants: Physiological implications and integration for stomatal function. J Membr Biol 124: 95–112 (1991).
13. Blatt MR: K^+ Channels of stomatal guard cells. Characteristics of the inward rectifier and its control by pH. J Gen Physiol 99: 615–644 (1992).
14. Bush DS, Hedrich R, Schoeder JI Jones RL: Channel-mediated K^+ flux in barley aleurone protoplasts. Planta 176: 368–377 (1988).
15. Cosgrove DJ, Hedrich R: Stretch-activated chloride, potassium, and calcium channels coexisting in the plasma membranes of guard cells of *Vicia faba* L.. Planta 186: 143–153 (1991).
16. Dietrich P, Hedrich R: Conversion of fast and slow

gating modes of GCAC1, a *g*uard *c*ell *a*nion *c*hannel. Planta, in press (1994).
17. Elzenga JTM, Van Volkenburg E: Characterization of ion channels in the plasma membrane of epidermal cells of expanding pea (*Pisum sativum* arg) leaves. J Membr Biol 137, in press (1994).
18. Fairley K, Laver D, Walker NA: Whole-cell and single-channel currents across the plasmalemma of corn shoot suspension cells. J Membr Biol 121: 11–22 (1991).
19. Fairley-Grenot KA, Assmann SM: Evidence for G-protein regulation of inward K^+ channel current in guard cells of fava bean. Plant Cell 3: 1037–1044 (1991).
20. Fairley-Grenot KA, Assmann SM: Permeation of Ca^{2+} through K^+ channels in the plasma membrane of *Vicia faba* guard cells. J Membr Biol 128: 103–113 (1992).
21. Falke L, Edwards KL, Pickard BG, Misler SA: A stretch-activated anion channel in tobacco protoplasts. FEBS Lett 237: 141–144 (1988).
22. Findlay GP, Tyerman SD, Garrill A, Skerrett M: Pump and K^+ inward rectifiers in the plasmalemma of wheat root protoplasts. J Membr Biol 139: 103–116 (1994).
23. Frommer W, Hummel S, Riesmeyer J: Expression cloning in yeast of a cDNA encoding a broad specifity amino acid permease from *Arabidopsis thaliana*. Proc Natl Acad Sci USA 90, 5944–5948 (1993).
24. Hedrich R, Neher R: Cytoplasmic calcium regulates voltage dependent ion channels in plant vacuoles. Nature 329: 833–835 (1987).
25. Hedrich R, Kurkdjian A: Characterization of an anion-permeable channel from sugar beet vacuoles: Effect of inhibitors. EMBO 7: 3661–3666 (1988).
26. Hedrich R, Barbier-Brygoo H, Felle H, Flügge UI: General mechanisms for solute transport across the tonoplast of plant vacuoles: a patch-clamp survey of ion channels and proton pumps. Bot Acta 101: 7–13 (1988).
27. Hedrich R, Schroeder JI: The physiology of ion channels and electrogenic pumps in higher plant. Annu Rev Plant Physiol 40: 539–569 (1989).
28. Hedrich R, Busch H, Raschke K: Ca^{2+} and nucleotide dependent regulation of voltage dependent anion channels in the plasma membrane of guard cells. EMBO J 9: 3889–3892 (1990).
29. Hedrich R, Marten I: Malate-induced feedback regulation of plasma membrane anion channels could provide a CO_2 sensor to guard cells. EMBO J 12: 897–901 (1993).
30. Hedrich R, Marten I, Lohse G, Dietrich P, Winter H, Lohaus G, Heldt HW: Malate-sensitive anion channels enable guard cells to sense changes in the ambient CO_2 concentration. Plant J 6: 741–748 (1994).
31. Hedrich R, Moran O, Conti F, Busch H, Becker D, Gambale F, Dreyer I, Küch A, Neuwinger K, Palme K: Voltage-dependence and high-affinity Cs^+ block of a cloned plant K^+ channel. Eur J Biophys in press (1994).
32. Hedrich R: Technical approaches to studying specific properties of ion channels in plants. In: Neher E, Sakmann B (eds) Single Channel Recordings II. Plenum Press, NY, in press (1994).
33. Hedrich R: Voltage-dependent chloride channels in plant cells: Identification, characterization, and regulation of a guard cell anion channel. In: Guggino WB (ed) Current Topics in Membranes 42, Chloride Channels, pp. 1–34. Academic Press, San Diego (1994).
34. Hille B: Ionic channels of excitable membranes. Sinauer Assoc., Sunderland, MA (1992).
35. Ho K, Nichols CG, Lederer WJ, Lytton J, Vassilev PM, Kanazirska MV, Herbert SC: Cloning and expression of an inwardly rectifying ATP-regulated potassium channel. Nature 362: 31–38 (1993).
36. Hodgkin HL, Huxley AF: A quantitative description of membrane current and its application to conduction and excitation in nerve. J Physiol 117: 500–544 (1952).
37. Hoffmann A, Laubinger W, Dimroth P: Sodium-coupled ATP synthesis in *Propionigenium modestum*: is it a unique system? Biochim Biophys Acta 1018: 206–210 (1990).
38. Iijima T, Hagiwara S: Voltage-dependent K^+ channels in protoplasts of trap-lobe cells *Dionaea muscipula*. J Membr Biol 100: 73–81 (1987).
39. Ilan N, Schwartz A, Moran N: External pH effects on the depolarization-activated K channels in guard cell protoplasts of *Vicia faba*. J Gen Physiol 103: 807–831 (1994).
40. Keller BU, Hedrich R, Raschke K: Voltage-dependent anion channels in the plasma membrane of guard cells. Nature 341: 450–453 (1989).
41. Ketchum KA, Shrier A, Poole RJ: Characterization of potassium dependent currents in protoplasts of corn suspension cells. Plant Physiol 89: 1184–1192 (1989).
42. Ketchum KA, Poole RJ: Cytosolic calcium regulates a potassium current in corn (Zea maize) protoplasts. J Membr Biol 119: 227–288 (1991).
43. Kolb HA, Marten I, Hedrich R: GCAC1 a guard cell anion channel with gating properties like the HH sodium channel. J Membr Biol in press (1994).
44. Kourie J, Goldsmith MHM: K^+ channels are responsible for an inwardly rectifying current in the plasma membrane of mesophyll protoplasts of *Avena sativa*. Plant Physiol 98: 1087–1097 (1994).
45. Kubo Y, Baldwin TJ, Jan YN, Jan LY: Primary structure and functional expression of a mouse inward rectifier potassium channel. Nature 362: 127–133 (1993).
46. Kwart M, Hirner B, Hummel S, Frommer WB: Differential expression of two related amino acid transporters with differing substrate specificity in *Arabidopsis thaliana*. Plant J 4: 993–1002 (1993).
47. Lew RR: Substrate regulation of single potassium and chloride ion channels in *Arabidopsis plasma* membrane. Plant Physiol 95: 642–647 (1991).
48. Linder B, Raschke K: A slow anion channel in guard cells, activating at large hyperpolarization, may be principal for stomatal closing. FEBS Lett 313: 27–31 (1994).
49. Lohse G, Hedrich R: Characterization of the plasma

membrane H^+-ATPase from *Vicia faba* guard cells. Modulation by extracellular factors and seasonal changes. Planta 188: 206–214 (1992).
50. Luan S, Lee W, Rusnack F, Assmann SM, Schreiber SL: Immunosuppressants implicate protein phsphatase regulation of K^+ channels in guard cells. Proc Natl Acad Sci USA 90: 2202–2206 (1993).
51. Maathuis FJM, Prins HBA: Patch clamp studies on root cell vacuoles of a salt-tolerant and a salt-sensitive *Plantago* species. Plant Physiol 92: 23–28 (1991).
52. Maathuis FJM, Prins HBA: Inhibition of inward rectifying tonoplast channels by a vacuolar factor, Physiological and kinetic implications. J Membr Biol 122: 251–258 (1991).
53. Marten I, Lohse G, Hedrich R: Plant growth hormones control voltage-dependent activity of anion channels in plasma membrane of guard cells. Nature 353: 758–762 (1991).
54. Moran N, Ehrenstein G, Iwasa K, Mischke C, Bare C, Satter RL: Potassium channels in motor cells of *Samanea saman*. A patch-clamp study. Plant Physiol 88: 643–648 (1988).
55. Moran N, Fox D, Sutter RL: Interaction of the depolarization-activated K^+ channel of *Samanea saman* with inorganic ions: A patch-clamp study. Plant Physiol 94: 424–431 (1990).
56. Müller-Röber B, Busch H, Ellenberg J, Becker D, Dietrich P, Provart N, Hedrich R, Willmitzer L: Cloning and electrophysiological characterisation of a voltage-dependent K^+ channel predominantly expressed in potato guard cells. EMBO J, submitted (1994).
57. Nelson N, Taiz L: The evolution of H^+-ATPases. Trends Biochem Sci 14: 113 (1989).
58. Ninnemann O, Jauniaux JC, Frommer WB: Identification of a high affinity ammonium transporter from plants. EMBO J 13: 3463–3471 (1994).
59. O'Rourke B, Ramza BM, Marban E: Oscillations of membrane current and excitability driven by metabolic oscillations in heart cells. Science 265: 962–966 (1994).
60. Pantoja O, Dainty J, Blumwald E: Cytoplasmic chloride regulates cation channels in the vacuolar membrane of plant cells. J Membr Biol 125: 219–229 (1992).
61. Pederson PL, Carafoli E: Ion motive ATPases. II Energy coupling and work output. Trends Biochem Sci 12 (5): 186–189 (1987).
62. Raschke K: Movements of stomata. In: Haupt W, Feinleib ME (eds) Encyclopedia of Plant Physiology, Bd. 7, Physiology of Movements. Springer Verlag, Berlin (1979).
63. Reifarth FW, Weiser T, Bentrum F-W: Voltage- and Ca^{2+}-dependence of the K^+ channel in the vacuolar membrane of *Chenopodium rubrum* L. suspension cells. Biochim Biophys Acta 1192: 79–87 (1994).
64. Riesmeier JW, Willmitzer L, Frommer WB: Isolation and characterization of a sucrose carrier cDNA from spinach by functional expression in yeast. EMBO J 11: 4705–4713 (1992).
65. Sauer N, Tanner W: Molecular biology of sugar transporters in plants. Bot Acta 106: 277–286 (1993).
66. Sauer N, Baier K, Gahrtz M, Stadler R, Stolz J, Truernit E: Sugar transport across the plasma membranes of higher plants. Plant Mol Biol 26: 1671–1679 (1994).
67. Schachtman DP, Tyerman SD, Terry BR: The K^+/Na^+ selectivity of a cation channel in the plasma membrane of root cells does not differ in salt-tolerant and salt-sensitive wheat species. Plant Physiol 97: 598–605 (1991).
68. Schachtmann DP, Schroeder JI, Lucas WJ, Anderson JA, Gaber RF: Expression of an inward-rectifying potassium channel by the Arabidopsis KAT1 cDNA. Science 258: 1654–1658 (1992).
69. Schachtmann DP, Schroeder JI: Structure and transport mechanism of a high-affinity potassium uptake transporter from higher plant. Nature 370: 655–658 (1994).
70. Schauf CL, Wilson KJ: Effects of ABA on K^+ channels in *Vicia faba* guard cell protoplasts. Biochem. Biophys Res Com 145: 284–290 (1987).
71. Schauf CL, Wilson KJ: Properties of single K^+ and Cl^- channels in *Asclepias tuberosa* protoplasts. Plant Physiol 85: 413–418 (1987).
72. Schönknecht G, Hedrich R, Junge W, Raschke K: A voltage-dependent chloride channel in the photosynthetic membrane of a higher plant. Nature 336: 589–592 (1988).
73. Schroeder JI, Hedrich R, Ferandez JM: Potassium-selective single channels in guard cell protoplasts of *Vicia faba*. Nature 312: 361–362 (1984).
74. Schroeder JI, Raschke K, Neher E: Voltage-dependence of K^+ channels in guard-cell protoplasts. Proc Natl Acad Sci USA 84: 4108–4112 (1987).
75. Schroeder JI: K^+ transport properties of K^+ channels in the plasma membrane of *Vicia faba* guard cells. J Gen Physiol 92: 667–683 (1988).
76. Schroeder JI, Hagiwara S: Cytosolic calcium regulates ion channels in the plasma membrane of *Vicia faba* guard cells. Nature 338: 427–430 (1989).
77. Schroeder JI, Hedrich R: Involvement of ion channels and active transport in osmoregulation and signalling of higher plant cells. Trends Biochem Sci 14: 187–192 (1989).
78. Schroeder JI: Quantitative analysis of outward rectifying K^+ currents in guard cell protoplasts from *Vicia faba*. J Membr Biol 107: 229–235 (1989).
79. Schroeder JI, Hagiwara S: Voltage-dependent activation of $Ca2^+$-regulated anion channels and K^+ uptake channels in *Vicia faba* guard cells. In: Leonard RT, Hepler PK (eds) Current Topics in Plant Physiology 4: Calcium in Plant Growth and Development, pp. 144–150. American Society of Plant Physiologists, Rockville, Maryland (1990).

80. Schroeder JI, Fang HH: Inward-rectifying K^+ channels in guard cells provide a mechanism for low-affinity K^+ uptake. Proc Natl Acad Sci USA 88: 11583–11587 (1991).
81. Schroeder JI, Keller B: Two types of anion channel currents in guard cells with distinct voltage regulation. Proc Natl Acad Sci USA 89: 5025–5029 (1992).
82. Schroeder JI, Schmidt C, Sheaffer J: Identification of high-affinity slow anion channels blockers and evidence for stomatal regulation by slow anion channels in guard cells. Plant Cell 5: 1831–1841 (1993).
83. Schulz-Lessdorf B, Hedrich R: Protons and calcium modulate SV-type channels in the vacuolar-lysosomal compartment – Interaction with calmodulin antagonists. J Gen Physiol, submitted (1994).
84. Schulz-Lessdorf B, Dietrich P, Marten I, Lohse G, Busch H, Hedrich R: Coordination of plasma membrane ion channels during stomatal movement. In: Leigh RA (ed) The SEB Symposium 48, Membrane Transport in Plants and Fungi. The Company of Biologists Limited, Cambridge (1994).
85. Schulz-Lessdorf B, Hedrich R: pH and Ca^{2+} modulate the activity of ion channels in the vacuolar membrane of guard cells – possible interaction with calmodulin. Poster Abstract [377], Botanikertagung Berlin, FRG (1992).
86. Sentenac H, Bonneaud N, Minet M, Lacroute F, Salmon J-M, Gaymard F, Grignon C: Cloning and expression in yeast of a plant potassium ion transport system. Science 256: 663–665 (1992).
87. Shimazaki K, Iino M, Zeiger E: Blue light-dependent proton extrusion by guard cell-protoplasts of *Vicia faba*. Nature 319: 324–326 (1986).
88. Skerrett M, Tyermann SD: A channels that allows inwardly directed fluxes of anions in protoplasts derived from wheat roots. Planta 192: 295–305 (1994).
89. Spalding EP, Goldsmith MHM: Activation of K^+ channels in the plasma membrane of *Arabidopsis* by ATP produced photosynthetically. Plant Cell 5: 477–484 (1993).
90. Spalding EP, Slayman CL, Goldsmith MHM, Gradmann D, Bertl A: Ion channels in *Arabidopsis plasma* membrane. Plant Physiol 99: 96–102 (1992).
91. Speer M, Kaiser WM: Ionic relations of symplastic and apoplastic space in leaves from *Spinatia oleracea* L. and *Pisum sativum* L. under salinity. Plant Physiol 97: 990–997 (1991).
92. Stoeckel H, Takeda K: Calcium-activated, voltage-dependent, non-selective cation currents in the endosperm plasma membrane from higher plants. Proc R Soc Lond B 237: 213–231 (1989).
93. Sussmann MR: Shaking *Arabidopsis thaliana*. Science 256: 619 (1992).
94. Sze H: H^+-translocating ATPases. Advances using plasma membrane vesicles. Annu Rev Plant Physiol 36: 175–208 (1985).
95. Terry BR, Tyerman SD, Findlay GP: Ion channels in the plasma membrane of Amaranthus protoplasts: One cation and one anion channel dominate the conductance. J Membr Biol 121: 223–236 (1991).
96. Thiel G, Homan U, Gradmann D: Microscopic elements of electrical excitation in chara: Transient activity of Cl^- channels in the plasma membrane. J Membr Biol 134: 53–66 (1993).
97. Thiel G, McRobbie EAC, Blatt MR: Membrane transport in stomatal guard cells. Importance of voltage control. J Membr Biol 126: 1–18 (1992).
98. Thuleau P, Ward JM, Ranjeva R, Schroeder JI: Voltage-dependent calcium-permeable channels in the plasma membrane of carrot suspension cells. EMBO J 13: 2970–2975 (1994).
99. Tsay Y-F, Schroeder JI, Feldmann KA, Crawford NM: The herbicide sensitivity gene *CHL1* of *Arabidopsis thaliana* enodes a nitrate-inducible nitrate transporter. Cell 72: 705–713 (1993).
100. Tyerman SD: Anion channels in plants. Annu Rev Plant Physiol Mol Biol 43: 351–373 (1992).
101. Vogelzang SA, Prins HBA: Patch clamp analysis of the dominant plasma membrane K^+ channel in root cell protoplasts of *Plantago media* L. Its significance for the P and K state. J Membr Biol 141: 113–122 (1994).
102. Ward JM, Schroeder JI: Ca-activated K^+ channels and Ca-induced Ca release by slow vacuolar ion channels in guard cells vacuoles implicated in the control of stomatal closure. Plant Cell 6: 669–683 (1994).
103. Wegner LH, Raschke K: Ion channels in the xylem parenchyma of barley roots. Plant Physiol 105: 799–813 (1994).
104. Wegner LH, DeBoer AH, Raschke K: Properties of the K^+ inward rectifier in the plasma membrane of xylem parenchyma cells from Barley roots: Effects of TEA^+, Ca^{2+}, Ba^{2+}, and La^{3+}. J Membr Biol, in press (1994).
105. Weiser T, Bentrup F-W: (+)-Tubocurarine is a potent inhibitor of cation channels in the vacuolar membrane of *Chenopodium rubrum* L. FEBS Lett 277: 220–222 (1990).
106. Weiser T, Bentrup F-W: Pharmacology of the SV channel in the vacuolar membrane of *Chenopodium rubrum* suspension cells. J Membr Biol 136: 43–54 (1993).
107. Weiser T, Blum W, Bentrup F-W: Calmodulin regulates the Ca^{2+}-dependent slow-vayacuolar ion channel in the tonoplast of *Chenopodium rubrum* suspension cells. Planta 185: 440–442 (1991).
108. Zimmermann S, Thomine S, Guern J, Barbier-Brygoo H: An anion current at the plasmamembrane of tobacco protoplast shows ATP-dependent voltage regulation and is modulated by auxin. Plant J 6: 707–716 (1994).

Plant Molecular Biology **26**: 1651–1670, 1994.

Transporters for nitrogenous compounds in plants

Wolf B. Frommer*, Marion Kwart, Brigitte Hirner, Wolf Nicolas Fischer, Sabine Hummel and Olaf Ninnemann
IGF, Ihnestrasse 63, D-14195 Berlin, Germany (author for correspondence)*

Received 23 March 1993; accepted 26 April 1994

Key words: amino acid transport, ammonium uptake, nitrate uptake

Introduction

Cells are surrounded by a membrane which separates intracellular from extracellular processes and allows controlled contact and exchange of the intracellular space with the environment. Communication is made possible by integral membrane proteins at the interface between intra- and extracellular space, thus establishing the contact to neighbouring cells and the surrounding medium. The membrane proteins can function either as signal transducers or as transporters. Transport proteins (permeases) transfer molecules across cell membranes, which can have nutritional or informational value for the cell. By using these transduction pathways, external signals can exert control on biochemical processes inside the cell. As a matter of fact, one molecule can be a nutrient and a regulatory signal at the same time. As has been demonstrated elegantly in many organisms, biochemical pathways are subject to control by metabolites, either by induction or repression mechanisms. The classical example is the induction of transcription of the lactose permease gene by its substrate [88]. In addition, substrate-inducible transport systems have been described also for other substrates and in other species [6, 31, 118]. Catabolite repression is another example where the presence of a metabolite in the medium can effect transport processes [138]. The combination of these two transport control mechanisms, positive and negative regulation, allows an optimal adaptation of metabolism to multiple sources of nutrients. Only those transport systems and their associated biochemical pathways are active that are actually required in a given environmental situation. If a substrate cannot enter the cell due to the absence of the respective carrier, the whole pathway will be silent. Cells can therefore respond flexibly and economically to different environmental requirements.

This review will describe one example of transport across cell membranes, the transport of nitrogenous compounds in plants. Nitrogen is an essential component of most biological molecules. The uptake of nitrogen is of special importance for plant development because, according to Liebig's hypothesis, the growth of a plant is limited by the nutrient that is present in limiting amounts. Like phosphate and potassium, nitrogen is one of the major limiting nutrients in the soil. Nitrogen metabolism and transport have been well characterized in unicellular organisms, and the regulatory networks have been worked out. However, relatively little is known in higher plants, especially regarding nitrogen transport processes. Main causes of this lack of knowledge are the difficulty to purify integral membrane proteins, to assay their activity and the complexity of their regulation. During the past few years, functional expression of heterologous genes in *Xenopus* oocytes and in yeast cells resulted in the identification and characterization of many transporter genes from a variety of higher organisms [5]. The availability of this approach has drawn the attention of many research groups to the important role of transport processes. This review will concentrate on transport across the plasma membrane and genes encoding transporters involved in the uptake and transfer of nitrogen, with an emphasis on amino acid transport and its regulation.

Three different levels of complexity are involved in nitrogen uptake and transfer.

1. Uptake. Nitrogen is available in the soil in different forms. The major sources of nitrogen can be nitrate, ammonium and, to a lower extent, also amino acids. The resorption of nutrients from the soil is a main function of the roots and is mediated only to a minor extent through other organs from the atmosphere. If the nitrogenous substance *per se* is not permeable, uptake has to occur across the membrane in the outer cell layers, i.e. the root hairs and in the cortex cells. Reduction, fixation and use may take place directly in the same or neighbouring cells or the material may be transported through the long-distance transport network to other organs. Apoplastic transport could be a means for translocation within the cortex, but the casparian stripe blocks further transfer to the stele via the apoplast. After symplastic transfer to the stele, the nutrients have to be transferred to the xylem by additional transport systems [84]. For those plants that live in symbiosis with nitrogen-fixing organisms such as bacteria or mycorrhiza, the use of atmospheric dinitrogen (N_2) is indirectly possible.

2. Translocation and processing. The location for the processing, i.e. the reduction and fixation into amino acids and the synthesis of proteins, depends on species and on conditions such as availability of precursors, energy, and other factors like pH of soil [4]. If nitrate is taken up, it may in certain species be reduced directly in the root. In case of ammonium uptake, the assimilation very probably takes place directly in the root. Reduced nitrogen is transported mainly in the form of amino acids, amides and ureides within the vascular system. Alternatively, nitrate is transported through the xylem to the leaves, where the photosynthetic energy can be used for the production of amino acids. Amino acids are distributed within the plant through both xylem and phloem. For ease of presentation we will refer only to the transport of amino acids, which is meant to cover also amides and ureides.

3. Allocation. Reduced nitrogen can be stored transiently in different vegetative organs, for example as vegetative storage proteins or in storage tissues of reproductive organs such as seeds. The nitrogen resources needed for the storage in reproductive organs only partially derive from direct uptake but are rather derived from nitrogen that has been taken up before. Therefore the plant needs to reallocate nitrogen during development. Exporting sources for transiently stored nitrogen can be roots or leaves or, in early stages of development, the endosperm and the cotyledons. The transport routes in this case are both xylem and phloem. Thus the situation is much more complex than for sucrose, which in many species represents the exclusive transport form for carbohydrates, whose transport is restricted to the phloem and for which in most cases leaves or cotyledons are the only sources of carbon [recent reviews by 13, 25].

Anatomical, physiological, biochemical and genetic evidence has been presented for the necessity and activity of carrier-mediated transport processes responsible for the uptake and transfer of nitrate, ammonium and amino acids.

Nitrate transport

In most species, the major compound absorbed from the soil is nitrate. Nitrate uptake seems to be mediated by specific transport systems [18, 27, 110]. The uptake is electrogenic and has been shown to act as a symport with at least two protons [26]. The uptake underlies a complex regulation in order to allow the plant to adapt to varying external availability of nitrate (up to 4 orders of magnitude [reviewed by 58, 86]). Further arguments for the involvement of carrier-mediated transport processes stem from genetic data. Mutants have been identified in *Aspergillus*, *Chlamydomonas* and *Arabidopsis*. The *Aspergillus crnA* mutant was used to isolate a nitrate transporter gene encoding for a hydrophobic membrane protein which is related to the sugar transporter superfamily [124]. Mutants in *Chlamydomonas* indicate that different transporters for nitrate and nitrite exist. The genes involved in nitrate uptake and reduction are organized in a gene cluster. The putative transporter genes are highly regulated and do not crosshybridize with the *CRNA* gene [74]. *Arabidopsis* mutants were selected for germination on media containing the toxic nitrate analogue chlorate. The gene for this transporter was cloned from a T-DNA-tagged allele of chlorate-resistant mutants [120]. The *CHL1* gene also encodes an integral membrane protein. The mutant (*chl1*) still retains residual nitrate uptake capacity and is impeded also in chloride and potassium uptake [18, 103]. The kinetic data and remaining uptake activity in the mutants indicate that additional systems must be present. These may be proteins related to the *Arabidopsis CHL1* or to the *Aspergillus CRNA* genes. Apart from nitrate uptake, there is also evidence for nitrite transport, that is not yet characterized in detail. In higher plants, the absorbed nitrate or nitrite is either reduced in the roots and exported in the form of amino acids or the nitrate is transported to the leaves, where it is reductively assimilated to produce amino acids. Thus theoretically at least three carrier systems are necessary in different tissues: one for uptake into cortex cells, one for uptake into the xylem and one for release in leaves.

Ammonium transport

Under normal field conditions plants also seem to be able to take up ammonium. This holds true especially in rice cultivation, forest ecosystems, in the Arctic tundra and even in winter varieties of cereals growing in cold soils, and in species growing in acidic or agriculturally extensively used soils. A common feature of these conditions is the inhibition of nitrification resulting in ammonium to become the prevalent source of nitrogen [cited in 134]. In contrast to nitrate, ammonium occurs mainly in bound forms in the soil, thus requiring efficient systems for release and ingestion. Ammonium transport has been a matter of debate for many years, since ammonia as an uncharged small molecule seems to be membrane-permeable. Already in the past century, ammonium uptake into plants was studied [17]. These investigations supported the conclusion that NH_3 passively diffuses across the membrane. Furthermore, multiple studies of organelles, i.e. chloroplasts, indicated that these membranes are permeable towards NH_3. The general conclusion was therefore that no specific transporters are required. However, genetic studies have shown that in a multitude of organisms mutations can strongly affect ammonium uptake [50]. These genes may either encode uptake systems [19] or essential constituents functioning as retrieval systems for ammonia that is passively leaking out of the cells [49, 50]. In plants, biphasic uptake kinetics were determined consisting of a saturable carrier-mediated system operating at low ammonium concentrations and a linear diffusive component at elevated ammonium concentrations [23, 123, 135]. The significance of the high affinity system from rice roots may be in allowing the plant to absorb sufficient nitrogen from the very low levels present in the rhizosphere. Despite the fact that in several organisms mutants have been identified that are affected in ammonium transport and despite extensive trials to identify the respective proteins, only genes encoding putative regulatory functions have been identified. Using heterologous complementation of a yeast mutant deficient in ammonium uptake, we have recently succeeded in isolating the first high-affinity ammonium transport system from the plasma membrane of plants [72]. The system operates in an energy-dependent manner and is probably driven by the membrane potential. Whether it constitutes a typical channel or a transporter remains to be shown. An interesting question concerns the number of systems involved. Under certain conditions significant amounts of ammonium can be found in the xylem. Therefore, similar to nitrate uptake, transporters are required for uptake into the cortex, transfer and release into/from the vessels, and, in addition, specific retrieval systems that prevent loss in the form of ammonia. Further studies will enable a better understanding of the role and function of such transporters. These results demonstrate the efficiency and elegance of molecular tools in studying even systems which are marginally understood at the physiological level.

In those plants which are able to use atmospheric nitrogen via symbiosis with dinitrogen-fixing organisms, dinitrogen reduced in the bacteria has to be transferred into the plant cytoplasm. As the rhizobial ammonium carriers are repressed in nodules, it is assumed that ammonia diffuses across the bacteroid and peribacteroid membranes and is trapped in the cell wall due to protonation in the more acidic milieu [for review see 51]. The assumption that ammonium is the major transport form for transfer across the two membranes is supported by the finding of only low capacities for the uptake of various amino acids [122]. So far the only high-affinity transport system found on the bacteroid membrane was specific for glutamate. However this system was absent from the peribacteroid membrane [121].

The ammonium that is taken up or produced during dinitrogen and nitrate reduction feeds into the biosynthesis of amino acids. Excess amino acids produced during assimilation can subsequently be exported via the vascular system to organs and cells that are dependent on external supply with reduced nitrogen due to the lack of assimilatory processes.

Amino acid transport

From the places of synthesis which can, depending on the species, be either predominantly roots or leaves, the amino acids are transported to other organs via both phloem and xylem. The long-distance translocation of amino acids can be studied physiologically by analysing the constituents of phloem and xylem sap. Different techniques for collecting efflux from incisions or detopped plants, from isolated cotyledons or from aphid stylets have been successfully used to isolate fluids [for review see 111]. The main constituents of the xylem sap are amino acids whereas in the phloem sap amino acids are the second prevalent compound beside sucrose [77, 78].

Physiological studies on amino acid transport in roots

In species where amino acid biosynthesis takes place mainly in roots, export has to occur via the xylem. Xylem transport is also found in leaves during the mobilization of stored nitrogen. Roots have been shown to possess uptake systems for amino acids. The involvement of active transport is supported by the finding that, compared to root cells, higher concentrations of amino acids are found in root exudate [101]. The major amino acids from root pressure exudates are glutamine in *Brassica oleraceae*, a species related to *Arabidopsis* [1], and asparagine and glutamine in *Lupinus* [109]. Amino acids may play a role in nitrogen supply of plants growing in soils where significant amounts of amino acids are present or where microorganisms provide amino acids to the plant. Data about the transfer of nitrate and amino acids from the roots into the xylem by using root pressure exudates and studies in cotyledons show significant differences, indicating that different carrier systems are involved [102].

Physiological studies on amino acid transport in leaves

Analysis of the composition of the phloem sap of spinach shows that amino acids such as glutamic and aspartic acid represent ca. 17% of the solutes in exudate whereas, for example, proline is only a marginal constituent [90]. Comparison of amino acid concentration in the cytosol of mesophyll cells to that of phloem sap from barley shows that the concentration of a number of amino acids is similar [140]. Concentration differences were only found for aspartate, glutamate, valine and lysine, that differed in light and darkness. Similar results were found in spinach where several amino acids (asparagine, glutamate, alanine, serine, glycine, valine and threonine) were found to be 1.1- to 2-fold higher in the phloem as compared to the cytosolic concentration in mesophyll cells [89]. The interdependence of the amino acids in the phloem sap and amino acid metabolism in the surrounding organs has been reported previously [22]. Thus the amino acid composition of mesophyll and phloem is similar, but the concentration gradient is clearly lower than sucrose gradients which can be more than tenfold. With few exceptions, the amino acid composition of phloem sap parallels that of the xylem, suggesting that the loading of amino acids is not very selective. In *Ricinus*, differential loading of amino acids into the phloem has been described. External application of glutamine to cotyledons led to large increases of amino acids in the phloem sap, whereas glutamate, alanine and arginine were loaded less efficient. Under these conditions, feeding of glutamine and alanine led to an up to tenfold accumulation of the respective amino acid in the phloem [101]. A strong interdependence with sucrose transport was observed, which could be due to effects on the mass flow in the phloem. A high sucrose content is related to low amino acid concentrations and vice versa. The phloem loading system of leaves was found to have a different specificity from that of the cotyledons.

The observation that the composition of the transport fluids is largely dependent on the actual concentration of respective amino acids can be interpreted in a way that a single transport system with a broad substrate specificity is involved in phloem loading.

Physiological studies on amino acid import into developing seeds

Removal of the developing seed in species producing large seeds has enabled to determine the composition of metabolites supplied to the developing embryo [reviewed by 119]. This system has allowed to determine factors that influence this 'unloading' process. Several studies have shown that amino acids arrive mainly in the phloem suggesting that a xylem-to-phloem transfer must occur [for review see 81]. Results obtained by studying the composition of phloem sap in rice are interpreted either as metabolic changes during translocation or selective unloading of aspartate and glutamate and an exchange between phloem and xylem along the translocation pathway [32].

Biochemical studies of amino acid transport

Regarding amino acid transport in plants, the best studied systems are green algae like *Chlorella*. The reason is that unicellular organisms are simpler to handle for transport studies. Inhibition studies indicated the presence of more than one transporter as glutamine transport is not completely inhibited by protonophores [57]. This system shows carrier-mediated kinetics and was named the 'unprotonated system'. Three inducible systems could be identified, one being specific for basic, one for neutral and one for a number of different amino acids which was therefore described as the general system. As the three systems do not cover the uptake of all amino acids, additional specific constitutive systems seem to be present. This assumption is further supported by an analysis of two *Chlorella* amino acid transport mutants that are still able to take up a number of different amino acids [109]. However the genes have not yet been isolated. The question is whether higher plants also contain such a multiplicity of systems.

In a variety of species the transport of radiolabelled amino acids has been measured either by feeding experiments, by determining the uptake into leaf discs, abraded leaves, protoplasts or cell cultures [87]. Since the development of efficient methods for isolating plasma membrane vesicles, comparatively few studies were carried out in this system, too. Most of the studies have concentrated on the phloem loading systems in leaves and were often carried out in parallel with sucrose transport studies. Those studies led to models for phloem loading with amino acids which are similar to what has been postulated previously for sucrose loading. It involves release of amino acids into the cell wall and subsequent loading of the phloem by proton-coupled symport that is energized by the H^+-ATPase (Fig. 1).

McNeil [68] fed valine to detached shoots of *Lupinus* and found evidence for carrier-mediated

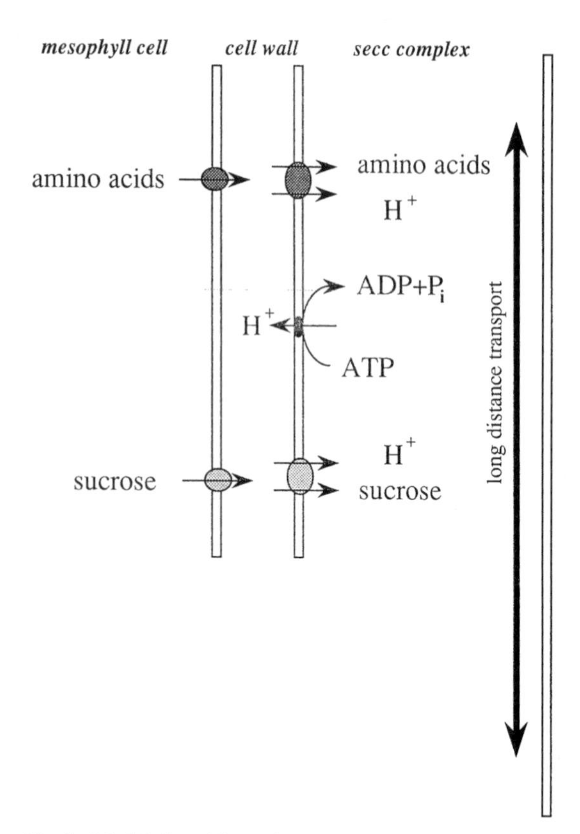

Fig. 1. Model for phloem loading. A broad specific amino acid permease catalysing proton symport is localized at the companion cell plasma membrane and is responsible for loading. Active transport is energized by the proton gradient generated by a H^+-ATPase. A second carrier system is postulated for efflux from the adjacent cells. The same principal mode has been put forward for loading with sucrose previously.

phloem loading. Studies on the export of exogenous leucine from abraded leaves still attached to the plant showed the existence of three saturable components [107]. The uptake kinetics are often complex biphasic or multiphasic isotherms like in barley roots [113]. Leaf discs from *Nicotiana* show uptake of valine by two systems with low and high affinity, 40 μM and 15 mM [7, 8]. There is evidence that the high-affinity system works with proton symport [16, 94]. Van Bel [126, 127] proposed that in *Commelina* the amino acids are maintained in the mesophyll by the high-affinity component, whereas the release near the conducting complex is mediated by the low-affinity system [125]. All tissues of tomato show biphasic kinetics for alanine uptake [128, 129]. Studies in more defined systems, i.e. in plasma membrane vesicles from *Ricinius*, also show biphasic kinetics. The ΔpH- and $\Delta\psi$-dependent uptake of glutamine was saturable with a K_m of 0.35 mM [139]. The uptake of several amino acids in vesicles from sugar beet also followed biphasic kinetics and was coupled to the cotransport of protons [59]. All strategies described up to now may be limited in their validation by involving multiple cell types with potentially different carriers. Therefore molecular studies with individual proteins are required for discrimination.

In general, the V_{max} for sucrose transport is much higher than for amino acids [107]. This could explain why the accumulation of amino acids in the phloem is lower than for sucrose. All amino acids seem to be translocated with the same velocities and mass transfer rates independent of their prevalence in the phloem. This has supported the assumption that only one general carrier system is present.

In summary, the number of transport systems present in the plant is dependent on two variables, i.e. cell and substrate specificity. Regarding tissue specificity, transport systems are involved in the uptake into roots, the mobilization from roots in the xylem, the unloading of the phloem in roots and possibly direct transfer between phloem and xylem. In leaves two systems are required for phloem loading (Fig. 1), and possibly one for uptake from the xylem. Sink organs such as seeds also require at least two systems, one in the maternal tissue for unloading of the phloem (in a broader sense) and one for the uptake into the developing embryo. The situation is thus very complex and if we take a high degree of regulation and tissue specificity into account, we may expect the presence of many genes or gene families in the plant.

Molecular studies on amino acid transport

Molecular studies of metabolite transport across the plasma membrane have been neglected for many years due to the problems associated with the identification and purification of the respective proteins. To circumvent the problems associated with a biochemical identification of transport proteins, yeast complementation systems were developed which seem to be efficient tools for isolating genes with specific functions also from heterologous organisms, i.e. potassium channels and ammonium and sucrose transporters from plants [2, 72, 90, 91, 106]. The complementation of different yeast mutants deficient in amino acid transport has allowed to isolate several gene families from *Arabidopsis* encoding amino acid transporters [24, 41, 56]. The *Saccharomyces cerevisiae* strain 22574d, which carries mutations in general amino acid, proline and γ-aminobutyric acid permease genes, is unable to grow on media containing citrulline, proline or γ-aminobutyric acid as the sole nitrogen source [43, 44]. To identify plant amino acid transporter genes, yeast strain 22574d was transformed with episomal plasmids containing a cDNA library derived from *Arabidopsis* seedlings under control of the phosphoglycerate kinase promoter [69]. The second cloning system is based on complementation of histidine auxotrophy. A strain lacking a functional histidine uptake and biosynthesis pathway can be used as a selection system on complex media containing histidine. Ammonium has to be present in order to suppress the general amino acid permease. For this purpose, the strain JT16 that is deficient in both histidine and arginine uptake has been chosen, as the selection is tight also on high concentrations of histidine and thus should be more sensitive than JT48 [117]. This system had also been used to try and identify human histidine transporters. However this resulted only in the isolation of a human gene that is able to derepress the general yeast amino acid permease GAP1 [105]. Complementation of the different yeast mutants with the *Arabidopsis* cDNA library enabled the isolation of at least two gene families encoding integral membrane proteins that are able to mediate amino acid transport. The first group of related genes consists of at least five members and was named amino acid permease (AAP) family [24, 56, 21a]. Two members of the family AAP1 and AAP2 were analysed in more detail. The two permeases share 56% identical amino acids. The yeast mutant JT16 has also allowed to isolate a cDNA clone from a different *Arabidopsis* ecotype named NAT2, which is 99% identical to AAP1 [41]. The predicted polypeptides of about 53 kDa are highly hydrophobic and contain 9–12 putative membrane-spanning regions. Several other plant cDNAs are also able to mediate amino acid uptake into the yeast mutants. Some of them also encode hydrophobic membrane proteins but are not closely related to the AAP family. Analysis of the genes and their function is in progress [25a, 25b].

In bacteria, fungi and animals, multiple transport systems with different specificities and transport modes have been identified. Many of the respective genes have been isolated and most of them seem to be distantly related to each other. In most cases genetic tools were used to clone the genes due to the difficult biochemistry of integral membrane proteins. In mammalian organisms the identification of amino acid transporter genes was made possible by expression cloning on *Xenopus* oocytes [29, 46, 137]. Amino acid permeases can be grouped into several subfamiles. The first family of mainly sodium dependent systems is conserved from bacteria to mammals and includes transporters specific for neutral and acidic amino acids and for proline [45, 54, 108]. The bacterial proton glutamate and aspartate cotransporters, but also the *Rhizobium* dicarboxylate transporter belong to this family [96, 133]. The second subfamily contains all *Saccharomyces* amino acid transporters identified so far and is related to the mammalian cationic amino acid transporters, the *Aspergillus* proline transporter and several bacterial transport systems such as AroP and PheP which transport aromatic amino acids, the GABA permease, the lysine-specific system LysP and the branched chain amino acid transporters BraZ [33, 37, 83, 114, 116, 131]. An amino acid transporter from *Arabidopsis* that is related to this family was recently identified, that has a low affinity and broad specificity [25b]. The third subfamily is constituted by the plant AAP family, the bacterial tyrosine and neutral amino acid permeases TyrP and LivI, LivII and LivIII and the neutral amino acid specific *Neurospora* Mtr permease [38, 55]. Several unusual neurotransmitter transporters also are distantly related to this family [62]. The fourth subfamily are the ABC proteins which contain ATP-binding cassettes [35]. Among this large family which contains the multidrug resistance proteins are also some amino acid transporters like the bacterial histidine permease.

Substrate specificity of plant amino acid transporters

The multiple transporters present in yeast have different substrate specificities. There is a general permease that is stereo-unspecific and recognizes all amino acids and some ureides, but also a multiplicity of more or less specific systems [28, 138]. *Chlorella* disposes of at least seven transport systems with differing substrate specificity (E. Komor, personal communication). In higher plants, the number of systems is controversial and reaches from one to several [87]. Three distinct transport systems, one for neutral including glutamine, asparagine and histidine, one for acidic and one for basic amino acids, were found in sugar-cane suspension cells [142]. In *Ricinus*, glutamine uptake is mediated through a broad specificity transport system which shows low or no competition by lysine and aspartic acid [139]. Furthermore, differences were observed between different tissues and different species [60, 139]. In sugar beet leaves, the complex results of competition data of uptake into vesicles of six different radiolabelled amino acids was interpreted as evidence for a minimum of four different permeases. At least one carrier each is specific for acidic, one for basic and two for neutral amino acids [59, 60]. A detailed competition analysis of the two neutral systems indicates that the α-amino group is the major determinant of the substrate specificity of the neutral systems, whereas the side chains play only a minor role [61]. As the kinetics are complex, however, a direct analysis of single carriers is necessary to unequivocally demonstrate the substrate specificity. A tool for this is the expression of the permease genes in heterologous cell types, i.e. oocytes or yeast. However these studies also have some limitations, because they are complicated by the presence of endogenous carriers and because competition does not necessarily mean that a compound is actually transported. Either direct uptake studies in a set of different yeast mutants, detailed kinetic studies or the use of purified and reconstituted proteins in synthetic vesicles will be necessary as a final proof for the actual transport specificity.

The transport activities of AAP1 and AAP2 were determined directly in yeast mutants deficient in four different amino-acid permeases, i.e. proline, citrulline, γ-amino butyric acid and histidine transport [56]. Both genes mediate uptake of proline, citrulline and histidine but not γ-amino-butyric acid. Proline uptake was studied in most detail and is saturable with a K_m in the range of 60–120 μM, therefore constituting the high-affinity component described [24, 56]. Since the endogenous activities of alanine uptake are very low, alanine uptake could also be demonstrated for NAT2/AAP1. The K_m of NAT2 for alanine was 300 μM [41]. Indications for the specificity were drawn from competition studies with the proteogenic amino acids. These data further support the assumption that AAPs transport a broad spectrum of amino acids. The strongest competitors for *L*-proline uptake by AAP1 were glutamic acid, cysteine, methionine and phenylalanine, intermediate inhibition was achieved by neutral amino acids such as glutamine and leucine, whereas the basic amino acids did not significantly compete for *L*-proline transport. Among the best competitors were glutamine and glutamate, major transport amino acids of both xylem and phloem. Citrulline, an important transport amino acid especially in nitrogen-fixing species, is also a good competitor. The two *Arabidopsis* carriers obviously differ in their substrate specificity with respect to basic amino acids. A detailed comparison of the competition data of AAP1 and AAP2 is shown in Fig. 2. Positive charges protruding into the lumen of the putative channel formed by the hydrophobic domains might reduce the passage of basic amino acids in case of AAP1. Taken together, the specificity of

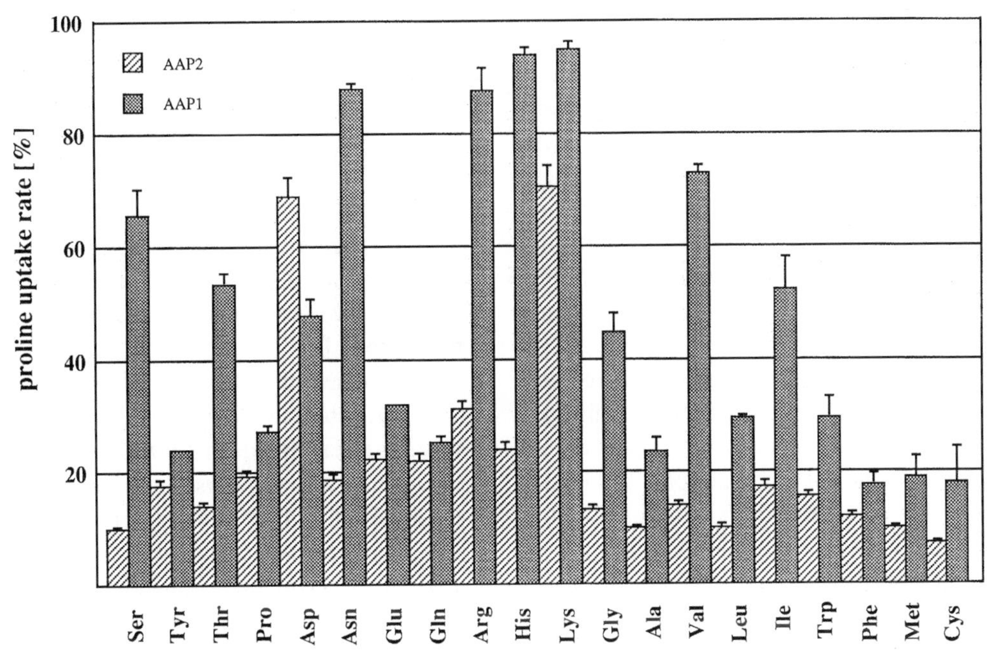

Fig. 2. Broad or general substrate specificity of plant amino acid permeases. The competition of different substrates for the uptake of radiolabelled proline into yeast cells expressing either of the two related amino acid permeases AAP1 and AAP2 from *Arabidopsis*. The analysis indicates that members of this protein family may be responsible for the transport of most proteogenic amino acids.

AAPs resembles the general or neutral systems identified in vesicles from leaves of different plant species [13].

The results obtained with D-amino acids in competition studies argue for a stereospecific transporter [24]. The stereospecificity in substrate recognition stresses the relevance of the amino and carboxy groups. This is further supported by the finding that γ-aminobutyric acid does not compete for uptake by the two *Arabidopsis* carriers. A detailed analysis of the substrate specificity of the neutral systems shows that β-alanine is not an effective competitor [61]. The two neutral systems measured in sugar beet differ in their affinity towards isoleucine, valine, threonine, and proline. This was interpreted in a way that branching at the β-carbon is responsible for steric hindrance. Due to the similarity in substrate specificity to the *Neutral System II* (as defined by [59–61]) from sugar beet leaves, the *Arabidopsis* gene was named NAT2 [41]. We do not follow this classification, first because of imperfect match (e.g. specificity for proline fits better to *Neutral System I*), but even more so because AAP1/NAT2 is mainly expressed in tissues other than leaves [see below and 56].

The broad or general specificity of the AAPs covers the transport of the major components found in phloem and xylem. An unresolved question is still whether the AAPs can also account for the basic and acidic systems determined in biochemical studies. Both basic and acidic amino acids can be transported by AAPs, though with different affinities. The presence of multiple systems in leaves with different but overlapping specificities might also explain the complex kinetics and competition data. We also cannot exclude that more specific systems belonging to the same or other gene families are present. In this context it is interesting to note that despite the high sequence homology, the members of the yeast amino acid transporter family discriminate between different substrates (side chains of amino acids). This is especially interesting because the GABA transporter from yeast also belongs to this family [3]. This means that the change in the primary structure is sufficient to account for the discrimination between substrates with α- and γ-amino groups. This might be explained by a change in the recognition domain of the transporter that displaces a hydrogen bond acceptor involved in binding of the substrate. The family also displays homology to transporters with different substrates such as choline [71].

The systems isolated so far all seem to be general systems that recognize a variety of amino acids though with differing affinity. Two related transporters can thus have different K_m values for a certain amino acid. If transport studies are carried out with plant tissues and the tissue contains more than one transporter and these differ strongly in their affinity, for example, for basic amino acids, bi- or multiphasic kinetics will be observed. Such a situation can explain the complex kinetics described above. Such a strategy could allow the plant to adapt each tissue or cell type to their own requirements. It allows to skim the excess of amino acids that is not needed for metabolism, for example by protein biosynthesis, and to export the surplus. Amino acid export has to occur in competition with metabolism. It has been calculated that the minimal requirement for the cell lies for leucine in the range of about 4 mM. Amino acids in excess of 4 mM could then be exported, especially under conditions of senescence [107]. The situation is thus more complicated than for sucrose transport, where sucrose represents mainly a transport metabolite and is not needed for basic cellular functions. The presence of a single broad specific permease makes sense in all cases where the cells export proteolysis products, which necessarily are a mixture of the different amino acids. The data presented here do not exclude that even more transporters are present in the plant that have not been or cannot be recognized by the complementation system, for example due to the requirement of multiple subunits or incorrect targetting in yeast.

Mechanism for amino acid transport

Several modes of transport are possible: passive diffusion, facilitated diffusion, secondary active and primary active transport. For active transport, several ways for energy coupling are used. Mammalian systems mainly use Na^+ coupling for active transport of metabolites. A family of proteins with related sequences is responsible for a variety of different substrates such as glucose, nucleosides, amino acids and other compounds. Relatives are also found in bacteria such as transporters for proline and panthotenate [41, 54, 70, 75]. The unifying property in this case is not the metabolite but the sodium cotransport mechanism. So far, no relatives of this family have been found in plants, possibly because in many plants sodium is excluded and plays only a minor role. However these systems might be present in halophytes [for further details see 85]. Both yeasts and plants mainly use the proton gradient generated by plasma membrane proton ATPases for secondary active transport. The best studied case is sucrose transport [for review see 13]. But also for amino acid transport evidence has been presented for proton symport [for review see 87]. Direct proof for proton-coupled amino acid transport has been obtained in plasma membrane vesicles [59].

Light inhibits the efflux of sugars and amino acids from the mesophyll but promotes a greater efflux from the veins with a net stimulation, indicating an energy dependence of transport [127]. In broad bean, light promotes the uptake of threonine and α-aminoisobutyric acid in both mesophyll and veins [16]. The translocation velocity of amino acids applied to abraded spots on leaves is similar to that of sucrose in the phloem of soybeans [39, 40, 107]. The close similarity of carbon and nitrogen transport mechanisms is also demonstrated by the fact that uncouplers effectively block both sucrose and amino acid loading and that thiol-group modifying agents inhibit sucrose transport by 90% and leucine transport by 65%. Application of potassium, which depolarizes the proton gradient between both mesophyll or phloem cells and the free space, leads to a comparable decrease in sucrose and leucine transport. These effects can either be independent or a block in sucrose transport also affects amino acid transport. This however seems improbable because different sucrose concentrations had no effect on amino acid transport [107]. McNeil [68] calculated that valine is transported into the phloem against a concentration gradient of about ten. A detailed analysis of the mesophyll and phloem sap concentration of a variety of species, indicating maximum differences of a factor of two, could not support this view [89, 136]. Therefore different mechanisms may be operating, the proton gradient may be used only to a lower extent or the concentration of amino acids in the apoplast is lower than that of sucrose. The transport studies of plant amino acid permeases in yeast show an increase of transport rates with a decrease in the pH. This may be interpreted as a low intrinsic pH dependence of the proteins, and thus may only reflect a difference in electron motive force. This and the finding that protonophores block uptake are taken as indications for a proton symport mechanism. Furthermore, inhibitors of glycolysis and respiration interfere with the transport and support the hypothesis of energy dependence for transport. However these data are no direct proof, as the pH dependence could also be due to pH optimum of the transporter and the inhibitors could directly act on the amino acid transporters. An interesting feature of the permeases is that, due to their broad specificity, they have to transport molecules differing in net charge, i.e. acidic, neutral and basic amino acids. Niven and Hamilton [73] proposed that acidic and neutral amino acids might be taken up with one proton, whereas basic amino acids may be transported in the positively charged form without an additional proton. Arginine transport in sugarcane suspension culture cells occurs in the fully ionized state and is mediated by electrogenic uniport [53]. Analysis of the neutral amino acid transport system from *Chlorella* indicates that the transport occurs in the zwitterionic form by proton symport [14]. Electrophysiological data have been presented for different mechanisms driving transport of amino acids. According to these data, neutral amino acids are cotransported with one proton, acidic ones with two cations (H^+ and K^+) and permanently protonated basic amino acids without protons along the electrochemical gradient [20, 48]. The low capacity of the AAPs to transport basic amino acids may therefore rely on the fact that only the rare unprotonated forms are recognized and cotransported with a proton. Direct electrophysiological studies should provide a possibility to unravel how one transport system is able to transport substrates with different charges. The proton symport of substrates such as amino acids depolarizes the membrane as long as the transport is electrogenic. Activation of the H^+-ATPase and compensation fluxes of potassium play an important role for charge compensation and thus for the overall transport of metabolites [53, 130, 142].

Amino acid transport mutants

An effective tool to assess the function of individual proteins and to study their control is the analysis of mutants. In general, there are two ways to obtain plants with modified transport activities: screening or selection. As changes in transport activity are difficult to measure and the phenotype to be expected is unclear, screening seems not feasible. A classical way to select for mutants is the use of toxic concentrations or toxic analogues of amino acids. Alternatively, reverse genetics can be applied to create specific mutants in transgenic organisms by overexpression, antisense inhibition or gene disruption if the respective genes are available. Gene disruption so far is not feasible in plants, since the efficiency of homologous recombination was found to be too low. Up to now, no transgenic plants have been described that show a decreased transport of nitrogenous compounds due to antisense inhibition, like in case of the sucrose transporter where an essential role of this protein in phloem loading could be demonstrated [93]. Experiments aiming at an inhibition of amino acid transport are in progress (M. Kwart and W.B. Frommer, unpublished results).

The simplest system for obtaining classical mutants are unicellular organisms. Two mutants were identified in *Chlorella* by selecting mutants on toxic amino acid analogues. The analysis of these mutants could clearly demonstrate the presence of more than three amino acid transport systems [99]. In a similar way, inhibition of seed germination or regeneration of protoplasts on media containing toxic amino acid analogues has been used to identify mutants. Some amino acids are also toxic when applied at elevated levels. A classical example is the selection on elevated levels of valine in the medium. A number of EMS mutagenized valine-resistant *Arabidopsis* lines, some of which might be mutants in transport have been identified and are waiting for further analysis [141]. After UV mutagenesis, haploid protoplasts from leaves of *Nicotiana plumbaginifolia* were selected for resistance on 5–10 mM valine. Besides mutants in acetolactacte synthase, some clones were isolated in which the resistance is due to decreased valine uptake [66]. It could be directly shown that the mutants are also affected in acidic and neutral amino acid uptake, but not in transport of basic amino acids. The trait is monogenic, whereas in the mutants of the amphidiploid *Nicotiana tabacum* two genes on the respective sets of chromosomes are present. The uptake rates for valine are reduced to 2–5% of the wild-type rates, whereas lysine uptake was only slightly affected (75% of wild type [9]). It could be shown that the high-affinity system is strongly reduced in the mutants [7, 8]. At first sight, growth and development of the tobacco mutant were unaltered, but the leaves often displayed necrotic spots. An unexplained observation is that the spots developed at 20 °C but not at 30 °C. However growth rates were also strongly reduced at 20 °C [10]. It is therefore probable that a transporter belonging to the AAP family is affected that is predominantly expressed in leaves. Surprisingly, the mutants were not further analysed. The phenotype of the tobacco mutants is unexplained, even more so as it seems to be temperature dependent. The nature of the remaining low-affinity system is still unclear.

In barley, a mutant in basic amino acid uptake into roots has been identified by selection on a toxic lysine analogue. Uptake measurements indicate that the high-affinity transport system for basic amino acids is affected as lysine, arginine and ornithine transport are strongly reduced, whereas the uptake of neutral and acidic amino acids is largely unaffected. The low-affinity transporter is still present and no effects were observed on transport characteristics in leaves, indicating that the mutated locus encodes a root-specific permease [11]. Also in this case, a detailed analysis of these mutants concerning, e.g. effects on metabolism has not been published. A high affinity amino acid transporter has been identified that efficiently transports basic amino acids and that is specifically expressed in *Arabidopsis roots* [21a]. A mutation in the corresponding gene of barley could be responsible for phenotype of the barley mutant.

Role of amino acid transporters in the plant

Due to the circulation of amino acids through the plant in phloem and xylem, the transport processes for amino acids as compared to sugars must necessarily be more complex. Uptake into roots has been observed, but the respective carrier genes have not yet been identified. The carriers may provide a means to take up amino acids from the soil. On the other hand the root has to be supplied with amino acids when nitrogen fixation is restricted to leaf tissues. As long as it is unclear whether carrier-mediated unloading occurs in sink organs, the presence of specific unloading carriers cannot be excluded.

Phloem and xylem transport

A controversy exists on how assimilates enter the phloem in exporting leaves. It is highly probable that similar mechanisms operate for amino acid and sucrose loading. The data obtained for sucrose should therefore also hold true for amino acids. As this area has been extensively reviewed, a short overview should be sufficient here [13, 25]. The distribution of plasmodesmata and microscopical studies with fluorescent dyes have provided evidence for symplastic transport at least between mesophyll cells. The concept of symplastic loading has been confined within limits to plants displaying a high degree of connectivity between mesophyll and the sieve-element companion cell complex [126]. The association of expression of the sucrose carrier gene with the development of active transport activity of maturing leaves suggests a role in phloem loading [91]. RNA *in situ* hybridization showed that the expression of the sucrose carrier was localized in the phloem supporting the model described in Fig. 1. A mutant, in which sucrose transporter gene expression has been inhibited partially by the presence of an antisense construct accumulates large amounts of carbohydrates in the leaves, transports less sugars through the phloem and is not able to fully sustain sink tissues. These results are interpreted as the best proof for an apoplastic loading of sucrose into the phloem at least in potato [92]. The fact that genes coding for amino acid permeases could be isolated and characterized as proton cotransports, indicates an apoplastic transport of amino acids. The massive transfer between phloem and xylem and the loading and unloading of the xylem also requires the presence of respective permeases. AAP2 due to its expression in the stem might be involved in this process [56].

Supply of reproductive organs with reduced nitrogen
The unloading of the phloem in seeds in general seems to be best understood. The amino acids are transported through both phloem and xylem, but due to xylem-to-phloem transfer along the translocation pathway, the assimilates arrive mainly through the phloem in the seeds. The embryo is symplastically isolated from the maternal tissue and assimilates have to pass the apoplastic space before entering the developing seeds [119]. Thus, at least two membranes have to be passed and possibly two different sets of transport systems are necessary. The two amino acid transporter genes that have been characterized in detail (AAP1 and 2) are highly expressed in developing pods, where they might be responsible for unloading of the phloem and translocation to the embryo [56]. The differential regulation of the two genes might indicate that AAP1 and AAP2 serve different functions and are specific for different cell types in the siliques. *In situ* hybridization using seeds at different developmental stages and comparison to the profile of storage protein accumulation is crucial to prove this hypothesis. Both genes are also expressed in the vascular system of the cotyledons from developing seedlings. The expression in the vascular system during this stage of development implies a role in loading of phloem or xylem to supply the growing tissues with reduced nitrogen. Thus both transporters seem to play a role in the supply of developing seeds with reduced nitrogen and lateron in the remobilization of stored nitrogen in the developing seedling. The expression patterns of the two carriers analysed so far are very specific and do not cover all expected transport activities, such as those in roots or mature leaves [11, 13, 102]. As described above, several other genes have already been isolated, some of which belong to the same gene family. These or others might be responsible for transport activities in leaves and other organs [25a, 25b].

Transport of other nitrogenous compounds

Peptide transport

Under certain conditions, the direct export of small peptides derived from proteolysis of storage proteins may increase the transport efficiency. Transport activities for peptides were found in a variety of plant tissues [34, 79]. Their distribution suggests that they might be involved in the transfer of proteolysis products from areas with a high level of protein hydrolysis, for example in germinating seedlings that mobilize storage proteins, during mobilization of vegetative storage proteins in soybean leaves for pod filling, and mobilization during senescence. Peptide transport activities were measured in barley embryos and in broad bean leaf plasma membrane vesicles [79; S. Delrot, pers. comm.]. Recently the first peptide transporter genes were isolated from yeast [19a]. Interestingly this protein shows significant homologies to the nitrate transporter Chl1 [82]. As random sequencing projects have identified several cDNAs related to *Chl1*, some of the respective proteins may be responsible for peptide transport. Complementation of the yeast peptide transport mutant has enabled the isolation and characterization of an *Arabidopsis* peptide transporter related to the peptide transporters described above [116a]. Furthermore, another member of this family was identified that mediates slow uptake of histidine and that might constitute another peptide transporter [25b].

Ureide transport

The complexity further increases if we look at other nitrogenous substrates that are found in the phloem or xylem sap of a number of species or under special conditions. A number of species has specialized in transporting ureides instead of amino acids and amides [104]. The ureide citrulline, which can serve as a nitrogen source for *Arabidopsis*, can also be transported by members of the AAP family [24, 64]. Whether allantoids are also recognized by the transporters isolated from *Arabidopsis* has not been tested so far. A compound found in the xylem sap of several species is γ-amino butyric acid which is not recognized by AAP1 and 2 [24, 56, 109 and references therein].

Regulation of nitrogen transport and allocation

Regulation of amino acid transport is well characterized in bacterial and fungal systems [138]. Several transporter genes are subject to nitrogen catabolite repression both on transcriptional and posttranslational level. The analysis of yeast mutants has indicated that a protein kinase *NPR1* is required for the activity of the ammonium-sensitive amino acid permeases [132]. All possible levels seem to be used, as even for the targeting to the plasma membrane, receptors specific for amino acid permeases could be identified [63]. Interestingly, a plant serine threonine kinase enables a yeast mutant deficient in the general amino acid permease GAP1 to grow on citrulline, indicating that a phosphorylation event may play a role in these processes (M. Kwart and W.B. Frommer, unpublished results). Correct glycosylation may not be necessary for the function of the cationic amino acid transporter from mouse [47]. In the majority of cases, a single polypeptide is sufficient for the function. However, oocyte expression cloning has led to the identification of a number of proteins that do not resemble integral membrane proteins. These polypeptides might represent regulatory subunits that activate the oocyte-endogenous permeases [137].

Since the plant transporters for nitrogenous compounds have been isolated only recently, very little is known about their regulation. The ammonium transporter and several amino acid transporters in *Chlorella* are regulated by glucose or by nitrogen starvation [15, 99, 104]. Furthermore, a general amino acid permease is inducible by a combination of glucose and ammonium [97]. The mechanism for this induction is not understood and it is not clear whether this is a general phenomenon or specific for *Chlorella*. Expression of the nitrate transporter in the roots of *Arabidopsis* is regulated at the RNA level by the pH of the medium and by nitrate [120]. Physiological control mechanisms such as control by phytohormones have been reviewed by Simpson [111]. The AAPs are under developmental control, and other modes of regulation have not been described. Yeast may serve as a system to study regulation events at the post-transcritional level, such as analysis of correct targeting in secretion pathway mutants.

Accumulation of storage protein: indication for regions of high transport activities

The two main storage forms of nitrogen in the plant are amino acids and proteins. To some extent, excess supply of nitrate may lead to accumulation of nitrate in the leaves. During plant development, two main periods of storage can be distinguished, one during vegetative growth and the other after reallocation of nitrogen to reproductive organs. Plants have developed special proteins for storage of nitrogen during the two periods. Before flowering, large quantities of fixed nitrogen have accumulated in leaves in the form of vegetative storage proteins. This ensures that sufficient nitrogen is already available before seed setting. Upon seed development, these proteins can serve as sources for amino acids which are mobilized and transported through both phloem and xylem to the storage organs [115]. Synthesis of specific storage proteins are induced during the development of seeds or vegetative storage organs such as tubers [36, 95]. The pattern of storage protein expression is best analysed in *Brassicaceae* [21, 76, 112]. In *Arabidopsis*, which has only a minor endosperm, nitrogen is mainly stored as 2S and 12S storage proteins in the cotyledons of the embryo. The synthesis of the storage proteins is regulated at the transcriptional level and follows a time-programmed gradient in the embryo starting from the outer layer of the cotyledons [21, 30, 76]. During germination, the storage proteins are degraded and the resulting amino acids are exported to supply the new developing plant in early stages. In potato, patatin which represents the major storage protein of the tubers is inducible by sucrose and amino acids [80, 95]. Thus, as in the case of starch synthesis, precursors can act as inducers. The simplest, though still very speculative, model is therefore that by turning on the import of metabolites in specific cells, metabolic pathways such as storage protein biosynthesis is induced. This process might even trigger the development of the sink organ. Thus amino acid transport is an essential component of the nitrogen partitioning and allocation of the plant. A better understanding of the regulation of amino acid transport and induction of storage proteins is necessary to assess such models.

Transport of nitrogen between subcellular compartments

The plant transport permeases complement yeast mutants deficient in transport across the plasma membrane and thus must be targeted correctly to the plasma membrane. In analogy, we conclude that the transporters in the plant are also responsible for plasma membrane transport. However, the data do not exclude the possibility that the proteins represent, for example, tonoplast carriers that are mistargeted in yeast. However, this is improbable, as the tonoplast of yeast seems to be the default compartment for membrane proteins [93]. To unequivocally demonstrate the subcellular distribution, immunolocalization studies will have to be conducted.

The question remains how to get a hand also on genes responsible for transport activities in other membranes, i.e. the tonoplast, but also envelopes of plastids and mitochondria. A multitude of transport activities for different substrates has been characterized, for example in the tonoplast [67]. The availability of plasma membrane protein genes might be a tool to search for homologous genes that might be responsible for such transport systems. Alternatively, yeast transport mutants, which so far have not been described, might be a more promising way to obtain the respective plant genes.

Conclusions

During the past few years, the combination of molecular tools with the analysis of mutants has allowed to get a first insight into the transport processes of higher plants. The way to proceed in studying the function of these proteins has been illustrated for sucrose transport. In that case the gene was cloned by complementation of an engeneered yeast mutant, expression and regulation in the plant was analysed and transgenic plants were constructed that are affected in phloem loading. Using the same rationale, similar studies will have to be devoted to the nitrogen transporters. This will help to assess the role of these transporters not only in nutrient transport but also in transfer of information about the nutritive state of certain cells or organs. Due to the much higher complexity, it seems feasible first to try and isolate all members of the different gene families and to study their expression profile. Alternatively, it may be useful to concentrate on specific processes such as the transport in roots or in seeds. The yeast expression system represents an excellent tool for studying structure-function relationships in order to understand the structure of the transporters and how the substrates are recognized [65]. Another important goal will be a comparison of transport characteristics of different plant species under varying conditions. Much more work will be necessary to elaborate the regulation and interaction of the different transport processes as has been done in simpler systems such as yeast.

Acknowledgements

We would like to express our gratitude for criticism and suggestions on this manuscript to E. Komor (Bayreuth), S. Delrot (Poitiers) and Frank R. Lauter (Berlin).

References

1. Allen S, Raven JA: Intracellular pH regulation in *Ricinus communis* grown with ammonium or nitrate as N source: the role of long distance transport. J Exp Bot 38: 580–596 (1987).
2. Anderson JA, Huprikar SS, Kochian LV, Lucas WJ, Gaber RF: Functional expression of a probable *Arabidopsis thaliana* potassium channel in *Saccharomyces cerevisiae*. Proc Natl Acad Sci USA 89: 3736–3740 (1992).
3. André B, Hein C, Grenson M, Jauniaux JC: Cloning and expression of the *UGA4* gene coding for the inducible GABA-specific transport protein of *Saccharomyces cerevisiae*. Mol Gen Genet 237: 17–25 (1993).
4. Andrews M: The partitioning of nitrate assimilation between the root and shoot of higher plants. Plant Cell Environ 9: 511–519: (1986).
5. Attwell D, Bouvier M: Neurotransmitter transporters: cloners quick on the uptake. Curr Biol 2: 541–543 (1992).
6. Boos W: Bacterial transport. Annu Rev Biochem 43: 123–146 (1974).
7. Borstlap AC: Tobacco mutants of amino acid membrane transport: uptake of *L*-valine in leaf discs from the double mutant Val^r-2 and its monogenic derivatives. In: Lambers H, Neeteson JJ, Stulen I (eds) Fundamental Ecological and Agricultural Aspects of Nitrogen Metabolism in Higher Plants, pp. 115–117. Martinus Nijhoff Publishers, Dordrecht (1986).
8. Borstlap AC, Schuurmans J: Kinetics of L-valine uptake in tobacco leaf discs; comparison of wild type, the digenic mutant Val^r-2 and its monogenic derivatives. Planta 176: 42–50 (1988).
9. Borstlap AC, Schuurmanns J, Bourgin JP: Amino acid transport mutant of *Nicotiana tabacum*. Planta 166: 141–144 (1985).
10. Bourgin JP, Goujaud J, Missionier C, Pethe C: Valine resistance, a potential marker in plant cell genetics. I. Distinction between two types of valine resistant tobacco mutants isolated from protoplast-derived cells. Genetics 109: 393–407 (1985).
11. Bright SWJ, Kueh JSH, Rognes SE: Lysine transport in two barley mutants with altered uptake of basic amino acids. Plant Physiol 72: 821–824 (1983).
12. Bush DR, Langstone-Unkefer PJ: Amino acid transport into membrane vesicles isolated from zucchini. Plant Physiol 88: 487–490 (1988).
13. Bush DR: Proton-coupled sugar and amino acid transporters in plants. Annu Rev Plant Physiol Plant Mol Biol 44: 513–542 (1993).
14. Cho BH, Komor E: Mechanism of proline uptake by *Chlorella vulgaris*. Biochim Biophys Acta 735: 361–366 (1983).
15. Cho BH, Sauer N, Komor E, Tanner W: Glucose induces two amino acid transport systems in *Chlorella*. Proc Natl Acad Sci USA 78: 3591–3594 (1981).
16. Despeghel JP, Delrot S: Energetics of amino acids uptake by *Vicia faba* leaf tissue. Plant Physiol 71: 1–6 (1983).
17. de Vries H: Sur la perméabilité du protoplasma des betteraves rouges. Arch Neerl Sci Exactes Nat 6: 118–126 (1871).
18. Doddema H, Telkamp GP: Uptake of nitrate by mutants of *Arabidopsis thaliana*, disturbed in uptake or reduction of nitrate. Physiol Plant 45: 332–338 (1979).
19. Dubois E, Grenson M: Methylamine/ammonia uptake systems in *Saccharomyces cerevisiae*: multiplicity and regulation. Mol Gen Genet 175: 67–76 (1979).
19a. Fei Y, Kanai Y, Nussberger S, Ganapathy V, Leibach FH, Romero MF, Singh SK, Boron WF, Hediger MA: Expression cloning of a mammalian proton-coupled oligopeptide transporter. Nature 368: 563–566 (1994).
20. Felle H: Stereospecificity and electrogenicity for amino acid transport in *Riccia fluitans*. Planta 152: 505–512 (1981).
21. Fernandez DE, Turner FR, Crouch M: *In situ* localization of storage protein mRNAs in developing meristems of *Brassica napus* embryos. Development 111: 299–313 (1991).
21a. Fischer WN, Kwart M, Hummel S, Frommer WB: Differential expression of amino acid transporters from *Arabidopsis*. Mol Gen Genet, submitted.
22. Fisher DB, Magnicol PK: Amino acid composition along the transport pathway during grain filling in wheat. Plant Physiol 82: 1019–1023 (1986).
23. Fried MF, Zsoldos F, Vose PB, Shatokhin IL: Characterizing the NO_3^- and NH_4^+ uptake process of rice roots by use of ^{15}N labelled NH_4NO_3. Physiol Plant 18: 313–320 (1965).
24. Frommer WB, Hummel S, Riesmeier JW: Expression cloning in yeast of a cDNA encoding a broad specificity amino acid permease from *Arabidopsis thaliana*. Proc Natl Acad Sci USA 90: 5944–5948 (1993).

25. Frommer WB, Hirner B, Harms K, Kühn C, Martin T, Riesmeier JW, Schulz B, Willmitzer L: Sugar transport in higher plants. In: Tartakoff A (ed) Membranes: Specialized Functions in Plants. JAI Press, USA, in press (1994).
25a. Frommer WB, Hummel S, Rentsch D: Cloning of an *Arabidopsis* histidine transporting protein related to nitrate and peptide transporters. FEBS Lett., in press (1994).
25b. Frommer WB, Hummel S, Unseld M, Ninneman O: An amino acid transporter from *Arabidopsis* with low affinity and broad specificity related to mammalian cationic amino acid transporters. Plant J, submitted (1994).
26. Glass ADM, Shaff JE, Kochian LV: Studies of the uptake of nitrate in barley. Plant Physiol 99: 456–463 (1992).
27. Goyal SS, Huffaker RC: A novel approach and a fully automated microcomputer-based system to study kinetics of NO_3^-, NO_2^-, and NH_4^+ transport simultaneously by intact wheat seedlings. Plant Cell Environ 9: 209–215 (1986).
28. Grenson M, Hou C, Crabeel M: Multiplicity of the amino acid permeases in *Saccharomyces cerevisiae*. IV. Evidence for a general amino acid permease. J Bact 103: 770–777 (1970).
29. Guastella J, Nelson N, Nelson H, Czyzyk C, Keynan S, Miedel MC, Davidson N, Lester HA, Kanner BI: Expression of a rat brain GABA transporter. Science 249: 1303–1306 (1990).
30. Guerche P, Tire C, Grossi de Sa F, De Clercq A, Van Montagu M, Krebbers E: Differential expression of the *Arabidopsis* 2S albumin genes and the effect of increasing gene family size. Plant Cell 2: 469–478 (1990).
31. Haskovec C, Kotyk A: Attempts at purifying the galactose carrier from galactose-induced baker's yeast. Eur J Biochem 9: 343–347 (1969).
32. Hayashi H, Chino M: Chemical composition of phloem sap from uppermost internode of the rice plant. Plant Cell Physiol 31: 247–251 (1990).
33. Heatwole VM, Somerville RL: Cloning, nucleotide sequence, and characterization of *mtr*, the structural gene for a tryptophan-specific permease of *Escherichia coli* K-12. J Bact 173: 108–115 (1991).
34. Higgins CF, Payne JW: Peptide transport by germinating barley embryos: evidence for a single common carrier for di- and oligopeptides. Planta 138: 217–221 (1978).
35. Higgins CF, Hyde SC, Mimmack MM, Gileadi U, Gill DR, Gallagher MP: Binding protein-dependent transport systems. J Bioenerg Biomembr 22: 571–592 (1990).
36. Higgins TJV: Synthesis and regulation of major proteins in seeds. Annu Rev Plant Physiol 35: 191–221 (1984).
37. Honoré N, Cole ST: Nucleotide sequence of the *aroP* gene encoding the general aromatic amino acid transport protein of *Escherichia coli* K-12: homology with yeast transport proteins. Nucl Acids Res 18: 653 (1990).
38. Hoshino T, Kose-Terai K, Uratani Y: Isolation of the *braZ* gene encoding the carrier for a novel branched-chain amino acid transport system in *Pseudomonas aeruginosa* PAO. J Bact 173: 1855–1861 (1991).
39. Housley TL, Peterson DM, Schrader LE: Long distance translocation of sucrose, serine, leucine, lysine, and CO_2 assimilates. I. Soybean. Plant Physiol 59: 217–220 (1977).
40. Housley TL, Schrader LE, Miller M, Setter TL: Partitioning of ^{14}C-photosynthate, and long distance translocation of amino acids in preflowering and flowering, nodulated and unnodulated soybeans. Plant Physiol 64: 94–98 (1979).
41. Hsu L, Chiou T, Chen L, Bush DR: Cloning a plant amino acid transporter by functional complementation of a yeast amino acid transport mutant. Proc Natl Acad Sci USA 90: 7441–7445 (1993).
42. Jackowski S, Alix JH: Cloning, sequence, and expression of the pantothenate permease (*panF*) gene of *Escherichia coli*. J Bact 172: 3842–3848 (1990).
43. Jauniaux JC, Grenson M: *GAP*1, the general amino acid permease gene of *Saccharomyces cerevisiae*: nucleotide sequence, protein homology with the other baker's yeast amino acid permeases, and nitrogen catabolite repression. Eur J Biochem 190: 39–44 (1990).
44. Jauniaux JC, Vandenbol M, Vissers S, Broman K, Grenson M: Nitrogen catabolite regulation of proline permease in *Saccharomyces cerevisiae*. Cloning of the *PUT*4 gene and study of *PUT*4 RNA levels in wild-type and mutant strains. Eur J Biochem 164: 601–606 (1987).
45. Kanai Y, Hediger MA: Primary structure and functional characterization of a high-affinity glutamate transporter. Nature 360: 467–471 (1992).
46. Kim JW, Closs EI, Albritton LM, Cunningham JM: Transport of cationic amino acids by the mouse ecotropic retrovirus receptor. Nature 352: 725–728 (1991).
47. Kim JW, Cunningham JM: N-linked glycosylation of the receptor for murine ecotropic retroviruses altered in virus-infected cells. J Biol Chem 268: 16316–16320 (1993).
48. Kinraide TB, Etherton B: Electrical evidence for different mechanisms of uptake for basic, neutral, and acidic amino acids in oat coleoptiles. Plant Physiol 65: 1085–1089 (1981).
49. Kleiner D: Transport of NH_3 and NH_4^+ across biological membranes. Biochim Biophys Acta 639: 41–52 (1981).
50. Kleiner D: Bacterial ammonia transport. FEMS Microbiol Rev 32: 87–100 (1985).
51. Kleiner D: NH_4^+ transport systems. In: Bakker EP (ed) Alkali Cation Transport Systems in Procaryotes, pp. 379–396. CRC Press, London (1993).
52. Komor E, Rotter M, Tanner W: A proton-cotransport system in a higher plant: sucrose transport in *Ricinus communis*. Plant Sci Lett 9: 153–162 (1977).
53. Komor E, Thom M, Maretzki A: Mechanism of uptake

of *L*-arginine by sugar-cane cells. Eur J Biochem 116: 527–533 (1981).

54. Kong C, Yet S, Lever JE: Cloning and expression of a mammalian Na^+/amino acid cotransporter with sequence similarity to Na^+/glucose cotransporters. J Biol Chem 268: 1509–1512 (1993).
55. Koo K, Stuart WD: Sequence and structure of *mtr*, an amino acid transport gene of *Neurospora crassa*. Genome 34: 644–651 (1991).
56. Kwart M, Hirner B, Hummel S, Frommer WB: Differential expression of two related amino acid transporters with differing substrate speceficity in *Arabidopsis thaliana*. Plant J 4: 993–1002 (1993).
57. Langmüller G, Springer-Lederer H: Membranpotential von *Chlorella fusca* in Abhängigkeit von pH Wert, Temperatur und Belichtung. Planta 120: 189–196 (1974).
58. Larsson CM, Ingemarsson B: Molecular aspects of nitrate uptake in higher plants. In: Wray J, Kinghorn J (eds) Molecular and Genetic Aspects of Nitrate Assimilation, pp. 3–14. Oxford Science Publishers, Oxford (1989).
59. Li Z, Bush DR: ΔpH-dependent amino acid transport into plasma membrane vesicles isolated from sugar beet (*Beta vulgaris* L.) leaves. Plant Physiol 94: 268–277 (1990).
60. Li Z, Bush DR: ΔpH-dependent amino acid transport into plasma membrane vesicles isolated from sugar beet (*Beta vulgaris* L.) leaves. Plant Physiol 96: 1338–1344 (1991).
61. Li Z, Bush DR: Structural determinants in substrate recognition by proton-amino acid symports in plasma membrane vesicles isolated from sugar beet leaves. Arch Biochem Biophys 294: 519–526 (1992).
62. Liu Q, Mandiyan S, López-Corcuera, Nelson H, Nelson N: A rat brain cDNA encoding the neurotransmitter transporter with an unusual structure. FEBS Lett 315: 114–118 (1993).
63. Ljungdahl PO, Gimeno CJ, Styles CA, Fink GR: *SHR*3: A novel component of the secretory pathway specifically required for localization of amino acid permeases in yeast. Cell 71: 463–478 (1992).
64. Ludwig RA: *Arabidopsis* chloroplasts dissimilate *L*-arginine and *L*-citrulline for use as N source. Plant Physiol 101: 429–434 (1993).
65. Hennessey EM, Broome-Smith JK: Gene-fusion techniques for determining membrane-protein topology. Curr Opin Struc Biol 3: 524–531 (1993).
66. Marion-Poll A, Missonier C, Goujard J, Caboche M: Isolation and characterization of valine-resistant mutants of *Nicotiana plumbaginifolia*. Theor Appl Genet 75: 272–277 (1988).
67. Martinoia E: Transport processes in vacuoles of higher plants. Bot Acta 105: 232–245 (1992).
68. McNeil DL: The kinetics of phloem loading of valine in the shoot of a nodulated legume (*Lupinus albus* L. cv. Ultra). J Exp Bot 118: 1003–1012 (1979).
69. Minet M, Dufour ME, Lacroute F: Complementation of *Saccharomyces cerevisiae* auxotrophic mutants by *Arabidopsis thaliana* cDNAs. Plant J 2: 417–422 (1992).
70. Nakao T, Yamato I, Anraku Y: Nucleotide sequence of the *putP*, the proline carrier gene of *Escherichia coli* K12. Mol Gen Genet 208: 70–75 (1987).
71. Nikawa JI, Hosaka K, Tsukagoshi Y, Yamashita S: Primary structure of the yeast choline transport gene and regulation of its expression. J Biol Chem 265: 15996–16003 (1990).
72. Ninnemann OW, Jauniauex JC, Frommer WB: Identification of a high affinity ammonium transporter from plants. EMBO J 13: 3464–3471 (1994).
73. Niven DF, Hamilton WA: Mechanisms of energy coupling to the transport of amino acids by *Staphylococcus aureus*. Eur J Biochem 44: 517–522 (1974).
74. Quesada A, Galván A, Schnell RA, Lefebvre PA, Fernández E: Five nitrate assimilation-related loci are clustered in *Chlamydomonas reinhardtii*. Mol Gen Genet 240: 387–394 (1993).
75. Pajor AM, Wright EM: Cloning and functional expression of a mammalian Na^+/nucleoside cotransporter. J Biol Chem 267: 3557–3560 (1992).
76. Pang PP, Pruitt RE, Meyerowitz EM: Molecular cloning, genomic organization, expression and evolution of 12S seed storage protein genes of *Arabidopsis thaliana*. Plant Mol Biol 11: 805–820 (1988).
77. Pate JS: Transport and partioning of nitrogenous solutes. Annu Rev Plant Physiol 31: 313–340 (1980).
78. Pate JS, Sharkey PJ, Atkins CA: Nutrition of a developing legume fruit. Plant Physiol 59: 506–510 (1977).
79. Payne JW, Hardy DJ: Characterization of the peptide transport system synthesized in germinating barley embryos. In: Dainty J *et al.* (ed) Plant Membrane Transport, pp. 507–508. Elsevier, Amsterdam (1989).
80. Peña-Cortés H, Liu X, Sanchez Serrano J, Schmid R, Willmitzer L: Factors affecting gene expression of patatin and proteinase-inhibitor II genes families in detached potato leaves. Planta 186: 495–502 (1991).
81. Peoples MB, Gifford RM: Long distance transport of nitrogen and carbon from sources to sinks in higher plants. In: Dennis DT, Turpin DH (eds) Plant Physiology, Biochemistry and Molecular Biology, pp. 442–455. Longman, Essex (1990).
82. Perry JR, Basrai MA, Steiner H, Naider F, Becker JM: Isolation and characterization of a *Saccharomyces cerevisiae* peptide transport gene. Mol Cell Biol 14: 104–115 (1994).
83. Pi J, Wookey PJ, Pittard AJ: Cloning and sequencing of the *pheP* gene, which encodes the phenylalanine-specific transport system of *Escherichia coli*. J Bact 173: 3622–3629 (1991).
84. Pitman MG: Ion transport into the xylem. Annu Rev Plant Physiol 28: 71–88 (1977).
85. Poole RJ: Energy coupling for membrane transport. Annu Rev Plant Physiol 29: 437–460 (1978).

86. Redinbaugh MG, Campbell WH: Higher plant responses to environmental nitrate. Physiol Plant 82: 640–650 (1991).
87. Reinhold L, Kaplan A: Membrane transport of sugars and amino acids. Annu Rev Plant Physiol 35: 45–83 (1984).
88. Rickenberg HV, Cohen GN, Buttin G, Monod J: La galactoside-perméase d'*Escherichia coli*. Ann Inst Pasteur 91: 829–857 (1956).
89. Riens B, Lohaus G, Heinke D, Heldt HW: Amino acid and sucrose content determined in the cytosolic, chloroplastic and vacular compartment and in the phloem sap of spinach leaves. Plant Physiol 97: 227–233 (1991).
90. Riesmeier JW, Willmitzer L, Frommer WB: Isolation and characterization of a sucrose carrier cDNA from spinach by functional expression in yeast. EMBO J 11: 4705–4713 (1992).
91. Riesmeier JW, Hirner B, Frommer WB: Potato sucrose transporter expression in minor veins indicates a role in phloem loading. Plant Cell 5: 1591–1598 (1993).
92. Riesmeier JW, Willmitzer L, Frommer WB: Evidence for an essential role of the sucrose transporter in phloem loading and assimilate partitioning. EMBO J 13: 1–7 (1994).
93. Roberts CJ, Nothwehr SF, Stevens TH: Membrane protein sorting in the yeast secretory pathway: evidence that the vacuole may be the default compartment. J Cell Biol 119: 69–83 (1992).
94. Robinson SP, Beevers H: Evidence for amino acid proton cotransport in *ricinus* cotyledon. Planta 152: 527–533 (1981).
95. Rocha-Sosa M, Sonnewald U, Frommer WB, Stratmann M, Willmitzer L: Both developmental and metabolic signals activate the promoter of a class I patatin gene. EMBO J 8: 23–29 (1989).
96. Ronson CW, Astwood PM, Downie JA: Molecular cloning and genetic organization of C4-dicarboxylate transport genes from *Rhizobium leguminosarum*. J Bact 160: 903–909 (1984).
97. Sauer N: A general amino acid permease is inducible in *Chorella vulgaris*. Planta 161: 425–431 (1984).
98. Sauer N, Komor E, Tanner W: Regulation and characterization of two inducible amino acid transport systems in *Chlorella vulgaris*. Planta 159: 404–410 (1983).
99. Sauer N, Tanner W: Selection and characterization of *Chlorella* mutants deficient in amino acid transport. Plant Physiol 79: 760–764 (1985).
100. Schlee J, Komor E: Ammonium uptake by *Chlorella*. Planta 168: 232–238 (1986).
101. Schobert C, Komor E: The differential transport of amino acids into the phloem of *Ricinus communis* L. seedlings as shown by the analysis of sieve-tube sap. Planta 177: 342–349 (1989).
102. Schobert C, Komor E: Transfer of amino acids and nitrate from the roots into the xylem of *Ricinus communis* seedlings. Planta 181: 85–90 (1990).
103. Scholten HJ, Feenstra WJ: Uptake of chlorate and other ions in seedlings of the nitrate uptake mutant B1 of *Arabidopsis thaliana*. Physiol Plant 66: 265–269 (1986).
104. Schubert KR, Boland MJ: The ureides. In: The biochemistry of Plants, vol. 16, pp. 197–281. Academic Press, New York (1990).
105. Segel GB, Boal TR, Cardillo TS, Murant FC, Lichtman MA, Sherman F: Isolation of a gene encoding a chaperonin-like protein by complementation of yeast amino acid transport mutants with human cDNA. Proc Natl Acad Sci USA 89: 6060–6064 (1992).
106. Sentenac H, Bonneaud N, Minet M, Lacroute F, Salmon JM, Gaynard F, Grignon C: Cloning and expression in yeast of a plant potassium ion transport system. Science 256: 663–665 (1992).
107. Servaites JC, Schrader LE, Jung DM: Energy-dependent loading of amino acids and sucrose into the phloem of soybean. Plant Physiol 64: 546–550 (1979).
108. Shafqat S, Tamarappoo BK, Kilberg MS, Puranam RS, McNamara JO, Guadano-Ferraz A, Fremeau RT Jr: Cloning and expression of novel NO^+-dependent neutral amino acid transporter structurally related to mammalian Na^+/glutamate cotransporters. J Biol Sci 268: 15351–15355 (1993).
109. Shelp BJ: The composition of phloem exudate and xylem sap from broccoli (*Brassica oleracea* var. *italica*) supplied with NH_4^+, NO_3^-. J Exp Bot 38: 1619–1636 (1987).
110. Siddiqi MY, Glass ADM, Ruth TJ, Rufty TW Jr: Studies on nitrate uptake in barley. Plant Physiol 93: 1426–1432 (1990).
111. Simpson RJ: Translocation and metabolism of nitrogen: whole plant aspects. In: Lambers H, Neetson JJ, Stulen I (eds) Fundamental, Ecological and Agricultural Aspects of Nitrogen Metabolism in Higher Plants, pp. 71–96. Martinus Nijhoff, Dordrecht (1986).
112. Sjödahl S, Gustavsson H, Rödin J, Lenman M, Höglund A, Rask L: Cruciferin gene familes are expressed coordinately but with tissue-specific differences during *Brassica napus* seed development. Plant Mol Biol 23: 1165–1176 (1993).
113. Soldal T, Nissen P: Multiphasic uptake of amino acids by barley roots. Physiol Plant 43: 181–188 (1978).
114. Sophianopoulou V, Scazzocchio C: The proline transport protein of *Aspergillus nidulans* is very similar to amino acid transporters of *Saccharomyces cerevisiae*. Mol Microbiol 3: 705–714 (1989).
115. Staswick PE: Novel regulation of vegetative storage protein genes. Plant Cell 2: 1–6 (1990).
116. Steffes C, Ellis J, Wu J, Rosen BP: The *lysP* gene encodes the lysine-specific permease. J Bact 174: 3242–3249 (1992).
116a. Steiner HY, Song W, Zhang L, Naider F, Becker JM, Stacey G: An *Arabidopsis* peptide transporter is a member of a novel family of membrane transport proteins. Plant Cell, in press (1994).

117. Tanaka J, Fink GR: The histidine permease gene (*HIP*1) of *Saccharomyces cerevisiae*. Gene 38: 205–214 (1985).
118. Tanner W: Light-driven uptake of 3-*O*-methylglucose via an inducible hexose uptake system of *Chlorella*. Biochem Biophys Res Commun 36: 278–283 (1969).
119. Thorne JH: Phloem unloading of C and N assimilates in developing seeds. Annu Rev Plant Physiol 36: 317–343 (1985).
120. Tsay Y, Schroeder JI, Feldmann KA, Crawford NM: The herbicide sensitivity gene *CHL*1 of *Arabidopsis* encodes a nitrate-inducible nitrate transporter. Cell 72: 705–713 (1993).
121. Udvardi MK, Salom CL, Day DA: Transport of L-glutamate across the bacteroid membrane but not the peribacteroid membrane from soybean root nodules. Mol Plant-Microbe Interact 1: 250–254 (1988).
122. Udvardi MK, Yang LO, Young S, Day DA: Sugar and amino acid transport across symbiotic membranes from soybean nodules. Mol Plant-Microbe Interact 3: 334–340 (1990).
123. Ullrich WR, Larsson M, Larsson CM, Lesch S, Novacky A: Ammonium uptake in *Lemna gibba* G1, related membrane potential changes, and inhibition of anion uptake. Physiol Plant 61: 369–376 (1984).
124. Unkles SE, Hawker KL, Grieve C, Campbell EI, Montague P, Kinghorn JR: *crnA* encodes a nitrate transporter in *Aspergillus nidulans*. Proc Natl Acad Sci USA 88: 204–208 (1991).
125. van Bel AJE, Borstlap AC, van Pinxteren-Bazuine A, Ammerlaan A: Analysis of valine uptake by *Commelina* mesophyll cells in a biphasic active and a diffusional component. Planta 155: 355–61 (1982).
126. van Bel AJE, Gamalei YV, Ammerlaan A, Bik LPM: Dissimilar phloem loading in leaves with symplastic or apoplastic minor vein configurations. Planta 186: 518–525 (1992).
127. van Bel AJE, Koops AJ, Dueck T: Does light promoted export from *Commelina benghalensis* leaves result from differential light sensitivity of the cells in the mesophyll to sieve tube path? Physiol Plant 71: 227–234 (1986).
128. van Bel AJE, van Leeuwenkamp P, van der Schoot C: Amino acid uptake by various tissues of the tomato plant. Effects of the external pH and light. Z Pflanzenphysiol Bot 104: 117–128 (1981).
129. van Bel AJE, van der Schoot C: Light-stimulated biphasic amino acid uptake by xylem parenchyma cells. Plant Sci Lett 19: 101–107 (1980).
130. van Bel AJE, van Erven AJ: Potassium co-transport and antiport during the uptake of sucrose and glutamic acid form the xylem vessels. Plant Sci Lett 15: 285–291 (1979).
131. Vandenbol M, Jauniaux JC, Grenson M: Nucleotide sequence of the *Saccharomyces cerevisiae PUT4* proline-permease-encoding gene: similarities between CAN1, HIP1, and PUT4 permeases. Gene 83: 153–159 (1989).
132. Vandenbol M, Jauniaux JC, Grenson M: The *Saccharomyces cerevisiae* NPR1 gene required for the activity of ammonia-sensitive amino acid permeases encodes a protein kinase homologue. Mol Gen Genet 222: 393–399 (1990).
133. Wallace B, Yang Y, Hong J, Lum D: cloning and sequencing of a gene encoding a glutamate and aspartate carrier of *Escherichia coli* K-12. J Bact 172: 3214–3220 (1990).
134. Wang MY, Siddiqi MY, Ruth TJ, Glass ADM: Ammonium uptake by rice roots I. Fluxes and subcellular distribution of $^{13}NH_4^+$. Plant Physiol 103: 1249–1258 (1993).
135. Wang MY, Siddiqi MY, Ruth TJ, Glass ADM: Ammonium uptake by rice roots. II. Kinetics of $^{13}NH_4^+$ influx across the plasmalemma. Plant Physiol 103: 1259–1267 (1993).
136. Weiner H, Heldt JW: Inter- and intracellular distribution of amino acids and other metabolites in maize (*Zea mays* L.) leaves. Planta 187: 242–246 (1992).
137. Wells RG, Hediger MA: Cloning of a rat kidney cDNA that stimulates dibasic and neutral amino acid transport and has sequence similarity to glucosidases. Proc Natl Acad Sci USA 89: 5596–5600 (1992).
138. Wiame JM, Grenson M, Arst HN Jr: Nitrogen catabolite repression in yeasts and filamentous fungi. Adv Micr Physiol 26: 1–87 (1985).
139. Williams LE, Nelson SJ, Hall JL: Characterization of solute transport in plasma membrane vesicles isolated from cotyledons of *Ricinus communis* L. Planta 182: 540–545 (1992).
140. Winter H, Lohaus G, Heldt HW: Phloem transport of amino acids and sucrose in correlation to the corresponding metabolite levels in barley leaves. Plant Physiol 99: 996–1004 (1992).
141. Wu K, Mourad G, King J: A valine resistant mutant of *Arabidopsis thaliana* displays an acetolactate synthase with altered feedback control. Planta 192: 249–255 (1994).
142. Wyse RE, Komor E: Mechanism of amino acid uptake by sugarcane suspension cells. Plant Physiol 76: 865–870 (1984).

Plant Molecular Biology **26**: 1671–1679, 1994.

Sugar transport across the plasma membranes of higher plants

Norbert Sauer*, Kerstin Baier, Manfred Gahrtz, Ruth Stadler, Jürgen Stolz and Elisabeth Truernit
Lehrstuhl für Zellbiologie und Pflanzenphysiologie, Universität Regensburg, D–93040 Regensburg, Germany
(author for correspondence)*

Received and accepted 20 July 1994

Key words: apoplastic space, glucose transport, heterologous expression, phloem loading, plasma membrane, sucrose transport

Abstract

The fluxes of carbohydrates across the plasma membranes of higher-plant cells are catalysed mainly by monosaccharide and disaccharide-H^+ symporters. cDNAs encoding these different transporters have been cloned recently and the functions and properties of the encoded proteins have been studied extensively in heterologous expression systems. Several of the proteins have been identified biochemically in these expression systems and their location in plants has been shown immunohistochemically or with transgenic plants which were transformed with reporter genes, expressed under the control of the promoters of individual transporter genes. In this paper we summarize the current knowledge on the molecular biology and biochemistry of higher-plant sugar transport proteins.

Introduction

Photosynthetic CO_2 fixation in the mature leaves of higher plants and the controlled partitioning of assimilated carbon between photosynthetically inactive organs, developing tissues or storage compartments are two highly interconnected processes. Photosynthetic tissues possess only a limited capacity for the storage of assimilates and most of the carbohydrates that are not needed for the leaf's own metabolism enter the sieve elements of the vascular system. The role of higher-plant long-distance phloem transport is therefore not only the supply of assimilates to all kinds of sink tissues but also the removal of fixation products from the source tissues to avoid accumulation of metabolites and feed back inhibition on the photosynthetic machinery. This allocation system seems to be carefully balanced, but mechanisms controlling its regulation are still poorly understood.

Two fundamentally different models are discussed for the way used by photoassimilates to get into or out of the sieve elements of the phloem. The symplastic model depends on the existence of symplastic connections between photosynthetically active mesophyll cells and sieve elements, the plasmodesmata, and similar connections would also be needed at the sink end of the phloem. According to this model carbohydrates synthesized in the mesophyll would diffuse from one cell type to the other through these plasmodesmata, never leaving the symplastic space and reaching their final destination without a single membrane translocation step. The apoplastic model of phloem loading and unloading, on the other hand, depends on the existence of specific transporters, catalysing the membrane penetration of sugars or amino acids, and several types of transporters might be necessary to allow this apoplastic pathway. The first type of transport protein is responsible for the efflux or export of assimilates, such as the efflux of sucrose from the mesophyll cells or the efflux of sucrose from the phloem at the sink end of the sieve elements. The second type of transporters is needed for the catalysis of substrate import, for example the loading of sucrose into the phloem or the transport of monosaccharides into the cells of a specific sink tissue.

Results have been published supporting either the apoplastic, carrier-mediated way or the symplastic transport via plasmodesmata. In recent years cDNAs and/or genes have been cloned for a number of higher-plant sugar transporters. The identification of DNA encoding such proteins not only provided strong evidence for the apoplastic model, but it also allowed studies of individual transporters and determination of their location *in planta*.

Identification of clones encoding sugar transporters and homology to other transport proteins

The first plant sugar transporter cDNA was cloned from the lower plant *Chlorella kessleri* by differential screening of a cDNA library from cells which were induced for monosaccharide transport [32]. The obtained cDNA clone (*HUP1*) was used for the screening of several cDNA and genomic libraries from higher plants and the cloning of the first higher-plant monosaccharide transporter from *Arabidopsis thaliana* was published in 1990 [28]. Both the *Chlorella* HUP1 and the *Arabidopsis* STP1 protein showed homology to transporters from mammals, yeast, and bacteria [19] and were unequivocally characterized as glucose-H^+ symporters by heterologous expression in *Schizosaccharomyces pombe* ([27, 28]; see below). The availability of probes for monosaccharide transporters and the fact that plant plasma membrane proteins could be functionally expressed were the basis for the cloning of many other transporters.

A sink-specific monosaccharide transporter (MST1) was cloned from *Nicotiana tabacum* [29], and families of at least 7 to 12 monosaccharide transporters (or monosaccharide transporter-like proteins) were shown to exist in *Ricinus communis* [45], *Chenopodium rubrum* [26], and *Arabidopsis thaliana* ([26], Baier and Sauer, unpublished data). Not all of these putative transporters have been fully characterized by heterologous expression and it may turn out that the substrate specificities of some of these proteins differ. In the lower plant *Chlorella kessleri*, for example, a second gene encoding a monosaccharide-H^+ symporter was cloned (*HUP2*) which gene turned out to be a galactose-H^+ symporter [37]. Up to now there is no indication, however, that any of the higher-plant transport proteins isolated by heterologous screening with monosaccharide transporter probes might transport disaccharides such as sucrose or maltose.

The cDNA of the first higher-plant sucrose transporter was cloned by Riesmeier *et al.* in 1992: a yeast strain which was deficient in both sucrose transport and invertase activity was transformed with a plant-derived sucrose synthase to allow sucrose degradation in the yeast cytoplasm. The resulting strain was then used for the expression of a spinach cDNA library and transformed yeast cells were screened for their capability to grow on sucrose as single carbon source which should only be possible when either a sucrose transporter or a secreted invertase is expressed. Shortly after the spinach sucrose transporter (*SoSUT1* [24]) a sucrose transporter from potato was isolated also by complementation cloning in *Saccharomyces cerevisiae* (*StSUT1* [23]) and both plants seem to have only one single gene encoding a sucrose transport protein. In contrast, cDNAs for two different sucrose transporters were identified by heterologous screening in cDNA libraries constructed from whole plants of *Arabidopsis thaliana* (*AtSUC1* and *AtSUC2* [30]) or from isolated vascular bundles of *Plantago major* (*PmSUC1* and *PmSUC2* [8], Gahrtz and Sauer, unpublished data).

The different higher-plant sucrose-H^+ symporters show a high degree of homology on the amino acid level with 63% (PmSUC1 versus AtSUC1) to 77% (AtSUC1 versus AtSUC2) of identical amino acids in their translated sequences and similar degrees of identity can be calculated for the monosaccharide-H^+ symporters (e.g. 79% for MST1 versus STP1 or 63% for STP1 versus STP4). A comparison of disaccharide transporters versus monosaccharide transporters, on the other hand, suggests that these two types of transport proteins are less closely related with only 20% of identical amino acids in the average. Despite this low homology between mono- and disaccharide transporters, however, the tertiary structures of the proteins in the membrane might not be so different. On the basis of hydrophilicity plots and in agreement with results from bacterial and mammalian sugar transporters [5, 21, 35], 12 putative transmembrane helices were predicted for all cloned plant sugar transporters; it is assumed that, as in these other transporters, N- and C-termini of the plant proteins are located on the cytoplasmic side of the membrane.

It may seem erroneous to use data on the structure of bacterial or mammalian sugar transporters to explain the structure of higher-plant sugar-H^+ symporters. However, as Marger and Saier [19] pointed out, the plant monosaccharide-H^+ symporters belong to a large superfamily of transmembrane facilitators which possibly evolved from one single ancestral transport protein. The structurally related members of this superfamily catalyse the transport of quite different substrates such as organic acids, drugs, organic phosphates or oligo- and monosaccharides. A typical feature of these 12 membrane transporters is a rather high degree of similarity between the first and the second half of the protein [12, 18, 33]. Such conserved sequences can also be found in the two halves of higher-plant mono- and disaccharide transporters [30], showing that the sucrose-H^+ symporters may also be members of this superfamily.

Heterologous expression and kinetic characterization of plant transporters

First hints towards an energy-dependent sugar-H^+ symport in higher plants came from the work of Sovonick *et al.* [36], Giaquinta [10] and Komor *et al.* [14] but the first unequivocal proof for sugar-H^+ symport in plants came again from *Chlorella kessleri* [13]. Since then proton-coupled mono- and disaccharide transporters have been identified and studied in many different tissues of higher plants (reviewed by Bush [4]). Only the cloning of the respective genes and cDNAs, and their expression in *Saccharomyces cerevisiae*, *Schizosaccharomyces pombe* and *Xenopus laevis* oocytes allowed the investigation of one specific higher-plant transporter at a time, separated from all other transporters possibly present in the same cell or plasmalemma preparation. Studies of sugar transport into tissue slices, cell suspensions, or even into lipid vesicles prepared from plant plasma membranes or reconstituted from detergent extracts thereof may always deal with the concerted reaction of a set of similar transporters. Thus the proton-coupled hexose carrier from *Chenopodium rubrum* [11] has turned out to consist of at least three different monosaccharide transporters [26] and also the inducible *Chlorella* glucose-H^+ symporter [39] has been shown to be composed of three different inducible proteins, with one of them being a galactose-H^+ symporter [37]. Since two sucrose-H^+ symporters have been identified in *Arabidopsis* [30] and *Plantago* [8] one cannot be sure whether the earlier results published on sucrose transporters in other plants represent the kinetic properties of only one or possibly two sucrose transporters.

Ideally, a transporter should be studied in a unicellular, stable expression system with no or only a negligible background of endogenous transporters. The system should be eukaryotic to allow possible modifications and it should be easy to handle. Fission yeast *Schizosaccharomyces pombe* and baker's yeast *Saccharomyces cerevisiae* fulfil all of these criteria: they can utilize several monosaccharides as sole carbon source and the uptake of these sugars is catalysed by sugar uniporters (or facilitators) with K_m values in the millimolar range. Furthermore there are yeast mutants deficient in sugar transporters, and the expression of many sugar uptake systems is repressed in the presence of glucose.

The *Arabidopsis thaliana* STP1 protein was the first higher-plant sugar transporter to be expressed in yeast cells [28, 38]. It could be shown by membrane fractionation and with immunohistochemical methods that 100% of the recombinant STP1 protein reach the yeast plasma membrane (Stadler and Stolz, personal communication). The protein was characterized as monosaccharide transporter with a K_m for *D*-glucose of 20 μM. Other sugars such as *D*-xylose or *D*-galactose, which are normally not transported by yeast cells, are also substrates of the STP1 protein allowing transport studies with no interference by yeast endogenous transporters. The STP1 protein is sensitive to incomplete uncouplers of the energy gradient, causes accumulation of non-metabolizable sugar analogs. This accumulation by STP1 depends both on the change in pH across the plasma membrane and on the energy status of the cells since starved yeast cells are unable to energize their plasma membranes. This suggests that STP1 is an energy-dependent monosaccharide-H^+ symporter. Several other monosaccharide transporters have been studied in yeast systems: the MST1 protein from tobacco [29], the HEX3 protein from *Ricinus* [45], and three transporters from *Arabidopsis* (STP2, STP3, and STP4; Baier and Sauer, unpublished data). All of them have similar kinetic properties as the STP1 protein described above.

For a better understanding of the energy dependence during active sugar transport by STP1-plasma membranes were isolated from transgenic, STP1-expressing yeast cells and fused with proteoliposomes containing cytochrome-*c*-oxidase from beef heart mitochondria [38]. In the presence of the electron donor ascorbate this system generates a proton motive force (*pmf*) which should be high enough to energize sugar accumulation inside the vesicles if STP1 is a monosaccharide-H^+ symporter. The same reconstitution system had already been used successfully for *in vitro* characterization of the *Chlorella* HUP1 protein [22] and, in fact, Stolz *et al.* [38] could demonstrate that a *pmf* generated by cytochrome-*c*-oxidase can drive the accumulation of *D*-glucose by STP1. Similar results could also be obtained with STP1 protein solubilized from yeast plasma membranes with octylglucoside after reconstitution in cytochrome-*c*-oxidase-containing proteoliposomes [38].

STP1 has also been studied in *Xenopus* oocytes [1] where it caused hexose-elicited depolarization of the oocyte membrane. Sugar uptake by STP1 was stimulated by low external pH and the determined K_m values and pH optima were in good agreement with results from the yeast system.

The cDNAs of the spinach SoSUT1 and the potato StSUT1 sucrose transporters were isolated by complementation cloning in yeast [23, 24]. The K_m values for sucrose uptake into transgenic yeast cells were shown to be 1.5 and 1 mM, respectively. These values correlate well with K_m values determined *in planta* (for review see [6]). Of all the substrates tested only maltose competed with sucrose for transport by SoSUT1 and StSUT1. The K_m values for maltose were about 5 and 10 mM, respectively. In contrast, Buckhout [2] and Bush [3] had previously published that sucrose transport in sugar beet is highly specific for sucrose with maltose causing no inhibition at a 10-fold higher concentration than sucrose. Sucrose uptake by SoSUT1 and StSUT1 was shown to depend on the energy status of the yeast cells, but a possible accumulation of the substrate inside the cells could not be shown, since sucrose was readily metabolized. This and the uncoupler sensitivity of sucrose uptake suggested that sucrose transport might be a H^+ symport.

Using *SoSUT1* cDNA as hybridization probe Sauer and Stolz [30] cloned two cDNAs from *Arabidopsis thaliana* (*AtSUC1* and *AtSUC2*) which were studied in a *S. cerevisiae* wild-type strain. Most kinetic parameters of the two *Arabidopsis* transporters were very similar to those determined for SoSUT1 and StSUT1 with one single exception: the activity of AtSUC1 was constant from pH 4 to 6 and decreased only to about 50% at pH 7, whereas the activity of AtSUC2, SoSUT1, and StSUT1 increased drastically with decreasing external pH values, suggesting different functions of the two *Arabidopsis* sucrose transporters. The same results were obtained when the two sucrose transporters PmSUC1 and PmSUC2 from *Plantago* were expressed in *S. cerevisiae*. PmSUC1 exhibited a pH dependence like AtSUC1, and the pH-dependence of PmSUC2 was like that of AtSUC2 ([8], Gahrtz and Sauer, unpublished data). Expression of the *Plantago* PmSUC2 transporter in an invertase-deficient strain of *Saccharomyces cerevisiae*, which is unable to split sucrose, revealed that sucrose can be accumulated more than 200-fold in this strain. Isolation of plasma membranes from this PmSUC2 expressing yeast strain and fusion of these membranes to cytochrome-*c*-oxidase-containing proteoliposomes (see above) proved unequivocally that *pmf* is the driving force for this accumulation and that sucrose transport by PmSUC2 is a H^+ symport [8].

Identification of the proteins

Only two sugar transport proteins had been identified biochemically: the *Chlorella* HUP1 protein [31] and a sucrose transporter from sugar beet [15]. Both proteins had an apparent molecular mass of about 42 kDa when separated in sodium dodecyl sulfate (SDS)-polyacrylamide gels. This molecular mass on SDS gels is clearly smaller than the molecular masses calculated from the cDNA sequences which are in the range of 55k Da for all transporters. Antibodies raised against C-terminal fusions of various monosaccharide transporters with *Escherichia coli* β-galactosidase showed that the *Arabidopsis* STP1 protein and the tobacco MST1 protein have apparent molecular masses of about 42 kDa in SDS gels when isolated from transgenic yeast cells [29, 38] and the same molecular mass was found for STP1 isolated from *Arabidopsis thaliana* plasma membranes [38]. This difference between apparent and calculated molecular mass is not due to N- or C-terminal modifications, since protein sequencing of the N-terminus of purified STP1 protein yielded the same sequence that had previously been determined by cDNA sequencing and a C-terminal histidine tag which had been used for purification was still present [38].

Histidine tagging has also been used for identification of the *Arabidopsis* AtSUC1 sucrose transporter protein in plasma membranes of transgenic yeast cells (45k Da including the tag of 6 histidines) and the subsequent purification of the protein [30]. Addition of a C-terminal histidine tag did not interfere with the transport activity of STP1 or AtSUC1 and may be an important tool for the large-scale purification of plant sugar transporters [30, 38]. The *Plantago* PmSUC2 sucrose-H^+ symporter is expressed at very high levels in *Saccharomyces cerevisiae*. About 10% of the protein in purified yeast plasma membranes is PmSUC2 protein. The apparent molecular mass of this transporter on SDS gels is only 35 kDa. For this transporter it has been shown that deletion of the cytoplasmic N-terminus has hardly any effect on the transport activity of PmSUC2 and on the sorting of the protein towards the yeast plasma membranes (Stadler and Sauer, unpublished data).

Plant sugar transporters do not seem to be N-glycosylated. Neither the *Chlorella* HUP1 glucose-H^+ symporter nor the *Arabidopsis* STP1 monosaccharide-H^+ symporter possess consensus sequences for N-glycosylation [28, 32]. The sucrose-H^+ symporters do have potential N-glycosylation sites. None of these sites, however, is at a conserved position in all sucrose transporters sequenced so far and in most of these transporters the consensus sequences are located on the putative cytoplasmic side of the protein (according to the 12-helix model) and would therefore never face the lumen of the endoplasmic reticulum (ER [8, 24, 30]). Despite the lack of N-glycosylation which would proof that these transporters use the secretory pathway in plants, it is very likely that the proteins reach the plasmalemma via this classical way, since similar transporters in yeast and mammals are glycosylated in their first extracellular loop which means that they have to be synthesized at the ER. Introduction of an artificial N-glycosylation consensus sequence into this loop of the STP1 protein might be a way to prove this hypothesis.

Localization in the plant

After the identification and characterization of mono- and disaccharide transporters it was important to determine the tissue and/or cell type specificity of expression of the respective proteins and, at least for one or two of them, the localization in the plasma membrane. Only recently it became obvious that sequence homology alone is not sufficient for the assignment of transporters to a specific membrane: GLUT7, a human glucose uniporter, shares 68% identical amino acids with the liver plasma membrane glucose transporter GLUT2 but it is located in liver microsomes, most likely the endoplasmic reticulum [44]. The fact that a protein is sorted to the plasma membrane in a heterologous expression system provides strong evidence but is no proof for the same localization *in planta*. Only for the *Chlorella* HUP1 protein the localization in the plasmalemma has unequivocally been determined using anti-HUP1 antibodies on *Chlorella* thin sections [37].

The existence of higher-plant sugar transporters, especially of monosaccharide transporters, has frequently been explained with the retrieval of effluxed solutes back into the cells by these proton-dependent carriers (for review see [17]). In view of the large numbers of monosaccharide transporters found in higher plants, however, this simple function is very unlikely. According to the apoplastic theory one would expect to find monosaccharide transporters mainly in the sink tissues where they catalyse the import of glucose and fructose into the respective cells after sucrose hydrolysis by cell wall bound invertases. Northern blot analysis of the organ-specific expression of the tobacco MST1 monosaccharide-H^+ symporter showed that this protein is expressed predominantly in roots, the probably strongest sink of these plants [29]. Lower levels of MST1 expression were also found in tobacco leaves with a clear decrease during leaf development and higher expression levels in young (sink) leaves than in mature (source) leaves.

A more precise answer on the location of a specific transporter *in planta* can be obtained using transgenic plants expressing the β-glucuronidase reporter gene under the control of the promoter of the corresponding gene. The expression of four monosaccharide-H^+ symporters from *Arabidopsis thaliana* (STP1 to STP4) has been studied with this method and the results were surprising: all of them seem to be expressed in a very restricted area of the plant with STP1 being found in the ovaries of *Arabidopsis* flowers, STP2 in the anthers, STP3 in the mesophyll cells of the green leaves and sepals, and STP4 in the anthers and the roots (Truernit and Sauer, unpublished). Thus, the expression of STP1, 2, and 4 seems to be strictly sink-specific, since the tissues of their expression are heterotrophic, non-green parts of the plant depending on carbon supply from the leaves. STP1 is the only transporter which is not expressed in a sink tissue and the function of STP3 in the green leaves of *Arabidopsis* is still unclear. Maybe STP1 has in fact a function in the retrieval of effluxed monosaccharides from the leaf apoplast as mentioned above.

Similar studies were performed with promoter/β-glucuronidase constructs of the *Arabidopsis thaliana SUC2* gene. Sucrose transporters would be expected in the phloem cells of the vascular system catalyzing the import of sucrose from the apoplastic space into the phloem companion cells and/or the sieve elements. The results obtained with transgenic *Arabidopsis plants* confirm these expectations showing that *SUC2* expression is found in the phloem practically all over the plant with one single exception, the petals [40]. Expression of *SUC2* is clearly limited to the vascular bundles and cross sections through *Arabidopsis* stems reveal that this expression is restricted to the phloem. Surprisingly *SUC2* expression was found also in the phloem of sink tissues such as the roots, the ovaries, and the filaments, all organs where phloem loading is not expected to occur. This observation may be taken as evidence that SUC2 is possibly not only responsible for phloem loading, but also for unloading. An almost identical expression pattern has been published for the *Arabidopsis thaliana* plasma membrane H^+-ATPase AHA3 [7] which therefore may represent the primary active pump for the AtSUC2 secondary active transporter. Expression of SUC2 in *Arabidopsis* leaves is regulated during growth starting at the tip of young rosette leaves und proceeding towards the leaf's basis during development finally reaching all the way down into the stem of fully developed leaves. This agrees with data published by Turgeon and Webb [42] showing that the sink/source transition of *Cucurbita* leaves goes from the tip towards the basis (for review see [41]). The tissue specificity of AtSUC1 expression has not yet been determined.

Using the method of *in situ* hybridization on thin sections of potato leaves, Riesmeier *et al.* [23] demonstrated that the potato StSUT1 sucrose transporter is expressed in the minor veins of potato which also provides strong evidence for phloem loading by StSUT1. These results were confirmed with potato plants expressing *StSUT1* antisense RNA. These plants have reduced levels of StSUT1 protein resulting in increased levels of soluble carbohydrates in the leaves and increased starch content [25]. Due to impaired sugar export from the leaves these plants show reduced root growth and tuber yield.

Gahrtz and coworkers isolated vascular bundles from *Plantago major* and northern blots and RNase protection experiments revealed that both PmSUC1 and PmSUC2 are predominantly expressed in the vascular bundles ([8]; Gahrtz and Sauer, personal communication). Antibodies raised in rabbits against PmSUC2 protein which were purified from transgenic *S. cerevisiae*, were used for a more precise analysis of PmSUC2 expression and thin sections through the basal part of *Plantago* leaves showed that PmSUC2 is in fact expressed in phloem of *Plantago* and in no other part of the vascular bundles (Stadler, Brandner and Sauer, unpublished data).

The presented data confirm the theory of apoplastic phloem loading and unloading in the investigated plants, with sucrose transporters being expressed only in the phloem and monosaccharide transporters being found mainly in the sinks of these plants.

Outlook

The results summarized in this article show that the field of plant sugar transport, carbohydrate partitioning and long-distance transport has quickly developed over the past years. With the help of molecular biology cDNAs and genes for many transporters have been cloned, the encoded proteins have been studied in detail in various expression systems, the location of some of these transporters has been determined in the plant, and mutants with reduced amounts of sucrose transporter have been generated. We are now starting to collect first insights into how plants manage to distribute the photosynthates to their different heterotrophic organs and storage tissues, but we are far from understanding the regulatory mechanisms controlling this important step.

On first sight it may seem that phloem loading is fully explained by the identification of sucrose-H^+ symporters and their localization in the phloem. But there are still many open questions. Where and how is sucrose leaving the mesophyll cells? Is sucrose entering the companion cells, the sieve elements, or both? How is the situation in plants transporting sorbitol, mannitol, or raffinose? And, last but not least, are there plants with symplastic phloem loading as suggested by data obtained mainly with members of the Cucurbitaceae [9, 16, 34]?

The situation in the sink tissues, where the phloem is unloaded, is comparable: the large number of monosaccharide-H^+ symporters identified in higher plants and the sink-specific expression of many of these proteins favour the idea of apoplastic unloading of sucrose from the phloem, followed by extracellular hydrolysis. Mutants with decreased or artificially increased levels of cell wall invertase undergo phenotypic changes which also supports an apoplastic step [20, 43]. But why do plants have so many monosaccharide transporters compared to only one or two disaccharide transporters, and how is sucrose leaving the phloem?

We are still far from understanding the actual step of membrane penetration, and we have nothing more than weak, preliminary models of how the sugars might pass the transport proteins, or which amino acid residues line the channel or pore. The possibility to express transporters in systems accessible to electrophysiological methods will be helpful in answering some of these questions; purification and three dimensional resolution of the structure of at least one of these proteins will be essential.

Acknowledgements

This work was supported by the Deutsche Forschungsgemeinschaft (SFB 43/C5) and a grant to NS from the Bundesministerium für Forschung und Technologie (BEO21-0310331A).

References

1. Boorer KJ, Forde BG, Leigh RA, Miller AJ: Functional expression of a plant plasma membrane transporter in *Xenopus* oocytes. FEBS Lett 302: 166–168 (1992).
2. Buckhout TJ: Sucrose transport in isolated plasma membrane vesicles from sugar beet. Planta 178: 393–399 (1989).
3. Bush DR: Proton-coupled sucrose transport in plasmalemma vesicles isolated from sugar beet (*Beta vulgaris* L. cv. Great Western) leaves. Plant Physiol 89: 1318–1323 (1989).
4. Bush DR: Proton-coupled sugar and amino acid transporters in plants. Annu Rev Plant Physiol Plant Mol Biol 44: 513–542 (1993).
5. Davies A, Meeran K, Cairns MT, Baldwin SA: Peptide specific antibodies as probes of the orientation of the glucose transporter in the human erythrocyte membrane. J Biol Chem 262: 9347–9352 (1987).
6. Delrot S: Loading of photoassimilates. In: Baker DA, Milburn JA (eds) Transport of Photoassimilates, pp. 167–205. Longman Scientific & Technical, New York (1989).
7. DeWitt ND, Harper JF, Sussman MR: Evidence for a plasma membrane proton pump in phloem cells of higher plants. Plant J 1: 121–128 (1991).
8. Gahrtz M, Stolz J, Sauer N: A phloem-specific sucrose H^+ symporter from *Plantago major* supports the model of apoplastic phloem loading. Plant J 6: 697–706 (1994).
9. Gamalei YV: Characteristics of phloem loading in woody and herbaceous plants. Fiziol Rastenii 32: 866–875 (1985).
10. Giaquinta WT: Possible role of pH gradient and membrane ATPase in the loading of sucrose into the sieve tubes. Nature 267: 369–370 (1977).
11. Gogarten JP, Bentrup F-W: The electrogenic proton/hexose carrier in the plasmalemma of *Chenopodium rubrum* suspension cells: effects of Δc, ΔpH and Δpsi on hexose exchange diffusion. Biochim Biophys Acta 978: 43–50 (1989).
12. Griffith JK, Baker ME, Rouch DA: Membrane transport proteins: implications of sequence comparisons. Curr Opinions Cell Biol 4: 684–695 (1992).
13. Komor E: Proton-coupled hexose transport in *Chlorella vulgaris*. FEBS Lett 38: 16–18 (1973).
14. Komor E, Rotter M, Tanner W: A proton-cotransport system in a higher plant: sucrose transport in *Ricinus communis*. Plant Sci Lett 9: 153–162 (1977).
15. Lemoine R, Delrot S, Gallet O, Larsson C: The sucrose carrier of the plant plasma membrane: II. Immunological characterization. Biochim Biophys Acta 978: 65–71 (1989).
16. Madore MA, Webb JA: Leaf free space analysis and vein loading in *Cucurbita pepo*. Can J Bot 59: 2550–2557 (1981).
17. Madore MA, Lucas WJ: Transport of photoassimilates between leaf cells. In: Baker DA, Milburn JA (eds) Transport of Photoassimilates, pp. 49–78. Longman Scientific & Technical, New York (1989).
18. Maiden MCJ, Davies EO, Baldwin SA, Moore DCM, Henderson PJE: Mammalian and bacterial sugar transport proteins are homologous. Nature 325: 641–643 (1987).
19. Marger MD, Saier MH Jr: A major superfamily of transmembrane facilitators that catalyze uniport, symport and antiport. Trends Biochem Sci 18: 13–20 (1993).
20. Miller ME, Chourey PS: The maize invertase-deficient miniature-1 seed mutation is associated with aberrant pedical and endosperm development. Plant Cell 4: 297–305 (1992).
21. Mueckler M, Caruso C, Baldwin SA, Panico M, Blanch I, Morris HR, Jaffrey W, Lienhard GE, Lodish HF: Sequence and structure of a human glucose transporter. Science 229: 941–945 (1985).
22. Opekarová M, Caspari T, Tanner W: The HUP1 gene product of *Chlorella kessleri*: H^+/glucose symport studied *in vitro*. Biochim Biophys Acta.
23. Riesmeier JW, Hirner B, Frommer WB: Potato sucrose transporter expression in minor veins indicates a role in phloem loading. Plant Cell 5: 1591–1598 (1993).
24. Riesmeier JW, Willmitzer L, Frommer W: Isolation and characterization of a sucrose carrier cDNA from spinach by functional expression in yeast. EMBO J 11: 4705–4713 (1992).
25. Riesmeier JW, Willmitzer L, Frommer WB: Evidence for an essential role of sucrose transport in phloem loading and assimilate partitioning. EMBO J 13: 1–7 (1994).
26. Roitsch T, Tanner W: Expression of a sugar-transporter gene family in a photoautotrophic suspension culture of *Chenopodium rubrum* L. Planta 193: 365–371 (1994).
27. Sauer N, Caspari T, Klebl F, Tanner W: Functional expression of the *Chlorella* hexose transporter in *Schizosaccharomyces pombe*. Proc Natl Acad Sci USA 87: 7949–7952 (1990).
28. Sauer N, Friedländer K, Gräml-Wicke U: Primary structure, genomic organization and heterologous expression

of a glucose transporter from *Arabidopsis thaliana*. EMBO J 9: 3045–3050 (1990).

29. Sauer N, Stadler R: A sink-specific H^+/monosaccharide co-transporter from *Nicotiana tabacum*: cloning and heterologous expression in baker's yeast. Plant J 4: 601–610 (1993).
30. Sauer N, Stolz J: SUC1 and SUC2: two sucrose transporters from *Arabidopsis thaliana*; expression and characterization in baker's yeast and identification of the histidine-tagged protein. Plant J 6: 67–77 (1994).
31. Sauer N, Tanner W: Partial purification and characterization of inducible transport proteins from *Chlorella*. Z Pflanzenphysiol 114: 367–375 (1984).
32. Sauer N, Tanner W: The hexose carrier from *Chlorella*. cDNA cloning of a eucaryotic H^+ cotransporter. FEBS Lett 259: 43–46 (1989).
33. Sauer N, Tanner W: Molecular biology of sugar transporters in plants. Bot Acta 106: 277–286 (1993).
34. Schmitz K, Cuypers B, Moll M: Pathway of assimilate transfer between mesophyll cells and minor veins in leaves of *Cucumis melo* L. Planta 171: 19–29 (1987).
35. Seckler R, Wright JK, Overath P: Peptide-specific antibody locates the COOH terminus of the lactose carrier of *E. coli* on the cytoplasmic side of the plasma membrane. J Biol Chem 258: 10817–10820 (1983).
36. Sovonick SA, Geiger DR, Fellows RJ: Evidence for active phloem loading in the minor veins of sugar beet. Plant Physiol 54: 886–891 (1974).
37. Stadler R, Wolf K, Hilgarth C, Tanner W, Sauer N: Subcellular localization of the inducible *Chlorella* HUP1 monosaccharide-H^+ symporter and cloning of a coinduced galactose-H^+ symporter. Plant Physiol, in press (1995).
38. Stolz J, Stadler R, Opekarová M, Sauer N: Functional reconstitution of the solubilized *Arabidopsis thaliana* STP1 monosaccharide-H^+ symporter in lipid vesicles and purification of the histidine tagged protein from transgenic *Saccharomyces cerevisiae*. Plant J 6: 225–233 (1994).
39. Tanner W: Light-driven active uptake of 3-*O*-methylglucose via an inducible hexose uptake system of *Chlorella*. Biochem Biophys Res Commun 36: 278–283 (1969).
40. Truernit E, Sauer N: The promoter of the *Arabidopsis thaliana SUC2* sucrose-H^+ symporter gene directs expression of β-glucuronidase to the phloem: evidence for phloem loading and unloading by SUC2. Plant J, in press (1994).
41. Turgeon R: The sink-source transition in leaves. Annu Rev Plant Physiol Plant Mol Biol 40: 119–138 (1989).
42. Turgeon R, Webb JA: Leaf development and phloem transport in *Cucurbita pepo*: transition from import to export. Planta 113: 179–191 (1973).
43. Von Schaewen A, Stitt M, Schmidt R, Sonnewald U, Willmitzer L: Expression of yeast-derived invertase in the cell wall of tobacco and *Arabidopsis* plants leads to accumulation of carbohydrate and inhibition of photosynthesis and strongly influences growth and phenotype of transgenic tobacco plants. EMBO J 9: 3033–3044 (1990).
44. Waddell ID, Zomerschoe AG, Voice MW, Burchell A: Cloning and expression of a hepatic microsomal glucose transport protein. Biochem J 286: 173–177 (1992).
45. Weig A, Franz J, Sauer N, Komor E: Isolation of a family of cDNA clones from *Ricinus communis* L. with close homology to the hexose carriers. J Plant Physiol 143: 178–183 (1994).

Subject index

A.t. = *Arabidopsis thaliana*
A.m. = *Antirrhinum majus*
P.h. = *Petunia hybrida*